Medical Radiation Dosimetry

Brian J. McParland

Medical Radiation Dosimetry

Theory of Charged Particle Collision Energy Loss

Springer

Brian J. McParland BASc, MSc,
PhD, CPhys, CSci, FCCPM, FIPEM, FInstP
Amersham, Buckinghamshire, UK

ISBN 978-1-4471-5402-0 ISBN 978-1-4471-5403-7 (eBook)
DOI 10.1007/978-1-4471-5403-7
Springer London Heidelberg New York Dordrecht

Library of Congress Control Number: 2013951430

Printed on acid-free paper

Springer is part of Springer Science+Business Media (www.springer.com)

For Sharon,
'Who ever loved that not loved at first sight?'
As You Like It
Act III, Scene 5

Preface

Background

Medical radiation dosimetry is the discipline of evaluating, through measurement or calculation, the absorbed dose received by a patient as a consequence of therapeutic or diagnostic exposures to ionising radiation. A high level of accuracy in the calculation and measurement of this dosimetry is essential for the optimisation of patient treatment and diagnostic efficacy, the assurance of patient safety and the continued use of medical exposures in retrospective epidemiological studies of the effects of ionising radiation. The achievement of this accuracy requires detailed theoretical knowledge of the means through which the radiation interacts with matter. As the absorbed dose is the immediate consequence of the transfer of kinetic energy by an energetic charged particle to a medium as it decelerates through collisions with atoms, the theoretical understanding of this physical mechanism of collisional energy loss is fundamental to all medical dosimetry evaluations and is the subject of this book.

The use of ionising radiations in medical practice is extensive. Throughout the world approximately 2.5 billion people are exposed annually for the detection or diagnosis of disease or abnormal pathology (IAEA 2002). That is, over one in three of all humans receive medical ionising radiation each year for diagnoses with the result of an associated annual worldwide per caput effective dose[1] reported by the International Atomic Energy Agency (IAEA) to be about 0.4 mSv,[2,3] In addition

[1] The effective dose is a measure of radiocarcinogenic and hereditary damage risk resulting from non-uniform exposures at low radiation absorbed doses (ICRP 1991) and has the unit of sievert (Sv). A typical annual effective dose resulting from the exposure to natural sources of radiation is about 2.5 mSv.

[2] This is estimated to be about 78 % due to non-dental x-ray procedures, 21 % due to dental radiological examinations and 1 % due to nuclear medicine examinations.

[3] There is a wide variation in this effective dose depending upon the degree and complexity of medical care (UNSCEAR 2000) with countries with the highest levels of health care being associated with a per caput annual effective dose of 1.3 mSv whereas those with the lowest being associated with 0.02 mSv.

(the energy loss per unit distance travelled in a medium and normalised to the physical density of that medium) for electrons as a function of kinetic energy in carbon (used here as an elemental radiological approximation to soft tissue). Electrons can be set into motion through photon interactions with the atoms in tissue through photoelectric absorption and Compton scatter or else released through the β-decay of a radionuclide. In diagnostic radiology and diagnostic nuclear medicine, the resulting electron energies are of only a few tens or hundreds of keV at most, and the radiative component of the total stopping power is several orders of magnitude less than the collision component and, hence, negligible. For electron energies of a few MeV, which are typical of Compton-scattered electrons set in motion by the megavoltage photons used in therapy, radiative energy losses are at least an order of magnitude less than those occurring through collisions. Perhaps for pure high-energy electron beams (with kinetic energies of about 25 MeV) does the *bremsstrahlung* component become appreciable (although the radiative stopping power is still only a quarter of that due to collisions). This contribution to the absorbed dose could be considered in patient radiotherapy planning, but only as a diffuse background component derived from measurement and not from explicit calculation (McParland et al. 1988). It is only at an electron kinetic energy of about 100 MeV that the collision and radiative energy losses in carbon equate. This is why x-ray production in x-ray tubes and medical linear accelerators is energetically inefficient, why extensive means of selecting the thermal capacity of the x-ray target is a critical engineering decising and why mechanically efficient cooling of the x-ray target are required as the collision energy loss dominates over the radiation production. As radiative energy losses do not play a significant role in the medical radiation dosimetry of low atomic number materials, such as tissue, they will not be considered in this book. The reader interested in fundamental *bremsstrahlung* theory should consult the classic textbooks by Heitler (1954) and Haug and Nakel (2004). For those interested in its applications to medical radiation dosimetry, the reader is referred to Attix (1986), Podgorsak (2010) and McParland (2010).

Development of the Book

As has been emphasised in the above discussion, knowledge of the collision stopping power[6] is essential to accurate medical radiation dosimetry. What makes it a unique dosimetric quantity is that it is an almost entirely theoretical quantity and not empirical.[7] This characteristic of the collision stopping power makes the

[6] The pedant may suggest that, first, the noun *collision* should be replaced by the adjective *collisional* and, second, that the units of the stopping *power* are of energy per distance travelled and, hence, the appropriate term should be *collisional stopping force*. This is, indeed, true but it would contradict common usage of the term over many decades.

[7] Perhaps with the exception of the mean ionisation potential which, although it can be calculated ab initio, can be determined through empirical measurements of, for example, the stopping power of protons through thin foils.

theoretical foundations of calculating charged particle energy loss the foci of patient care and safety. The practising medical physicist can extract stopping power data for elements and compounds of radiological interest from, for example, tables provided in various reports of the International Commission on Radiological Units and Measurements (ICRU) or, for example, from the website of the National Institute of Science and Technology. One can use these data without understanding or appreciation of the theoretical developments which led to their calculation. This, in this author's view, is not desirable. Moreover, the research medical physicist designing dosimetry apparatus or using the numerous Monte Carlo codes available for medical radiation transport calculations must understand the theory of collision energy loss in order to select the appropriate modelling to use and to be able to interpret the results. This book is not intended to replace the copious sources of numerical data that are readily available to the medical physicist; in fact, these sources of data are cited throughout. Instead, the book is written so as to be used as an adjunct to these data compilations and to provide the theoretical bases for understanding the data, their behaviours as functions of particle energy and medium physical properties and the limitations of their accuracies.

Most medical physics textbooks derive a simple expression for the collision stopping power based upon classical electrodynamics largely through the Rutherford formula or another classical electrodynamic approach. Then, the Bethe equation, derived from quantum theory, is given usually without derivation or explanation of the limitations to the theory and the corrections required to compensate for these. On the other hand, textbooks that discuss the quantum theory of charged particle interactions with matter tend to tailor the discussion for the theorist audience. This book attempts to bridge this divide between the two approaches and derives and explores the theory to its full depth and appreciation of the beauty inherent to it, but always within the context of medical applications.

At one level, this book emerged from a chapter in a book previously written by me on the same topic, but which was limited to applications in nuclear medicine radiation dosimetry (McParland 2010). Here, the subject matter has been expanded to cover the charged particle radiations of interest to all current diagnostic and therapeutic applications and to elucidate the theory of charged particle energy loss to a more complete depth. The intended readership of this book is the research medical physicist, the radiation metrologist, the medical physics graduate student and the medical physicist practising in radiation oncology and diagnostic imaging (including nuclear medicine). It is unlikely to yield benefit for the physician community, including radiation oncologists, radiologists or nuclear medicine physicians. The reader is expected to be familiar with nonrelativistic and relativistic quantum mechanics and kinematics, electromagnetism theory, special relativity and advanced mathematics. The last item would include familiarity with complex variable theory (including integration) and statistics. An extensive Appendix is provided to act as a convenient refresher and reference for the reader. Quantum perturbation theory and quantum scattering theory are the core of this book and are developed in a separate chapter.

At the end of each chapter are two sections, that of the Bibliography and Further Reading and that of the References. The former cites books, reports and scientific literature that were not explicitly cited in the chapter, but helped form the background of the development of that chapter and from which the reader can obtain further details of the topics developed and discussed. The latter cites those sources that were directly acknowledged within that chapter.

The construction of this book is as follows:

- The starting material provides the Glossary, a historical introduction to the field and a detailed review of quantum perturbation theory which will be drawn extensively on in later chapters.
- **Part II** provides an extensive examination of the Coulomb scatter of charged particles based upon the Rutherford theory which is derived through two classical means and the quantum theory to demonstrate the consistencies between the approaches. The effects of finite target size and screening due to atomic electrons upon scatter are examined and eventually lead us to the exploration of multiple scatter theory which dominates the transport of charged particles.
- **Part III** begins our examination of the energy loss through atomic electron excitation and ionisation. We isolate the interactions leading to this energy loss into *soft* and *hard* components. The soft component is examined through both classical (Bohr) and quantum (Bethe) means, whereas the hard component is explored within solely the quantum theory. Higher-order effects are then presented, along with estimations of the range of the charged particle as it decelerates and the practical considerations of media which are molecular and not elemental.
- **Part IV** recognises that the results of the calculations derived in Part III are of the mean energy loss and that we must not ignore the fact that the energy losses through collisions is stochastic. After developing the foundations through which we can calculate the probability distribution functions describing stochastic energy loss, we focus on the general Vavilov theory and demonstrate how, in the cases of thin and thick absorbers, the Landau and Gaussian theories are derived.
- Included is an Appendix, which covers much of the detailed mathematics used in the book and may be treated by the experienced reader as a refresher or, by others, a tutorial.

Additional Resources

For the medical physicist interested in further additional academic resources to supplement his or her reading of the topic of this textbook, the following may be of use. These are, in alphabetical order:

- **Attix FH (1986) Introduction to Radiological Physics and Radiation Dosimetry (John Wiley & Sons, New York)**: This is one of the classical medical physics textbooks. It will enable the reader to extend from the theory of charged particle energy deposition in matter to practical dosimetry calculations.

- **Johns HE and Cunningham JR (1983) The Physics of Radiology, 4th Edition (Charles C. Thomas, Springfield)**: This, in the author's opinion, is *the* classic of medical radiation physics textbooks. Despite it having been 30 years since the last edition was last published, it remains the foundation of the training of any medical radiation physicist.
- **Kase KR and Nelson WR (1978) Concepts of Radiation Dosimetry (Pergamon Press, New York)**: One must ignore its age as this compilation of lecture notes provides still one of the most comprehensive reviews of the physics of radiation dosimetry, in particular that of the interactions of charged particles with matter.
- **McParland BJ (2010) Nuclear Medicine Radiation Dosimetry: Advanced Theoretical Principles (Springer-Verlag, London)**: An overview of the theories that are developed within this book were provided in Chapter 7 of that book. The focus of the former were, however, on the internal radiation dosimetry associated with radionuclides distributed within the body, whereas this current book provides both a broader and deeper aspect of the theory of charged particle interactions with matter in medical applications.
- **Podgorsak EB (2010) Radiation Physics for Medical Physicists (Springer-Verlag, Berlin 2010)**: This textbook provides perhaps the most comprehensive current overview of radiation physics as applied to medicine.
- **Rossi B (1952) High-Energy Particles (Prentice-Hall, New York)**: Even though it is over 60 years of age, this classic textbook still generates use through its unequalled compilation of the calculations of charged particle interactions with matter.
- **Sigmund P (2006) Particle penetration and radiation effects (Springer-Verlag, Berlin)**: This is *the* definitive modern-day textbook on the physics of charged particle interactions with matter. It is a generalist book, but the medical physicist with an interest in radiation dosimetry will benefit from it.

Amersham, England Brian J. McParland

References

Attix FH. Introduction to radiological physics and radiation dosimetry. New York: Wiley; 1986.

Hall EJ, Giaccia AJ. Radiobiology for the radiologist. 6th ed. Philadelphia: Lippincott, Williams & Wilkins; 2006.

Haug E, Nakel W. The elementary process of *bremsstrahlung*. Singapore: World Scientific; 2004.

Heitler W. The quantum theory of radiation. 3rd ed. London: Oxford University Press; 1954.

IAEA. Radiological protection for medical exposure to ionizing radiation. Safety Standards Series No. RS-G-15. Vienna: International Atomic Energy Agency; 2002.

ICRP. Recommendations of the International Commission on Radiological Protection. ICRP Publication 60. Oxford: Pergamon; 1991.

McParland BJ. Nuclear radiation dosimetry: advanced theoretical principles. London: Springer; 2010.

McParland BJ, Cunningham JR, Woo MK. The optimization of pencil beam widths for use in an electron pencil beam algorithm. Med Phys. 1988;15:489–97.

Podgorsak EB. Radiation physics for medical physicists. Berlin: Springer; 2010.

UNSCEAR. Sources and effects of ionizing radiation – Vol. I, United Nations Scientific Committee on the effects of atomic radiation 2000 report to the general assembly. New York: United Nations; 2000.

Wagner LK, McNeese MD, Marx MV, Siegel EL. Severe skin reactions from interventional fluoroscopy: case report and review of literature. Radiology 1999;213:773–6.

Acknowledgements

While the topic of this book was first introduced to me in the context of several electromagnetic theory courses as an undergraduate electrical engineering student, it was as a graduate student in Medical Physics that I learnt of its medical applications and the need to employ it in my experimental research. My theoretical interest in the topic grew further through training in quantum scattering theory. This was in parallel to the stimulation of my experimental interest in the topic via the practical needs of particle detector design and construction in my experimental investigations of the nuclear production and absorption of π-mesons at TRIUMF. But my specific interest in the medical applications of understanding charged particle transport in tissue was ignited when I was a Terry Fox Fellow at the Department of Medical Biophysics of the University of Toronto and was encouraged to develop further the use of electron beams in the treatment of neoplasia through theoretical and experimental evaluations of the associated radiation dosimetry. This was then followed by my investigations required for a planned proton therapy programme at the Princess Margaret Hospital in Toronto.

The results of over three decades of work in the medical applications of the theory of charged particle interactions with tissue eventually led to a significant exploration of the topic of charged particle dosimetry in one of the chapters of my previous book, 'Nuclear Medicine Radiation Dosimetry: Advanced Theoretical Principles'. After completion of that chapter, I realised that a broader and more collective theoretical review of the interactions of moving charged particles with matter in the context of medical radiation dosimetry was required. This necessity is owed to its fundamental importance as the mechanism through which radiation absorbed dose is provided to tissue in both diagnostic and therapeutic medicine. I hope that this book satisfies sufficiently that need.

I wish to thank the support of Springer Publishing, especially through Melissa Morton, during my authorships with it. I appreciate the many sources that have provided me with photographs which, I believe, add to the history, and appreciation, of the richness of the topic of this book. In particular, I wish to acknowledge and express my thanks to The Alexander Turnbull Library (Wellington, New Zealand), the Niels Bohr Institute (Copenhagen, Denmark), Cornell University,

CERN, Prof. Stamatios M Krimigis of the Applied Physics Laboratory of The Johns Hopkins University, Prof. Bruce A. Faddegon of the University of California San Francisco, Spire Biomedical Inc. and TRIUMF.

My interest in the topic of this book has been the culmination of the experimental and theoretical stimuli provided by the mentorships and teachings of the late Prof. Edward G. Auld, Prof. John R. Cunningham, Prof. Garth Jones, Dr. Richard O. Kornelsen, Prof. Malcolm MacMillan and Prof. J. Alan Rawlinson. I hope that this book reflects the knowledge that they have conveyed to me. Of course, any errors within this book are solely attributable to me.

Finally, as always, I acknowledge my wife, Sharon, and our daughters, Siobhán and Áine. Words continue to remain inadequate in the expression of my love for them and my humility at their continued tolerance of my writing and their support throughout it.

Amersham, England Brian J. McParland

Contents

Part I Introduction to Charged Particles

Units, Glossary and Nomenclature

> And the Lord said, 'Behold they are one people, and they have all one language; and this is only the beginning of what they will do; and nothing that they propose to do will now be impossible for them. Come, let us go down, and there confuse their language, that they may not understand one another's speech'.
> **Genesis 11, 6–7**

Physical Units

The above Divine punishment has been clearly inflicted upon science, medicine and engineering as the variation of units and nomenclature is annoyingly vast throughout the disciplines and their sub-specialties. The topic of this book lies within medical radiation physics, and because that field has embraced, although not entirely, the *Système International d'Unités* (SI) system,[1] the SI system will be used almost universally throughout this book. However, there will be reservations. For example, it would be inappropriate and confusing to adopt the macroscopic units of the *metre* and *kilogram* where, in the microscopic realm, nuclear and atomic dimensions are most conveniently and commonly described in the unit of the formal *femtometre*, and the colloquial *fermi* (10^{-15} m), and where the rest-mass of a particle is quantified in terms of its energy equivalent (given by $E = mc^2$) of the MeV (10^6 eV or 1.6×10^{-13} J). Hence, SI units will be used where required and other units will be used where clarity, convenience and convention so require. Where potential confusion could arise, the associated text and context will explain.

Unfortunately, this general use of the SI system will perhaps introduce problems to those readers who are accustomed to Gaussian or "natural" units. Whilst these units can assist in making a mathematical derivation of a physical solution somewhat more transparent to some, the transport of these to the world of where calculations and measurements must be made can lead to error and requires the

[1] Which may be considered as an attempt to counter the aforementioned Divine intervention and may be conveniently found at http://www.bipm.org/en/si/si_brochure/

injections of the factors of the reduced Planck constant $\hbar$, the speed of light c and 4π. As the topic of this book has clear applications to the clinical world, I will tend to the SI units, although there will be practical cases where clinical quantities are perhaps best expressed in *cgs* units rather than *mks* units. However, throughout all of this text, considerable effort has been made to ensure the utmost clarity.

However, I will exercise some exceptions in the microscopic theoretical derivations in order to simplify mathematical expressions. For example, in general, I will frequently normalise units so that $c = 1$, thus allowing rest masses to be expressed in units of MeV and not MeV/c^2 and units of the magnitude of the 3-vector momentum to be expressed in units of MeV and not MeV/c. This will reflect the common vernacular and it does, indeed, simplify, for example, the expression of the total energy of a relativistic particle, $E = \sqrt{(pc)^2 + (mc^2)^2}$, to that of $E = \sqrt{p^2 + m^2}$. Such usage will limit the number of factors and powers of c that can overgrow expressions: for example, the rest mass of the electron will be given simply as $m_\mathrm{e} = 0.511$ MeV.

However, this simplification comes at a cost of requiring the use of the conversion factor, $\hbar c = 197.33$ MeV fm. This will be accompanied by the frequent use of the definitions of the fine structure constant $\alpha = e^2/4\pi\varepsilon_0\hbar c$ and the classical radius of the electron $r_0 = e^2/4\pi\varepsilon_0 m_e c^2$ throughout the text in order to simplify expressions and still retain transparency. However, this will also present advantages: for example, as a result of the use of these factors, the magnitude of the electric field due to a point charge e will be written in the succinct form as $E(r) = \alpha\hbar c/er$.

As discussed in the Preface, radiative interactions of charged particles with matter are not of dosimetric interest for the charged particles and kinetic energies of medical interest in tissue. Hence, the expressions for the linear and mass collision stopping powers $\mathrm{d}E/\mathrm{d}x$ and $\mathrm{d}E/\rho\,\mathrm{d}x$, respectively, are provided without any modifier, such as a subscript *Coll*, to indicate that they refer to collision energy losses.

Vector Notation

Three-vectors are denoted in bold. If $\mathbf{x}$ is a three-vector, then the corresponding unit vector is,

$$\hat{\mathbf{x}} = \frac{\mathbf{x}}{|\mathbf{x}|}.$$

A four-vector is denoted in bold italics, e.g., $\boldsymbol{x}$. The contravariant components are denoted by superscripts whereas the covariant components are denoted by subscripts,

$$\mathsf{x}^\mu \equiv \left(\mathsf{x}^0, \mathsf{x}^1, \mathsf{x}^2, \mathsf{x}^3\right) \equiv \left(\mathsf{x}^0, \mathbf{x}\right).$$

The covariant and contravariant components are related through,

$$\mathsf{x}_\mu = \sum_{\nu=0}^{3} g_{\mu\nu}\mathsf{x}^\nu$$

where the metric tensor is,

$$g_{\mu\nu} = g^{\mu\nu} = \begin{pmatrix} 1 & 0 & 0 & 0 \\ 0 & -1 & 0 & 0 \\ 0 & 0 & -1 & 0 \\ 0 & 0 & 0 & -1 \end{pmatrix}.$$

Hence,

$$\mathsf{x}_\mu = \left(\mathsf{x}^0,\ -\mathrm{x}\right).$$

We will use the summation convention that when an identical superscript and subscript occur, the components are summed over that superscript/subscript. That is,

$$\mathsf{p}_\mu \mathsf{x}^\mu \equiv \sum_{\mu=0}^{3} \mathsf{p}_\mu \mathsf{x}^\mu.$$

In the event that no ambiguity can arise, we will simplify this summation by omitting the indices so as to write $\mathsf{p}_\mu \mathsf{x}^\mu \equiv \mathsf{px}$. Note also that,

$$\mathsf{p}_\mu \mathsf{x}^\mu = \mathsf{p}^\mu \mathsf{x}_\mu = \mathsf{p}^0 \mathsf{x}^0 - \mathbf{p} \cdot \mathbf{x}.$$

The four-vector momentum for a particle of rest mass m is,

$$\mathsf{p}^\mu = (E, \mathbf{p})$$

Where E is the total energy (sum of kinetic energy and rest mass) and $\mathbf{p}$ is the three-vector momentum. From the above, we see that,

$$\begin{aligned} \mathsf{p} \cdot \mathsf{p} &= E^2 - \mathbf{p} \cdot \mathbf{p} \\ &= m^2 \end{aligned}$$

and, hence, is an invariant. The product of the four-vector momenta of two distinguishable particles A and B is,

$$\mathsf{p}_A \cdot \mathsf{p}_B = E_A E_B - \mathbf{p}_A \cdot \mathbf{p}_B.$$

Complex Conjugation

Let $z = x + iy$ be a complex variable. Its complex conjugate is denoted by

$$z^* = x - iy$$

Hermitian Conjugation

Let $\mathbf{M}$ be a matrix with complex elements $(M)_{ij} = m_{ij}$. The Hermitian conjugate of this matrix is denoted by $\mathbf{M}^\dagger$ where $(M^\dagger)_{ij} = m^*_{ij}$. That is, the Hermitian conjugate is the transpose of the original matrix and with the elements replaced by their complex conjugates. If $\mathbf{M} = \mathbf{M}^\dagger$, the matrix is referred to as being Hermitian or self-adjoint.

Adjoint Operator

The adjoint of the spinor $\psi(\mathbf{r}, t)$ is,[2]

$$\overline{\psi}(\mathbf{r}, t) = \psi^\dagger(\mathbf{r}, t)\gamma^0$$

where γ_0 is a Dirac matrix.

Media

For most of our calculations, special consideration must be given to the medium that the charged particle is traversing. Although we will assign an atomic number and atomic mass number to the medium, in many cases we treat the medium as a cold and dilute gas in that there are no interactions between the atoms of the medium. Chapters 8 and 12 deal with realistic cases of condensed media, such as tissue, where inter-atomic interactions and medium polarisation can occur. The foundational calculations will be for elemental and monatomic media, in particular, carbon and calcium as representations of soft tissue and bone. More advanced calculations will consider more complex and condensed media such as lead, liquid water, striatal muscle and compact bone. The compositions of the latter are provided in Chap. 12.

[2] Refer to Appendix F for the definition of the spinor and the Dirac matrices.

Symbols

As there are only 26 letters in the Latin alphabet and 24 letters in the Greek alphabet, there will inevitably be duplication in the use of a particular letter as a symbol for a physical or mathematical quantity. The Table of symbols following will provide the meanings of the symbols used in this book. Where a symbol has multiple applications, the application for which it is being used will be both described in the book and evident from the context of discussion.

Table of Symbols

α	Fine structure constant, $\alpha = e^2/4\pi\varepsilon 0\hbar c \approx 1/137$; The ^{4}He nucleus
α_e	Electronic polarisability
$\alpha\hbar c$	Conversion factor, $\alpha\hbar c$ = 1.44 MeV fm
$\boldsymbol{\alpha}_i$	Dirac matrix, $\boldsymbol{\alpha}_i = \begin{pmatrix} 0 & \boldsymbol{\sigma}_i \\ \boldsymbol{\sigma}_i & 0 \end{pmatrix}$
β	Speed normalised to the speed of light, c Beta particle (electron or positron) produced through beta decay
$\boldsymbol{\beta}_\text{i}$	Dirac matrix, $\boldsymbol{\beta} = \begin{pmatrix} \mathbf{1} & 0 \\ 0 & -\mathbf{1} \end{pmatrix}$
χ_0	Screening angle for charged particle-atom interactions
χ_a	Molière characteristic screening angle
χ_c	Molière angle
χ_e	Electric susceptibility
δ	Density effect of a medium to the collision stopping power δ particle (recoil electron)
$\delta(x)$	Dirac δ-function
δ_{jk}	Kronecker δ-function
Δ^2	$\equiv \hbar^2\partial^2/\partial t^2 - (\hbar c)^2\nabla^2$
ΔE	Energy loss
$\overline{\Delta E}$	Mean energy loss
ε_0	Permittivity of free space, 8.854×10^{-12} F/m
ε_R	Relative dielectric constant
$\hat{\boldsymbol{\varepsilon}}_m$	Photon unit polarisation vector (m = 1, 2)
$\overline{\varepsilon}$	Mean energy of secondary electrons resulting from ionisation but which are insufficiently energetic to cause further ionisations
γ	Relativistic contraction factor, $\gamma = 1/\sqrt{1-\beta^2}$ Photon

(continued)

$\boldsymbol{\gamma}^0$	Dirac matrix, $\boldsymbol{\gamma}^0 = \boldsymbol{\beta}$
$\boldsymbol{\gamma}^i$, i = 1, 2, 3	Dirac matrix, $\boldsymbol{\gamma}^i \equiv \begin{pmatrix} 0 & \boldsymbol{\sigma}_i \\ -\boldsymbol{\sigma}_i & 0 \end{pmatrix} i = 1,2,3$
γ_5	Dirac matrix, $\gamma_5 = i\gamma^0\gamma^1\gamma^2\gamma^3 = \begin{pmatrix} 0 & \mathbf{1} \\ \mathbf{1} & 0 \end{pmatrix}$
γ_{EM}	Euler-Mascheroni constant
Γ	Damping coefficient
$\Gamma(z)$	Gamma function
$\eta(E)$	Bohr screening condition parameter
κ	Atomic electron screening factor
	Stochastic scaling parameter
λ_{fi}	Transition rate (per unit time) from initial state i to final state f
λ_{fo}	Transition rate (per unit time) from ground state o to final state f
λ_L	Landau parameter
λ_V	Vavilov parameter
$\lambdabar$	Reduced de Broglie wavelength, $\lambdabar = \hbar c/p$
$\boldsymbol{\mu}$	Laplace-Runge-Lanz vector
μ	Eccentricity
μ_O	Permeability of free space, $4\pi \times 10^{-7}$ N/A^2
$\nu_{e,\mu,\tau}$	Neutrinos associated with the electron (e), muon (μ) and tau (τ) lepton
$\theta_{\mathbf{1}}$	Scattering angle for a single scattering event
$\overline{\boldsymbol{\theta}_1}$	Mean scattering angle for a single scattering event
$\overline{\boldsymbol{\theta}_1^2}$	Mean-square scattering angle for a single scattering event
θ_A	Molière characteristic screening angle
$\theta_A^{'}$	Modified Molière characteristic screening angle
θ_S	Single/multiple scattering angle cutoff
$\overline{\Theta^2}$	Variance of Gaussian multiple scatter probability distribution function
ρ	Physical density (mass per unit volume)
$\rho_B(E)$	Bohr collision diameter for kinetic energy, E
ρ_e	Electron number density (number of electrons per unit volume)
ρ_f	Phase space factor (density of final states per unit energy level)
ρ_P	Proton number density (number of protons per unit volume)
σ	Total cross section
	Conductivity
σ_{Inel}	Total inelastic cross section
σ_{Ion}	Total ionisation cross section
ϑ	Reduced Molière scattering angle
$\boldsymbol{\sigma}_i$; i = 1, 2, 3	Pauli matrices
$\mathrm{d}\sigma/d\Omega$	Differential cross section in solid angle
$\mathrm{d}\sigma/\mathrm{d}Q$	Differential cross section in energy transfer
$\xi(E)$	Bohr screening condition parameter

(continued)

ω	Angular frequency
$\omega(Q)$	Differential probability of energy loss Q
ω_0	Natural (resonant) frequency
ω_D	Damped frequency
ω_P	Plasma frequency
Ω	Solid angle
	Molière parameter
$\psi_{i,f}$	Wavefunctions of (i)nitial and (f)inal state of interaction
ψ	Logarithmic derivative of the gamma function, Γ
a_∞	Bohr radius, $a_\infty = \hbar c/\alpha m_e$
A	Atomic mass number
	Activity
$\mathbf{A}$	Three-vector electromagnetic potential
A_0	Scalar electromagnetic potential
$A^\mu(x)$; $A_\mu(x)$	Four-vector electromagnetic potential
$Ai(t)$	Airy function, $Ai(t) = \frac{1}{\pi}\int_0^\infty dy \cos\left(yt + \frac{y^3}{3}\right)$
b	Barn (1b $= 10^{-24}$ cm^2)
	Impact parameter
	Molière parameter
$\boldsymbol{b}$	Vectorial impact parameter
$b_{\min}$	Minimum impact parameter
$b_{\max}$	Maximum (adiabatic) impact parameter
B	Molière parameter (measure of the width of the Gaussian probability distribution function for multiple Coulomb scatter)
$\mathbf{B}$	Magnetic field
c	Speed of light
C	Coulomb
	Numerical constant, 0.1535 MeV atom cm^2/mole
$C_e(\beta)$	Term describing effect of atomic electron shells on collision stopping power
$Ci(x)$	Cosine integral, $Ci(x) = \gamma_{EM} + \ln x + \int_0^x dt \frac{\cos t - 1}{t}$
CSDA	Continuous slowing-down approximation
$d^3\mathbf{r}$	Differential spherical element, $d^3\mathbf{r} \equiv r^2 dr\, d\varphi\, d(\cos\theta)$
D	Absorbed dose
e	Unit of electric charge
eV	Electron volt
E	Total energy (sum of kinetic and rest mass energies)
E_B	Atomic electron binding energy
$\mathbf{E}_{rad}$	Electric radiation field
dE/dx	Linear collision stopping power
$dE/dx\vert_H$	Linear hard-collision stopping power

(continued)

$dE/dx\|_S$	Linear soft-collision stopping power
$dE/dx\|_\Delta$	Restricted linear collision stopping power
$dE/\rho\,dx$	Mass collision stopping power
$dE/\rho\,dx\|_H$	Mass hard-collision stopping power
$dE/\rho\,dx\|_S$	Mass soft-collision stopping power
$dE/\rho\,dx\|_\Delta$	Restricted mass collision stopping power
D	Electric flux density
E	Electric field
E0	Initial kinetic energy
ET	Total kinetic energy
$Ei(x)$	Exponential integral, $Ei(x) = \int_{-x}^{\infty} dt\,\frac{e^{-t}}{t}$
e	Fundamental unit of electric charge
$f_n(q, Z)$	Generalised oscillator strength (GOS)
$F(q)$	Scattering form factor for three-vector momentum transfer, q
$F(q, Z)$	Atomic form factor
$F(Z_Y, E_e)$	Fermi β-decay nuclear Coulomb correction factor for daughter nucleus Z_Y and β particle energy, E_e
$f(\mathbf{q})$	Scattering amplitude as a function of three-vector momentum transfer, $\mathbf{q}$
$f_{GS}(\theta;s)$	Goudsmit-Saunderson multiple scattering probability distribution function
$f_M(\theta;s)$	Molière multiple scattering probability distribution function
g	gram
	Fraction of liberated charged particles' initial kinetic energies that is irradiated as *bremsstrahlung*
$G(\mathbf{r},\mathbf{r}')$	Green's function
Gy	Gray
$h_l^{\binom{1}{2}}(x) = \sqrt{\frac{\pi}{2x}}\left(J_{l+½}(x) \pm iN_{l+½}(x)\right)$	Hankel functions of first and second type of order l
$\hbar$	Reduced Planck constant, $\hbar = h/2\pi$
$\hbar c$	Conversion factor, $\hbar c$ = 197.3 MeV · fm
H	Total Hamiltonian
	Magnitude of the magnetic field strength
$Ha(x)$	Hankel function
$H_i(x)$	ith Hermite polynomial
H_0	Free Hamiltonian
$H(t)$	Heaviside function: $H(t) = 0 t<0$ $= 1 t>0$
$H_{EM}(\mathsf{x})$	Electromagnetic Hamiltonian
$\boldsymbol{H}$	Magnetic field strength
$\boldsymbol{H}_{rad}$	Magnetic radiation field
Im	Imaginary component
$\overline{I}$	Mean excitation energy

(continued)

$\boldsymbol{J}$	Total angular momentum
	Electric current density
$J_l(x)$	Bessel function of the first kind of order l
$j_l(x)$	Spherical Bessel function of the first kind of order l
$K_\nu(x)$	Modified Besself function of the second kind of order ν
l	Quantum angular momentum number
L_∞	Unrestricted linear energy transfer
$L(\beta)$	Stopping number for charged particle with speed normalised to c, β
LET	Linear energy transfer
$\boldsymbol{L}$	Total angular momentum
m_e	Rest mass of the electron
m_N	Rest mass of a nucleon
m_p	Rest mass of the proton
M_{fi}	Matrix element, $M_{fi} = \int d^3\mathbf{r}\, \psi_f^* U \psi$, for transition from initial state i to final state f
$M(a,b,c)$	Confluent hypergeometric function
n	Index of refraction
N_A	Avogadro's number
$N_l(x)$	Neumann function of order l
p	Magnitude of three-vector momentum vector, $\mathbf{p}$
$p_{T,j}$	Spectral distribution of particle radiance for particle species j
pdf	Probability distribution function
p_F	Fermi momentum
$\boldsymbol{p}$	Three-vector momentum
	Electric dipole moment
p	Four-vector momentum, $\mathsf{p} = (\mathrm{E}, \mathbf{p})$
$\mathbf{P}$	Poynting vector
	Polarisation (number of electric dipole moments per unit volume)
$P_{lm}(\mu)$	Associated Legendre polynomial of the first kind
$P_l(\mu)$	Legendre polynomial
q	Electric charge
	Magnitude of three-vector momentum transfer
Q	Electric charge
	Kinetic energy transfer
Q_C	Energy transfer demarcating the difference between soft and hard charged-particle collisions
$\boldsymbol{q}$	Three-vector momentum transfer
$\Re_{CSDA}$	Charged particle range in the continuous slowing down approximation
r_e	Classical electron radius, $r_e = e^2/4\pi\varepsilon_0 m_e c^2$
r_{TF}	Thomas-Fermi radius

(continued)

$Si(x)$	Sine integral, $Si(x) = \int_0^x dt \frac{\sin t}{t}$
$Sinc(x)$	Sine cardinal function
$S_F(\mathsf{x} - \mathsf{y})$	Feynman propagator, $S_F(\mathsf{x} - \mathsf{y}) = \frac{1}{(2\pi)^4} \int d^4\mathsf{p} \frac{e^{-\mathsf{p}\cdot(\mathsf{x}-\mathsf{y})}}{\mathsf{p} - m}$
SI	Système International
Tr	Trace of a matrix
$\mathfrak{I}$	Linear scattering power
$\mathfrak{I}/\rho$	Mass scattering power
$\mathbf{v}$	Velocity
v	Speed (magnitude of velocity vector)
V	Electric potential
U	Potential energy
x	Three-vector position
x	Four-vector position
$y_l(x)$	Spherical Bessel function of the second kind of order l (also known as a Neumann function)
$Y(t,t')$	Time evolution operator
$Y_l(x)$	Bessel function of the second kind of order l
$Y_m^l(\theta,\phi)$	Spherical harmonic
z	Electric charge, in units of e, of a projectile
Z	Atomic number Electric charge, in units of e, of a target

Part I
Introduction to Charged Particles

Chapter 1
Introduction

Contents

Abstract This chapter provides an historical overview of non-radiative interactions of charged particles with matter.

B.J. McParland, *Medical Radiation Dosimetry*,
DOI 10.1007/978-1-4471-5403-7_1,

1.1 Introduction

This chapter introduces the charged particles and their applications that are relevant to the uses of ionising radiation in diagnostic and therapeutic medicine. This will provide the background of the variety of applications of the theories of charged particle energy loss that are developed in the book. The chapter is divided into three sections following this Introduction. The first reviews the physical characteristics of these charged particles and their interactions and the histories of their discoveries which were dependent upon the detection of the transfer of kinetic energy to matter. The next reviews the early history of the development of theories describing this transfer, and the third section reviews the applications of these theories.

1.2 History and Discovery

1.2.1 Introduction

The organisation of a discussion of the history of the science of the interactions of charged particles with matter is surprisingly tricky. A simple chronological approach has the benefit in that it can perhaps elucidate the development of thought and experiment over time with the gaining of new knowledge and techniques. Yet it can suffer from those cases where either the theoretical explanation of an experimental result or the experimental validation of a theoretical prediction is separated considerably in time and can thus yield incoherence and complication in explanation and association. Similarly, the sharp division of the discussion into two components, those of theory and of experiment, can suffer from the same lack of coherence that chronology can offer. The approach taken here is to isolate the discussion in a manner so as to link chronology, theory and experiment together in as coherent a manner as possible. Hence, the introduction and review of the history of charged particle interactions with matter will be based upon particles with applications in medicine: cathode rays (electrons), α- and β-particles, internal conversion and Auger electrons, protons, heavy ions (in their initial presentation in the form of fission fragments), π^--mesons and the neutral indirectly ionising particles, photons and neutrons, which generate secondary fluxes of charged particles through interactions with the atoms of the medium. Clearly these are not all of the possible charged particles that could be considered, but are reflective of those that are associated with diagnostic and therapeutic medicine. This will follow a scientific current which will be roughly parallel to the chronology and the developments of theory and experiment. Congruity of the three specifications will not be complete as this will be impossible to achieve. However, it will provide, it is hoped, an appreciation of how our understanding of how charged particles interact with matter in discovery and experiment evolved.

While the genesis of the modern science and engineering of charged particle interactions with matter lies within the past 150 years, the general problem of the rate at which a moving object loses energy as it passes through a resistive medium has long been of practical interest. Perhaps the earliest modern development can be traced back to Book II of *The Principia* in which Newton (1687) described mathematically the resistance felt by objects moving through a fluid. Other investigators of this field, which evolved into fluid dynamics and aerodynamics, included Euler, d'Alembert and Bernoulli. Of peculiar interest to us is the treatise published by Robins (1742) on the kinematics of artillery and firearm projectiles through air, in which he described his invention of the ballistic pendulum as a means of measuring the velocity of a projectile. Robins wrote, ... *that if a globe sets out in a resisting medium with a velocity much exceeding that with which the particles of the medium would rush into a void space ... the resistance of this medium to the globe will be ... in proportion to its velocity. We may further conclude, that the resisting power of the medium will gradually diminish, as the velocity of the globe decreases...*, a conjecture that could be considered prescient of the theory of a charged particle moving through a Fermi gas of electrons.

However, our interest will lie in more recent times, primarily in those following the triad of the years of 1895, 1896 and 1897, the years during which x-rays, radioactivity and the nature of the electron were discovered – results which each culminated in the awarding to the discoverers of the then newly presented Nobel Prize in Physics. But before we provide the historical foundations to the medical applications of the theory of charged particle interactions with matter described in this book, we first review the modern understanding of the constitution and organisation of these particles.

1.2.2 The Standard Model

"Who ordered that?"

II Rabi on the announcement of the discovery of the muon

As the intent of this book is to present, within the context of medical applications, the means by which moving charged particles interact with matter and yield a radiation absorbed dose, it is appropriate that we first review the nature of these particles, their compositions and their properties as described by the Standard Model. This model reflects our current understanding of nature and three of its four known forces (electromagnetic, weak and strong), and it represents one of the great pinnacles of physics and intellectual development in the latter part of the twentieth century. These four known forces have widely differing levels of magnitude or 'strength'. These are reflected in Table 1.1.

The periodicity of Mendeleev's periodic table of the elements can be explained on the basis of systematic groupings of only three particles required to form atoms – protons, neutrons and electrons. An analogous situation requiring the simplification of complexity through the introduction of systematic arrangements of fundamental particles arose in the 1940s and 1950s with the discoveries of copious particles

Table 1.1 The four known forces of nature

Force	Mediating boson	Relative magnitude	Range of interaction
Strong	Gluon	1	1 fm
Electromagnetic	Photon	1/137	∞
Weak	$W^{\pm}, Z^0$	10^{-13}	10^{-3} fm
Gravity	Graviton	10^{-38}	∞

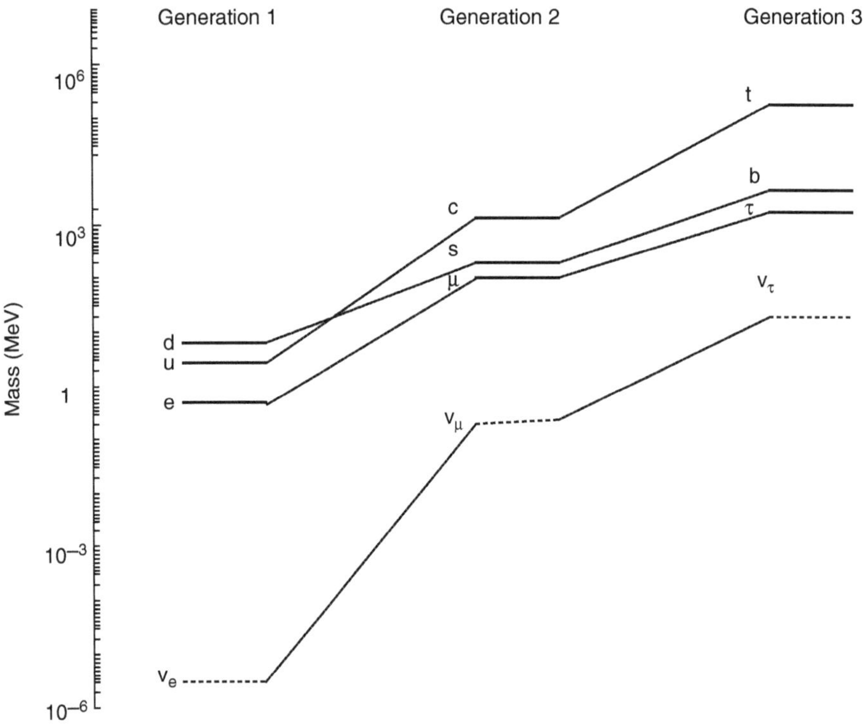

Fig. 1.1 The Standard Model of the known elementary particles. Each generation contains a pair of quarks and a pair of leptons, one of which is the electrically neutral neutrino. The neutrino masses shown are the current estimates of their upper limits

which required a similarly systematic means of explanation. From analyses of the systematics of the physical attributes of these particles (such as mass, charge, intrinsic spin, isotopic spin, lifetime), the Standard Model has evolved as a means of describing these particles.[1]

The Standard Model considers known matter to be made up of 12 elementary particles (and their antimatter partners) which are divided over three generations. Figure 1.1 shows the rest mass distribution of these particles. Each are fermions as

[1] An excellent review of the development of experiment and theory leading to the Standard Model can be found in the compilation of lectures, given at a 1985 symposium held at Fermilab, provided by Brown et al. (1989).

they each have an intrinsic spin of ½ (in fundamental units of the reduced Planck constant, $\hbar$), making them subject to Fermi–Dirac statistics and the Pauli exclusion principle. Of the four particles assigned to each generation, the two with the greater masses are the quarks (the name having been coined by Murray Gell-Mann and derived from a quotation in Joyce's Finnegan's Wake, '*Three quarks for Muster Mark*') and two are leptons (from the Greek *λεπτοζ* or *leptos*, meaning 'small and thin'). The quarks have a unique quantum entity associated with them, that of flavour, and the states of the flavour quantum number are *up* (u), *down* (d), *charm* (c), *strange* (s), *top* (t) and *bottom* (b),[2] with two flavours being assigned to each generation. Similarly, a pair of leptons is assigned to each generation, one which carries an electric charge (the electron (e), the μ-meson or muon (μ) and the tau lepton (τ)) and an electrically neutral neutrino that is associated with the charged lepton of that generation (i.e. the neutrino has a lepton flavour). The neutrino has, until recent times, been considered as having zero mass. Experimental evidence, however, has established a non-zero mass, the existence of which is reflected by *neutrino oscillations* in which neutrinos can change from one lepton flavour to another.

With the exception of the neutrinos, all of the fermions carry an electric charge. The quarks have fractional charge (e.g. the up quark carries the charge $+\frac{2}{3}e$ and the down quark carries the charge $-\frac{1}{3}e$), whereas the charged leptons have an integral charge, $\pm\ 1e$. Of course, the antiparticle counterparts carry the opposite electric charge. As the quarks are subject to the strong force, they, unlike the leptons, carry the additional quantum number of colour which has been assigned as the three states of red, green and blue. While only the quarks are subject to the strong force, all of the elementary particles of Fig. 1.1 are subject to the weak force,[3] and all of the charged particles are subject to the electromagnetic force. The interactions via the three forces are mediated through bosons as shown in Fig. 1.2. Bosons have zero or integral spin (in units of $\hbar$) and are thus described by Bose–Einstein statistics for which the Pauli exclusion principle is not applicable.

Quarks have not been detected as separate and individual entities, and all experimental evidence to date is that they form composite colour-neutral[4] particles known as hadrons (from the Greek *αδροζ* or *hadrós*, meaning 'thick'). A quark–antiquark pair forms a meson which has integral spin due to the coupling of their intrinsic spin-½ and any integral orbital angular momentum between the pair. For example, the positively charged π-meson consists of an up and an anti-down quark in the s state (i.e. with zero orbital angular momentum). Mesons are physically unstable and decay through weak interactions via the $W^{\pm}$ and Z^0

---

[2] And also referred to as 'beauty'.

[3] They are also all subject to the gravitational force, but, as demonstrated by Table 1.1, this force is inconsequential and will be neglected.

[4] To the 'outside world', a composite of quarks does not exhibit colour. This is achieved, in the case of the meson which is made up of two quarks, by a colour/anti-colour pair and, in the case of the baryon made up of three quarks, by quarks with red, green and blue colour.

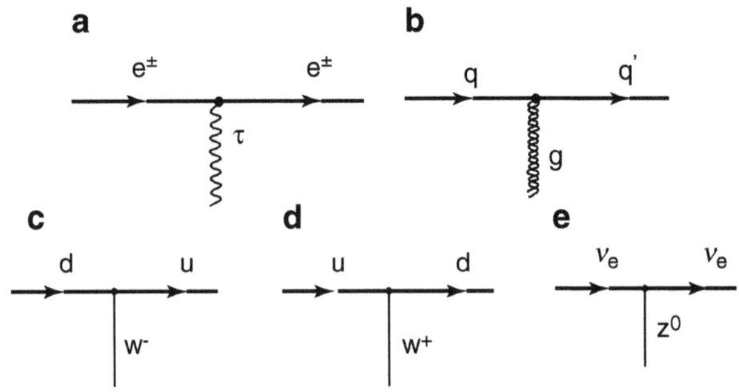

Fig. 1.2 Feynman diagrams of electromagnetic, strong and weak interaction vertices. (**a**) The charged electron/positron interacts with another charged particle through the mediation of a photon; (**b**) the strong force that a quark is subject to is mediated by a gluon; (**c**) and (**d**) the weak force-associated flavour changes of a quark are due to the exchange of the W^- boson with the appropriate electric charge; and (**e**) the neutrino interacts weakly through the exchange of the neutral Z^0 boson. Charged particle energy loss is mediated through the electromagnetic process (**a**)

intermediate vector bosons.[5,6] The π^+ consists of an up and anti-down quark pair and decays as $\pi^+ \rightarrow \mu^+ + \nu_\mu$, mediated through a W^+ boson as shown in Fig. 1.3.

A quark triplet forms a baryon (from the Greek *βαρυζ* or *barys*, meaning 'heavy'). For example, the proton consists of two up quarks and one anti-down quark and the neutron consists of two down quarks and one anti-up quark. As with the mesons, the baryons are colour neutral (e.g. consist of a red quark, a green quark and a blue quark so that the net combination of colours cancel with the result being referred to as 'white'), and in fact, it was the existence of a baryon, the Δ^{++} resonance at 1,232 MeV, that first provided the argument for the existence of the colour quantum number. Analysis demonstrated that the Δ^{++} was a particle with electric charge $+2e$, intrinsic spin-$\frac{3}{2}$ and isotopic spin-$\frac{3}{2}$[7] – features that were all consistent with three up quarks in the spin-0*s* state, which is a wavefunction symmetric upon exchange. However, a composite of three fermions must have an antisymmetric wavefunction with respect to the exchange of any two quarks. As it was not possible to explain this anomaly through the existence of any other angular momenta, it was the introduction

---

[5] The 'W' refers to the 'weak force' and the 'Z' refers to 'zero' charge. The $W^\pm$ and Z^0 bosons have masses of about 80 GeV and 91 GeV, respectively, and hence, the weak force has an extremely short range. In contrast, the photon has zero mass leading to the electromagnetic force having infinite range.

[6] The neutral pion decays predominantly through the electromagnetic process $\pi^0 \rightarrow 2\gamma$ and, less frequently, through $\pi^0 \rightarrow \gamma + e^+ + e^-$.

[7] Isotopic spin (or 'isospin') was proposed as a new quantum number by Heisenberg in 1932 due to the near equality of the rest masses of the proton and neutron, allowing consideration of them as being two quantum states of the same particle, the nucleon. It is an additive quantum number and is conserved in strong interactions. It is not conserved in weak interactions as shown in the β-decay of the neutron of Fig. 1.3.

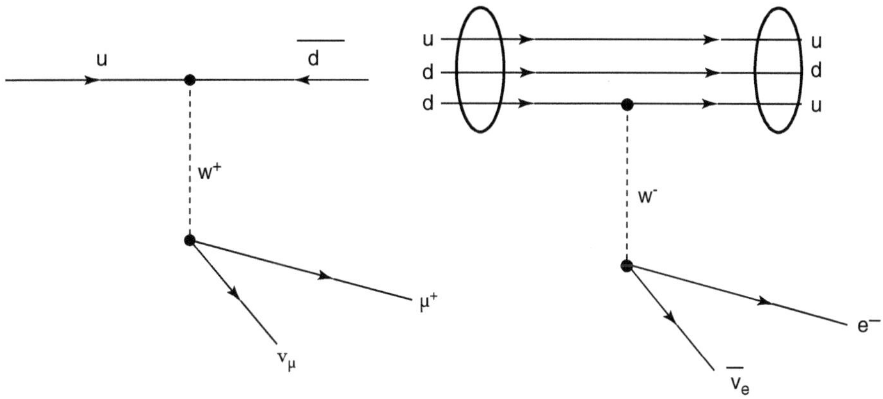

Fig. 1.3 Feynman diagrams of the $\pi^+ \rightarrow \mu^+ + \nu_\mu$ decay and the β-decay of a neutron to a proton. The former decay is a weak process where the up and anti-down quarks of the pion annihilate to form a W^+ boson which then decays to create a positively charged muon and a neutrino. In the latter decay, a down quark transits to an up quark through the emission of a W^- boson which subsequently decays to create an electron and an electron antineutrino

of a new quantum number, that of colour, that was necessary in order to differentiate between the quarks in order to resolve this anomaly.[8]

Of the three generations, only the first comprises stable matter: the electron has an experimentally established lower bound on its lifetime of 4.6×10^{26} years and that for the proton is in excess of 6.6×10^{33} years. As the discussions of this book focus on medical applications, we will consider only the first generation of elementary particles and their composites.

1.2.3 Energy Loss/Transfer, Linear Energy Transfer and Range

In anticipation of the discussion to follow in this chapter, four quantities associated with energy loss will be introduced (although they will be defined in much greater detail in later chapters).

[8] On its own, this would not be a conclusive argument for the existence of the colour quantum number. However, measurements of high-energy electron–positron annihilations demonstrate that the ratio of the cross section at which these annihilations produce hadrons to that at which they produce muon pairs is consistent with the existence of the colour quantum number with three states. Similar confirmation is obtained from measurements of the decay of the neutral pion to two photons.

1.2.3.1 Collision Stopping Power

The transfer of the kinetic energy of a charged particle moving through a medium to the atoms of that medium is primarily through Coulomb interactions with atomic electrons. The linear collision stopping power, $\frac{\mathrm{d}E}{\mathrm{d}x}$, is the rate of kinetic energy loss per unit distance travelled through such collisions. It is a mean quantity. A complete description is necessarily stochastic as the energy loss is through multiple interactions. Although the stopping power describes an energy *loss* and is, hence, a negative quantity, the negative sign is largely omitted in the literature and is taken to be understood. In the medical physics radiation community, the most commonly used unit for the linear collision stopping power is MeV/cm. The mass collision stopping power, $\frac{\mathrm{d}E}{\rho\,\mathrm{d}x}$, is the linear collision stopping power normalised to the physical density of the medium. In medical applications, it has units of MeV cm^2/g. The use of the mass collision stopping power removes much of the density dependence of the stopping power, except for the polarisation of the medium.

The collision stopping power is customarily divided into components. For electrons and positrons, these components are the collision stopping power, which is due to Coulomb collisions with bound electrons, and the radiative stopping power, which is a consequence of the emission of *bremsstrahlung* in the Coulomb fields of the atomic nucleus and atomic electrons. For hadrons and heavy ions, the collision stopping power is dominated by the electronic stopping power due to collisions with atomic electrons and the nuclear stopping power which is due to elastic Coulomb interactions with the atom as a whole. As *bremsstrahlung* is the result of the deflection of the projectile with the result that the radiation yield is inversely proportional to the square of the projectile mass, the radiative stopping power is negligible for heavy particles at kinetic energies typical of medical applications. In addition, as the recoil kinetic energy resulting from a collision is proportional to the ratio of the projectile to target masses, the nuclear stopping power is insignificant for electrons and positrons.

In medical radiation physics, it is customary to further separate the electronic stopping power into soft- and hard-collision stopping powers. The former describes the energy transfer to the atomic electrons as an aggregate, and the latter reflects the energy loss resulting from a direct collision between projectile and a single-atomic electron. The demarcation between soft and hard collisions is defined in the Bohr classical theory in configuration space and, in the Bethe quantum theory, in energy/momentum space.

We will be required to be specific in our use of the terms of 'energy transfer', 'energy loss' and 'energy deposited', although all three are used interchangeably. 'Energy transfer' refers to the kinetic energy lost by a moving charged particle and transferred to an atom (or its atomic electrons) in the medium it is traversing in a single interaction. In this book, this energy will be denoted by Q. 'Energy loss' is the reduction in the kinetic energy of the charged particle, regardless of the energy channel used, through multiple interactions over the distance of interest. It includes radiative energy loss which, as described in the Preface, is largely negligible in medical radiation dosimetry calculations and those cases of the transfer of sufficient kinetic energy to atomic electrons for them to be ejected and deposit energy at points away from the interaction site. This energy will be denoted by ΔE. Note that in those cases where the distance is sufficiently small, only a single interaction can occur $Q = \Delta E$. The 'energy deposited' is that kinetic energy lost by the projectile but which is contained within the region of the interaction site. This is characterised by the restricted collision stopping power, $\left.\frac{\mathrm{d}E}{\rho\,\mathrm{d}x}\right|_{\Delta}$, which excludes the contributions of those interactions with atomic electrons with energy transfers exceeding the amount Δ as such transfers result in energy being transported to a significant distance away from the interaction site.

1.2.3.2 Linear Energy Transfer

The linear energy transfer (LET) is the linear collision stopping power but rescaled to different units, keV/μm. The use of this scale is of interest to radiobiology, microdosimetry and radiation protection.

1.2.3.3 Continuous Slowing Down Approximation (CSDA) Range

The CSDA range of a charged particle of kinetic energy E_0 is the distance over which the particle's kinetic energy goes to zero,[9] $\Re_{\mathrm{CSDA}} = \int_0^{E_0} \frac{\mathrm{d}E}{\left(\frac{\mathrm{d}E}{\mathrm{d}x}\right)}$. The upper and lower limits of integration are E_0 and *0*, respectively, as the collision stopping power is treated as a positive quantity. This expression excludes radiative energy losses. Should these be significant, the linear radiative stopping power is added to the term in the denominator. It is assumed that the energy loss is continuous and that no scatter occurs to deviate the particle from its linear trajectory. The range of a charged particle is also a mean value.

1.2.4 Cathode Rays

The cathode ray was discovered in 1859 by Julius Plücker through his observations of the phosphorescence that occurred on the glass wall in the vicinity of the cathode of an evacuated glass tube across which an electric potential had been applied. As he noted that the position of the site of phosphorescence could be altered should a magnet be brought near it, Plücker concluded that this region of phosphorescence was due to an electric current travelling from the cathode to the glass wall. Further

---

[9] The particle's minimum energy will in fact equalise to the thermal energy at equilibrium which will be greater than zero.

study of this phenomenon was continued by Plücker's student, Johann Hittorf, one of whose observations of note was that the placement of a solid body on the inside of the glass wall of the discharge tube formed a shadow, suggesting that the cause of the phosphorescence had been blocked. This particular observation continued to be studied by others and it was in 1876 that Eugen Goldstein reported that by placing a solid body with an area much smaller than that of the cathode near the cathode, a distinct shadow was cast. This simple geometric observation was proof that the rays producing the phosphorescence were coming from the direction of the cathode. While it was Goldstein who coined the name *Kathodenstrahlen* (cathode rays) for these rays, he regarded them as being waves in the aether rather than a particulate. Goldstein also observed fluorescence in the tube's gas from a perforated cathode on the side opposite to the anode and attributed these to rays moving in the direction opposite to the cathode rays. These were named by him as *Kanalstrahlen* (canal rays) as they were produced by the holes, or canals, in the cathode.[10]

Goldstein's wave hypothesis of the nature of cathode rays was challenged by both Cromwell Varley and William Crookes who proposed, on the basis of experiment, that the cathode rays were particulate in nature and were ejected orthogonally from the cathode at high velocity. In a highly effective and transparent approach, Crookes placed a small turbine into the trajectory of these cathode rays and observed that it rotated in a direction opposite from the direction of the cathode. This demonstrated that these rays carried momentum, which implied strongly their particulate nature. Jean Perrin demonstrated unequivocally that the cathode rays had a negative electric charge by assessing their deflection by a magnetic field. He allowed a cathode ray beam to pass through two holes in two coaxial cylinders, with the outer grounded and the inner connected to an electroscope. The resulting signal demonstrated the negative electric charge of the cathode ray. No signal from the electroscope was found when a magnetic field was used to deflect the cathode rays away from the cylinders.

The particulate hypothesis of the cathode ray had assumed that the associated particle was of molecular or atomic dimensions. However, the experiments of Hertz, which demonstrated that these cathode rays could traverse thin foils of aluminium and gold, suggested that this assumption was unlikely to be correct. The nature of this work was continued by Lenard who studied, in detail, the penetration of these cathode rays in absorbers. Finally, it was JJ Thomson at Cambridge (Fig. 1.4) who concluded in 1897 that the data of Lenard showing the depth of penetration of cathode rays in air was evidence that the size of the particles comprising the cathode rays was much

[10] As these moved in the direction opposite to the cathode rays, we now recognise that these have a positive electric charge. While Goldstein was unable to observe a deflection of these canal rays by a magnetic field, Wien was subsequently able to do so by using a very strong magnetic field and thus demonstrated their positive charge. Further, using combined electrostatic and magnetic fields, he concluded that some of these canal rays had a mass comparable to that of hydrogen; hence, this was one of the first pieces of experimental evidence of the existence of the proton in motion. Being what we now know as positive ions, canal rays had a far more complex composition than the cathode rays which we now know as electrons (Thomson 1913).

Fig. 1.4 JJ Thomson

smaller than that of atomic or molecular dimensions. Clearly, this was a profound statement to have been made in so far as it proclaimed for the first time that there existed a physical entity of subatomic dimensions.

Thomson's investigations led eventually to the definitive proof of the particulate nature of cathode rays.[11] Prior experiments of attempting to deflect the cathode rays by an electrostatic field were imprecise due to the imperfect vacuum of the tubes manufactured for these experiments. Such imperfections meant that the positive ions which had been created by cathode rays colliding with remnant gas atoms within the tube would be attracted to the cathode and create a shielding electric field of charge to reduce the net intensity of the field. But Thomson had created tubes with exquisitely high vacuums so that this field-masking problem did not occur and he could measure accurately the electric deflection of the cathode ray. The combination of electrostatic magnetic fields allowed the measurement of the electron velocity, $\frac{\mathrm{d}\mathbf{r}}{\mathrm{d}t}$, through $m\frac{\mathrm{d}^2\mathbf{r}}{\mathrm{d}t^2} = e\mathbf{E} + e\left(\frac{\mathrm{d}\mathbf{r}}{\mathrm{d}t} \times \mathbf{B}\right)$ where e is the electric charge of the

[11] This was a refutation of the hypothesis that the cathode ray (or electron) was a wave in the aether, not the de Broglie wave–particle duality hypothesis.

electron, **E** is the electric field and **B** is the magnetic field (Thomson 1897). The velocity was found to be of the order of 1.5×10^7 m s^{-1} which, again, implied that the cathode ray mass was likely to be much less than that of an atom. Thomson determined the ratio of the cathode ray mass to charge through two particularly elegant experiments. In the first, he bombarded a target with a known thermal capacity, C, with a beam of cathode rays over a fixed period of time. The total charge carried by the beam is

$$Q = Ne \tag{1.1}$$

where N is the number of the particles making up the cathode ray beam during the time of irradiation and e is the electric charge carried by each. For an anode that is much thicker than the range of the cathode rays, all of the kinetic energy of this bolus of cathode rays will be absorbed by the target. The temperature of the target will, as a result, be raised by an amount ΔT determined through

$$\frac{1}{2}Nmv_{||}^2 = C\,\Delta T \tag{1.2}$$

where m is the mass of the individual cathode ray particle and $v_{||}$ is the velocity (the need for the || subscript to identify that the velocity is parallel to the cathode–anode axis will become soon apparent). As the cathode ray was known to be charged and deflected in a magnetic field B, the product of its radius of curvature in the field and B is given by

$$Br = \left(\frac{m}{e}\right) v_{||}. \tag{1.3}$$

This quantity, the particle momentum per unit charge, is also known as the magnetic rigidity. From these equations, the mass-to-charge ratio of the elementary particle making up the cathode ray is given in terms of measurable quantities:

$$\frac{m}{e} = \frac{Q\,(Br)^2}{2C\,\Delta T}. \tag{1.4}$$

Thomson's second experiment used coterminous electrostatic and magnetic fields. Consider the cathode ray (particle) of velocity $v_{||}$ traversing a length l across which an electric field $\mathcal{E}$ orthogonal to the direction of travel exists. The time that the particle spends within the field is $\frac{l}{v_{||}}$ during which it will experience a force $e\mathcal{E}$ perpendicular to the direction of travel. As the force is constant over this distance, the net change in velocity normal to the direction of travel is

$$\Delta v_{\perp,E} = \left(\frac{e}{m}\right)\mathsf{E}\left(\frac{l}{v_{||}}\right) \tag{1.5}$$

The angle of deflection caused by this electric field is

$$\begin{aligned}\theta_E &= \frac{\Delta v_{\perp,\mathsf{E}}}{v_{||}} \\ &= \frac{\left(\frac{e}{m}\right)\mathsf{E}l}{v_{||}^2}\end{aligned} \tag{1.6}$$

A magnetic field B was then introduced and directed along the axis mutually perpendicular to both the electric field and the cathode–anode axis. The net change in the velocity orthogonal to the cathode ray's direction (and antiparallel to that caused by the electron field) as a result of the magnetic field is

$$\Delta v_{\perp,B} = \left(\frac{e}{m}\right) Bl \tag{1.7}$$

with an associated angle of deflection:

$$\begin{aligned}\theta_B &= -\frac{\Delta v_{\perp,B}}{v_{||}} \\ &= -\left(\frac{e}{m}\right)\frac{Bl}{v_{||}}.\end{aligned} \tag{1.8}$$

From (1.6) and (1.8), an expression for $\frac{m}{e}$ is obtained:

$$\frac{m}{e} = \frac{B^2 l}{\mathsf{E}}\frac{\theta_E}{\theta_B^2} \tag{1.9}$$

By adjusting the magnetic field B, Thomson was able to obtain the force-equalising condition of $-\theta_E = \theta_B$ (i.e. cancelling the deflection caused by the electric field) so that (1.9) yields

$$\frac{m}{e} = \frac{B^2 l}{\mathsf{E}\theta_E} \tag{1.10}$$

which also provides the mass-to-charge ratio in terms of measurable quantities. In both experiments, Thomson obtained a value for $\frac{m}{e}$ equivalent to about 5.7×10^{-12} kg/C for the cathode ray, a value which is some three orders of magnitude less than that of the 1.6×10^{-8} kg/C value that had been prior known as the smallest $\frac{m}{e}$ and which had been obtained for the hydrogen ion (proton) in electrolysis.

Finally, some 38 years after its discovery, Thomson was able to present the fact that the cathode ray was a particulate, with a negative electric charge and a mass of about 1/2,000th that of the hydrogen atom. He labelled this particulate a 'negative corpuscle', but this was subsequently renamed as the 'electron', a name coined earlier in 1894 by George Stoney as the fundamental unit of electric charge. The first subatomic particle and its basic characteristics of mass and electric charge had been identified and characterised. The electron, and its rate of energy loss as it moves through matter, will be very much a focus of this book. But we shall now look at the creation of other moving charged particles of relevance to medicine.

1.2.5 α- and β-Particles

> Radioactivity is shown to be accompanied by chemical changes in which new types of matter are being continually produced – The conclusion is drawn that these chemical changes must be sub-atomic in character.
>
> —Ernest Rutherford

Although the α- and β-particles are two entirely different physical entities, a doubly ionised helium atom and an electron (or positron), respectively, their coincident discoveries and investigations require that they be examined in parallel.[12]

Whereas the electron was the first subatomic particle to have been discovered, the manner of its production was artificial in that it required the thermionic emission of electrons from a cathode and their acceleration within an evacuated tube towards an anode. On the other hand, natural radioactivity, discovered serendipitously by Henri Becquerel in 1896 by noting the effects of the emissions of a uranium salt on photographic emulsion (Becquerel 1896), is a natural source of subatomic particles which has played and continues to play significant roles in medical radiation physics. Radioactive decay of nuclei leads, in the main, to the production of α- and β-particles. The medical application of the decelerated α-particle resulting from radioactive decay, which is the ^{4}He nucleus, is exclusively therapeutic due to its short range in tissue. Over its short range, it deposits its kinetic energy tissue within a small volume consisting typically of only a few cells and yields a high and localised absorbed dose. As the short range of the α-particle does not enable it to exit tissue and be detected externally in order to form a diagnostic image, its use is relegated to the treatment of malignant disease in radionuclide therapy, as will be discussed.

---

[12] A review of the theory and phenomenology of α-decay is provided by Chapter 4 of McParland (2010); that of β-decay can be found in that same chapter and in the comprehensive review by Grotz and Klapdor (1990) and the classic text by Lipkin (1962).

Fig. 1.5 Ernest Rutherford (*right*) and Hans Geiger (*left*) (Courtesy and © of Alexander Turnbull Library, Wellington, New Zealand)

On the other hand, the β-particle, which is either an electron or a positron that is emitted through nuclear decay, has both diagnostic and therapeutic applications in medicine. In its modern application, the positively charged β-particle, the positron, can annihilate with a nearby electron, resulting in the production of a pair of collinear γ-rays each with an energy equal to the rest masses of the electron and positron and which can be detected externally in order to form a diagnostic image in what is referred to as positron emission tomography (PET). On the other hand, the negatively charged β-particle, the electron, decelerates and produces an absorbed dose. The electron will tend to have a greater range than α-particles and, hence, the associated absorbed dose spatial distribution will be less localised. However, unlike the α-particle in traversing a medium, the electron can also undergo radiative loss through the production of *bremsstrahlung* which, in special circumstances, can be imaged in soft tissue, although this is rarely performed in nuclear medicine clinical practice.

The discovery of, and the assignation of the nomenclature to, α- and β-radiation from radioactive decay is attributed to Ernest Rutherford (Fig. 1.5), although he had initially misunderstood the physical nature of both.

The discoveries of these two radiations evolved out of his experimental work through which he measured the differential penetrations through aluminium foils of the emitted radiations resulting from the nuclear decay of uranium. Even though the partitioning of the emitted radiation into two components was Rutherford's discovery, Becquerel had earlier observed that the radiation emitted by the uranium salt appeared to consist of components which had differential absorption properties.

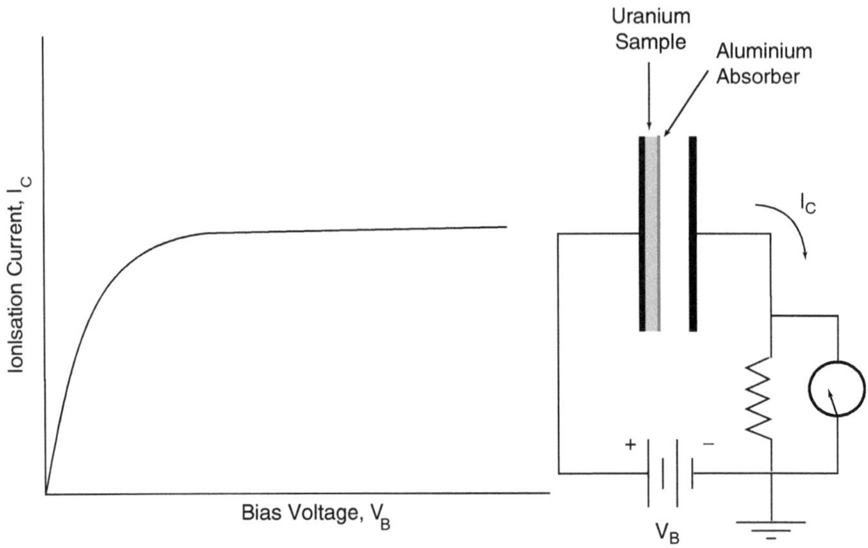

Fig. 1.6 Conceptual diagram of the ionisation detection apparatus used by Rutherford to identify the α- and β-particle components of the radiation emitted by uranium. As the two components had different electric charges, their detections required that the bias voltage be reversed for each (Refer to text for details)

A significant feature of this experimental work leading to the discovery of the α- and β-radiations was the nature of the radiation detector used by Rutherford. Prior work in the nascent field of radioactivity research had been dominated by the use of the photographic emulsion as a detector. Emulsion consists of silver bromide grains suspended in a gelatine matrix which are sensitised when radiation deposits energy in them through ionisation. These grains remain in this sensitised state, thus storing a latent image of the ionisation spatial pattern. During the development process, these sensitised grains are converted to metallic silver which affects surrounding silver molecules and increases the visibility of an ionisation, affecting a single silver bromide grain. Rutherford, however, considered emulsion as a slow and inefficient means of detecting radiation. In order to achieve the radiation detection sensitivity he required, Rutherford instead opted for gaseous ionisation detectors following his and Thomson's earlier studies of the ionisation of gases exposed to x-rays (Thomson and Rutherford 1896; Rutherford 1897). Such observed ionisation was the product of photoelectric absorption and Compton scattering of photons by air molecules which resulted in the production of electrons and ions. The basic structure of the experiment performed by Rutherford in identifying α- and β-particles is shown in Fig. 1.6. This detector was an air ionisation chamber operated in the saturation mode as shown by the voltage–current curve in that figure. At low bias voltages, the electrons and positive ions that are produced following ionisation of the air molecules by the radiation have low drift velocities which result in a high probability of remaining within physical proximity and recombining, leading to a low ionisation current, I_C. With an increasing bias voltage, the electrostatic field becomes sufficient to separate the electrons and positive ions prior to recombination. The current I_C thus increases with bias voltage to reach a plateau (the 'saturation mode') reflecting the condition for which all of the ionisation products are collected.

In Rutherford's experiment, a thin layer of a uranium salt placed on one electrode provided the source of ionisation and aluminium absorbers of various thicknesses were placed over the uranium salt, and the changes in I_C were then measured as a function of the absorber thickness. Rutherford used an exponential model of the variation of the ionisation current with absorber thickness, $I_C = e^{-\mu t}$, where t is the thickness of absorber, and analysed the variation of the absorption parameter μ as a function of t. Charged particle radiations were stopped when the collision energy losses exceeded the particle's incident kinetic energy. He determined that uranium emitted two types of radiations of different penetrability which he referred to as α and β in increasing order of penetration of the aluminium foil (Rutherford 1899). Despite what is commonly thought, Rutherford did not realise these to be the charged particles we now know them to be but, instead, considered them to be x-rays of two different energies. The differential penetration of the two particles in the foil is an immediate reflection of the collision stopping power, which is proportional to the square of the particle's charge and the reciprocal of its mass.

Rutherford's hypothesis of α- and β-radiations being photons of different energies began to succumb through the analysis using a magnetic field by Pierre Curie. The curvature of the trajectories of the β-radiations in the magnetic field demonstrated that this entity had an electric charge and, hence, was not a photon. Moreover, the direction of the trajectory revealed that it had a negative electric charge, and subsequent measurements of the mass to charge, $\frac{m}{q}$, proved unequivocally that the β-radiation was made up of fast electrons.[13] On the other hand, the α-radiation could not be detected as deflecting in the magnetic field. This was the result of a combination of a weak magnetic field generated by the apparatus and the heavy mass of the α-particle. As the orbital radius of a particle with mass m, charge q and velocity v in a magnetic field B is $r = \left(\frac{m}{q}\right)\left(\frac{v}{B}\right)$, this combination of low magnetic field strength and high particle mass led to high values of r and the resulting imperceptible curvatures. Thus, further experimental investigation was required in order to identify the nature of the α-radiation.

It was the Curies who continued the investigation by using intense radioactive sources of radium and polonium which produced copious fluxes of α-particles. Of

[13] Whereas in our previous discussion of Thomson's discovery of the electron we had used e as the symbol of the electric charge of the entity discussed – and which was entirely appropriate as that is the electric charge carried by the electron – we will now use q as the generic symbol for electric charge. In our nomenclature of the interactions of charged particles with matter, the projectile will have an electric charge ze where z is an integer equal to 1 or greater and the target will be assigned an electric charge Ze where, again, Z is an integer equal to 1 or greater.

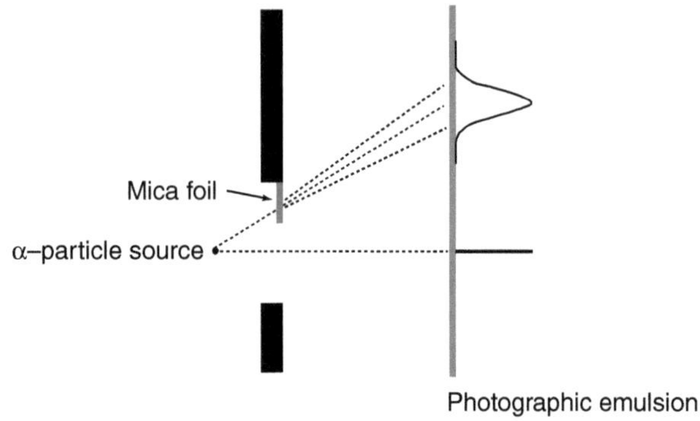

Fig. 1.7 Conceptualisation of the Rutherford experiment demonstrating that α-particles are scattered and spatially distributed by matter and the evaluation of the electric field strength required to achieve this (See text for details)

particular interest to our subject of charged particle energy loss, Pierre Curie measured a finite range of 6.7 cm in air for the α-radiation emitted by radium salts. Beyond that distance, the α-radiation failed to produce any ionisation in the detector. He found that this range was independent of the activity of the radium or of its chemical properties (Curie 1900a). We now know that this finite range is a function of the kinetic energy of the α-radiation and the properties of the absorbing material (in this case, air). The activity and chemical composition of the source of α-particles could in no way bear any influence on the result. Independently, Marie Curie measured the relative differential loss in ionisation of the α-radiation, $\frac{\Delta I}{I}$, and found that this ratio of ionisation loss *increased* with penetration distance (Curie 1900b). Unknowingly, she had discovered that the collision stopping power is inversely proportional to the square of the particle's velocity – that is, the slower a particle moves through a medium due to previous energy transfers to the medium, the more rapidly it loses kinetic energy. While at the Adelaide University, William H Bragg (Bragg and Kleeman 1905) continued this work and showed that this energy deposition increased momentously near the end of the α-particle's range. This feature is referred to as the *Bragg peak* and is one of the important elements of hadron therapy, as to be discussed.

In a further series of experiments, Becquerel was able to demonstrate that the α-radiation was not readily scattered in air and made diffuse. Using polonium as a source of α-radiation, he was able to image the sharp shadow of a wire on a photographic plate. We now recognise this inability to detect scattering in air as a consequence of the combination of the α-particle's heavy mass and the low physical density of air.

However, Rutherford was able to demonstrate that α-particles could be scattered by materials and, in so doing, determined the large electric fields involved by using the experimental arrangement of Fig. 1.7.

A sharply defined source of α-radiation was created using a wire coated with radium and a photographic emulsion was used as the detector. The α-particles were allowed to pass through a slit in a collimator and impinge on the emulsion. A thin sheet of mica with a thickness equivalent to that of 3.5 cm of air covered part of the slit. α-particles passing through the air and the foil would undergo scattering as a consequence of Coulomb interactions with the atoms and which would result in the deflection of the α-particles from their original trajectories. As to be derived in Chap. 4, the probability of Coulomb scatter through the angle θ is proportional to $\frac{1}{\sin^4\left(\frac{\theta}{2}\right)}$. Hence, the scattering will be predominantly forward directed. Along the extended pathway of air, the α-particles will undergo multiple scatter with a net zero scattering angle integrated over the entire path length at the position of the detector. Those that pass through the mica foil will also be scattered but, due to the lever-arm effect and the mean zero scattering through air as shown in Fig. 1.7, will demonstrate an angular deviation at the position of the emulsion. By following the same route of development of (1.6), it can be readily shown that the electric field within the atoms of the mica to cause such a deflection is given by the expression

$$\mathsf{E} = \frac{E_0\theta}{el} \tag{1.11}$$

where E_0 is the kinetic energy of the α-particle, θ is the mean angle of scattering, e is the fundamental unit of electric charge and l is the linear thickness of the electric foil. An order-of-magnitude estimate of the strength of this electric field is obtained readily from (1.11). For example, the α-particle's kinetic energy will be of the order of 5 MeV and the measured mean scattering angle was reported to be 2° or 35 mrad. The air-equivalent thickness of the mica yields an areal density of about 4.2 mg/cm^2 which, for a density of 3 g/cm^3 for mica, yields a thickness of 14 μm for the mica foil. Using these values in (1.11) yields an electric field E value of 1.25×10^8 V/cm, thus demonstrating how a simple experiment can yield the magnitude of the Coulomb interaction causing the scatter and which will dominate the discussions of this book.

Rutherford had earlier resumed his examination of the problem of identifying the nature of the α-particle with the use of an intense magnetic field and a particularly intense radium source that had been provided him by the Curies. He demonstrated that the α-radiation was made up of positively charged particles and led to the suggestion, from the charge-to-mass ratio, that the α-particle could be the ^{4}He nucleus (Rutherford 1903). This proof arose from the experimental results of Rutherford and Royds (1909). By collecting the α-particles from a glass tube containing radon gas that had traversed the extremely thin wall of the tube (and collected electrons from the glass in the process to be neutralised), compressing the resultant gas, and then by observing the optical spectrum resulting from an electric discharge through the captured gas, they demonstrated that the collected gas was helium. In conclusion, the α-particle was a cluster composed of, as we now know, two neutrons and two protons.

Experimental investigations of the rates at which α- and β-particles lost velocity while traversing material began in earnest at around this time at the Cavendish Laboratory in Cambridge and at Manchester. This work can probably be considered to have begun with Rutherford in 1906 using α-particles, with further refinements by Darwin and Geiger. The latter had determined that the α-particle velocity v as it

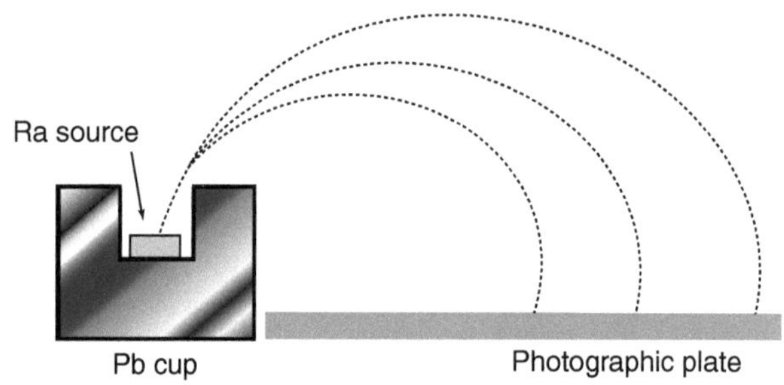

Fig. 1.8 Experiment used by Becquerel to study the spectrum of β-particle velocities

traverses air could be described by the empirical relationship $v_0^3 - v^3 = bt$, where v_0 is the original velocity, t is the thickness of air and b is a constant of proportionality. Experimental study of the β-particle proved to be more difficult than those of the α-particle. This difficulty was a consequence of the greater amount of scatter due to the β-particle's much lower mass, the confounding effects of characteristic x-ray production caused by atomic relaxation resulting from its ionisation of the medium, *bremsstrahlung*, and the fact that the β-particles exhibited a velocity spectrum due to what we now know as the consequence of the three-body final state of nuclear β-decay, ${}^{A}Z \rightarrow {}^{A}(Z+1) + e^{-} + \bar{\nu}_e$. But even at this early time, elegant experiments were capable of extracting much detail of this phenomenon. Strutt's measurements of the assumed exponential absorption of β-particles failed to find an absorption parameter μ that was independent of absorber thickness (Strutt 1900). This suggested a heterogeneity in the energies of the β-particles emitted from β-decay. Perhaps one of the simplest, yet most powerful, experiments to demonstrate the fact that the β-particles had a momentum spectrum was that of Becquerel as shown in Fig. 1.8. A radium source was placed within a lead cup and a magnetic field applied. The β-particles emitted during the radium's decay were then bent into orbits impinging on a photographic plate so that the physical spread of the darkening of the film was reflective of the spread of β-particle velocities (Becquerel 1903).

1.2.6 Internal Conversion and Coster–Krönig Electrons

Nuclear β-decay is not the only means by which electrons are emitted in a radioactive process. Non-radiative transitions and atomic relaxation processes can also result in the production of moving electrons. Unlike β-particles, these electrons are monoenergetic, while internal conversion electrons can have energies and ranges comparable to those typical of nuclear β-decay; Auger and Coster–Krönig electrons have, in general, much reduced energies and ranges.

Internal conversion (IC) is a non-radiative decay process which may be made available to an excited nucleus. As low-orbital (*K* and, to a lesser degree, *L*) electrons have wavefunctions that can overlap that of the nucleus, it is possible for the nucleus to transfer energy to the electron directly without the presence of an

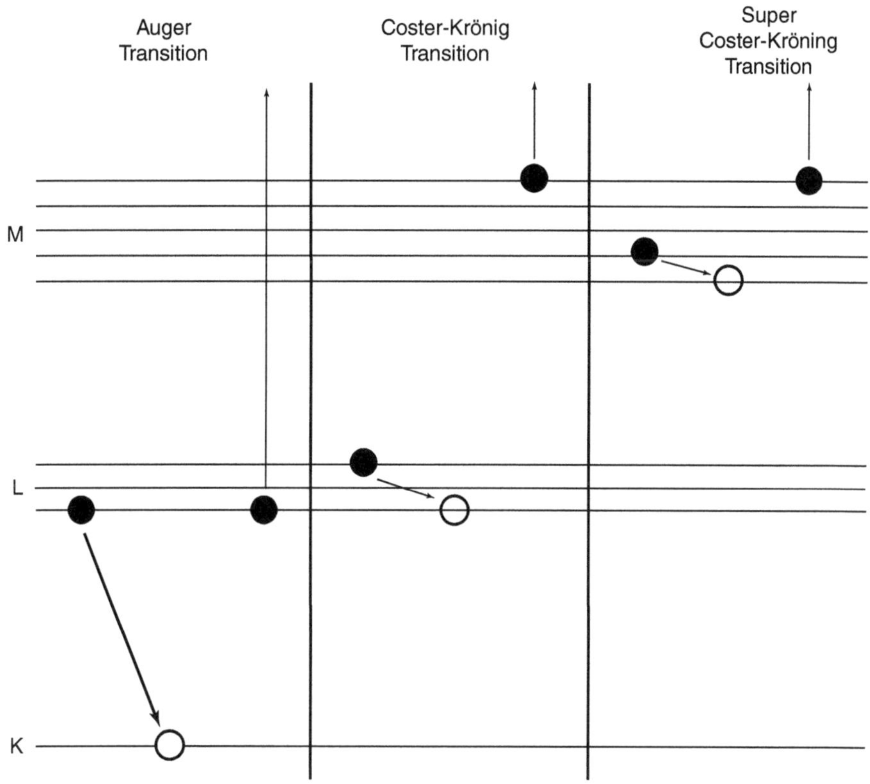

Fig. 1.9 Non-radiative Auger and Coster–Krönig transitions

intermediate γ-ray, but through a virtual photon. In other words, IC is not a photoelectric absorption event in which a γ-ray is emitted by the de-exciting nucleus and absorbed by an atomic electron.[14] The electron is subsequently ejected with energy equal to the difference between the nuclear excitation energy and the atomic electron's binding energy. The IC electron plays a significant role in internal radiation dosimetry calculation (Smith et al. 1965).

A key consequence of the ejection of an IC electron is that of a hole in the orbital from which the electron was removed. This will result in the transition of an electron from a higher orbital to fill this hole (Fig. 1.9). The transition can be radiative with the emission of a characteristic x-ray with an energy equal to the difference between the two orbital binding energies. However, this energy can also

[14] That this is the case demonstrated by the existence of transitions between nuclear states with zero angular momentum ($0 \rightarrow 0$) resulting in IC, whereas electromagnetic ($0 \rightarrow 0$) transitions without the involvement of an electron are impossible since there are no multipoles with zero angular momentum in the radiation field. See Section 4.4.3 of McParland (2010) for a proof of this derived using perturbation theory.

Table 1.2 Total Auger-electron yield for radionuclides commonly used in nuclear medicine

Radionuclide	Mean yield of Auger electrons per decay
^{67}Ga	4.7
^{99m}Tc	4.0
^{111}In	14.7
^{123}I	14.9
^{201}Tl	36.9

Data are from Howell (1992)

be directed towards another electron which is then ejected to create another hole. Another transition can occur to fill this new hole and the difference in energy between the two orbitals transferred to yet another electron which is ejected, etc. This process is known as the Auger cascade and results in the ejection of multiple low-energy electrons. Auger electrons were discovered and explained by Lise Meitner in 1923, but only in the course of a longer treatise on the β-decay spectrum of ^{234}Th which, as a result, this discovery received limited attention. Two years later, Pierre Auger discovered these eponymous electrons independently by observing their tracks in a Wilson cloud chamber.[15]

If the electron vacancy is in a different orbital, the process is referred to as an Auger transition. Coster–Krönig and super Coster–Krönig transitions occur within suborbital levels. A Coster–Krönig transition is one in which the transiting electron and vacancy are in the same main orbit, but the ejected electron is from a different main orbit. A super Coster–Krönig transition is one in which the transiting electron, vacancy and ejected electron are all from the same main orbit.[16]

The absorbed doses associated with Auger transitions can be significant and, if in the vicinity of a radiosensitive biological target, lead to significant biological effects (Boswell and Brechbiel 2005). Auger electrons will tend to have low kinetic energies and, subsequently, low ranges of penetration into tissue. Should an Auger electron-emitting radionuclide be brought into close proximity of the radiosensitive centre of a cell (the nucleus), significant damage leading to cell death can occur. This could be a desirable effect for therapy but could also be of radiation protection concern in diagnostic applications if the diagnostic radiopharmaceutical uses a radionuclide which is an Auger-electron emitter. Moreover, another result of an Auger cascade is that of a highly charged ion which, if in the vicinity of radiosensitive cell structures, can also cause biological damage. Hence, the roles played by Auger electrons in therapeutic nuclear medicine and in the radiation protection considerations of diagnostic nuclear medicine are of considerable interest due to their low energy and the $\frac{1}{\beta^2}$ dependence of the collision energy loss. Table 1.2 summarises the Auger-electron emission characteristics of a number of radionuclides used in nuclear medicine.

[15] This was not the only time that Meitner had been supplanted or excluded in a major physics discovery, as will be discussed later in the synopsis of the discovery of nuclear fission.

[16] For convenience, we will group Auger, Coster–Krönig and super Coster–Krönig non-radiative transitions as Auger transitions in later discussions.

1.2.7 Protons

The proton is the lightest baryon and its existence had been inferred as early as 1808 through Dalton's gravimetric weight measurements which had implied that the elements were made up of integral constituents. This was followed by Proust's hypothesis in 1815 that all elements were integral multiples of the hydrogen atom, which he had termed as the *protyle*. The proton had been recognised as being a hydrogen ion through, for example, electrolysis. It was inferred as a nuclear constituent gravimetrically and, to some degree, by Rutherford's interpretation of the results of Geiger's and Marsden's experiment of scattering α-particles from gold foils. The definitive proof of the existence of the proton as a subnuclear particle can be traced to Rutherford's investigations of the scintillations observed on a zinc sulphide screen at one end of a glass tube and a radium source of α-particles at the other. The tube could be filled with a variety of gases. Critically, the scintillation screen was placed at a distance from the radium source beyond the α-particle range, and a transverse magnetic field was applied to deflect β-particles emitted by the radium decay from reaching the screen (Rutherford 1919). As a consequence, any scintillations observed would be due to any new particles that had been created by interactions between the α-particles and the gas and which had a range greater than the α-particle and a magnetic rigidity exceeding that of a β-particle. When the tube was filled with either dry air or nitrogen, Rutherford observed scintillations which satisfied the criteria of the range and mass of the secondary particle sought. As the former is dependent on electric charge, this result indicated that these scintillations were due to a charged particle with a unit charge of e and a mass intermediate to that of an electron and that of a $^{4}He^{++}$ ion, in other words, the proton. As the rate of scintillations occurring was 1.25 times greater for pure nitrogen than for air, it was deduced that these scintillations were caused by the interactions between the α-particles and the nitrogen nuclei. On the basis of this observation, Rutherford hypothesised that the nitrogen nucleus was composed of three helium nuclei and two hydrogen nuclei (recall that this experiment was performed before the discovery of the neutron by Chadwick). In this case, he conjectured that the α-particle collided with the nitrogen nucleus and imparted sufficient momentum to a valence proton to eject it from the nucleus.

In medical applications, protons have figured significantly in hadron radiotherapy over the past 60 years and are the most common radiotherapeutic modality after photons and electrons. This is a result of their heavy mass which leads to the Bragg peak which has a number of therapeutic advantages. We will return to discuss proton therapy in detail later in the chapter.

1.2.8 Heavy Ions (Fission Fragments)

Our discussion of heavy ions will exclude the low-energy α-particles emitted in radioactive decay of heavy nuclei but will instead consider more massive ions which, in modern medical research practice, are generated by accelerators. Even though this is the case, we can begin our historical consideration of the interactions of heavy ions with matter through the historical development of the theory in the context of fission fragments. Although being both densely ionising and heavy ions, fission fragments have not been used in medicine but the theoretical analyses of their interactions with matter have. Heavy ions, produced through accelerators, have the therapeutic advantage of the associated Bragg peak and dense ionisation and have been under medical investigation since the late 1950s.

The first high-energy heavy ions to be discovered were the fragments resulting from the fission of uranium nuclei, discovered in 1938 through, at first, radiochemical means and then confirmed through ionisation measurements. Up to that time, it had been considered possible to cleave off only small components of the nucleus such as individual nucleons and, at most, α-particle clusters. Heavy nuclei, such as that of uranium, when bombarded by a projectile (most notably a neutron), can fission into two fragments. The fragments are massive and asymmetric in mass, e.g. ^{97}Kr and ^{139}Ba being the most probable pair of post-fission fragments from the fission of ^{238}U. As can be seen, there is a difference between the sum of the atomic mass numbers of the products and the original nucleus which reflects the evaporation of neutrons. Having come from a heavy nucleus, the fission fragments are themselves neutron rich and undergo a subsequent series of β^--decays along chains of isobars to reach the stable nucleus at the minimum of the mass parabola for the relevant atomic mass number.[17] However, our interest lies in the facts that these fragments are massive and highly charged, and why this interest should be so and should be relevant to medical applications is revealed below.

The discovery of nuclear fission was presaged by the neutron absorption measurements of Fermi and his group in Rome during the mid-1930s. Their discovery of radioactive products following neutron irradiation of uranium was attributed to new transuranic elements which underwent subsequent α- and β-decays. As the Fermi group's analysis had terminated at the element lead, the German chemist Ira Noddack challenged their conclusion that a transuranic element had been created (Noddack 1934) and suggested that the heavy uranium nucleus had been fragmented into two or more heavy products. This suggestion by a chemist was largely dismissed by physicists. However, the German radiochemists Otto Hahn and Fritz Strassmann and their Austrian physicist collaborator Lise Meitner determined, on the basis of radiochemistry, that Noddack's hypothesis of fission was correct.[18]

---

[17] Refer to Chapter 3 of McParland (2010) for the derivation of the mass parabolae and the $\beta^{\pm}$-decays that reach the stable nucleus at the minimum of the parabola.

[18] Being an Austrian Jew, Meitner had been forced to leave Germany for Sweden following the *Anschluss* in 1938. She continued to collaborate with Hahn and Strassmann and during Christmas of 1938, with her nephew Otto Frisch, modelled fission on the basis of Bohr's liquid-drop model of the nucleus and accurately calculated that the resulting fission fragments would carry a net kinetic energy of about 200 MeV due to both mass defect and Coulomb repulsion. Hahn received the 1945 Nobel Prize in Chemistry for the discovery, but Meitner did not. Controversy over that decision remains.

The experimental confirmation of fission by the measurement of ionisation in air caused by the fission fragments followed soon afterwards. In a conceptually similar experiment to that of Rutherford (Fig. 1.6), Frisch placed a uranium-coated foil within an air ionisation chamber, the ionisation current of which was amplified and displayed on an oscilloscope. He exposed the foil to a neutron source and observed the high pulses created by the densely ionising fission fragments almost immediately. The amplitudes of the ionisations corresponded to energies of between 100 and 200 MeV, in agreement with the Meitner's calculation.[19]

As these fragments are more massive, highly charged and energetic than the α- and β-particles and nucleons which had been studied up to that time, considerable interest was developed in evaluating their stopping properties within a medium. Because of their high electric charge and their ability to carry atomic electrons with them, these ions are able to shed electrons or capture further atomic electrons from the medium with the result that their net electric charge can vary as they move along their trajectory. The first measurements of the effective charge of fission fragments using magnetic fields suggested that the value of the effective charge was about 20. As the energy loss of a moving charged particle is proportional to the square of the particle's electric charge, the energy deposition by these fission fragments will be rapid and within a small volume at the end of the Bragg peak.

Quickly following the above experimental work, initial calculations of the collision energy losses of heavy ions were presented by Bohr (1940), Bohr et al. (1940), Lamb (1940) and by Knipp and Teller (1941). These analyses of fission fragment energy losses remain the foundations of the dosimetry-related calculations for heavy-ion therapy. However, the definition of what is a *heavy ion* will be dependent upon the application and varies between the disciplines. While a $^{4}He^{++}$ ion would likely not be considered as being *heavy*, a radiation oncologist would term a ^{12}C ion as being *heavy*, whereas a nuclear physicist would not (certainly in comparison to a fission fragment). Because the energy deposited by a charged particle of mass m and charge z per unit distance travelled within a medium is proportional to the factor $\frac{mz^2}{E}$, where E is the particle's kinetic energy, heavy ions have long appeared particularly attractive in radiation oncology.

Much of the pioneering work of the medical use of heavy ions which began in the 1950s was performed at the Bevatron, which was a proton synchrotron at the Lawrence Berkeley Laboratory of the University of California. Beginning in 1971, it was also used to accelerate heavy ions, but in 1974, it was combined with the SuperHILAC linear accelerator as an injector to produce beams of relativistic heavy ions. The combination was referred to as the Bevalac. Ions as massive as ^{40}Ar were accelerated to kinetic energies between 50 MeV/nucleon to 2.1 GeV/nucleon for

[19] Frisch was staying at the Bohr Institute when he performed these measurements there. Also staying there was an American biologist, William A Arnold, who was approached by Frisch to describe the division of cell. Arnold replied that this was 'binary fission' and Frisch adopted the term 'fission' to describe the process he had just measured.

research into therapy and radiobiology. The Bevalac was decommissioned in 1993. In modern practice, heavy-ion therapy is limited to ^{12}C ions. These applications are discussed later in this chapter.

1.2.9 *π-Mesons*

The existence of the π-meson as the exchange boson of the strong nuclear force was predicted by Hideki Yukawa in 1935 (Yukawa 1935) and was based upon the empirical evidence of the increase in nuclear binding energy with atomic mass number, A. It had been suggested by Wigner (1933) that this behaviour was indicative of an attractive nuclear potential with a range of the order of about 1–2 fm. Yukawa concluded that the attractive nuclear potential was mediated by a particle which was an analogue to the photon in electromagnetic interactions. However, whereas the electromagnetic force has infinite range due to the photon being massless, the mediating particle of the nuclear force must have a non-zero mass as demonstrated by the finite range of that force. It is possible to infer the mass of this particle on the basis of the magnitude of this range and the Heisenberg uncertainty principle. For a nuclear range R of the order of 1.5 fm, the mass of this mediating particle would be approximately

$$\begin{aligned} m &= \frac{\hbar c}{R} \\ &\approx \frac{197.3\,MeV\,\text{fm}}{1.5\,\text{fm}} \\ &\approx 130\,\text{MeV} \end{aligned}$$

The particle was referred to as a meson in that its predicted mass was intermediate between that of an electron (the lightest particle) and the nucleon. Experimental evidence of this meson, known as the π-meson (or *pion*), was obtained in balloon-borne emulsion measurements (Lattes et al. 1947). The pion is a trinity in that it comes in two charged forms and a neutral form, as described earlier in terms of combinations of quark pairs. The two charged pions, each having a mass of about 139.6 MeV, undergo decay to muons and (anti)neutrinos via a weak interaction as shown in Fig. 1.3, whereas the neutral pion, with a mass of 135 MeV, decays through the electromagnetic interaction.

1.2.10 *Neutral Radiations: Photons and Neutrons*

The photon is ubiquitous in its use in diagnostic and therapeutic medicine and it, and the neutron (which is used only in therapy), sets into motion within tissue secondary fluxes of charged particles. It is these particles which then create directly an absorbed dose through collision losses with the atoms in the medium. In the case of the photon, this secondary flux is composed primarily of electrons created through photoelectric absorption and Compton scatter. At photon energies above 1.022 MeV, positrons can also form part of this flux through the pair production process; and, at very high photon energies, protons can be a small component due to photonuclear effects. The secondary charged particle flux in the case of the neutron is composed of protons generated through (n,p) elastic scatter in the hydrogenous tissue. We review now the history of the discoveries of the photon and the neutron.

1.2.10.1 Photons

> I have been for a long time interested in the problem of cathode rays from a vacuum tube as studied by Hertz and Lenard. I had followed theirs and other researches with great interest, and determined, as soon as I had time, to make some researches of my own. This time I found at the close of October [1895].
>
> **WK Röntgen recounting on what led to his discovery of x-rays on 8 November 1895 – some 8 days after he found 'the time'.**

In medical radiation physics, high-energy photons are defined as being either x- or γ-rays on the basis of the means of their production, rather than of their energy as is common in other areas of physics. X-rays are considered to be either the high-energy *bremsstrahlung* photons caused by the interactions of a moving electron with the Coulomb field of an atomic nucleus and which has a continuous energy spectrum dominated by lower-energy photons or are the monoenergetic characteristic photon radiations which are the consequence of atomic electron transitions. γ-rays are photons produced either by nuclear transitions and, hence, are monoenergetic or are the result of positron–electron annihilation. In nuclear transitions, the probabilities and angular distributions of the emitted γ-radiation are functions of the quantum-mechanical states between which the transition is made.

The serendipitous discovery of x-rays by Röntgen and the almost immediate effect it had upon the medical community will almost certainly be well known to the reader. Röntgen had been Professor of Physics and Director of the New Physical Institute at the University of Wurzburg for 7 years prior to when he made this discovery on the evening of 8 November 1895 (Brailsford 1946). He had connected a Crookes tube[20] across a Ruhmkorff induction coil[21] with a 20-ampere current through the primary. The tube had been encased within an opaque paper box. Röntgen noted a fluorescence from a piece of barium platinocyanide-coated paper

[20] The Crookes tube is a partially evacuated tube in which electrons are created through the ionisation of residual air by a high-tension supply. These are accelerated by this potential between that cathode and anode.

[21] This is a step-up transformer in which the primary coil is connected to an interruptible DC voltage. The change in the primary voltage causes a voltage in the secondary coil through induction.

in the vicinity of the tube. Because of the distance and positioning of the paper from the tube, it was not possible that this was due to the cathode rays exiting from the Crookes tube. By interposing various materials between the tube and the paper, Röntgen was able to characterise systematically the degree of absorption of these rays by matter. It was also during this time that he interposed his hand between the Crookes tube and the barium platinocyanide paper and observed the bones within his hand on the paper, thus creating the first fluoroscope. Following an intense degree of experimental work lasting up to 28 December 1895, he presented his work on 6 January 1896.[22] The diagnostic applications of x-rays were immediately apparent, especially those of the identification of bone fractures and the location of dense foreign bodies. On the other hand, therapeutic applications were considered only after the observations of radiation effects upon diagnostic patients and x-ray apparatus operators alike. Due to the primitive apparatus and photographic materials, extremely long exposures were typically required, especially in imaging thick anatomy (it has been reported that a skull radiograph would take more than an hour to acquire with the x-ray tube placed at about 5 cm from the head). Desquamation and epilation would frequently result, appearing about 2 weeks following exposure. While certainly an undesired sequelae in diagnosis, these observations led directly to the use of x-rays in the treatment of *tinea capitis*.

The discovery of γ-rays resulting from nuclear decays is attributable to Paul Villard who worked as a researcher in the chemistry department of the *École Normale* in Paris (Gerward 1999). Becquerel's serendipitous discovery of radioactivity in February 1896, through the fogging of photographic film by a uranium salt, led to a series of papers presented by him, between February and May 1896, to the Paris *Académie des Sciences*. The subsequent identification of the α and β components by Rutherford, described previously in Sect. 1.2.5, followed. Research into radioactivity accelerated after the Curies' discovery of the elements of polonium and radium as, in particular, the latter had a specific activity several orders of magnitude greater than that of uranium. In 1900, Villard obtained a small sample of radium and began a series of experiments comparing cathode rays and β-rays using a photographic plate as a detector. However, he noted on the plate traces of a beam from the radium source that was not scattered nor deflected by a magnetic field unlike the electrons. Moreover, the radiation was penetrating in being capable of traversing a 0.2-mm-thick lead foil. On 18 May 1900, Villard announced at the *Société francaise de physique* meeting that he had discovered a third radiation from radium that was similar to x-rays in that it was not charged and that it was highly penetrating. Surprisingly, there was little interest in this discovery and it was not until 3 years later the first incidence of this radiation being referred to as γ-radiation was made, and this was by Rutherford (Rutherford 1903).

[22] There has been a suggestion that Nikola Tesla may have detected x-rays prior to Röntgen (Hrabak et al. 2008).

1.2.10.2 Neutrons

'Up to the present, all the evidence is in favour of the neutron, while the quantum hypothesis can only be upheld if the conservation of energy and momentum be relinquished at some point'.

—James Chadwick in proposing that the neutral radiation resulting from α-particle bombardment of beryllium was not photon in nature.

The proposition of the neutron as an uncharged nuclear particle with a mass almost equal to that of the proton was made by Rutherford in 1920 on the basis of the observation that the ratio of the atomic mass number to the atomic number, $\frac{A}{Z}$, was very nearly equal to 2 for the light elements (excluding, obviously, hydrogen). This observation implied that a nucleus was composed of Z protons and $A - Z$ electrically neutral particles, each with a mass similar to that of the proton. However, this was not the only viable explanation of the ratio $\frac{A}{Z} \approx 2$ that could be based solely upon the combination of mass and charge. This ratio could also be achieved where the nucleus was made up of A protons and Z electrons. However, quantum-mechanical considerations, based upon the Pauli exclusion principle, the de Broglie particle–wave hypothesis and Heisenberg's uncertainty principle, led to the rejection of this proton–electron nuclear model. The first consideration is based upon the angular momentum of the diatomic homonuclear $^{14}N_2$ molecule (Ehrenfest and Oppenheimer 1931). If the ^{14}N nucleus were composed of 14 protons and 7 electrons, as proposed in the proton–electron model, the combination of an odd number of fermions would lead necessarily to a non-integer angular momentum of the nucleus. Raman scattering measurements, however, had shown that the ^{14}N nuclear angular momentum was, in fact, equal to $1\hbar$. If, on the other hand, the ^{14}N nucleus were to be considered as a combination of 7 protons and 7 of the hypothesised electrically neutral particle, with the latter also being a fermion, then the ^{14}N nucleus problems was resolved by considering it as three pairs of protons and three pairs of the neutral particles with their intrinsic spins coupling to cancel and produce a net zero spin, with the remaining proton and neutral coupling to form a nuclear spin of $1\hbar$. The second consideration against the proton–electron nuclear model was through application of the de Broglie hypothesis and Heisenberg's uncertainty principle. Consider an electron to be confined to the spatial dimensions of a hydrogen atom, which we will take to be of the order of 0.2 nm or 2×10^5 fm. From the uncertainty principle, we have $\Delta p\, \Delta x \approx \frac{\hbar c}{2} \approx 100\,\text{MeV} \bullet \text{fm}$. This leads to $\Delta p \approx 500$ eV which corresponds to an energy of just a few eV – certainly the correct order of magnitude for the energy of an atomic electron. Applying this to the case of an electron confined to the volume of a ^{14}N nucleus (radius of about 3 fm), we obtain $\Delta p \approx 33$ MeV or a kinetic energy of about 32 MeV. However, the nuclear Coulomb confinement energy would be of the order of $\frac{Z\alpha\hbar c}{r} \approx 3.4\,\text{MeV}$. Clearly, this inconsistency indicated that electrons could not be bound within the nucleus.

The first experimental evidence of the existence of the neutron arose from the work in 1930 of Walther Bothe and Herbert Becker which demonstrated the presence of a neutral radiation arising from the bombardment of a beryllium target by α-particles. The neutral radiation they had detected had a significant penetration in lead with an attenuation coefficient of 0.3 cm^{-1}. Initial speculation was that this neutral radiation was due to photons generated through interactions between the α-particles and the beryllium. The Joliot-Curies continued this work and observed that when this radiation was incident to a hydrogenous material, protons were generated as a tertiary beam.[23] They attributed this to being the result of a photon–proton Compton scatter resulting from an incident photon energy of about 50 MeV.

At Cambridge, Chadwick performed a series of similar experiments and measured the ionisation produced by the tertiary radiation (Chadwick 1932). He replicated the result that the tertiary radiation from a hydrogenous material was that of protons and further expanded his study to consider various other elements. On the basis of his measurements of the proton energy and recoil nuclei, he concluded that the assumption of the neutral radiation being made up of photons was inconsistent with the simultaneous conservations of momentum and energy. In particular, he noted that an α-particle incident to a ^{9}Be nucleus would produce a compound ^{13}C nucleus. By knowing the mass defect of ^{13}C, Chadwick concluded that any photon resulting from a de-excitation could not have an energy greater than about 14 MeV, a result which was inconsistent with the photon energy requirement for the observation to be the result of photon–proton Compton scatter. His conclusion was that the neutral radiation was made up of the hypothesised neutron and was created through the ^{9}Be (α,n)^{12}C reaction.

1.3 Early Theories of Charged Particle Energy Loss in Matter

1.3.1 Introduction

Theoretical descriptions of the transfer of charged particle kinetic energy to the atoms of the medium it traverses are fundamental to medical applications such as radiotherapy planning, radiation protection and equipment design. Theory is the basis of determining this energy loss rather than through empirical experimental measurement.[24] This section reviews the three pillars of the theory of charged particle energy loss, Bohr, Bethe and Bloch.

---

[23] The primary beam being the α-particles incident to the beryllium and the secondary beam being the neutral radiation incident to the medium.

[24] Experiment does, however, play a role in determining parameters used in modifications to the theoretical derivation.

This early work was of the interactions between a charged particle and a single atom. In such a case, these can result in a loss of some of the particle's kinetic energy through five mechanisms:

- Transfer of energy to the target atom as a whole
- Transfer of energy to the target atom's electrons resulting in excitation or ionisation
- In the case of a complex projectile such as an ion, changes in internal degrees of freedom within the projectile
- Should the energy transfer to the atomic nucleus be sufficient, nuclear excitation and disintegration
- *Bremsstrahlung*

1.3.2 Classical Theory: Niels Bohr

In the main, the classical theory describing how charged particles slow down and scatter in matter is reliant upon Rutherford's explanation of the large-angle scattering of α-particles by a thin gold foil as observed by Geiger and Marsden (1909). Rutherford's deduction of these data (Rutherford 1911) was that the atom was mainly empty space with a point-like, massive, positively charged nucleus linked with less massive, negatively charged electrons surrounding it.[25] Even so, what is likely the first attempt at describing theoretically charged particle energy loss is that of Thomson (1906) and which preceded Rutherford's deduction of the existence of the nucleus. Thomson's theory was based upon the assumption that energy loss was the result of collisions with the atomic electrons, which were considered at rest. He extended that theory to estimate the mean number of ionisations with the requirement that the energy transferred by the charged particle during a collision with an atomic electron exceeded the electron's binding energy.

Following the presentation of Rutherford's nucleus-based theory of the atom, Darwin (1912) shortly afterwards developed a theoretical study of the scattering and absorption of α-particles. The intent of that work, which was begun at Rutherford's instigation, was to provide a basis for Geiger's empirically derived formula of $v_0^3 - v^3 = bt$ describing the reduction in α-particle velocity as it passes through air. Darwin's calculation was based upon four main assumptions. First, as had Thomson, Darwin assumed that the α-particle energy loss was a consequence of collisions with atomic electrons, which was reasonable as it was consistent with the fact that α-particles were known to be readily capable of ionisation. Second, it was assumed that the atomic electrons were at rest prior to the collision. As the α-particles from radioactive decay had velocities of the order of 10^7 m/s, this was also a reasonable assumption. Third, Darwin assumed that the net negative charge of the

[25] The notion of electrons 'orbiting' the nucleus only became considered with the development by Bohr of his model of the atom.

Fig. 1.10 Niels Bohr (Courtesy and © of Niels Bohr Archive, Copenhagen)

atomic electrons was uniformly distributed throughout the atom. Fourth, and most critically, Darwin's calculation was based upon the assumption that the energy loss resulted only if the α-particle penetrated the interior of the atom. That paper and its assumptions were the kernel of the classical theory developed by Niels Bohr that paved the way to our modern understanding of charged particle energy loss in matter.

Niels Bohr (Fig. 1.10) obtained his doctorate in Copenhagen in the study of the motion of electrons in metals. Following this, in 1911, he then moved to the Cavendish Laboratory at Cambridge to work under JJ Thomson. However, after only a few months, he asked Rutherford if he could join his group at Manchester. This request was accepted and Bohr moved there in the spring of 1912. Following his arrival, as Rutherford was on holiday, Bohr attended an experimental course on radioactivity whose instructors included Geiger and Marsden. Interestingly, the course included measurements of the penetration of α- and β-particles through aluminium and tin absorbers. Afterwards, Bohr became aware of the publication by Darwin described in the previous paragraph and queried the validities of some of its assumptions. He quickly developed an alternative theory of how charged particles lose energy through soft collisions while traversing a medium. This work led to his two seminal papers on the topic (Bohr 1913, 1915).

Bohr's theory of soft-collision energy loss will be derived and discussed in Chap. 8. Here, we shall summarise its basic foundations through its definition of what is a soft collision. The theory is based upon a description of interactions within the context of classical mechanics in which the atomic electrons within a quasi-static Coulomb field are modelled by an ensemble of harmonic oscillators. Using this model, let us begin by considering a single harmonic oscillator with a resonant frequency of ω_0. The oscillator is acted upon by an external force F over the finite time interval T (this is the Coulomb force of a passing charged particle). If the time interval of the interaction is sufficiently abrupt, i.e. $\frac{2\pi}{\omega_0} \gg T$, then the oscillator will receive an impulse of FT and respond to this force as if it were a free particle with an energy transfer to it. On the other hand, should the time interval be much longer than the oscillator's response time so that $\frac{2\pi}{\omega_0} \ll T$, then the response of the oscillator will be adiabatic and the energy transfer can be considered negligible.

These two extrema of the oscillator response provide the basis of Bohr's concept of the impact parameter, b, which is the closest distance between a moving projectile and a target that would occur if there were no interaction between them. For an impact parameter greater than a given value, the charged particle would be unable to transfer sufficient energy to ionise or to excite the atom. On the other hand, for an impact parameter less than some specified value, the charged particle would interact directly with an individual electron in a singular and 'hard' collision. Thus, these two boundaries of minimum and maximum impact parameters define the Bohr region of soft collisions within which the charged particle interacts with the entire ensemble of atomic electrons. A 'hard' collision, which is that between the projectile and an atomic electron, will be quantal in description in that it must include not only the relativistic kinematics of the collision (as described in the Appendix D) but also the inherent quantum-mechanical attributes, such as Pauli's exclusion principle for an electron projectile.

Bohr retained a long-standing interest in developing the theoretical descriptions of the energy losses of energetic charged particles moving through matter. As noted earlier, he was one of the first to address the problems associated with the scattering and stopping of the heavy ions of fission fragments and their ability to shed and capture atomic electrons. This work helped provide the foundations for heavy-ion therapy planning. Even some 60 or more years later after its publication, his monograph describing charged particle energy loss (Bohr 1948) remains a classic in the field with its elucidating derivations.

1.3.3 Quantum Theory: Max Born and Hans Bethe

Extending from Bohr's classical theory to one based upon quantum mechanics was of intense interest during the 1920s. However, practical calculational success of a quantum theory of collision energy loss of charged particles came through the development of quantum perturbation theory which can be used to derive expressions describing interaction cross sections and decay rates for a wide variety of processes. Being based upon perturbation theory, this calculation of charged particle energy loss does, of course, have restrictions enforced upon it which limit the conditions of validity of its results, and this is especially true when it is applied to the Coulomb interaction as a result of the potential's infinite range. This effect will be investigated in considerable detail in Chaps. 2, 4 and 8. Max Born, while at Göttingen, published in 1926 the derivation of his eponymous iterative solution of the Schrödinger equation as applied to the scattering problem (Born 1926). In later life, Born reflected upon this derivation as having had its genesis in his observation of the wake of a docking ship incident to a cylindrical post of a pier. The wake was a model of the quantum-mechanical plane-wave representation of the charged projectile incident to a scattering centre as represented by the pier. These waves were then reflected from the pier to form emanating circular waves. Born's result evolved from his previous linking of the matrix and wave interpretations of quantum mechanics so as to provide the scattering model in which the scattered wavefunction is a spherical wave with an angular dependence modulated by a scattering amplitude which, in turn, is a function of the scattering potential (see, e.g. Fig. 2.1). The solution of the scattered wavefunction in configuration space is an iterative integral equation which contains the scattering potential within the integrand.[26] For a 'weak' potential (i.e. one smaller in magnitude than the free projectile's kinetic energy), the Born series calculation is terminated generally at first order.

The most successful foundation of quantum energy loss theory, and one which is still used today, is based upon Born's development of quantum perturbation theory, in particular, its application to the description of the influence of a potential generated by a scattering centre upon an incident wavefunction. The introduction of the Born expansion to the calculation of charged particle energy loss was made by Hans Bethe (Fig. 1.11). He had studied chemistry at Frankfurt initially but, purportedly due to a personal dislike of white laboratory coats, switched to studying for his doctorate in physics under Arnold Sommerfeld in Munich which he received in 1928. Afterwards, Bethe worked in Stuttgart as an assistant to Paul Ewald. It would seem that this was a particularly happy time for Bethe where he had become renowned for his knowledge of the rapidly developing field of quantum mechanics. However, he was recalled to Munich by Sommerfeld to become *Privatdozent* in 1929. It was here that Bethe used the Born iterative solution of the Coulomb interaction (to first order[27]) to derive a quantum-mechanical solution for the soft-collision losses of a charged particle passing through matter. His resulting paper, 'Theory of the Passage of Fast Corpuscular Rays Through Matter', was written during the winter of 1929 and was received by *Annalen der Physik* on 3 April 1930 (Bethe 1930). It is said that Bethe was to have regarded this publication as his best (Lee and Brown 2007).[28]

Bethe's theory is developed in Chap. 8 and will remain the basis of discussion and derivation for the remainder of the book. As it is the result of the first term of the

---

[26] The Born series is derived in Chap. 2 as a means of solving the Lippmann–Schwinger equation.

[27] The problems associated with the second-order Born series for the Coulomb interaction are discussed in Chap. 4. Correction terms to account for conditions unmanageable by the first-order term of the Born expansion are derived in Chap. 11.

[28] Upon reflection, this is a rather profound judgment on Bethe's part considering his having been awarded the 1967 Nobel Prize in Physics for his discovery of the stellar energy production mechanism, his significance in the development of nuclear weaponry and his later political challenges to their use and his deep and wide work in nuclear physics.

Fig. 1.11 Hans Bethe (Courtesy of Cornell University Photography)

Born expansion of the Coulomb interaction, a number of factors to account for the associated limitations will also be derived. Unlike Bohr's classical theory which characterises soft- and hard-collision types in configuration space, the quantum theory of Bethe is based upon a classification within momentum/energy space. That is, small momentum transfers are associated with large impact parameters and vice versa. An energy transfer threshold is used to demarcate between the soft collisions of the Bethe theory and the head-on 'hard' collisions with singular atomic electrons. When the expressions for 'soft-' and 'hard-'collision powers are summed to yield the 'total' collision stopping power resulting from the combinations of both collisions, this energy threshold disappears from the final expression. The gap between Bohr's classical and Bethe's quantum theories required resolution, however. And that was provided by Felix Bloch.

1.3.4 Bridging Classical and Quantum Theories: Felix Bloch

From the prior two sections, by the 1930s there existed classical and quantum-mechanical descriptions of the collision energy loss of charged particles moving through matter. The Bohr and Bethe theories, while providing mathematically similar expressions for the collision stopping power, also yielded different numerical results. These differences required resolution and this was achieved through the work of Felix Bloch (Fig. 1.12). Bloch, in fact, is associated mostly with magnetism

Fig. 1.12 Felix Bloch (Courtesy and © of CERN)

(as a result of his having shared the Nobel Prize in Physics of 1952 with Purcell for the discovery of nuclear magnetic resonance) and solid-state physics, rather than that of the energy loss of charged particles traversing a medium. His interests were extended through the study of charged particle energy loss during the autumn of 1931 when he was Oersted Fellow working with Niels Bohr at the latter's institute in Copenhagen. Bohr had encouraged Bloch to investigate the differences between the Bohr classical and Bethe quantum theories that had become apparent. Bloch (1933) was able to demonstrate that the two theories were, in fact, opposite limiting approaches arising from the different ways in which the phase of the projectile wavefunction varies as it passes by the atom. Bloch's modification of the Bethe stopping power equation to account for this fact is studied in Chap. 12.

1.4 Applications

While the subject of this book is the development and exploration of the theory of charged particle collision energy loss and deposition in matter in medical applications, the theory has much broader applications. To consider this, this Section presents the following. First, available databases of stopping power data and software for evaluating these are documented. The customer desiring such data will be, in most cases, the medical or radiation protection physicist. Second, Monte Carlo simulations of the transport of radiation and its interaction with matter have become immensely useful in medical physics, in particular the evaluation of radiation absorbed doses. While the intent is not to summarise Monte Carlo applications in medical radiation physics, a very brief review (emphasising the methods of calculating the collision stopping power) is provided. Third, the diverse range of non-medical applications of collision stopping power calculations is provided through a variety of examples. Fourthly, and finally, the medical applications of these calculations are summarised in therapeutic and diagnostic considerations.

1.4.1 Collision Stopping Power Databases

The Bethe theory provides the means by which collision stopping powers are calculated in both tabular form and in Monte Carlo code modules, the latter discussed in the following section. Tables for stopping powers for electrons and positrons in various media are provided in Report 37 of the International Commission on Radiological Units and Measurements (ICRU) (ICRU 1984) and those for protons and α-particles are given in ICRU Report 49 (ICRU 1993). These tabulated data are the basis of calculations provided by the web-based databases ESTAR, PSTAR and ASTAR for electrons/positrons, protons and α-particles, respectively, and which are made available to the scientific community by the National Institute of Science and Technology (NIST 2012). The results of these calculations are provided for elements and for compounds and mixtures of radiological interest. In ESTAR, data are provided for 72 materials on a standard grid covering 81 kinetic energies ranging from 10 keV to 10 GeV. In PSTAR and ASTAR, data are provided for 74 materials on a standard grid which covers 133 proton kinetic energies ranging from 1 keV to 10 GeV and on a standard grid covering 122 α-particle kinetic energies ranging from 1 keV to 1 GeV.

ESTAR provides, in addition to the collision stopping power, the radiative stopping power and the CSDA range. PSTAR and ASTAR provide, in addition, the nuclear stopping power, which is reflected in the kinetic energy of the recoiling atom following a collision (this is important only for heavy projectiles), and the CSDA range.

Also available on the Internet are software codes for the evaluation of the collision stopping powers and ranges of heavy ions in matter (SRIM 2012). These are based upon the work by Ziegler and others at IBM Research (Ziegler 1980, 1999). In addition, the *S*topping *P*ower *A*nd *R*anges (SPAR) code is available for evaluations for muons, pions, protons and heavy ions (RSICC 1985).

1.4.2 Monte Carlo Simulations

While it is perhaps commonly thought that the Monte Carlo method was originally developed in modelling radiation transport in early nuclear weapon design (Metropolis and Ulam 1949), the sampling method that is inherent to Monte Carlo simulations was first mentioned in the Old Testament (1 Samuel 28 6) as the *Urim*, a stochastic means of divining, through random selection, a decision from a deity. (*When Saul saw the army of the Philistines, he was afraid, and his heart trembled greatly, And when Saul inquired of the Lord, the Lord did not answer him, either by dreams, or by Urim, or by prophets.*) The applications of the Monte Carlo method to the radiological sciences are not reviewed here in detail, and the discussion will be restricted to the methods used for calculating charged particle collision energy deposition. Contemporary descriptions of the Monte Carlo method in medical radiation physics may be found in Andreo (1991), Ljungberg et al. (1998) and Zaidi and Sgouros (2003), and the interested reader is referred to these for an introduction to the field.

As recognised by Bohr, the interaction of a charged particle with matter is an intrinsically stochastic process described by a probability distribution function (pdf). The single-valued collision stopping power and CSDA range introduced in Sect. 1.2.3 represent the mean values associated with these statistical variations. In many radiological dosimetry calculations, the uses of these mean values are adequate. However, there are calculations which require accounting of the stochastic nature of the interactions of radiation with matter in order to accurately replicate the physical case. While this can be done through analytical means for simple scenarios, e.g. by use of the diffusion equation or Boltzmann's transport equation, the Monte Carlo method of simulating radiation transport is the most common means used for modelling radiation transport in modern medical radiation physics.

Radiological applications of the Monte Carlo method are dominated by the calculations of the transport and interaction with matter of photons, electrons and positrons. Energetic electrons and positrons are generated through radioactive decay or photon interactions with matter (photoelectric absorption, Compton scatter and pair/triplet production). At energies typical of diagnostic radiology or nuclear medicine, the kinetic energies of these particles are sufficiently small that their ranges can often be neglected and energy deposition may be considered to be local to the photon interaction and not transported away. That is, the absorbed dose and kerma are assumed to be the same in both magnitude and location. Investigators have modelled the individual interactions of electrons in microscopic calculations at low electron energies (Andreo 1991). At higher energies or for those cases where the charged particle range is comparable of the physical dimensions of the problem, the increased number of scattering events to be modelled as the electron slows down makes such a detailed calculation computationally burdensome. The algorithms for evaluating charged particle energy deposition are based upon the Bethe quantum theory and the statistical models derived by Landau or Vavilov (Chapter 18). These can be used to evaluate the collision stopping power for high-energy and complex environments through Monte Carlo simulations of charged particle transport and which are based upon the theory of charged particle interactions developed in this book (e.g. Faddegon et al. 1992). The three 'off-the-shelf' Monte Carlo codes which are perhaps most commonly used in medical

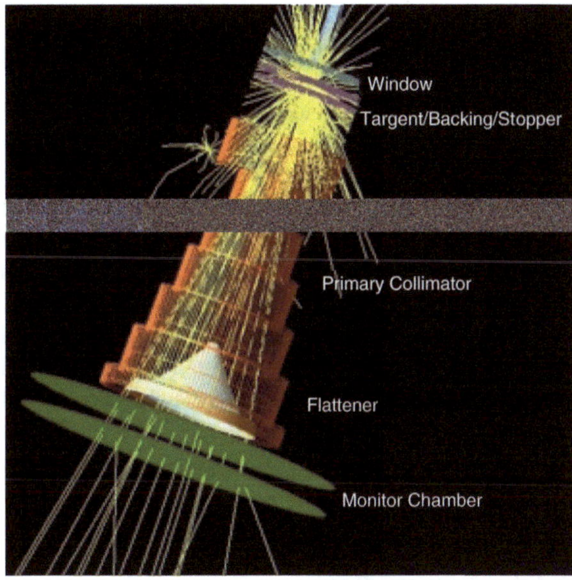

Fig. 1.13 Photon, electron and positron transport and interactions in the head of a therapeutic electron linear accelerator simulated with the EGSnrc Monte Carlo system of the National Research Council of Canada (Courtesy of Bruce Faddegon, University of California San Francisco)

radiation physics are MCNP, GEANT4 and EGS.[29] The MCNP (Monte Carlo N-Particle) code was developed at, and is supported by, the Los Alamos National Laboratory, and GEANT4 (GEometry ANd Tracking) was developed and continues to be supported by CERN. While EGS was developed originally at the Stanford Linear Accelerator Center, it is maintained and updated currently by the National Research Council of Canada. Please refer to their user manuals via the websites provided in the footnote for further information as to how charged particle energy losses are calculated and accounted for in the codes.

Figure 1.13 presents an example of such a Monte Carlo simulation of radiation transport and production by showing the trajectory of an electron beam incident to a thick tungsten target in a medical linear accelerator and the resultant photons and secondary electrons and positrons. The electrons in the incident beam undergo both *bremsstrahlung* (required to produce the therapeutic x-rays required) and collision energy losses which will require the target to be cooled. In order to design the apparatus to ensure that the exiting photon beam does not contain an unacceptable level of electron contamination (which, due to their short depth of penetration, can lead to a high radiation absorbed dose to the skin), the Monte Carlo code must be

[29] The current Internet sites for these codes are:

MCNP	https://laws.lanl.gov/vhosts/mcnp.lanl.gov/
GEANT4	http://geant4.cern.ch/
EGS	http://irs.inms.nrc.ca/software/egsnrc/

able to calculate accurately the energy losses of the electrons as they traverse an electron shield within the accelerator.

In this Figure, the electron beam exits the vacuum of the accelerator waveguide at the top centre and is incident to an electron-opaque tungsten target. Photons are generated through *bremsstrahlung* as are secondary electrons which must be shielded from the patient as these impart an absorbed dose to the skin and superficial tissues, but not to the tumour at depth due to their finite range. The primary collimator shown restricts the lateral extent of the x-ray beam. As the spatial distribution of this beam is highly forward directed, a beam flattener is used to make this distribution uniform across the lateral extent of the field. The x-rays then pass through a monitor ionisation chamber which is used to measure the x-ray flux incident to the patient. This is performed by collecting, through a static electric field, the electron and positive ions generated within the chamber gas through photon- or electron-caused ionisations of gas molecules. Not shown are the variable collimators after this chamber which are used to restrict the x-ray flux to a specific cross-sectional distribution.

1.4.3 Non-medical

1.4.3.1 High-Energy Physics

Medical physicists will be familiar with the small radiation detectors used in therapeutic, diagnostic and radiation safety applications, such as the Geiger–Müller counter or the scintillation counter. On the other hand, modern-day high-energy physics research uses massive detectors in order to extract physical data (mass, momentum, energy, etc) of interactions between particles. These detectors include the behemoths such as the ATLAS (*A Torodial LHC ApparatuS*) detector at the Large Hadron Collider at CERN (Fig. 1.14). ATLAS detects the particle products of collisions between counter-circulating proton beams. The consequent charged particles are detected through energy transfer to matter resulting in ionisation and analysed in terms of charge, momentum and mass in order to identify them and to deduce the mechanisms by which they were produced.

Regardless of the size or the absorber used, charged particle identification and detection is frequently based upon the measurement of the energy deposited within the medium of the detector. An example of the principle behind this charged particle-identification technique is shown in Fig. 1.15. Particles with a rest mass m, velocity v, kinetic energy E and electric charge ze are incident to a thin detector (here, considered to be a 0.5 g/cm^2 slab of carbon). The particles have a kinetic energy E sufficient to transit the detector and deposit an energy ΔE through collisions with atoms in the detector. This energy loss is given by

$$\Delta E = \frac{\mathrm{d}E}{\mathrm{d}x}\Delta x \qquad (1.12)$$

Fig. 1.14 The ATLAS (A Toroidal LHC ApparatuS) detector under installation at the Large Hadron Collider in CERN in 2006. The physical dimensions of the detector can be sensed by comparison with the people standing to the left of the image's centre (Courtesy and © CERN)

where Δx is the thickness of the detector. The collision stopping power will be shown later to have the proportionality

$$\frac{\mathrm{d}E}{\mathrm{d}x} \propto \left(\frac{z}{v}\right)^2. \tag{1.13}$$

This proportionality can be rewritten by using the relativistic expression for the velocity as

$$\frac{\mathrm{d}E}{\mathrm{d}x} \propto z^2 \frac{\left(1 + \frac{m}{E}\right)^2}{1 - \left(\frac{m}{E}\right)^2}. \tag{1.14}$$

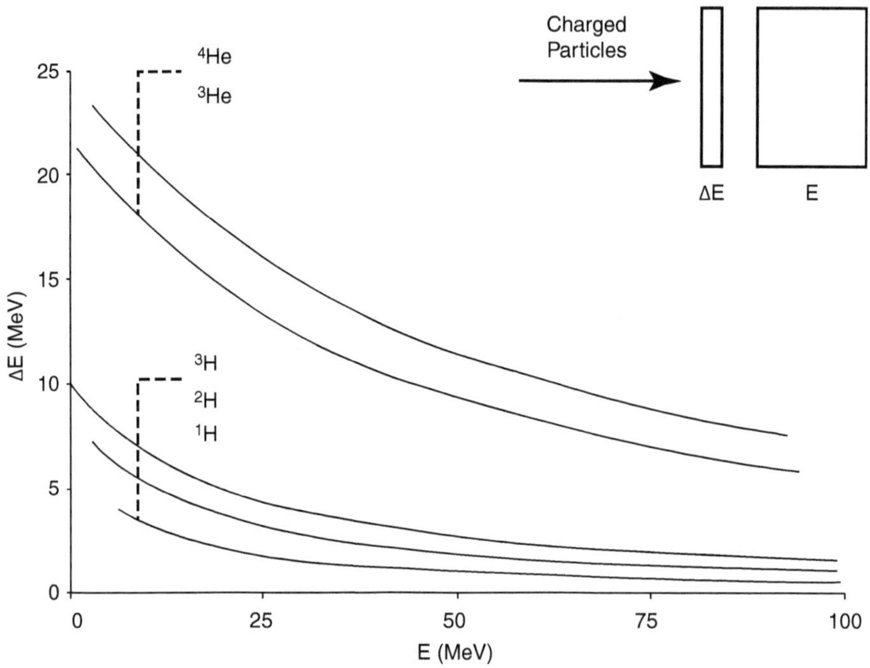

Fig. 1.15 Concept of charged particle identification by measuring simultaneously the collision energy loss as they traverse a thin (non-opaque) detector in which the energy loss (ΔE) is measured and are then incident to an opaque detector in which the particle is stopped and the remnant energy (E) is measured. As charged particle energy loss is a function of the mass and electric charge of the particle, cross-plotting both measurements can identify the nature of the particle. Examples are shown for three isotopes of hydrogen and two isotopes of helium incident to plastic scintillator detectors with the thickness of the thin detector equal to 0.5 cm^2/g

Expanding the numerator, ignoring all terms of the order of $\left(\frac{m}{E}\right)^2$ and assuming that $m >> \frac{E}{2}$ (both actions being equivalent to considering the incident kinetic energy much less than the particle rest mass), this yields a proportional relationship between the energy loss and the particle characteristics:

$$\Delta E \propto \left(\frac{2z^2 m}{E}\right) \Delta x. \tag{1.15}$$

This result shows the three features of a ΔE measurement that allow differentiation between charged particles. First, ΔE is inversely proportional to E (i.e. a plot of the logarithm of E versus the logarithm of ΔE will be linear with a negative slope). Second, for a fixed combination of mass and kinetic energy, the energy loss increases with the square of the charge of the particle. Third, for a fixed combination of z and E, ΔE increases with particle mass. The kinetic energy can be determined by having the particles, after having traversed the thin detector, incident to another detector sufficiently thick to stop them and deposit all of the remaining energy, $E - \Delta E$. As shown in Fig. 1.15, a plot of E versus ΔE for a range of particles and energies generates loci (tending to a $\Delta E \propto E^{-1}$ dependence with the constant of proportionality being mz^2 where m is the projectile's rest mass and ze is its electric charge) which clearly differentiate charged particles in terms of mass and electric charge. Energy exclusion is possible by incorporating a third detector following the thick detector and analysing only those events (energy depositions) that occur in the first two detectors but not the third. That is, the third detector is operated in anticoincidence such that if there is any particle that penetrates to the third detector, that event is excluded from analysis.

While the numerical example shown by Fig. 1.15 considered carbon-based detectors, the detectors could also be silicon in which the amount of collected charge generated by the interaction of the charged particles with the silicon atoms is the measure of the deposited energies. This arrangement of multiple axial detectors is colloquially referred to as a 'telescope' and need not be limited to only a single 'transparent' thin detector as will be shown in the next section in the discussion of a particular piece of apparatus.

1.4.3.2 Space Physics and Astrophysics

At the other extreme of both size and distance of a charged particle detector is the LECP (Low-Energy Charged Particle) experiment on the *Voyager 1* and *Voyager 2* unmanned spacecraft shown in Fig. 1.16, some 16 billion km distant from Earth at the time of this writing and still operational a third of a century after their launch (Krimigis et al. 1977). These detectors are designed to characterise the charged particle environment within interplanetary and interstellar space.

The LECP experiment comprises two solid-state detector configurations, one of which is the Low-Energy Particle Telescope (LEPT) which is designed to measure the charge and energy distributions of low- and medium-energy charged particles. A schematic of the LEPT is shown in Fig. 1.17. The basic principle of the LEPT's ability to identify charged particles is the same as that of the arrangement of Fig. 1.15. All of the detectors are solid state and so the signal is the collected charge of the electron–hole pairs created when a charged particle transfers kinetic energy to the atoms of the medium contained within the detectors and thus creates electron–ion pairs. The LEPT was made to be a binary axial detector in order to detect charged particles with different energies in two directions. As the spacecraft has a fixed orientation, the detectors are placed on a rotating turntable in order to achieve a total azimuthal detection coverage.

For particles incident from the left-hand side of Fig. 1.17, the D_1 detector (segmented into three discs, as can be seen in Fig. 1.16) provides the ΔE element and D_2 the E element (D_3 can be used in anticoincidence). For particles from the

Fig. 1.16 The LECP (Low-Energy Charged Particle Detector) carried on the Voyagers 1 and 2 spacecraft. The *upper aperture* is for the LEMPA (Low-Energy Magnetosphere Particle Analyzer) detector and the *lower aperture* is for the LEPT (Low-Energy Particle Telescope) detector, which is described in the text. Both detectors are on a rotatable platform. The dome at the *top* is stationary and used for detecting low-energy electrons. For a sense of dimension, the diameter of the LEPT aperture is 7.1 cm. The electronic package is contained in the base of the LECP device (Courtesy of Stamatios M Krimigis, Applied Physics Laboratory, The Johns Hopkins University)

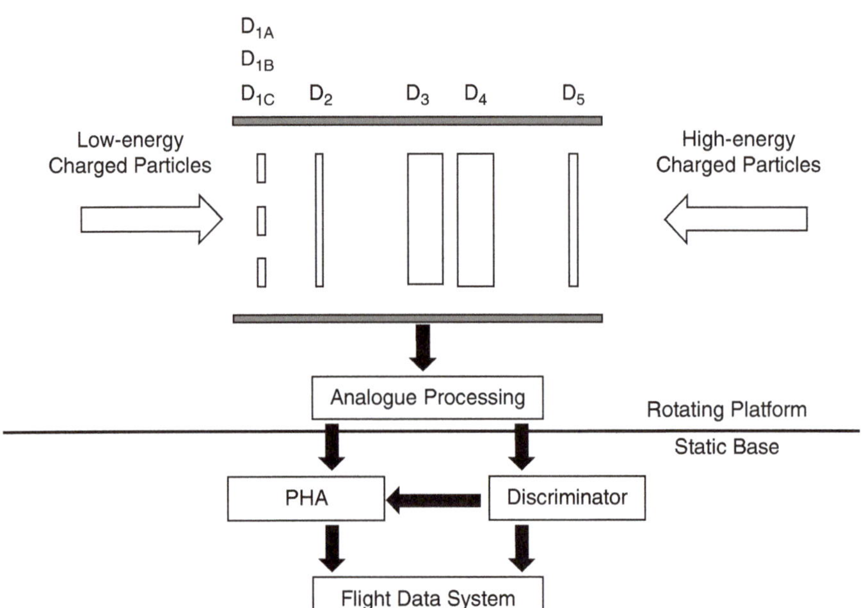

Fig. 1.17 Schematic diagram of the bilateral Low-Energy Particle Telescope carried on the *Voyager 1* and *Voyager 2* spacecraft (See text for details)

right-hand side, the D_5 detector in combination with the D_3 and D_4 detectors provides a telescope for particles with higher energies. When a detection logic condition is satisfied (e.g. $D_1D_2\bar{D}_3$ which signifies a particle having been detected by D_1 and stopped in D_2 without having reached D_3), the analogue signals (i.e. the charges of the collected electron–hole pairs in the detectors caused by the ionisations generated by the passing charged particles) are amplified logarithmically because of the $\Delta E = k_1E^{k_2}$ dependence of (1.15). These signals are then transported to a discriminator and a pulse height analyser (PHA) in order to digitise them prior transfer to the Flight Data System and radio transmission to Earth.

1.4.3.3 Engineering and Manufacturing

The interactions of charged particles with matter, resulting in the transfer of kinetic energy to a medium as they decelerate, play many roles from the mundane to the profound within modern society and often with little realisation. For example, the common household smoke detector is reliant upon the detection of such interactions through the ionisations of air by an α-emitting radionuclide source.

Semiconductor Manufacturing

Fabrication of the ubiquitous semiconductor can require the implantation of ions into a material in order to change its electric properties, a process referred to as 'doping'. One means of doing so is to generate an ion beam and bombard the substrate with the charged ions which penetrate the substrate and stop at the end of their range in the medium. This precise placement of the dopant requires calibration of the necessarily precise submicron penetration into a semiconductor substrate (Fig. 1.18).

The ions are electrostatically accelerated, momentum analysed, focussed and then scanned over a substrate. The selection of the beam kinetic energy is important for the application. Low kinetic energies (<1 keV) deposit the ions on the surface whereas kinetic energies of up to 10 keV will result in the deposition of ions to depths of the order of 10^{-9} m. Higher energies result, clearly, in greater penetrations. The typical dopants used in semiconductor manufacturing are arsenic, boron and phosphorus. Accurate modelling of energy loss and penetration of the ions is required in this application.

Materials Engineering

In addition to its uses in semiconductor manufacturing, ion implantation is also used in the strengthening of materials used in orthopaedic prosthetic implants. For example, orthopaedic prosthetic implants can employ a titanium alloy femoral head

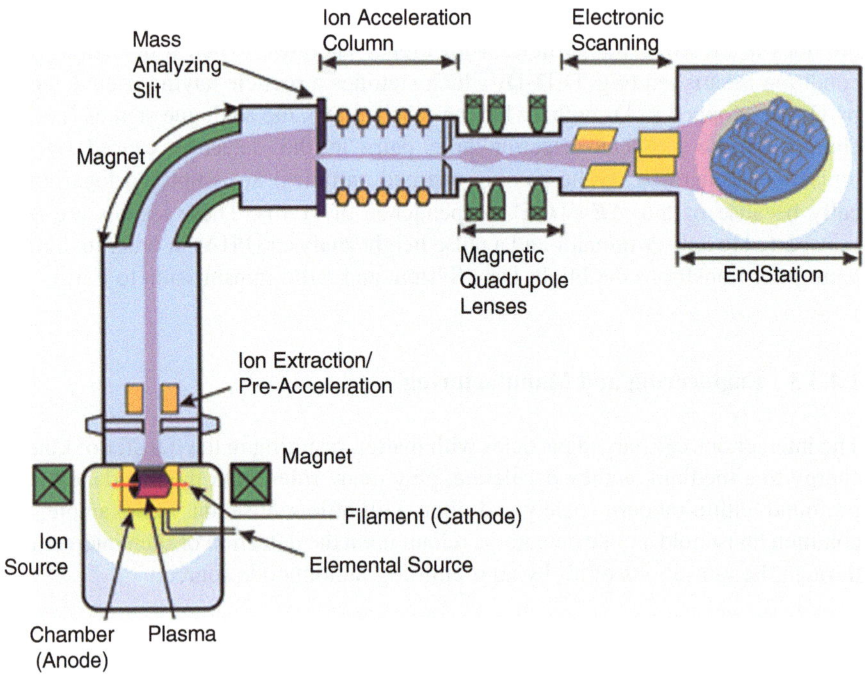

Fig. 1.18 Ion implantation of a semiconductor substrate. Ions are produced and pre-accelerated prior to entering a dipole magnet. A slit is used to select ions of a given momentum and which are then further accelerated and undergo focusing and scanning over the substrate (Courtesy of Spire Biomedical, Inc.)

in contact with an ultrahigh molecular weight polyethylene (UMWPE). Irradiation by beams of nitrogen ions is used to improve both the wear resistance of the titanium and the same used to improve the sliding wear resistance of the UMWPE (Allen et al. 1995). Again, calculation of energetic ion energy loss and the CSDA range is essential to this modification of materials.

1.4.3.4 Sensitivity of Semiconductor Electronics to Cosmic Rays

Terrestrial cosmic rays are energetic charged particles at ground level created by cosmic ray interactions in the upper atmosphere. Cosmic rays are about 90 % protons and 9 % α-particles with the rest being made up of heavy nuclei and which impact the upper atmosphere with a flux of about $10^3/m^2/s$ (Gaisser 1990). These produce showers of energetic charged particles, such as pions, which decay into muons (Fig. 1.3), and neutrons which reach the level of the Earth's surface. Modern semiconductor electronics can be sensitive to the ionisations caused by these cosmic rays, and this sensitivity is categorised in terms of the Soft Error Rate, which is the rate of a spontaneous change in data or logic. An important component of the assessment of this sensitivity and in the design of integrated circuits is the modelling of the interactions of the charged particles with the material (Ziegler and Srinivasin 1996). This requires the ability to accurately calculate the energy loss of charged particles in the semiconductor material.

1.4.4 Medical

The demand of medical applications for accurate calculations of charged particle energy deposition in matter is greatest for radiation dosimetry. In radiotherapy, an accurate calculation of this energy deposition which results in an absorbed dose to tissue is needed for the minimisation of the absorbed dose to uninvolved healthy tissue and for the maximisation of a tumouricidal dose (i.e. the maximisation of the therapeutic ratio). This required accuracy has been quantified on the basis of clinical data of tumour control. The ICRU has recommended that the net uncertainty in the absorbed dose to a target volume be less than 5 % at the two standard-deviation level (ICRU 1976). More recently, the IAEA has suggested that the uncertainty in the physical measurement of the absorbed dose in a reference phantom be less than 3–5 % at the one standard-deviation level (IAEA 2000). Assuming that all of the sources of uncertainty are statistically uncorrelated, the net uncertainty is described by their quadrature sum of uncertainties, leading to the conclusion that these requirements would indicate that the uncertainty in the dose calculation should be no greater than 1 %. As both the measurement and the calculation are reliant upon numerical data derived from theoretical calculations of charged particle energy loss in matter, the accuracy of these theories is fundamental. Although radiogenic biological effects are not sought in diagnostic radiology and nuclear medicine, because the associated absorbed doses are at least some four orders of magnitude less and a larger number of patients undergo diagnostic procedures than therapeutic, the *per caput* probabilistic radiation risk is the focus of dosimetry accuracy, as discussed in the Preface. That is, accurate evaluation of the energy deposition of secondary charged particles in diagnostic examinations is primarily for the protection of a population rather than of the individual patient. As in therapy, accurate dosimetry required in diagnosis will be also dependent upon accurate stopping power calculations.

1.4.4.1 Therapy: External Photon and Electron Beams and Sealed Sources

Modern conventional external beam radiotherapy is based largely upon photons and electrons. Throughout much of the world, ^{60}Co teletherapy devices producing γ-rays have been largely supplanted by the electron linear accelerator which generates an intense electron beam with a kinetic energy of up to about 25 MeV to bombard a high atomic number target to create an intense flux of x-rays through *bremsstrahlung*. The engineering design and the evaluation of the radiation transport in tissue and consequential absorbed dose deposition require accurate calculation of the electron and photon interactions with materials, including tissue (Fig. 1.18).

Electron beam therapy is that if the electron beam is not intercepted by an electron-opaque target but is instead allowed to impinge upon the patient directly.[30] As charged particles have a finite range due to the repeated transfer of kinetic energy to the medium, electron therapy is suitable for the treatment of superficial tumours or those at shallow depth in order to protect uninvolved tissue at depths beyond the tumour. Knowledge of changes in the electron stopping power in tissue with energy (or, equivalently, with depth) is of significant importance in electron beam therapy planning (Attix 1986). Because of their light mass, electrons are more subject to multiple scattering than are the heavier hadrons (protons, α-particles, etc.) and do not exhibit the Bragg peak of heavy charged particles.

But the delivery of photons or electrons for the radiation treatment of neoplasia need not be external. Internal radiotherapy is performed using sealed or unsealed radioactive sources. Brachytherapy is the permanent or temporary implantation or intracavitary insertion of sealed radioactive sources. Examples include the use of ^{137}Cs sources in the treatment of gynaecological cancers by the temporary insertion of such sources into the vagina or uterus (Mould and Hobday 1983) and the use of ^{125}I and ^{103}Pd sealed sources permanently implanted into the prostate gland for the treatment of neoplasia. The use of unsealed radioactive sources, also known as radionuclide therapy (RNT), is deferred to a later point in this chapter.

The evaluation of the radiation dosimetry associated with medical external beam irradiations is described by Johns and Cunningham (1983) and Attix (1986) and that associated with internal irradiations is provided by McParland (2010). An example of the importance of accurate theoretical evaluations of the collision stopping power is given by the Bragg–Gray relationship which is used to calculate the absorbed dose to a photon-irradiated medium through the measurement of the ionisation caused within the gas volume of an ionisation chamber placed within the medium. The absorbed dose to the medium as measured with an ionisation chamber of a wall material different from the gas and medium is proportional to the factor:

$$f\frac{\left.\frac{dE}{\rho\,dx}\right|_{\text{Wall}}}{\left.\frac{dE}{\rho\,dx}\right|_{\text{Gas}}}\frac{\left.\frac{\mu_{en}}{\rho}\right|_{\text{Medium}}}{\left.\frac{\mu_{en}}{\rho}\right|_{\text{Wall}}}+(1-f)\frac{\left.\frac{dE}{\rho\,dx}\right|_{\text{Medium}}}{\left.\frac{dE}{\rho\,dx}\right|_{\text{Gas}}}$$

[30] In such a case, the electron beam current within the accelerator must be reduced for patient safety from that used for x-ray therapy by a factor of about 10^3, as the latter current is higher due to the low efficiency of *bremsstrahlung* production at electron energies used for radiotherapy.

where *f* is the fraction of the total ionisation within the chamber gas volume caused by electrons generated by photon interactions within the chamber wall and where $\left.\frac{\mu_{en}}{\rho}\right|_a$ is the photon mass–energy absorption coefficient for material *a* averaged over the photon energy spectrum as it exists at the point of measurement in the medium. The first term describes the contribution to the total ionisation due to the secondary electrons generated within the ionisation chamber wall, and the second term accounts for those generated within the medium and which cross the wall. It is clearly evident from this expression that the mass collision stopping power is essential to the determination of the absorbed dose delivered to the radiation oncology patient.

1.4.4.2 Therapy: Hadrons

Charged hadrons (protons, π^--mesons and heavy ions) and the neutron have long been of interest in radiotherapy. For the charged hadrons, this has been due to the attribute of the Bragg peak which enables increased ability to optimise the spatial distribution of the absorbed dose within tissue. Because of the denser distribution of absorbed dose within the position of the Bragg peak, there are also radiobiological advantages, as discussed below. This discussion of hadron therapy is limited to external beam irradiation by accelerator-produced protons, neutrons, heavy ions and π^--mesons. Internal irradiation by α-particles created through radioactive decay is discussed later in the context of therapeutic nuclear medicine. The first accelerator-produced hadron used in radiotherapy was the neutron where neutron beams were typically generated by the (^{2}H,n) stripping reaction with a cyclotron-generated deuteron beam incident to a beryllium target. However, in modern times, the proton is the prevalent hadron in use in radiotherapy with ^{12}C ion therapy becoming of increasing interest.

There are three main reasons why hadrons have a long-generated interest in radiation treatment of cancer. Two are based upon the physics of the radiation transport and one is based upon radiobiological considerations. The first is that charged hadrons demonstrate a Bragg peak which, through modulation of the incident hadron beam energy and associated depth of penetration, allows both the minimisation of the absorbed dose to overlying tissue in regions between the beam entrance point and the target and the maximisation of absorbed dose at the target (Fig. 1.19). There is no absorbed dose beyond the Bragg peak to affect underlying normal tissue.

The second reason is that the large masses of the charged hadrons minimise the degree of multiple scatter, thus retaining a well-defined lateral definition of the beam. The third, and radiobiological reason, is common to both charged hadrons and neutrons. The highly localised deposition of energy in tissue, described by a high LET, gives the hadrons a greater biological effect (proportion of cells killed per unit energy deposited) than other radiations (photons and electrons) which do not deposit energy into tissue in such a localised fashion. Thus, per unit absorbed

Fig. 1.19 The Bragg peak

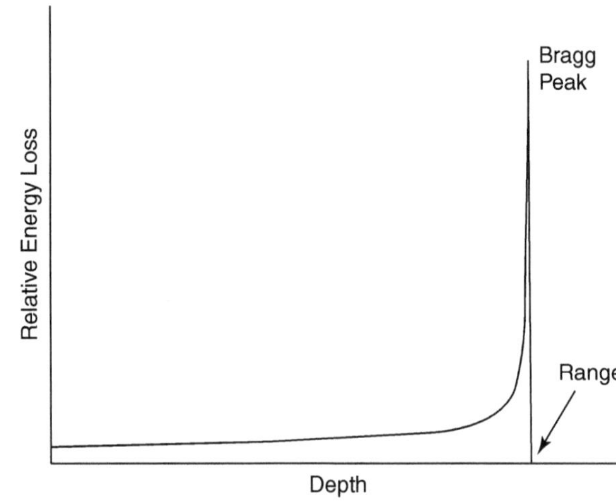

dose, hadrons have a higher biological effect than photons or electrons or, in other words, a greater relative biological effectiveness (RBE). Another important radiobiological attribute of hadrons is the oxygen enhancement ratio (OER) which reflects the fact that tumours tend to be hypoxic and the reduced oxygen tension increases their radioresistance.[31] This can be overcome with the localised energy depositions of hadron therapy with the expectation that a tumour's OER is reduced (Fig. 1.20) which are expressed in terms of the RBE and the OER. For conventional photon and electron therapy (excluding Auger and internal conversion electrons), the RBE is defined as being equal to 1.0. A value of 1.1 is assigned to protons and a range of 2–4 is adopted for heavy ions (Jäkel et al. 2001). This radiobiological effect is accounted for using the iso-effective dose, defined as the product of the radiation absorbed dose and the RBE. A contemporary review of the precision required in hadron therapy dosimetry and the means of achieving this is provided by Karger et al. (2010).

1.4.4.3 Hadron Therapy: Protons

Protons are the most commonly used radiation in therapeutic medicine after photons and electrons. By the end of 2010, approximately 90,000 patients worldwide had undergone proton therapy (PTCOG 2011). The use of protons as a modality for the treatment of malignant disease was first proposed by Wilson (1946). But it was not until the mid-1950s when proton therapy was first applied in a collaboration between the Harvard Cyclotron Laboratory and the Massachusetts General Hospital and at Uppsala University (the latter still continuing). Up to the late 1980s, proton therapy was a tool limited primarily to research facilities with proton accelerators that were

[31] Details of the roles of RBE and OER can be found in Chapters 6 and 7 of Hall and Giaccia (2006).

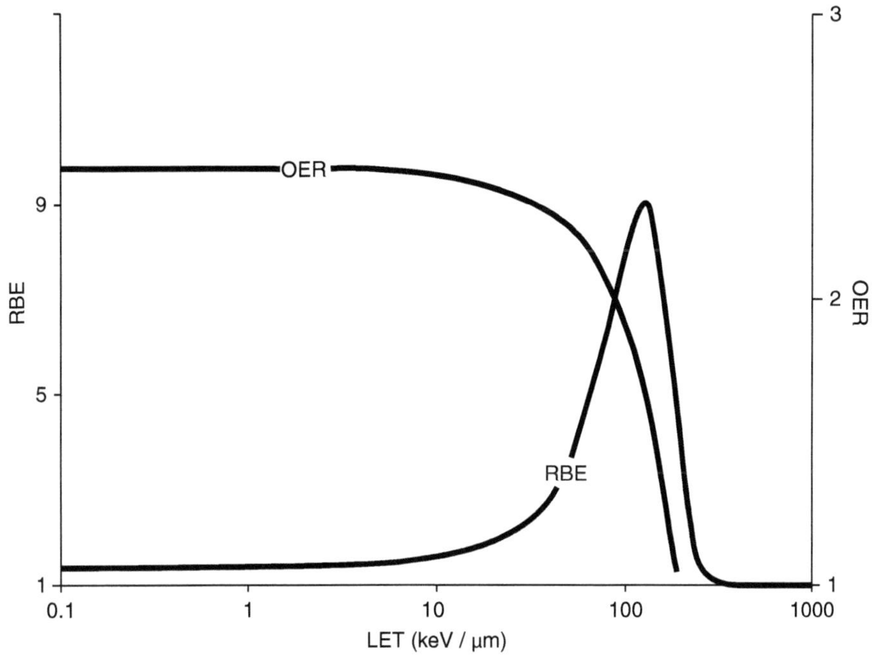

Fig. 1.20 The relative biological effectiveness (*RBE*) and oxygen enhancement ratio (*OER*) as a function of linear energy transfer (*LET*). The high *LET* offers the advantages of elevated *RBE* and reduced *OER*

built and used in basic science applications. It was not until 1990 that the first purpose-built facility for proton therapy was opened and this was at Loma Linda University in California (Slater 2006).

Isocentric techniques with protons and other accelerator-produced hadrons require complex beam transport systems in order for the treatment gantry to rotate about the patient.[32] As a result, some proton facilities are restricted to a horizontal beam which restricts the types of anatomical sites that can be treated. An example of the horizontal proton therapy beam as used at TRIUMF in Vancouver is shown in Fig. 1.21 and which is used for the treatment of ophthalmological neoplasia. The depth of the proton Bragg peak can be adjusted in order to be coincident with the depth of the tumour, by varying the incident proton beam. This is done by using a fixed proton beam energy (selected so that the undegraded Bragg peak is placed at the distal depth of the volume to be irradiated) and a variable thickness energy degrader that the beam passes through prior to being incident to the patient. Collision energy losses from the protons resulting from interactions with the atoms in

[32] One solution is the mounting of the cyclotron on the rotating gantry. In practice, this would require the cyclotron to have an especially intense magnetic field in order to reduce its size and, hence, the use of superconducting magnetic coils.

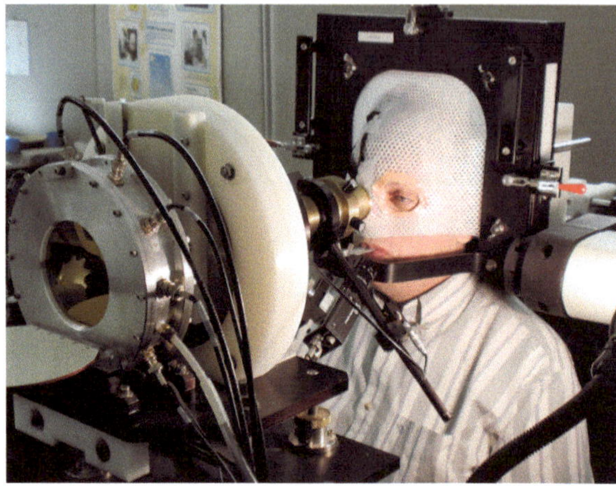

Fig. 1.21 Proton therapy at TRIUMF, Vancouver (Courtesy and © of TRIUMF)

the degrader result in the exiting proton beam having a lower kinetic energy and a shallower Bragg peak depth. The lateral extent of the beam can be adjusted by using a passive scatterer or by the use of scanning magnets. The latter method is considered generally to be more desirable as the former approach creates secondary neutrons which can deposit absorbed doses in unintended volumes of tissue.

1.4.4.4 Hadron Therapy: Heavy Ions

As with protons, heavy ions are of interest to radiation therapy due to the Bragg peak and the dense ionisations that lead to improved radiobiological properties as a result of a particularly elevated RBE and reduced OER. However, these are expected to be more significant due to the greater masses and charges of the heavy ions.

Accelerator-based heavy-ion radiotherapy was begun at Berkeley with eventually some 2,500 oncology patients treated with carbon, neon, silicon and argon ions between 1957 and 1992. At the time of writing, the only active heavy-ion therapy facilities are in Europe and Japan, and these are dominated by the use of carbon ions (plans are reported to be considered for heavy-ion therapy facilities in China). As of 2011, about 6,600 patients throughout the world have been treated with carbon ions, 90 % or more of them in Japan at the Heavy-Ion Medical Accelerator in Chiba and the Hyogo Ion Beam Medical Center (Cirrilli 2011). Due to the fragmentation of the carbon ion which, through direct nuclear interactions, results in lighter fragments which have a greater degree of penetration into tissue, the distal extent of the carbon ion Bragg peak is greater than that of the proton. One advantage of the therapeutic use of carbon ions is that as a result of the nuclear interactions in tissue, neutrons can be stripped from the ^{12}C nucleus to leave a ^{11}C nucleus and which is a positron-emitting nucleus. As the range of this nucleus is very nearly that of the ^{12}C nucleus, it provides a visualisation of the treatment volume in posttreatment PET imaging.

As a result of the Bragg peak and reduced scatter, carbon ion therapy is used primarily for those cases of where the tumour to be treated is in close physical proximity to radiosensitive normal tissue or in paediatric patients where the incidences of radiation-induced sequelae are sought to be reduced. As noted earlier, the calculation of heavy-ion energy loss has its genesis in the theoretical studies of the energy losses of nuclear fission fragments. These have indicated the complexities resulting from the heavy ion's ability to capture atomic electrons and thus vary its effective electric charge.

1.4.4.5 Hadron Therapy: π^--Mesons

Due to their rest mass of 139 MeV, charged pions will exhibit a Bragg peak as they decelerate in tissue. But, in addition to this therapeutic advantage, the capture of the negatively charged pion by the nucleus will lead to the deposition of a significant of energy within the nucleus.[33] These features led to negatively charged pions being explored as a means of selective radiotherapy as a result of the proposal, in 1961, of Perkins and Fowler of the use of the negatively charged π-meson as a means of delivering a highly localised radiation absorbed dose (Perkins and Fowler 1961). Being negatively charged and subject to the strong nuclear force, the decelerated π^--meson will be captured by the positively charged nucleus of an atom and cascade to lower orbitals through a chain of x-ray emissions. As the Bohr atomic radius is inversely proportional to the particle mass, the pion will be much closer to the atomic nucleus than any of the atomic electrons, and the pion wavefunction will overlap that of the nucleus. Much of the rest mass of the pion will be within the nucleus and thus interacts directly with the nucleons. Clearly, this deposition of energy leads to an unstable situation and the nucleus will break apart to create multiple densely ionising fragments. This final state is referred to as a 'star' due to the appearance of the ionisation tracks within a film emulsion (Fig. 1.22). These densely ionising products will, due to their combination of high mass and electric charge, slow down quickly and create a highly localised radiation absorbed dose.

Clinical trials of π^- therapy in oncology were initiated in the 1970s and 1980s at TRIUMF in Vancouver, SIN (now the Paul Scherrer Institute) in Villigen and at LAMPF in Los Alamos. At each centre, pions were produced through the bombardment of targets by protons generated in high-beam current proton accelerators (e.g. Fig. 1.23). Some of these trials continued until the 1980s and 1990s using conventional external megavoltage photon beam therapy as a control (von Essen et al. 1987; Goodman et al. 1990). In general, the clinical results were not decisive which, combined with the need of advanced large and expensive proton

[33] In this case, the pion (or its wavefunction) entering the nucleus is 'real' and the effects of its rest mass are cogent. Pions mediating the strong nuclear force within the nucleus are 'virtual' and do not contribute to the same result.

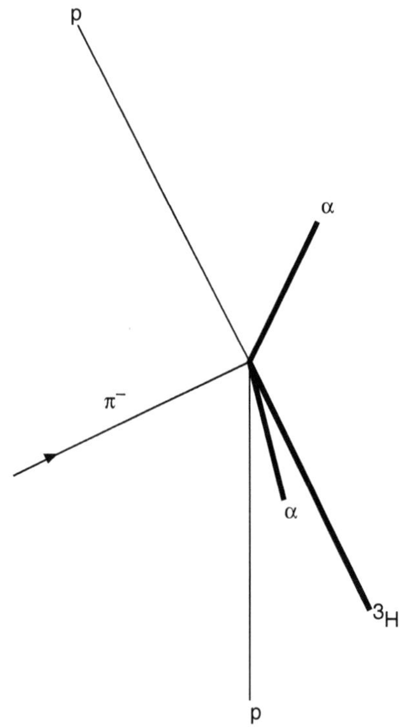

Fig. 1.22 'Star' production resulting from the capture of a π^- by an oxygen nucleus in a film emulsion. The oxygen nucleus disintegrates following the absorption of the π^- into energetic fragments: two protons, two α-particles and a ^{3}H nucleus (Redrawn from the emulsion photograph Powell et al. (1959))

accelerators to produce pion beams of sufficient energy and flux, has led to the abandonment of π^- as a therapeutic modality.

1.4.4.6 Hadron Therapy: Neutrons

As the neutron has no electric charge, it does not exhibit a Bragg peak but, instead, has a depth-dose profile (i.e. the absorbed dose along the central axis of the beam) similar to that of photons in that both exponentially decrease with depth. Neutrons deposit absorbed dose primarily through a secondary proton flux created through the (n,p) elastic scattering reaction on hydrogen in tissue. These protons have a LET, which is closely related to the collision stopping power. This results in a highly localised dose deposition and a greater RBE and OER.

Neutron therapy was the first externally administered hadron modality, beginning with the use of fast neutrons[34] in the late 1930s at Berkeley University in California. In fact, it would appear that the first patient treated with neutrons at

[34] 'Fast' neutrons are defined as those with an energy of the order of 1 MeV. These are moderated as they pass through a hydrogenous medium (such as tissue) through (n,p) elastic scatters until they 'thermalise'.

Fig. 1.23 The TRIUMF cyclotron in Vancouver during construction in January 1972. It remains the world's largest. TRIUMF was one of three sites worldwide used for pioneering studies of π^- therapy up to the 1990s, and it is currently used for proton therapy (Courtesy and © of TRIUMF)

Berkeley was the mother of Ernest O Lawrence (who won the Nobel Prize in Physics in 1939 for the invention of the cyclotron) and John Lawrence (often referred to as one of the first nuclear medicine physicians). This was likely performed in 1938 using the cyclotron that the former had designed and had built, and it was an act that was intended to be kept confidential.[35] The initial interest in neutron beam therapy was not rigorously based upon radiobiological considerations, although a differential in effect upon normal tissue versus tumour had been noted. Rather, it was more of a result of interest in the availability of a new radiotherapy modality.

Neutron therapy has had a mixed history of results as demonstrated in a variety of clinical trials, with the most fundamental ones coming from the UK's Medical Research Council's 60-inch-diameter cyclotron at the Hammersmith Hospital throughout the 1950s and into the 1970s. In more recent times, neutron therapy

[35] The Lawrences, mother had been diagnosed with a pelvic tumour with the prognosis of a 3-month survival. Following the neutron therapy, she survived for sixteen years.

has evolved into the use of the capture of thermal neutrons by boron which has been taken up by tumour. Boron neutron-capture therapy (BCNT) is based upon the high thermal neutron absorption cross section of boron through the $^{10}B(n, {}^{4}He)^{7}Li$ reaction (Hatanaka and Nagakawa 1994). Both of the reaction products, due to their high mass and electric charge, have a short range and are densely ionising, thus yielding a highly localised absorbed dose. This requires a vector to deliver the boron atom with high specificity to the tumour. In addition, the boron must be uniformly distributed throughout the tumour in order to compensate for the limited ranges of the ^{4}He and ^{7}Li nuclei.

In addition to external neutron beam irradiation, neutron brachytherapy has been investigated using sealed sources of ^{252}Cf. In the main, ^{252}Cf undergoes α-decay (97 %) with the remainder of the decays being of spontaneous fission. During the latter decay, it emits fast neutrons with a mean kinetic energy of about 2.2 MeV.

1.4.4.7 Therapy: Nuclear Medicine

Radionuclide therapy is the administration of unsealed radioactive substances for the treatment of disease, with administration either oral or intravenous. The first application of RNT (as we would consider it) was the use of ^{32}P by Lawrence in the treatment of leukaemia and, then, polycythaemia vera (a myeloproliferative disease) in the 1930s. However, the most prevalent radioisotope in RNT is reactor-produced ^{131}I used in the treatment of hyperthyroidism and carcinoma of the thyroid gland with orally administered Na^{131}I. Thyroid tissue preferentially takes up the ^{131}I and is consequently irradiated by the β-particle component of the radioactive decay.

RNT is based upon the need of a suitable vector to deliver an appropriate radionuclide to the target. The nature of the vector is not within the scope of this book whereas the emissions of the radionuclide are. Because of the greater mean free path length, a photon-emitting radionuclide is not preferred in therapeutic applications as the density of ionisations and energy deposition is limited. Charged particle-emitting radionuclides are desirable due to their much reduced mean free path lengths.[36] These will limit the irradiation of normal tissues – but at the demand of requiring a high specificity (targeting ability) of the vector carrying the radionuclide to the target tissue. Therapeutic radionuclides include those that emit β-particles, low-energy Auger electrons and α-particles. These have limited physical ranges which enable a higher therapeutic ratio (ratio of absorbed dose to tumour to that of normal tissue). The selection of the type of particle to use is dependent upon the vector profile. If a large volume is to be treated or there is limited specificity, longer-ranged β-particles may be preferable due to 'cross fire' across

[36] The mean free path length is derived in Chap. 6.

the target volume. Otherwise, shorter-ranged α-particles may be desirable if they can be delivered to the target with high specificity (Salvatori et al. 2008).

Unlike external beam radiotherapy, where highly accurate patient-specific dosimetry is the norm, RNT is not accompanied frequently with pretreatment planning. The amount of activity of the radionuclide to be administered can often be selected on the basis of clinical indication and experience, body weight or body surface area, in a fashion similar to conventional nonradioactive therapeutics. The ability to determine the distribution of the activity in vivo over time through external radiation detection enables more sophisticated means of RNT planning (Fleming and Perkins 2000; McParland 2010). The Medical Internal Radiation Dose (MIRD) schema (Loevinger et al. 1988) presents a means of calculating the absorbed dose to target regions due to the emissions from a radionuclide contained within multiple source regions through calculation of the transport of radiation through tissue and the energy deposition by charged particles. Because of the computational burden in evaluating this transport, the MIRD approach is most often used in simple stylised cases using reference anthropomorphic phantoms which are, in general, not suitable for therapeutic applications where the magnitudes of the absorbed doses are high.

1.4.4.8 Diagnosis: Radiology

In a diagnostic radiology examination, the patient is exposed to an external beam of x-rays which transits the body to a detector and the resulting attenuation information used to form an image. As the radiation source is external to the body, the anatomy irradiated is known from the relative geometric placements of the beam and the patient. Unlike therapy, patient-specific dosimetry calculations are rarely performed for diagnostic procedures due to both the relatively low absorbed doses involved (and the general restriction to stochastic effects) and the large number of patients who undergo such examinations. On the other hand, patient-specific measurements which are indirect metrics of dosimetry are frequently done. These would include the dose–area product for planar x-ray beam exposures (the product of the in-air collision kerma incident to the patient and the cross-sectional area of the beam) and the dose–length product for a CT scanner. It is also possible to infer dosimetry from technical exposure data such as x-ray tube potential, tube current or current–time product, filtration and the distance between the x-ray tube focus and that patient. These physical measurements can then be used to calculate the corresponding radiation dosimetry of a reference anthropomorphic phantom using pre-calculated Monte Carlo simulation-derived data. In diagnostic radiology, procedure-specific dose reference levels are frequently assessed by audit and published in order to attempt to standardise imaging practice. In fact, within the UK, this is mandated by law through the Ionising Radiation (Medical Exposures) Regulations (IR(ME)R) 2000.

While the radiation measurement is patient specific, the dosimetry calculation is not as it is done for one of a number of reference anthropomorphic phantoms which act as surrogates for the patient (Stabin et al. 1995; Xu and Eckerman 2010).

1.4.4.9 Diagnosis: Nuclear Medicine

Whereas radiology (imaging with x-rays) is based upon anatomical imaging in that it attempts to reproduce the morphology, architecture or structure of anatomy, diagnostic nuclear medicine is based upon functional imaging in that the image produced replicates underlying pathology or function rather than anatomy. In diagnostic nuclear medicine, a radiopharmaceutical is administered to the patient, in most cases intravenously. The radionuclide attached to the radiopharmaceutical may emit photons directly as a result of radioactive decay or may emit positrons which annihilate with electrons present in tissue to produce γ-rays. The photons are detected externally and used to create either a planar or tomographic image.

The administration of a radioactive substance necessarily involves the exposure of the patient to a radiation absorbed dose. As in diagnostic radiology, patient-specific dosimetry calculations are not performed. However, as part of the characterisation of the safety aspects of the radiopharmaceutical, a radiation absorbed dose profile must be evaluated. This is performed using the Medical Internal Radiation Dose (MIRD) equations for reference anthropomorphic phantoms rather than for an individual patient. Details of how this calculation is performed may be found in Stabin (2008) and McParland (2010). In this dosimetry evaluation, the mean absorbed dose to the target region $\mathbf{r}_T$ due to radiation from a source region $\mathbf{r}_S$ containing activity is given by the expression

$$\bar{D}_{\mathbf{r}_T} = \tilde{A}_{\mathbf{r}_S} \frac{\sum_i w_{R,i} Y_i E_i \phi(\mathbf{r}_T \leftarrow \mathbf{r}_S; E_i)}{m_{\mathbf{r}_T}} \tag{1.16}$$

where $\tilde{A}_{\mathbf{r}_S}$ is the cumulated activity in $\mathbf{r}_S$, which is the total number of decays of the radionuclide in the source region, and $m_{\mathbf{r}_T}$ is the mass of the target region. The summation in the numerator is over all available nuclear decay channels. $w_{R,i}$ is the radiation weighting factor for the i-th emission channel and is a quantity related to the RBE: for photons and electrons/positrons (excluding Auger electrons) of energies typical of nuclear medicine, it is equal to unity and is assigned a value of 20 for α-particles (ICRP 2007). Y_i and E_i are the yield and energy released per decay of the i-th emission channel; the product $Y_i E_i$ is the mean energy released by the i-th emission channel. The quantity $\phi(\mathbf{r}_T \leftarrow \mathbf{r}_S; E_i)$ is the absorbed fraction for the source and target region combination for the energy E_i and is the fraction of that energy released in $\mathbf{r}_S$ that is absorbed in $\mathbf{r}_T$. The absorbed fraction is calculated by Monte Carlo simulation of radiation transport in a phantom. In most current clinical applications, this phantom is one of the series designed by Cristy and Eckerman (1987) and by Stabin et al. (1995).

Bibliography and Further Reading

Bohr N, Thorsen J. Niels Bohr – Collected works, vol 8: The penetration of charged particles through matter (1912–1948). Amsterdam: North-Holland Physics Publishing; 1987.
Davis NP. Lawrence and Oppenheimer. New York: Da Capo Press; 1986.
Harbert JC, Eckelman WC, Neumann RD, editors. Nuclear medicine – diagnosis and therapy. New York: Thieme Medical Publishers; 1996.
Haug E, Nakel W. The elementary process of *bremsstrahlung*. Singapore: World Scientific Publishing Co; 2004.
Heitler W. The quantum theory of radiation. 3rd ed. Oxford: Clarendon Press; 1954.
Heilbron JL. The scattering of α and β particles and Rutherford's atom. Arch Hist Exact Sci. 1967;17:247–307.
Moore R. Niels Bohr. Cambridge, MA: The MIT Press; 1985.
Rhodes R. The making of the atomic bomb. New York: Simon and Schuster; 1988.
Segrè E. Enrico Fermi – Physicist. Chicago: University of Chicago Press; 1970.
Sigmund P. Particle penetration and radiation effects. Berlin: Springer; 2008.
Sime RL. Lise Meitner – a life in physics. Berkeley: University of California Press; 1996.
Thomson JJ, Thomson GP. Conduction of electricity through gases, vol II: Ionisation by collision and the gaseous discharge. 3rd ed. Cambridge: University Press; 1933.
Wilson D. Rutherford – simple genius. London: Hodder and Stoughton; 1983.

References

Allen C, Bloyce A, Bell T. The effect of ion implantation on the sliding wear behaviour of ultrahigh molecular weight polyethylene against an oxidized titanium alloy Ti-6A1–4V. Tribol Lett. 1995;1:35–38.
Andreo P. Monte Carlo techniques in medical radiation physics. Phys Med Biol. 1991;36:861–920.
Attix FH. Introduction to radiological physics and radiation dosimetry. New York: Wiley; 1986.
Becquerel H. Sur les propriétés différentes des radiations invisibles émises par ltes sels de l'uranium et du rayonnement de la paroi anticathodique d'un tube de Crookes. Comp Rend. 1896;122:702–7.
Becquerel H. Sur la déviabilité et la nature de certains rayons émis par le radium et le polonium. Comp Rend. 1903;136:199–204.
Bethe HA. Zur Theorie des Durchgangs schneller Korpuskularstrahlen durch Materie. Ann Phys. 1930;5:324–400.
Bloch F. Zur Bremsung rasch bewegter Teilchen beim Durchgang durch Materie. Ann Phys. 1933;16:285–320.
Bohr N. On the theory of the decrease of velocity of moving electrified particles on passing through matter. Philos Mag. 1913;25:10–31.
Bohr N. On the decrease of velocity of swiftly moving electrified particles in passing through matter. Philos Mag. 1915;30:581–612.
Bohr N. Scattering and stopping of fission fragments. Phys Rev. 1940;58:654–5.
Bohr N. The penetration of atomic particles through matter. Mat Fys Medd Dan Vid Selsk. 1948;18(8):1–144.
Bohr N, Bøggild JK, Brostrøm KH, Lauritsen T. Velocity-range relation for fission fragments. Phys Rev. 1940;58:839–40.
Born M. Zur quantenmechanik der Stovorgänge. Z Physik. 1926;37:863–72.
Boswell CA, Brechbiel MW. Auger electrons: lethal, low energy, and coming to a tumor cell nucleus near you. J Nucl Med. 2005;46:1946–7.

Bragg WH, Kleeman R. On the alpha particles of radium, and their loss of range in passing through various atoms and molecules. Philos Mag. 1905;10:318–41.

Brailsford JF. Roentgen's discovery of x rays. Br J Radiol. 1946;19:453–61.

Brown LM, Dresden M, Hoddesson L, editors. Pions to quarks. Cambridge: Cambridge University Press; 1989.

Chadwick J. Possible existence of a neutron. Nature. 1932;129:312.

Cirrilli M. Hadron therapy: collaborating for the future. CERN Cour. 2011;51:37–40.

Cristy M, Eckerman KF. Specific absorbed fractions of energy at various energies from internal photon sources, vols I–VII. Report ORNL/TM-8381. Oak Ridge: Oak Ridge National Laboratory; 1987.

Curie P. Action du champ magnétique sure les rayons de Becquerel. Rayons déviés et rayons non déviés. Comp Rend. 1900a;130:73–6.

Curie P. Sur la pénétration des rayons de Becquerel non déviables par le champ magzlétique. Comp Rend. 1900b;130:76–9.

Darwin CG. A theory of the absorption and scattering of the α rays. Philos Mag. 1912;23:901–21.

Ehrenfest P, Oppenheimer JR. Note on the statistics of nuclei. Phys Rev. 1931;37:333–8.

Faddegon B, Ross CK, Rogers DWO. Measurements of collision stopping powers of graphite, aluminium and copper for 10 and 20 MeV electrons. Phys Med Biol. 1992;37:1561–78.

Fleming JS, Perkins AC. Targeted radiotherapy. IPEM Report 83. York: Institute of Physics and Engineering in Medicine; 2000.

Gaisser TK. Cosmic rays and particle physics. Cambridge: Cambridge University Press; 1990.

Geiger H, Marsden E. On a diffuse reflection of the α-particles. Proc R Soc. 1909;A82:495–500.

Gerward L. Paul Villard and his discovery of gamma rays. Phys Perspect. 1999;1:367–83.

Goodman GB, Skarsgard LD, Thompson GB, Harrison R, Lam GJ, Ludgate C. Pion therapy at TRIUMF. Treatment results for astrocytoma grades 3 and 4: a pilot study. Radiother Oncol. 1990;17:21–8.

Grotz K, Klapdor HV. The weak interaction in nuclear, particle and astrophysics. Bristol: IOP Publishing; 1990.

Hall EJ, Giaccia AJ. Radiobiology for the radiologist. 6th ed. Philadelphia: Lippincott, Williams and Wilkins; 2006.

Hatanaka H, Nagakawa Y. Clinical results of long surviving brain tumor patients who underwent boron neutron capture therapy. Int J Radiat Oncol Biol Phys. 1994;28:1061–6.

Howell RW. Radiation spectra for Auger-emitting radionuclides. Med Phys. 1992;19:1371–83.

Hrabak M, Padovan RS, Kralik M, Ozretic D, Potocki K. Nikola Tesla and the discovery of x-rays. Radiographics. 2008;28:1189–992.

IAEA. Absorbed dose determination in external beam radiotherapy – an international code of practice for dosimetry based on standards of absorbed dose to water. Technical Report Series No 398. Vienna: International Atomic Energy Agency; 2000.

ICRP. The 2007 recommendations of the International Commission on Radiological Protection. ICRP publication 103. Oxford: Pergamon; 2007.

ICRU. Determination of absorbed dose in a patient irradiated by beams of x or gamma rays in radiotherapy procedures. ICRU Report No 24. Bethesda: International Commission on Radiological Units and Measurements; 1976

ICRU. International Commission on Radiation Units and Measurements. ICRU Report 37 – Stopping Powers for Electrons and Positrons. Bethesda: ICRU; 1984.

ICRU. International Commission on Radiation Units and Measurements. ICRU Report 49 – Stopping Powers and Ranges for Protons and Alpha Particles. Bethesda: ICRU; 1993.

Jäkel O, Krämer M, Karger CP, Debus J. Treatment planning for heavy ion radiotherapy: clinical implementation and application. Phys Med Biol. 2001;46:1101–16.

Johns HE, Cunningham JR. The physics of radiology. 4th ed. Springfield: CC Thomas; 1983.

Karger CP, Jäkel O, Palmans H, Kanai T. Dosimetry for ion beam radiotherapy. Phys Med Biol. 2010;55:R193–234.

Knipp J, Teller E. On the energy loss of ions. Phys Rev. 1941;59:659–69.

Krimigis SM, Armstrong TP, Axford WI, Bostrom CO, Fan CY, Gloeckler G, Lanzerotti LJ. The low energy charged particle (LECP) experiment on the Voyager spacecraft. Space Sci Rev. 1977;21:329.

Lamb WE. Passage of uranium fission fragments through matter. Phys Rev. 1940;58:696–702.

Lattes CMG, Muirhead H, Occhialini GPS, Powell CF. Processes involving charged mesons. Nature. 1947;159:694–7.

Lee S, Brown GE. Hans Albrecht Bethe. 2 July 1906 – 6 March 2005: Elected ForMemRS 1957. Biogr Mems Fell R Soc. 2007;53:1–20.

Lipkin HJ. Beta decay for pedestrians. Amsterdam: North-Holland Publishing; 1962.

Ljungberg M, Strand S-E, King MA, editors. Monte Carlo calculations in nuclear medicine – applications in diagnostic imaging. Bristol: Institute of Physics; 1998.

Loevinger R, Budinger T, Watson E. MIRD primer for absorbed dose calculations. New York: Society of Nuclear Medicine; 1988.

McParland BJ. Nuclear medicine radiation dosimetry: advanced theoretical principles. London: Springer; 2010.

Metropolis N, Ulam S. The Monte Carlo method. J Am Stat Assoc. 1949;44:335–41.

Mould RF, Hobday PA. Radiation dosimetry for the Amersham caesium 137 manual afterloading system for gynaecological brachytherapy. Amersham: Amersham International plc; 1983.

Newton I. The Principia. Translated by Andrew Motte (1995). Amherst: Prometheus Books; 1687.

NIST (National Institute of Science and Technology). http://www.nist.gov/pml/data/star/index.cfm (2012). Accessed February 2012.

Noddack I. Über das element 93. Z fur Ange Chemie. 1934;47:653.

Perkins PH, Fowler DH. The possibility of the therapeutic applications of beams of negative pi-mesons. Nature. 1961;189:524–8.

Powell CF, Fowler PH, Perkins DH. Study of elementary particles by the photographic method. Oxford: Pergamon; 1959.

PTCOG (Particle Therapy Co-operative Group). http://ptcog.web.psi.ch/Archive/Patientenzahlen-updateMay2011.pdf (2011). Accessed November 2011.

Robins B. New principles of gunnery. London; 1742.

RSICC (Radiation Safety Information Computational Center). C00228 SPAR, Calculation of stopping powers and ranges for muons, charged pions, protons and heavy ions. http://rsicc.ornl.gov (1985). Accessed March 2012.

Rutherford E. On the electrification of gases exposed to Röntgen rays, and the absorption of Röntgen radiation by gases and vapours. Philos Mag. 1897;43:241–55.

Rutherford E. Uranium radiation and the electrical conduction produced by it. Philos Mag. 1899;47:109–63.

Rutherford E. The magnetic and electric deviation of the easily absorbed rays from radium. Philos Mag. 1903;5:177–87.

Rutherford E. The scattering of α and β particles by matter and the structure of the atom. Philos Mag. 1911;21:669–88.

Rutherford E. Collisions of α particles with light atoms. IV. An anomalous effect in nitrogen. Philos Mag. 1919;37:581.

Rutherford E, Royds T. The nature of the α-particle from radioactive substances. Philos Mag. 1909;17:281–6.

Salvatori M, Indovina L, Mansi L. Targeted α-particle therapy: a clinical overview. Clin Radiopharm. 2008;1:251–3.

Slater JD. Clinical applications of proton radiation treatment at Loma Linda University: review of a fifteen-year experience. Tech Cancer Res Treat. 2006;5:81–9.

Smith EM, Harris CC, Rohrer RH. Calculation of local energy deposition due to electron capture and internal conversion. J Nucl Med. 1965;7:23–31.

SRIM (The Stopping and Range of Ions in Matter). http://www.srim.org/ (2012). Accessed February 2012

Stabin MG. Fundamentals of nuclear medicine dosimetry. New York: Springer; 2008.

Stabin MG, Watson EE, Cristy M, Ryman JC, Eckerman KF, Davis JL, Marshall D, Gehlen MK. Mathematical models and specific absorbed fractions of photon energy in the nonpregnant

female and at the end of each trimester of pregnancy. Report ORNL/TM-12907. Oak Ridge: Oak Ridge National Laboratory; 1995.
Strutt RJ. The absorption of the Becquerel rays by solid and gaseous bodies. Nature. 1900;61:539–40.
Thomson JJ. Cathode rays. Philos Mag. 1897;44:293.
Thomson JJ. Conduction of electricity through gases. 2nd ed. Cambridge: University Press; 1906.
Thomson JJ. Rays of positive electricity. Proc R Soc. 1913;A89:1–20.
Thomson JJ, Rutherford E. On the passage of electricity through gases exposed to Röntgen rays. Philos Mag. 1896;42:392–407.
Von Essen CF, Bagshaw MA, Bush SE, Smith AR, Kigerman MM. Long-term results of pion therapy at Los Alamos. Int J Radiat Oncol Biol Phys. 1987;13:1389–98.
Wigner E. On the mass defect of helium. Phys Rev. 1933;43:252–7.
Wilson RR. Radiological use of fast protons. Radiology. 1946;47:487–91.
Xu XG, Eckerman KF. Handbook of anatomical models for radiation dosimetry. Boca Raton: CRC Press; 2010.
Yukawa H. On the interaction of elementary particles. I. Proc Phys Math Soc Jpn. 1935;17:48.
Zaidi H, Sgouros G, editors. Therapeutic applications of Monte Carlo calculations in nuclear medicine. Bristol: Institute of Physics; 2003.
Ziegler JF. Handbook of stopping cross-sections for energetic ions in all elements, The stopping and ranges of ions in matter, vol. 5. New York: Pergamon Press; 1980.
Ziegler JF. The stopping of energetic light ions in matter. J Appl Phys. 1999;85:1249–72.
Ziegler JF, Srinivasin GR (eds). Terrestrial cosmic rays and soft errors. IBM J Res Dev. 1996;40:1–128.

Chapter 2
Elements of Quantum Scattering Theory

Contents

Abstract This chapter provides the elements of quantum scattering theory which shall be used throughout this book. The general scattering problem is reviewed in the context of scattering and time evolution operators and recognises that the observables are the pre- and post-scattering asymptotic states. Quantum perturbation theory is developed with the two linked goals of deriving an expression for the transition probability rate between two quantum states (Fermi's Golden Rule) and the scattering amplitude and differential cross section (first Born approximation). The Born approximation is derived through the Lippmann–Schwinger equation. The chapter concludes with a phase shift analysis of the scattering problem.

B.J. McParland, *Medical Radiation Dosimetry*,
DOI 10.1007/978-1-4471-5403-7_2,

2.1 Introduction

The topic of this book is that of the interactions of a moving charged particle as it slows down within a medium which result in kinetic energy loss through energy transfers to the medium. These interactions are referred to as 'scatters' or 'collisions'[1] as the projectile will be scattered from its original trajectory into a different direction, a process which may or may not involve a significant transfer of energy from the projectile to the target. A light projectile, such as an electron, scattered from a heavy nucleus will not, due to the requirement of the simultaneous conservations of energy and momentum, impart significant recoil energy to the nucleus, and little energy is transferred to the medium containing that nucleus. On the other hand, if the projectile and target have masses which are comparable or of the same order of magnitude (e.g. as in electron–electron scatter or the scatter of an α-particle from a light nucleus), then significant energy transfer can occur (see Appendix D). These interactions are probabilistic in nature and are described by differential cross sections. On the other hand, the kinematics of the scatter is not probabilistic and is well defined. For radiation dosimetry calculations, we are most interested in the probabilities of given amounts of kinetic energy being transferred through a scatter. This chapter provides the basic components of the quantum 'toolkit' required to calculate these interactions and their consequences. It does not intend to be a full and detailed exposition of quantum scattering theory: it is expected that the reader will already be at least aware of much of the material summarised here, and that, in a sense, this chapter provides a review of those aspects relevant to the calculation of the transfer of the kinetic energy of a moving charged particle to the medium it transverses. More specific attributes of the theory that we make singular or infrequent use of will be introduced at the time of the relevant discussion elsewhere in this book. The reader who wishes further detailed description of quantum scattering theory may refer to the texts cited in the references to this chapter.

Depending upon the physics application, there can be a variety of definitions of 'elastic' and 'inelastic' scatter. In this book, an elastic scatter is defined as one in which the total kinetic energy is conserved: the sum of the kinetic energies of the projectile and target in the initial state is equal to the sum of the kinetic energies of the products in motion in the final state. On the other hand, an inelastic scatter is defined as one in which the sum of the post-scatter kinetic energies is less than that of the pre-scatter state. This would be due to the diversion of the initial state kinetic energies to internal degrees of freedom provided by the scattering system. Such channels would include atomic or nuclear excitation and ionisation.[2] In general, we will consider the interaction to be that of the scattering from the Coulomb potential energy of $\frac{\alpha\hbar c z Z}{r}$, where z and Z are the electric charges of the projectile and target in

[1] Collision and scatter have been differentiated by some authors so that the former correspond to multichannel final states, and the latter was linked to a single channel final state. Such a distinction will not be made here.

[2] *Bremsstrahlung* is sometimes referred to as being inelastic. However, it is an elastic process by our definition as the pre- and post-scatter energies of the electron and of the electron/photon/recoiling nucleus set are the same.

integer multiples of the fundamental electric charge, e, or a Yukawa-type modification of the Coulomb potential as a model of electrostatic screening. Should the scattering centre not be well modelled as infinite mass or itself be in motion, then the scattering problem can be solved in the centre-of-mass system in which the projectile mass in the calculation is replaced by the reduced mass of the projectile and target and the target considered to be a potential. Even the inelastic scatter from an atomic electron which is ejected as a consequence can be well approximated as an elastic interaction if the kinetic energies involved greatly exceed the electron binding energy. Indeed, it will often be the case that the atomic electron can be considered to be at rest if the projectile velocity greatly exceeds the orbital velocity of the electron.[3]

Our study of scatter, however, will not consider the pure multi-body state.[4] There is no need to introduce this complication as the flux of projectiles and the density of targets are such that, at the microscopic level, a binary interaction need only be considered.[5] Hence, we will restrict ourselves to the problems of a single projectile incident to a single target or potential. As the scattering problem is fundamental to many diverse fields of physics in addition to our interest in radiation dosimetry, we will model our derivations of the calculated tools provided by this chapter upon the most general aspects of particle scattering.

2.2 The General Scattering Problem

2.2.1 *Initial and Final Observable States*

The general scattering problem must be thought of as follows. Before and after the interaction in those regions of space where the scattering potential can be assumed to be effectively equal to zero, the particles are considered to be described by wave packets which are solutions to the homogeneous Schrödinger equation. The Heisenberg uncertainty principle $\Delta p\, \Delta x \approx \hbar c$ states that one cannot precisely measure the position of the particle without a corresponding uncertainty in determining its momentum. In this region of space, the Hamiltonian is the free representation, H_0, with the eigenvalue of the particle's kinetic energy, E. This implicitly requires the static potential U which resulted in the scattering of the potential to be

---

[3] Although this will not always be the case, particularly for high-Z atomic targets. The effect of a non-negligible atomic electron velocity is to reduce the rate at which energy is lost. The correction factor for this effect is derived in Chap. 12.

[4] When we do consider the multi-body problem in which the target is made up of multiple charges (e.g. the scatter of a projectile due to the combined Coulomb interactions with the nucleus and electrons of an atom), the multi-body problem will be simplified and reduced to the sum of independent two-body scattering calculations in which it is assumed that the interactions between individual targets are uncorrelated.

[5] In the case of a 'dressed ion' projectile (i.e. one carrying atomic electrons), this cannot be the case as the electrons themselves can participate in the scattering.

non-zero only within a finite range a. More precise definitions of this requirement include, first, that the potential be locally square integrable

$$\int_0^a d^3\mathbf{r}' \, |U(\mathbf{r}')|^2 < \infty \tag{2.1}$$

and, second, that the magnitude of the potential be limited at infinite distances,

$$|U(\mathbf{r})| < U_{\text{Max}} \quad \text{for} \quad |r| \geq a \quad \text{and} \quad a < \infty. \tag{2.2}$$

Within the range a of the potential, the total Hamiltonian which includes the potential $H = H_0 + U$ must be considered as to be discussed. Note that we have used the general form of the potential as having a directional dependence through the use of the vector $\mathbf{r}$ as the argument rather than the scalar quantity, r. While a well-known potential having a directional dependence is that of the tensor nuclear force associated with the deuteron, such potentials are of no interest to the scattering problems examined in this book, although the initial steps of derivations will use $\mathbf{r}$ for generality before adopting r when it becomes appropriate. The potentials that we will consider (the attractive/barrier well, the Yukawa and the Coulomb) are all central. However, in particular, it should be noted that the $\frac{1}{r}$ nature of the Coulomb potential means that that potential cannot meet the conditions of either (2.1) or (2.2). This necessitates either the careful use of perturbation theory for the Yukawa potential $\left(\propto \frac{e^{-\kappa r}}{r}\right)$ where, following integration, the screening parameter κ is set to zero or else a full and precise calculation of the Coulomb scattering problem is performed, as shall be shown in Chap. 4.

Continuing from these requirements, let us next consider the macroscopic fundamentals of particle scattering from the perspective of the experimentalist who can observe only the macroscopic initial state and the macroscopic consequences of the scatter. It is from these observations that the microscopic details of the scattering process (e.g. the form of the scattering potential) are deduced. The experimentalist must first create or determine a well-defined initial state consisting of a projectile incident to a target. Having done so, he will then detect and measure the final state achieved as a consequence of the scatter. In practical medical applications, the initial state can be that of a flux of charged particles external to the body directed towards tissue. This flux may be generated by an accelerator (e.g. a linear accelerator or a cyclotron) in which charged particles are set into motion, filtered through a momentum analyser and then collimated to form a beam which is, macroscopically, well defined in terms of momentum and spatial distribution. This is also the result of the external beam being of x- or γ-rays which produces a secondary charged particle flux within the medium through Compton scatter, photoelectric absorption or, if highly energetic, photonuclear reactions. The energy spectra of these charged particles will be defined by the combination of the photon energy spectrum and the kinematics of the physics of these photon–matter interactions. Another scenario arises from neutron irradiation in which a secondary charged particle flux arises from neutron–proton elastic scatter. Charged particle fluxes can also be generated by the nuclear decays of a distributed source of a radionuclide within tissue resulting in the emission of α-particles, electrons or positrons. In this case, the kinetic energies of the charged particles are dictated by the kinematics of the nuclear decay and are generally constrained to a spectrum.[6] Further, the emission of these charged particles will be isotropic with respect to the site of emission. A photon-emitting internally distributed radionuclide also generates secondary electrons in the same manner as an external beam of photons.[7] In all cases, the initial state can be defined by the experimentalist through measurement or calculation who, after the scattering, will detect the scattered particle with an apparatus.[8] Depending upon the application, this apparatus could provide information on, for example, the momentum, scatter direction and spin orientation of the scattered particle and/or of the recoiling target. Hence, to repeat, the experimentalist will know only the initial state of the projectile and its final state following the interaction, at distances far from the scattering centre and at a time which may be treated as being infinite in comparison to the actual duration of the scattering interaction linking the initial and final observed states. In other words, the experimentalist has information on the asymptotic forms of the initial and final states only. Clearly, what is of interest is the determination of the probability that the known initial state will yield the observed final state. This leads to the question of, *what is the nature of the interaction that bridged the initial and final states*? Of more practical and immediate interest to our application of this theory to the dosimetry problem is the question of, what is the probability of a given final state, following the transfer of energy to the medium, occurring for a given initial state?

To place the above into a more quantitative representation, the initial observed state is characterised by the wavefunction ψ_i and that of the final observed state is by a wavefunction ψ_f with both states existing far from the scattering centre and, thus, being solutions to the homogeneous Schrödinger equation. An example of this is provided by the geometry of Fig. 2.1 in which ψ_i is incident to the potential centred at the origin, $U(r)$. Following the interaction with this potential, the particle is deflected from its original trajectory and is described asymptotically by the wavefunction ψ_f. This deflection may or may not be accompanied by the loss of energy from the projectile and its transfer to the scattering centre. In this problem,

[6] In the case of α-decay, as a result of its two-body nature, the energy spectrum of the α-particle is discrete with levels corresponding to the energy channels generated by the excited states of the daughter nucleus. In the case of β-decay, as a result of its three-body final state, the energy spectrum of the electron/positron is continuous.

[7] With the exception of photonuclear reactions which have a photon energy threshold of about 10 MeV, well above the energies of photons emitted by radionuclides in medical use.

[8] However, in the case of internal radiation dosimetry, the charged particle is rarely detected, but its transport can be calculated from the known physics of its interactions with matter, such as that described in this book.

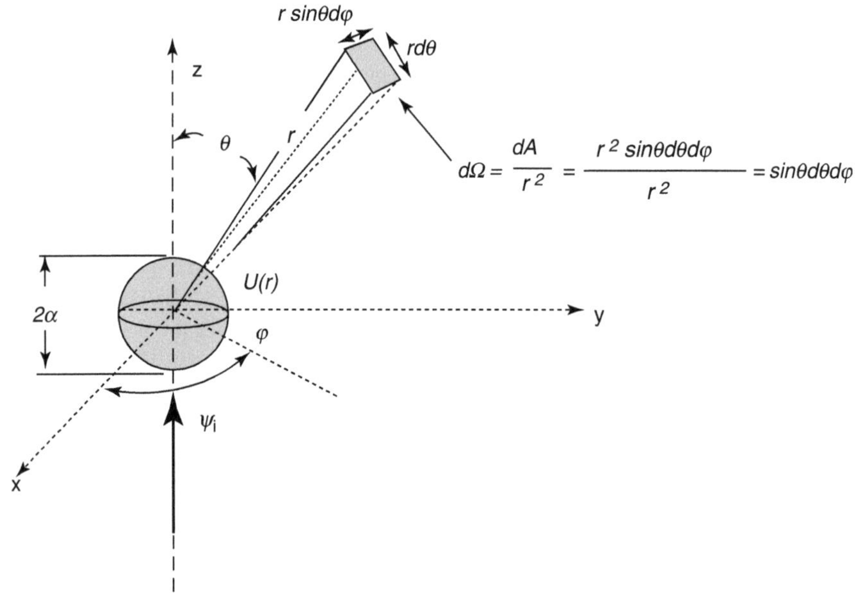

Fig. 2.1 Scattering geometry. The wavefunction is incident along the z-axis to a central and isotropic potential, $U(r)$, with an effective range a centred at the origin. The wave is scattered to yield a quasi-spherical outgoing wave representing the probability of the scattered particle being at a certain point in space, a fraction of which traverses the differential solid angle element, dΩ, at a distance r from the scattering centre

we wish to determine the probability of the particle being scattered through the angle θ into the differential solid angle element dΩ at a distance r from the scattering centre. Note that for a central and isotropic potential $U(r)$, there is no dependence upon the azimuthal scattering angle, φ. In many practical cases in dosimetry calculations, we are not interested directly in the directional changes caused by the scatter but, rather, the changes in energy (i.e. energy loss) that occur as a result. This is often obtained by first calculating the directional dependence of the scatter and then incorporating the kinematics of two-body scatter.

With the exception of the influence of the Coulomb potential (which has infinite range), outside the range a of the potential, the particles are in free motion and are not influenced by the potential that gave rise to the scattering. Further, as the time of the actual interaction is very short (of the orders of 10^{-15} to 10^{-12} s or less), we can consider the incident and final states to be those of the distant past and remote future, $t \to \mp\infty$. That is, at these limits of time, there are no interactions between the projectile and the scattering force, and the scattering potential does not affect the projectile. These free and final states are described, in state vector form, by the time-dependent Schrödinger equation:

$$\begin{aligned} H_0\psi_i(t) &= i\hbar\frac{\partial\psi_i(t)}{\partial t} \\ H_0\psi_f(t) &= i\hbar\frac{\partial\psi_f(t)}{\partial t}. \end{aligned} \tag{2.3}$$

When the particle is within the range of the potential U and is under its influence, the total Hamiltonian is the sum of the free Hamiltonian and the perturbing potential:

$$H = H_0 + U \tag{2.4}$$

with the corresponding time-dependent Schrödinger equation

$$H\psi(t) = i\hbar \frac{\partial \psi(t)}{\partial t} \tag{2.5}$$

for the total state wavefunction, $\psi(t)$. The asymptotic requirement of the scattering problem is that, as $t \to \mp\infty$, the total wavefunction described by (2.5) asymptotically approaches the unperturbed wavefunctions, $\psi_i(t)$ and $\psi_f(t)$. We now examine operators that describe the scattering process and which lead to the physical observables seen by the experimentalist. The relevant operator algebra is summarised in Appendix C.8.

2.2.2 Scattering and Time Evolution Operators

As the experimentalist can only measure the initial and final states, any information regarding the interaction must be determined indirectly through comparison of these two states, one occurring at an earlier time (pre-scatter) and the other at a later time (post-scatter). Hence, what needs to be understood by the theorist (i.e. us, who are calculating the radiation dosimetry resulting from such an interaction) is the descriptor of the scattering process in which the quantum mechanical operator S links these initial and final states of the projectile:

$$\psi_f(t) = S\psi_i(t). \tag{2.6}$$

As the experimentalist can extract only the physical observables associated with the wavefunctions $\psi_i(t)$ and $\psi_f(t)$, such as the fraction of the incident particle flux scattered into a solid angle set about a given direction, it will be the scattering operator that contains all of the information that links the two systems of the initial and final states and which must be determined from these experimental observables.

Upon inspection, this scattering operator must have two fundamental properties. First, it must be unitary:

$$SS^\dagger = S^\dagger S = 1. \tag{2.7}$$

This property means that the sum of the probabilities of all of the final states that can possibly arise from a given initial state must equal unity. It follows from (2.7) that the adjoint of S must equal the inverse of S:

$$S^{\dagger} = S^{-1}. \tag{2.8}$$

The third property of the scattering operator is that it must be stationary (i.e. it has no time dependence).

The S-matrix elements provide the physical observables presented to the experimentalist. That is, by using the Dirac *bra–ket* notation (described in Appendix E), the squared modulus $\left|\langle\psi_f|S|\psi_i\rangle\right|^2$ is proportional to differential cross sections which describe the measurable probabilities of the transition from the initial state $|\psi_i\rangle$ to the final state $|\psi_f\rangle$ occurring. In general, the state variables affected by charged particle scattering that are of interest are usually energy or momentum, scattering direction and intrinsic spin. The effect of scattering upon the energy variable is clearly of the greatest interest in a dosimetry calculation. This requires the integration of the scattering cross section over all scattering angles as the directions of the particles in the post-scattering state are generally of no interest in a dosimetry calculation, except for the case of regions in the vicinity of an interface between two radiologically differing media. Further, should it be necessary to account explicitly for the effects of intrinsic spin, this is done by first recognising that the pre- and post-scatter spin orientations are not known nor are of interest in medical radiation applications. In this case, the incident beam of particles and the targets are unpolarised, and so every possible state of spin orientation has the same a priori probability of existing. Hence, one first averages the calculation over the initial spin states as the contributions of the initial spin states are incoherent. One then sums over the final spin states. We will return to this specific problem, for example, in our derivation of the Mott scatter differential cross section for a Dirac spin-½ projectile, such as the electron.

The time-dependent Schrödinger equation of (2.5) yields a determinism in that by knowing the state of the system described by the wavefunction $\psi(t')$ at a prior time t', the state will be known for all time. In other words, solving the scattering problem is to solve the time evolution of the quantum system (or the two subsystems of incident and scattered particles in the case of scattering). For example, this time evolution can be described by

$$|\psi(t')\rangle = Y(t,t')\,|\psi(t)\rangle \tag{2.9}$$

where $Y(t,t')$ is the time evolution operator and which clearly must be linear. It follows from (2.9) that the inverse operator $Y^{-1}(t,t')$ must also exist due to time invariance. Substituting (2.9) into the time-dependent Schrödinger equation of (2.5) yields the result

$$HY(t,t') = i\hbar\frac{\partial Y(t,t')}{\partial t}. \tag{2.10}$$

The solution to the differential equation of (2.10) yields an expression for the time evolution operator $Y(t,t')$:

$$Y(t,t') = e^{-i\frac{H(t-t')}{\hbar}}. \tag{2.11}$$

Further properties of the operator follow from the above equations. Clearly, $Y(t',t') = 1$ follows from (2.11), as expected. Also, as the conservation of probability demands that the scalar product of the wavefunction at different times must remain constant (as demonstrated below), or

$$\langle\psi(t')|\,\psi(t')\rangle = \langle\psi(t)|\,\psi(t)\rangle, \tag{2.12}$$

then the insertion of (2.9) into (2.12) gives

$$\langle\psi(t)|Y^{\dagger}(t',t)\,Y(t',t)|\psi(t)\rangle = \langle\psi(t)|\psi(t)\rangle \tag{2.13}$$

which subsequently demonstrates the unitarity of the time evolution operator:

$$Y^{\dagger}(t',t)\,Y(t',t) = 1. \tag{2.14}$$

For example, as the inverse of the time evolution operator $Y^{-1}(t,t')$ exists, then

$$\begin{aligned} Y^{-1}(t,t') &= \mathrm{Y}^{\dagger}(\mathrm{t},t') \\ &= Y(t',t) \end{aligned} \tag{2.15}$$

which follows from the conservation of probability. In order to ensure the above, $||\psi(t)\rangle\,|^2$ must remain independent of time which requires that the Hamiltonian be Hermitian (i.e. $H^{\dagger} = H$). To assure ourselves that this condition is met, let us consider the complex conjugate (adjoint) of the Schrödinger equation:

$$H\psi^*(t) = -i\hbar\frac{\partial\psi^*(t)}{\partial t}, \tag{2.16}$$

which we then multiply by $\psi(t)$ to give

$$\psi(t)\,H\psi^*(t) = -i\hbar\psi(t)\frac{\partial\psi^*(t)}{\partial t}. \tag{2.17}$$

Subtracting this result from its complex conjugate results in

$$\psi(t)\,H\psi^*(t) - \psi^*(t)H\psi(t) = -i\hbar\left(\psi(t)\frac{\partial\psi^*(t)}{\partial t} - \psi^*(t)\frac{\partial\psi(t)}{\partial t}\right). \tag{2.18}$$

As $\psi(t)H\psi^*(t) = \psi^*(t)H\psi(t)$, it then follows from the right-hand side of (2.18) that

$$\frac{\partial(\psi^*(t)\psi(t))}{\partial t} = 0, \tag{2.19}$$

and hence, $\psi^*(t)\,\psi(t) = |\psi(t)\,|^2$ is a constant for all time, thus justifying (2.15).

2.3 Presentation of Calculation Tools

In the remainder of this chapter, we present the tools for calculating the physical observables associated with the scattering of a charged particle. In nearly all cases, the chapter will be restricted to nonrelativistic theory. The rationale for this is that in medical applications, the relativistic theory for a Dirac spin-½ particle is only applicable to fast electrons and positrons under the influence of the Coulomb potential. This discussion is deferred to Chap. 4 where the discussions of relativistic quantum scattering theory are presented in the context of Mott scatter and to Chaps. 8 and 9 for the topics of the Bethe, Møller, Bhabha and Bloch theories.

2.4 Quantum Perturbation Theory

2.4.1 *Fermi's Golden Rule*

As noted earlier, the scattering of a particle can be described by the transition from the initial state denoted by the incident particle's kinetic energy E and three-vector momentum $\mathbf{p}$ to a final state of kinetic energy E' and momentum $\mathbf{p}'$. The inclusion of intrinsic spin as an additional variable to be considered in the scatter is discussed in later chapters in relation to electrons and positrons and is not considered here. The probability of a transition occurring between two quantum states can be described through time-dependent perturbation theory. Its most practical appearance is that of Fermi's Golden Rule which we will make substantial use of and will derive in this section.

We begin the derivation by considering a quantum system in steady state and described by the free Hamiltonian H_0. As noted before, this Hamiltonian describes the scattering system in which the projectile, either before or after scattering, is outside of the range of the influence of the perturbing potential. Schrödinger's equation for this particular state is

$$i\hbar\frac{\partial\Psi_0}{\partial t} = H_0\Psi_0. \tag{2.20}$$

A solution to this equation can be written in terms of a series of eigenstates which accounts for the evolution of the system over time:

$$\Psi_0 = \sum_k a_{k,0}\psi_k e^{-i\frac{E_k t}{\hbar}} \tag{2.21}$$

where the coefficients $a_{k,0}$ are constants. We next consider the case when this system is perturbed by a time-dependent potential, $U(t)$, and attempt to determine the probability of a particular final state being achieved through a transition from the initial state. This state during the perturbation reflects the brief time of the actual scatter when the particle is under the influence of the scattering potential. We begin the calculation by writing the total Hamiltonian of the interacting system as

$$H = H_0 + \lambda U(t). \tag{2.22}$$

A dimensionless weighting parameter, λ, has been included to scale the influence of the potential for the purpose of enabling a perturbative calculation. The Schrödinger equation of this perturbed system is

$$\begin{aligned} i\hbar\frac{\partial\Psi}{\partial t} &= H\Psi \\ &= (H_0 + \lambda U(t))\,\Psi. \end{aligned} \tag{2.23}$$

As the eigenstates ψ_k must form a complete set, we can write the wavefunction of the perturbed state as

$$\Psi = \sum_k a_k(t)\,\psi_k e^{-i\frac{E_k t}{\hbar}} \tag{2.24}$$

where the coefficients, unlike those of (2.21), have a time dependence due to the potential. These coefficients are such that we define them at time $t = 0$ as $a_k(0) \equiv a_{k,0}$, and we interpret the squared modulus $|a_k(t)|^2$ as representing the probability of finding the perturbed system in the k-th state at the time t. These time-dependent coefficients are determined by substituting this series representation of the wavefunction into the Schrödinger equation:

$$\begin{aligned} & i\hbar\left[\sum_k\left(\frac{\mathrm{d}a_k(t)}{\mathrm{d}t}\psi_k e^{-i\frac{E_k t}{\hbar}} - i\frac{E_k}{\hbar}a_k(t)\,\psi_k e^{-i\frac{E_k t}{\hbar}}\right)\right] \\ & = (H_0 + \lambda U(t))\sum_k a_k(t)\,\psi_k e^{-i\frac{E_k t}{\hbar}}. \end{aligned} \tag{2.25}$$

Separating the free Hamiltonian and perturbative terms in (2.25) leads to the two equalities:

$$\sum_k a_k(t) E_k \psi_k e^{-i\frac{E_k t}{\hbar}} = H_0 \sum_k a_k(t)\, \psi_k e^{-i\frac{E_k t}{\hbar}} \tag{2.26}$$

$$i\hbar \sum_k \frac{\mathrm{d}a_k(t)}{\mathrm{d}t} \psi_k e^{-i\frac{E_k t}{\hbar}} = \lambda \sum_k U(t)\, a_k(t) \psi_k e^{-i\frac{E_k t}{\hbar}}. \tag{2.27}$$

The result given by (2.26) is trivial and is ignored. The effect of the perturbative potential alone upon the system is described by (2.27). By using the Dirac *bra–ket* formalism, (2.27) is rewritten as

$$i\hbar \sum_k \frac{\mathrm{d}a_k(t)}{\mathrm{d}t} |\psi_k\rangle e^{-i\frac{E_k t}{\hbar}} = \lambda \sum_k U(t) a_k(t)\, |\psi_k\rangle e^{-i\frac{E_k t}{\hbar}}. \tag{2.28}$$

We next multiply both sides of (2.28) by the *bra* $\langle \psi_f|$ of the final state (i.e. that following the perturbation) and integrate the wavefunctions over all space:

$$i\hbar \sum_k \int \mathrm{d}^3\mathbf{r} \frac{\mathrm{d}a_k(t)}{\mathrm{d}t} \langle \psi_f|\psi_k\rangle e^{-i\frac{E_k t}{\hbar}} = \lambda \sum_k \int d^3\mathbf{r}\, \langle \psi_f|U(t)\, |\psi_k\rangle a_k(t)\, e^{-i\frac{E_k t}{\hbar}}. \tag{2.29}$$

From the definition of the scalar product and using the orthonormality of states (which follows from the definition of the scalar product), $\langle \psi_n \mid \psi_m\rangle = \delta_{mn}$, where δ_{mn} is the Kronecker δ-function, (2.29) becomes

$$i\hbar \frac{\mathrm{d}a_f}{\mathrm{d}t} e^{-i\frac{E_f t}{\hbar}} = \lambda \sum_k U_{fk}(t)\, a_k(t)\, e^{-i\frac{E_k t}{\hbar}} \tag{2.30}$$

where the matrix element relating the final state f to the state k is defined as $U_{fk}(t) \equiv \langle \psi_f|U(t)\, |\psi_k\rangle$. Rearranging (2.30) gives the derivatives of the coefficients as a series:

$$\frac{\mathrm{d}a_f}{\mathrm{d}t} = -i\frac{\lambda}{\hbar} \sum_k U_{fk}(t) a_k(t) e^{-i\omega_{fk} t} \tag{2.31}$$

where we have defined the frequency of the transition from the kth state to the final f state as

$$\omega_{fk} \equiv \frac{E_f - E_k}{\hbar}. \tag{2.32}$$

It is now that we make use the coefficient λ that weights the influence of the perturbative potential. We demand that the strength of the potential relative to the eigenvalues of the steady-state Hamiltonian be weak. Consequently, the coefficients $a_k(t)$ are then expanded through a power series of the coefficient λ:

$$a_k(t) = \sum_{m=0}^{\infty} \lambda^m a_{k,m}(t). \tag{2.33}$$

Substituting this expansion into Eq. (2.31) gives

$$\sum_{m=0}^{\infty} \lambda^m \frac{\mathrm{d}a_{f,m}(t)}{\mathrm{d}t} = -\frac{i}{\hbar} \sum_{m=0}^{\infty} \sum_{k} U_{fk}(t) \lambda^{m+1} a_{k,m}(t) e^{i\omega_{fk}t}. \tag{2.34}$$

Then, by equating the coefficients of equal powers of λ, we obtain the following expressions for the derivatives of the coefficients:

$$\frac{\mathrm{d}a_{f,0}(t)}{\mathrm{d}t} = 0 \tag{2.35}$$

and

$$\frac{\mathrm{d}a_{f,m}(t)}{\mathrm{d}t} = -\frac{i}{\hbar} \sum_{k} U_{fk}(t) e^{i\omega_{fk}t} a_{k,m-1}; \qquad m \geq 1. \tag{2.36}$$

In order to use these results to calculate the coefficients, let us assume that the perturbing potential is 'switched on' at time $t = 0$ and remains constant for $t > 0$:

$$U(t) = U_0 H(t) \tag{2.37}$$

where $H(t)$ is Heaviside's function defined here as

$$\begin{aligned} H(t) &= 1 \quad t > 0 \\ &= 0 \quad t < 0. \end{aligned} \tag{2.38}$$

In this model, we further assume that the system is initially in a single and well-defined state $|\psi_i\rangle$ for $t \leq 0$. Thus, $|a_{i,0}|^2 = 1$ and the remaining coefficients at time t are then found by integrating (2.36) where, as λ has been previously defined as being small, the calculation need only be limited to

$$\begin{aligned} a_{f,1}(t) &= -\frac{i}{\hbar}\int_0^t \mathrm{d}t' M_{fi} e^{i\omega_{fi}t'} \\ &= -i\frac{M_{fi}}{\hbar}\int_0^t \mathrm{d}t' e^{i\omega_{fi}t'} \\ &= \frac{M_{fi}}{\hbar}\left(\frac{1-e^{i\omega_{fi}t}}{\omega_{fi}}\right) \end{aligned} \tag{2.39}$$

where $M_{fi} \equiv \langle \psi_f | \psi_i \rangle U_0$. The probability that a transition from the state $|\psi_i\rangle$ to the state $|\psi_f\rangle$ occurring at time t is

$$\begin{aligned} P_{fi}(t) &= \left|a_{f,1}(t)\right|^2 \\ &= 2\frac{\left|M_{fi}\right|^2}{\hbar^2}\frac{\left(1-\cos\,\omega_{fi}t\right)}{\omega_{fi}^2} \\ &= 2\frac{\left|M_{fi}\right|^2}{\hbar^2}B\left(\omega_{fi},t\right). \end{aligned} \tag{2.40}$$

As shown in Fig. 2.2, the function $B(\omega,t)$ is at a maximum for $\omega = 0$, and the sharpness of this maximum increases with increasing t. This observation

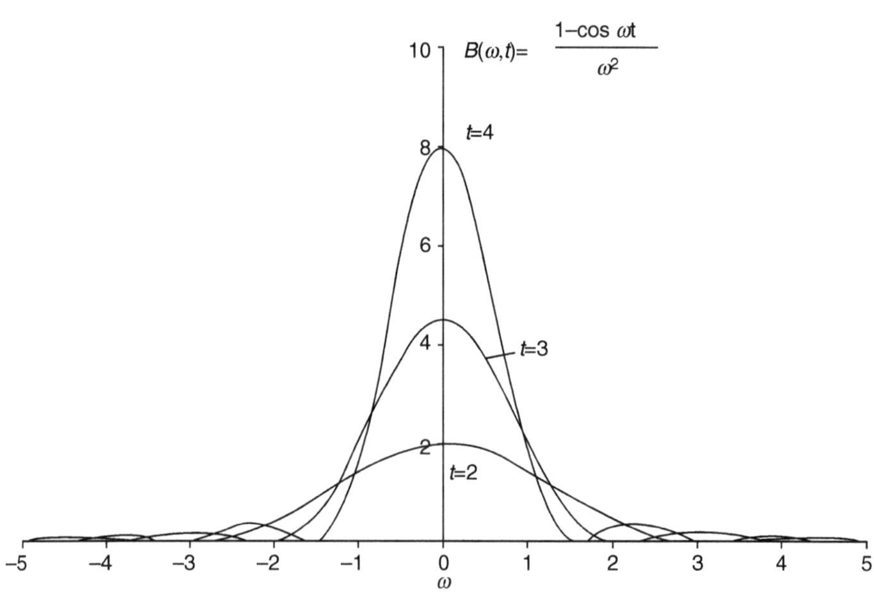

Fig. 2.2 The function $B(\omega,t)$ of (2.40)

demonstrates that the probability of a given transition occurring between two states is greater when the energy difference between the initial and final states is smaller.

However, in reality, in a quantum transition, it is unlikely that the final state will have a singular energy associated with it. The most likely case will be, in fact, that of an ensemble of neighbouring states centred around $|\psi_f\rangle$ with different (energy) eigenvalues around the final-state energy, E_f. As a consequence, the transition probability must then be integrated over this ensemble of energy eigenvalues:

$$P_{fi}(t) = \frac{2}{\hbar^2} \int_{-\infty}^{\infty} \mathrm{d}E_{f'} \left|M_{f'i}\right|^2 B\left(\omega_{f'i}, t\right) \rho_{f'} \tag{2.41}$$

where the number density of final states is

$$\rho_{f'} = \frac{\mathrm{d}N\left(E_{f'}\right)}{\mathrm{d}E_{f'}}. \tag{2.42}$$

The calculation of the integral of (2.41) can be simplified if it is assumed that both the matrix element $M_{f'i}$ and the number density of states are constant over the region where the magnitude of $B\left(\omega_{f'i}, t\right)$ is significant. This assumption allows these quantities to be extracted from the integrand:

$$\begin{aligned} P_{fi}(t) &= \frac{2}{\hbar^2} \left|M_{fi}\right|^2 \rho_f \int_{-\infty}^{\infty} \mathrm{d}E_{f'} B\left(\omega_{f'i}, t\right) \\ &= \frac{2}{\hbar} \left|M_{fi}\right|^2 \rho_f \int_{-\infty}^{\infty} \mathrm{d}\omega_{f'} \frac{\left(1 - \cos\, \omega_{f'i} t\right)}{\omega_{f'i}^2}. \end{aligned} \tag{2.43}$$

The resulting integral is solved by the substitution of the variable $y = \omega_{f'} - \omega_i$ to give

$$\int_{-\infty}^{\infty} \mathrm{d}\omega_{f'} \frac{\left(1 - \cos\, \omega_{f'i} t\right)}{\omega_{f'i}^2} = \int_{-\infty}^{\infty} \mathrm{d}y \frac{\left(1 - \cos\, yt\right)}{y^2}. \tag{2.44}$$

It is then straightforward to solve this integral by integration by parts to yield,

$$\int_{-\infty}^{\infty} \mathrm{d}y \frac{\left(1 - \cos\, yt\right)}{y^2} = 2t \int_{0}^{\infty} \mathrm{d}x \frac{\sin\, x}{x}. \tag{2.45}$$

This latter integral is solved by inserting the exponential $e^{-\alpha x}$ into the integrand:

$$f(\alpha) = \int_0^\infty \mathrm{d}x\, e^{-\alpha x} \frac{\sin x}{x} \tag{2.46}$$

integrating and then letting $\alpha \to 0$ following integration. By first differentiating this integral, we obtain

$$\begin{aligned} \frac{\mathrm{d}f}{\mathrm{d}\alpha} &= \frac{\mathrm{d}}{\mathrm{d}\alpha} \int_0^\infty \mathrm{d}x\, e^{-\alpha x} \frac{\sin x}{x} \\ &= -\int_0^\infty \mathrm{d}x\, e^{-\alpha x} \sin x. \end{aligned} \tag{2.47}$$

As $\sin x = \mathrm{Im}\, e^{ix}$, (2.47) can be rewritten as

$$\begin{aligned} \frac{\mathrm{d}f}{\mathrm{d}\alpha} &= -\mathrm{Im} \int_0^\infty \mathrm{d}x\, e^{-(\alpha - i)x} \\ &= \mathrm{Im} \frac{1}{(\alpha - i)} e^{-\alpha x} e^{ix} \Big|_0^{x=\infty} \\ &= \mathrm{Im} \frac{1}{i - \alpha} \\ &= -\mathrm{Im} \frac{i + \alpha}{1 + \alpha^2} \\ &= -\frac{1}{1 + \alpha^2}. \end{aligned} \tag{2.48}$$

Integrating (2.48),

$$f(\alpha) = -\int \frac{\mathrm{d}\alpha}{1 + \alpha^2}, \tag{2.49}$$

and by solving this integral, we achieve

$$f(\alpha) = -\tan^{-1}\alpha + C \tag{2.50}$$

where C is the integration constant. As, from (2.46), $f(+\infty) = 0$, then (2.50) leads to

$$C = \tan^{-1}(\infty) = \frac{\pi}{2} + n\pi \tag{2.51}$$

where n is either zero or an integer. The exact value of n is determined from the limits:

$$\pi > \int_0^{\pi} \mathrm{d}t \frac{\sin t}{t} > \int_0^{\infty} \mathrm{d}t \frac{\sin t}{t} > 0. \tag{2.52}$$

Hence, n must be equal to zero, giving

$$\int_0^{\infty} dx \frac{\sin x}{x} = \frac{\pi}{2}, \tag{2.53}$$

and (2.45) becomes

$$\int_{-\infty}^{\infty} \mathrm{d}\omega_{f'} \frac{\left(1 - \cos \omega_{f'i} t\right)}{\omega_{f'i}^2} = \pi t. \tag{2.54}$$

Finally, the probability of the transition can now be written as

$$P_{fi}(t) = \frac{2\pi t}{\hbar} \left|M_{fi}\right|^2 \rho_f. \tag{2.55}$$

However, what we are seeking is not the probability of the transition occurring but the transition rate between the initial state i and the final state f. This rate is equal simply to the transition probability per unit time:

$$\lambda_{fi} = \frac{\mathrm{d}P_{fi}}{\mathrm{d}t} \tag{2.56}$$

or

$$\lambda_{fi} = \frac{2\pi}{\hbar} \left|M_{fi}\right|^2 \rho_f. \tag{2.57}$$

This final expression of the transition rate between initial and final quantum states is known as Fermi's Golden Rule and has a profound number of applications including the determination of radioactivity decay rates and scattering cross sections.

2.4.2 Lippmann–Schwinger Equation and the Born Expansion

Introduction

Our first study of perturbation theory was to evaluate the general case of the rate at which a transition from an initial quantum state i to a final quantum state f. This was achieved through a series expansion of the perturbing potential that induces the transition. A requirement of this perturbation calculation was that the potential be much weaker than the eigenvalues of the free Hamiltonian. Other expansion methods based upon perturbation theory are also possible, although they are generally manifestations of the same theory.

The foundation of this analysis is the Lippmann–Schwinger equation which, in turn, provides an iterative solution to the scattered wavefunction which yields the Born series. We will derive this equation initially in its integral form in configuration space. In a later section, we will extend this derivation to provide the Lippmann–Schwinger equation in operator form in which the propagation of a wavefunction and its interaction with a potential are described in terms of operators.

Although this was not the historical route of the derivation, the Born expansion follows from the Lippmann–Schwinger equation by the invocation of two main assumptions: first, the scattering potential is weak relative to the eigenvalues of the free-particle Hamiltonian and, second, the observation of the final-state wavefunction occurs at a great enough distance from the scattering centre that it is, in fact, the asymptotic form that is measured. What is commonly called the *Born approximation* is, in fact, the first term of the *Born series* which follows from the iterative solution to the Lippmann–Schwinger equation. This result describes the final-state wavefunction as an infinite series in both the time evolution of the initial to final states and of the perturbing potential that causes this quantum transition. Ignoring the zeroth term, the first term accounts for a single channel of that from the initial state to the final state. Additional terms account for additional channels that occur through intermediate states. In most practical calculations, the first perturbative term is adequate. However, we will derive the Born series up to second order as we will show, in Chap. 4, that while the first term of the Born series can replicate exactly the classical result for the Coulomb scatter cross section, the second term diverges as a consequence of the infinite-range nature of the Coulomb potential. The fact that the Born series cannot provide a complete picture of elastic Coulomb scatter leads inevitably to the need of the Bloch correction term in the complete expression for the collision stopping power.

In this section, we apply the Born series to the calculation of the *scattering amplitude*. This quantity, as will be shown, leads us directly to a physical observable; the scattering differential cross section in solid angle, which, when coupled with the kinematics of the scatter, yields the probability of energy transfer. The scattering amplitude modulates spatially the outgoing wavefunction following the scatter and, hence, will contain all of the information pertaining to the scattering interaction. As a consequence, the scattering amplitude will be much of the focus of this section.

Scattering Amplitude and the Differential Cross Section

Unlike the classical Hamiltonian which has only continuous eigenvalues, a quantum-mechanical Hamiltonian can have both discrete and continuous eigenvalues depending upon whether or not the system it is describing is bound. Discrete eigenvalues correspond to the bound states of a system which are the consequences of an attractive interaction potential confining the system of particles to within a finite volume of space. In such a case, the Schrödinger equation of the bound system is

$$H\Psi_n = E_n\Psi_n \tag{2.58}$$

with energy eigenvalues E_n and eigenfunctions Ψ_n for the Hermitian Hamiltonian. The magnitudes of the wavefunctions of the bound particles will decrease rapidly with growing interparticle spacing. The strength of the binding potential must thus exceed the kinetic energies of the bound particles.

Continuous eigenvalues correspond to unbound, or, in our case of interest, scattering, states in which the wavefunctions extend asymptotically to infinity. It is these asymptotic states that are defined by the initial and final states of our energy transfer calculation. In scattering, a particle is incident to a potential and is then scattered by that potential so that the final observed state is the sum of the incident (unscattered) and scattered wavefunctions. Whereas the discrete eigenfunction problem corresponds to a single and confined system, the continuous eigenfunction problem describes the interactions between two subsystems (here, the projectile and the target) which are not mutually bound states. In Appendix D, the kinematics of a collision between a projectile and a target is derived. Those results provide only the energy and momentum of the two particles in the final state once a scattering or recoil angle has been specified – the probability of such a final state occurring is not answered by the kinematical solution. The scattering problem is that of interacting systems representing asymptotically free states which must exist not just during the time interval about the collision but also for all time beyond the instant of scatter (or, in other words, $t \to \pm\infty$). Following a scatter, the observer can detect only the outgoing waves from the point of interaction which corresponds to the asymptotic limit of time $t \to +\infty$. An approach to solving the scattering problem is through the solution of the Schrödinger equation. In particular, the time-dependent version can be an initial focus for obtaining a solution. The incident beam of projectiles in such a case would be represented by identical wave packets. However, the mechanics of obtaining the solution to the time-dependent Schrödinger equation are laborious. Instead, one can make a number of assumptions in order to simplify this time-dependent scattering problem to a time-independent one in which the time dependence remains implicit and need not enter the calculation. For example, let us first assume that the wave packets have a well-defined momentum **p** so that each packet has an extended length over many wavelengths as a consequence of the Heisenberg uncertainty principle. Thus, each particle can be considered as a plane wave:

$$\psi_i(\mathbf{r}) = e^{i\mathbf{k}\bullet\mathbf{r}} \tag{2.59}$$

where the wavevector is defined as

$$\mathbf{k} \equiv \frac{\mathbf{p}}{\hbar c} \tag{2.60}$$

and has units of inverse length. Let us next assume that the beam of projectile particles is incident to the potential for a period of time much longer than the duration of the interaction itself. Under these two assumptions, steady-state conditions can be taken to be valid with the result that the problem reduces to finding the stationary solutions to the time-independent Schrödinger equation. In this case, the outgoing wavefunction following the scatter is the sum of the incident plane wave and a scattered (i.e. modulated) spherical wave:

$$\psi_f(\mathbf{r}) = e^{i\mathbf{k}\bullet\mathbf{r}} + f(\mathbf{q})\,\frac{e^{ikr}}{r}. \tag{2.61}$$

The quantity $f(\mathbf{q})$, where $\mathbf{q}$ is the momentum transfer resulting from the collision, which modulates the intensity of the scattered wave is known as the *scattering amplitude* and has units of length. In terms of perturbation theory, the scattering amplitude can be represented by the series

$$f(\mathbf{q}) = \sum_{m=1}^{\infty} f_{(m)}(\mathbf{q}), \tag{2.62}$$

a fact that we will take advantage of in Chap. 4 in the development of the second term of the Born expansion for Coulomb scatter.

The argument $\mathbf{q}$ is frequently replaced by the scattering angle, θ.

Recall the general scattering problem of Fig. 2.1 in which a particle interacts with a potential and, with a certain probability, is scattered through the angle θ into the differential solid angle element dΩ. The potential is assumed to be isotropic and central, i.e. there is no dependence upon the azimuthal angle φ. In order to calculate the probability of this scatter occurring, we first consider the free particle with mass m and the corresponding nonrelativistic Hamiltonian:

$$H_0 \Leftrightarrow \frac{p^2}{2m} \tag{2.63}$$

and the associated Schrödinger equation

$$H_0|\psi_0\rangle = E_0|\psi_0\rangle \tag{2.64}$$

where $|\psi_0\rangle$ is the energy eigenfunction of the free Hamiltonian operator H_0 and E_0 is the eigenvalue corresponding to the kinetic energy of (2.63). The wavefunction corresponding to the particle is given by the scalar product of the Dirac combination of *bra* and *ket*:

$$\psi_0(\mathbf{r}) = \langle \mathbf{r} | \psi_0 \rangle. \tag{2.65}$$

The time-independent (stationary) Schrödinger equation for the particle, when the influence of the potential is included, is

$$(H_0 + U - E)\,|\psi\rangle = 0. \tag{2.66}$$

By using the correspondence rule of $\mathbf{p} = -i\hbar c\nabla$ (where the factor of $\hbar c$ is incorporated to ensure that $|\mathbf{p}|$ has the units of energy in the system followed here), we write the free Hamiltonian in operator form as $-\left(\frac{(\hbar c)^2}{2m}\right)\nabla^2$ and rearrange the Schrödinger equation to the form

$$\left(-\nabla^2 + k^2\right)\,|\psi\rangle = \Xi(\mathbf{r})\,|\psi\rangle \tag{2.67}$$

where we have defined

$$\Xi(\mathbf{r}) \equiv \frac{2mU(\mathbf{r})}{(\hbar c)^2} \tag{2.68}$$

and where the magnitude of the wavevector is

$$k = \frac{\sqrt{2mE}}{\hbar c}. \tag{2.69}$$

Note that k is the reciprocal of the reduce de Broglie wavelength, $\lambda\!\!\!^{-}$. Note that the homogeneous solution is for the equation without the perturbing potential (i.e. $\Xi(\mathbf{r}) = 0$), $(-\nabla^2 + k^2)|\psi_0\rangle = 0$. Solving for the *ket* $|\psi\rangle$ and including the homogeneous solution to this differential equation give

$$|\psi\rangle = |\psi_0\rangle + \frac{1}{k^2 - \nabla^2}\Xi(\mathbf{r})\,|\psi\rangle, \tag{2.70}$$

a result which clearly returns us with the incoming wave for the case $U = 0$. The necessity for adding the homogeneous solution is the consequence of the conditions required when dividing through by an operator (as shown in the appendices). Multiplying both sides of (2.70) with the *bra* $\langle\mathbf{r}|$ and using the closure relationship of the eigenstates of momentum rather than kinetic energy (which are also those of the free Hamiltonian),

$$\langle \mathbf{r}|\psi\rangle = \langle \mathbf{r}|\psi_0\rangle - \frac{1}{(2\pi)^{\frac{3}{2}}} \int \mathrm{d}^3\mathbf{k}' \langle \mathbf{r} \mid \mathbf{k}'\rangle \frac{1}{k^2 - k'^2} \langle \mathbf{k}'|\Xi(\mathbf{r})|\psi\rangle. \tag{2.71}$$

Repeating with the integration over the eigenstates of position,

$$\begin{aligned}\langle \mathbf{r}|\psi\rangle &= \langle \mathbf{r}|\psi_0\rangle - \frac{1}{(2\pi)^3} \int \mathrm{d}^3\mathbf{r}' \int \mathrm{d}^3\mathbf{k}' \langle \mathbf{r} \mid \mathbf{k}'\rangle \langle \mathbf{k}' \mid \mathbf{r}'\rangle \frac{1}{k^2 - k'^2} \langle \mathbf{r}'|\Xi(\mathbf{r})|\psi\rangle \\ &= \langle \mathbf{r}|\psi_0\rangle - \frac{1}{(2\pi)^3} \int \mathrm{d}^3\mathbf{r}' \int \mathrm{d}^3\mathbf{k}' \frac{|\mathbf{k}'\rangle\langle \mathbf{k}'|}{k^2 - k'^2} \Xi(\mathbf{r}')\psi(\mathbf{r}').\end{aligned} \tag{2.72}$$

This result will be recognised as being of the general form

$$\psi(\mathbf{r}) = \psi_0(\mathbf{r}) - \int \mathrm{d}^3\mathbf{r}' G(\mathbf{r}, \mathbf{r}')\Xi(\mathbf{r}')\psi(\mathbf{r}') \tag{2.73}$$

where $G(\mathbf{r}, \mathbf{r}')$ is the Green's function or operator which obeys

$$\left(-\nabla^2 + k^2\right) G(\mathbf{r}, \mathbf{r}') = \delta(\mathbf{r} - \mathbf{r}') \tag{2.74}$$

as shown in Appendix C.11.

This result is not a mathematically unique solution as one can add the solution of the homogeneous equation to the Green's function. That this Green's function yields a solution to the Schrödinger equation is readily demonstrated by direct substitution:

$$\begin{aligned}\left(-\nabla^2 + k^2\right) \psi(\mathbf{r}) = &\left(-\nabla^2 + k^2\right) \psi_0(\mathbf{r}) \\ &+ \int \mathrm{d}^3\mathbf{r}' \left(-\nabla^2 + k^2\right) G(\mathbf{r} - \mathbf{r}')\Xi(\mathbf{r}')\, \psi(\mathbf{r}').\end{aligned}$$

The first term on the right-hand side is equal to zero as it is the homogeneous solution of the Schrödinger equation and, as $(-\nabla^2 + k^2)$ acts only upon the variable $\boldsymbol{r}$,

$$\int \mathrm{d}^3\mathbf{r}' \left(\left(-\nabla^2 + k^2\right) G(\mathbf{r} - \mathbf{r}')\right) \Xi(\mathbf{r}')\, \psi(\mathbf{r}') = \Xi(\mathbf{r})\, \psi(\mathbf{r}),$$

through (2.74).

The Green's function, as defined, does not specify fully the physical nature of the scattering problem but only the propagation of the wavefunction. Recall that, in reality, we have a single incoming plane wave and an outgoing quasi-spherical wave, with the latter going in all directions with an intensity modulated with scattering angle θ. However, the Schrödinger equation has an equally valid solution in which there are incoming waves arriving from all directions. Thus, the Green's function must be selected so as to generate spherical outgoing waves only in order to provide a physically realistic solution. As a result, we must next determine an expression for such a Green's function:

$$
\begin{aligned}
G(\mathbf{r}) &= \frac{1}{(2\pi)^3}\int \mathrm{d}^3\mathbf{k}'\,\frac{|\mathbf{k}'\rangle\langle\mathbf{k}'|}{k^2-k'^2} \\
&= \frac{1}{(2\pi)^3}\int \mathrm{d}^3\mathbf{k}'\,\frac{e^{i\mathbf{k}'\bullet\mathbf{r}}}{k^2-k'^2} \\
&= \frac{1}{4\pi^2}\int_{-1}^{1}\mathrm{d}(\cos\,\theta)\int_{0}^{\infty}\mathrm{d}k'\,k'^2\,\frac{e^{ik'|\mathbf{r}|\cos\theta}}{k^2-k'^2} \\
&= -\frac{i}{4\pi^2|\mathbf{r}|}\left(\int_{0}^{\infty}\mathrm{d}k'\,k'\,\frac{e^{ik'|\mathbf{r}|}}{k^2-k'^2}-\int_{0}^{\infty}\mathrm{d}k'\,k'\,\frac{e^{-ik'|\mathbf{r}|}}{k^2-k'^2}\right) \\
&= -\frac{i}{4\pi^2|\mathbf{r}|}\int_{-\infty}^{\infty}\mathrm{d}k'\,k'\,\frac{e^{ik'|\mathbf{r}|}}{k^2-k'^2} \\
&= \frac{i}{4\pi^2|\mathbf{r}|}\int_{-\infty}^{\infty}\mathrm{d}k'\,k'\,\frac{e^{ik'|\mathbf{r}|}}{(k'-k)\,(k'+k)}.
\end{aligned}
\tag{2.75}
$$

The conversion to the single integral to yield the penultimate step above was accomplished by exchanging the sign of the scattered wavevector k' in the second integral to allow integration from $-\infty$ to 0 and then to sum the two integrals to yield a single integral over k' from $-\infty$ to $+\infty$.

Singularities in the integrand exist clearly at momentum values of $k' = \pm k$, and these must be dealt with. It is at this point that we can exclude the non-physical case of incoming waves from all direction, i.e. the Green's function is to be an integration over all outgoing waves exclusively. In order to both solve this integral using the residue theorem described in Appendix A.4 and to ensure a physical solution, we let the two singularities be adjusted along the imaginary axis so that they exist at $k' = \pm k \pm i\varepsilon$ (as shown in Fig. 2.3). Following integration, we will let $\varepsilon \to 0$. Integration will be taken over the semi-circular contour on the positive imaginary side of the complex k' plane with radius R, following which we allow $R \to \infty$, through the justification that $e^{ik'r}$ represents the outgoing wave due to the suppressed time dependence factor of $e^{-i\frac{Et}{\hbar}}$, and so the contour must contain the 'adjusted' singularity of $k' = k + i\varepsilon$ alone. The solution to the integral of (2.75) is, where Res $(f(z))$ represents the residue of $f(z)$,

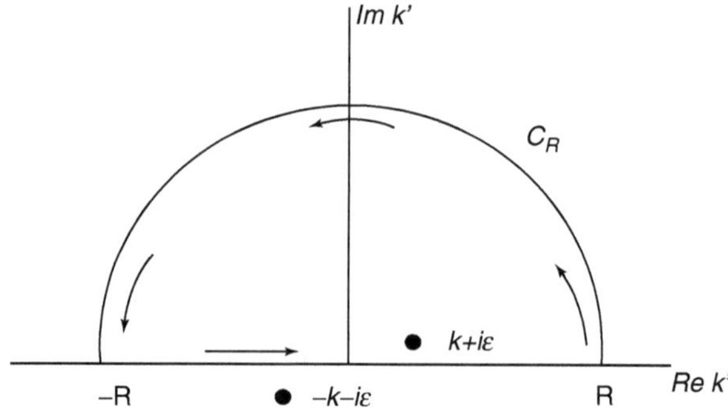

Fig. 2.3 Contour for solving the integral of (2.75). The poles at $k' = \pm k$ are displaced by the incremental amounts of $\pm i\varepsilon$ to allow the contour integration for $R \to \infty$ using residue theory, following which the result is given by taking the limit of $\varepsilon \to 0$

$$
\begin{aligned}
&\oint_C \mathrm{d}k' k' \frac{e^{ik'|\mathbf{r}|}}{(k' - (k + i\varepsilon))\,(k' - (-k - i\varepsilon))} \\
&= 2\pi i \,\mathrm{Res}\left(k' \frac{e^{ik'|\mathbf{r}|}}{(k' - (k + i\varepsilon))\,(k' - (-k - i\varepsilon))}\right) \\
&= 2\pi i \lim_{k' \to (k+i\varepsilon)} \left((k' - (k + i\varepsilon))\,k' \frac{e^{ik'|\mathbf{r}|}}{(k' - (k + i\varepsilon))\,(k' - (-k - i\varepsilon))}\right) \\
&= 2\pi i \lim_{k' \to (k+i\varepsilon)} \left(k' \frac{e^{ik'|\mathbf{r}|}}{(k' - (-k - i\varepsilon))}\right) \\
&= 2\pi i (k + i\varepsilon) \frac{e^{i(k+i\varepsilon)|\mathbf{r}|}}{(k + i\varepsilon - (-k - i\varepsilon))} \\
&= i\pi\, e^{ik|\mathbf{r}|} (\text{for} \quad \varepsilon \to 0).
\end{aligned} \tag{2.76}
$$

Substituting this result into (2.75) gives us a final expression for the Green's function:

$$
G(\mathbf{r})^{+} = -\frac{1}{4\pi} \frac{e^{ik|\mathbf{r}|}}{|\mathbf{r}|}. \tag{2.77}
$$

The + superscript indicates that this particular Green's function leads to the physical case of outgoing waves in all directions (i.e. $t \to +\infty$). We then place this result into (2.73) to yield the post-scatter wavefunction:

$$
\begin{aligned}
\psi(\mathbf{r}) &= \psi_0(\mathbf{r}) - \frac{1}{4\pi} \int \mathrm{d}^3\mathbf{r}' \frac{e^{ik|\mathbf{r}-\mathbf{r}'|}}{|\mathbf{r}-\mathbf{r}'|} \Xi(\mathbf{r}') \psi(\mathbf{r}') \\
&= \psi_0(\mathbf{r}) - \frac{m}{2\pi(\hbar c)^2} \int \mathrm{d}^3\mathbf{r}' \frac{e^{ik|\mathbf{r}-\mathbf{r}'|}}{|\mathbf{r}-\mathbf{r}'|} U(r')\, \psi(\mathbf{r}').
\end{aligned} \tag{2.78}
$$

This is known as the *Lippmann–Schwinger* equation in configuration space. It is obvious that (2.78) is an iterative result due to the placement of $\psi(\mathbf{r}')$ within the integrand and will thus lead inevitably to a series solution. This will be demonstrated now in terms of the Born expansion.

To develop the Born series from (2.78), we recall that throughout our consideration of the scattering problem, we have always assumed that the potential is weak relative to the energy eigenvalue of the free Hamiltonian and that the scattered particle detector is far away from the scattering centre. This latter assumption is reflected by the inequality of $|\mathbf{r}| >> |\mathbf{r}'|$, where $|\mathbf{r}|$ indicates the relative distance of the detector from the scattering centre and $|\mathbf{r}'|$ indicates the relative dimensions of the spatial region over which the scattering potential acts. In this case, the Taylor expansion of $|\mathbf{r} - \mathbf{r}'|$ to first order gives

$$\begin{aligned}|\mathbf{r} - \mathbf{r}'| &= r\sqrt{1 + \left(\frac{r'}{r}\right)^2 - 2\left(\frac{r'}{r}\right)\cos\,\theta} \\ &\approx r\left(1 - \frac{r'}{r}\cos\,\theta\right) \\ &\approx r - \hat{\mathbf{r}} \bullet \mathbf{r}'.\end{aligned} \tag{2.79}$$

While the denominator of the integrand can be simplified to $|\mathbf{r} - \mathbf{r}'| \approx r$, such a simplification cannot be applied properly to the exponent of the numerator of the integrand as, from (2.79),

$$\begin{aligned}k|\mathbf{r} - \mathbf{r}'| &\cong kr - k\hat{\mathbf{r}} \bullet \mathbf{r}' \\ &\equiv kr - \mathbf{k}_f \bullet \mathbf{r}'\end{aligned} \tag{2.80}$$

where $\mathbf{k}_f$ is the wavevector of the scattered wavefunction. This result introduces a non-negligible phase factor that must be accounted for. As a result, $e^{ik|\mathbf{r}-\mathbf{r}'|} = e^{ikr}$ $e^{-i\mathbf{k}_f \bullet \mathbf{r}'}$ where $e^{-i\mathbf{k}_f \bullet \mathbf{r}'}$ represents the phase term of the scattered wavefunction. Hence, for $|\mathbf{r}| >> |\mathbf{r}'|$, the expression of (2.78) simplifies to

$$\psi(\mathbf{r}) = \psi_0(\mathbf{r}) - \frac{m}{2\pi(\hbar c)^2}\frac{e^{ikr}}{r}\int \mathrm{d}^3\mathbf{r}' e^{-i\mathbf{k}_f \bullet \mathbf{r}'} U(r')\,\psi(\mathbf{r}'). \tag{2.81}$$

A first-order approximation description of the scattering, which is the first-order term of the Born series, can be obtained by recalling the above description of the magnitude of the potential relative to the energy eigenvalue of the free Hamiltonian. That is, the potential is sufficiently weak that we can approximate the scattered wavefunction by the incident one (i.e. an unperturbed plane wave). This allows us to replace the scattered wavefunction $\psi(\mathbf{r}')$ in the integrand with the incident wavefunction which we have modelled by $e^{i\frac{\mathbf{p} \bullet \mathbf{r}}{\hbar c}}$:

$$\psi_{\text{Born 1}}(\mathbf{r}) = e^{i\mathbf{k}\bullet\mathbf{r}} - \frac{m}{2\pi(\hbar c)^2}\frac{e^{ikr}}{r}\int d^3\mathbf{r}' e^{-i\frac{\mathbf{q}\bullet\mathbf{r}'}{\hbar c}}U(r') \tag{2.82}$$

where $\mathbf{q} = \mathbf{p}_f - \mathbf{p}$ is the previously presented three-vector momentum transfer as a consequence of the scatter. This result of (2.82) is known as the *first Born approximation* from which we obtain the associated scattering amplitude:

$$f_{(1)}(\mathbf{q}) = -\frac{m}{2\pi(\hbar c)^2}\int d^3\mathbf{r}' e^{-i\frac{\mathbf{q}\bullet\mathbf{r}'}{\hbar c}}U(r') \tag{2.83}$$

which modulates the outgoing spherical wave. Note that the scattering amplitude in the first Born approximation is proportional to the Fourier transform of the scattering potential which, in such cases, will reduce to

$$\begin{aligned}
f_{(1)}(q) &= -\frac{m}{2\pi(\hbar c)^2}\int d^3\mathbf{r}' e^{-i\frac{\mathbf{q}\bullet\mathbf{r}'}{\hbar c}}U(r') \\
&= -\frac{m}{2\pi(\hbar c)^2}\int_0^{2\pi} d\varphi \int_{-1}^{1} d(\cos\,\theta)\int_0^{\infty} dr'\, r'^2\, e^{-i\frac{qr'\cos\theta}{\hbar c}}\, U(r') \\
&= -\left(\frac{2m}{\hbar c q}\right)\int_0^{\infty} dr'\, r' \sin\left(\frac{qr'}{\hbar c}\right) U(r').
\end{aligned} \tag{2.84}$$

Thus, we write in the first Born approximation, assuming an incident plane wave and a potential without an azimuthal dependence as

$$\psi(\mathbf{r}) = e^{i\mathbf{k}\bullet\mathbf{r}} + f(\theta)\frac{e^{ikr}}{r} \tag{2.85}$$

where the original argument of the scattering amplitude of the momentum transfer $\mathbf{q}$ has been replaced by the scattering angle. This exchange of variable follows from

$$\begin{aligned}
|\mathbf{q}| &= |\mathbf{p}_f - \mathbf{p}| \\
&= \sqrt{p_f^2 + p^2 - 2p_f p\cos\,\theta} \\
&= p\sqrt{2(1-\cos\,\theta)} \\
&= 2p\sin\frac{\theta}{2}
\end{aligned} \tag{2.86}$$

where the elasticity of the scatter, $|\mathbf{p}| = |\mathbf{p}_f|$, has been accounted for.

As the scattering amplitude modifies the outgoing spherical wave, it provides the physical observable of the differential cross section as shall now be demonstrated.

This outcome follows from the development of the time-dependent Schrödinger equation and its complex conjugate:

$$\begin{aligned} i\hbar\frac{\partial\psi}{\partial t} &= -\frac{(\hbar c)^2}{2m}\nabla^2\psi \\ -i\hbar\frac{\partial\psi^*}{\partial t} &= -\frac{(\hbar c)^2}{2m}\nabla^2\psi^*. \end{aligned} \tag{2.87}$$

Premultiplying the Schrödinger equation with ψ^* and its complex conjugate with ψ and then summing, we have

$$i\hbar\left(\psi^*\frac{\partial\psi}{\partial t}+\psi\frac{\partial\psi^*}{\partial t}\right)=\frac{(\hbar c)^2}{2m}\left(\psi\nabla^2\psi^*-\psi^*\nabla^2\psi\right). \tag{2.88}$$

This is equivalent to

$$\begin{aligned} \frac{\mathrm{d}|\psi|^2}{\mathrm{d}t} &= \frac{(\hbar c)^2}{i2m\hbar}\left(\psi\nabla^2\psi^*-\psi^*\nabla^2\psi\right) \\ &= \frac{(\hbar c)^2}{i2m\hbar}\nabla\bullet\left(\psi\nabla\psi^*-\psi^*\nabla\psi\right). \end{aligned} \tag{2.89}$$

Now, if we were to consider the right-hand side of the (2.89) $\frac{(\hbar c)^2}{i2m\hbar}\left(\psi\nabla\psi^*-\psi^*\nabla\psi\right)$ as being a vector current, **J** then (2.89) becomes the continuity equation:

$$\frac{\mathrm{d}|\psi|^2}{\mathrm{d}t}+\nabla\bullet\mathbf{J}=0. \tag{2.90}$$

In order to link the scattering amplitude to an experimental observable, we begin by determining the vector current, or vector particle fluence, of the incoming beam described by the plane wave, $\psi_0(\mathbf{r})=e^{i\mathbf{k}\bullet\mathbf{r}}$. It follows that this is

$$\mathbf{J}_0(\mathbf{r})=\frac{(\hbar c)^2 k}{m\hbar}\,\hat{\mathbf{r}}. \tag{2.91}$$

The scattered wave in the radial direction of the spherical co-ordinate system has the current:

$$\begin{aligned} \mathbf{J}_S(\mathbf{r}) &= \frac{(\hbar c)^2}{i2m\hbar}\left(f^*(\theta)\frac{e^{ikr}}{r}\frac{\partial}{\partial r}\left(f(\theta)\frac{e^{ikr}}{r}\right)\hat{\mathbf{r}}-f(\theta)\frac{e^{ikr}}{r}\frac{\partial}{\partial r}\left(f^*(\theta)\frac{e^{-ikr}}{r}\right)\hat{\mathbf{r}}\right) \\ &= \frac{(\hbar c)^2}{i2m\hbar}|f(\theta)|^2\frac{e^{-ikr}}{r}\left(\left(ik\frac{e^{ikr}}{r}-\frac{e^{ikr}}{r^2}\right)-\frac{e^{ikr}}{r}\left(-ik\frac{e^{-ikr}}{r}-\frac{e^{i-kr}}{r^2}\right)\right)\hat{\mathbf{r}} \\ &= \frac{(\hbar c)^2 k}{m\hbar r^2}|f(\theta)|^2\,\hat{\mathbf{r}}. \end{aligned} \tag{2.92}$$

The flux of scattered particles that cross the area $r^2\,\mathrm{d}\Omega$ is given by

$$N\,\mathrm{d}\Omega = \mathbf{J}_S(r) \bullet \hat{\mathbf{r}}\, r^2\,\mathrm{d}\Omega$$
$$= \frac{(\hbar c)^2 k}{m\hbar}\,|f(\theta)|^2\mathrm{d}\Omega. \quad (2.93)$$

As the incoming flux is, from (2.91),

$$|\mathbf{J}_0|\,\mathrm{d}\sigma = \frac{(\hbar c)^2 k}{m\hbar},$$

then the differential cross section in solid angle is

$$\frac{\mathrm{d}\sigma}{\mathrm{d}\Omega} = \frac{N}{|\mathbf{J}_0|}$$
$$= |f(\theta)|^2. \quad (2.94)$$

Thus, the measurable quantity of the differential cross section can be predicted by the calculated scattering amplitude which, in turn, is derived from knowledge of the scattering potential.

Recall that the Lippmann–Schwinger equation provided an iterative result for the final-state wavefunction and that the first Born approximation evolved from it through the truncation of the Born series at first order and the use of a number of approximations. We can repeat this exercise to bring the expansion to second order:

$$\psi(\mathbf{r}) = \psi_0(\mathbf{r}) - \frac{m}{2\pi(\hbar c)^2}\int \mathrm{d}^3\mathbf{r}'\,\frac{e^{ik|\mathbf{r}-\mathbf{r}'|}}{|\mathbf{r}-\mathbf{r}'|}U(r')\,\psi_0(\mathbf{r}')$$
$$+\left(\frac{m}{2\pi(\hbar c)^2}\right)^2 \int \mathrm{d}^3\mathbf{r}'\,\frac{e^{ik|\mathbf{r}-\mathbf{r}'|}}{|\mathbf{r}-\mathbf{r}'|}U(r')\int \mathrm{d}^3\mathbf{r}''\,\frac{e^{ik|\mathbf{r}'-\mathbf{r}''|}}{|\mathbf{r}'-\mathbf{r}''|}U(r'')\psi(\mathbf{r}''). \quad (2.95)$$

We will shortly return the iterative nature of the Lippmann–Schwinger equation (in operator form) and will look at (2.95) in the specific context of the Coulomb potential in Chap. 4.

2.4.3 *Differential Cross Section Derived from Fermi's Golden Rule*

We deviate slightly at this point to demonstrate that the result of (2.94) can be also achieved by using Fermi's Golden Rule. The result of (2.94) was obtained on the basis of the fundamental assumption of perturbation theory in that the magnitude of the potential is much weaker than the Hamiltonian's energy eigenfunction, thus allowing us to replace the scattered wavefunction in the integrand of (2.81) with the incident wavefunction. To obtain the same result using Fermi's Golden Rule, we modify (2.57) slightly by incorporating the Dirac δ-function to enforce the elasticity of the scatter within the calculation:

$$\lambda_{fi} = \frac{2\pi}{\hbar} \left| \langle \psi_f | U \mid \psi_i \rangle \right|^2 \rho_f \, \delta\left(E_f - E_i\right).$$

The derivation of the elastic differential cross section can be obtained in a straightforward, but detailed, manner. Here, a summary of that derivation is provided.

The normalised incident wavefunction is

$$\psi_i = \frac{1}{L^{\frac{3}{2}}} e^{i\frac{\mathbf{pr}}{\hbar c}} \tag{2.96}$$

where L^3 is a volume used for normalisation (and which is eliminated eventually through the course of the derivation) and $\mathbf{p}$ is the three-vector momentum of the projectile. As the exiting spherical wave will be expressed as the sum of weighted plane waves, the resulting matrix element is

$$\begin{aligned} \langle \psi_f | U \mid \psi_i \rangle &= \frac{1}{L^3} \int \mathrm{d}^3\mathbf{r} \, e^{-i\frac{\mathbf{p}_f \bullet \mathbf{r}}{\hbar c}} U(\mathbf{r}) e^{i\frac{\mathbf{p}_i \bullet \mathbf{r}}{\hbar c}} \\ &= \frac{1}{L^3} \int \mathrm{d}^3\mathbf{r} \, U(\mathbf{r}) e^{-i\frac{\mathbf{q} \bullet \mathbf{r}}{\hbar c}}. \end{aligned} \tag{2.97}$$

The phase space factor is

$$\begin{aligned} \rho_f \, \mathrm{d}E_f &= \left(\frac{L}{2\pi \, \hbar^3} \right) \mathrm{d}^3\mathbf{p}_f \\ &= \left(\frac{L}{2\pi \, \hbar^3} \right) p_f^2 \, \mathrm{d}p_f \, \mathrm{d}\Omega \end{aligned} \tag{2.98}$$

Note that here E_f represents the total energy, the sum of the kinetic energy and rest mass, of the scattered projectile. As

$$\begin{aligned} \mathrm{d}p_f &= \frac{\mathrm{d}p_f}{\mathrm{d}E_f} \mathrm{d}E_f \\ &= \frac{E_f}{p_f} \, \mathrm{d}E_f, \end{aligned}$$

we can write

$$\rho_f\,\mathrm{d}E_f = \left(\frac{L}{2\pi\,\hbar^3}\right) p_f\, m\,\mathrm{d}E_f\mathrm{d}\Omega \tag{2.99}$$

where $m \approx Ef$ for nonrelativistic energies. The transition rate between the two states is, thus,

$$\lambda_{fi} = \frac{m\,p_f\,\mathrm{d}\Omega\,\delta\left(E_f - E_i\right)}{\left(2\pi\hbar^2\right)L^3}\left|\int \mathrm{d}^3\mathbf{r}\,U(\mathbf{r})e^{-i\frac{\mathbf{q}\bullet\mathbf{r}}{\hbar c}}\right|^2. \tag{2.100}$$

Normalising to the incident flux leads to

$$\begin{aligned}\mathrm{d}\sigma &= \frac{L^3}{\beta c}\,\lambda_{fi}\\ &= \frac{m\,\mathrm{d}\Omega}{\left(2\pi\hbar^2\right)}\left|\int \mathrm{d}^3\mathbf{r}\,U(\mathbf{r})e^{-i\frac{\mathbf{q}\bullet\mathbf{r}}{\hbar c}}\right|^2\end{aligned}$$

to give

$$\begin{aligned}\frac{\mathrm{d}\sigma}{\mathrm{d}\Omega} &= \left(\frac{m}{2\pi\hbar^2}\right)\left|\int \mathrm{d}^3\mathbf{r}\,U(\mathbf{r})e^{-i\frac{\mathbf{q}\bullet\mathbf{r}}{\hbar c}}\right|^2\\ &= |f(\theta)\,|^2\end{aligned} \tag{2.101}$$

which is the same result derived from the quantum current approach.

2.4.4 Lippmann–Schwinger Equation in Operator Form

We now return to a discussion of the Lippmann–Schwinger equation, but now within the context of operators, eigenfunctions and eigenvectors. The Schrödinger equation is thus written as

$$(H_0 + U)\,|\psi\rangle = E|\psi\rangle \tag{2.102}$$

to yield

$$\begin{aligned}|\psi\rangle &= |\psi_0\rangle + \frac{1}{E - H_0 + i\varepsilon}U|\psi\rangle\\ &= |\psi_0\rangle + G_+U|\psi\rangle.\end{aligned} \tag{2.103}$$

The introduction of the imaginary infinitesimal $i\varepsilon$ is necessary in order to satisfy the conditions of dividing through by an operator (see Appendix C.8.3). The + subscript for the Green function operator indicates that these are outgoing waves. Further iterations yields the series

$$|\psi\rangle = |\psi_0\rangle + G_+U|\psi_0\rangle + G_+UG_+U|\psi_0\rangle + G_+UG_+UG_+U|\psi_0\rangle + \ldots. \quad (2.104)$$

By treating the potential as an operator and operating both sides of (2.103) with it result in

$$\begin{aligned} U|\psi\rangle &= U|\psi_0\rangle + UG_+U|\psi_0\rangle + UG_+UG_+U|\psi_0\rangle + UG_+UG_+UG_+U|\psi_0\rangle + \ldots \\ &= T|\psi_0\rangle \end{aligned} \quad (2.105)$$

where T is the transition matrix defined by the sequence

$$T = U + UG_+U + UG_+UG_+U + UG_+UG_+UG_+U + \ldots. \quad (2.106)$$

The T-operator $T(E)$ can be written in the form

$$T(E) = U + U\frac{1}{E - H_0 + i\varepsilon}T(E). \quad (2.107)$$

The matrix element $\langle \mathbf{k}'|T(E)|\mathbf{k}\rangle$ contains all of the information that describes the scattering process. If $k'^2 = k^2$, the corresponding matrix elements are referred to as being *on the energy shell* as E is conserved in a two-body scattering process. Should there be a third particle available to take away part of the energy during the scatter (e.g. as in electron–electron *bremsstrahlung*, $e^- + e^- \rightarrow e^- + e^- + \gamma$), then $k'^2 \neq k^2$ and the two-body matrix element $\langle \mathbf{k}'|T(E)|\mathbf{k}\rangle$ is referred to as being *off the energy shell.*

This picture shows that the Born series of (2.104) can be described graphically as the sum of multiple interactions within a potential, as shown in Fig. 2.4.

2.4.5 Conditions of Validity of the Born Expansion

As the Born expansion is the outcome of perturbation theory, it will be subject to criteria that define its range of validity. Clearly, such validity criteria will demand that the perturbation be small in order that the number of necessary terms in the Born series be limited in its extent for a practical calculation. Here, we shall consider the first term of the Born approximation only. As this is a function of the projectile momentum, we examine the extrema of low and high momenta. In the modified Lippmann–Schwinger equation,

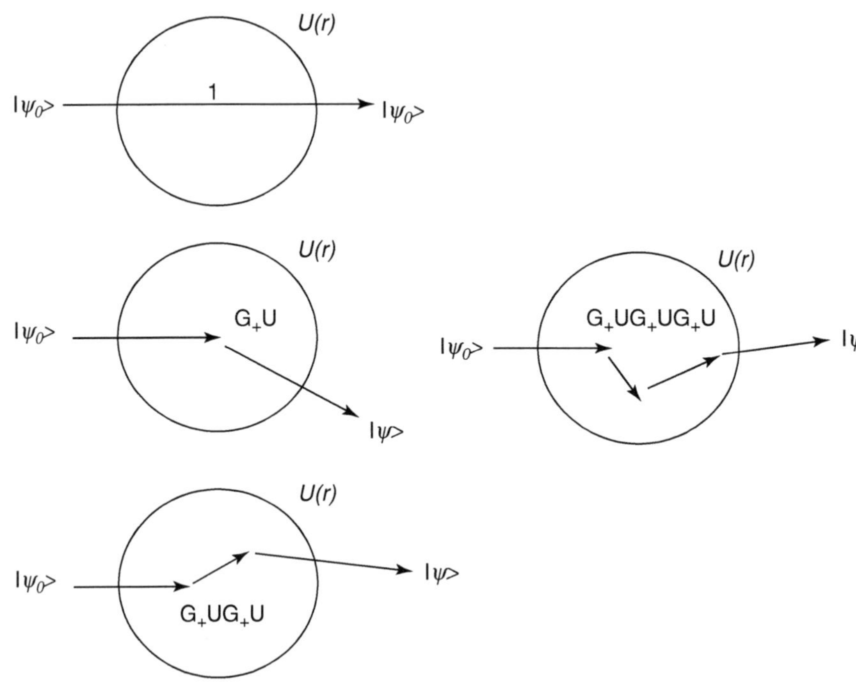

Fig. 2.4 Graphical representation of the first four terms of the Born series described by (2.104) in which each term is treated as propagations of the wavefunction within the potential and interactions with the potential

$$\psi(\mathbf{r}) = \psi_0(\mathbf{r}) - \frac{m}{2\pi(\hbar c)^2}\frac{e^{ikr}}{r}\int \mathrm{d}^3\mathbf{r}' e^{-i\mathbf{k}_f\bullet\mathbf{r}'}U(\mathbf{r}')\,\psi(\mathbf{r}'),$$

we require that the difference in magnitude between the scattered and incident wavefunctions described by the approximation be less than the magnitude of the incident wavefunction:

$$|\psi(\mathbf{r}) - \psi_0(\mathbf{r})\,| << |\psi_0(\mathbf{r})|$$

or

$$\frac{m}{2\pi(\hbar c)^2}\left|\int \mathrm{d}^3\mathbf{r}'\frac{e^{-ik|\mathbf{r}-\mathbf{r}'|}}{|\mathbf{r}-\mathbf{r}'|}U(\mathbf{r}')\,\psi(\mathbf{r}')\right| << |\psi_0(\mathbf{r})\,|. \tag{2.108}$$

Because of the presence of k within the integrand, the validity of the first Born approximation will be explicitly dependent upon the projectile momentum. (2.108) leads to the limiting case of

$$\frac{m}{2\pi(\hbar c)^2}\left|\int d^3\mathbf{r}'\,\frac{e^{-ik|\mathbf{r}-\mathbf{r}'|}}{|\mathbf{r}-\mathbf{r}'|}U(\mathbf{r}')\,e^{i\mathbf{k}\bullet\mathbf{r}'}\right| << 1. \tag{2.109}$$

To derive the validity conditions, let us apply the simplest case in which the potential is approximated by having a constant magnitude, U_0, and a finite range, a. We calculate within the vicinity of the potential (i.e. $r \approx 0$) and, at first, the low-momentum case of $ka \ll 1$. In this case, (2.109) provides

$$\frac{mU_0}{2\pi(\hbar c)^2}\left|\int d^3\mathbf{r}'\,\frac{e^{-ikr'}}{r'}\,e^{i\mathbf{k}\bullet\mathbf{r}'}\right| << 1. \tag{2.110}$$

The integral is

$$\begin{aligned}\int d^3\mathbf{r}'\,\frac{e^{-ikr'}}{r'}\,e^{i\mathbf{k}\bullet\mathbf{r}'} &= 2\pi\int_0^a dr'\,r'\,e^{-ikr'}\int_{-1}^{1} d(\cos\,\theta)\,e^{ikr'\,\cos\,\theta}\\ &= \frac{2\pi}{ik}\int_0^a dr'\left(e^{2ikr'}-1\right)\\ &= \frac{\pi}{k^2}\left((1-\cos\,2ka)+i(2ka-\sin\,2ka)\right).\end{aligned} \tag{2.111}$$

Substituting this result into (2.110) gives, at first,

$$\frac{mU_0}{2(k\hbar c)^2}\left|\left((1-\cos\,2ka)+i(2ka-\sin\,2ka)\right)\right| << 1. \tag{2.112}$$

The modulus of (2.112) is then calculated for $ka \ll 1$ by expanding the cosine and sine functions to the third term to demonstrate that the modulus is equal to $2(ka)^2$. Thus the low-momentum validity criterion for the first Born approximation is

$$U_0 << \frac{(\hbar c)^2}{ma^2}\quad ka << 1. \tag{2.113}$$

On the other hand, for the high-momentum state dictated by the inequality of $ka \gg 1$, the phase factor within the integral of (2.109) will rapidly oscillate, and the integral averages to zero as $k \to \infty$. Hence, the first Born approximation should be good at high energies (an expectation we will verify shortly). The exact behaviour of the integral for large k can be determined by using the stationary phase approximation. The main contributions to the integral will arise from those regions surrounding those points where the oscillatory factor is stationary. This yields the high-momentum validity criterion for the first Born approximation:

$$U_0 << \frac{2(\hbar c)^2 k}{ma} \quad ka \gg 1. \tag{2.114}$$

As the magnitude of the wavevector is $k \equiv \frac{p}{\hbar c}$, (2.114) can also be written as

$$U_0 << \frac{2\hbar c\beta}{a} \quad ka \gg 1. \tag{2.115}$$

The dependence of the Born expansion upon the incident momentum (energy) requires that the projectile's energy exceed that of the scattering potential, as has just been demonstrated. This requirement is, in fact, obvious in that the fundamental assumption leading to the Born series is that the influence of the scattering potential upon the incident wavefunction can be neglected and that $\psi(\mathbf{r})$ within the integrand can be replaced by $\psi_0(\mathbf{r})$. It is further clear that if the potential were attractive and exceeded the incident kinetic energy, a bound state would arise.

We now consider the sensitivity of the first term of the Born series to the projectile momentum by evaluating the differential cross section amplitude for the 'hard sphere' potential defined by

$$\begin{aligned} U(r) &= U_0; \qquad r < a \\ &= 0; \qquad r \geq a \end{aligned} \tag{2.116}$$

From (2.84), the first Born approximation of the scattering amplitude is

$$\begin{aligned} f_{(1)}(q) &= -\left(\frac{2m}{\hbar c q}\right) \int_0^\infty \mathrm{d}r'\, r' \sin\left(\frac{qr'}{\hbar c}\right) U(r') \\ &= -\left(\frac{2m\, U_0}{\hbar c q}\right) \int_0^a \mathrm{d}r'\, r' \sin\left(\frac{qr'}{\hbar c}\right) \end{aligned} \tag{2.117}$$

This integral is solved by a change in variable

$$y = \frac{qr'}{\hbar c}$$

and then an integration by parts

$$f_{(1)}(\theta) = -\frac{mU_0}{(2\hbar c)^2 k^3} \left(\frac{\sin(\zeta(\theta)) - \zeta(\theta)\cos(\zeta(\theta))}{\sin^3\left(\frac{\theta}{2}\right)}\right) \tag{2.118}$$

where k is the wavevector and (2.86) have been used. The dimensionless variable is

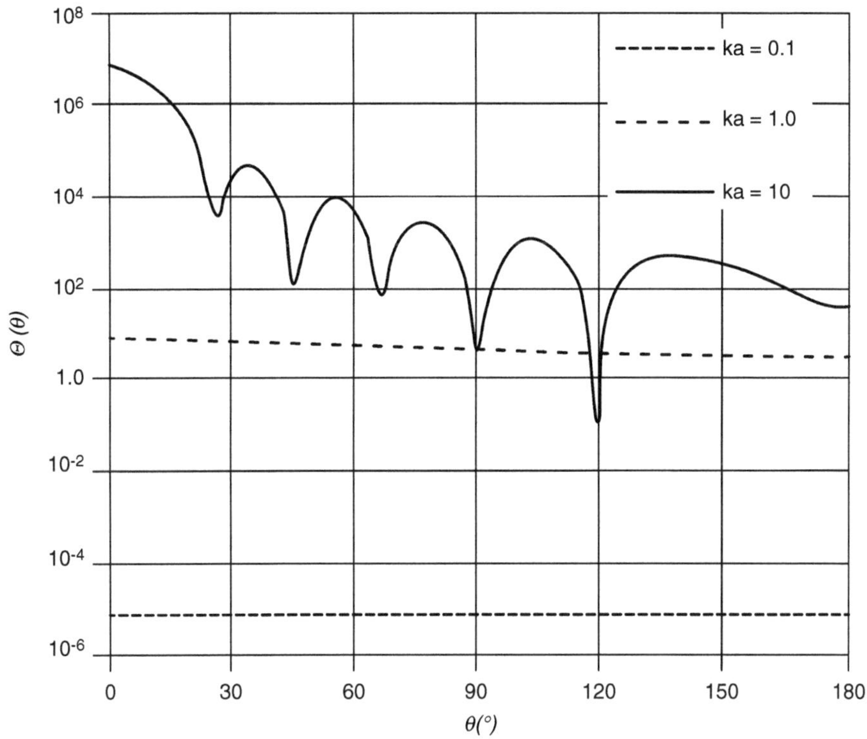

Fig. 2.5 The angular function $\Theta(\theta)$ as a function of scattering angle θ from (2.120) for three different values of the product ka

$$\zeta(\theta) = 2ka \sin\left(\frac{\theta}{2}\right),$$

and the resulting differential cross section in solid angle is, from (2.94),

$$\begin{aligned}\frac{\mathrm{d}\sigma}{\mathrm{d}\Omega} &= |f(\theta)|^2 \\ &= \left(\frac{U_0}{8Ek}\right)^2 \Theta(\theta)\end{aligned} \qquad (2.119)$$

and where the angular function is

$$\Theta(\theta) = \frac{\zeta(\theta)^2 + \left(1 - \zeta(\theta)^2\right)\sin^2 \zeta(\theta) - 2\zeta(\theta) \sin\ \zeta(\theta) \cos\ \zeta(\theta)}{\sin^6\left(\frac{\theta}{2}\right)}. \qquad (2.120)$$

An initial inspection of this result reveals an interesting feature: the differential cross section is proportional to the square of the ratio of the potential to the kinetic energy. There is no discontinuity for the case of the projectile's kinetic energy being less than the magnitude of the potential. Further, we consider the angular function $\Theta(\theta)$ which is shown in Fig. 2.5 as a function of the scattering angle for various values of the product ka. For the cases of the de Broglie wavelength being smaller or comparable to the range of the potential ($ka = 0.1,\ 1$), the angular distribution shows little dependence upon the scattering angle. The magnitude of $\Theta(\theta)$, however, increases with ka. For the condition of the de Broglie wavelength being smaller than the potential's range ($ka = 10$), diffraction effects occur, and local maxima and minima occur with scattering angle. As this result is derived from the Born approximation, it affirms our above derivations that the Born approximation is insufficient at low momenta or, in other words, those cases where the projectile's kinetic energy is weak relative to the scattering potential.

2.5 Phase Shift Analysis

2.5.1 *Introduction*

Further examination of the quantum scattering process can be had by now explicitly accounting for the role of angular momentum. Why this should be done is demonstrated by considering the scattering case of Fig. 2.6 of a projectile of mass m and speed v incident to an infinitely massive scattering centre. The trajectory of the projectile is offset from the position of the scattering centre by the perpendicular distance b, which is known as the impact parameter and which represents the closest distance between projectile and target if there were no interaction between the two.

Adopting a mixture of classical and quantum mechanics, we consider the projectile to be well localised. Classically, the magnitude of the angular momentum of the projectile about the scattering centre is

$$L = pb. \tag{2.121}$$

In anticipation of our bridge to a quantum-mechanical description, we write

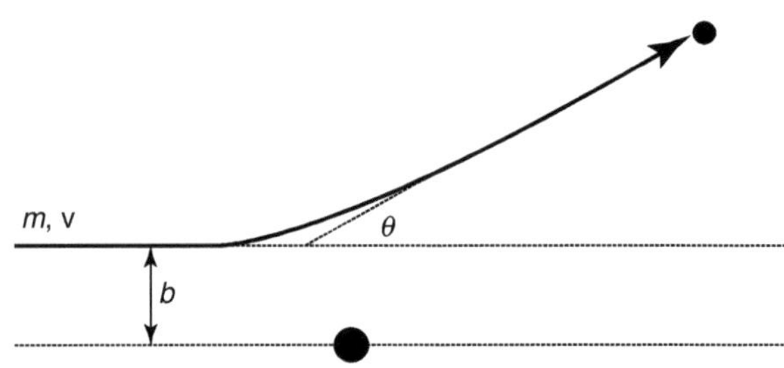

Fig. 2.6 Nonrelativistic scattering of a particle of mass m and speed v at an impact parameter b through an angle θ from an infinitely massive scattering centre

$$L = \hbar ckb \tag{2.122}$$

accommodating our use of the units of energy for momentum. However, in quantum mechanics angular momentum is quantised:

$$L = \hbar \sqrt{l(l+1)} \tag{2.123}$$

where l is an integer The proportionality between the angular momentum quantum number l and the impact parameter is obtained from (2.122) and (2.123) to yield

$$(l(l+1)) \propto (bk)^2. \tag{2.124}$$

It thus follows that a beam of projectiles of finite width will consist of many angular momenta. Hence, this incident beam of momentum **p** and wavenumber **k**, which is described by a plane wave, can be expanded into a series of all of the wavefunctions appropriate for the range of angular momenta. The derivation of this expansion is involved and will not be provided here, but is

$$\begin{aligned}\psi_0(\mathbf{r}) &= e^{i\mathbf{k}\bullet\mathbf{r}} = e^{ikr\cos\,\theta} \\ &= \sum_{l=0}^{\infty} (2l+1)\, i^l\, j_l(kr)\, P_l(\cos\ \theta)\end{aligned} \tag{2.125}$$

where *jl*(*kr*) is the spherical Bessel function and *Pl*(cos θ) is the Legendre polynomial, both of order l (see Appendices C.1 and C.2). The summation is taken to infinity reflecting infinite angular momenta corresponding to an infinitely broad beam of projectiles.

One can crudely interpret, from (2.125), that the quantity $|(2l+1)\, j_l(kr)\,|^2$ represents the probability that a projectile with the quantum angular momentum number l can be found at a distance r from the scattering centre. The use of angular momentum in describing the scattering problem thus leads to the development of the phase shift in which the scatter is reflected by a shifting of the wavefunction phase, in addition to a modulation of its amplitude.

Subsequent to the scatter, the exiting waveform is, as described before, the sum of the incident wave and a spherical wave, with the latter weighted by the scattering amplitude. Hence, following from the result of (2.125), this exiting wave can also be expanded into a series by first considering an expansion in spherical harmonics (Appendix C.3):

$$\psi(\mathbf{r}) = \sum_{l=0}^{\infty} \sum_{m=-l}^{+l} R_{lm}(r) Y_{lm}(\theta, \varphi) \tag{2.126}$$

where $Rlm(r)$ is the radial solution to the Schrödinger wave equation. As our interest is solely of central scattering potentials with no azimuthal symmetry, there is azimuthal symmetry and $m = 0$. The spherical harmonics can thus be replaced with Legendre polynomials to give the series

$$\psi(\mathbf{r}) = \sum_{l=0}^{\infty} (2l + 1)\, i^l\, R_l(r)\, P_l(\cos\,\theta) \tag{2.127}$$

so as to keep a form similar to (2.125).

In this section, the description of the scattering problem in terms of angular momenta (as suggested by the two prior equations) and, subsequently, phase shifts will be shown to be useful in deriving further expressions for the scattering amplitude and, by extension, the differential cross section for the scatter.

2.5.2 Scattering Amplitude

As the solution to the scattering problem is obtained by determining the asymptotic form of the scattered wavefunction at a distance from the scattering centre, it can be demonstrated that scattering results in a shift in the wavefunction phase in addition to modulating the intensity and angular distribution of the wavefunction. Recall that the scattering amplitude is determined from

$$\psi(\mathbf{r}) - \psi_0(\mathbf{r}) = \frac{e^{ikr}}{r} f(\mathbf{q}). \tag{2.128}$$

In order to determine $f(\mathbf{q})$, we adopt our usual case of examining the results of the scatter at an extended distance from the scattering centre. We can write the radial form of the solution to the Schrödinger equation given by (2.127) as the weighted sum of the spherical Bessel functions:

$$R_l(r) = A_l j_l(kr) + B_l n_l(kr). \tag{2.129}$$

As we are interested in the asymptotic forms for $r \to \infty$, the asymptotic forms of the Bessel functions are used so that (2.129) is written as

$$\begin{aligned} R_l(r) &= \frac{1}{kr}\left(A_l \sin\left(kr - \frac{l\pi}{2}\right) - B_l \cos\left(kr - \frac{l\pi}{2}\right)\right) \\ &= \frac{\sqrt{A_l^2 + B_l^2}}{kr}\left(\frac{A_l}{\sqrt{A_l^2 + B_l^2}} \sin\left(kr - \frac{l\pi}{2}\right) - \frac{B_l}{\sqrt{A_l^2 + B_l^2}} \cos\left(kr - \frac{l\pi}{2}\right)\right). \end{aligned} \tag{2.130}$$

Let us first consider the quantity within the parentheses of this result. If we were to interpret the weighting factors of the sine and cosine functions as being trigonometric functions of an angle δ_l

$$\begin{aligned} \frac{A_l}{\sqrt{A_l^2 + B_l^2}} &= \cos\ \delta_l \\ \frac{B_l}{\sqrt{A_l^2 + B_l^2}} &= \sin\ \delta_l, \end{aligned} \tag{2.131}$$

it follows from the trigonometric identity

$$\sin(\alpha - \beta) = \sin\ \alpha\ \sin\ \beta - \cos\ \alpha\ \cos\ \beta$$

that we can write that quantity as

$$\cos\ \delta_l \sin\left(kr - \frac{l\pi}{2}\right) + \sin\ \delta_l \cos\left(kr - \frac{l\pi}{2}\right) = \sin\left(kr - \frac{l\pi}{2} + \delta_l\right). \tag{2.132}$$

From (2.131), we can write the phase shift angle as

$$\delta_l = -\tan^{-1}\frac{B_l}{A_l}. \tag{2.133}$$

δl represents the phase shift of the wavefunction component of the l-th order. By defining

$$C_l = \sqrt{A_l^2 + B_l^2} \tag{2.134}$$

the radial function can then be written as

$$\begin{aligned} R_l(r) &= \frac{C_l}{kr} \sin\left(kr - \frac{l\pi}{2} + \delta_l\right) \\ &= \frac{C_l}{2ikr}\left(e^{i\left(kr - \frac{l\pi}{2} + \delta_l\right)} - e^{-i\left(kr - \frac{l\pi}{2} + \delta_l\right)}\right). \end{aligned} \tag{2.135}$$

The wavefunction described by (2.127) is

$$\psi(\mathbf{r}) = \frac{1}{2ikr} \sum_{l=0}^{\infty} (2l+1)\, i^l\, C_l \Big(e^{i\left(kr - \frac{l\pi}{2} + \delta_l\right)} - e^{-i\left(kr - \frac{l\pi}{2} + \delta_l\right)} \Big) P_l(\cos\,\theta). \qquad (2.136)$$

We wish to link the coefficients Cl with the phase shifts δl. This is accomplished through the use of asymptotic form of the spherical Bessel function:

$$\begin{aligned} \psi_0(\mathbf{r}) &= \frac{1}{kr} \sum_{l=0}^{\infty} (2l+1)\, i^l \sin\left(kr - \frac{l\pi}{2} \right) P_l(\cos\,\theta) \\ &= \frac{1}{2ikr} \sum_{l=0}^{\infty} (2l+1)\, i^l \left(e^{i\left(kr - \frac{l\pi}{2}\right)} - e^{-i\left(kr - \frac{l\pi}{2}\right)} \right) P_l(\cos\,\theta). \end{aligned} \qquad (2.137)$$

By substituting (2.136) and (2.137) into (2.128) and equating the coefficients of the exponentials, we obtain, first,

$$\frac{1}{2ikr} \sum_{l=0}^{\infty} (2l+1)\, i^l\, P_l(\cos\,\theta)\, e^{i\frac{l\pi}{2}} \left(1 - C_l e^{-i\delta_l}\right) = 0 \qquad (2.138)$$

and, second,

$$\frac{1}{2ik} \sum_{l=0}^{\infty} (2l+1)\, i^l\, P_l(\cos\,\theta)\, e^{-i\frac{l\pi}{2}} \left(C_l e^{i\delta_l} - 1\right) = f(\theta). \qquad (2.139)$$

(2.138) gives the result

$$C_l = e^{i\delta_l}$$

directly and which, when substituted into (2.139), yields, after further algebraic manipulation, an initial expression for the scattering amplitude:

$$\frac{1}{2ik} \sum_{l=0}^{\infty} (2l+1)\; P_l(\cos\,\theta)\, e^{i\delta_l} \left(e^{i\delta_l} - e^{-i\delta_l}\right) = f(\theta)$$

and a final expression in terms of partial wave shifts

$$f(\theta) = \frac{1}{k} \sum_{l=0}^{\infty} (2l+1)\, e^{i\delta_l} \sin\,\delta_l P_l(\cos\,\theta). \qquad (2.140)$$

This result is a consequence of the conservation of angular momentum (rotational invariance) and the conservation of probability. Using (2.94) and this expression for the scattering amplitude, we can then calculate the differential cross section in terms of the partial waves:

$$\frac{\mathrm{d}\sigma}{\mathrm{d}\Omega} = \frac{1}{k^2} \sum_{l=0}^{\infty} \sum_{l'=0}^{\infty} (2l+1)\,(2l'+1)\, e^{i\delta_l} e^{-i\delta_{l'}} \sin\,\delta_l\, \sin\,\delta_{l'}\, P_l(\cos\,\theta) P_{l'}(\cos\,\theta). \tag{2.141}$$

The total cross section is given by the integration of (2.141) over all solid angle:

$$\begin{aligned}
\sigma &= \frac{1}{k^2} \int \mathrm{d}\Omega\,(2l+1)\,(2l'+1)\, e^{i\delta_l} e^{-i\delta_{l'}} \sin\,\delta_l\, \sin\,\delta_{l'}\, P_l(\cos\,\theta) P_{l'}(\cos\,\theta) \\
&= \frac{1}{k^2} \sum_{l=0}^{\infty} \sum_{l'=0}^{\infty} (2l+1)\,(2l'+1)\, e^{i\delta_l} e^{-i\delta_{l'}} \sin\,\delta_l\, \sin\,\delta_{l'} \\
&\quad \times \int \mathrm{d}\Omega P_l(\cos\,\theta) P_{l'}(\cos\,\theta) \\
&= \frac{2\pi}{k^2} \sum_{l=0}^{\infty} \sum_{l'=0}^{\infty} (2l+1)\,(2l'+1)\, e^{i(\delta_l-\delta_{l'})} e^{-i\delta_{l'}} \sin\,\delta_l\, \sin\,\delta_{l'} \\
&\quad \times \int_{-1}^{1} \mathrm{d}(\cos\,\theta) P_l(\cos\,\theta) P_{l'}(\cos\,\theta) \\
&= \frac{4\pi}{k^2} \sum_{l=0}^{\infty} \sum_{l'=0}^{\infty} (2l+1)\,(2l'+1)\, e^{i(\delta_l-\delta_{l'})} e^{-i\delta_{l'}} \sin\,\delta_l\, \sin\,\delta_{l'} \frac{\delta_{ll'}}{(2l+1)} \\
&= \frac{4\pi}{k^2} \sum_{l=0}^{\infty} (2l+1) \sin^2\,\delta_l.
\end{aligned} \tag{2.142}$$

where the orthogonality property of the Legendre polynomials (Appendix C.2) has been used. $\delta_{ll'}$ is the Kronecker delta function. Note that, from (2.140),

$$f(0) = \frac{1}{k} \sum_{l=0}^{\infty} (2l+1) e^{i\delta_l} \sin\,\delta_l \tag{2.143}$$

from which it follows that

$$\operatorname{Im} f(0) = \frac{1}{k} \sum_{l=0}^{\infty} (2l+1) \sin^2\,\delta_l. \tag{2.144}$$

From this result and the expression for the total cross section given by (2.142),

$$\sigma = \frac{4\pi}{k} \operatorname{Im} f(0). \tag{2.145}$$

This is known as the *optical theorem* which demonstrates that the total cross section is a measure of the reduction in flux which results from the destructive interference between the incident and scattered wavefunctions for the scattering angle $\theta = 0$.

Bibliography and Further Reading

Abramowitz M, Stegun IA (eds). Handbook of mathematical functions. New York: Dover Publications; 1972.

Belkic D. Principles of quantum scattering theory. Bristol: Institute of Physics Publishing; 2004.

Churchill RV, Brown JW, Verhey RF. Complex variables and applications. 3rd ed. New York: McGraw-Hill Inc; 1974.

Dyson F. Advanced quantum mechanics. Singapore: World Scientific Publishing; 2007.

Goldberger ML, Watson KM. Collision theory. Mineola: Dover Publications; 2004.

Gribov VN, Nyiri J. Quantum electrodynamics. Cambridge: Cambridge University Press; 2001.

Jackson JD. Classical electrodynamics. 3rd ed. Hoboken: Wiley; 1999.

McParland BJ. Nuclear medicine dosimetry: advanced theoretical principles. Berlin: Springer; 2010.

Messiah A. Quantum mechanics. New York: Wiley.

Sigmund P. Particle penetration and radiation effects. Berlin: Springer; 2008.

Part II
Elastic Coulomb Scatter

Chapter 3
Introduction to Part II

Contents

Abstract Part II presents topics of the elastic Coulomb scatter of charged particles. This chapter summarises these topics.

3.1 Introduction

By our, now well known, definition, the Coulomb elastic scatter between two charged objects is that in which the sum of the pre-scatter kinetic energies is equal to that of the post-scatter state. In such a case, the projectile and target entities are to be regarded as singular entities without any internal degrees of freedom. Excitation or ionisation channels, which would redirect this energy elsewhere and away from the post-scattering kinetic energies, are either not present or else the magnitudes of the energies extracted through these channels are sufficiently low that they can be ignored.

In Chap. 2, we developed the quantum theory of scattering and have emphasised the fact that the observer (i.e. the experimentalist) can only measure the asymptotic forms of the pre- and post-scattering states. This will prove the foundation of our derivations in the following three chapters of Part II.

The Coulomb elastic scatter phenomenon is important to medical radiation dosimetry in many facets. It dictates the transport of charged particles within the medium, a mechanism which is fundamental to the calculation of how charged particles transfer kinetic energy to the medium resulting in absorbed radiation dose and subsequent biological effects which the medical practitioner wishes either to achieve or to avoid. At one level, we recognise that the ability to calculate this transport, either directly through Monte Carlo simulations or else through approximations such as through the Molière theory of multiple scatter, will reveal

B.J. McParland, *Medical Radiation Dosimetry*,
DOI 10.1007/978-1-4471-5403-7_3,

to use the spatial distribution of the charged particle fluence and, subsequently, the absorbed dose spatial distribution.

We must also recognise that the transfer of energy to the medium is also via elastic interactions. In some cases, this is the recoil energy of the atom as a whole, but in most practical cases, it is the recoil energy of the electron which may be significant to be ejected as a δ-ray to transport energy to an extended distance from the interaction site. This is the subject of Chap. 10 in Part III: the projectile's velocity far exceeds that of the orbiting atomic electron to such an extent that the latter is considered to be free and kinematically at rest. There are other interaction avenues: the atom may absorb energy through inelastic channels; in other cases, such as the adiabatic response, the atomic electrons are perturbed, but there is an insignificant transfer of energy from the moving charged projectile to the medium.

Part II considers the elastic interactions of the charged particle with the medium. This does not mean, as is sometimes commonly thought, that kinetic energy is not transferred to the medium via an elastic collision. Energy can be transferred, but perhaps what is of most immediate interest to a medical radiation physics' description of the process is an understanding of the transport of the charged particle through the medium. This is of particular interest to the multitude of Monte Carlo software codes simulating charged particle transport that is used in medical applications.

Part II begins the exposition of this book in Chap. 4 through the fundamental understandings of elastic charged particle interactions with an unscreened (bare) point charge through both classical and quantum theory, the latter dependent upon the Born series developed in Chap. 2. But, in medical radiation physics, the absorbed dose-producing interactions between the charged particle and tissue are through atomic interactions, and hence, the foundations of Chap. 4 are expanded through Chap. 5 to consider the more realistic conditions of a distributed and screened target, the effects upon the scattering cross section and how they can be estimated through the appropriate use of atomic models, such as the Thomas–Fermi statistical model and the discrete atomic electron model. A consequence of Chap. 5 is our appreciation that elastic scatter is dominated by the interaction between the charged projectile and the atomic nucleus (whereas, as will be demonstrated in Part III, energy transfer is dominated by the interaction between the projectile and the atomic electrons). Chapters 4 and 5 emphasise that elastic Coulomb scatter is highly forward directed with a $sin^{-4}\frac{\theta}{2}$ scattering angle dependence. Part II concludes with Chap. 6 which investigates the consequence of this angular dependence which is manifested as multiple elastic Coulomb scatter. Four theories which are, or have been, widely used in medical radiation physics are derived and discussed – those of Goudsmit and Saunderson, Williams, Fermi and Eyges and that of Molière.

Chapter 4
Elastic Coulomb Scatter from an Unscreened Point Charge

Contents

Abstract This chapter begins Part II of this book with an exploration of the physics of a process which will be fundamental throughout our future discussions, that of the Coulomb interaction between a moving charged particle and a stationary charged body or, equivalently, a Coulomb potential. The result is scatter, or the deviation of the moving particle from its original trajectory. Energy can be transferred through the recoil of the target. Here, the target is assumed to be a bare (unscreened) point charge. We begin with two independent classical descriptions of the problem: one based upon our development of the scattering integral and the other based upon the Laplace–Runge–Lenz vector. Both yield the same result in that the Coulomb scatter differential cross section with solid angle is proportional to $\frac{1}{\sin^4\left(\frac{\theta}{2}\right)}$, where θ is the angle through which the projectile is scattered from its original trajectory. The validity conditions for such a classical description are then examined. We then proceed to derivations of this differential cross section through

B.J. McParland, *Medical Radiation Dosimetry*,
DOI 10.1007/978-1-4471-5403-7_4, © Springer-Verlag London 2014

quantum-mechanical means. We first use the first term of the Born expansion to yield an expression for the scattering amplitude and, thus, the differential cross section which is in exact agreement with that derived classically. But as the Coulomb potential does not satisfy the requirements of this approximation, we extend the Born expansion to second order for the Coulomb potential and determine that this second-order term will diverge, but can be made convergent for a screened charged target. An exact nonrelativistic quantum-mechanical solution for the Coulomb scattering differential cross section is then derived from the Schrödinger equation and compared to the classical results and that based upon the Born expansion. The problem of electron (Mott) scattering, in which the intrinsic spin of the electron is not neglected, is examined and a differential cross section derived from Dirac theory. Finally, the case of electron–electron (Møller) scatter and the inherent need to account for spin statistics and the indistinguishability between the electrons in the final state is derived.

4.1 Introduction

In this chapter, we explore the Coulomb scatter of a moving projectile of charge ze from a bare point target of charge Ze. Depending upon the masses of the projectile and target (or if the target is moving relative to the projectile), the calculation is performed in either the laboratory or centre-of-mass reference systems using the actual or reduced masses, accordingly. This chapter will deal exclusively with elastic scatter as it is concerned with a point projectile undergoing Coulomb scatter with a point target which is not screened. No degrees of freedom which can channel away kinetic energy are considered to be present.

Collisions with an atom will be differentiated between those with an atom as a whole (which Bohr (1948) termed, perhaps rather confusingly, as 'nuclear' collisions) and with those with individual atomic electrons (known as 'electronic' collisions). In later discussion, we will categorise collisions as 'soft' and as 'hard', with the differentiation between these being defined by the theoretical model (spatial distance from the atom or energy transfer). The former is, to a reasonable approximation, considered elastic in that the atomic recoils as a whole with a complete transfer of kinetic energy and momentum. The latter is frequently inelastic in that additional energy channels exist which remove kinetic energy and momentum from the projectile. These channels include atomic excitations and, should the energy transfer to the atomic electron exceed its binding energy, ionisation and the ejection of the electron.[1] These collisions are primarily responsible for the stopping of a charged particle in a medium.

As the Coulomb interaction is the foundation of the rest of the discussions to follow, it is the focus of this and the following two chapters. We begin by considering the description of Coulomb scatter using classical mechanics.

[1] As noted earlier, when the projectile's kinetic energy is much greater than the atomic binding energy, the latter can be ignored and the interaction considered as elastic for all practical purposes.

The goal is to calculate the differential cross section in solid angle of the charged projectile being scattered from its trajectory by an angle θ. We will consider two separate ways of evaluating this cross section through classical methods. The first is based upon the scattering integral which relates the scattering angle to both the kinematics of the scatter and the spatial separation between the projectile and target. We will consider this solution in both nonrelativistic and relativistic cases. The use of classical perturbation theory as a means of evaluating the scattering integral concludes the discussion of the scattering integral. The second classical mechanical method investigated of evaluating the Coulomb cross section is based upon the Laplace–Runge–Lenz vector which not only reproduces the differential cross section calculated using the scattering integral but also describes the shape of the trajectory followed by the scattered projectile. We then conclude this section with a detailed examination of the conditions under which descriptions of Coulomb scatter using classical mechanics are valid and the uncertainties that can arise.

Quantum-mechanical descriptions of Coulomb scatter will then follow with the nonrelativistic case considered first. The Born expansion of Chap. 2 is applied to the unscreened Coulomb potential and investigated in detail. The first term of the expansion for Coulomb scatter is evaluated and the differential cross section calculated – a result shown to be the same as that derived from the classical mechanical calculations. However, as a result of the perturbation theory which is the foundation of the Born expansion as an iterative solution to the Lippmann–Schwinger equation, the expansion is strictly valid only for a short-range potential that is weaker than the projectile's kinetic energy. Because the range of the Coulomb potential is infinite, we will extend the Born expansion to second order in order to more fully investigate its utility in describing the Coulomb interaction.

Going beyond the use of perturbation theory in calculating for Coulomb scatter, the exact solution is derived from the Schrödinger equation. The chapter then concludes with discussion of relativistic Coulomb scatter.

4.2 Classical Mechanical Solutions

4.2.1 Scattering Integral

4.2.1.1 Nonrelativistic Solution

Figure 4.1 describes the basic scattering geometry that we will consider in this discussion.

In this figure, a particle of mass m, 3-vector momentum $\mathbf{p}$ (and kinetic energy $E = \frac{p^2}{2m}$) and electric charge ze is incident from the left at an impact parameter b from the infinitely massive scattering centre of charge Ze' which is at the origin, and is scattered through the angle θ with a momentum $\mathbf{p}'$. Note that $|\mathbf{p}| = |\mathbf{p}'|$. In this calculation, the target is assumed implicitly to have infinite mass, but should

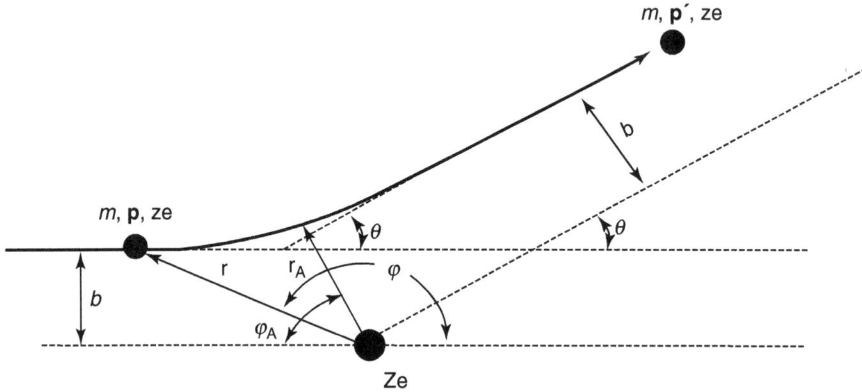

Fig. 4.1 Scattering geometry from an unscreened charged point target of infinite mass (See text for description)

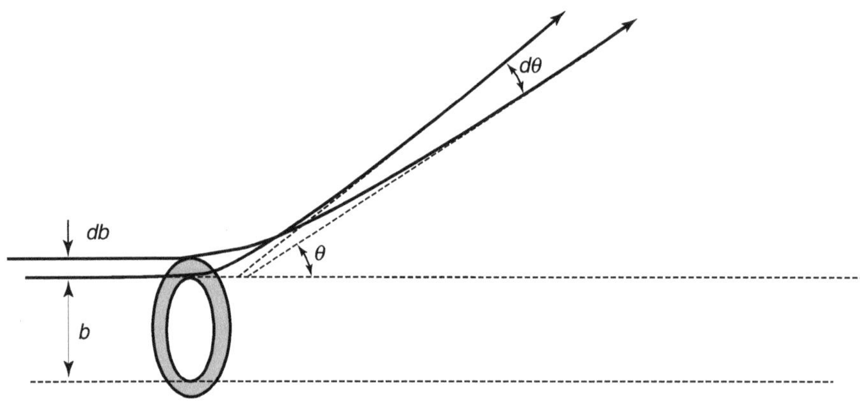

Fig. 4.2 Scattering geometry for calculating the differential cross section. Azimuthal symmetry exists in Coulomb scattering, and the differential solid angle is $\mathrm{d}\Omega = 2\pi \sin\theta \, \mathrm{d}\theta$

it not, the calculation can be performed in the centre-of-mass reference system and the mass taken to be the reduced mass (see Appendix D). The potential affecting the projectile is the Coulomb, $U(r) = \frac{\alpha\hbar c z Z}{r}\cdot$ As the differential cross section for the scatter is the annular area afforded by a differential element of the impact parameter, as shown in Fig. 4.2, we wish to determine the relationship between the scattering angle and the impact parameter. This differential cross section is

$$\begin{aligned} \mathrm{d}\sigma &= \pi\left(|b + \mathrm{d}b|^2 - |b|^2\right) \\ &= 2\pi|b\,\mathrm{d}b| + \pi|\mathrm{d}b|^2 \\ &\cong 2\pi|b\,\mathrm{d}b| \end{aligned} \tag{4.1}$$

as we will neglect terms in the differential of second order or higher.

Recall from our discussions of Chap. 2 that we are only able to observe the asymptotic forms of the projectile pre- and post-scatter or, equivalently, at infinite times in the past and in the future. Hence, the angle φ will have the properties

$$\begin{aligned}\varphi &\to \pi \text{ as } t \to -\infty \\ \varphi &\to \theta \text{ as } t \to +\infty\end{aligned} \tag{4.2}$$

as is immediately evident from Fig. 4.1. The vector $\mathbf{r}$ is the radial vector from the scattering centre to the projectile, and the vector $\mathbf{r}_\mathbf{A}$ is the value of r at the point of closest approach (i.e. the periapsis) between the projectile and the scattering centre. Of course, the magnitudes of the radial vector and the polar angle are both functions of time.

Let us consider conserved mechanical quantities of the scatter. The first will be that of energy. At infinite time in the past (and at infinite distance where there is no influence from the potential), the total energy is kinetic only and given by $\frac{p^2}{2m}$. The conservation of energy is, in cylindrical co-ordinates,

$$\frac{m}{2}\left(\left(\frac{\mathrm{d}r}{\mathrm{d}t}\right)^2 + r^2\left(\frac{\mathrm{d}\varphi}{\mathrm{d}t}\right)^2\right) + \frac{\alpha\hbar c}{r} zZ = \frac{p^2}{2m}. \tag{4.3}$$

The second conserved mechanical quantity considered is that of angular momentum, the magnitude of which is

$$|\mathbf{L}| = pb \text{ as } t \to -\infty. \tag{4.4}$$

From the geometry of Fig. 4.1, the conservation of angular momentum is described by

$$mr^2\frac{\mathrm{d}\varphi}{\mathrm{d}t} + pb = 0. \tag{4.5}$$

Also from Fig. 4.1, we see that the relationship between the angle φ_A and the scattering angle θ is

$$\theta = \pi - 2\varphi_\mathrm{A}. \tag{4.6}$$

The angle φ_A is given by the integral

$$\begin{aligned}\varphi_\mathrm{A} &= \int_{r_\mathrm{A}}^{\infty} \mathrm{d}r\,\frac{\mathrm{d}\varphi}{\mathrm{d}r} \\ &= \int_{r_\mathrm{A}}^{\infty} \mathrm{d}r\,\frac{\left(\frac{\mathrm{d}\varphi}{\mathrm{d}t}\right)}{\left(\frac{\mathrm{d}r}{\mathrm{d}t}\right)}.\end{aligned} \tag{4.7}$$

The goal of the calculation reduces now to determining the scattering angle, θ, as a function of the impact parameter b. (4.1) showed the relationship between the physically measurable cross section and this impact parameter. We begin by, as suggested by (4.7), determining the rate of change in φ with r. This requires that the derivatives in the numerator and denominator of the integrand be solved for. That of the numerator is determined by extracting the time dependence of the angle φ from (4.5):

$$\frac{\mathrm{d}\varphi}{\mathrm{d}t} = -\frac{pb}{mr^2}. \tag{4.8}$$

The denominator is calculated by substituting this result into (4.3) to yield the change in the magnitude of the radial vector with time:

$$\frac{\mathrm{d}r}{\mathrm{d}t} = \pm\left(\frac{p}{m}\right)\sqrt{1-\left(\frac{2m\alpha\hbar czZ}{rp^2}\right)-\left(\frac{b}{r}\right)^2}. \tag{4.9}$$

Substituting the two time dependencies of (4.8) and (4.9) into the integral of (4.7) gives the expression for the angle:

$$\begin{aligned}\varphi_{\mathrm{A}} &= \pm b\int_{r_{\mathrm{A}}}^{\infty}\frac{\mathrm{d}r}{r^2\sqrt{1-\left(\frac{2m\alpha\hbar czZ}{rp^2}\right)-\left(\frac{b}{r}\right)^2}}\\ &= \pm b\int_{r_{\mathrm{A}}}^{\infty}\frac{\mathrm{d}r}{r^2\sqrt{1-\left(\frac{\rho_B(E)}{r}\right)-\left(\frac{b}{r}\right)^2}}\end{aligned} \tag{4.10}$$

where the spatial parameter

$$\begin{aligned}\rho_{\mathrm{B}}(E) &\equiv \frac{2m\alpha\hbar czZ}{p^2}\\ &= \frac{\alpha\hbar czZ}{E}\end{aligned} \tag{4.11}$$

has been defined. This parameter will feature throughout much of our derivations and will be recognised as being the closest distance of approach between the projectile and target if a repulsive Coulomb force exists between the two entities. More subtly, Bohr (1948) termed $\rho_{\mathrm{B}}(E)$ as the 'collision diameter' for a reason that is to be established following our derivation of the Coulomb scatter differential cross section. An expression for r_{A} can now be obtained by noting that the derivative $\frac{dr}{dt} = 0$ when $r = r_A$. This is equivalent to the denominator of (4.10) being equal to zero for $r = r_A$ or

$$r_A^2 - \rho_B(E) r_A - b^2 = 0. \tag{4.12}$$

Solving for r_A and allowing for only the physically valid solution, we have

$$r_A = \frac{\rho_B(E)}{2}\left(1 + \sqrt{1 + \left(\frac{2b}{\rho_B(E)}\right)^2}\right). \tag{4.13}$$

As a result of the conservation of angular momentum, it can be seen that the vector $\mathbf{r}_A$ defines the axis of symmetry of the projectile's trajectory. The sign used in (4.10) is positive when the projectile is in the approach phase and negative in the departure phase. Hence, the scattering angle is, from (4.6), given by

$$\theta = \pi - 2b \int_{r_A}^{\infty} \frac{dr}{r^2 \sqrt{1 - \left(\frac{\rho_B(E)}{r}\right) - \left(\frac{b}{r}\right)^2}}. \tag{4.14}$$

The integral of (4.14), when multiplied by the impact parameter, shall be referred to as the scattering integral. By making the substitution of variable, $x = \frac{1}{r}$, this expression for the scattering angle becomes

$$\theta = \pi - 2b \int_{0}^{\frac{1}{r_A}} \frac{dx}{\sqrt{1 - \rho_B(E)\,x - b^2 x^2}}. \tag{4.15}$$

This integral is of the form with the solution

$$\int \frac{dx}{\sqrt{\alpha + \beta x + \gamma x^2}} = \frac{1}{\sqrt{-\gamma}} \cos^{-1}\left(-\frac{\beta + 2\gamma x}{\sqrt{\beta^2 - 4\alpha\gamma}}\right) \quad \text{for} \quad \gamma < 0. \tag{4.16}$$

Using this, the solution to the scattering integral is

$$b \int_{0}^{\frac{1}{r_A}} \frac{dx}{\sqrt{1 - \rho_B(E)x - b^2x^2}} = \cos^{-1}\left(\frac{\rho_B(E) + 2b^2 x}{\sqrt{\rho_B^2(E) + 4b^2}}\right)\Bigg|_{0}^{\frac{1}{r_A}}$$

$$= \cos^{-1}\left(\frac{1 + 2\frac{b^2}{\rho_B(E)\, r_A}}{\sqrt{1 + \left(2\frac{b}{\rho_B(E)}\right)^2}}\right) - \cos^{-1}\left(\frac{1}{\sqrt{1 + \left(2\frac{b}{\rho_B(E)}\right)^2}}\right). \tag{4.17}$$

Use of (4.13) followed by tedious algebraic manipulation of the argument of the first $\cos^{-1}$ function (which we shall omit) shows that

$$\cos^{-1}\left(\frac{1+2\dfrac{b^2}{\rho_{\mathrm{B}}(E)r_{\mathrm{A}}}}{\sqrt{1+\left(2\dfrac{b}{\rho_{\mathrm{B}}(E)}\right)^2}}\right)=\cos^{-1}(1).$$

$$=0 \tag{4.18}$$

Thus, the scattering angle is given by the expression

$$\sin\frac{\theta}{2}=\frac{1}{\sqrt{1+\left(2\dfrac{b}{\rho_{\mathrm{B}}(E)}\right)^2}}. \tag{4.19}$$

We can then use this result to solve for the impact parameter as a function of scattering angle:

$$b=\frac{\rho_{\mathrm{B}}(E)}{2}\cot\frac{\theta}{2}. \tag{4.20}$$

The result of (4.20), which was derived originally via a similar approach taken by Thomson (1906), is the functional link between scattering angle and impact parameter that we sought. More importantly, it leads directly to the physical observable of the differential cross section in solid angle. We can show this by first differentiating (4.20) with respect to scattering angle:

$$\frac{\mathrm{d}b}{\mathrm{d}\theta}=-\frac{\rho_{\mathrm{B}}(E)}{4}\frac{1}{\sin^2\dfrac{\theta}{2}} \tag{4.21}$$

and then substituting this result and (4.20) into the expression for the differential cross section given by (4.1) to arrive at

$$\mathrm{d}\sigma=\pi\left(\frac{\rho_{\mathrm{B}}(E)}{2}\right)^2\frac{\cos\dfrac{\theta}{2}}{\sin^3\dfrac{\theta}{2}}\,\mathrm{d}\theta. \tag{4.22}$$

Using

$$\mathrm{d}\Omega=2\pi\sin\theta\,\mathrm{d}\theta$$

and the trigonometric identity

$$\sin\,\theta = 2\sin\frac{\theta}{2}\,\cos\frac{\theta}{2},$$

can be rewritten as

$$\frac{\mathrm{d}\sigma}{\mathrm{d}\Omega} = \left(\frac{\rho_{\mathrm{B}}(E)}{4}\right)^2 \frac{1}{\sin^4\dfrac{\theta}{2}}. \tag{4.23}$$

This result is frequently referred to as the Rutherford, or unscreened Coulomb, cross section. We will provide a discussion of the various features of this result which will be of significant later interest in greater detail following the associated derivation using quantum theory.

Why Bohr described the spatial parameter $\rho_{\mathrm{B}}(E)$ as the 'collision diameter' can now be demonstrated following the above derivation through the use of (4.23). From (4.20), it follows that the condition of backscatter (i.e. that of the scattering angle θ exceeding $\pi/2$) corresponds to the inequality of $\rho_{\mathrm{B}}(E) \geq 2b$. That is,

$$\begin{aligned}\sigma\left(\theta \geq \frac{\pi}{2}\right) &= 2\pi \int_{-1}^{0} \mathrm{d}(\cos\,\theta)\,\frac{\mathrm{d}\sigma}{\mathrm{d}\Omega} \\ &= 2\pi\left(\frac{\rho_{\mathrm{B}}(E)}{4}\right)^2 \int_{-1}^{0} \frac{\mathrm{d}(\cos\,\theta)}{\sin^4\frac{\theta}{2}}.\end{aligned} \tag{4.24}$$

The integral is solved using the trigonometric identity, $\sin\frac{\theta}{2} = \sqrt{\frac{(1-\cos\theta)}{2}}$, and the substitution of variable $y = 1 - \cos\theta$ to give the result

$$\sigma\left(\theta \geq \frac{\pi}{2}\right) = \frac{\pi\,\rho_{\mathrm{B}}^2(E)}{4} \tag{4.25}$$

which is equal to the area of a circle of diameter ρ_{B} and, hence, leading to Bohr's description of it as the collision diameter. Although ρ_{B} is a classically derived quantity, its functional form will appear throughout our discussions due to its convenience.

Figure 4.3 shows the dependence of the Bohr scattering diameter, normalised to the zZ product of the projectile and target charges, as a function of the projectile's kinetic energy. This shows the reciprocal dependence of the Bohr scattering diameter with kinetic energy for a fixed zZ product and implies the decrease for fixed kinetic energy and increasing Coulomb potential. This also leads to the consideration of the conditions of validity over which the classical Coulomb scattering calculation is valid, which are to be discussed in Sect. 4.2.3.

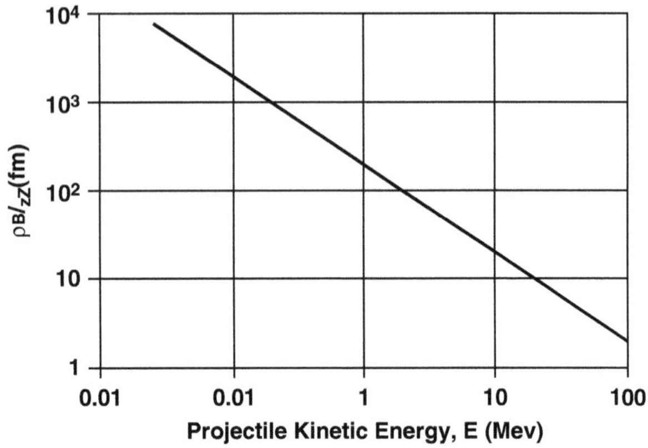

Fig. 4.3 The Bohr scattering diameter normalised to the product of the projectile and charge zZ values as a function of projectile energy

4.2.1.2 Relativistic Solution

The extension of the above derivation to relativistic energies begins through the expression of the conservation of energy in the form given by

$$E_{\mathrm{T}} = \sqrt{p^2 + m^2} + \frac{\alpha\hbar czZ}{r} \tag{4.26}$$

where E_{T} is the total energy of the projectile (rather than only the kinetic energy we have considered previously). Using the conservation of angular momentum in cylindrical co-ordinates, we have

$$L = m\gamma r^2 \frac{\mathrm{d}\phi}{\mathrm{d}t} \tag{4.27}$$

where $\gamma = \frac{1}{\sqrt{1-\beta^2}}$ and β is the velocity normalised to the speed of light. Thus, we have

$$p^2 = \left(\frac{L}{r}\right)^2 + \left(m\gamma \frac{\mathrm{d}r}{\mathrm{d}t}\right)^2. \tag{4.28}$$

By calculating p^2 from (4.26) and equating this with (4.28), we achieve the derivative of the radial distance with respect to time:

$$\frac{\mathrm{d}r}{\mathrm{d}t} = \frac{1}{m\gamma}\sqrt{\left(E - \frac{\alpha\hbar czZ}{r}\right)^2 - m^2 - \left(\frac{L}{r}\right)^2}. \tag{4.29}$$

The derivative of the angle φ with respect to time is, from (4.27),

$$\frac{\mathrm{d}\varphi}{\mathrm{d}t} = \frac{L}{m\gamma r^2}. \tag{4.30}$$

From (4.29) and (4.30), we then have the result:

$$\frac{\mathrm{d}\varphi}{\mathrm{d}r} = \pm \frac{L}{r^2\sqrt{\left(E - \frac{\alpha\hbar czZ}{r}\right)^2 - m^2 - \left(\frac{L}{r}\right)^2}}. \tag{4.31}$$

We follow our earlier procedure of the substitution of variable $x = \frac{1}{r}$ to arrive at

$$\varphi_{\mathrm{A}} = L \int_0^{\frac{1}{r_{\mathrm{A}}}} \frac{\mathrm{d}x}{\sqrt{E^2 - 2\alpha\hbar czZEx - m^2 + \left((\alpha\hbar czZ)^2 - L^2\right)x^2}}. \tag{4.32}$$

Recalling that $\theta = \pi - 2\varphi_{\mathrm{A}}$ and using a result from the tables of Abramowitz and Stegun (1972), the scattering angle is

$$\theta = \pi - \frac{2L}{\sqrt{L^2 - (\alpha\hbar czZ)^2}} \tan^{-1}\left(\beta \frac{\sqrt{L^2 - (\alpha\hbar czZ)^2}}{\alpha\hbar czZ}\right) \tag{4.33}$$

where β is the projectile velocity normalised to that of light. Clearly, this classical result of (4.33) can be valid only for large values of angular momentum, i.e. $L > \alpha\hbar czZ$. In most practical cases, the product zZ will tend to be small, and then, $\sqrt{L^2 - (\alpha\hbar czZ)^2} \approx L$ and (4.33) will simplify to the expression

$$\theta = \pi - 2\tan^{-1}\left(\frac{\beta L}{\alpha\hbar czZ}\right). \tag{4.34}$$

Further simplification can be achieved by writing the angular momentum as the product of projectile momentum and impact parameter $(L = pb)$, allowing us to write an expression for the scattering angle in the simpler form

$$\theta = \pi - 2\tan^{-1}\left(\frac{2b}{\gamma\,\rho_B(E)}\right). \tag{4.35}$$

It is straightforward to show that inversion of (4.35) leads us to the relationship between impact parameter, Bohr collision diameter and scattering angle of (4.20), but with the Bohr collision diameter scaled by the relativistic factor, γ.

4.2.1.3 Classical Perturbation Theory

In Chap. 2, we developed quantum perturbation theory from first principles and, in particular, derived the Born expansion in terms of the summation of powers of the perturbing potential subject to the 'weakness' of the perturbing potential relative to the projectile's kinetic energy. While these quantum results will be examined later in this chapter for the specific case of the Coulomb potential, we investigate the use of classical perturbation theory in calculating the scattering integral. Classical perturbation theory will also play a role in our later discussion of energy transfer during a collision.

In order to advance the general nature of the method, we will not impose the requirement that the potential $U(r)$ be Coulombic. The form of the scattering integral is, in this general case,

$$\varphi_{\mathrm{A}} = b \int_{r_{\mathrm{A}}}^{\infty} \frac{\mathrm{d}r}{r^2 \sqrt{1 - \left(\frac{U(r)}{E}\right) - \left(\frac{b}{r}\right)^2}} \tag{4.36}$$

where, as we are considering nonrelativistic kinematics, $E = \frac{p^2}{2m}$ is the kinetic energy of the projectile. If an analytical solution of (4.36) does not exist or else is too laborious to derive, the integral may be solved numerically in order to yield the angle φ_{A}. As a result, one may question what utility of the iterative approach is provided by perturbation theory in solving this equation. Depending upon the nature of the potential, the first-order term of the perturbative expansion may be useful and easier to use than either a numerical solution to the integral or the inclusion of further terms obtained by iteration. Hence, examination of a higher-order term may indicate that the lower-order term expansion is sufficient and practical.

In our use of classical perturbation theory in evaluating the scattering integral, we will follow the approaches previously described by Lehmann and Leibfried (1963), Andersen and Sigmund (1966) and Sigmund and Schinner (2000). Why the use of perturbation theory should be preferable to a straightforward Taylor's expansion is made immediately apparent upon expansion of the square root in the denominator of the integrand of (4.36) in terms of the potential:

$$\sqrt{1 - \left(\frac{U(r)}{E}\right) - \left(\frac{b}{r}\right)^2} \cong \frac{1}{1 - \left(\frac{b}{r}\right)^2} + \frac{\left(\frac{U(r)}{E}\right)}{2\left(1 - \left(\frac{b}{r}\right)^2\right)^{\frac{3}{2}}} + \frac{3\left(\frac{U(r)}{E}\right)^2}{8\left(1 - \left(\frac{b}{r}\right)^2\right)^{\frac{5}{2}}} + \cdots . \tag{4.37}$$

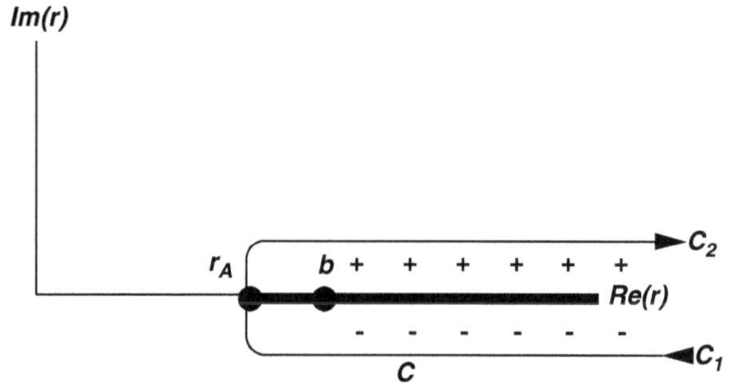

Fig. 4.4 Integration contour in the complex r-plane to be used to solve (4.36) using perturbation theory

We see immediately the singularities that occur when $r = b$.

The perturbation solution to the scattering integral of (4.36) is achieved by using contour integration in the complex r-plane as shown in Fig. 4.4.[2]

The integrand of (4.36) is multi-valued in the complex r-plane. However, it is possible to make it single valued if a cut is made along the real axis for $r > r_A$, as shown in Fig. 4.4. Thus, the positive value is approached from the +Im(r) plane, whereas the negative value is approached from the −Im(r) plane. We then write the scattering integral over this contour around the branch cut as

$$\varphi_A = b \int_C \frac{dr}{r^2\sqrt{1 - \left(\frac{U(r)}{E}\right) - \left(\frac{b}{r}\right)^2}}. \tag{4.38}$$

The contour bypasses the cut which, as the integrand is regular[3] external to the cut, makes use of the Cauchy–Goursat theorem. Thus,

$$\theta = \pi - \varphi_A. \tag{4.39}$$

[2] And is described in Appendix (A.7).

[3] A 'regular' function is one which is analytical and single valued throughout a region R.

We now apply the Taylor expansion, in this example, to the first three terms in $U(r)$:

$$\begin{aligned}\varphi_{\mathrm{A}} &= b\int_C \frac{\mathrm{d}r}{r^2\sqrt{1-\left(\frac{U(r)}{E}\right)-\left(\frac{b}{r}\right)^2}} \\ &= b\left(\int_C \frac{\mathrm{d}r}{r^2\sqrt{1-\left(\frac{b}{r}\right)^2}}+\int_C \frac{\mathrm{d}r}{2\,r^2\left(1-\left(\frac{b}{r}\right)^2\right)^{\frac{3}{2}}}\frac{U(r)}{E}+\right. \\ &\qquad \left.\int_C \frac{\mathrm{d}r}{r^2\left(1-\left(\frac{b}{r}\right)^2\right)^{\frac{5}{2}}}\frac{3}{8}\left(\frac{U(r)}{E}\right)^2+\cdots\right) \\ &= \varphi_{\mathrm{A}}^{(0)}+\varphi_{\mathrm{A}}^{(1)}+\varphi_{\mathrm{A}}^{(2)}+\cdots. \end{aligned} \tag{4.40}$$

In this example, we evaluate each of these three terms in turn. The first (zeroth-order) term is

$$\begin{aligned}\varphi_{\mathrm{A}}^{(0)} &= b\int_C \frac{\mathrm{d}r}{r^2\sqrt{1-\left(\frac{b}{r}\right)^2}} \\ &= b\int_b^{\infty} \frac{\mathrm{d}r}{r^2\sqrt{1-\left(\frac{b}{r}\right)^2}} \end{aligned} \tag{4.41}$$

where the path of integration has been manipulated towards the real axis. In this case, the zeroth-order approximation to the scattering angle is $\theta^{(0)} = \pi - 2\varphi_{\mathrm{A}}^{(0)}$. By using the exchange of variable $x = \frac{b}{r}$, the zeroth-order term of the expansion of the scattering integral is

$$\begin{aligned}\varphi_{\mathrm{A}}^{(0)} &= b\int_b^{\infty} \frac{\mathrm{d}r}{r^2\sqrt{1-\left(\frac{b}{r}\right)^2}} \\ &= \int_0^1 \frac{\mathrm{d}x}{\sqrt{1-x^2}} \\ &= \frac{\pi}{2}. \end{aligned} \tag{4.42}$$

Thus, the scattering angle corresponding to this term is

$$\begin{aligned}\theta^{(0)} &= \pi - 2\varphi_{\mathrm{A}}^{(0)} \\ &= 0.\end{aligned}$$

That this result of a zero scattering angle should be the case is intuitive: the perturbing potential does not appear in the zeroth-order term of the integrand of (4.41); hence, there can be no interaction between projectile and target, and the scattering angle must be zero.

The second term of the scattering integral expansion is

$$\varphi_{\mathrm{A}}^{(1)} = \frac{b}{2E} \int_C \mathrm{d}r \frac{U(r)}{r^2 \left(1 - \left(\frac{b}{r}\right)^2\right)^{\frac{3}{2}}}. \tag{4.43}$$

This integral can be solved by parts by first noting the derivative:

$$\frac{\mathrm{d}}{\mathrm{d}r} \frac{1}{\sqrt{1 - \left(\frac{b}{r}\right)^2}} = -\frac{b^2}{r^3 \left(1 - \left(\frac{b}{r}\right)^2\right)^{\frac{3}{2}}}$$

which enables us to write (4.43) in the form

$$\varphi_{\mathrm{A}}^{(1)} = -\frac{1}{2bE} \int_C \mathrm{d}r \,(r\,U(r))\, \frac{\mathrm{d}}{\mathrm{d}r} \frac{1}{\sqrt{1 - \left(\frac{b}{r}\right)^2}}. \tag{4.44}$$

Partial integration is then used to yield

$$\varphi_{A}^{(1)} = -\frac{1}{2bE} \left. \frac{rU(r)}{\sqrt{1 - \left(\frac{b}{r}\right)^2}} \right|_{C_1}^{C_2} + \frac{1}{2bE} \int_C \frac{\mathrm{d}r}{\sqrt{1 - \left(\frac{b}{r}\right)^2}} \frac{\mathrm{d}(r\,U(r))}{\mathrm{d}r}. \tag{4.45}$$

This result can be left as is for any arbitrary scattering potential. However, to continue further this examination for a Coulomb-like potential, we treat $U(r)$ as a Yukawa or Wentzel potential modelling a screened charged source with the screening parameter, κ, $U(r) = U_0 \frac{e^{-\kappa r}}{r}$. The integrated term of (4.45) is, then,

$$
\begin{aligned}
-\frac{1}{2bE}\frac{rU(r)}{\sqrt{1-\left(\frac{b}{r}\right)^2}}\Bigg|_{C_1}^{C_2} &= -\frac{1}{2bE}\frac{r^2U(r)}{\sqrt{r^2-b^2}}\Bigg|_{C_1}^{C_2} \\
&= -\frac{U_0}{2bE}\frac{re^{-\kappa r}}{\sqrt{r^2-b^2}}\Bigg|_{C_1}^{C_2} \\
&= 0; \quad \text{as } C_1, C_2 \to \infty.
\end{aligned}
\tag{4.46}
$$

Thus (4.45) reduces to, by returning the integration contour to the $Re(r)$ axis,

$$
\begin{aligned}
\varphi_{\mathrm{A}}^{(1)} &= \frac{1}{2bE}\int_b^\infty \frac{\mathrm{d}r}{\sqrt{1-\left(\frac{b}{r}\right)^2}}\frac{\mathrm{d}(r\,U(r))}{\mathrm{d}r} \\
&= -\left(\frac{U_0\,\kappa}{2Eb}\right)\int_b^\infty \mathrm{d}r\,\frac{r\,e^{-\kappa r}}{\sqrt{r^2-b^2}}.
\end{aligned}
\tag{4.47}
$$

We solve for this by using the exchange of variables within the integral, $x=\frac{r}{b}$, to give

$$
\varphi_{\mathrm{A}}^{(1)} = -\frac{U_0\kappa}{2E}\int_1^\infty \mathrm{d}x\,\frac{x\,e^{-\kappa bx}}{\sqrt{x^2-1}}. \tag{4.48}
$$

From Appendix C.1, the modified zeroth-order Bessel function in integral form is $K_0(y)=\int_1^\infty \mathrm{d}x\,\frac{e^{-yx}}{\sqrt{x^2-1}}$ with the relation, $\frac{\mathrm{d}K_0(y)}{\mathrm{d}y}=-K_1(y)$, where $K_1(y)$ is the modified first-order Bessel function. Thus, the first term of this series is given by

$$
\varphi_{\mathrm{A}}^{(1)} = -\frac{U_0\kappa}{2E}K_1(\kappa b). \tag{4.49}
$$

The next term of the scattering integral expansion is solved in the same manner, but now by using two integrations by parts to give

$$
\varphi_{\mathrm{A}}^{(2)} = -\left(\frac{U_0\kappa}{2E}\right)^2 K_1(2\kappa b). \tag{4.50}
$$

The ratio of the two results of (4.50) and (4.49) is

$$
\frac{\varphi_{\mathrm{A}}^{(2)}}{\varphi_{\mathrm{A}}^{(1)}} = \frac{U_0\kappa}{2E}\frac{K_1(2\kappa b)}{K_1(\kappa b)}. \tag{4.51}
$$

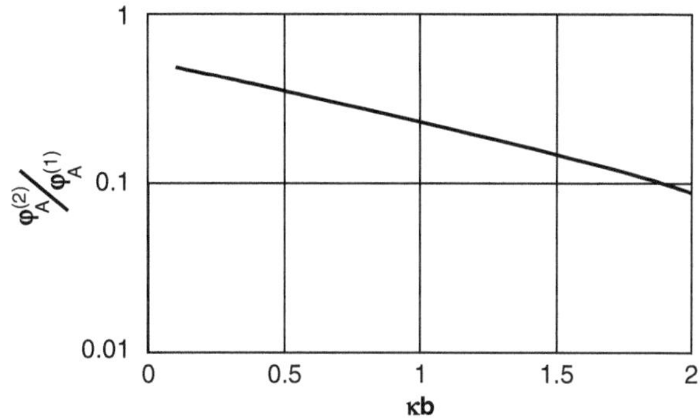

Fig. 4.5 Ratio of the contributions of the first two non-zero terms of the classical perturbative expansion of the scattering integral for a screened Coulomb potential as a function of the product of the screening and impact parameters

Figure 4.5 shows this ratio as a function of the product of the screening and impact parameters. From both (4.51) and this figure, we observe two dependencies of the ratio of these two terms. First, the $\varphi_{\mathrm{A}}^{(1)}$ term will dominate for the combination of a weak scattering potential and high projectile kinetic energies which suggests that such conditions do not require extended terms in the perturbation expansion. Second, for a fixed screening parameter κ, the contribution of $\varphi_{\mathrm{A}}^{(2)}$ relative to the dominant $\varphi_{\mathrm{A}}^{(1)}$ decreases exponentially as a function of increasing impact parameter.

4.2.2 Laplace–Runge–Lenz Vector

The discussions of the previous section were the results of the simultaneous conservation of energy and angular momentum. However, these are not the only quantities of motion that are conserved during the interaction between projectile and target. The Laplace–Runge–Lenz vector (here referred to as the *LRL* vector for convenience) is another quantity that is conserved throughout Coulomb scatter (Aguiar and Barroso 1996).[4] There are a variety of scalings used in the definition of the *LRL* vector, but here, we shall take it to be the dimensionless vector defined as

$$\mu = \hat{\mathbf{r}} - \frac{1}{zZm\alpha\hbar c}\mathbf{L} \times \mathbf{p}, \tag{4.52}$$

which is equivalent to the eccentricity vector of celestial mechanics. Note that the *LRL* vector lies within the scattering plane. Historically, the *LRL* vector has been used to determine the shape of Keplerian orbits and in early quantum-mechanical descriptions of the hydrogen atom, prior to Schrödinger's development of his eponymous equation. We will demonstrate now how using the *LRL* vector can

[4] The *LRL* vector is conserved only if the interaction force varies with the inverse square of the distance between the two interacting bodies, such as the Coulomb and gravitational forces.

provide the relationship between the impact parameter, the Bohr collision diameter and the scattering angle given by (4.20) and show that the projectile trajectory becomes hyperbolic as a result of the scatter.

First, however, we must demonstrate that the *LRL* vector is indeed a conserved quantity of motion. Differentiating (4.52) with respect to time gives

$$\begin{aligned}\frac{\mathrm{d}\boldsymbol{\mu}}{\mathrm{d}t} &= \frac{\mathrm{d}}{\mathrm{d}t}\left(\frac{\mathbf{r}}{r}\right) - \frac{1}{zZm\alpha\hbar c}\left(\frac{\mathrm{d}\mathbf{L}}{\mathrm{d}t}\times\mathbf{p} + \mathbf{L}\times\frac{\mathrm{d}\mathbf{p}}{\mathrm{d}t}\right)\\ &= \frac{1}{r}\frac{\mathrm{d}\mathbf{r}}{dt} - \mathbf{r}\frac{\frac{\mathrm{d}r}{\mathrm{d}t}}{r^2} - \frac{1}{zZm\alpha\hbar c}\mathbf{L}\times\frac{\mathrm{d}\mathbf{p}}{\mathrm{d}t}\\ &= \frac{1}{r}\mathbf{v} - \frac{v}{r^2}\mathbf{r} - \frac{1}{zZm\alpha\hbar c}\mathbf{L}\times\frac{\mathrm{d}\mathbf{p}}{\mathrm{d}t}\end{aligned} \tag{4.53}$$

where the conservation of angular momentum (i.e. $\frac{\mathrm{d}\mathbf{L}}{\mathrm{d}t} = \mathbf{0}$) has been included. From the definition of the angular momentum and using the vector triple product, we find that the cross product of (4.53) reduces to

$$\begin{aligned}\mathbf{L}\times\frac{\mathrm{d}\mathbf{p}}{\mathrm{d}t} &= \mathbf{r}\times\mathbf{p}\times\frac{\mathrm{d}\mathbf{p}}{\mathrm{d}t}\\ &= m^2\left(\mathbf{r}\times\mathbf{v}\times\frac{\mathrm{d}\mathbf{v}}{\mathrm{d}t}\right)\\ &= m^2\left(\left(\mathbf{r}\bullet\frac{\mathrm{d}\mathbf{v}}{\mathrm{d}t}\right)\mathbf{v} - (\mathbf{r}\bullet\mathbf{v})\frac{\mathrm{d}\mathbf{v}}{\mathrm{d}t}\right).\end{aligned} \tag{4.54}$$

The acceleration is determined from Newton's Third Law incorporating the Coulomb force:

$$\frac{\mathrm{d}\mathbf{v}}{\mathrm{d}t} = \alpha\hbar c\frac{zZ}{m}\frac{\mathbf{r}}{r^3}. \tag{4.55}$$

Using this, (4.55) can be written as

$$\begin{aligned}\mathbf{L}\times\frac{\mathrm{d}\mathbf{p}}{\mathrm{d}t} &= m^2\left(\left(\mathbf{r}\bullet\alpha\hbar c\frac{zZ}{m}\frac{\mathbf{r}}{r^3}\right)\mathbf{v} - (\mathbf{r}\bullet\mathbf{v})\alpha\hbar c\frac{zZ}{m}\frac{\mathbf{r}}{r^3}\right)\\ &= mzZ\frac{\alpha\hbar c}{r^3}\left(r^2\mathbf{v} - (\mathbf{r}\bullet\mathbf{v})\,\mathbf{r}\right)\end{aligned} \tag{4.56}$$

Substituting this expression of the cross product of angular momentum and acceleration into (4.53) then provides the proof that the *LRL* vector is a conserved quantity:

$$\begin{aligned}\frac{\mathrm{d}\boldsymbol{\mu}}{\mathrm{d}t} &= \frac{1}{r}\mathbf{v} - \frac{v}{r^2}\mathbf{r} - \frac{1}{r^3}\left(r^2\mathbf{v} - (\mathbf{r}\bullet\mathbf{v})\,\mathbf{r}\right)\\ &= \left(\frac{v}{r}\right)\hat{\mathbf{r}} - \left(\frac{v}{r}\right)\hat{\mathbf{r}} - \frac{1}{r^3}\left(r^2\mathbf{v} - (\mathbf{r}\bullet\mathbf{v})\,\mathbf{r}\right)\\ &= \mathbf{0}.\end{aligned} \tag{4.57}$$

As noted earlier, $\boldsymbol{\mu}$ lies within the plane of motion so we can now take the dot product:

$$\begin{aligned}\boldsymbol{\mu}\bullet\mathbf{r} &= \hat{\mathbf{r}}\bullet\mathbf{r} - \frac{1}{zZm\alpha\hbar c}\mathbf{L}\times\mathbf{p}\bullet\mathbf{r}\\ &= r - \frac{1}{zZm\alpha\hbar c}\mathbf{L}\times\mathbf{p}\bullet\mathbf{r}\\ &= r + \frac{L^2}{zZm\alpha\hbar c}.\end{aligned} \tag{4.58}$$

Now, by again using the polar co-ordinate system, we get

$$\mu r\cos\varphi = r + \frac{L^2}{zZm\alpha\hbar c}. \tag{4.59}$$

Solving for the radial distance between the projectile and the target centre yields the equation

$$r = -\frac{\left(\dfrac{L^2}{zZm\alpha\hbar c}\right)}{1-\mu\cos\varphi} \tag{4.60}$$

which is the equation of a hyperbola with eccentricity $\mu > 1$, thus demonstrating that the trajectory resulting from Coulomb scatter is hyperbolic.

Let us consider, now, the two components of the *LRL* vector parallel, and perpendicular, to the projectile trajectory. As time $t \to -\infty$, these are

$$\begin{aligned}\mu_{||} &= 1\\ \mu_{\perp} &= -\frac{mv^2b}{\alpha\hbar czZ}\\ &= -\frac{2b}{\rho_{\mathrm{B}}(E)}\end{aligned} \tag{4.61}$$

where the Bohr collision diameter has been used. Then,

$$\begin{aligned}\tan\varphi &= \left|\frac{\mu_{||}}{\mu_{\perp}}\right|\\ &= \frac{2b}{\rho_{\mathrm{B}}(E)}\end{aligned} \tag{4.62}$$

which leads to (4.20) and, as before, the expression for the Rutherford cross section of Coulomb scatter.

4.2.3 Validity Conditions for Classical Descriptions

To conclude our discussion of the Coulomb scattering process as described by using classical mechanics, we review those conditions under which such a calculation is valid. This presents an appropriate segue to those calculations based upon quantum mechanics and which follow.

We can look at this validation on the basis of a comparison between the spatial dimensions of the wavepacket representing the projectile and the dimensions of the scatterer. Although the scatterer is defined as being a point, the Bohr collision diameter, ρ_B, can be used as a cross-sectional measure of the scatterer. A classical mechanical description of Coulomb scatter is acceptable if the reduced de Broglie wavelength is much smaller than the Bohr collision diameter:

$$\lambda\!\!\!^{-} << \rho_B(E) \quad \text{classical mechanical solution valid.} \tag{4.63}$$

Recalling that $\lambda\!\!\!^{-}$ is the reciprocal of $k = \frac{p}{\hbar c}$ and by using the definition of $\rho_B(E)$, we can specify that the classical solution is valid if the criterion

$$\frac{2mzZ}{p} >> \frac{1}{\alpha} \approx 137 \tag{4.64}$$

is met. This, as expected, reflects how the combinations of a low incident momentum, high projectile mass and a large Coulomb interaction are requirements for the validity of the classical description. For decreasing values of $\frac{mzZ}{p}$, the classical model of hyperbolic orbits loses validity, and once $\frac{2mzZ}{p} \approx \frac{1}{\alpha}$ all validity is lost. For an α-particle incident on a ^{12}C nucleus (a scenario typical of α-particle therapy), this occurs for a kinetic energy of about 14 MeV.

We can further investigate the validity of the classical solution and infer the degree of uncertainty that it can lead to within the results of the calculation. The approach of Bohr (1948) is followed as a guide. Consider Fig. 4.6 in which a beam of charged particles of momentum **p** is incident to an opaque collimator with a circular aperture of diameter *d*. Classically, we consider this to be a stream of charged particles exiting through the aperture and incident to a charged scattering centre for an impact parameter *b*. The particles are scattered into a hyperbolic trajectory through the angle θ. In the quantum-mechanical picture, the beam of incident particles is considered a wave with a wave vector magnitude of $k = \frac{1}{\lambda\!\!\!^{-}}$. Diffraction occurs when the wave passes through the aperture, producing an interference pattern with a lateral extent described by σ_ξ. If the lateral extent of the interference pattern is less than the scattering angle, then the classical mechanical description remains valid and appropriate. Should it be comparable to, or exceed, θ, then one must allow for quantum effects and the resulting combined uncertainties arising from the beam definition caused by the aperture and the scatter itself.

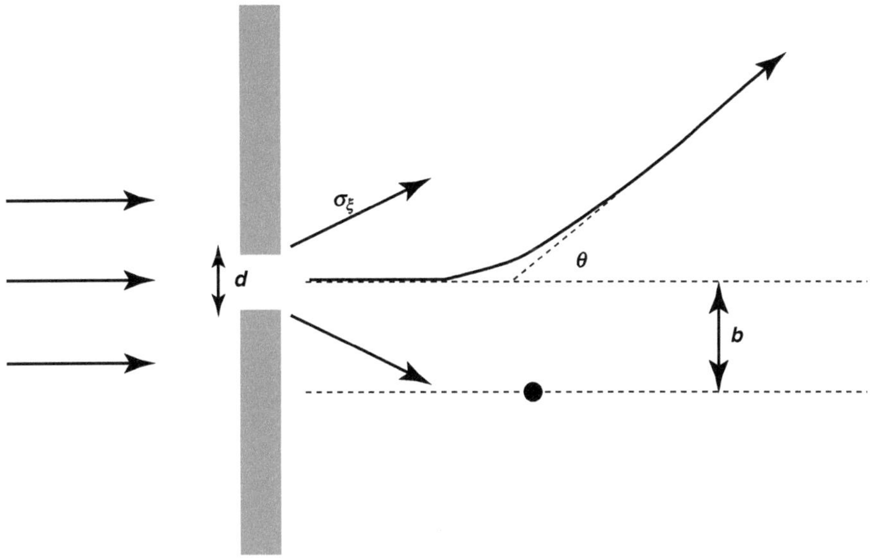

Fig. 4.6 Geometry for specifying the validity of the classical mechanical solution of Coulomb scatter

Let us assume that the condition of Fraunhofer diffraction given by $\frac{kd^2}{2\pi} << L$, where L is the distance from the aperture, is applicable to this aperture condition. Then, the intensity profile orthogonal to the central axis and at an extended distance from the aperture will be

$$I(\xi) = \text{sinc}^2\left(\frac{\pi d}{\lambda\!\!\!^{-}} \sin\xi\right) \tag{4.65}$$

where

$$\begin{aligned} \text{sinc}\,x &= \frac{\sin x}{x}; \quad x \neq 0 \\ &= 1; \qquad x = 0. \end{aligned} \tag{4.66}$$

The sine cardinal function $\text{sinc}\,x$ is, of course, also the zeroth-order spherical Bessel function of the first kind. Figure 4.7 shows $\sin c^2 x$.

We now approximate the central peak, which is bound by the local minima of $\pm\sin^{-1}\left(\frac{\lambda\!\!\!^{-}}{d}\right)$, with a Gaussian function with an approximate standard deviation:

$$\sigma_\xi \equiv \frac{\lambda\!\!\!^{-}}{d}. \tag{4.67}$$

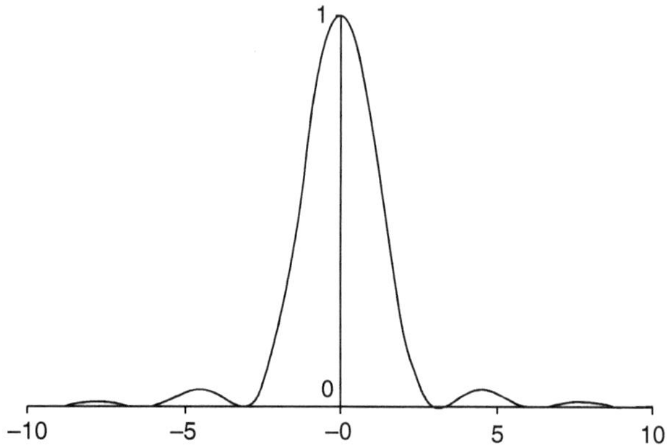

Fig. 4.7 The function $\mathrm{sinc}^2\, x$

In the small scattering angle approximation to (4.20), the scattering angle can be approximated by

$$\theta \approx \frac{\rho_\mathrm{B}(E)}{b}. \tag{4.68}$$

For the diffraction width of (4.67) to be much smaller than the scattering angle, then the inequality

$$\frac{\lambda\!\!\!^{-}}{d} << \frac{\rho_\mathrm{B}(E)}{b} \tag{4.69}$$

must be satisfied. This can be achieved by the use of a sufficiently large aperture diameter d. However, a consequence of a large aperture diameter is the associated uncertainty in the impact parameter. This uncertainty will be limited by the radius of the aperture or by setting it by an upper limit:

$$\delta b = \frac{d}{2}. \tag{4.70}$$

The uncertainty in the scattering angle is, from (4.68),

$$\begin{aligned}\sigma_\theta &= \left|\frac{\mathrm{d}\theta}{\mathrm{d}b}\right| \mathrm{d}b \\ &= \frac{\rho_B(E)\,\mathrm{d}}{2b^2} \\ &= \frac{\mathrm{d}}{2\rho_B(E)}\,\theta^2.\end{aligned} \tag{4.71}$$

The combined effects of the uncertainties of (4.67) and (4.71) can be estimated from their quadrature sum:

$$\Delta\theta = \sqrt{\sigma_\xi^2 + \sigma_\theta^2} \\ \geq \sqrt{\frac{\lambda\!\!\!^-}{\rho_B(E)}}\theta. \tag{4.72}$$

4.3 Quantum-Mechanical Solutions

4.3.1 Introduction

We have now developed, through a number of methods based upon classical mechanics, relationships between scattering angle, impact parameter, the Bohr collision diameter and the differential cross section for the Coulomb scattering of a charged projectile from a charged point target. The derivations we have explored were based upon the conservation of kinematic quantities and a static Coulomb potential. The discussions of the relativistic considerations were limited to the scattering integral and the modification of Bohr's collision diameter. More importantly, the intrinsic spins of projectile and target and the indistinguishability between an identical fermion projectile and target pair were not considered as these entities are inherently quantum features. We now make use of the quantum scattering tools developed in Chap. 2 in great detail in order to explore the quantum descriptions of Coulomb scatter.

4.3.2 Born Expansion: First Term

Recall from (2.83) that the first term of the Born series is proportional to the Fourier transform of the scattering potential:

$$f_{(1)}(\mathbf{q}) = -\frac{m}{2\pi(\hbar c)^2}\int d^3\mathbf{r}' e^{-i\frac{\mathbf{q}\bullet\mathbf{r}'}{\hbar c}} U(r') \tag{4.73}$$

where $\mathbf{q}$ is the three-vector momentum transfer and m is the rest mass of the projectile. Should the scattering potential be central and isotropic, it was shown in Chap. 2 that the scattering amplitude in the first Born approximation reduces to

$$f_{(1)}(q) = -\left(\frac{2m}{\hbar c q}\right)\int_0^\infty dr'\, r' \sin\left(\frac{qr'}{\hbar c}\right) U(r'). \tag{4.74}$$

For the Coulomb potential between a projectile and target with electric charges ze and Ze, respectively, this scattering amplitude is

$$
\begin{aligned}
f_{(1)}(q) &= -\left(\frac{2m}{\hbar c q}\right) \int_0^\infty \mathrm{d}r'\, r' \sin\left(\frac{qr'}{\hbar c}\right) \left(\frac{\alpha \hbar c z Z}{r'}\right) \\
&= -\left(\frac{2m\alpha z Z}{\hbar c}\right) \int_0^\infty \mathrm{d}r'\, r' \frac{\sin\left(\frac{qr'}{\hbar c}\right)}{\left(\frac{qr'}{\hbar c}\right)} \\
&= -\left(\frac{2m\alpha z Z}{\hbar c}\right) \int_0^\infty \mathrm{d}r'\, r'\, j_0\left(\frac{qr'}{\hbar c}\right)
\end{aligned} \tag{4.75}
$$

where $j_0(x)$ is the zeroth-order spherical Bessel function of the first kind. For an unscreened target, the integral of (4.75) can be solved by first changing the Coulomb potential into a Yukawa form, integrating and then letting the screening parameter κ go to zero:

$$
\begin{aligned}
\int_0^\infty \mathrm{d}r'\, r'\, j_0\left(\frac{qr'}{\hbar c}\right) &= \lim_{\kappa \to 0} \int_0^\infty \mathrm{d}r'\, r'\, e^{-\kappa r'} j_0\left(\frac{qr'}{\hbar c}\right) \\
&= \frac{\hbar c}{q} \lim_{\kappa \to 0} \int_0^\infty \mathrm{d}r'\, e^{-\kappa r'} \sin\left(\frac{qr'}{\hbar c}\right) \\
&= \frac{\hbar c}{2iq} \lim_{\kappa \to 0} \int_0^\infty \mathrm{d}r'\, e^{-\kappa r'} \left(e^{i\left(\frac{qr'}{\hbar c}\right)} - e^{-i\left(\frac{qr'}{\hbar c}\right)}\right) \\
&= \frac{\hbar c}{2iq} \lim_{\kappa \to 0} \int_0^\infty \mathrm{d}r'\, \left(e^{-r'\left(\kappa - i\frac{q}{\hbar c}\right)} - e^{-r'\left(\kappa + i\frac{q}{\hbar c}\right)}\right) \\
&= \frac{\hbar c}{2iq} \lim_{\kappa \to 0} \frac{\left(\frac{2iq}{\hbar c}\right)}{\kappa^2 + \left(\frac{q}{\hbar c}\right)^2} \\
&= \lim_{\kappa \to 0} \frac{1}{\kappa^2 + \left(\frac{q}{\hbar c}\right)^2} \\
&= \left(\frac{\hbar c}{q}\right)^2.
\end{aligned} \tag{4.76}
$$

The penultimate step in the derivation of (4.76) which included a non-zero κ will be investigated in the next chapter when the screened Coulomb potential is discussed. Returning this result to (4.75) yields the scattering amplitude for the Coulomb potential in the first term of the Born series:

$$f_{(1)}(q) = -\left(\frac{2m\alpha\, zZ}{\hbar c}\right)\left(\frac{\hbar c}{q}\right)^2$$
$$= -2\alpha\hbar c\left(\frac{mzZ}{q^2}\right). \tag{4.77}$$

The Coulomb differential cross section in the first Born approximation is, consequently,

$$\frac{d\sigma}{d\Omega} = |f_{(1)}(q)|^2$$
$$= 4(\alpha\hbar c)^2\frac{(mzZ)^2}{q^4}$$
$$= \left(\frac{\rho_B(E)}{4}\right)^2\frac{1}{\sin^4\left(\frac{\theta}{2}\right)} \tag{4.78}$$

where the relationship between scattering angle and projectile momentum/energy of (2.86) has been used. The result is exactly that derived from each of our previous explorations of classical theory. Such a consequence of an exact agreement may appear surprising when we recall that the Born expansion assumes an unperturbed (or, at least, a weakly perturbed) wavefunction for which a plane wave is considered an acceptable model at an extended distance from the scattering potential. As the Coulomb potential has an infinite range, the incident wave will be perturbed or distorted by the potential and may not necessarily be well modelled by a plane wave. We will demonstrate the effects of this consequence in two ways. First, we extend the Born expansion to second order for the Coulomb potential to show that the result, with its assumption of an infinite-range potential, diverges and that convergence requires the screening of the potential. Second, we then derive the exact result for nonrelativistic Coulomb scattering by solving the Schrödinger equation for that potential and thus demonstrate the distortion to the wavefunction caused by the potential. In the interim, we explore some aspects of the differential cross section of (4.78) which we will return to in later discussions. For small-scattering angles, this expression of the differential cross section reduces to a very simple result

$$\frac{d\sigma}{d\Omega} = \left(\frac{\rho_B(E)}{\theta^2}\right)^2 \quad \text{(small angle approximation)}. \tag{4.79}$$

It is immediately evident that two points should be noted from these expressions for the differential cross section:

- The cross section for an unscreened point charge target diverges as the scattering angle $\theta \to 0$.
- As the cross section is independent of the signs of the electric charges of the probe and the target, the attractive or repulsive nature of the Coulomb potential is irrelevant.

The first point yields a tremendous impact upon the results of multiple Coulomb scatter, which we will consider in Chap. 6. To anticipate that discussion, the predominance of forward scatter means that the spatial divergence of a charged particle beam traversing a medium can be described statistically through a probability distribution function. However, as demonstrated originally by the experimental results of Geiger and Marsden, a scatter through large angles can be due only to a single interaction.

4.3.3 Born Expansion: Second Term

We now investigate the Born expansion applied to Coulomb scatter to second order.

Continuing on from the Lippmann–Schwinger equation of (2.95), the Born expansion of the total wavefunction post-scatter to second order (with the zeroth-order being the original and unscattered wavefunction) is

$$\psi(\mathbf{r}) = \psi_0(\mathbf{r}) - \frac{m}{2\pi(\hbar c)^2} \int d^3\mathbf{r}' \frac{e^{ik|\mathbf{r}-\mathbf{r}'|}}{|\mathbf{r}-\mathbf{r}'|} U(r')\psi_0(\mathbf{r}') + \left(\frac{m}{2\pi(\hbar c)^2}\right)^2 \int d^3\mathbf{r}' \frac{e^{ik|\mathbf{r}-\mathbf{r}'|}}{|\mathbf{r}-\mathbf{r}'|} U(r') \int d^3\mathbf{r}'' \frac{e^{ik|\mathbf{r}'-r''|}}{|\mathbf{r}'-\mathbf{r}''|} U(r'')\,\psi(r''). \tag{4.80}$$

We consider this second term of the expansion for the Coulomb potential following the approach given by Dalitz (1951). In order to simplify this, we transform the problem from configuration space to momentum space through the use of Fourier transforms, as shown in Appendix B.1. In so doing, the total wavefunction post-scattering is given by

$$\phi(\mathbf{p}') = \delta(\mathbf{p}' - \mathbf{p}) + \tilde{f}(\mathbf{p}')\,\delta_+\left(p'^2 - p^2\right) \tag{4.81}$$

where $\tilde{f}(\mathbf{p}')$ is the Fourier transform of the scattering amplitude into momentum space. The first term of (4.81), described by the Dirac δ-function, describes the incident plane wave. The function $\delta_+(x)$ is defined as

$$\delta_+(x) = \delta(x) + \frac{i}{\pi x}. \tag{4.82}$$

The total scattering amplitude in momentum space is a summation as following from the Born expansion:

$$\tilde{f}(\mathbf{p}') = \sum_{n=1}^{\infty} \tilde{f}_{(n)}(\mathbf{p}'). \tag{4.83}$$

The first term of the expansion, $\tilde{f}_{(1)}(\mathbf{p}')$, has been previously derived, and here we rewrite it as the proportionality to the Fourier transform of the potential from configuration space to momentum space:

$$\tilde{f}_{(1)}(\mathbf{p}') = \frac{m}{2\pi(\hbar c)^2} \tilde{U}(\mathbf{p}', \mathbf{p}) \tag{4.84}$$

where

$$\tilde{U}(\mathbf{p}', \mathbf{p}) = \int \mathrm{d}^3\mathbf{r}\, e^{-i\frac{(\mathbf{p}'-\mathbf{p})\bullet\mathbf{r}}{\hbar c}} U(r). \tag{4.85}$$

The second term of the Born expansion of the scattering amplitude is

$$\tilde{f}_{(2)}(\mathbf{p}') = \left(\frac{m}{2\pi(\hbar c)^2}\right)^2 \int \mathrm{d}^3\mathbf{p}''\, \tilde{U}(\mathbf{p}', \mathbf{p}'')\tilde{f}_{(1)}(\mathbf{p}'')\delta_+\left(p''^2 - p'^2\right). \tag{4.86}$$

We consider a screened Coulomb potential with screening parameter κ for which, from our earlier derivations, yields

$$\tilde{f}_{(1)}(\mathbf{p}') = -\left(\frac{2m\alpha zZ}{\hbar c}\right) \frac{1}{\kappa^2 + \left(\frac{p'^2}{\hbar c}\right)}. \tag{4.87}$$

The integral of (4.86) is solved using the methods of complex integration described in Appendix A and a number of 'tricks' attributable to Feynman (1949) and recounted by Dalitz. This long and tedious solution, which does not add materially to our understanding, is avoided here, and we refer the interested reader directly to the original results of Dalitz. We can write the sum of the first two terms of the Born expansion of the Coulomb scattering amplitude as (including the conversion of the scattering amplitude's argument from momentum transfer to scattering angle).

$$\begin{aligned}\tilde{f}_{(1)}(\theta)+\tilde{f}_{(2)}(\theta) &= -\frac{2\alpha\hbar cmzZ}{\kappa^2+4\left(\frac{p}{\hbar c}\right)^2\sin^2\frac{\theta}{2}}e^{-i\left(\frac{zZ\alpha}{\beta}\right)\ln\left(\frac{2\left(\frac{p}{\hbar c}\right)\sin\frac{\theta}{2}}{\kappa}\right)}\\ &\cong -\frac{2\alpha\hbar cmzZ}{\kappa^2+4\left(\frac{p}{\hbar c}\right)^2\sin^2\frac{\theta}{2}}\times\left(1-i\left(\frac{zZ\alpha}{\beta}\right)\ln\left(\frac{2\left(\frac{p}{\hbar c}\right)\sin\frac{\theta}{2}}{\kappa}\right)\right)\end{aligned} \tag{4.88}$$

where β is the particle velocity normalised to that of light. It is obvious from the result of (4.88) that for a pure Coulomb potential ($\kappa = 0$), the first term of the Born expansion remains finite and will replicate the classical Coulomb scattering cross section, whereas it diverges at the second term. In the practical scattering case from an atom, where the nucleus is screened by atomic electrons (i.e. $\kappa \neq 0$), the Born expansion is applicable, and the first term of the expression is used in the Bethe quantum theory of charged particle energy loss, as shown in Chap. 9.

4.3.4 Exact Nonrelativistic Solution

Instead of using perturbation theory, an exact solution describing Coulomb scatter can be obtained by solving the Schrödinger equation directly for the Coulomb potential. Consider the centre-of-mass system for a projectile of charge ze incident to a target particle of charge Ze where m is the reduced mass and E is the kinetic energy in the centre-of-mass reference frame. The corresponding stationary Schrödinger equation is

$$\left(-\frac{(\hbar c)^2}{2m}\nabla^2+\frac{\alpha\hbar czZ}{r}\right)\psi(\mathbf{r}) = E\psi(\mathbf{r}) \tag{4.89}$$

which, after algebraic manipulation, gives

$$\left(\nabla^2+k^2\left(1-\frac{\rho_{\mathrm{B}}(E)}{r}\right)\right)\psi(\mathbf{r}) = 0. \tag{4.90}$$

A regular solution to (4.90) is of the form $e^{ik\zeta}f(r-\zeta)$ (where we have substituted the symbol ζ for the cylindrical spatial co-ordinate z in order to avoid confusion with the electric charge of the projectile, ze). By defining the variable $u = r-\zeta$ and substituting this into (4.90), we obtain a second-order differential equation in the radial variable of the wavefunction:

$$u\frac{\mathrm{d}^2f(u)}{\mathrm{d}u^2}+(1-iku)\frac{\mathrm{d}f(u)}{\mathrm{d}u}-\frac{\rho_{\mathrm{B}}(E)k^2}{2}f(u) = 0. \tag{4.91}$$

By further defining the variable $v = iku = ik(r - \zeta)$, this equation becomes

$$\frac{\mathrm{d}^2 f(v)}{\mathrm{d}v^2} + \left(\frac{1}{v} - 1\right)\frac{\mathrm{d}f(v)}{\mathrm{d}v} + i\frac{\left(\frac{k\rho_\mathrm{B}(E)}{2}\right)}{v} f(v) = 0. \tag{4.92}$$

In Appendix C.6, it is shown that a regular solution to this equation is of the form

$$\psi_\mathrm{c} = Ae^{ik\zeta} M\left(-i\frac{k\rho_\mathrm{B}(E)}{2}, 1, ik(r - \zeta)\right) \tag{4.93}$$

where A is a normalisation constant and $M(a,b,x)$ is the confluent hypergeometric function. As always in our consideration of the scattering problem, we are interested in only the asymptotic solution at large distances from the scattering centre in order to link the microscopic scattering process to physical observables (i.e. measurable quantities). In Appendix C.6, it is shown that the confluent hypergeometric functions can be written as the sum of the functions:

$$M\left(-i\frac{k\rho_\mathrm{B}(E)}{2}, 1, ik(r - \zeta)\right) = W_1\left(-i\frac{k\rho_\mathrm{B}(E)}{2}, 1, ik(r - \zeta)\right) + W_2\left(-i\frac{k\rho_\mathrm{B}(E)}{2}, 1, ik(r - \zeta)\right) \tag{4.94}$$

where the W_1 and W_2 functions are alternate solutions and can be interpreted as the incoming and outgoing wavefunction. They are particularly attractive in that although they are complex integrals, they have relatively simple asymptotic forms for $|ik(r - \zeta)| \rightarrow \infty$ which, of course, is the condition that what we are interested in the scattering problem. These asymptotic forms are

$$W_1\left(-i\frac{k\rho_\mathrm{B}(E)}{2}, 1, ik(r - \zeta)\right) \approx \frac{\Gamma(1)}{\Gamma(1 + i\tau)}(-(ik(r - \zeta)))^{i\left(\frac{k\rho_\mathrm{B}(E)}{2}\right)} \times \sum_{n=0}^{\infty}\left(\frac{\Gamma\left(n - i\frac{k\rho_\mathrm{B}(E)}{2}\right)}{\Gamma\left(i\frac{k\rho_\mathrm{B}(E)}{2}\right)}\right)^2 \frac{(ik(r - \zeta))^{-n}}{n!} \tag{4.95}$$

and

$$W_2\left(-i\frac{k\rho_B(E)}{2}, 1, ik(r-\zeta)\right) \approx \frac{\Gamma(1)}{\Gamma\left(-i\frac{k\rho_B(E)}{2}\right)} e^{ik(r-\zeta)} (ik(r-\zeta))^{-\left(1+i\frac{k\rho_B(E)}{2}\right)} \times \sum_{n=0}^{\infty} \left(\frac{\Gamma\left(n+1+i\frac{k\rho_B(E)}{2}\right)}{\Gamma\left(1+i\frac{k\rho_B(E)}{2}\right)}\right)^2 \frac{(ik(r-\zeta))^{-n}}{n!} \tag{4.96}$$

As the complete wavefunction after the scatter is the sum of the incident (ψ_i) and scattered (ψ_S) wavefunctions, we then write the incident wavefunction as

$$\psi_i = Ae^{ik\zeta} W_1\left(-i\frac{k\rho_B(E)}{2}, 1, ik(r-\zeta)\right)^{i\frac{k\rho_B}{2}} \tag{4.97}$$

and the scattered wavefunction as

$$\psi_S = Ae^{ik\zeta} W_2\left(-i\frac{k\rho_B(E)}{2}, 1, ik(r-\zeta)\right). \tag{4.98}$$

If we were to select the arbitrary normalisation constant A to be equal to $\Gamma(1+i\tau)\, e^{-\frac{\pi k\rho_B(E)}{4}}$, then the asymptotic forms of the two solutions appear in the convenient forms

$$\psi_i \approx e^{i\left(kz+\left(\frac{k\rho_B(E)}{2}\right)\ln k(r-\zeta)\right)} \left(1+\frac{\left(\frac{k\rho_B(E)}{2}\right)^2}{ik(r-\zeta)}+\cdots\right) \tag{4.99}$$

and

$$\psi_S \approx -\frac{\left(\frac{k\rho_B(E)}{2}\right)}{k(r-\zeta)} \times \frac{\Gamma\left(1+i\left(\frac{k\rho_B(E)}{2}\right)\right)}{\Gamma\left(1-i\left(\frac{k\rho_B(E)}{2}\right)\right)} e^{i\left(kr-\left(\frac{k\rho_B(E)}{2}\right)\ln\, k(r-\zeta)\right)} \left(1+\frac{\left(1+i\left(\frac{k\rho_B(E)}{2}\right)\right)^2}{ik(r-\zeta)}+\cdots\right). \tag{4.100}$$

In their asymptotic forms described by (4.99) and (4.100), ψ_i and ψ_S resemble superficially a plane wave and an outgoing spherical wave, respectively. However, as a result of the long-range nature of the Coulomb force, the effects of which are mediated through the Bohr scattering diameter, $\rho_\mathrm{B}(E)$, the correspondence is not exact, and both waves will, as a result, be distorted by the Coulomb potential through the $e^{\pm i\left(\frac{k\rho_\mathrm{B}(E)}{2}\right)\ln k(r-\zeta)}$ terms that appear in both equations. The expression for the scattered wavefunction ψ_S given by (4.100) can be further simplified by noting that $\zeta = r\cos\theta$, thus allowing us to write it in a somewhat more familiar format:

$$\psi_\mathrm{S} \approx \frac{e^{i\left(kr-\left(\frac{k\rho_\mathrm{B}(E)}{2}\right)\ln 2kr\right)}}{r} f_\mathrm{Coul}(\theta) \tag{4.101}$$

where the Coulomb scattering amplitude is

$$f_\mathrm{Coul}(\theta) = -\frac{\rho_\mathrm{B}(E)}{4\sin^2\frac{\theta}{2}}\, e^{-i\left(\left(\frac{k\rho_\mathrm{B}(E)}{2}\right)\ln\left(\sin^2\frac{\theta}{2}\right)-2\arg\Gamma\left(1+i\left(\frac{k\rho_\mathrm{B}(E)}{2}\right)\right)\right)}. \tag{4.102}$$

It is of interest to rewrite the first Born approximation of the Coulomb scattering amplitude given by (4.78)

$$f_{(1)}(\theta) = -\frac{\rho_\mathrm{B}(E)}{4\sin^2\frac{\theta}{2}}$$

and compare this with (4.102). It is seen that the exact Coulomb scattering amplitude is that of the first term of the Born series but now including a multiplicative exponential term which yields a phase. However, this term will not affect the resulting calculation of the differential cross section in solid angle as the latter is equal to the modulus squared of the scattering amplitude:

$$\begin{aligned}\frac{\mathrm{d}\sigma}{\mathrm{d}\Omega} &= |f_\mathrm{Coul}(\theta)|^2\\ &= \left(\frac{\rho_\mathrm{B}(E)}{4}\right)^2 \frac{1}{\sin^4\frac{\theta}{2}}.\end{aligned} \tag{4.103}$$

Comparing this with the expression for the scattered wavefunction given by (4.101), we can see that the wavefunction-distorting factor is

$$e^{-i\left(\frac{k\rho_\mathrm{B}(E)}{2}\right)\ln 2kr}$$

which will not completely disappear. However, the kr term in the exponent of (4.101) will dominate over the logarithm of $2kr$ as $r\to\infty$, so that the effect of this effect becomes negligible.

4.3.5 Partial-Wave Decomposition

In Chap. 2, it was shown how accounting for the angular momentum of a scatter can lead to a partial wave solution to the Schrödinger equation. The radial component of Equation (4.90) is

$$\frac{\mathrm{d}^2 R_l(r)}{\mathrm{d}r^2} + \left(k^2\left(1 - \frac{\rho_{\mathrm{B}}(E)}{r}\right) - \frac{l(l+1)}{r^2}\right) R_l(r) = 0. \tag{4.104}$$

Because of the infinite-range nature of the Coulomb potential (the effects of which are mediated in (4.104) through the $\frac{\rho_{\mathrm{B}}(E)}{r}$ term), the expansion into spherical harmonics (see Appendix C.3) leads to a slowly converging solution which limits the usefulness of a partial-wave decomposition applied to Coulomb scatter. Moreover, an exact solution has already been derived in the previous section.

A partial-wave expansion of Coulomb scatter does, however, have utility if the Coulomb potential were to be combined with a short-range potential (i.e. one that decreases faster than the $\frac{1}{r}$ -dependence of the Coulomb potential). An example would be the scatter of an α-particle from a nucleus through which it would experience both the Coulomb potential arising from the protons within the nucleus and the strong nuclear force from all of the nucleons within the target if the α-particle were able to be within the range of that force. The latter would correspond to the Yukawa potential, $\frac{e^{-\kappa r}}{r}$. Because of the additional potential's short range, only the first few terms of the expansion of the scattering amplitude for the combined potentials, $f_{\mathrm{Comb}}(\theta)$, will be affected. Hence, the difference between the expansions of $f_{\mathrm{Comb}}(\theta)$ and $f_{\mathrm{Coul}}(\theta)$ which is given by $f_{\mathrm{Comb}}(\theta) - f_{\mathrm{Coul}}(\theta)$ will converge rapidly.

The solution to (4.104) is derived by using the following changes of both function and variable:

$$\begin{aligned} R_l(r) &= e^{ikr}\,(kr)^{l+1}\,v_l \\ \xi &= -2ikr \end{aligned} \tag{4.105}$$

to yield the differential equation

$$\xi\frac{\mathrm{d}^2 v_l}{\mathrm{d}\xi^2} + (2l + 2 - \xi)\frac{\mathrm{d}v_l}{\mathrm{d}\xi} + \left(l + 1 + i\frac{k\rho_{\mathrm{B}}(E)}{2}\right) v_l = 0. \tag{4.106}$$

This result is, in fact, Kummer's equation, described earlier in this chapter and derived in Appendix C. This equation has the solution $M\left(l + 1 + i\frac{k\rho_{\mathrm{B}}(E)}{2}, 2l + 2, -2ikr\right)$. As usual, we evaluate the asymptotic form of that solution for large kr. There will be thus both a regular solution and an irregular solution to (4.106):

$$\begin{aligned} F_l\left(\frac{k\rho_{\mathrm{B}}(E)}{2}, kr\right) &\approx g\cos\theta_l + f\sin\theta_l \\ G_l\left(\frac{k\rho_{\mathrm{B}}(E)}{2}, kr\right) &\approx f\cos\theta_l - g\sin\theta_l, \end{aligned} \tag{4.107}$$

where

$$\theta_l = kr - \frac{k\rho_{\mathrm{B}}(E)}{2}\ln 2kr - \frac{l\pi}{2} + \delta_l^{(\mathrm{Coul})}. \tag{4.108}$$

The phase shift angle for the Coulomb potential is

$$\delta_l^{(\mathrm{Coul})} = \arg\,\Gamma\left(l + 1 + i\frac{k\rho_{\mathrm{B}}(E)}{2}\right). \tag{4.109}$$

We have previously seen the gamma function with a complex argument in our development of the exact solution to the Coulomb scatter problem. In the complex plane, it has the following property for the complex variable $z = x + iy$:

$$\begin{aligned}\arg\,\Gamma(1+z) &= \arg\,\Gamma(z) + \tan^{-1}\left(\frac{y}{x}\right)\\ &= y\psi(x) + \sum_{n=0}^{\infty}\left(\frac{y}{x+n} - \tan^{-1}\left(\frac{y}{x+n}\right)\right)\end{aligned} \tag{4.110}$$

where $\psi(x)$ is the logarithmic derivative of the gamma function. This is also known as the *digamma function* (Appendix C.13),

$$\psi(x) = \frac{\mathrm{d}\ln\Gamma(z)}{\mathrm{d}z}\,\frac{\dfrac{\mathrm{d}\Gamma(x)}{\mathrm{d}x}}{\Gamma(x)}. \tag{4.111}$$

As $G_l\left(\frac{k\rho_{\mathrm{B}}(E)}{2}, kr\right)$ is an irregular solution, it is ignored. Further as $f \to 1$ and $g \to \frac{1}{kr}$ as $kr \to \infty$ (Abramowitz and Stegun (1972)), then the radial component of the wavefunction becomes

$$R_l(kr) \approx \frac{\cos\left(kr - (l+1)\dfrac{\pi}{2} + \arg\,\Gamma\left(l + 1 + i\dfrac{k\rho_{\mathrm{B}}(E)}{2}\right) - \left(\dfrac{k\rho_{\mathrm{B}}(E)}{4}\right)\ln 2kr\right)}{kr}. \tag{4.112}$$

We can now employ the phase shift analysis methods of Chap. 2 using the Coulomb phase shift angle given by (4.109). The partial-wave expansion of the Coulomb scattering amplitude is

$$f(\theta) = \frac{1}{k}\sum_{l=0}^{\infty}(2l+1)\,e^{i\arg\,\Gamma\left(l+1+i\left(\frac{k\rho_{\mathrm{B}}(E)}{2}\right)\right)}\sin\left(\arg\,\Gamma\left(l + 1 + i\left(\frac{k\rho_{\mathrm{B}}(E)}{2}\right)\right)\right)P_l(\cos\,\theta). \tag{4.113}$$

Then, the differential cross section in solid angle is given by the following expression:

$$\frac{\mathrm{d}\sigma}{\mathrm{d}\Omega} = \frac{1}{k^2}\sum_{l=0}^{\infty}\sum_{l'=0}^{\infty}(2l+1)(2l'+1)e^{i\arg\Gamma\left(l+1+i\left(\frac{k\rho_{\mathrm{B}}(E)}{2}\right)\right)}\left(e^{i\arg\Gamma\left(l'+1+i\left(\frac{k\rho_{\mathrm{B}}(E)}{2}\right)\right)}\right)^*$$
$$\times\left(\sin\left(\arg\Gamma\left(l+1+i\left(\frac{k\rho_{\mathrm{B}}(E)}{2}\right)\right)\right)P_l(\cos\theta)\right)$$
$$\left(\sin\left(\arg\Gamma\left(l'+1+i\left(\frac{k\rho_{\mathrm{B}}(E)}{2}\right)\right)\right)P_{l'}(\cos\theta)\right)^*. \tag{4.114}$$

4.3.6 Mott Scatter

The intrinsic spin of a particle is not a quantity inherent to classical mechanics and hence was not considered during our discussion of classical Coulomb scatter. In fact, even our quantum-mechanical development of Coulomb scatter theory has not, up to this point, considered the role that the intrinsic spin plays in the scattering process. In order to understand this role, we consider a Dirac particle with intrinsic spin-½ as this will have the greatest relevance in medical radiation physics calculations due to the prominences of electrons and positrons.

Here, in this section, we derive the differential cross section in solid angle of an electron scattered from a static Coulomb field of an electric charge $+Ze$. This was originally derived by Mott (1929). Here, however, in order to invoke the intrinsic spin-½ nature of the electron, the Dirac theory is used with the relativistic propagator formalism described by, for example, Bjorken and Drell (1964) and Dyson (2007). As the result obtained here for the electron is also that of the positron (to the lowest order of the fine-structure constant, α), which can be deduced by charge conjugation invariance or else shown explicitly, an explicit calculation for the positron cross section is not given, although this can be found in Mott (1932).

The derivation begins with the calculation of the transition matrix element for the elastic Coulomb scatter of an electron which describes the interaction of the electron with the electromagnetic field:

$$S_{\mathrm{fi}} = -ie\int \mathrm{d}^3\mathbf{r}\,\mathrm{d}t\,\bar{\psi}_f(\mathbf{r},t)\ \not{A}(\mathbf{r},t)\,\psi_i(\mathbf{r},t). \tag{4.115}$$

The Feynman 'dagger' notation used in (4.115) is

$$\not{A} = \gamma^0 A_0 - \gamma \bullet \mathbf{A} \tag{4.116}$$

where $(\gamma^\circ,\boldsymbol{\gamma})$ are the 4×4 Dirac matrices described in Appendix F and A^0 and $\mathbf{A}$ are the scalar and vector electromagnetic potentials, respectively. As the Coulomb gauge ($\gamma\bullet\mathbf{A} = 0$) is to be used, we have

$$\gamma^0 A_0 = -\frac{(Z\alpha\hbar c)}{e}\frac{\gamma^0}{|\mathbf{r}|}. \tag{4.117}$$

Because the calculation is performed to first order, plane-wave approximations are assumed for the incident and scattered electron:

$$\psi_{\mathrm{i}}(\mathbf{r},t) = \sqrt{\frac{m_e}{EL^3}}\, u(\mathbf{p},s)\, e^{\left(i\frac{\mathbf{p}\bullet\mathbf{r}}{\hbar c} - \frac{Et}{\hbar}\right)} \tag{4.118}$$

$$\psi_{\mathrm{f}}(\mathbf{r},t) = \sqrt{\frac{m_e}{E'L^3}}\, u(\mathbf{p}',s')\, e^{\left(i\frac{\mathbf{p}'\bullet\mathbf{r}}{\hbar c} - \frac{E't}{\hbar}\right)}, \tag{4.119}$$

where s and s' are the pre- and post-scattering electron spin orientations and E and E' are the corresponding total energies. The $\frac{m_e}{E}$ factor is required in the normalisation in order to both compensate for the Lorentz contraction of the volume element along the projectile's motion and to ensure that the spinor probability density remains invariant. The bar over $\psi_{\mathrm{f}}(\mathbf{r},t)$ denotes it as an adjoint spinor, and L^3 is the usual volume of normalisation and is a quantity that will disappear at the end of the calculation. The transition matrix is then

$$S_{\mathrm{fi}} = iZ(\alpha\hbar c)\, \frac{m_e}{L^3\sqrt{E'E}}\, \bar{u}(\mathbf{p}',s)\gamma^0 u(\mathbf{p},s) \int \mathrm{d}^3\mathbf{r}\, \frac{e^{i\frac{\mathbf{q}\bullet\mathbf{r}}{\hbar c}}}{|\mathbf{r}|} \int_0^\infty \mathrm{d}t\, e^{-i\frac{(E-E')\,t}{\hbar}}. \tag{4.120}$$

The space integral is known as the Bethe integral and was solved previously in our calculation of the first term of the Born expansion of Coulomb scatter. The time integral is calculated on the basis of the definition of the δ-function, $\delta(k) = \frac{1}{2\pi} \int_{-\infty}^{\infty} \mathrm{d}\omega\, e^{ik\omega}$, given by (C.164):

$$\int_0^\infty \mathrm{d}t\, e^{-i\frac{(E-E')\,t}{\hbar}} = 2\pi\, \delta\left(\frac{E'-E}{\hbar}\right). \tag{4.121}$$

Substituting these two results of these integrations into the expression for the S-matrix element gives

$$S_{\mathrm{fi}} = i\, 8\pi^2 Z(\alpha\hbar c)\, \frac{m_e}{L^3\sqrt{E'E}} \left(\frac{\hbar c}{q}\right)^3 \bar{u}(\mathbf{p}',s)\gamma^0 u(\mathbf{p},s)\, \delta\left(\frac{E'-E}{\hbar}\right). \tag{4.122}$$

Following from our discussions of quantum perturbation theory, the transition probability of the scatter occurring per incident electron is

$$p_{\mathrm{fi}} = |S_{\mathrm{fi}}|^2 \rho_{\mathrm{f}} \tag{4.123}$$

where the phase-space factor ρ_{f} is the density of final states per unit energy level $\mathrm{d}E'$:

$$\rho_{\mathrm{f}}\,\mathrm{d}E' = \left(\frac{L}{2\pi\hbar c}\right)^3 \mathrm{d}^3\mathbf{p}'. \tag{4.124}$$

By expanding the differential of (4.124), the phase-space factor is

$$\begin{aligned}\rho_{\mathrm{f}} &= \frac{1}{4\pi^2}\left(\frac{L}{\hbar c}\right)^3 p'^2\,\frac{\mathrm{d}p'}{\mathrm{d}E'}\,\mathrm{d}(\cos\,\theta)\\ &= \frac{1}{4\pi^2\beta'}\left(\frac{L}{\hbar c}\right)^3 p'^2\,\mathrm{d}(\cos\,\theta).\end{aligned} \tag{4.125}$$

Then, as

$$|\mathrm{d}(\cos\,\theta)| = \frac{q}{p^2}\,\mathrm{d}q,$$

the phase-space factor is

$$\rho_{\mathrm{f}} = \frac{1}{4\pi^2\beta}\left(\frac{L}{\hbar c}\right)^3 q\,\mathrm{d}q \tag{4.126}$$

where $\beta = \beta'$ and $p = p'$ due to the elasticity of the scatter and the negligible recoil of the scatterer.[5] The squared magnitude of the S-matrix element is

$$|S_{\mathrm{fi}}|^2 = 64\pi^4 Z^2(\alpha\hbar c)^2\,\frac{m_e^2\hbar^2}{L^3E'E}\left(\frac{\hbar c}{q}\right)^4 |\bar{u}(\mathbf{p}',s)\gamma^0 u(\mathbf{p},s)|^2 \left|\delta\left(\frac{E'-E}{\hbar}\right)\right|^2 \tag{4.127}$$

of the square of the δ-function is determined by assuming that the scattering takes place over a finite time interval bounded by $-\frac{T}{2}$ and by $+\frac{T}{2}$. Over that time interval, the δ-function will be broadened or

$$\begin{aligned}\delta\left(\frac{E'-E}{\hbar}\right) &\approx \frac{1}{2\pi}\int_{-\frac{T}{2}}^{\frac{T}{2}} \mathrm{d}t\, e^{i\frac{(E'-E)\,t}{\hbar}}\\ &\approx \frac{1}{\pi}\,\frac{\sin\left(\left(\frac{T}{2}\right)\left(\frac{E'-E}{\hbar}\right)\right)}{\left(\frac{E'-E}{\hbar}\right)}.\end{aligned} \tag{4.128}$$

[5] We are considering the electron being scattered from a potential or, equivalently, from another charged particle using the centre-of-mass reference system.

Thus,

$$\left|\delta\left(\frac{E'-E}{\hbar}\right)\right|^2 \approx \frac{1}{\pi^2}\frac{\sin^2\left(\left(\frac{T}{2}\right)\left(\frac{E'-E}{\hbar}\right)\right)}{\left(\frac{E'-E}{\hbar}\right)^2}. \tag{4.129}$$

As a function of $\frac{E'}{\hbar}$, the area under this curve is equal to $2\pi T$, leading to

$$\left|\delta\left(\frac{E'-E}{\hbar}\right)\right|^2 = \frac{T}{2\pi}\,\delta\left(\frac{E'-E}{\hbar}\right). \tag{4.130}$$

Returning this result to (4.127), we have

$$|S_{\rm fi}|^2 = 32\pi^3 Z^2(\alpha\hbar c)^2\,\frac{Tm_e^2}{L^3E'E}\left(\frac{\hbar c}{q}\right)^4\left|\bar{u}(\mathbf{p}',s)\gamma^0 u(\mathbf{p},s)\right|^2\delta\left(\frac{E'-E}{\hbar}\right). \tag{4.131}$$

The transition rate corresponding to the transition probability per incident electron is

$$\lambda_{\rm fi} = \frac{p_{\rm fi}}{T} \tag{4.132}$$

which, as a result of the above, is

$$\lambda_{\rm fi} = 8\pi Z^2(\alpha\hbar c)^2\,\frac{m_e^2(\hbar c)}{\beta q^3L^3E'E}\left|\bar{u}(\mathbf{p}',s)\gamma^0 u(\mathbf{p},s)\right|^2 {\rm d}q\,\delta\left(\frac{E'-E}{\hbar}\right). \tag{4.133}$$

We now consider the evaluation of the quantity $\left|\bar{u}(\mathbf{p}',s)\gamma^0 u(\mathbf{p},s)\right|^2$ which is a function of the electron spin orientations before and after the scatter. Before doing so, it should be recognised that we have neither knowledge of the spin orientation of the incident electron prior to the scatter nor interest in it following the scatter. In other words, as discussed in Chap. 2, one first averages over the initial spin states and then sums over the final spin states. This process is denoted by an overline over the quantity, $\overline{|\bar{u}(\mathbf{p}',s)\gamma^0 u(\mathbf{p},s)|^2}$, thus giving an expression for the transition rate:

$$\lambda_{\rm fi} = 8\pi Z^2(\alpha\hbar c)^2\,\frac{m_e^2(\hbar c)}{\beta q^3L^3E'E}\,\overline{|\bar{u}(\mathbf{p}',s)\gamma^0 u(\mathbf{p},s)|^2}\,{\rm d}q\,\delta\left(\frac{E'-E}{\hbar}\right). \tag{4.134}$$

where,

$$\overline{|\bar{u}(\mathbf{p}',s)\gamma^0 u(\mathbf{p},s)|^2} = \frac{1}{2}\sum_{\pm s,s'}\left|\bar{u}(\mathbf{p}',s)\gamma^0 u(\mathbf{p},s)\right|^2. \tag{4.135}$$

This sum over spin orientations is next expanded by first reducing it to traces and then using the trace properties of the Dirac matrices. As the trace algebra of Dirac matrices applied to this solution is long and tedious and a review of it will not add to our understanding of the result of the derivation,[6] the result of this summation is presented without proof,

$$\sum_{\pm s,s'} \bar{u}(\mathbf{p}',s')\,\gamma^0\,u(\mathbf{p},s) = Tr\left(\left(\frac{\not{p}' + m_e}{2m_e}\right)\gamma^0\left(\frac{\not{p}' + m_e}{2m_e}\right)\gamma^0\right). \tag{4.136}$$

The resultant trace of (4.136) is

$$\begin{aligned} Tr\left(\left(\frac{\not{p}' + m_e}{2m_e}\right)\gamma^0\left(\frac{\not{p}' + m_e}{2m_e}\right)\gamma^0\right) &= \frac{1}{m_e^2}\left(2EE' - \mathbf{p}\bullet\mathbf{p}' + m_e^2\right) \\ &= \frac{2EE'}{m_e^2}\left(1 - \beta^2\sin^2\frac{\theta}{2}\right) \end{aligned} \tag{4.137}$$

leading to

$$\overline{\left|\bar{u}(\mathbf{p}',s)\gamma^0 u(\mathbf{p},s)\right|^2} = \frac{EE'}{m_e^2}\left(1 - \beta^2\sin^2\frac{\theta}{2}\right). \tag{4.138}$$

Applying this result to the calculation of the transition rate gives the expression

$$\lambda_{\mathrm{fi}} = \frac{8\pi}{q^3}\,Z^2(\alpha\hbar c)^2\,\frac{(\hbar c)}{\beta L^3}\left(1 - \beta^2\,\sin^2\frac{\theta}{2}\right)\mathrm{d}q\,\delta\left(\frac{E' - E}{\hbar}\right). \tag{4.139}$$

Finally, the differential cross section in momentum transfer is

$$\begin{aligned} \frac{\mathrm{d}\sigma}{\mathrm{d}q} &= \frac{L^3}{\beta c}\frac{\lambda_{\mathrm{f}i}}{dq} \\ &= \frac{8\pi}{q^3}\left(\frac{Z\alpha\hbar c}{\beta}\right)^2\left(1 - \beta^2\,\sin^2\frac{\theta}{2}\right). \end{aligned} \tag{4.140}$$

[6] The interested reader can refer to Section (7.1) of Bjorken and Drell (1964) for this trace calculation.

We can now convert this result to the differential cross section in solid angle:

$$\begin{aligned}
\frac{d\sigma}{d\Omega} &= \frac{d\sigma}{dq}\frac{dq}{d\Omega} \\
&= \frac{1}{2\pi}\frac{d\sigma}{dq}\frac{dq}{d(\cos\theta)} \\
&= \frac{1}{4\pi}\frac{d\sigma}{dq}\frac{p}{\sin\left(\frac{\theta}{2}\right)} \\
&= \frac{2}{q^3}\left(\frac{Z\alpha\hbar c}{\beta}\right)^2 \frac{p}{\sin\left(\frac{\theta}{2}\right)}\left(1-\beta^2\sin^2\frac{\theta}{2}\right) \\
&= \frac{1}{4}\left(\frac{Z\alpha\hbar c}{p\beta}\right)^2 \frac{\left(1-\beta^2\sin^2\frac{\theta}{2}\right)}{\sin^4\left(\frac{\theta}{2}\right)} \\
&= \left(\frac{\rho_B(E)}{4}\right)^2 \frac{\left(1-\beta^2\sin^2\frac{\theta}{2}\right)}{\sin^4\left(\frac{\theta}{2}\right)} \\
&= \left.\frac{d\sigma}{d\Omega}\right|_{\text{Spin}-0}\left(1-\beta^2\sin^2\frac{\theta}{2}\right)
\end{aligned} \tag{4.141}$$

where $\left.\frac{d\sigma}{d\Omega}\right|_{\text{Spin}-0}$ is the elastic Coulomb (Rutherford) scatter differential cross section for which intrinsic spin is ignored.

$$\frac{d\sigma_{\text{Mott}}}{d\Omega} = \frac{d\sigma_{\text{Ruth}}}{d\Omega}\left(1-\beta^2\sin^2\frac{\theta}{2}\right). \tag{4.142}$$

Inclusion of the electron's intrinsic spin into the calculation thus modifies the Rutherford spin-0 cross section by the multiplicative factor of $\left(1-\beta^2\sin^2\frac{\theta}{2}\right)$. This factor is shown as a function of electron kinetic energy for scattering angles of 15°, 45° and 90° in Fig. 4.8. This reduction in electron scattering cross section due to the projectile's intrinsic spin becomes significant only for highly energetic electrons and large scattering angles.

In the relativistic limit of $\beta \to 1$, the cross section becomes

$$\frac{d\sigma}{d\Omega} \to \frac{d\sigma_{\text{Ruth}}}{d\Omega}\cos^2\frac{\theta}{2}. \tag{4.143}$$

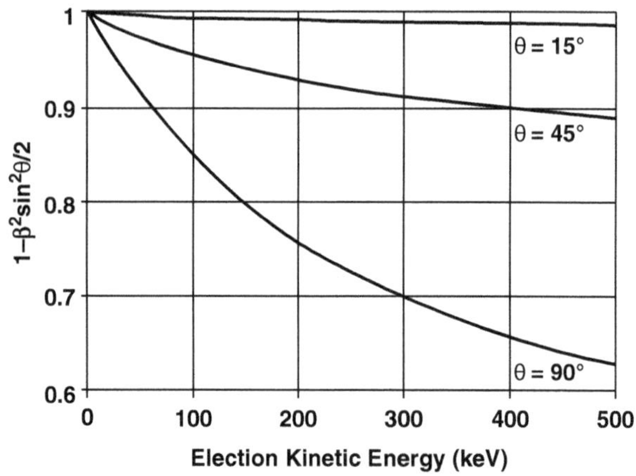

Fig. 4.8 Reduction in the Coulomb scatter cross section for electrons due to consideration of the intrinsic electron spin

4.3.7 Møller (Electron–Electron) Scatter

In our derivation of the Mott scattering cross section, we considered the electron as having been scattered by a Coulomb potential without any consideration of what was the source of that potential. Should that source be another electron, we must then account for two issues. First, the system of two electrons will be subject to the Pauli exclusion principle and spin statistics. Second, the final state will consist of two moving electrons, and it will not be possible to differentiate between the scattered and recoil electrons by observation, thus requiring an a priori definition of which electron is which following the scatter. We now consider the case of the projectile being an electron with kinetic energy, *E*. The Feynman diagrams of electron–electron (Møller 1932) scatter are shown in Fig. 4.9. Two graphs must arise as a result of the inability to distinguish between the two exiting electrons of which was the projectile or the target. From Appendix D, the maximum energy transferred to the target electron is equal to the kinetic energy of the incident. Because of the indistinguishability between the exiting electrons, the exiting electron with the highest energy is assumed to be the primary by convention.

To demonstrate the effect of the exchange interaction, we derive the nonrelativistic expression of the Møller differential cross section. As the electron is a fermion, the wavefunction of an electron pair system must therefore be antisymmetric through the interchange of the two electrons. Thus, if the electron spins are parallel (i.e. the system is in the spin-triplet state), the system will be symmetric under the exchange of spins. This will require the spatial wavefunctions to be antisymmetric under the exchange of the electron's relative co-ordinates. Hence, the scattering amplitude associated with the spin-triplet state will be

$$f_T(\theta) = f(\theta) - f(\pi - \theta) \tag{4.144}$$

where $f(\theta)$ is the scattering amplitude of elastic Coulomb scatter and the transformation from θ to $\pi - \theta$ is the equivalent of the interchange of the two electrons.

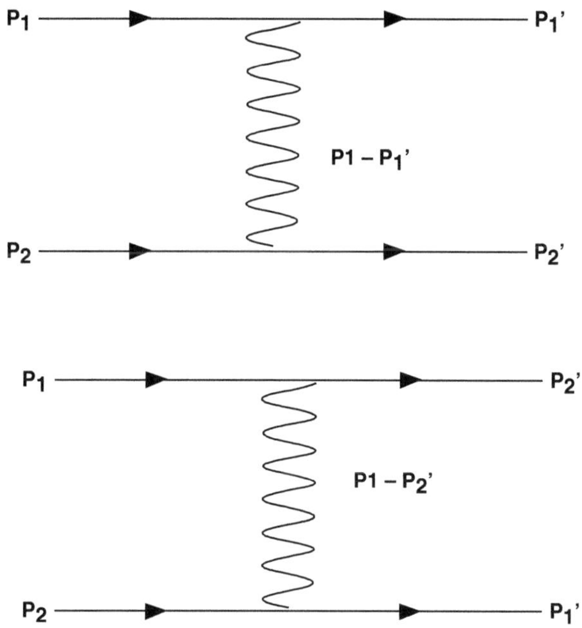

Fig. 4.9 Feynman diagrams for electron–electron Coulomb scatter. *Top*: direct interaction; *bottom*: exchange interaction

For the condition of antiparallel electron spins (i.e. the spin-singlet state), the system will be antisymmetric under the interchange of spins, and as a result, the spatial wavefunctions must thus be symmetric:

$$f_S(\theta) = f(\theta) + f(\pi - \theta). \tag{4.145}$$

However, it must be recognised that in dosimetry calculations, the projectile and target electrons are considered to be unpolarised (as we have no knowledge of their spin orientations), and e^-e^- scatter is thus taken to have a random distribution of spins. Hence, spin-singlet and spin-triplet states will exist with a ratio of relative probabilities of 1:3. The resulting differential cross section, then, is the statistically weighted sum of the squared moduli of the scattering amplitudes:

$$\begin{aligned}
\frac{d\sigma}{d\Omega} &= \frac{1}{4}|f_S(\theta)|^2 + \frac{3}{4}|f_T(\theta)|^2 \\
&= \frac{1}{4}|f(\theta) + f(\pi - \theta)|^2 + \frac{3}{4}|f(\theta) - f(\pi - \theta)|^2 \\
&= |f(\theta)|^2 + |f(\pi - \theta)|^2 - |f(\theta)|\,|f(\pi - \theta)| \\
&= \left(\frac{\rho_B}{4}\right)^2 \left(\frac{1}{\sin^4\frac{\theta}{2}} + \frac{1}{\cos^4\frac{\theta}{2}} - \frac{1}{\sin^2\frac{\theta}{2}\cos^2\frac{\theta}{2}} \right) \\
&= \left(\frac{\rho_B}{4}\right)^2 \left(\frac{1}{\sin^4\frac{\theta}{2}} + \frac{1}{\cos^4\frac{\theta}{2}} - \frac{1}{\sin^2\frac{\theta}{2}\cos^2\frac{\theta}{2}} \right)
\end{aligned} \tag{4.146}$$

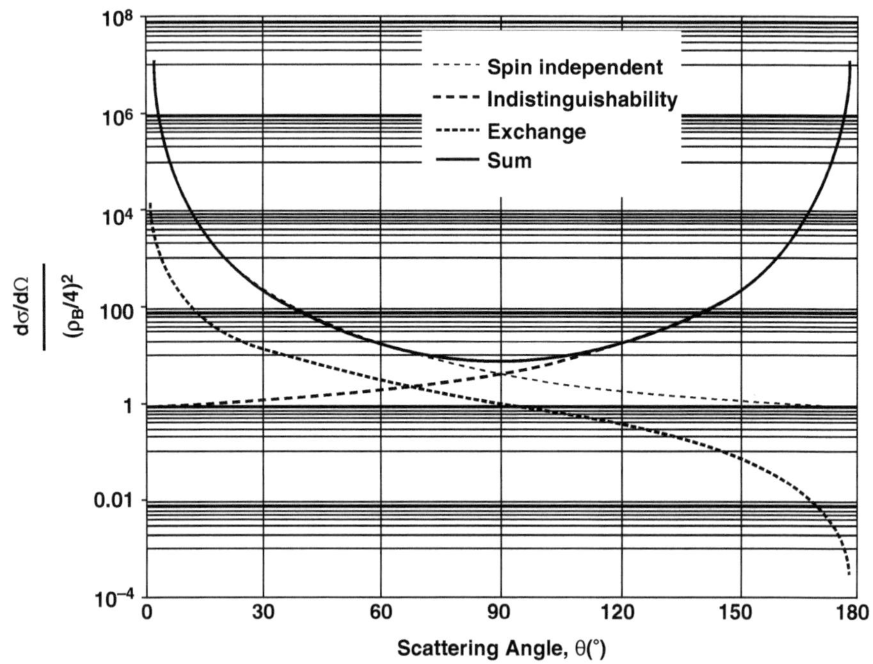

Fig. 4.10 The angular dependence of the three terms of the Møller electron–electron scatter of (4.146). The spin-independent term is the conventional Rutherford dependence, the indistinguishability term reflects the inability to differentiate in the final state between the incident and target electron and the exchange term is the contribution of the second Feynman diagram of Fig. 4.9

The first term of (4.146) is that of the original spin-independent elastic Coulomb scatter cross section, whereas the second reflects the inability to distinguish between the incident and scattered electrons. The third 'cross' term is the exchange term. Figure 4.10 shows the dependencies of each of these terms upon scattering angle. The spin-independent contribution to the differential cross section decreases with scattering angle to reach a value equal to 1 for backscatter ($\theta = 180°$). The angular dependence of the exchange term has a symmetry about ($\theta = 90°$) and thus diverges for both $\theta \to 0°$ and $\theta \to 180°$. Finally, the combined terms yield an angular distribution symmetric about $\theta = 90°$ diverging for both of the conditions of forward and backscatter.

Møller scatter, and the associated Bhabha (electron–positron) scatter, will be revisited in our later discussions in energy loss resulting from charged particle interactions with matter.

4.3.8 Spin-1 Projectile

For completeness, we present the differential cross section for a spin-1 projectile without derivation (Massey and Corben 1939):

$$\frac{\mathrm{d}\sigma_{\mathrm{Mott}}}{\mathrm{d}\Omega} = \frac{\mathrm{d}\sigma_{Ruth}}{\mathrm{d}\Omega}\left(1 + \frac{1}{6}\left(\frac{p\beta}{m}\right)^2 \sin^2\frac{\theta}{2}\right). \tag{4.147}$$

As with (4.142), the effects of intrinsic spin are negligible for small deflection angles and low projectile kinetic energies.

Bibliography and Further Reading

Belkić D. Principles of quantum scattering theory. Bristol: Institute of Physics Publishing; 2004.
McParland BJ. Nuclear medicine radiation dosimetry: advanced theoretical concepts. London: Springer; 2010.
Podgoršak EB. Radiation physics for medical physicists. 2nd ed. Berlin: Springer; 2010.
Sigmund P. Particle penetration and radiation effects. Berlin: Springer; 2006.

References

Abramowitz M, Stegun IA, editors. Handbook of mathematical functions. New York: Dover Publications; 1972.
Aguiar CE, Barroso MF. The Runge-Lenz vector and perturbed Rutherford scattering. Am J Phys. 1996;64:1042–8.
Andersen HH, Sigmund P. A simple nonbinary scattering model applicable to atomic collisions in crystals at low energies. Mat Fys Medd Dan Vid Selsk. 1966;34:1–50.
Bjorken JD, Drell SD. Relativistic quantum mechanics. New York: McGraw-Hill; 1964.
Bohr N. The penetration of atomic particles through matter. Matt Fys Medd Dan Vild Selsk. 1948;18:1–144.
Dalitz RH. On higher Born approximations in potential scattering. Proc R Soc. 1951;206:509–20.
Dyson F. Advanced quantum mechanics. Singapore: World Scientific Publishing; 2007.
Feynman RP. Space-time approach to quantum electrodynamics. Phys Rev. 1949;76:769–89.
Lehmann C, Leibfried G. Higher order momentum approximations in classical collision theory. Z Physik. 1963;172:465–87.
Massey HJ, Corben HC. Elastic collisions of mesons with electrons and protons. Proc Camb Philos Soc. 1939;35:463–73.
Møller C. Zur theorie des durchgangs schneller elektronen durch materie. Ann Physik. 1932;14:531–85.
Mott NF. Scattering of fast electrons by atomic nuclei. Proc R Soc. 1929;A124:425–42.
Mott NF. The elastic scattering of fast positrons by heavy nuclei. Proc R Soc. 1932;A135:429.
Sigmund P, Schinner A. Binary stopping theory for swift heavy ions. Eur Phys J. 2000;D 12:425–34.
Thomson JJ. Conduction of electricity through gases. 2nd ed. Cambridge: University Press; 1906.

Chapter 5
Elastic Coulomb Scatter from Distributed and Screened Charges

Contents

Abstract With the exception of point targets, such as a free electron, the target in the elastic Coulomb scatter of a charged particle is typically distributed and screened, such as the atomic nucleus. The degree of spatial variation and Coulomb screening perceived by the projectile will be dependent upon its de Broglie wavelength of both the atom and the nucleus. In this chapter, we extend the previous discussion of elastic

B.J. McParland, *Medical Radiation Dosimetry*,
DOI 10.1007/978-1-4471-5403-7_5, © Springer-Verlag London 2014

Coulomb scatter from a point charge to the more realistic case of a distributed and screened target. This will begin with the definition of the form factor for a spatially distributed charge that the projectile is incident to. Examples such as a hard-edge homogeneous sphere and a sphere with a charge distribution with a Gaussian radial charge distribution are considered in the calculation of these form factors.

Screened target are next considered and in much detail. We begin with the simple model of continuous screening using a Yukawa-type potential to represent that of the screened nuclear potential. A key parameter used in this model is the atomic electron screening parameter, which is the coefficient of the spatial variable in the exponent of the Yukawa form. The Thomas–Fermi statistical model of the atom is developed and its results used to derive a functional form of the atomic electron screening parameter.

We then proceed to the model of discrete screening in which the potential perceived by the projectile is the summation of those due to the atomic electrons and that due to the atomic nucleus. From this calculation is extracted the atomic elastic scattering form factor.

The outcomes of the continuous and discrete atomic electron screening models are differential cross sections which display modifications from the Rutherford (unscreened) Coulomb differential cross section which avoid the divergence at zero scattering angle.

We conclude with a numerical analysis of these results.

5.1 Introduction

In the previous chapter, the elastic Coulomb scatter differential cross sections for a point charged projectile incident to a point bare charged target (i.e. Rutherford scatter) were evaluated as functions of momentum transfer and solid angle in both classical and quantum-mechanical models. The result in solid angle has a $\frac{1}{\sin^4\left(\frac{\theta}{2}\right)}$ dependence which leads to the divergence $\frac{d\sigma}{d\Omega} \to \infty$ as $\theta \to 0$.

In practice, the target is not a bare point. It is the nucleus which, in the Coulomb picture, is a spatial distribution of protons and, hence, cannot be considered a point if the projectile's de Broglie wavelength is sufficiently small. Further, the nucleus resides within a cloud of atomic electrons bound to the nucleus. The nuclear Coulomb field experienced by the projectile will thus vary with projectile–nucleus separation in a manner different from the inverse square law due to the shielding of the nuclear charge by these electrons.

In this chapter, we will investigate these effects upon the elastic Coulomb differential cross section. Atomic electron screening will, in particular, remove the divergence noted above and the differential cross section will reach a finite value as $\theta \to 0$. On the other hand, the tendency of the finite nuclear spatial distribution will be to diminish the cross section at larger angles or, in some cases, introduce a diffraction-like behaviour with angle. The combination of both cases leads to elastic Coulomb scatter being predominantly forward directed and to the phenomenon of multiple scatter (which is the subject of the following chapter).

We begin this chapter by considering the nuclear target as a spatial distribution $\rho_P(r)$ of protons. The result of this analysis will be that the (unscreened) Rutherford cross section is modified by the modulus squared of the nuclear form factor, which will be shown to be equal to the Fourier transform of $\rho_P(r)$.[1] Examples of this quantity are derived for a variety of simple spherically symmetric charge distributions. We must also recognise, however, that with positrons and electrons (i.e. charged leptons), the effect of the finite nuclear size is purely electromagnetic in character through their interactions with the nuclear protons. On the other hand, for hadrons (protons, π^--mesons, α-particles and ions) that are used as medical radiations, there is the additional contribution of the attractive strong nuclear force. As the range of this force is of only the order of 1–2 fm, its effect would not be evident in most interactions of medical radiation interest unless there were sufficient overlap between the hadron projectile and nuclear wavefunctions, such as in the formation of a pionic atom and the resulting production of star products as described in Chap. 1. We will, however, limit the discussion of this chapter to electromagnetic interactions.

Following our derivation of the modification of the Rutherford cross section by a finite nuclear size, we then proceed to an investigation of the screened atomic target and develop it in two ways. We note that this requires our determination of the spatial distribution of the atomic electrons, $\rho_e(r)$.

In the first, the screening is considered to result in a continuous and exponentially decreasing Coulomb potential with decreasing distance r from the nucleus (i.e. a Wentzel type of potential, i.e. $\propto \frac{e^{-\kappa r}}{r}$, where κ is the screening parameter). The scattering differential cross section for this screened potential is calculated from the first-order term of the Born series for this screened potential. Following an exposition given by Bohr (1948), we then proceed to analyse the various conditions of scattering arising from the exponentially screened Coulomb potential and the underlying roles of kinematics and the dimensions of the electron cloud surrounding the nucleus.[2]

We next consider the means of evaluating numerically the effects of screening. This will be largely based upon the Thomas–Fermi statistical model of multi-electron atoms which is then derived. There are three reasons for our focus on the Thomas–Fermi approach. First, it provides a number of analytical expressions of the atomic electron screening parameter used in the exponentially screened potential and which are readily scaled between particular atoms. Second, the model itself provides an expression of an effective potential that can be used to calculate for the screened Coulomb scatter differential cross section. Third, it will be referred to in later discussions of charged particle energy loss. However, the Thomas–Fermi result is not analytical, and thus its practical implementation requires either a numerical calculation in order for it to be evaluated or else to be approximated by an analytical formula (as given in the multiexponential approximation of Molière).

[1] This feature makes the measurement of elastic scattering cross sections as a useful means of discerning the internal structure of a target.

[2] The discussions of the screening provided in this chapter are of the target only. The projectile is considered to be 'bare', i.e. it is not an ion with its own atomic electrons. This particular case is discussed later within the context of energy loss.

In the second approach of accounting for screening, the scattering centre is treated explicitly as a nucleus surrounded by a discrete number of noninteracting electrons such that the Coulomb scattering potential felt by the projectile is the sum of the individual contributions of these sources. A differential cross section is calculated from these contributions using quantum perturbation theory and the results compared with the continuous-screening model. The chapter concludes with numerical examples of calculations for carbon and calcium targets, because their presences within tissue and their large differences in atomic number Z will be used as the elemental targets for these calculations.

Throughout, where numerical data are introduced in order to assist the reader in understanding the orders of magnitude of the quantities discussed, we will cite examples of 100-keV electrons and 250-MeV protons incident to carbon and calcium. The particles and kinetic energies are typical of many medical applications of these projectiles, and the target selection covers the range of atomic numbers relevant to tissue.[3]

5.2 Spatially Distributed Targets

5.2.1 *Form Factor: Definition*

Instead of the point charge discussed throughout Chap. 4, let us now consider the target to be a spatially distributed electric charge, such as the nucleus, with a charge density function $\rho_e(\mathbf{r})$ with unit normalisation:

$$\int \mathrm{d}\mathbf{r}\,\rho_e(\mathbf{r}) = 1 \tag{5.1}$$

$\rho_e(\mathbf{r})$ is the number density/spatial distribution of protons within the nucleus normalised to Z. This unit normalisation is selected for this specific discussion as it leads to form factors such that $|F(q)|^2 \in [0, 1]$. Clearly, we could change the normalisation of (5.1) such that the integral is equal to the atomic number of the target, Z, and such non-unity normalisations are used in general in later discussions within the context of the atomic electron spatial distribution. This normalisation (which gives the dimension of the density function as that of inverse volume) is also used as it allows the Z charge of the target to be explicitly noted in the expression of the distributed Coulomb potential:

$$V(\mathbf{r}) = \left(\frac{\alpha\hbar c Z}{e}\right) \int \mathrm{d}^3\mathbf{r}'\, \frac{\rho_e(\mathbf{r}')}{|\mathbf{r} - \mathbf{r}'|} \tag{5.2}$$

---

[3] Carbon, at $Z = 6$, is selected as it represents the smallest atomic number atom that can be justifiably modelled using the Thomas–Fermi method from which we derive a screening parameter.

where the integration is over the nuclear volume. This extraction of Z will be consistent with our nomenclature so far, as to be shown now. The potential energy for the projectile due to this electrostatic potential is

$$\begin{aligned} U(r) &= \pm ze\, V(r)/r \\ &= \pm(\alpha\hbar c\, zZ) \int \mathrm{d}^3\mathbf{r}' \, \frac{\rho_\mathrm{e}(\mathbf{r}')}{|\mathbf{r}-\mathbf{r}'|}. \end{aligned} \tag{5.3}$$

Spherical symmetry will soon be applied (i.e. $\mathrm{d}^3\mathbf{r} = 4\pi\,\mathrm{d}r\,r^2$) so as to simplify the calculation and to reflect the azimuthal independence of Coulomb scatter.

In order to calculate the differential cross section of the scattering from this distributed charge, we use the scattering amplitude evaluated from the first term of the Born expansion as derived in Chap. 2. This is a function of the three-vector momentum transfer $\mathbf{q} = \mathbf{p}' - \mathbf{p}$, where $\mathbf{p}'$ is the post-scatter momentum and $\mathbf{p}$ is the incident momentum (for elastic scatter, $|\mathbf{p}'| = |\mathbf{p}|$). The scattering amplitude is

$$\begin{aligned} f(\mathbf{q}) &= -\frac{m}{2\pi(\hbar c)^2} \int \mathrm{d}^3\mathbf{r}' e^{-i\frac{\mathbf{q}\bullet\mathbf{r}'}{\hbar c}} U(\mathbf{r}') \\ &= \mp \frac{zZm\alpha}{2\pi\hbar c} \int \mathrm{d}^3\mathbf{r}' e^{-i\frac{\mathbf{q}\bullet\mathbf{r}'}{\hbar c}} \frac{\rho_e(\mathbf{r}')}{|\mathbf{r}-\mathbf{r}'|}. \end{aligned} \tag{5.4}$$

By using the substitution of variable of $\mathbf{R} = \mathbf{r} - \mathbf{r}'$, (5.4) can be written in the form

$$f(\mathbf{q}) = \mp \frac{zZm\alpha}{2\pi\hbar c} \int \mathrm{d}^3\mathbf{R} \frac{e^{-i\frac{\mathbf{q}\bullet\mathbf{R}}{\hbar c}}}{|\mathbf{R}|} \int \mathrm{d}^3\mathbf{r}\, e^{-\frac{\mathbf{q}\bullet\mathbf{r}}{\hbar c}} \rho_\mathrm{e}(\mathbf{r}). \tag{5.5}$$

Let us now compare this scattering amplitude for a distributed charge with that derived previously for a point charge (2.83). We see that the latter has now been scaled by the integral, $\int \mathrm{d}^3\mathbf{r}\, e^{-\frac{i\mathbf{q}\bullet\mathbf{r}}{\hbar c}} \rho_\mathrm{e}(\mathbf{r})$, which is the Fourier transform into $\mathbf{q}$-space of the nuclear proton spatial distribution. Thus, the unscreened point target (Rutherford) differential cross section is modified by the modulus squared of this Fourier transform to yield the differential cross section for a distributed target:

$$\frac{\mathrm{d}\sigma_\mathrm{Dist}}{\mathrm{d}\Omega} = \frac{\mathrm{d}\sigma_\mathrm{Ruth}}{\mathrm{d}\Omega} |F(\mathbf{q})|^2, \tag{5.6}$$

where the form factor is given by the Fourier transform of the electric charge distribution of the target

$$F(\mathbf{q}) = \int \mathrm{d}^3\mathbf{r}\, e^{-\frac{i\mathbf{q}\bullet\mathbf{r}}{\hbar c}} \rho_\mathrm{e}(\mathbf{r}). \tag{5.7}$$

The dependence of the form factor upon the scattering angle arises through the magnitude of the momentum transfer $|\mathbf{q}| = 2p\sin\frac{\theta}{2}$ that is within the exponent of the integrand. Clearly, as noted earlier, from the normalisation of the electric charge distribution given by (5.1), $|F(\mathbf{q})|^2 \in [0, 1]$. At high momentum transfers, the phase term in (5.5) rapidly oscillates and the integral consequently averages to zero, i.e. $|F(\mathbf{q})|^2 \to 0$ as $q \to \infty$. It is also immediately evident from (5.1) and (5.6) that $|F(\mathbf{q})|^2 \to 1$ as $q \to 0$.

5.2.2 *Form Factors of Specific Charge Spatial Distributions*

5.2.2.1 Spherically Symmetric Charge Distributions

Examples of charge distributions that are spherically symmetric will be considered here. The assumption of continuous electric proton charge distribution makes ease of the calculation of the effects of a finite charge distribution, such as the nucleus, upon the scattering cross section. A calculation for a discrete distribution of protons within the nucleus is also possible, but this requires knowledge of the nuclear wavefunction which may not always be known to great precision (Glauber and Matthiae 1970). However, such evaluations are not of practical interest to medical radiation dosimetry calculations.

For a continuous and spherically symmetric charge distribution, the form factor can be simplified to the following integral which can be solved to yield a final expression in terms of the charge spatial distribution:

$$\begin{aligned} F(q) &= \int \mathrm{d}^3\mathbf{r}\, e^{-\frac{i\mathbf{q}\bullet\mathbf{r}}{\hbar c}}\, \rho_e(r) \\ &= 2\pi \int_0^\infty \mathrm{d}r\, r^2\, \rho_e(r) \int_{-1}^{1} \mathrm{d}(\cos\,\theta)\, e^{-i\frac{qr\,\cos\,\theta}{\hbar c}} \\ &= -4\pi \left(\frac{\hbar c}{q}\right) \int_0^\infty \mathrm{d}r\, r \sin\left(\frac{qr}{\hbar c}\right) \rho_e(r). \end{aligned} \tag{5.8}$$

5.2.2.2 Point Charge Distribution

Before progressing, it must be verified first that we can recover the result for a point charge target from (5.8) (which is equivalent to stating that $|F(q)|^2 = 1$ for a point charge). In this case the spatial distribution of the target electric charge is described by the Dirac δ-function:

$$\rho_e(\mathbf{r}) = \rho_0 \delta(\mathbf{r}) \tag{5.9}$$

where the normalisation constant is determined from (5.1),

$$4\pi \rho_0 \int \mathrm{d}r\, r^2\, \delta(r) = 1, \tag{5.10}$$

giving $\rho_e(r) = \frac{\delta(r)}{4\pi}$. Substituting this into (5.8) yields the expected constant form factor modulus squared, $|F(q)|^2 = 1$.

5.2.2.3 Hard-Edge and Homogeneous Spherical Charge Distribution

This is the case of the proton charge distribution uniformly distributed within a spherical nucleus with a well-defined and finite radius R_N. This is described by

$$\rho_e(r) = \rho_0 H(R_N - r) \tag{5.11}$$

where ρ_0 is a constant and $H(x)$ is the Heaviside function. The parameter R_N will be set equal to the root mean square nuclear radius given by the empirical formula $R_N = 0.94\, A^{\frac{1}{3}}$ fm (McParland 2010). This reflects a crude model of the nucleus whereas, in reality, the nuclear edge demonstrates a smooth and continuous decrease in electric charge with radius. ρ_0 is again determined from the normalisation condition of (5.1):

$$4\pi \rho_0 \int_0^{R_N} \mathrm{d}r\, r^2 = 1 \tag{5.12}$$

yielding

$$\rho_0 = \frac{3}{4\pi R_N^3}. \tag{5.13}$$

Substituting this into (5.11) and solving for the form factor,

$$\begin{aligned} F(q) &= -\frac{3}{R_N^3}\left(\frac{\hbar c}{q}\right)\int_0^{R_N} \mathrm{d}r\, r \sin\left(\frac{qr}{\hbar c}\right) \\ &= -3\frac{j_1\left(\frac{qR_N}{\hbar c}\right)}{\left(\frac{qR_N}{\hbar c}\right)} \end{aligned} \tag{5.14}$$

where $j_1(x)$ is the spherical Bessel function of the first kind, as derived in Appendix C. Figure 5.1 shows the hard-edge radial distribution and the modulus squared of the associated form factor.

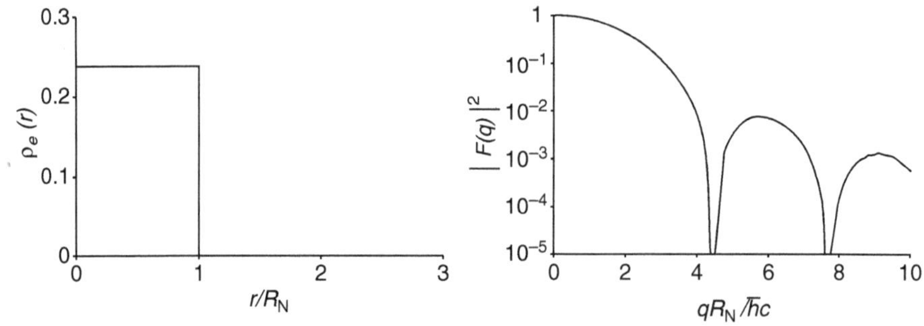

Fig. 5.1 Charge spatial distribution for spherical homogeneous hard-edge target and the corresponding modulus squared of the form factor

This charge distribution yields an oscillatory form factor as a consequence of the diffraction of the incident projectile's wavefunction by a target with a finite dimension. The dimensionless variable $\frac{qR_N}{\hbar c} = 2\left(\frac{R_N}{\lambda\!\!\!^{-}}\right)\sin\frac{\theta}{2}$, where $\lambda\!\!\!^{-}$ is the reduced de Broglie wavelength, provides the dependence of $|F(q)|^2$ upon the projectile momentum and scattering angle. For $\lambda\!\!\!^{-}$ greater than the scattering centre dimensions (i.e. low incident momentum), $\frac{qR_N}{\hbar c}$ is small and $|F(q)|^2$ is close to unity such that the modified differential cross section is essentially equal to that of a point target. This reflects the fact that the target is perceived by the projectile as a point for as long as $R_N << \lambda\!\!\!^{-}$. In general, $|F(q)|^2$ reduces rapidly with decreasing $\lambda\!\!\!^{-}$ by (ignoring the multiple local minima caused by diffraction) a factor of about 10^{-3} for $\frac{qR_N}{\hbar c} \approx 10$.

5.2.2.4 Gaussian Charge Distribution

The modelling of the nuclear charge distribution as having a finite radius previously discussed is a rather simplistic and harsh one. Evidence derived from scattering measurements has demonstrated that the nuclear radial charge distribution is smooth, that the nuclear 'edge' is diffuse and that, consequently, $\rho_e(r)$ is a continuous function. Here, we consider the use of a Gaussian distribution to describe this radial variation of the charge distribution:

$$\rho_e(r) = \rho_0\, e^{-\left(\frac{r}{R_N}\right)^2} \tag{5.15}$$

where R_N is the radial distance at which the charge distribution is equal to e^{-1} that of the central value. Experimental evidence indicates that (5.15) is a reasonable model for light nuclei. The example provided in the introduction of the Cauchy–Goursat theorem in Appendix A.3 is useful here. As before, we evaluate ρ_0 from the normalisation condition of (5.1):

$$4\pi\rho_0 \int_0^\infty dr\, r^2\, e^{-\left(\frac{r}{R_N}\right)^2} = 1. \tag{5.16}$$

This integral is solved by noting that

$$\int_0^\infty \mathrm{d}r\, r^2\, e^{-\left(\frac{r}{R_\mathrm{N}}\right)^2} = \frac{d}{\mathrm{d}\left(\frac{1}{R_\mathrm{N}^2}\right)} \int_0^\infty \mathrm{d}r\; e^{-\left(\frac{r}{R_\mathrm{N}}\right)^2}. \tag{5.17}$$

The integral on the right-hand side of (5.17) is solved by first squaring it and creating the double integral:

$$\begin{aligned}\left(\int_0^\infty \mathrm{d}r\; e^{-\left(\frac{r}{R_\mathrm{N}}\right)^2}\right)^2 &= \int_0^\infty \mathrm{d}s\; e^{-\left(\frac{s}{R_\mathrm{N}}\right)^2} \int_0^\infty \mathrm{d}t\, e^{-\left(\frac{t}{R_\mathrm{N}}\right)^2} \\ &= \int_0^\infty \mathrm{d}s \int_0^\infty \mathrm{d}t\, e^{-\frac{(s^2+t^2)}{R_\mathrm{N}^2}}.\end{aligned} \tag{5.18}$$

This double integral is solved, in turn, by converting it from Cartesian co-ordinates into polar co-ordinates:

$$\int_0^\infty \mathrm{d}s \int_0^\infty \mathrm{d}t\, e^{-\frac{(s^2+t^2)}{R_\mathrm{N}^2}} = \frac{1}{4}\left(2\pi \int_0^\infty \mathrm{d}r\, r\, e^{-\left(\frac{r}{R_\mathrm{N}}\right)^2}\right). \tag{5.19}$$

As $\mathrm{d}(r^2) = 2r\,\mathrm{d}r$,

$$\begin{aligned}2\pi \int_0^\infty \mathrm{d}r\, r\, e^{-\left(\frac{r}{R_\mathrm{N}}\right)^2} &= \pi R_\mathrm{N}^2 \int_0^\infty \mathrm{d}t\, e^{-t} \\ &= \pi R_\mathrm{N}^2.\end{aligned} \tag{5.20}$$

This gives

$$\int_0^\infty \mathrm{d}r\; e^{-\left(\frac{r}{R_\mathrm{N}}\right)^2} = \frac{\sqrt{\pi}\, R_\mathrm{N}}{2} \tag{5.21}$$

and, hence, from (5.16),

$$\int_0^\infty \mathrm{d}r\, r^2\, e^{-\left(\frac{r}{R_\mathrm{N}}\right)^2} = \frac{\sqrt{\pi}}{4}\, R_\mathrm{N}^3 \tag{5.22}$$

which leads to the normalised Gaussian charge distribution:

$$\rho_e(r) = \frac{e^{-\left(\frac{r}{R_N}\right)^2}}{\pi^{\frac{3}{2}} R_N^3}. \tag{5.23}$$

Having obtained this normalised charge distribution, we now evaluate the corresponding form factor

$$F(q) = -\left(\frac{\hbar c}{q}\right) \frac{4}{\sqrt{\pi} R_N^3} \int_0^\infty \mathrm{d}r\, r \sin\left(\frac{qr}{\hbar c}\right) e^{-\left(\frac{r}{R_N}\right)^2}. \tag{5.24}$$

This integral is solved using the complex integration techniques presented in Appendix A. We begin by noting that the integrand is an even function and can thus write the integral as

$$\begin{aligned}\int_0^\infty dr\, r\, \sin\left(\frac{qr}{\hbar c}\right) e^{-\left(\frac{r}{R_n}\right)^2} &= \frac{1}{2}\int_{-\infty}^\infty dr\, r\, \sin\left(\frac{qr}{\hbar c}\right) e^{-\left(\frac{r}{R_n}\right)^2} \\ &= \frac{1}{2}\mathrm{Im}\int_{-\infty}^\infty dr\, r\, e^{-\left(\left(\frac{r^2}{R_n^2}\right)-i\frac{qr}{\hbar c}\right)}.\end{aligned} \tag{5.25}$$

Symmetrising this integral will be shown shortly to be necessary in order to simplify the contour integration that is to follow. The exponent of (5.25) is rewritten as the square

$$\frac{r^2}{R_N^2} - i\frac{qr}{\hbar c} = \left(\frac{r}{R_N} - i\frac{qR_N}{2\hbar c}\right)^2 + \left(\frac{qR_N}{2\hbar c}\right)^2 \tag{5.26}$$

and using the substitution of variable $z = \left(\frac{r}{R_N}\right) - i\left(\frac{qR_N}{2\hbar c}\right)$, the integral of (5.25) can be separated into the sum of two simpler integrals:

$$\begin{aligned}\int_0^\infty \mathrm{d}r\, r\, e^{-\left(\left(\frac{r^2}{R_N^2}\right)-i\frac{qr}{\hbar c}\right)} = R_N^2\, e^{-\left(\frac{qR_N}{2\hbar c}\right)^2} \Bigg(&\int_{-\infty-i\frac{qR_N}{2\hbar c}}^{\infty-i\frac{qR_N}{2\hbar c}} \mathrm{d}z\, z\, e^{-z^2} \\ &+ i\left(\frac{qR_N}{2\hbar c}\right) \int_{-\infty-i\frac{qR_N}{2\hbar c}}^{\infty-i\frac{qR_N}{2\hbar c}} \mathrm{d}z\; e^{-z^2}\Bigg).\end{aligned} \tag{5.27}$$

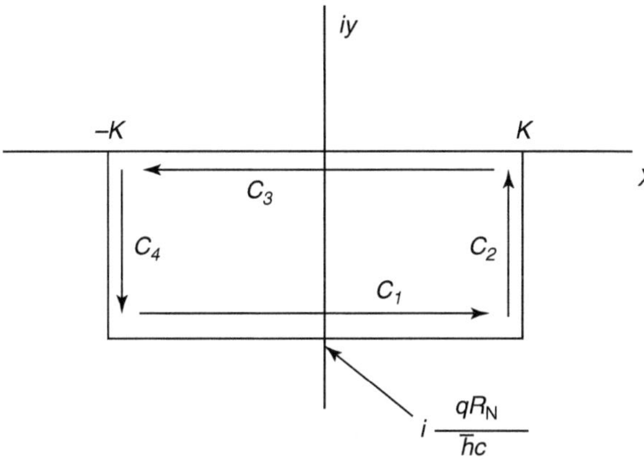

Fig. 5.2 Contour for solving for integrals of Eq. (5.27)

Both integrals are solved by using the Cauchy–Goursat integral theorem, given in Appendix A.3, and which states that the contour integral $\oint_C \mathrm{d}z f(z) = 0$, provided that the integrand is holomorphic (analytic) within a region containing the closed contour C. The contour to be used for both integrals is shown in Fig. 5.2, $C = C_1 \cup C_2 \cup C_3 \cup C_4$, and where $z = x + iy$.

We solve the second integral of (5.27) first by writing it in the form of

$$\int_{-\infty - i\frac{qR_N}{2\hbar c}}^{\infty - i\frac{qR_N}{2\hbar c}} dz\, e^{-z^2} = \lim_{K \to \infty} \int_{-K - i\frac{qR_N}{2\hbar c}}^{K - i\frac{qR_N}{2\hbar c}} \mathrm{d}z\, e^{-z^2}. \tag{5.28}$$

Then, as the function e^{-z^2} is holomorphic for this contour, the Cauchy–Goursat theorem gives

$$\begin{aligned} \int_{-K - i\frac{qR_N}{2\hbar c}}^{K - i\frac{qR_N}{2\hbar c}} \mathrm{d}z\, e^{-z^2} &= \int_{C_1} \mathrm{d}z\, e^{-z^2} \\ &= -\int_{C_2} \mathrm{d}z\, e^{-z^2} - \int_{C_3} \mathrm{d}z\, e^{-z^2} - \int_{C_4} \mathrm{d}z\, e^{-z^2}. \end{aligned} \tag{5.29}$$

The integration over the contour C_2 is

$$\begin{aligned} \int_{C_2} \mathrm{d}z\, e^{-z^2} &= \int_{K - i\frac{qR_N}{2\hbar c}}^{K} \mathrm{d}z\, e^{-z^2} = i \int_{0 - i\frac{qR_N}{2\hbar c}}^{0} \mathrm{d}y\, e^{-(K+iy)^2} \\ &= 0 \quad \text{as} \quad K \to \infty. \end{aligned} \tag{5.30}$$

The integration over C_3 is

$$\int_{C_3} \mathrm{d}z\, e^{-z^2} = \int_{K}^{-K} \mathrm{d}x\; e^{-x^2}$$
$$= -\sqrt{\pi} \quad \text{as} \quad K \to \infty \tag{5.31}$$

and that over C_4 is

$$\int_{C_4} \mathrm{d}z\, e^{-z^2} = \int_{-K}^{-K-i\frac{qR_\mathrm{N}}{2\hbar c}} \mathrm{d}z\; e^{-z^2}$$
$$= 0 \quad \text{as} \quad K \to \infty. \tag{5.32}$$

Combining these results gives

$$\int_{-\infty-i\frac{qR_\mathrm{N}}{2\hbar c}}^{\infty-i\frac{qR_\mathrm{N}}{2\hbar c}} \mathrm{d}z\; e^{-z^2} = \sqrt{\pi}. \tag{5.33}$$

The other integral of (5.27) is solved in the same manner:

$$\int_{-K-i\frac{qR_\mathrm{N}}{2\hbar c}}^{K-i\frac{qR_\mathrm{N}}{2\hbar c}} \mathrm{d}z\; z\, e^{-z^2} = \int_{C_1} \mathrm{d}z\, z\, e^{-z^2}$$
$$= -\int_{C_2} \mathrm{d}z\, z\, e^{-z^2} - \int_{C_3} \mathrm{d}z\, z\, e^{-z^2} - \int_{C_4} \mathrm{d}z\, z\, e^{-z^2}. \tag{5.34}$$

As before, the integrals over C_2 and C_4 will go to zero as $K \to \infty$. The remaining integral (that over C_3) is solved by using the substitution of variable $s = x^2$ which yields $\mathrm{d}s = 2x\,\mathrm{d}x$. Thus,

$$\int_{C_3} \mathrm{d}z\, z\, e^{-z^2} = \int_{K}^{-K} \mathrm{d}x\; x\, e^{-x^2} = \frac{1}{2}\int_{K}^{-K} \mathrm{d}s\; e^{-s^2}$$
$$= 0 \quad \text{as} \quad K \to \infty. \tag{5.35}$$

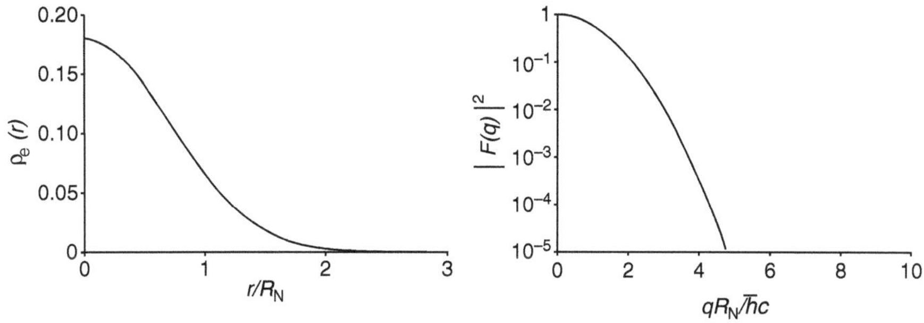

Fig. 5.3 Gaussian charge spatial distribution and corresponding squared modulus of the form factor

Hence,

$$\int_{-K-i\frac{qR_N}{2\hbar c}}^{K-i\frac{qR_N}{2\hbar c}} \mathrm{d}z\; z\, e^{-z^2} = 0. \tag{5.36}$$

Returning the results of (5.33) and (5.36) to (5.27), we have

$$\int_0^\infty \mathrm{d}r\, r\, e^{-\left(\left(\frac{r^2}{R_N^2}\right)-i\frac{qr}{\hbar c}\right)} = i\frac{\sqrt{\pi}}{2}\frac{qR_N^3}{\hbar c}e^{-\left(\frac{qR_N}{2\hbar c}\right)^2} \tag{5.37}$$

and, from (5.25),

$$\int_0^\infty \mathrm{d}r\, r \sin\left(\frac{qr}{\hbar c}\right) e^{-\left(\frac{r}{R_N}\right)^2} = \frac{\sqrt{\pi}}{4}\frac{qR_N^3}{\hbar c}e^{-\left(\frac{qR_N}{2\hbar c}\right)^2}. \tag{5.38}$$

Finally, the substitution of this result into (5.24) yields the form factor for a Gaussian charge spatial distribution:

$$F(q) = -e^{-\left(\frac{qR_N}{2\hbar c}\right)^2} \tag{5.39}$$

which is also Gaussian in q-space. Figure 5.3 shows the Gaussian spatial distribution and the modulus squared of the form factor. The smooth drop-off of $\rho_e(r)$ with r avoids diffraction effects in $|F(q)|^2$ and a similarly smooth and monotonic drop-off is observed in momentum transfer space.

5.2.2.5 Exponential Charge Distribution

Another smoothly decreasing distribution with radius is that of the exponential charge distribution and which is described by

$$\rho_e(r) = \rho_0 e^{-\left(\frac{r}{R_N}\right)}. \tag{5.40}$$

This distribution is an accurate representation of that of the proton. The normalisation constant is determined in the usual way:

$$4\pi \rho_0 \int_0^\infty \mathrm{d}r\, r^2\, e^{-\frac{r}{R_N}} = 1. \tag{5.41}$$

The integral is solved easily through two sequential integrations by parts to yield

$$\rho_0 = \frac{1}{8\pi R_N^3} \tag{5.42}$$

and the normalised exponential spatial charge distribution is thus

$$\rho_e(r) = \frac{e^{-\left(\frac{r}{R_N}\right)}}{8\pi R_N^3}. \tag{5.43}$$

Substituting this result into the form factor expression of (5.8) gives

$$F(q) = -\frac{1}{2R_N^3}\left(\frac{\hbar c}{q}\right)\int_0^\infty \mathrm{d}r\, r \sin\left(\frac{qr}{\hbar c}\right) e^{-\left(\frac{r}{R_N}\right)}. \tag{5.44}$$

The integral of (5.44) is solved be rewriting it as

$$\begin{aligned}\int_0^\infty \mathrm{d}r\, r \sin\left(\frac{qr}{\hbar c}\right) e^{-\left(\frac{r}{R_N}\right)} &= \mathrm{Im}\int_0^\infty \mathrm{d}r\, r\, e^{-\left(\left(\frac{1}{R_N}\right)-i\left(\frac{q}{\hbar c}\right)\right)r} \\ &= \mathrm{Im}\int_0^\infty \mathrm{d}r\, r\, e^{-wr}\end{aligned} \tag{5.45}$$

where $w = \frac{1}{R_N} - i\left(\frac{q}{\hbar c}\right)$. Using the substitution of variable $x = wr$ and following a subsequent integration by parts, the integration is

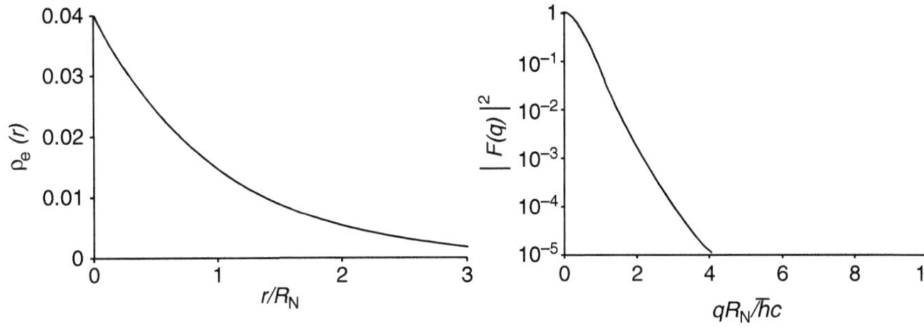

Fig. 5.4 Exponential charge spatial distribution and corresponding modulus squared of the form factor

$$\int_0^\infty \mathrm{d}r\, r\, e^{-wr} = \frac{1}{w^2}. \tag{5.46}$$

Substituting this into (5.45) gives

$$\int_0^\infty \mathrm{d}r\, r \sin\left(\frac{qr}{\hbar c}\right) e^{-\left(\frac{r}{R_N}\right)} = 2R_N^2 \frac{\left(\frac{qR_N}{\hbar c}\right)}{\left(1+\left(\frac{qR_N}{\hbar c}\right)^2\right)^2}. \tag{5.47}$$

The resulting form factor for the exponential charge spatial distribution is

$$F(q) = -\frac{1}{\left(1+\left(\frac{qR_N}{\hbar c}\right)^2\right)^2}. \tag{5.48}$$

Figure 5.4 shows the exponential spatial distribution and the modulus squared of the form factor. As with the Gaussian distribution, the smooth drop-off of the exponential distribution yields a smooth drop-off in momentum transfer by $|F(q)|^2$.

5.2.3 Summary of Nuclear Form Factors

It should be noted that the ordinates in all of the plots of $|F(q)|^2$ are logarithmic. This indicates, from (5.6), that the Coulomb scattering cross section for an extended target decreases more rapidly with momentum transfer/scattering angle than it does for a point target. As the abscissa of these plots is the dimensionless quantity, $\left(\frac{qR_N}{\hbar c}\right) = \left(\frac{2R_N}{\lambda\!\!\!^-}\right)\sin\left(\frac{\theta}{2}\right)$, small values of $|F(q)|^2$ will correspond to forward scattering angles and reduced de Broglie wavelengths relative to scattering centre dimensions (i.e. high momentum transfer). Another target-dependent parameter to be specified is R_N, which is the radius reflective of nuclear dimensions to be used in the evaluation of the form factor.

In addition to the analytical forms shown above, the spatial distribution of the protons within the nucleus can be described by the Woods–Saxon, or Fermi, distribution:

$$\rho_e(r) = \frac{\rho_0}{1 + e^{k(r-R_N)}} \tag{5.49}$$

where $k = 1.832\ \mathrm{fm}^{-1}$. The form factor for this distribution is a complicated expression given in terms of poly-logarithms of various orders (McParland 2010) and must be evaluated numerically. Another model of the nuclear charge distribution was presented by Helm (1956) and which treats the nuclear charge distribution as the convolution of two uniform distributions (5.11) or

$$\rho_e(r) = Z \int d^3\mathbf{r}'\, \rho_1(\mathbf{r}')\, \rho_2(\mathbf{r} - \mathbf{r}') \tag{5.50}$$

where the distributions are over two spheres of radii R_1 and R_2.

5.3 Screened Charges

5.3.1 *Introduction*

The previous consideration of a finite scattering centre introduced only one component of a realistic model of the atom to be used in the calculation of the elastic Coulomb scattering of a particle with high momentum and which has penetrated the atomic electrons surrounding the nucleus. The effect of 'screening', or the reduction of the effective electric charge of the nucleus by this electron cloud, as perceived by the projectile, is now to be reviewed.

In the discussions to follow, we will always consider the target atom to be both neutral (i.e. of atomic number Z and with Z bound electrons) and in its ground state. In this case, the effective interaction energy sensed by the projectile of charge ze incident to an atomic target with a nuclear charge Ze and an electron charge Ze is given by

$$U(r) = ze\,(V_{\mathrm{nuc}}(r) + V_{\mathrm{elec}}(r)) \tag{5.51}$$

where $V_{\mathrm{nuc}}(r)$ and $V_{\mathrm{elec}}(r)$ are the Coulomb potentials associated with the atomic nucleus and the distributed atomic electrons, respectively. The nuclear potential is given by the distributed-charge result of (5.2). For an electron density $\rho_e(r)$ of atomic electrons per unit volume, we have the non-unity normalisation:

$$\int d^3\mathbf{r}\,\rho_e(r) = 4\pi \int_0^\infty dr\, r^2 \rho_e(r) \\ = Z. \tag{5.52}$$

The electronic Coulomb potential can thus be written as

$$V_{\text{elec}}(r) = -4\pi e \left(\frac{1}{r} \int_0^r dr'\, r'^2 \rho_e(r') + \int_r^\infty dr'\, r' \rho_e(r') \right). \tag{5.53}$$

In order to focus this discussion on the screening by atomic electrons, we will treat the atomic nucleus as a point charge, i.e.

$$V_{\text{nuc}}(r) = \frac{Ze}{4\pi\varepsilon_0 r}. \tag{5.54}$$

The combination of a finite nuclear charge distribution and screening is discussed in the next chapter.

Accounting for the effects of atomic electron screening, the effective Coulomb potential affecting the projectile is, in its most general form,

$$U(r) = \frac{\alpha\hbar c\, zZ}{r} \Phi(r) \tag{5.55}$$

where the screening function $\Phi(r)$ has the property $\Phi(r) \in [0, 1]$ and the limiting properties of $\Phi(r) \to 1$ as $r \to 0$ and $\Phi(r) \to 0$ as $r \to \infty$. The latter properties reflect the two cases of the projectile having penetrated the screening electrons and experiencing the nuclear Coulomb field alone or of being at an extended distance from the atom and feeling a zero Coulomb potential. Recall that these conditions are implicit to the target being a neutral atom. Hence, the effective nuclear charge beyond screening that is seen by the projectile is $Ze\Phi(r)$, i.e. it has an explicit dependence upon distance from the nucleus.

In this section, we examine two approaches of accounting for the effects of atomic electron screening upon the elastic Coulomb scattering cross section for atomic targets:

- The first assumes that the atomic electrons provide a continuous screen of the nuclear electrostatic potential so that the effective Coulomb potential can be modelled by the analytical form of the Wentzel type of potential (or, similarly, that of Debye and Hückel), which is $\propto \frac{e^{-\kappa r}}{r}$, or by a numerical result derived from the Thomas–Fermi statistical model of the atom. Both of these results yield expressions for the Coulomb differential cross section, which provide a finite screening angle limiting the cross section to a finite value as the scattering angle $\theta \to 0$.

- The second calculation instead treats the atomic electrons as being discrete electric charges and evaluates their effect upon the scattering by using quantum perturbation theory from a net Coulomb potential given by the sum of that of a point nucleus and the individual contributions of the Z electrons. The result is an elastic form factor expression which also maintains a finite result for the Coulomb elastic scatter cross section as $\theta \to 0$ and replicates that of the continuous-screening calculation.

Using our results derived for continuous screening, we examine the various conditions of screening and the effects that they have upon the angular distribution of the screened Coulomb elastic scatter differential cross section.

5.3.2 Continuous Screening Part I: Exponential Screening

The derivation of the unscreened cross section given by (4.78) began with the calculation of the scattering amplitude (to the first order of the Born expansion) for an exponential modification of the Coulomb potential between a projectile of charge ze and a target of charge Ze, such that the interaction energy is

$$U(r) = \frac{\alpha\hbar c\, zZ}{r} e^{-\kappa r}. \tag{5.56}$$

Using this as an approximate model of atomic electron screening, from (5.55), the screening function is

$$\Phi(r) = e^{-\kappa r}. \tag{5.57}$$

In the original evaluation of the unscreened cross section, κ was treated as a calculation tool in the evaluation of the scattering amplitude and then allowed to go to zero, following the integration of (4.76). Here, we shall instead treat $e^{-\kappa r}$ as a continuous-screening term which approximates the moderation of the nuclear Coulomb potential by the surrounding atomic electrons and retain it within the calculation as a result. Thus, the spatial variation of the effect of screening is described by a single-atomic electron screening parameter, κ. Equation (4.76) is rewritten as

$$\int_0^\infty \mathrm{d}r'\, r'\, e^{-\kappa r'} j_0\left(\frac{qr'}{\hbar c}\right) = \frac{1}{\kappa^2 + \left(\frac{q}{\hbar c}\right)^2} \tag{5.58}$$

and the scattering amplitude (to the first order of the Born expansion) is

$$f_{(1)}(q) = -\left(\frac{2m\,\alpha z Z}{\hbar c}\right)\left(\frac{1}{\kappa^2 + \left(\frac{q}{\hbar c}\right)^2}\right). \tag{5.59}$$

The screened differential cross section in solid angle, which we recall is proportional to the modulus squared of the scattering amplitude, can be written as, using the definitions of the Bohr collision diameter $\rho_{\mathrm{B}}(E) = \frac{\alpha\hbar c\, zZ}{E}$, the reduced de Broglie wavelength $\lambda\!\!\!^{-} = \frac{\hbar c}{p}$ and the relationship between the elastic momentum transfer and scattering angle, $q = 2p\sin\frac{\theta}{2}$:

$$\frac{\mathrm{d}\sigma_{\mathrm{Scr}}}{\mathrm{d}\Omega} = \left(\frac{\rho_{\mathrm{B}}(E)}{4}\right)^2 \frac{1}{\left(\left(\frac{\lambda\!\!\!^{-}\kappa}{2}\right)^2 + \sin^2\frac{\theta}{2}\right)^2}. \tag{5.60}$$

As our interest of the effects of atomic electron screening will be in the divergence of the cross section at small scattering angles, we can invoke the small-angle approximation of $\sin^2\frac{\theta}{2} \approx \frac{\theta^2}{4}$ to give[4]

$$\begin{aligned}\frac{\mathrm{d}\sigma_{\mathrm{Scr}}}{\mathrm{d}\Omega} &\approx \frac{\rho_{\mathrm{B}}^2(E)}{\left((\lambda\!\!\!^{-}\kappa)^2 + \theta^2\right)^2} \\ &\equiv \left(\frac{\rho_{\mathrm{B}}(E)}{\chi_0^2 + \theta^2}\right)^2 \quad \text{for} \quad \theta \to 0\end{aligned} \tag{5.61}$$

where the screening angle χ_0 is defined as the product of the atomic electron screening parameter and the de Broglie wavelength:

$$\chi_0 \equiv \kappa\lambda\!\!\!^{-}. \tag{5.62}$$

Clearly, the divergence of the differential cross section at $\theta = 0$ has now been avoided through the inclusion of screening and the resulting calculated cross section at $\theta = 0$ is finite and given by

$$\frac{\mathrm{d}\sigma_{\mathrm{Scr}}}{\mathrm{d}\Omega} = \frac{\rho_{\mathrm{B}}^2(E)}{\chi_0^4} \quad \text{for} \quad \theta = 0. \tag{5.63}$$

We can also see that the screened differential cross section of (5.60) can be related to the (unscreened) Rutherford cross section through

[4] For practical calculational purposes, this assumption is valid over a surprisingly long range. For example, $\sin\theta \approx \theta$ to within 10 % for $0^\circ < \theta < 45^\circ$.

$$\frac{d\sigma_{Scr}}{d\Omega} = \frac{d\sigma_{Ruth}}{d\Omega} \frac{1}{\left(1 + \left(\frac{\chi_0}{2\sin\frac{\theta}{2}}\right)^2\right)^2}. \tag{5.64}$$

We rewrite this in the small-angle approximation:

$$\frac{d\sigma_{Scr}}{d\Omega} = \frac{d\sigma_{Ruth}}{d\Omega} \frac{1}{\left(1 + \left(\frac{\chi_0}{\theta}\right)^2\right)^2}. \tag{5.65}$$

The multiplicative term in angle in both (5.64) and (5.65) is a screening term. For the conditions of where screening becomes negligible (i.e. $\chi_0 \to 0$),

$$\frac{d\sigma_{Scr}}{d\Omega} \to \frac{d\sigma_{Ruth}}{d\Omega}; \quad \chi_0 \to 0. \tag{5.66}$$

That is, the scattering cross section goes over to the Rutherford form. On the other hand, for significant screening (i.e. χ_0 large), we have

$$\frac{d\sigma_{Scr}}{d\Omega} < \frac{d\sigma_{Ruth}}{d\Omega}; \quad \chi_0 \quad \text{large}. \tag{5.67}$$

We can also interpret these two results in terms of the ratio of the screening angle to the scattering angle, $\frac{\chi_0}{\theta}$. Equation (5.66) corresponds to $\chi_0 < \theta$; at scattering angles far beyond the screening angle, the Rutherford cross section is valid. Equation (5.67) corresponds to $\chi_0 > \theta$; at scattering angles of the order of, or less than, the screening angle, the actual cross section is less than the Rutherford unscreened result. These characterisations will be revisited a number of times later, especially in our discussion of the Molière theory of multiple scattering in Chap. 6.

As (5.64) was determined using the first Born approximation, its validity is based upon the assumption of a non-appreciable perturbation of the incident wavefunction. This can be considered by integrating (5.64) over all solid angle in order to yield the total screened scattering cross section

$$\sigma = \left(\frac{\pi}{\kappa^2}\right) \frac{\eta^2(E)}{1 + \left(\frac{\chi_0}{2}\right)^2} \tag{5.68}$$

where the ratio of the Bohr collision diameter to the reduced de Broglie wavelength has been written as the dimensionless parameter:

$$\eta(E) \equiv \frac{\rho_B(E)}{\lambda\!\!\!^-} = \frac{\alpha\hbar c\, zZ}{E\left(\frac{\hbar c}{p}\right)} = \frac{2\alpha\, zZ}{\beta}. \tag{5.69}$$

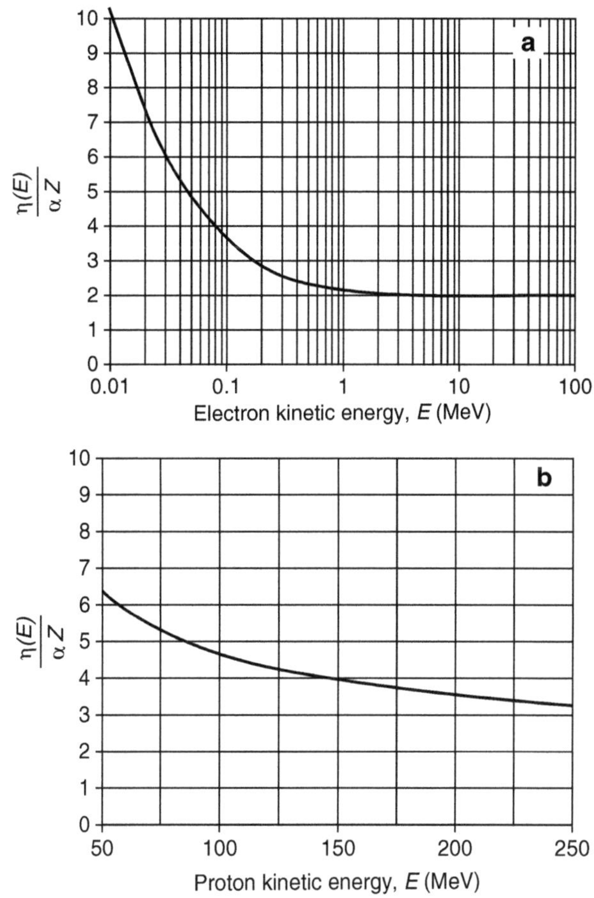

Fig. 5.5 The ratio $\frac{\eta(E)}{\alpha Z}$ from Eq. (5.69) as a function of kinetic energy for (**a**) electrons and (**b**) protons

Clearly, $\eta(E)$ decreases with increasing kinetic energy to approach an asymptotic value of $2\alpha zZ$. This ratio is a kinematic quantity that is shown as a function of kinetic energy E for electrons and protons in Fig. 5.5.

The factor of $\frac{\pi}{\kappa^2}$ of (5.68) can be considered as a geometric cross section of a sphere of radius $\frac{1}{\kappa}$ or, somewhat crudely, the cross-sectional area of the screened field. If $\eta^2(E) << 1$, which corresponds to a high β and a low zZ product (i.e. a weak Coulomb potential) and is analogous to (2.115), then the screened total cross section will be much smaller than this geometric cross section. That is, for that condition, only a small fraction of the projectiles will be scattered and the assumption of a weak Coulomb interaction is validated.

5.3.3 Screening Conditions

In this section, we consider the various conditions of screening, in particular those under which the unscreened (Rutherford) differential cross section of (4.23) can be considered to hold. This will be based upon the analysis provided by Bohr (1948).

From (4.20), the scattering angle can be written in terms of the Bohr collision diameter and the impact parameter, b:

$$\theta = 2\tan^{-1}\left(\frac{\rho_{\mathrm{B}}(E)}{2b}\right). \tag{5.70}$$

Following the discussion of (5.66) and (5.67), we would expect the Rutherford differential cross section to be accurate for scattering angles greater than that given by (5.70) for the case of the impact parameter being of the same dimension as that of the reciprocal of the atomic electron screening parameter, $b = \frac{1}{\kappa}$. We define this angle as

$$\begin{aligned}\theta_{(\mathrm{a})} &\equiv 2\tan^{-1}\left(\frac{\kappa\rho_{\mathrm{B}}(E)}{2}\right)\\ &= 2\tan^{-1}\left(\frac{\zeta(E)}{2}\right)\end{aligned} \tag{5.71}$$

where the dimensionless parameter

$$\begin{aligned}\zeta(E) &\equiv \kappa\,\rho_{\mathrm{B}}(E)\\ &= \frac{\kappa\,\alpha\hbar c\, zZ}{E}\end{aligned} \tag{5.72}$$

has been defined. The two dimensionless parameters of $\eta(E)$ and $\zeta(E)$ will prove useful in our imminent investigations of various screening conditions.

The condition of $\zeta(E) << 1$ (i.e. small κ) reflects the cases of either very limited screening or for which screening is restricted to very forward angles (coming from $\chi_0 = \kappa\lambda\!\!\!^{-}$). In this case, the small-angle approximation can be applied to (5.71) and leads to this parameter being equal to the scattering angle beyond which the Rutherford cross section is valid:

$$\theta_{(\mathrm{a})} \approx \zeta(E). \tag{5.73}$$

The consequence of this condition is that the Rutherford differential cross section will be valid over most of the range of scattering angles, as indicated previously by (5.66).

From the definitions of (5.69) and (5.72), we can write the screening angle as

$$\chi_0 = \frac{\zeta(E)}{\eta(E)}. \tag{5.74}$$

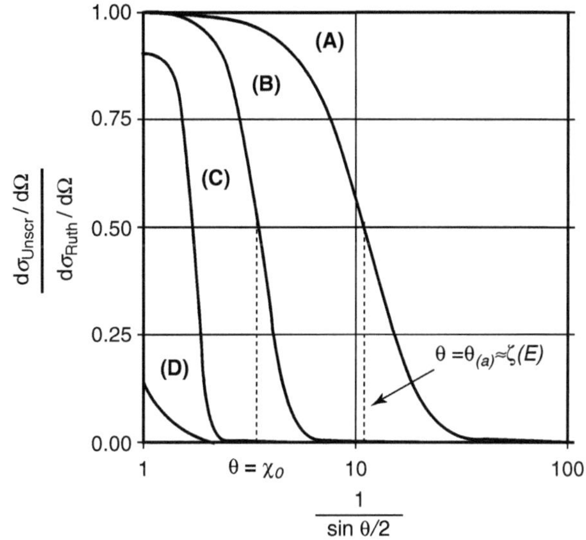

Fig. 5.6 Conceptual curves of the ratio of the exponentially screened to unscreened elastic Coulomb cross sections (See text for details. Figure follows from one provided by Bohr (1948))

Substituting this expression into (5.64) and rearranging the result to yield the ratio of the differential cross sections results in

$$\frac{\frac{d\sigma_{Scr}}{d\Omega}}{\frac{d\sigma_{Ruth}}{d\Omega}} = \frac{1}{\left(1 + \left(\frac{\left(\frac{\zeta(E)}{\eta(E)}\right)}{2\sin\frac{\theta}{2}}\right)^2\right)^2}. \tag{5.75}$$

We now examine the conditions of atomic electron screening, aided by the definitions of $\eta(E)$ and $\zeta(E)$, and their effects upon the scattering angular distribution.

Figure 5.6 shows the ratio of the screened to the Rutherford elastic Coulomb cross sections, (5.75), as a function of $\frac{1}{\left(\sin\frac{\theta}{2}\right)}$ for the small and constant value of $\zeta(E)$ and for various values of $\eta(E)$. Note that the abscissa is logarithmic and that decreasing scattering angle corresponds to movement to the right along the axis.

Curve (A): $\eta(E) \gg 1$ In the small-angle limit, we write the ratio of cross sections as

$$\frac{\frac{d\sigma_{Scr}}{d\Omega}}{\frac{d\sigma_{Ruth}}{d\Omega}} \approx \frac{1}{\left(1 + \frac{1}{\eta^2(E)}\left(\frac{\theta_{(a)}}{\theta}\right)^2\right)^2}. \tag{5.76}$$

For large θ, this ratio is approximately unity as shown in Fig. 5.6. In the region of $\theta_{(a)}$, the ratio decreases to

$$\frac{\frac{d\sigma_{\mathrm{Scr}}}{d\Omega}}{\frac{d\sigma_{\mathrm{Ruth}}}{d\Omega}} \approx \frac{1}{\left(1+\frac{1}{\eta^2(E)}\right)^2}; \quad \theta = \theta_{(a)} \approx \zeta(E). \tag{5.77}$$

And as the scattering angle is reduced,

$$\frac{\frac{d\sigma_{\mathrm{Scr}}}{d\Omega}}{\frac{d\sigma_{\mathrm{Ruth}}}{d\Omega}} \to 0 \quad \text{as} \quad \theta \to 0. \tag{5.78}$$

Curve (B): *$\eta(E) \ll 1$; $\eta(E) \approx \zeta^{1/2}(E)$* For the combination of these conditions, (5.75) becomes, in the small-angle limit,

$$\frac{\frac{d\sigma_{\mathrm{Scr}}}{d\Omega}}{\frac{d\sigma_{\mathrm{Ruth}}}{d\Omega}} \approx \frac{1}{\left(1+\left(\frac{\theta_{(a)}}{\theta^2}\right)\right)^2}. \tag{5.79}$$

Again, for large scattering angle, the ratio is equal to unity. Unlike Curve (A), this curve will begin to decrease at a lesser angle. While Curve (A) decreases in the region of $\theta_{(a)}$, Curve (B) drops off to zero in the region of $\sqrt{\theta_{(a)}} = \chi_0$.

Curve (C): *$\eta(E) \ll 1$; $\eta(E) \approx \zeta(E)$* For the combination of conditions, (5.75) is

$$\frac{\frac{d\sigma_{\mathrm{Scr}}}{d\Omega}}{\frac{d\sigma_{\mathrm{Ruth}}}{d\Omega}} \approx \frac{1}{\left(1+\frac{1}{\theta^2}\right)^2}. \tag{5.80}$$

This represents the region of transition where the deviation of the screened cross section from the Rutherford cross section becomes increasingly significant for large θ.

Curve (D): *$\eta(E) \ll 1$; $\eta(E) \ll \zeta(E)$* For this combination of conditions, one approaches a spherically uniform distribution.

In summary, for those scattering problems in which the condition $\zeta(E) << 1$ is valid, the scattering cross section agrees with the Rutherford unscreened form over a large range of scattering angles. However, once the condition $\zeta(E) \approx 1$ is reached, the angular distribution will differ from the unscreened result for all values of $\eta(E)$. Eventually, the high probability of forward-directed scatters will be sufficiently suppressed such that the differential cross section with solid angle approaches a constant value:

$$\frac{d\sigma}{d\Omega} \approx \frac{\sigma}{4\pi} \tag{5.81}$$

where σ is the total cross section. This latter case of excessive atomic electron screening can be examined by using the conditions of the validity of the classical

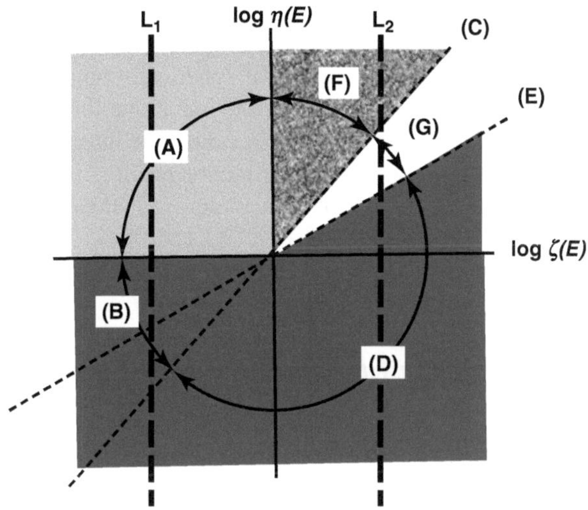

Fig. 5.7 log $\zeta(E)$ versus log $\eta(E)$ plot representing the various conditions of screened Coulomb scattering (After Bohr (1948). Refer to text for details)

mechanical description of Coulomb scatter discussed earlier. It follows from (4.67) that the selection of a diaphragm diameter d in order to restrict the impact parameter b to be less than $\frac{1}{\kappa}$ will make impossible the detection of scatters at an angle of less than χ_0. For excessive screening, large-angle scatters play an increasingly significant role, and an analysis based upon classical mechanics is acceptable for a small screening angle, $\chi_0 << 1$ or $\eta(E) >> \zeta(E)$. In the case of $\lambda\!\!\!^{-} > \frac{1}{\kappa}$, the requirement of $\eta(E) << 1$ ceases to be a necessary condition for the above wave diffraction-based analysis. From the total cross section expression (5.68), we can see that $\sigma < \frac{\pi}{\kappa^2}$ if $\eta(E) < \frac{\kappa\lambda\!\!\!^{-}}{2}$ or, equivalently, $\eta(E) << \sqrt{\zeta(E)}$ which, for $\zeta(E) >> 1$, is a somewhat less restrictive condition than that of $\eta(E) << 1$. Thus, for the condition of $\sqrt{\zeta(E)} < \eta(E) < \zeta(E)$, an analysis of the scattering problem based upon classical mechanics would no longer be accurate and quantum-mechanical considerations must be had.

The above discussions are summarised by the schematic of Fig. 5.7, for which every point in the graph represents a combination of values of $(\eta(E), \zeta(E))$. The components of this graph are described now:

Line (C): $\eta(E) = \zeta(E)$ This represents the condition of $\chi_0 = 1$.
Line (E): $\eta(E) = \zeta^{1/2}(E))$ This represents the condition of $\chi_0 = \sqrt{\theta_{(a)}}$.
Region (A): $\eta(E) > 1; \zeta(E) < 1$
Region (B): $\zeta^{1/2}(E) < \eta(E) < 1; \zeta(E) < 1$
Region (D): $\eta(E) < \zeta^{1/2}(E)$
Region (F): $1 < \zeta(E) < \eta(E) < 1$
Region (G): $\zeta^{1/2}(E) < \eta(E) < \zeta(E)$.

The vertical line L_1, which is for a constant small $\zeta(E)$, transits the regions (A) and (B). Its intersections with lines (C) and (E) corresponds to the curves

(C) and (B), respectively, of Fig. 5.7. The other vertical line L_2, which is for a constant large $\zeta(E)$, transits the regions (D), (G) and (F). The section below line (E) represents the conditions of spherically uniform scattering (corresponding to curve (D) of Fig. 5.6) and the section above line (C) represents the region where classical mechanical analysis is valid. The intermediate region (G) is where such analysis fails and where the diffraction result of region (D) is invalid and represents the conditions for which a quantum-mechanical description of the screening is required.

5.3.4 Atomic Electron Screening Parameter: Part I

In our discussion of atomic electron screening of the nuclear Coulomb field, we have described the screening using a single parameter, κ, on the basis of an assumption of a continuous exponential screening function, $\Phi(r) = e^{-\kappa r}$. We now consider the means of evaluating numerically this parameter. These will be based upon atomic models of the spatial distribution of the atomic electrons and the effective atomic Coulomb potential resulting from the electrons' screening of the nuclear Coulomb potential.

In principle, this spatial distribution of electrons can be derived from a determination of the wavefunction of a multiple-electron atomic system of Z electrons which can be described by the time-independent Schrödinger equation:

$$\left(-\frac{(\hbar c)^2}{2m_\mathrm{e}} \sum_{i=1}^{Z} \nabla_i^2 + \frac{1}{2} \sum_{\substack{i=1 \\ i \neq j}}^{Z} v(\mathbf{r}_i, \mathbf{r}_j) \right) \psi(\mathbf{r}_1 \ldots \mathbf{r}_Z) = E_i \psi(\mathbf{r}_1 \ldots \mathbf{r}_Z) \tag{5.82}$$

or the Dirac equation for those cases of the electrons having relativistic orbital velocities. Clearly, the solution of (5.82) will be extraordinarily complicated and impractical to obtain directly for the atoms of interest to medical radiation physics (with the exception of hydrogen). Another approach to deriving this spatial distribution of atomic electrons is to instead focus explicitly upon the electron density which, for a system of Z electrons, is given by

$$\rho_\mathrm{e}(\mathbf{r}) = Z \int \mathrm{d}^3\mathbf{r}_2 \ldots \mathrm{d}^3\mathbf{r}_Z \, \psi^*(\mathbf{r}, \mathbf{r}_2 \ldots \mathbf{r}_Z) \, \psi(\mathbf{r}, \mathbf{r}_2 \ldots \mathbf{r}_Z) \tag{5.83}$$

which is equal to the probability of finding any electron at the position $\mathbf{r}$. Solving (5.82) using (5.83) in tandem could be done iteratively but this would still remain a complicated and involved solution. We briefly describe such approaches later, but do not employ them.

On the other hand, a statistical approximation of the atomic electron energy/momentum distribution can provide a model of the spatial distribution of the charge due to the atomic electrons, from which the degree of screening of the nuclear Coulomb charge can be estimated. We now consider the semi-classical atomic model of Thomas and Fermi and use this to derive an expression for the screened potential observed by the projectile, from which values of κ will be estimated.

5.3.5 Thomas–Fermi Model of the Multi-electron Atom

Rather than attempting an exact solution of the spatial distribution of atomic electrons, a statistical solution for $\rho_e(r)$ can be obtained for the condition of $Z >> 1$. This treats the atom as a locally homogeneous electron gas that is bound by the electrostatic field of the nucleus. A semi-classical statistical method can then be employed to yield a solution for this distribution from which a screening potential can be derived. This is equivalent to stating that an expression of $\rho_e(r)$ is derived and used to define a 'radius' of the atom from which we can establish a description of the atomic electron screening parameter, κ.

Such a statistical atomic model was derived independently by Thomas (1927) and by Fermi (1927). In our calculations, the target for the Coulomb scattering of a projectile is considered to be a neutral atom in its ground state and the atomic electrons assumed to be subject to a central Coulomb potential and to the interactive Coulomb potentials with each other.[5] As before, the co-ordinate system of the calculation is one for which the nucleus is centred at the origin. The intrinsic spins of the electrons and the Pauli exclusion principle (beyond the assumption that no more than two electrons can reside in each phase-space unit volume) are not considered in this derivation. In order to invoke this statistical model, it is also necessary to assume that potential is static and that the spatial distance over which it varies is small relative to the de Broglie wavelength of the electrons,

$$\left|\frac{\mathrm{d}\lambda\!\!\!^{-}}{\mathrm{d}x}\right| << \frac{1}{2\pi}. \tag{5.84}$$

This inequality leads to the necessary homogeneity of the electron distribution. In the Thomas–Fermi model of a neutral atom in its ground state, all atomic electron levels up to the Fermi energy will be occupied. Thus, considering the phase-space volume, we can write the number of atomic electrons in the form

$$Z = \frac{4\pi}{3} p_F^3 L^3 \frac{2}{(2\pi\hbar c)^3} = \frac{\left(\frac{p_F L}{\hbar c}\right)^3}{3\pi^2} \tag{5.85}$$

[5] Ions will be considered inter alia in the derivation, but are not a focus as they are not relevant targets in medical radiation physics considerations.

where p_F is the Fermi momentum, L^3 is the volume of phase space and the additional multiplicative factor of 2 accounts for electron spin degeneracy. As the mean electron density is

$$\rho_e = \frac{Z}{L^3}, \tag{5.86}$$

then the Fermi momentum is related to the electron density through

$$p_F = \hbar c \left(3\pi^2 \rho_e\right)^{\frac{1}{3}}. \tag{5.87}$$

Let us further assume that the atomic electrons move as classical particles subject to a common potential so that the total energy of the most energetic electron is given by

$$E_{max} = \frac{p_F^2}{2m_e} + U(r). \tag{5.88}$$

E_{max} cannot have a spatial dependence as, otherwise, electrons would transit to that point in space where E_{max} is a minimum. For a bound system, $E_{max} \leq 0$. By substituting (5.87) into (5.88) and solving for the electron charge density distribution $\rho_e(r)$, we obtain an expression for the electron density in terms of the maximum energy and the potential energy:

$$\rho_e(r) = \frac{1}{3\pi^2}\left(\frac{2m_e}{(\hbar c)^2}\right)^{\frac{3}{2}} (E_{max} - U(r))^{\frac{3}{2}}. \tag{5.89}$$

As the inequality $U(r) > E_{max}$ will be forbidden classically, it will be necessary for us to define $\rho_e(r) = 0$ within that region. Should this condition not be imposed then the Fermi momentum would become imaginary and represent a non-physical state. The effective electrostatic potential can be calculated by the introduction of a Lagrange multiplier and the subsequent minimisation of the total energy of the system (see, e.g. Parr and Ghosh (1986)). Here, we instead relate the energies to the electrostatic potentials and assume spherical symmetry. We have the potential energy as $U(r) = -eV(r)$. Since

$$E_{max} = -eV_0 \tag{5.90}$$

and $E_{max} \leq 0$, then $V_0 \geq 0$ with the latter remains a constant to be determined. Substituting the relationship between potential energy and the potential and (5.90) into (5.89) yields an expression of the electron spatial density:

$$\begin{aligned}\rho_e(r) &= \frac{1}{3\pi^2}\left(\frac{2m_e e}{(\hbar c)^2}\right)^{\frac{3}{2}} (V_{eff}(r))^{\frac{3}{2}} \quad & V_{eff} \geq 0 \\ &= 0 & V_{eff} < 0\end{aligned} \tag{5.91}$$

where the effective electrostatic potential has been defined by

$$V_{\text{eff}}(r) = V(r) - V_0. \tag{5.92}$$

Recall that our calculation is for a neutral atom. It is assumed that a finite boundary at a radius r_0 such that and the potential at that point is $V(r_0) = 0$. As $V_{\text{eff}}(r_0) = 0$, then $V_0 = 0$ with the result that $V_{\text{eff}}(r) = V(r)$. Should the calculation be for an ionic target, then $V_0 > 0$.

While (5.91) relates the electron density to the effective potential, an additional relationship can be obtained. We first note that the sources creating $V(r)$ (or $V_{\text{eff}}(r)$) experienced by an atomic electron are the combined point Coulomb charges of the nucleus and the other $Z - 1$ electrons. As the spatial distribution of atomic electrons is assumed to be continuous, Poisson's equation can be used to provide an additional relationship between $\rho_e(r)$ and $V_{\text{eff}}(r)$:

$$\nabla^2 V_{\text{eff}}(r) = \frac{e\rho_e(r)}{\varepsilon_0} \tag{5.93}$$

where ε_0 is the permittivity of free space. Both (5.91) and (5.93) present separate relationships between the potential and the atomic electron charge distribution. Using spherical co-ordinates and eliminating the charge density $\rho_e(r)$ from both, we find that, for $V_{\text{eff}}(r) \geq 0$,

$$\frac{1}{r^2}\frac{\mathrm{d}^2}{\mathrm{d}r^2}(rV_{\text{eff}}(r)) = \frac{e}{3\pi^2\,\varepsilon_0}\left(\frac{2m_e}{(\hbar c)^2}\right)^{\frac{3}{2}}(eV_{\text{eff}}(r))^{\frac{3}{2}}. \tag{5.94}$$

We now specify the boundary condition and normalisation of the problem. As $r \to 0$, the Coulomb potential becomes solely that of the point nucleus or

$$\begin{aligned}\lim_{r \to 0} rV_{\text{eff}}(r) &= \frac{Ze}{4\pi\,\varepsilon_0}\\ &= \frac{\alpha\hbar c Z}{e}\end{aligned} \tag{5.95}$$

and, from the definition of the electron density, the normalisation condition is

$$\int \mathrm{d}^3\mathbf{r}\,\rho_e(r) = Z. \tag{5.96}$$

We next simplify (5.94) by now introducing the dimensionless variables x and $\chi(x)$. The variable x is the normalised radius, defined as

$$x \equiv \frac{r}{r_{\text{TF}}} \tag{5.97}$$

where r_{TF} is known as the Thomas–Fermi radius and is to be determined shortly. The potential is also normalised to what is referred to as the Thomas–Fermi atomic electron screening function:

$$\chi(x) \equiv \frac{V_{\mathrm{eff}}(r_{\mathrm{TF}}x)}{\left(\dfrac{Ze}{4\pi\,\varepsilon_o r_{\mathrm{TF}}x}\right)} = \frac{V_{\mathrm{eff}}(r_{\mathrm{TF}}x)}{\left(\dfrac{\alpha\hbar c\,Z}{e r_{\mathrm{TF}}x}\right)}. \tag{5.98}$$

These normalisations lead to an expression for the electric charge density function:

$$\rho_{\mathrm{e}}(r) = \frac{Z}{4\pi r_{\mathrm{TF}}^3}\left(\frac{\chi}{x}\right)^{\frac{3}{2}} \quad \chi(x) \geq 0$$
$$= 0 \qquad\qquad \chi(x) < 0. \tag{5.99}$$

Substituting (5.98) and (5.99) into (5.94) gives, after rearrangement, the differential equation

$$\frac{\mathrm{d}^2\chi(x)}{\mathrm{d}x^2} = \frac{e^2}{3\pi^2\,\varepsilon_0}\left(\frac{2r_{\mathrm{TF}}}{a_\infty}\right)^{\frac{3}{2}}\frac{\sqrt{Z}}{\alpha\hbar c}\frac{\chi(x)^{\frac{3}{2}}}{\sqrt{x}} = \frac{\chi(x)^{\frac{3}{2}}}{\sqrt{x}} \tag{5.100}$$

where the Thomas–Fermi radius has been defined:

$$r_{\mathrm{TF}} \equiv \frac{(3\pi)^{\frac{2}{3}}}{2^{\frac{7}{3}}}\frac{a_\infty}{Z^{\frac{1}{3}}} \approx 0.885\,\frac{a_\infty}{Z^{\frac{1}{3}}} \tag{5.101}$$

where $a_\infty = \frac{\hbar c}{\alpha m_{\mathrm{e}}}$ is the Bohr radius. Note that the Thomas–Fermi radius of (5.101) varies as $Z^{-\frac{1}{3}}$ which, from (5.99), results in the atomic electron density increasing with Z^2.

The result of (5.100) is known as the Thomas–Fermi equation and is valid only for a positively valued Thomas–Fermi function, $\chi(x) > 0$; (5.100) thus tells us that $\chi(x)$ is concave up. It follows from (5.95) (for $V_{\mathrm{eff}} < 0$) and (5.98) that

$$\frac{\mathrm{d}^2\chi(x)}{\mathrm{d}x^2} = 0 \quad \chi(x) < 0 \tag{5.102}$$

(5.96) specifies the boundary condition

$$\chi(0) = 1 \tag{5.103}$$

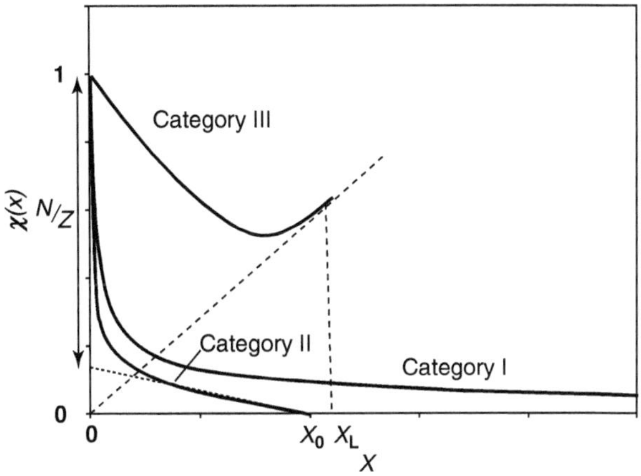

Fig. 5.8 The three categories of solutions to the Thomas–Fermi equation (See text for description. Redrawn from Bransden and Joachain (2003))

which provides a single constraint to $\chi(x)$. As such, multiple solutions to (5.100) will exist and which will differ from each other by the value of the initial slope, $\frac{d\chi(0)}{dx}$.

Before discussing these solutions, we note that the Thomas–Fermi function $\chi(x)$ can, at most, have only one zero in the interval $(0, +\infty)$, the position of which is defined as x_0 corresponding to the boundary of the (neutral) atomic system. Hence, (5.102) leads to

$$\chi(x) = \frac{d\chi(x_0)}{dx}(x - x_0) \tag{5.104}$$

where the derivative $\frac{d\chi(x_0)}{dx}$ must be both negative (as $\chi(x) > 0$ and $(x - x_0) < 0$) and a non-zero constant. If this latter condition were not met, then we would be led to the trivial solution of $\chi(x) = 0$.

The solutions to the Thomas–Fermi equation can be separated into three categories, recognising that $\chi(x)$ is concave up:

Category I. A single solution with the Fermi–Thomas function $\chi(x) \to 0$ as $x \to \infty$ (i.e. the solution is asymptotic with the abscissa).

Category II. The solutions are for the Fermi–Thomas function $\chi(x) = 0$ for $x \geq x_0$.

Category III. The solutions are for the Fermi–Thomas function $\chi(x) \neq 0$ and which diverge as $x \to \infty$.

These three categories are shown schematically in Fig. 5.8.

The physical natures of these categories of solutions are now explored. First, we consider Categories I and II for which $\chi(x)$ goes eventually to zero. Although our interest is in the neutral atom, let us consider an ion with N_e electrons for the purpose of discussion. The normalisation condition of (5.96) is now

$$\begin{aligned}
N_\mathrm{e} &= \int \mathrm{d}^3\mathbf{r}\,\rho_\mathrm{e}(\mathbf{r}) \\
&= Z \int_0^{x_0} \mathrm{d}x\,\sqrt{x}\,(\chi(x))^{\frac{3}{2}} \\
&= Z \int_0^{x_0} \mathrm{d}x\, x\,\frac{\mathrm{d}^2\chi(x)}{\mathrm{d}x^2} \\
&= Z\left(x\frac{\mathrm{d}\chi(x)}{\mathrm{d}x} - \chi(x)\right)\Bigg|_{x=0}^{x_0} \\
&= Z\left(x_0\frac{\mathrm{d}\chi(x_0)}{\mathrm{d}x} + 1\right)
\end{aligned} \tag{5.105}$$

where the boundary condition of (5.103) has been used. It follows from (5.106) that

$$x_0\frac{\mathrm{d}\chi(x_0)}{\mathrm{d}x} = \frac{N_\mathrm{e} - Z}{Z}. \tag{5.106}$$

For the case of a neutral atom ($N_\mathrm{e} = Z$), the derivative at x_0 is

$$\frac{\mathrm{d}\chi(x_0)}{\mathrm{d}x} = 0. \tag{5.107}$$

The facts that $\chi(x_0) = 0$ and that the derivative of the function is also equal to zero at that point lead to x_0 being at infinity, i.e. $\chi(x) \to 0$ as $x \to \infty$. Thus, the solution for the neutral atom is the unique solution of Category I above. As $\chi(x)$ is now extended to infinity, the neutral Thomas–Fermi atom will not have a boundary.

Consider, now, Category II of the solutions of $\chi(x)$ which vanish for $x \geq x_0$. Such solutions correspond to $N_\mathrm{e} \neq Z$ and, hence, are associated with ions with a finite radius given by $r_0 = x_0 r_\mathrm{TF}$. As $\frac{\mathrm{d}\chi(x_0)}{\mathrm{d}x} < 0$ for Category II solutions, (5.106) shows that $Z > N_\mathrm{e}$. In other words, these solutions correspond to positive ions. As $\chi(x_0) = 0$, then, from (5.99), the electric charge density will be zero for $x \geq x_0$. However, accounting for (5.90), (5.92) and (5.98) and recognising that, for an ion, $V_0 > 0$, it is seen that the potential energy $U(r)$ remains finite at $r_0 = x_0 r_\mathrm{TF}$.

The physical interpretations of the Category III solutions are more difficult to make as the electric charge density $\rho_\mathrm{e}(r)$ does not go to zero with increasing r. First consider the electric charge within a sphere of radius $r = x\,r_\mathrm{TF}$:

$$\begin{aligned}
Ze - e\int \mathrm{d}^3\mathbf{r}'\,\rho_\mathrm{e}(r') &= Ze - 4\pi e\int_0^r \mathrm{d}r'\,r'^2\,\rho_\mathrm{e}(r') \\
&= Ze - Ze\int_0^x \mathrm{d}x'\sqrt{x'}\chi(x') = Ze - Ze\int_0^x \mathrm{d}x'\,x'\,\frac{\mathrm{d}^2\chi(x')}{\mathrm{d}x'^2}
\end{aligned} \tag{5.108}$$

where (5.99) and (5.100) have been used. The integral is straightforward to solve by parts and by incorporating the boundary condition of $\chi(0) = 1$, we obtain the result:

$$Ze - e\int d^3\mathbf{r}'\,\rho_e(r') = Ze\left(\chi(x) - x\frac{d\chi(x)}{dx}\right). \tag{5.109}$$

At some radius $r_L = x_L r_{TF}$, the quantity within the parentheses of (5.109) will necessarily vanish:

$$\chi(x_L) - x_L\frac{d\chi(x_L)}{dx} = 0. \tag{5.110}$$

That is, the total charge within the sphere of radius r_L will be zero. From (5.110), it is seen that the tangent to the locus of $\chi(x)$ at x_L passes through the origin (as shown by Fig. 5.8). The solutions of Category III can thus be interpreted by noting that for $x \leq x_L$, the Thomas–Fermi function can be considered as a neutral atom with a finite boundary specified by r_L for which $\rho_e(r)$ is non-zero. Thus, Category III solutions can be considered to represent neutral ions under pressure.

For the remainder of our discussion of atomic electron screening, we will restrict ourselves to the Category I solution of the Thomas–Fermi equation which represents the neutral atom. It is this model of the multi-electron atom through which we will investigate the roles of atomic electron screening in elastic Coulomb scatter.

A non-trivial solution to (5.100) can be determined only numerically, such as through the Runge–Kutta method of solving ordinary differential equations.[6] Figure 5.9 shows the Category I solution to (5.100).

From (5.99), it is possible to calculate the spatial distribution of the electron charge density. The screened potential energy for the Thomas–Fermi model is

$$U(r) = \frac{\alpha\hbar c\, zZ}{r}\chi(\kappa r). \tag{5.111}$$

Comparing this result to our original definition of the screening function in (5.55), we see that the Thomas–Fermi function is the screening function, $\Phi(r)$. It is clear from Fig. 5.9 that $\chi(\kappa r)$ and, hence, $U(r)$, is negligible for $\kappa r > 15$.

These results of the Thomas–Fermi model will, as shown in the next section, allow us to define expressions for the atomic electron screening parameter which have some basis upon the atomic electron charge distribution and the effective potential seen by the projectile. But it should be noted that this atomic model is not accurate in some important details as are now summarised.

The most obvious deficiency is that it does not account for the quantum orbital structure of the atom – but this is not a significant issue in calculating for screening

[6] See, for example, Hoffman (1993)

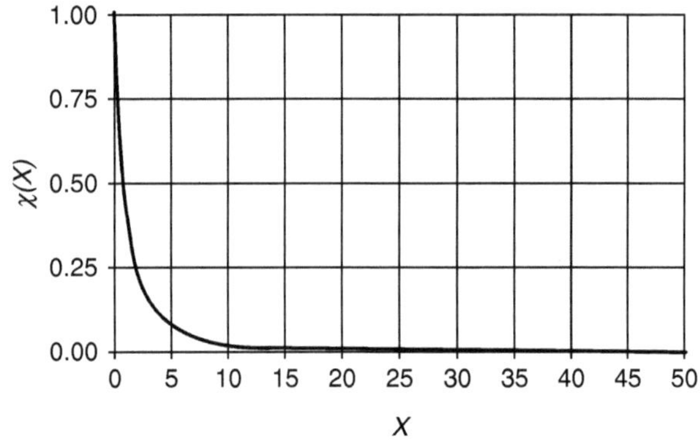

Fig. 5.9 The Thomas-Fermi function, $\chi(x)$, given by the Category I solution to Eq. (5.100)

effects in dosimetry evaluations. And, as we have seen, this model cannot provide a bounded solution for negative ions (Category III solution). Perhaps of major interest for our screening calculation is knowing the limitations of the asymptotic behaviours of the effective potential and the atomic electron charge distribution calculated from the Thomas–Fermi model. If we were to assume that the effective potential approaches an inverse-power radial dependence,

$$V_{\text{eff}}(r) \propto \frac{1}{r^n} \quad \text{as} \quad r \to \infty \tag{5.112}$$

then, from the left-hand side (5.94),

$$-\frac{1}{r^2}\frac{\mathrm{d}^2}{\mathrm{d}r^2}(rV_{\text{eff}}(r)) \propto \frac{1}{r^{n+2}} \quad \text{as} \quad r \to \infty \tag{5.113}$$

and, from the right-hand side,

$$(-V_{\text{eff}}(r))^{\frac{3}{2}} \propto \frac{1}{r^{\frac{3n}{2}}}. \tag{5.114}$$

It then follows from these results that the exponent of (5.112) is $n = 6$. Thus, for $r \to \infty$, one can conclude that the radial variation of the Thomas–Fermi charge density is

$$\rho_{\text{e}}(r) \propto \nabla^2 V_{\text{eff}}(r) \propto (-V_{\text{eff}}(r))^{\frac{3}{2}} \propto \frac{1}{r^6}$$

whereas it is, in fact, $\propto e^{-r}$. It also follows that, for $r \to 0$,

$$V_{\text{eff}}(r) \propto \frac{Z}{r}, \tag{5.115}$$

leading to

$$\rho_e(r) \propto \frac{1}{r^{\frac{3}{2}}}. \tag{5.116}$$

That is, the expression for the atomic electron charge density diverges as $r \rightarrow 0$ whereas it is, in reality, finite.

5.3.6 Other Models of Multi-electron Atoms

While the Thomas–Fermi model of the atom will allow a simple modelling of the atomic electron screening of the nuclear charge, it is not the only model applicable.

For electron projectiles, consideration must be given to electron exchange effects (e.g. for an electron projectile). Dirac proposed a modification (leading to the Thomas–Fermi–Dirac atomic model) to address these (Dirac 1930). As with the Thomas–Fermi model, the screening function is given by a differential equation that requires numerical solution (Lieb 1981; Condon and Odabasi 1980). However, the Thomas–Fermi–Dirac model does not have the $Z^{-\frac{1}{3}}$ scaling attribute of the Thomas–Fermi model, thus requiring that the screening function be evaluated separately for each particular atom. Practical calculations require parameterisations of these results. One provided by Bonham and Strand (1963) is

$$\Phi(r) = \sum_{j=1}^{3} B_j \, e^{-b_j r} \tag{5.117}$$

where the coefficients b_j and B_j are fourth-order polynomials in ln Z.

More accurate models of the spatial distributions of atomic electrons and which can lead to expressions of screening can be obtained through the Hartree–Fock self-consistent method in which a trial wavefunction for the Z bound electrons is created through an antisymmetric product of central orbitals. The radial components are then adjusted so as to minimise the total binding energy (Hartree 1957).

The use of these models in evaluating the elastic Coulomb scattering cross sections for electrons is described by Jablonski et al. (2004).

5.3.7 Atomic Electron Screening Parameter: Part II

Having derived the Thomas–Fermi model of the multi-electron atom, we now explore how it can yield estimates of the atomic electron screening parameter, κ.

Here, the Thomas–Fermi function is approximated by the exponential screening function:

$$\chi(\kappa r) = e^{-\kappa r}. \tag{5.118}$$

The simplest form is that of the reciprocal of the Thomas–Fermi radius (5.83) (see, e.g. McParland 2010):

$$\begin{aligned}\kappa &\equiv \frac{1}{r_{\mathrm{TF}}} \\ &= 1.13\,\frac{Z^{\frac{1}{3}}}{a_\infty}.\end{aligned} \tag{5.119}$$

Jackson (1999) has also suggested the fractional-power dependence upon atomic number and the inverse Bohr radius, but with a different multiplier:

$$\kappa \approx 0.71\,\frac{Z^{\frac{1}{3}}}{a_\infty} \tag{5.120}$$

which he stated was derived from an approximate fit to the Thomas–Fermi potential.

These two expressions neglect any charge characteristics of the projectile which, of course, would be important should the projectile be an α particle or a heavy ion (i.e. $z > 1$). Bohr (1948) proposed a symmetrical form for the atomic electron screening parameter (but without derivation):

$$\kappa = \frac{\sqrt{z^{\frac{2}{3}} + Z^{\frac{2}{3}}}}{a_\infty}. \tag{5.121}$$

Firsov (1957) expanded upon this result of Bohr and analysed numerically the collision between two Thomas–Fermi atoms from which he suggested another expression for the atomic electron screening parameter:

$$\kappa = \frac{\left(z^{\frac{1}{3}} + Z^{\frac{1}{3}}\right)^{\frac{2}{3}}}{a_\infty}. \tag{5.122}$$

Figure 5.10 shows the variations of these calculated atomic electron screening parameters (expressed as the dimensionless product κa_∞) as a function of Z between carbon and calcium.

All of the six curves shown in Fig. 5.10 demonstrate the same general behaviour of a slow increase with the target atomic number, Z. An increase in the Z by nearly a factor of 8 yields a corresponding increase of only about a factor of 2 or less in κ. Curves (A) and (B) correspond to (5.119) and (5.120) and are proportional to the

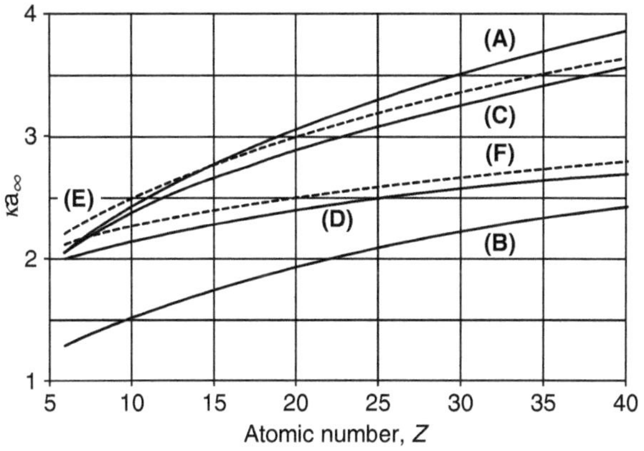

Fig. 5.10 The product of the atomic electron screening parameter and the Bohr radius, κa_∞, as a function of target atomic number Z between carbon and calcium for the different models presented: *solid lines* are for a projectile with $z = 1$ (e.g. electron or proton) and the dashed lines are for a projectile with $z = 2$ (e.g. α-particle). Curves (A) and (B) are for the reciprocal proportionalities of the Thomas–Fermi radius of Eqs. (5.119) and (5.120), respectively, and, hence, are valid for any value of projectile z. Curves (C) and (D) are the Bohr and Firsov forms of Eqs. (5.121) and (5.122) for $z = 1$, respectively. Curves (E) and (F) are the Bohr and Firsov forms for $z = 2$, respectively. The calculations do not extend below $Z = 6$ as the validity of the Thomas–Fermi statistical model of the atom fails in this region

reciprocal of the Thomas–Fermi radius; their only difference is in the constant of proportionality. As (5.119) and (5.120) have no dependence upon the ze of the projectile, they are valid for any charged projectile. Curves (C) and (D) are those of the Bohr (5.121) and Firsov (5.122) calculations for $z = 1$. Curves (A) and (B), which ignore the nature of the projectile, span a range within which curves (C) and (D), which do account for the projectile's charge, lie.

Curves (E) and (F) are the Bohr and Firsov results for $z = 2$; curves (A) and (B) are, obviously, also applicable to this case. The differences between Curves (C) and (E) and between (D) and (F) are quite reflecting the effect of the fractional powers applied to the projectile's charge.

5.3.8 Continuous Screening Part II: The Thomas–Fermi Model

Instead of assuming an exponentially screened potential with an atomic electron screening parameter extracted from an exponential approximation to the Thomas–Fermi function, we can use the latter itself to describe the effects of screening directly. As this function is nonanalytical, a calculation for the screened Coulomb differential cross section would be numerical in character if (5.93) were to be used. This, clearly, is not a convenient means of calculation. A more practical

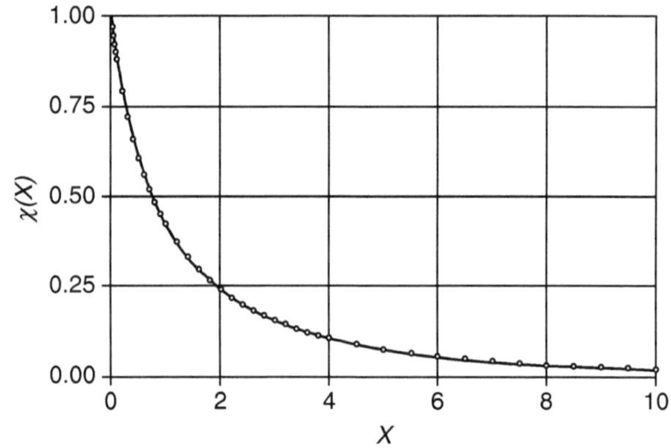

Fig. 5.11 The Thomas–Fermi function calculated from a numerical solution to the Thomas–Fermi equation (*solid line*) and from Molière's empirical fit

means would be to use an analytical approximation to $\chi(x)$, of which a common one is the multiexponential form provided by Molière (1947),

$$\chi(x) = \sum_{j=1}^{3} c_j e^{-k_j x} \tag{5.123}$$

with the coefficients having been derived from experimental data (Robinson 1969):

$$\begin{aligned} c_1 &= 0.35 \qquad & k_1 &= 0.3 \\ c_2 &= 0.55 \qquad & k_2 &= 1.2 \\ c_3 &= 0.10 \qquad & k_3 &= 6.0. \end{aligned}$$

The form of (5.123) reproduces the limiting requirements of the Thomas–Fermi function of $\chi(0) = 1$ and $\chi(x) \to 0$ as $x \to \infty$. Clearly, it also returns a form of the Wentzel-type potential if the summation were to be limited to the first term. In fact, if the summation were truncated at the first term only with c_1 fixed at 1.0, the resulting expression for $\chi(x)$ will agree with the actual value to within 13 % for x up to 1.0. Figure 5.11 shows a comparison of the Thomas–Fermi function and the empirical fit of Molière.

The elastic Coulomb scattering differential cross section for an atom with screening modelled by the Molière expression of (5.123) is now derived. Recall that the scattering amplitude to the first term of the Born expansion for a central and spherically symmetric potential of (2.84) is

$$f(q) = -\left(\frac{2m}{\hbar c q}\right) \int_0^\infty \mathrm{d}r'\, r' \sin\left(\frac{qr'}{\hbar c}\right) U(r').$$

Recalling the definition of the Bohr collision diameter and the relationship between momentum transfer and scattering angle, we obtain, through our usual means of derivation, an expression for the scattering amplitude,

$$f(q) = -\left(\frac{\rho_{\mathrm{B}}(E)}{4}\right) \sum_{j=1}^{3} \frac{c_j}{\left(\frac{k_j \kappa \lambda\!\!\!^{-}}{2}\right)^2 + \sin^2\frac{\theta}{2}} \tag{5.124}$$

with the corresponding differential cross section

$$\frac{\mathrm{d}\sigma}{\mathrm{d}\Omega} = \left(\frac{\rho_{\mathrm{B}}(E)}{4}\right)^2 \left(\sum_{j=1}^{3} \frac{c_j}{\left(\frac{k_j \kappa \lambda\!\!\!^{-}}{2}\right)^2 + \sin^2\frac{\theta}{2}}\right)^2 . \tag{5.125}$$

By taking the small-angle approximation, we obtain the expression

$$\frac{\mathrm{d}\sigma}{\mathrm{d}\Omega} \approx \left(\rho_{\mathrm{B}}(E) \sum_{j=1}^{3} \frac{c_j}{\chi_{0,j}^2 + \theta^2}\right)^2 \tag{5.126}$$

with the j-th Molière screening angle being given by

$$\chi_{0,j} = k_j \kappa \lambda\!\!\!^{-}. \tag{5.127}$$

5.3.9 *Discrete Screening: Elastic Scatter Form Factor*

A continuous-screening function for the nuclear potential can model the screening effects of the atomic electrons' charges by calculating their individual contributions to the scattering process. We will develop such a model and calculate a differential cross section through the use of quantum perturbation theory. Although this is a multi-body problem consisting of Z atomic electrons, the nucleus and the projectile, it is made solvable through the simplifying assumption that the atomic electrons are uncorrelated and do not interact. The spins of the electrons and projectile are ignored.

As usual, the origin of the scattering system is fixed at the centre of the nucleus. The net Coulomb potential between the projectile of charge ze and the atom is the sum of those between the projectile and the nucleus and the Z atomic electrons:

$$U(R, r_1 \ldots r_Z) = \alpha \hbar c\, z \left(\frac{Z}{R} - \sum_{j=1}^{Z} \frac{1}{|\mathbf{R} - \mathbf{r}_j|}\right) \tag{5.128}$$

where $\mathbf{R}$ is the position vector of the projectile and $\mathbf{r_j}$ is that for the j-th atomic electron. The total system vector is given by $\mathbf{r} = (\mathbf{R}, \mathbf{r}_1 \ldots \mathbf{r}_Z)$. As before, the atom is considered to be neutral and in its ground state and will remain so throughout the scatter as dictated by our assumption of elasticity. By treating the projectile as a plane wave, the projectile–atom system prior to scattering is (in *bra–ket* notation)

$$\langle \mathbf{r} \mid \mathbf{p}, 0\rangle = \frac{1}{\sqrt{L^3}} e^{i\frac{\mathbf{p}\bullet\mathbf{r}}{\hbar c}} |0\rangle \tag{5.129}$$

where L^3 is the volume used for normalisation and the system after scattering is described by

$$\langle \mathbf{r} \mid \mathbf{p}', 0\rangle = \frac{1}{\sqrt{L^3}} e^{i\frac{\mathbf{p}'\bullet\mathbf{r}}{\hbar c}} |0\rangle. \tag{5.130}$$

The phase-space factor, or the density of final states per final energy interval dE, is obtained from

$$\rho_f \, \mathrm{d}E' = \left(\frac{L}{2\pi\hbar c}\right)^3 \mathrm{d}^3\mathbf{p}'. \tag{5.131}$$

Expanding the vector differential $\mathrm{d}^3\mathbf{p}' = 2\pi p'^2 \, \mathrm{d}p' \, \mathrm{d}(\cos\,\theta)$ and recalling that $\frac{\mathrm{d}p'}{\mathrm{d}E'} = \beta'$ and that $|\mathrm{d}(\cos\theta)\,| = \left(\frac{q}{p^2}\right) \mathrm{d}q$, the phase factor is

$$\rho_{\mathrm{f}} = \frac{1}{4\pi^2\beta} \left(\frac{L}{\hbar c}\right)^3 q \, dq \tag{5.132}$$

where the primes are redundant and have now been removed due to the elasticity of the scattering.

The matrix element is calculated by first separating the contributions of the nuclear and atomic potentials. That due to the nuclear Coulomb potential is quickly solved by expansion into co-ordinate space:

$$\begin{aligned}
\langle \mathbf{p}', 0|U_{\mathrm{nuc}}|\mathbf{p}, 0\rangle &= \int \mathrm{d}^3\mathbf{r}' \int \mathrm{d}^3\mathbf{r} \, \langle \mathbf{p}', 0 \mid \mathbf{r}\rangle\langle \mathbf{r}|U_{\mathrm{nuc}}|\mathbf{r}'\rangle\langle \mathbf{r}' \mid \mathbf{p}, 0\rangle \\
&= \frac{zZ\,\alpha\hbar c}{L^3} \int \mathrm{d}^3\mathbf{R} \, \frac{e^{i\frac{\mathbf{q}\bullet\mathbf{R}}{\hbar c}}}{\mathbf{R}} \\
&= 4\pi \, \frac{zZ\,\alpha\hbar c}{L^3} \left(\frac{\hbar c}{q}\right)^2
\end{aligned} \tag{5.133}$$

where the solution of the Bethe integral previously derived has been used. The matrix element due to the combined electronic Coulomb potentials is determined also through expansion in co-ordinate space:

$$\begin{aligned}
\langle \mathbf{p}',0|U_{\rm elec}|\mathbf{p},0\rangle &= \int d^3\mathbf{r}' \int d^3\mathbf{r}\, \langle \mathbf{p}',0 \mid \mathbf{r}\rangle\langle \mathbf{r}|U_{\rm elec}|\mathbf{r}'\rangle\langle \mathbf{r}' \mid \mathbf{p},0\rangle \\
&= \int d^3\mathbf{r}' \int d^3\mathbf{r}\, \langle \mathbf{p}',0 \mid \mathbf{r}\rangle U_{\rm elec}(\mathbf{r})\langle \mathbf{r}' \mid \mathbf{p},0\rangle \\
&= \frac{1}{L^3}\int d^3\mathbf{R}\, e^{i\frac{\mathbf{q}\bullet\mathbf{R}}{\hbar c}} \prod_{j=1}^{Z}\int d^3\mathbf{r}_j \psi_0^*(\mathbf{r}_1\ldots\mathbf{r}_Z)\, U(\mathbf{r})\, \psi_0(\mathbf{r}_1\ldots\mathbf{r}_Z) \\
&= -\frac{\alpha\hbar c\, z}{L^3}\prod_{j=1}^{Z}\int d^3\mathbf{r}_j \psi_0^*(\mathbf{r}_1\ldots\mathbf{r}_Z)\, \psi_0(\mathbf{r}_1\ldots\mathbf{r}_Z)\int d^3\mathbf{R}\,\frac{e^{i\frac{\mathbf{q}\bullet\mathbf{R}}{\hbar c}}}{|\mathbf{R}-\mathbf{r}_j|} \\
&= -\frac{\alpha\hbar c\, z}{L^3}\prod_{j=1}^{Z}\int d^3\mathbf{r}_j \psi_0^*(\mathbf{r}_1\ldots\mathbf{r}_Z)\, \psi_0(\mathbf{r}_1\ldots\mathbf{r}_Z)\int d^3\mathbf{R}'\,\frac{e^{i\frac{\mathbf{q}\bullet\left(\mathbf{R}'+\mathbf{r}_j\right)}{\hbar c}}}{|\mathbf{R}'|} \\
&= -\frac{\alpha\hbar c\, z}{L^3}\prod_{j=1}^{Z}\int d^3\mathbf{r}_j \psi_0^*(\mathbf{r}_1\ldots\mathbf{r}_Z)\, \psi_0(\mathbf{r}_1\ldots\mathbf{r}_Z)\, e^{i\frac{\mathbf{q}\bullet\mathbf{r}_j}{\hbar c}}\int d^3\mathbf{R}'\,\frac{e^{i\frac{\mathbf{q}\bullet\mathbf{R}'}{\hbar c}}}{|\mathbf{R}'|} \\
&= -4\pi\frac{\alpha\hbar c\, z}{L^3}\left(\frac{\hbar c}{q}\right)^2\prod_{j=1}^{Z}\int d^3\mathbf{r}_j \psi_0^*(\mathbf{r}_1\ldots\mathbf{r}_Z)\, e^{i\frac{\mathbf{q}\bullet\mathbf{r}_j}{\hbar c}}\, \psi_0(\mathbf{r}_1\ldots\mathbf{r}_Z) \\
&= -4\pi\frac{\alpha\hbar c\, z}{L^3}\left(\frac{\hbar c}{q}\right)^2 \langle 0|\sum_{j=1}^{Z} e^{i\frac{\mathbf{q}\bullet\mathbf{r}_j}{\hbar c}}|0\rangle \\
&= -4\pi\frac{\alpha\hbar c\, z}{L^3}\left(\frac{\hbar c}{q}\right)^2 \langle 0|\sum_{j=1}^{Z} e^{i\frac{\mathbf{q}\bullet\mathbf{r}_j}{\hbar c}}|0\rangle \\
&= -4\pi\frac{\alpha\hbar c\, z}{L^3}\left(\frac{\hbar c}{q}\right)^2 F_0(\mathbf{q},Z)
\end{aligned} \tag{5.134}$$

where the elastic scattering form factor, $F_0(\mathbf{q},Z)$, has been defined in terms of the matrix element. This same expression for the atomic form factor is obtained through a derivation of Rayleigh scattering of photons (McParland 2010). An explicit calculation of $F_0(\mathbf{q},Z)$ requires knowledge of the atom's ground state wavefunction which, for non-hydrogen atoms, may not be known accurately. We can simplify the calculation by assuming a continuous atomic electron spatial distribution, from which it can be seen that the form factor is proportional to the Fourier transform of that distribution into momentum space. From the normalisation of the electron charge density over all space,

$$\begin{aligned}
\int d^3\mathbf{r}\,\rho_e(r) &= F_0(0,Z) \\
&= Z.
\end{aligned} \tag{5.135}$$

We will assume the radial distribution of the atomic electron density here to be exponential and, using the normalisation arguments previously discussed, write it as

$$\rho_{\mathrm{e}}(r) = \frac{Z\kappa^3}{4\pi}\, e^{-\kappa r}. \tag{5.136}$$

Using these two expressions and following our prior paths in derivations, we obtain an explicit expression for the form factor:

$$\begin{aligned} F_0(\mathbf{q}, Z) &= \frac{Z\kappa^3}{4\pi} \int \mathrm{d}^3\mathbf{r}\, e^{-\left(\kappa r + i\frac{\mathbf{q}\bullet\mathbf{r}}{\hbar c}\right)} \\ &= Z\left(\frac{\kappa^2}{\kappa^2 + \left(\frac{q}{\hbar c}\right)^2}\right)^2. \end{aligned} \tag{5.137}$$

The result of (5.137) is recast as a function of the scattering angle θ by using the previously derived relationship between elastic momentum transfer and scattering angle:

$$F_0(\theta, Z) = \frac{Z}{\left(1 + \left(\left(\frac{2}{\chi_0}\right)\sin\frac{\theta}{2}\right)^2\right)^2}. \tag{5.138}$$

The total matrix element for the elastic Coulomb scattering from the atom is obtained by summing (5.133) and (5.134):

$$\langle \mathbf{p}', 0|U|\mathbf{p}, 0\rangle = -4\pi\frac{\alpha\hbar c\, z}{L^3}\left(\frac{\hbar c}{q}\right)^3 (Z - F_0(\mathbf{q}, Z)). \tag{5.139}$$

Combining (5.132) and (5.139), we obtain the transition rate of the scattering:

$$\begin{aligned} \lambda_{\mathrm{fi}} &= \frac{2\pi}{\hbar} \mid \langle \mathbf{p}', 0|U|\mathbf{p}, 0\rangle \mid^2 \rho_{\mathrm{f}} \\ &= \left(\frac{8\pi c}{\beta}\right)\frac{(\alpha\hbar c\, z)^2}{(qL)^3} \mid Z - F_0(\mathbf{q}, Z) \mid^2 \\ &= \left(\frac{8\pi c}{\beta}\right)\frac{(\alpha\hbar c\, zZ)^2}{(qL)^3} \left|1 - \frac{F_0(\mathbf{q}, Z)}{Z}\right|^2. \end{aligned} \tag{5.140}$$

From our discussions of quantum perturbation theory provided in Chap. 2, we can use this to derive the differential cross section in momentum transfer:

$$\frac{\mathrm{d}\sigma}{\mathrm{d}q} = \frac{8\pi}{q^3}\left(\frac{\alpha\hbar c\, zZ}{\beta}\right)^2 \left|1 - \frac{F_0(\mathbf{q}, Z)}{Z}\right|^2. \tag{5.141}$$

From the relationship between momentum transfer and scattering angle, we can recast (5.141) as the differential cross section in solid angle:

$$\frac{\mathrm{d}\sigma}{\mathrm{d}\Omega}=\left(\frac{\alpha\hbar c\, zZ}{2p\beta}\right)^2\frac{\left|1-\dfrac{F_0(\theta,\,Z)}{Z}\right|^2}{\sin^4\dfrac{\theta}{2}}. \tag{5.142}$$

We did not use the Bohr collision diameter as a simplifying parameter in (5.122) as the contribution of Z has been extracted and placed within the modulus squared of the difference between it and the elastic scattering form factor. Comparing (5.142) with the unscreened potential's results of (4.78), we see that the inclusion of screening has replaced the Z^2 factor with $|Z-F_0(\mathbf{q},Z)|^2$ resulting in an expected reduction in the cross section as a result of electronic screening.

5.3.10 Differences Between Electron and Positron Scattering from Atoms

In Chap. 4, it was noted that, to first order in α, the electron and positron differential cross sections for Coulomb scatter from a central charge were equal. We now briefly consider the differences between electron and positron Coulomb scatter from an atom.

An electron scattered from a high-Z atom can fall into the atom's potential well with sufficient acceleration to reach relativistic energies and thus require a description by the Dirac equation. A positron, however, will be repelled by the nuclear Coulomb field and will not enter the atomic interior. The subsequent deceleration of the positron allows it to be described by the Schrödinger equation.

The interactions of a positron with matter are markedly different from those of an electron. This is suggested by the Dirac theory in which the positron is considered to be a negative-energy solution to the Dirac equation. As the electron and positron are different particles with opposing electric charge signs, there is no exchange interaction between the two and the Coulomb interaction will be attractive. This latter case allows for the formation of a bound electron–positron system known as positronium.

5.4 Numerical Examples

5.4.1 Introduction

A number of expressions have been derived in this chapter from various models in order to describe the differences between elastic Coulomb scattering from atoms from a bare point charge, as investigated in Chap. 4, due to the finite size of the nucleus and of the screening of the nuclear Coulomb potential by the atomic electrons. In this section, the magnitudes of these effects for conditions typical of medical irradiation are evaluated in order to assess under what specific set of conditions (e.g. energy, scattering angle and projectile/target combination) these effects are significant. Numerical results are calculated for the nonrelativistic cases of a 100-keV electron and a 250-MeV proton, with respective reduced de Broglie wavelengths of 589 fm and 0.27 fm, respectively. Hence, we immediately expect to see significant differences in screening effects between these two projectiles due to the large difference between the associated values of $\lambda\!\!\!^{-}$. The elements of carbon ($Z = 6$) and calcium ($Z = 20$) will be used as two examples of elemental targets as this range of atomic numbers spans that to be expected in tissue.[7] We will calculate the effects upon elastic Coulomb scatter due to finite nuclear size and atomic screening for the above cohort of projectiles, incident kinetic energies and targets for the following four models:

$$\frac{d\sigma}{d\Omega}(\theta) = \left(\frac{\rho_B(E)}{2\sin^2\frac{\theta}{2}}\right)^2 \quad \text{Unscreened point nuclear target}$$

$$\frac{d\sigma}{d\Omega}(\theta) = \left(\frac{\rho_B(E)}{2\sin^2\frac{\theta}{2}}\right)^2 |F(\theta)|^2 \quad \text{Unscreened finite nuclear target}$$

$$\frac{d\sigma}{d\Omega}(\theta) = \left(\frac{\rho_B(E)}{2\left(\left(\frac{\lambda\!\!\!^{-}\kappa}{2}\right)^2 + \sin^2\frac{\theta}{2}\right)}\right)^2 |F(\theta)|^2 \quad \text{Exponentially screened finite target.}$$

5.4.2 Nuclear Form Factor

For the ease of calculation, we will instead use the simpler Gaussian form factor with the squared modulus:

$$|F(\theta)|^2 = \left(e^{-\left(\frac{qR_N}{2\hbar c}\right)^2}\right)^2.$$

[7] In addition, carbon would reflect the lowest atomic number that could be expected to be treated as valid in the Thomas–Fermi statistical atomic model which requires $Z \gg 1$.

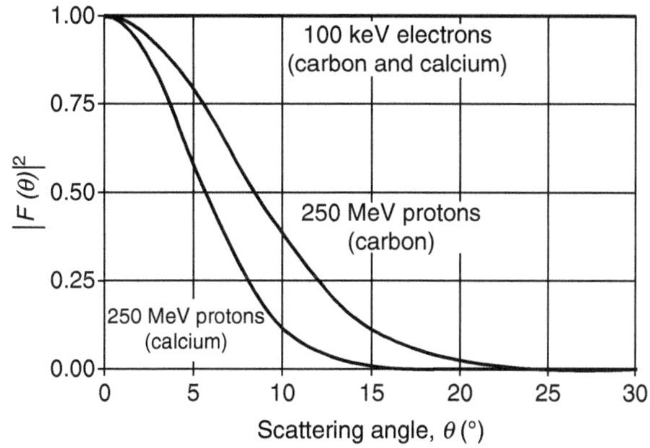

Fig. 5.12 Squared moduli of the nuclear form factors for 100-keV electrons and 250-MeV protons incident to carbon and calcium

The exponent of the Gaussian form factor is

$$\begin{aligned}\frac{qR_\mathrm{N}}{2\hbar c} &= \frac{p\sin\frac{\theta}{2}R_\mathrm{N}}{\hbar c} \\ &= \left(\frac{R_\mathrm{N}}{\lambda\!\!\!^{-}}\right)\sin\frac{\theta}{2} \\ &= \left(\frac{0.94A^{\frac{1}{3}}}{\lambda\!\!\!^{-}}\right)\sin\frac{\theta}{2}\end{aligned} \tag{5.143}$$

where the reduced de Broglie wavelength of the projectile is given in units of fm. Figure 5.12 shows the squared moduli of the nuclear form factors for a Gaussian spatial distribution of proton charge calculated for 100-keV electrons and 250-MeV protons incident to carbon ($A = 12$) and calcium ($A = 40$) as functions of the scattering angle, θ.

As shown in Fig. 5.12, the nuclear form factors for 100-keV electrons incident to carbon and calcium do not differ significantly from unity. This is due to the ratio of the nuclear radii (2.2 fm and 3.2 fm for carbon and calcium, respectively) and the reduced de Broglie wavelength of an electron of this energy of 589 fm. Even at the typically highest electron kinetic energy used in radiotherapy, 25 MeV, the minimum value of $|F_0(\theta)|^2$ is only 0.86. On the other hand, for 250-MeV protons, the nuclear form factor shows a strong dependence upon scattering angle (i.e. momentum transfer) as a result of the corresponding reduced de Broglie wavelength of 3.7 fm being comparable to the nuclear dimensions.

5.4.3 Atomic Electron Screening Parameter

In each of these models, the atomic screening parameter κ is to be defined. Table 5.1 shows various such parameters calculated for a unit-charge projectile incident to carbon and calcium using the three examples of the approximations of the reciprocal of the Thomas–Fermi radius and the expressions of Bohr and Firsov.

Table 5.1 Values of the atomic electron screening parameter for carbon and calcium calculated from different models

		Atomic electron screening parameter, κ (for projectile $z = 1$)		
Element	Z	$1.13\,\frac{Z^{\frac{1}{3}}}{a_\infty}$	$\frac{\sqrt{1+Z^{\frac{2}{3}}}}{a_\infty}$	$\frac{\left(1+Z^{\frac{1}{3}}\right)^{\frac{2}{3}}}{a_\infty}$
Carbon	6	$3.88 \times 10^{-5}\ \text{fm}^{-1}$	$3.92 \times 10^{-5}\ \text{fm}^{-1}$	$3.77 \times 10^{-5}\ \text{fm}^{-1}$
Calcium	20	$5.80 \times 10^{-5}\ \text{fm}^{-1}$	$5.47 \times 10^{-5}\ \text{fm}^{-1}$	$4.53 \times 10^{-5}\ \text{fm}^{-1}$

The values of these atomic electron screening parameters about the mean do not vary significantly between the three models. There is less than a 9 % variation for the carbon target and less than 14 % for the calcium target.

For the numerical calculations to follow, we will use the Bohr screening parameter, $\frac{\sqrt{1+Z^{\frac{2}{3}}}}{a_\infty}$, for electron and proton projectiles.

5.4.4 Atomic Electron Screening Angle

The value of the screening angle $\chi_0 \equiv \lambda\!\!\!^{-}\kappa$ dictates the limiting magnitude of the elastic Coulomb differential cross section as $\theta \to 0$.

The screening angles calculated for the Bohr screening parameter are shown in Fig. 5.13 for electrons and protons incident to carbon and calcium targets over ranges of kinetic energies typical in medical applications. Both exhibit a decrease in χ_0 with kinetic energy through the $\frac{\hbar c}{p}$ dependence of the reduced de Broglie wavelength. As expected through the Z-dependence of κ, the screening angle is greater for calcium than for carbon.

5.4.5 Differential Cross Sections

The theoretical unscreened and screened differential cross sections for 100-keV electrons incident to carbon and calcium are shown as functions of scattering angle Fig. 5.14. The deviations between the screened and unscreened results become most apparent for scattering angles of less than about 2° where the screened differential cross sections approach finite values whereas the unscreened differential cross sections continue to diverge. This is a practical example demonstrating aspects of the Bohr conditions for screening discussed in Sect. 5.3.3. It will also be relevant to our derivation of multiple-scattering probability distribution functions in Chap. 6.

Similarly, the unscreened differential cross sections for 250-MeV protons and those for the combined Yukawa screening and Gaussian form factor are shown as functions of scattering angle incident to carbon and calcium atomic targets in Fig. 5.15.

Fig. 5.13 Screening angles for (**a**) electrons and (**b**) protons impinging on carbon and calcium for kinetic energies typical of medical applications. Note that the ordinate scale for the electron plot is logarithmic

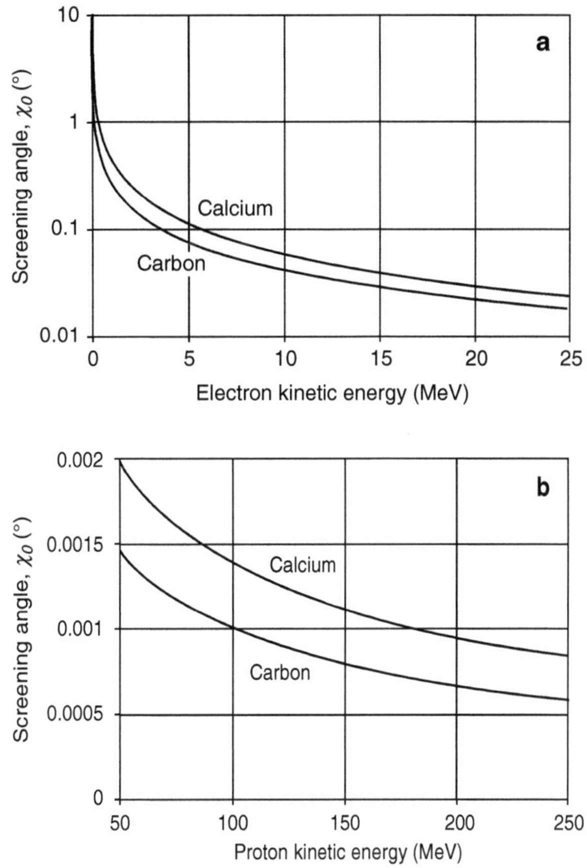

5.5 Combined Effects of Finite Nuclear Size and Atomic Electron Screening

5.5.1 *Qualitative Differences*

We have evaluated the effects of finite nuclear size and atomic electron screening upon elastic Coulomb scatter. The two effects have opposing dependencies upon scattering angle and projectile momentum.

The differential cross section is modified by the squared modulus of the form factor which is the Fourier transform of the charge distribution within the target, $|F(\mathbf{q})|^2$. This quantity decreases (although not necessarily monotonically as shown in Fig. 5.1) with scattering angle through $\frac{qR_N}{\hbar c} = \left(\frac{2pR_N}{\hbar c}\right)\sin\frac{\theta}{2}$ and as shown in Fig. 5.13. That is, the effect of the finite size of the target is to suppress the Rutherford differential cross section at large scattering angles. From this dependence, the suppression at a fixed scattering angle increases with projectile momentum.

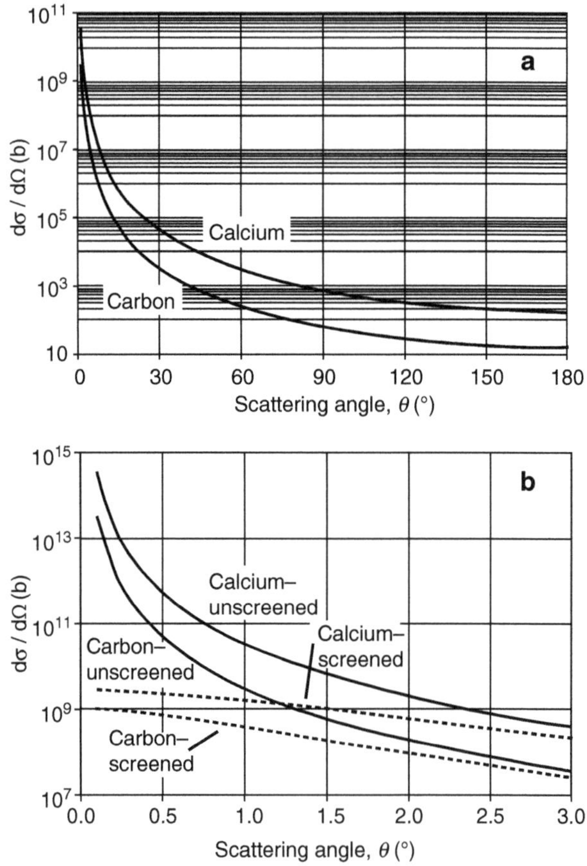

Fig. 5.14 100-keV electrons: elastic Coulomb scatter differential cross sections, calculated from the unscreened and the exponential–potential screened expressions (the latter including the Gaussian form factor term) as functions of scattering angle for carbon and calcium atomic targets. (**a**) shows results over all angles, whereas (**b**) zooms in on those small scattering angles where the screening effects that maintain a finite cross section become evident

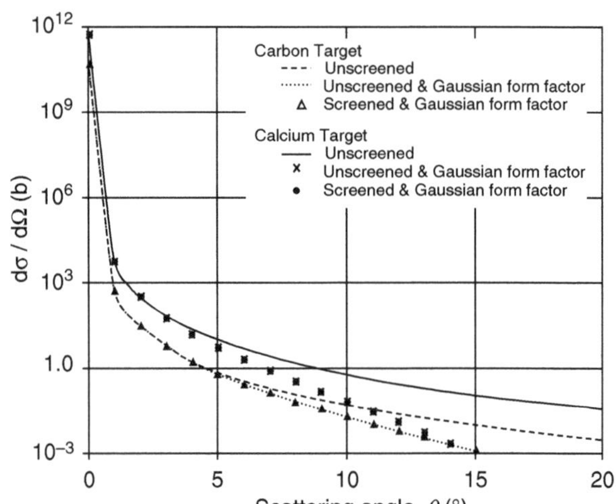

Fig. 5.15 250-MeV protons: elastic Coulomb scatter differential cross sections calculated from the unscreened and the screened expressions with and without the Gaussian form factor term as functions of scattering angle for carbon and calcium atomic targets

On the other hand, atomic electron screening led to a modification of the Rutherford differential cross section's angular dependence from that of $\frac{1}{\sin^4\frac{\theta}{2}}$ to that of $\frac{1}{\left(\left(\frac{\chi_0}{2}\right)^2+\sin^2\frac{\theta}{2}\right)^2}$. This leads to a suppression of the Rutherford differential cross section at small scattering angles. Further, this effect is more prominent at lower projectile momenta through $\chi_0 = \kappa\lambda\!\!\!^{-} = \frac{\kappa\hbar c}{p}$.

5.5.2 *Combined Effects Upon Elastic Coulomb Scatter Total Cross Section*

We now evaluate the total elastic Coulomb cross section with the combined effects of finite target size and atomic electron screening upon the elastic Coulomb total cross section. We begin with the formal extension of the differential cross section in solid angle:

$$\frac{\mathrm{d}\sigma}{\mathrm{d}\Omega} = \left(\frac{\rho_{\mathrm{B}}(E)}{4}\right)^2 \frac{|F(\theta)|^2}{\left(\left(\frac{\chi_0}{2}\right)^2 + \sin^2\frac{\theta}{2}\right)^2}. \tag{5.144}$$

For the sake of simplicity of this derivation, we will consider the nuclear form factor to be modelled by the Gaussian form of (5.39):

$$|F(\theta_1)| = e^{-\left(\frac{qR_{\mathrm{N}}}{2\hbar c}\right)^2} \tag{5.145}$$

with its modulus squared being

$$\begin{aligned} |F(\theta_1)|^2 &= e^{-2\left(\frac{qR_{\mathrm{N}}}{\hbar c}\right)^2} \\ &= e^{-8\left(\frac{R_{\mathrm{N}}}{\lambda\!\!\!^{-}}\right)^2 \sin^2\frac{\theta_1}{2}} \\ &= e^{-\left(\frac{2R_{\mathrm{N}}}{\lambda\!\!\!^{-}}\right)^2 (1-\cos\theta_1)}. \end{aligned} \tag{5.146}$$

Substituting this result into (5.144), applying the same trigonometric identity to the denominator and then integrating, we find the initial expression for the total elastic Coulomb scattering cross section, including the effects of atomic electron screening and a finite nuclear size:

$$\begin{aligned} \sigma &= 2\pi \left(\frac{\rho_{\mathrm{B}}(E)}{4}\right)^2 e^{-\left(\frac{2R_{\mathrm{N}}}{\lambda\!\!\!^{-}}\right)^2} \int_{-1}^{1} \mathrm{d}(\cos\theta_1) \frac{e^{\left(\frac{2R_{\mathrm{N}}}{\lambda\!\!\!^{-}}\right)^2 \cos\theta_1}}{\left(\left(\frac{\chi_0}{2}\right)^2 + \frac{1}{2} - \frac{\cos(\theta_1)}{2}\right)^2} \\ &= \frac{\pi\rho_{\mathrm{B}}^2(E)}{2} e^{-\left(\frac{2R_{\mathrm{N}}}{\lambda\!\!\!^{-}}\right)^2} \int_{-1}^{1} \mathrm{d}(\cos\theta_1) \frac{e^{\left(\frac{2R_{\mathrm{N}}}{\lambda\!\!\!^{-}}\right)^2 \cos\theta_1}}{\left(\left(\frac{\chi_0^2}{2} + 1\right) - \cos\theta_1\right)^2}. \end{aligned} \tag{5.147}$$

We now evaluate the integral of (5.147) which is of the general form

$$\int_{-1}^{1} \mathrm{d}x \, \frac{e^{ax}}{(b-x)^2}.$$

Using two sequential changes of variables beginning with $y = b - x$,

$$\int_{-1}^{1} \mathrm{d}x \, \frac{e^{ax}}{(b-x)^2} = e^{ab} \int_{b-1}^{b+1} \mathrm{d}y \, \frac{e^{-ay}}{y^2} \tag{5.148}$$

and, then, $t = ay$, we arrive at

$$e^{ab} \int_{b-1}^{b+1} \mathrm{d}y \, \frac{e^{-ay}}{y^2} = ae^{ab} \int_{a(b-1)}^{a(b+1)} \mathrm{d}u \, \frac{e^{-u}}{u^2}. \tag{5.149}$$

By noting that

$$\frac{\mathrm{d}}{\mathrm{d}u} \frac{e^{-u}}{u} = -\frac{e^{-u}}{u} - \frac{e^{-u}}{u^2} \tag{5.150}$$

the integral of (5.149) is

$$ae^{ab} \int_{a(b-1)}^{a(b+1)} \mathrm{d}u \, \frac{e^{-u}}{u^2} = -ae^{ab} \int_{a(b-1)}^{a(b+1)} \mathrm{d}u \frac{e^{-u}}{u} - ae^{ab} \frac{e^{-u}}{u} \bigg|_{a(b-1)}^{a(b+1)}. \tag{5.151}$$

From the definition of the exponential integral provided in Appendix C.6

$$E_1(x) = \int_{x}^{\infty} \mathrm{d}u \, \frac{e^{-u}}{u}, \tag{5.152}$$

we can write the integral of the right-hand side of (5.152) as

$$\int_{a(b-1)}^{a(b+1)} \mathrm{d}u \frac{e^{-u}}{u} = E_1(a(b-1)) - E_1(a(b+1)). \tag{5.153}$$

Thus, the integral of the left-hand side of (5.151) is

$$
\begin{aligned}
ae^{ab}\int\limits_{a(b-1)}^{a(b+1)} \mathrm{d}u\,\frac{e^{-u}}{u^2} &= -ae^{ab}\left(E_1(a(b-1)) - E_1(a(b+1))\right) - ae^{ab}\left.\frac{e^{-u}}{u}\right|_{a(b-1)}^{a(b+1)} \\
&= -ae^{ab}\left(\left(E_1(a(b-1)) - E_1(a(b+1))\right) + \left.\frac{e^{-u}}{u}\right|_{a(b-1)}^{a(b+1)}\right) \\
&= -ae^{ab}\left(\left(E_1(a(b-1)) - E_1(a(b+1))\right) + \frac{e^{-a(b+1)}}{a(b+1)} - \frac{e^{-a(b-1)}}{a(b-1)}\right) \\
&= ae^{ab}\left(E_1(a(b+1)) - E_1(a(b-1))\right) + \frac{e^{a}}{b-1} - \frac{e^{-a}}{b+1}. \qquad (5.154)
\end{aligned}
$$

Having obtained this result, we return to the integral of (5.147) and write it as

$$
\begin{aligned}
&\int\limits_{-1}^{1} \mathrm{d}(\cos\theta)\,\frac{e^{\left(\frac{2R_{\mathrm{N}}}{\lambda\!\!\!^{-}}\right)^2\cos\theta}}{\left(\left(\frac{\chi_0^2}{2}+1\right) - \cos\theta\right)^2} = \\
&\left(\frac{2R_{\mathrm{N}}}{\lambda\!\!\!^{-}}\right)^2 e^{\left(\frac{2R_{\mathrm{N}}}{\lambda\!\!\!^{-}}\right)^2\left(1+\frac{\chi_0^2}{2}\right)}\left(\left(E_1\left(\left(\frac{2R_{\mathrm{N}}}{\lambda\!\!\!^{-}}\right)^2\left(2+\frac{\chi_0^2}{2}\right)\right) - E_1\left(\left(\frac{2R_{\mathrm{N}}}{\lambda\!\!\!^{-}}\right)^2\frac{\chi_0^2}{2}\right)\right)\right. \\
&\left. + \frac{e^{\left(\frac{2R_{\mathrm{N}}}{\lambda\!\!\!^{-}}\right)^2}}{\left(\frac{\chi_0^2}{2}\right)} - \frac{e^{-\left(\frac{2R_{\mathrm{N}}}{\lambda\!\!\!^{-}}\right)^2}}{\left(2+\frac{\chi_0^2}{2}\right)}\right).
\end{aligned}
\qquad (5.155)
$$

This leads us to a final expression for the total elastic Coulomb cross section which includes the combined effects of atomic electron screening and a finite nuclear size:

$$
\begin{aligned}
\sigma = &\frac{\pi\rho_{\mathrm{B}}^2(E)}{2}\left(\left(\frac{2R_{\mathrm{N}}}{\lambda\!\!\!^{-}}\right)^2 e^{\left(\frac{2R_{\mathrm{N}}}{\lambda\!\!\!^{-}}\right)^2\left(\frac{\chi_0^2}{2}\right)}\left(E_1\left(\left(\frac{2R_{\mathrm{N}}}{\lambda\!\!\!^{-}}\right)^2\left(2+\frac{\chi_0^2}{2}\right)\right) - E_1\left(\left(\frac{2R_{\mathrm{N}}}{\lambda\!\!\!^{-}}\right)^2\frac{\chi_0^2}{2}\right)\right)\right) \\
&+\frac{\pi\rho_{\mathrm{B}}^2(E)}{2}\left(\frac{1}{\left(\frac{\chi_0^2}{2}\right)} - \frac{e^{-2\left(\frac{2R_{\mathrm{N}}}{\lambda\!\!\!^{-}}\right)^2}}{\left(2+\frac{\chi_0^2}{2}\right)}\right)
\end{aligned}
\qquad (5.156)
$$

where the term containing the exponential integrals has been isolated deliberately. This allows us to see clearly that in the case of a point nucleus (i.e. $\frac{R_{\mathrm{N}}}{\lambda\!\!\!^{-}} \to 0$), the first term of (5.156) goes to zero and the second term remains finite to give the resulting screened elastic Coulomb scattering cross section:

$$
\sigma = \frac{4\pi\rho_{\mathrm{B}}^2(E)}{\chi_0^2(\chi_0^2+4)}, \qquad (5.157)
$$

a result which is also obtained directly by integrating (5.60).

Let us again consider our practical medical radiation examples of 100-keV electrons and 250-MeV protons incident to carbon and calcium. From Fig. 5.12 of the nuclear form factors for the electron examples, it is clear that the first term of (5.156) is negligible and $e^{-2\left(\frac{R_N}{\lambda}\right)^2} \to 1$, and the total elastic Coulomb cross sections for the 100-keV electrons incident to carbon and calcium are given by (5.157). As Fig. 5.12 shows that the nuclear form factor is significant for 250-MeV protons incident to these two targets, it is of interest to quantify the effect of the finite size of the nucleus upon the total elastic Coulomb cross section. In this case, we consider the example of 250-MeV protons incident to carbon. In this case, $\frac{R_N}{\lambda} = 7.96$ and $\chi_0 = 1.1 \times 10^{-5}$, from which $E_1\left(\left(\frac{2R_N}{\lambda}\right)^2\left(2+\frac{\chi_0^2}{2}\right)\right) \approx 0$ and $E_1\left(\left(\frac{2R_N}{\lambda}\right)^2\frac{\chi_0^2}{2}\right) \approx 17.5$. The inclusion of the effects of the finite nuclear size yields a result with a relative difference of less than 3×10^{-7} from the result of (5.157). Hence, despite the effort taken to include the role of the nuclear form factor in the derivation of the total elastic Coulomb scatter cross section, we conclude that, for particles and energies typical of medical applications, the result of (5.157), which arises from the second term of (5.156) and the integration of (5.60) with the nuclear form factor excluded, is adequate for our further discussions.

Bibliography and Further Readings

Dearnaley G. Ion bombardment and implantation. Rep Prog Phys. 1969;32:405–91.

ICRU. Elastic scattering of electrons and positrons. ICRU Report No. 77. International Commission on Radiation Units and Measurements. Oxford: Oxford University Press; 2007.

Messiah A. Quantum mechanics, vol. II. New York: Wiley; 1958.

Monceau P, Szydlo T, Valent G. Screened Coulomb scattering versus Thomson scattering. Eur J Phys. 2003;24:47–65.

Preston MA, Bhaduri RK. Structure of the nucleus. Reading: Addison-Wesley; 1975.

Rossi B. High-energy particles. New York: Prentice-Hall; 1952.

Semon MD, Taylor JR. Screened Coulomb scattering in the Eikonal approximation. Phys Rev A. 1977;16:33–40.

References

Bohr N. The penetration of atomic particles through matter. Matt Fys Medd Dan Vild Selsk. 1948;18(8):1–144.

Bonham RA, Strand TG. Analytical expressions for potentials of neutral Thomas-Fermi-Dirac atoms and for the corresponding atomic scattering factors for x-rays and electrons. J Chem Phys. 1963;39:2200–4.

Bransden BH, Joachain CJ. Physics of atoms and molecules. 2nd ed. Harlow: Pearson Education; 2003.

Condon EU, Odabasi H. Atomic structure. Cambridge: Cambridge University Press; 1980.

Dirac PAM. Note on exchange phenomena in the Thomas-Fermi atom. Proc Camb Philos Soc. 1930;26:376–85.

Fermi E. Un metodo statistic per la determinazione di alcune proprietà dell-atomo. Rend Acad Maz Lancei. 1927;6:602–7.

Firsov OB. Calculations of the interaction potential of atoms for small nuclear separations. Sov Phys JETP. 1957;5:1192–6.

Glauber RJ, Matthiae G. High-energy scattering of protons by nuclei. Nucl Phys 1970; B21:135–157

Hartree DR. The calculation of atomic structures. New York: Wiley; 1957.

Helm RH. Inelastic and elastic scattering of 187-MeV electrons from selected even-even nuclei. Phys Rev. 1956;104:1466–75.

Hoffman JD. Numerical methods for engineers and scientists. Singapore: McGraw-Hill; 1993.

Jablonski A, Salvat F, Powell CJ. Comparison of electron elastic-scattering cross sections calculated from two commonly used atomic potentials. J Phys Chem Ref Data. 2004;33:409–51.

Jackson JD. Classical electrodynamics. 3rd ed. Hoboken: Wiley; 1999.

Lieb EH. Thomas-Fermi and related theories of atoms and molecules. Rev Mod Phys. 1981;53:603–41.

McParland BJ. Nuclear medicine radiation dosimetry: advanced theoretical principles. London: Springer; 2010.

Molière VG. Theorie der Streung schneller geladener Teilchen. I. Einzelstreuung am ambegeschirmten Coulomb-field. Z Naturforschg. 1947;2a:133–45.

Parr RG, Ghosh SK. Thomas-Fermi theory for atomic systems. Proc Natl Acad Sci USA. 1986;83:3577–9.

Robinson MT. Interpretation of channelled ion energy loss spectra. Phys Rev. 1969;179:327–42.

Thomas LH. The calculation of atomic fields. Proc Camb Phil Soc. 1927;23:542–8.

Chapter 6
Multiple Elastic Coulomb Scatter

Contents

Abstract In the two previous chapters, we have considered singular scatter events only. However, the results of our derivations from these, whether for unscreened or screened charges, have shown that scatter is preferentially forward directed. As a consequence, when we consider the transport of a charged particle within a medium made up of multiple-scattering centres, we realise that a parallel beam of charged particles entering such a medium will generate an angular spectrum upon exiting. This is a consequence of multiple and primarily forward-directed scatters. We can, up to a point, assume that all of the scatters are forward directed and, providing a sufficient number of elastic scatters have occurred, apply the central-limit theorem and conclude that the spatial/angular distribution of projectiles after traversing a finite distance of medium is Gaussian. However, we cannot neglect the small, but non-negligible, contributions of large-angle multiple scatter which will add 'tails' to the Gaussian.

DOI 10.1007/978-1-4471-5403-7_6, © Springer-Verlag London 2014

In this chapter, we consider the statistics of multiple elastic Coulomb scatter, introduce the concept of the scattering power and begin investigations of the contributions of atomic electrons to the multiple scattering of charged projectiles from atoms. The focus on the chapter will be upon our detailed derivation and examination of four theories which derive multiple scatter probability distribution functions: those of Goudsmit and Saunderson, Williams, Fermi and Eyges and the most advanced, that of Molière.

Note For this chapter only, we will alter the nomenclature for the scattering angle: θ_1 signifies the scattering angle after a single-scattering event, and θ signifies the cumulative scattering angle after multiple single-scattering events.

6.1 Introduction

We have demonstrated that Coulomb elastic scatter from a bare, or unscreened, target leads to a $\sin^{-4}\left(\frac{\theta_1}{2}\right)$ angular dependence which diverges as $\theta_1 \to 0$. This divergence was removed when we considered the more realistic case of the target being screened by a surrounding charge, as with the nucleus of the atom. This led to a $\left(\left(\frac{\chi_0}{2}\right)^2 + \sin^2\left(\frac{\theta_1}{2}\right)\right)^{-2}$ angular dependence, where χ_0 is the atomic electron screening angle. Despite the finite result of the latter, in both cases, the differential cross section is preferentially forward peaked, although it is the screened cross section that remains finite as $\theta_1 \to 0$ or, from (4.20), a large impact parameter. We shall now progress to the case of scattering within a medium containing multiple-scattering centres. These multiple interactions, combined with the forward-directed angular dependence of the scattering cross section, will lead to the overall tendency for charged particles traversing a medium to be scattered through repeated small scattering angles with the net scattering angle being described by a probability distribution function (*pdf*). This is shown conceptually in Fig. 6.1.

A charged particle observed to have been scattered through a large angle will have been almost exclusively due to a single interaction as the product of the probabilities of individual backscatters through smaller angles becomes insignificant. Hence, particles with a small net angular deviation from the initial direction are considered to have undergone *multiple scatter*. The *pdf* of the net scattering angle for multiple scatter is to be shown to be Gaussian, as would be expected from the central-limit theorem, with a width that is derivable from a fundamental single-scattering theory. On the other hand, those events observed at large scattering angles are considered to have undergone *single scatter*. The intermediate state is that of *plural scatter* which while consisting of multiple-scattering events, the number of events being insufficient to yield the Gaussian *pdf*. Thus, the transitions from multiple to plural to single scatter follow with increasing scattering angles.

This chapter investigates the phenomenon of multiple elastic Coulomb scatter and its role in charged particle transport. The nature of multiple interactions

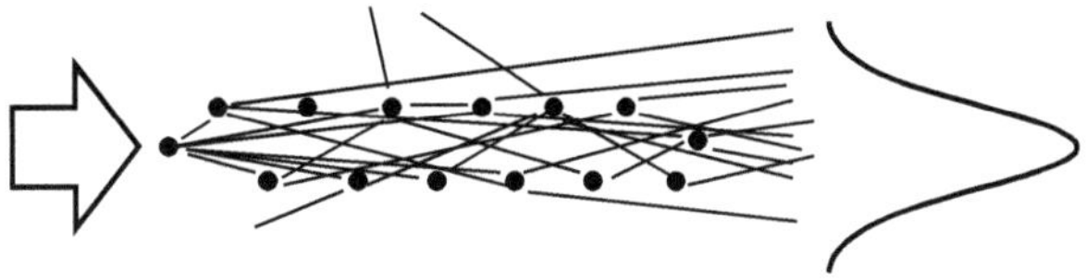

Fig. 6.1 Conceptual representation of multiple Coulomb scatter for a beam of charged projectiles incident from the left to an array of scattering centres. As a consequence of the angular distribution of the scattering, the tendency will be for multiple small-angle scatters for which the overall effect can be described by a probability distribution function as shown. Coulomb scatters through large angles are almost certainly, as a result of the low probability of large-angle scatter, due to a singular interaction between the charged particle and the scattering centre rather than a series of scatters of increasingly greater scatter angles with a negligible net probability of occurring

between a moving charged particle and the atoms of a medium leads inevitably to a statistical description of the resultant distribution of charged particle directions. These interactions will occur at various distances along the paths and which are dependent upon the properties of the medium and particle and the probability of interaction, the latter described by the total cross section. The mean value of these distances, defined as the mean free path length is derived followed by the derivation of the descriptive statistics of multiple scatter. The mean-square scattering angle following a single scatter is calculated and then extended to that for multiple scatter on the basis of the central-limit theorem which assumes a Gaussian distribution function describing multiple scatter. This is then followed by the presentation of the scattering power, which describes the change in mean-square scattering angle per unit distance travelled in a medium.

Upon these foundations, the theory of multiple elastic Coulomb scatter is investigated in detail. Although there have been many theoretical descriptions of multiple scatter, four which have played roles in medical radiation dosimetry calculations are developed. The first is the Goudsmit–Saunderson theory which describes the scattering probability distribution function as a Legendre polynomial expansion. The second is that of Williams which is based upon the central-limit theorem as a descriptor of multiple scatter and incorporates a single-scatter *pdf* to model the 'tails' reflecting large-angle scattering events. The third, and one which has been used most prominently in radiotherapy dose calculations, is the Fermi–Eyges theory. This is based upon an explicit solution by Fermi to the diffusion equation describing multiple scatter, but which excluded the effects of the energy lost by the charged particles through collisions and which yielded a Gaussian *pdf*. Eyges subsequently modified this result to account for the effects of energy transfer at each scattering event and which affects the multiple scatter distribution as a function of depth. Finally, we conclude with the comprehensive Molière theory which describes in a single form the combination of multiple, plural and single scattering of charged particles.

As previously performed, numerical examples of 100 keV electrons and 250 MeV protons incident to carbon and calcium will be provided so as to establish the ranges of magnitudes of the discussed effects of relevance to medical applications.

6.2 Mean Free Path Length for Elastic Coulomb Scatter

6.2.1 Definition of the Mean Free Path Length

Consider a charged particle traversing a medium of physical density ρ and atomic mass number A. We define the number density of scattering centres, i.e. the number of scattering centres (i.e. atoms) per unit volume, as

$$N \equiv \frac{\rho N_{\mathrm{A}}}{A} \tag{6.1}$$

where N_{A} is Avogadro's number. The probability per unit path length of an interaction occurring between a charged particle and an atom of the medium is proportional to the number of scattering centres, σN, where the constant of proportionality, σ, is the total cross section of the interaction. For our discussion, the total cross section will be that for elastic Coulomb scatter only. There will be a statistical distribution of path lengths due to the stochastic nature of these interactions and the mean free path length is the reciprocal of the probability per unit path length:

$$\lambda_{\mathrm{mfp}} = \frac{1}{N\sigma}. \tag{6.2}$$

6.2.2 Mean Free Path Length Between Elastic Coulomb Scatters

The total elastic Coulomb scattering cross section is given by

$$\begin{aligned}\sigma &= \frac{4\pi\rho_{\mathrm{B}}^2(E)}{\chi_0^2(\chi_0^2+4)} \\ &\approx \frac{\pi\rho_{\mathrm{B}}^2(E)}{\chi_0^2}.\end{aligned} \tag{6.3}$$

Substituting (6.3) into (6.2) yields the expression for the mean free path length between elastic Coulomb scatters from atoms (including the effects of atomic electron screening),

$$\begin{aligned}\lambda_{\mathrm{mfp}} &= \frac{1}{N\sigma} \\ &= \frac{\chi_0^2}{\pi N\rho_{\mathrm{B}}^2(E)}\end{aligned} \tag{6.4}$$

Fig. 6.2 Mean free path lengths between elastic Coulomb scatters from atoms for electrons and protons in carbon and calcium

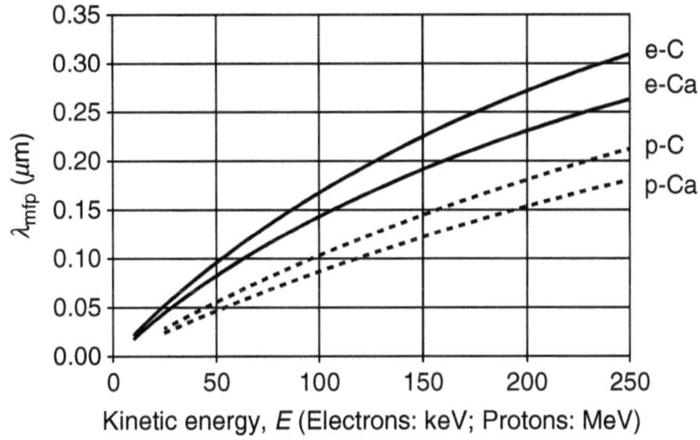

or by expanding the definitions of the atomic electron screening angle, scattering centre density and the Bohr collision diameter, we obtain the nonrelativistic result

$$\lambda_{\text{mfp}} = \left(\frac{A}{4\pi N_{\text{A}}\rho}\right)\left(\frac{\kappa\beta}{\alpha zZ}\right)^2 \tag{6.5}$$

where β is the projectile velocity normalised to that of light. Looking at the functional dependence of (6.5), we see that the mean free path length increases with increasing screening and decreases with increased Coulomb interaction with the scattering centre. Figure 6.2 shows the variation of λ_{mfp} as a function of kinetic energy for nonrelativistic electrons and protons in carbon and calcium.

It can be seen from Fig. 6.2 that the traversing of even 1 cm of tissue will result in several thousands of Coulomb interactions. This mean number of interactions will be calculated explicitly shortly.

It will be noted from (6.5) that for a given medium and projectile electric charge, λ_{mfp} scales as β^2.

6.3 Multiple Elastic Coulomb Scatter

6.3.1 *Introduction*

The transport of charged particles within a medium is, as a result of the forward-directed elastic Coulomb scatter and the variation of the mean free path length with energy, described stochastically due to multiple scattering from atoms. Hence, a *pdf* in scattering angle (or its projection onto orthogonal planes) is used to characterise the nature of this scattering. In this section, we consider four statistical multiple-scattering theories that have played, and some continue to play, significant roles in medical radiation practice. Their foundations all lie within the Rutherford cross section with additional considerations limited to atomic electron screening.

As shown in our previous derivation of the total elastic Coulomb scatter cross section, the role of finite nuclear size is often limited for medical radiation applications. The interested reader can find an examination of its effect in the paper by Cooper and Rainwater (1955).

6.3.2 *Scattering Angle Statistics*

6.3.2.1 Normalised Single-Scatter Distribution

As multiple scatter is the result of large numbers of single-scattering events, we look first at the descriptive statistics of single scatter. Two major quantities of interest in the description of multiple scatter will be the mean-square scattering angle, $\overline{\theta_1^2}$, and the root-mean-square scattering angle, $\sqrt{\overline{\theta_1^2}}$. In order to evaluate these, we first define the normalised single-scatter angular distribution,

$$f_1(\theta_1) = \frac{\frac{d\sigma}{d\Omega}}{\sigma}. \tag{6.6}$$

6.3.2.2 Mean-Square, Root-Mean Square and Mean Scattering Angles

The mean-square single-scattering angle is given by the integral of the square of the scattering angle weighted by the single-scatter distribution function,

$$\begin{aligned} \overline{\theta_1^2} &= \int d\Omega\, \theta_1^2 f_1(\theta_1) \\ &= \frac{\int d\Omega\, \theta_1^2 \frac{d\sigma}{d\Omega}}{\int d\Omega \frac{d\sigma}{d\Omega}}. \end{aligned} \tag{6.7}$$

We will use the unscreened (Rutherford) differential cross section and account for atomic electron screening through the imposition of the atomic electron screening angle χ_0 as the lower limit to the integration. Further, as the scattering cross section is very highly peaked for $\theta_1 \to 0$, we can use the small-angle approximation[1] (i.e. the differential cross section is proportional to $\frac{1}{\theta_1^4}$ and $d(\cos\theta_1) = -d\theta_1 \sin\theta_1 \approx -d\theta_1\, \theta_1$). We now consider an expression for the upper integration limit which will be dictated by the finite nuclear size as suggested in our earlier discussion of the nuclear form factor in Chap. 5. For all natures of distributed charges, $|F(\theta_1)|^2$ drops off rapidly with the dimensionless variable which is

[1] Recall that this small-angle approximation is reasonable over a large angular range with an agreement between $\sin\theta$ and θ being within 10 % for $\theta < \pi/4$.

proportional to the 3-vector momentum transfer, $\frac{qR_N}{2\hbar c} = \left(\frac{pR_N}{\hbar c}\right)\sin\frac{\theta_1}{2}$, so one could approximate the upper limit of the integral with the ratio $\frac{\hbar c}{pR_N}$. For low-energy electrons, the value of $\frac{\hbar c}{pR_N}$ can be very large, e.g. $\frac{\hbar c}{pR_N} \approx 273$ for 100 keV electrons incident to carbon. On the other hand, for 250 MeV protons incident to carbon, this upper limit angle is three orders-of-magnitude smaller, $\frac{\hbar c}{pR_N} \approx 0.13$. As an alternative, we can set the upper limit of the integration over θ as equal to π, reflecting the maximum scattering angle. Rossi (1952) has even suggested that one can set the upper limit to 1. In other calculations, the upper limit of other angular integrals has been set equal to ∞ for mathematical convenience. Then, the mean-square single-scattering angle in the small-angle approximation is

$$\overline{\theta_1^2} = \frac{\int_{\chi_0}^{\pi} d\theta_1\, \theta_1^3 \frac{d\sigma}{d\Omega}}{\int_{\chi_0}^{\pi} d\theta_1\, \theta_1 \frac{d\sigma}{d\Omega}} = \frac{\int_{\chi_0}^{\pi} \frac{d\theta_1}{\theta}}{\int_{\chi_0}^{\pi} \frac{d\theta_1}{\theta_1^3}} = \frac{2\ln\left(\frac{\pi}{\chi_0}\right)}{\left(\frac{1}{\chi_0^2} - \frac{1}{\pi^2}\right)} \approx 2\chi_0^2 \ln\left(\frac{\pi}{\chi_0}\right). \tag{6.8}$$

Using the Bohr form of the screening parameter from Table 5.1, the screening angle is, for the atomic number of the projectile much less than that of the target.[2]

$$\chi_0 = \frac{\sqrt{z^{\frac{2}{3}} + Z^{\frac{2}{3}}}}{a_\infty}\frac{\hbar c}{p}, \approx \frac{Z^{\frac{1}{3}}}{a_\infty}\frac{\hbar c}{p}. \tag{6.9}$$

Substituting this into (6.8) gives

$$\overline{\theta_1^2} \approx 2Z^{\frac{2}{3}}\left(\frac{\hbar c}{pa_\infty}\right)^2 \ln\left(\frac{\pi p a_\infty}{\hbar c} Z^{-\frac{1}{3}}\right) \approx 2Z^{\frac{2}{3}}\left(\frac{\hbar c}{pa_\infty}\right)^2 \ln\left(842 p Z^{-\frac{1}{3}}\right) \tag{6.10}$$

[2] This dependence upon Z would have also arisen had we defined the screening angle χ_0 in terms of the reciprocal of the Bohr radius, a_∞.

where the 3-vector momentum p is given in units of MeV. This result contains an explicit momentum dependence within the argument of the logarithm. It would be useful if this dependence were not present. If, instead of using π as the upper limit of integration, we were to replace it with $\frac{\hbar c}{pR_N}$, then these momentum terms cancel

$$\overline{\theta_1^2} \approx 2Z^{\frac{2}{3}} \left(\frac{\hbar c}{pa_\infty}\right)^2 \ln\left(\frac{\left(\frac{r_\infty}{R_N}\right)}{Z^{\frac{1}{3}}}\right). \tag{6.11}$$

We can simplify the argument of the logarithm further by noting that the nuclear radius can be approximated by $R_N \approx 0.94\,A^{\frac{1}{3}} \approx 1.18\,Z^{\frac{1}{3}}$ (in units of fm), where we have used the approximation of $A \approx 2Z$ for light nuclei. In this case, (6.11) will reduce to

$$\overline{\theta_1^2} \approx 2Z^{\frac{2}{3}} \left(\frac{\hbar c}{pa_\infty}\right)^2 \ln\left(0.84 r_0 Z^{-\frac{2}{3}}\right), \tag{6.12}$$

for which the Bohr radius is in units of fm. Then, ln $\left(0.84 a_\infty Z^{-\frac{2}{3}}\right) =$ ln $\left(4.44 \times 10^4\, Z^{-\frac{2}{3}}\right) \approx 2$ ln $\left(211\, Z^{-\frac{1}{3}}\right)$, which leads to

$$\begin{aligned}\overline{\theta_1^2} &\approx 4Z^{\frac{2}{3}} \left(\frac{\hbar c}{pa_\infty}\right)^2 \ln\left(211\, Z^{-\frac{1}{3}}\right) \\ &\approx 4\chi_0^2 \ln\left(211\, Z^{-\frac{1}{3}}\right).\end{aligned} \tag{6.13}$$

As has been noted by Jackson (1999), the multiplier within the logarithm of (6.13) varies between the derivations provided by different authors. Whereas our result has a value of 211, he derived a value of 204, and Rossi (1952) quoted a value of 196 $\left(\frac{Z}{A}\right)^{\frac{1}{6}}$ ≈ 175 also with the approximation of $A \approx 2Z$. However, as all of these multipliers reside within the argument of the logarithm, the differences between the magnitudes of the mean-square angles calculated with these multipliers are about only 4 % for elements with the low atomic numbers typical of tissue.

Figure 6.3 shows the root-mean-square (RMS) single-scattering angle ($\sqrt{\overline{\theta_1^2}}$) as a function of electron and proton kinetic energies typical of medical use in carbon and calcium as calculated from (6.13).[3] There are a number of factors to note from this Figure. First, $\sqrt{\overline{\theta_1^2}}$ for both projectiles is higher for calcium than carbon, reflecting the atomic number dependence of the scattering cross section. Second, $\sqrt{\overline{\theta_1^2}}$ decreases with increasing energy as slower particles are more easily deflected.

[3] Of course, the intrinsic spin of the electron is ignored in this calculation. But as noted in our derivation of the Mott scattering differential cross section, this effect will be restricted to high projectile velocities and large scattering angles.

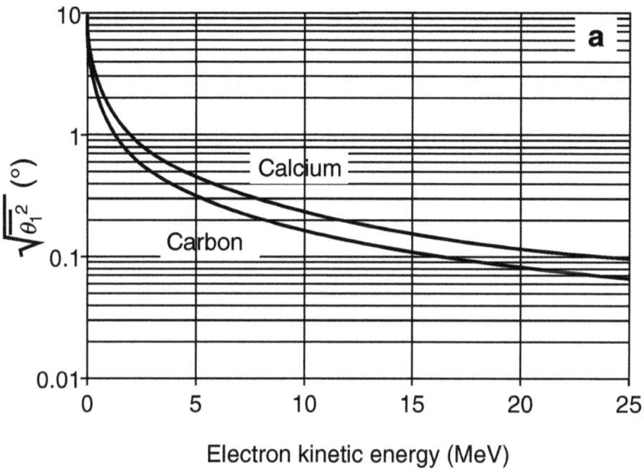

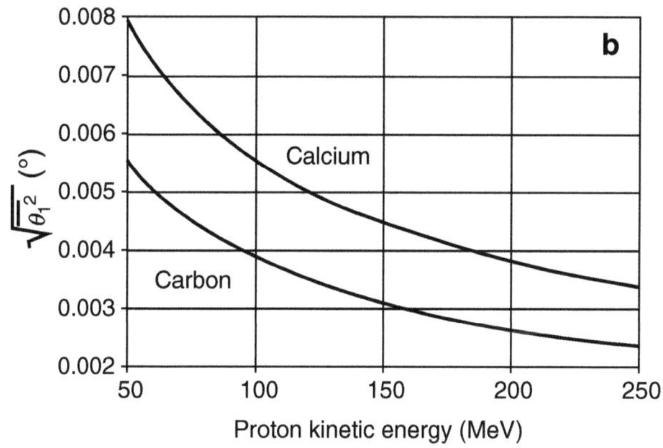

Fig. 6.3 Root-mean-square scattering angles for elastic Coulomb single scatter calculated for (**a**) electrons and (**b**) protons in carbon and calcium at kinetic energies typical of medical applications. Note that the ordinate of the plot in (**a**) is logarithmic

Third, $\sqrt{\overline{\theta_1^2}}$ is markedly higher for electrons than for protons as lighter particles are more easily deflected.

The mean single-scattering angle is

$$\overline{\theta_1} = \frac{\int \mathrm{d}\Omega\, \theta_1 \dfrac{\mathrm{d}\sigma}{\mathrm{d}\Omega}}{\int \mathrm{d}\Omega \dfrac{\mathrm{d}\sigma}{\mathrm{d}\Omega}} \approx \frac{\int_{\chi_0}^{\pi} \mathrm{d}\theta_1\, \theta_1^2 \dfrac{\mathrm{d}\sigma}{\mathrm{d}\Omega}}{\int_{\chi_0}^{\pi} \mathrm{d}\theta_1\; \theta_1 \dfrac{\mathrm{d}\sigma}{\mathrm{d}\Omega}}$$

$$\approx \frac{\int_{\chi_0}^{\pi} \frac{\mathrm{d}\theta_1}{\theta_1^2}}{\int_{\chi_0}^{\pi} \frac{\mathrm{d}\theta_1}{\theta_1^3}} \approx 2\frac{\left(\frac{1}{\chi_0} - \frac{1}{\pi}\right)}{\left(\frac{1}{\chi_0^2} - \frac{1}{\pi^2}\right)} \approx 2\chi_0. \qquad (6.14)$$

We will now anticipate our later derivations of the multiple-scattering theories and note that as the successive collisions are statistically independent,[4] following n scatters with n large, the central-limit theorem leads to the result that the distribution in scattering angle approaches the Gaussian with a variance given by $\overline{\Theta^2} = \bar{n}\overline{\theta_1^2}$, where $\bar{n}$ is the mean number of scatterings. From the definition of the probability of an elastic Coulomb scatter occurring per unit path length, one can calculate the mean number of such scatters, while traversing a medium of thickness s and physical density ρ is

$$\bar{n} = N\,\sigma\,s. \tag{6.15}$$

We now derive two expressions for $\overline{\Theta^2}$ which will depend upon the nature of the $\overline{\theta_1^2}$ selected. For the case of $\frac{\hbar c}{pR_{\mathrm{N}}} > \pi$, we would use the expression for $\overline{\theta_1^2}$ given by (6.8):

$$\begin{aligned}\overline{\Theta^2} &= 2\pi N \rho_{\mathrm{B}}^2(E)\, s \ln\left(\frac{\pi}{\chi_0}\right) \\ &= 2\pi\left(\frac{\rho N_{\mathrm{A}} s}{A}\right)\left(\frac{\alpha\hbar c\, zZ}{E}\right)^2 \ln\left(\frac{\pi\, a_\infty\, p}{\hbar c\,\sqrt{z^{\frac{2}{3}} + Z^{\frac{2}{3}}}}\right).\end{aligned} \tag{6.16}$$

On the other hand, for the case of $\frac{\hbar c}{pR_{\mathrm{N}}} < \pi$, one can use the form of $\overline{\theta_1^2}$ given by (6.13):

$$\begin{aligned}\overline{\Theta^2} &= 4\pi N_{\mathrm{S}}\, \rho_{\mathrm{B}}^2(E)\, s \ln\left(211\, Z^{-\frac{1}{3}}\right) \\ &= 4\pi\left(\frac{\rho N_{\mathrm{A}} s}{A}\right)\left(\frac{\alpha\hbar c\, zZ}{E}\right)^2 \ln\left(211\, Z^{-\frac{1}{3}}\right).\end{aligned} \tag{6.17}$$

Although we will return to this discussion later, there are two functional dependencies that should be noted from these results:

- $\overline{\Theta^2}$ increases linearly with s. This observation shall need to be considered in the context of the fact that as the charged particle moves through the medium, it loses energy along its path and this linearity will be lost eventually along large path lengths as E in the denominator will decrease. For short path lengths and non-negligible energy loss, one can calculate for the arithmetic mean of the kinetic energy along the path length. For thicker path lengths, integration accounting for energy loss is required. This is performed in the Fermi–Eyges multiple-scattering theory to be considered later in this chapter.

---

[4] Which, by the definition of elastic scatter in which there is no energy loss in a collision, is true.

- By ignoring the small contribution from the logarithmic term, $\overline{\Theta^2}$ increases linearly with Z^2 as high atomic number media induce greater degrees of multiple scattering. This dependence upon the target's atomic number will be investigated shortly as it neglects the separate contributions of the nucleus and of the atomic electrons to the scattering interaction. A more thorough examination is provided at the end of this chapter.

6.3.3 *Scattering Power*

6.3.3.1 Introduction

The above discussion considered the mean-square scattering angle associated with a single scatter and the resultant cumulative mean-square angle after traversing a given thickness of medium. What is often of practical interest in radiation transport and dosimetry calculations is the rate of change in the mean-square scattering angle with distance. As noted before, this will be the consequence of large numbers of single-scattering events which are each considered to be statistically independent.

There are two metrics that are useful in describing this change in mean-square scattering angle over distance travelled. The first is the linear scattering power defined as the derivative of the mean-square angle with total distance travelled, s,

$$\Im \equiv \frac{\mathrm{d}\overline{\Theta^2}}{\mathrm{d}s}. \tag{6.18}$$

The second metric arises as (6.18) will vary with the physical density of the medium as seen in (6.16) and (6.17). This dependency can be removed by normalising the linear scattering power to the physical density so as to define the mass scattering power,

$$\frac{\Im}{\rho} \equiv \frac{\mathrm{d}\overline{\Theta^2}}{\rho\,\mathrm{d}s}. \tag{6.19}$$

6.3.3.2 Spin-0 Projectiles

From (6.16), the mass scattering power for spin-0 projectiles is

$$\frac{\Im}{\rho} = 2\pi\left(\frac{N_{\mathrm{A}}}{A}\right)\left(\frac{\alpha\hbar c\,zZ}{E}\right)^2 \ln\left(\frac{\pi\, r_\infty\, p}{\hbar c\,\sqrt{z^{\frac{2}{3}}+Z^{\frac{2}{3}}}}\right). \tag{6.20}$$

and that from (6.17) is

$$\frac{\Im}{\rho} \approx 4\pi \, \frac{N_{\mathrm{A}}}{A} \left(\frac{\alpha\hbar c \, zZ}{E} \right)^2 \ln\left(211 \, Z^{-\frac{1}{3}} \right). \tag{6.21}$$

6.3.3.3 Spin-½ I: Mott Scattering of Electrons

In Chap. 4, we derived the differential cross section for the electric Coulomb scatter of an electron from an electrostatic field and which accounted for the intrinsic spin-½ of the electron. It was noted that the derivation would be the same for the positron.[5] The resulting differential cross section in the small-angle limit is

$$\frac{\mathrm{d}\sigma}{\mathrm{d}\Omega} \approx \rho_{\mathrm{B}}^2(E) \frac{\left(1 - \left(\frac{\beta\theta_1}{2}\right)^2\right)}{\theta_1^4}. \tag{6.22}$$

The Mott mean-square single-scattering angle is

$$\begin{aligned}
\overline{\theta_1^2} &= \frac{\int\limits_{\chi_0}^{\pi} \mathrm{d}\theta_1 \, \frac{1 - \left(\frac{\beta\theta_1}{2}\right)^2}{\theta_1}}{\int\limits_{\chi_0}^{\pi} \mathrm{d}\theta_1 \, \frac{1 - \left(\frac{\beta\theta_1}{2}\right)^2}{\theta_1^3}} \\
&= \frac{\int\limits_{\chi_0}^{\pi} \frac{\mathrm{d}\theta_1}{\theta_1} - \left(\frac{\beta}{2}\right)^2 \int\limits_{\chi_0}^{\pi} \mathrm{d}\theta_1 \, \theta_1}{\int\limits_{\chi_0}^{\pi} \frac{\mathrm{d}\theta_1}{\theta_1^3} - \left(\frac{\beta}{2}\right)^2 \int\limits_{\chi_0}^{\pi} \frac{\mathrm{d}\theta_1}{\theta_1}} \\
&\approx \frac{\ln\left(\frac{\pi}{\chi_0}\right) - \frac{(\beta\pi)^2}{8}}{\left(\frac{1}{2\chi_0^2}\right) - \left(\frac{\beta}{2}\right)^2 \ln\left(\frac{\pi}{\chi_0}\right)}.
\end{aligned} \tag{6.23}$$

As expected, this result reduces to the spin-0 result of (6.8) for $\beta \to 0$.

The Mott total cross section is straightforward to derive and only the highlights are presented here,[6]

[5] To the lowest order of expansion in powers of α.

[6] As the integration is to be over all angles greater than the atomic electron screening angle, we decide to not use the small-angle approximation in the calculation of the total Mott cross section.

$$\sigma = 2\pi \left(\frac{\rho_B(E)}{4}\right)^2 \int_{\chi_0}^{\pi} d\theta_1 \sin\theta_1 \left(\frac{1}{\sin^4\frac{\theta_1}{2}} - \frac{\beta^2}{\sin^2\frac{\theta_1}{2}}\right)$$
$$= \pi \frac{\rho_B^2(E)}{8} \left(I_1 - \beta^2 I_2\right) \tag{6.24}$$

where

$$I_1 = \int_{\chi_0}^{\pi} d\theta_1 \frac{\sin\ \theta_1}{\sin^4\frac{\theta_1}{2}}$$
$$= 2\frac{1}{\sin^2\left(\frac{\chi_0}{2}\right) - 1} \tag{6.25}$$

and

$$I_2 = \int_{\chi_0}^{\pi} d\theta_1 \frac{\sin\ \theta_1}{\sin^2\frac{\theta_1}{2}}$$
$$= -4\ln\ \sin\frac{\chi_0}{2}. \tag{6.26}$$

Inserting these results into (6.24) and noting that $\chi_0 << 1$, we obtain

$$\sigma \approx \pi \frac{\rho_B^2(E)}{2} \left(\frac{2}{\chi_0^2} + \beta^2 \ln\chi_0\right). \tag{6.27}$$

Note that in the nonrelativistic case, the Mott total cross section of (6.27) is in agreement with that of the spinless projectile given by (6.3). The Mott mass scattering power is

$$\left.\frac{\Im}{\rho}\right|_{\text{Mott}} = 2\pi \frac{N_A}{A} \rho_B^2(E) \left(\ln\left(\frac{\pi}{\chi_0}\right) - \frac{(\beta\pi)^2}{8}\right). \tag{6.28}$$

We can see that this result reduces to the previously derived spinless case of (6.20) as $\beta \to 0$. Using the previous arguments regarding the argument of the logarithm, we can rewrite this result as

$$\left.\frac{\Im}{\rho}\right|_{\text{Mott}} = 2\pi \frac{N_A}{A} \rho_B^2(E) \left(2\ln\left(211\, Z^{-\frac{1}{3}}\right) - \frac{(\beta\pi)^2}{8}\right)$$
$$= 4\pi \frac{N_A}{A} \left(\frac{\alpha\hbar c Z}{E}\right)^2 \left(\ln\left(211\, Z^{-\frac{1}{3}}\right) - \frac{(\beta\pi)^2}{16}\right) \tag{6.29}$$

where, as the projectile is an electron, $z = 1$.

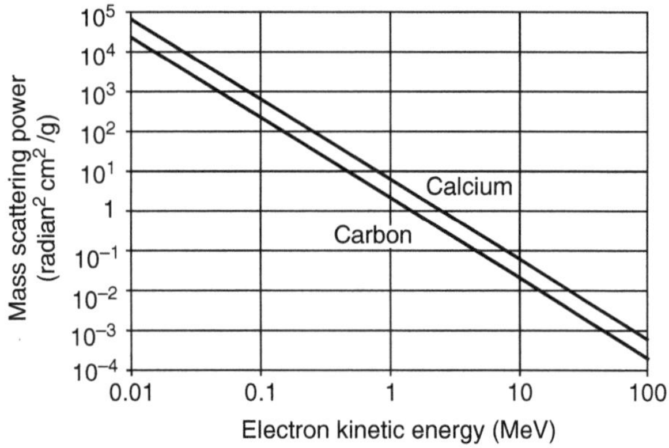

Fig. 6.4 Electron mass scattering power for carbon and calcium calculated from Eq. (6.29). Note that the contributions of the target's atomic electrons are not included

The mass scattering power for electrons in carbon and calcium as a function of electron kinetic energy, as calculated from (6.29), is plotted in Fig. 6.4. As the ordinate and abscissa scales are both logarithmic, the linear appearances and negative slopes show that the $\frac{(\beta\pi)^2}{16}$ term, due to the electron's intrinsic spin-½, has quite little influence upon the magnitude of the mass scattering power. This is evident by noting that $\frac{(\beta\pi)^2}{8} < 0.62$ and that the magnitude of the term $\ln\left(211\,Z^{-\frac{1}{3}}\right)$ is equal to 4.75 and 4.35 for carbon and calcium, respectively.

A comparison of the magnitudes of the mass scattering powers of Fig. 6.4 with those calculated by the author (McParland 2010) using the exact expressions (i.e. without assumptions of $Z \approx \frac{A}{2}$, small-angle approximation) and a different screening angle expression will show that there are differences in the numerical results of the scattering power calculations and that these are significant only at low energies.

6.3.4 Contributions of Atomic Electrons I

The above derivations have not considered the contribution to the elastic Coulomb scatter cross section that can be made by the atomic electrons (this is not the same as the screening of the nuclear electrostatic field by these electrons). These will increase the net electric charge and, hence, 'effective' atomic number seen by the projectile. An approximate method of allowing for this effect was suggested by Williams (1940) and Kulchitsky and Latyshev (1942) through the replacement of Z by $\sqrt{Z(Z+1)}$ which reflects a coherent summation of Z charges each with a single electric charge, e (i.e. the electron). This approximation will increase the scattering powers of carbon and calcium by 17 % and 5 %, respectively. This approach, however, assumes implicitly that the scattering cross sections for nuclear and electronic targets are the same which, clearly, they are not. Moreover, it implicitly assumes that the recoil kinetic energy far exceeds the atomic electron's binding energy. A refinement proposed by Fano (1954), which allows for these effects, will be discussed later in this chapter in the context of the Molière multiple-scattering theory.

6.3.5 *Goudsmit–Saunderson Theory*

6.3.5.1 Introduction

The charged particle of significant interest in medical dosimetry is the electron as it is set into motion through photon–matter interactions (i.e. photoelectric effect, Compton scatter, pair/triplet production) in tissue and is itself used directly as a means of therapy. As the electron has little mass, it will be subject to considerable magnitudes of scatter as was demonstrated in Fig. 6.3. Hence, it is appropriate that we begin our discussion of multiple-scattering theories with that by Goudsmit and Saunderson (1940) as it is exclusive to electrons and extends from our previous discussion of multiple, statistically independent scatters.

In the Goudsmit–Saunderson theory, the scattering angle is assumed to be sufficiently small so that the electron's path length is equal to the thickness of the scatterer (in other words, the result is strictly valid only for thin foils or short discrete steps in a Monte Carlo simulation). The energy losses occurring through these collisions are neglected as the scatters are assumed to be perfectly elastic. One significant property of the Goudsmit–Saunderson theory is that it is not restricted to any underlying single-scattering differential cross section. As such, it uses the normalised single-scatter angle distribution of (6.6) from where we begin. Our derivation of the Goudsmit–Saunderson multiple-scattering *pdf* will follow the following three steps:

- The *pdf* will be written as an expansion (infinite series) of Legendre polynomials with the series coefficients incorporate the Poisson distribution describing the probability of a given number of scatters occurring within a given path length (thus, using our previous derivation of λ_{mfp}).
- The coefficients of this series are then determined using the screened Coulomb elastic single-scattering cross section as input.
- The Goudsmit–Saunderson result is shown to be well represented by a Gaussian *pdf* with a variance evaluated in terms of the series.

6.3.5.2 Legendre Polynomial Expansions

Prior to the derivation, we first recall a property of the Legendre polynomials derived in Appendix C.2. As these polynomials are orthogonal, we can expand the scattering distribution into a series of Legendre polynomials. Consider the electron to have been scattered n times, after which it is following a net direction given by an angle θ to its original trajectory. An additional property of the Legendre polynomial is that, if the scattering has azimuthal symmetry, the mean value of the Legendre polynomial after n scatters will be equal to the nth power of its mean value following a single scatter. This can be shown readily by considering two sequential scatters. Let the electron be first scattered through an angle of deflection θ_a with a corresponding azimuthal angle, ϕ_a. The electron then undergoes a second scatter with an atom through the corresponding pair of angles (θ_b, ϕ_b). The net scattering angle is thus $\theta_a + \theta_b$ and

$$\begin{aligned} P_l(\cos(\theta_a+\theta_b)) &= P_l(\cos\theta_a)\,P_l(\cos\theta_b) \\ &\quad + \sum_{m=-l}^{l} P_l^m(\cos\theta_a)P_l^m(\cos\theta_b)\sin(m(\phi_b-\phi_a)). \end{aligned} \tag{6.30}$$

Averaging both sides gives

$$\overline{P_l(\cos(\theta_a+\theta_b))} = \overline{P_l(\cos\theta_a)}\,\overline{P_l(\cos\theta_b)} \tag{6.31}$$

where, because of the axial symmetry, averaging over $\sin(m(\phi_b - \phi_a))$ yields zero. As this result is for two sequential scatters, it can be extended to n sequential scatters to give the general result,

$$\overline{P_l(\cos(\theta))} = \left(\overline{P_l(\cos\theta_1)}\right)^n. \tag{6.32}$$

Recall our original introduction of the normalised single-scatter angular distribution of $f_1(\theta_1)$. This distribution may be conveniently expanded in a series of Legendre polynomials:

$$f_1(\theta_1) = \frac{1}{4\pi}\sum_{l=0}^{\infty} F_l\,(2l+1)\,P_l(\cos\theta_1) \tag{6.33}$$

where the coefficients are given by

$$\begin{aligned} F_l &= 2\pi\int_{-1}^{1} \mathrm{d}(\cos\theta_1)\,P_l(\cos\theta_1)f_1(\theta_1) \\ &= \overline{P_l(\cos\theta_1)}. \end{aligned} \tag{6.34}$$

6.3.5.3 Multiple-Scattering Probability Distribution Function

Following the above expansion, one is then able to write the Goudsmit–Saunderson multiple scatter *pdf* as

$$f_{GS}(\theta; s) = \frac{1}{4\pi} \sum_{l=0}^{\infty} (2l+1)\, G_l\, P_l(\cos\theta) \tag{6.35}$$

where G_l is the mean of the lth-order Legendre polynomial after the case of all possible values of n scatters, leading to the net scattering angle of θ.[7] This is denoted as

$$G_l = \sum_{n=0}^{\infty} p(n; s) \left(\overline{P_l(\cos\theta_1)}\right)^n \tag{6.36}$$

where $p(n; s)$ is the Poisson probability of n scatters occurring while the electron traverses a thickness s of the medium. (6.35) is normalised such that

$$\begin{aligned} \int \mathrm{d}\Omega f_{GS}(\theta; s) &= 2\pi \int_{-1}^{1} \mathrm{d}(\cos\theta) f_{GS}(\theta; s) \\ &= 1. \end{aligned} \tag{6.37}$$

The mean of the lth-order Legendre polynomial term is

$$\begin{aligned} \int \mathrm{d}\Omega f_{GS}(\theta; s)\, P_l(\cos\theta) &= 2\pi \int_{-1}^{1} \mathrm{d}(\cos\theta) \left(\frac{1}{4\pi} \sum_{l'=0}^{\infty} (2l'+1)\, G_{l'}\, P_{l'}(\cos\theta) \right) P_l(\cos\theta) \\ &= \frac{1}{2} \sum_{l'=0}^{\infty} (2l'+1) G_{l'} \int_{-1}^{1} \mathrm{d}(\cos\theta)\, P_{l'}(\cos\theta)\, P_l(\cos\theta) \\ &= \frac{1}{2} \sum_{l'=0}^{\infty} (2l'+1) G_{l'} \left(\frac{2}{2l+1} \right) \delta_{l'l} \\ &= G_l. \end{aligned} \tag{6.38}$$

Hence, once the parameters G_l are known, all angular averages of the scattering distribution can be evaluated. Two examples of this result are

$$\begin{aligned} G_1 &= \int \mathrm{d}\Omega f_{GS}(\theta; s)\, P_1(\cos\theta) \\ &= \overline{\cos\theta} \end{aligned}$$

and

[7] It should be noted that s is the total distance travelled, which, due to scatter, will in general be greater than the maximum penetration, x, of the electron into the medium. However, the Goudsmit–Saunderson derivation considers sufficiently thin media such that s and x are equal.

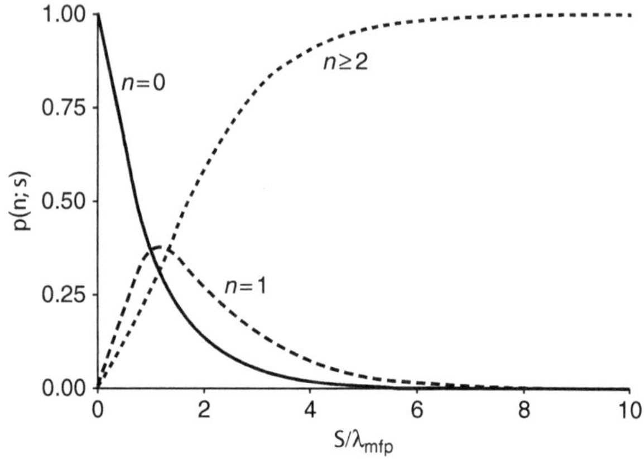

Fig. 6.5 The probability of electron elastic scatter occurring for $n = 0$, 1 and ≥ 2 scatters as a function of the ratio of the path length s to the elastic Coulomb scatter mean free path, λ_{mfp}

$$G_2 = \int d\Omega f_{GS}(\theta; s) P_2(\cos\theta)$$
$$= \frac{1}{2} \int d\Omega f_{GS}(\theta; s) \left(\overline{3\cos^2\theta} - 1\right)$$

from which one obtains

$$\overline{\cos^2\theta} = \frac{1}{3}(2G_2 + 1).$$

As a result, the evaluation of the scattering distribution$f_{GS}(\theta; s)$ has been reduced to the evaluation of the mean values of the Legendre polynomials. However, it is necessary to recall that the definition of G_l incorporates the probability of the electron undergoing n collisions, which is described by the associated Poisson distribution, $p(n; s)$. In order to evaluate this probability, we recall the definition of the mean number of collisions, $\overline{n}$, over the thickness s given by (6.15).

$\overline{n}$ must be large in order to have the condition of multiple scattering be valid through the central-limit theorem. The probability that an electron will undergo n collisions while traversing s is

$$p(n; s) = \frac{e^{-\overline{n}} (\overline{n})^n}{n!}$$
$$= \frac{e^{-\left(\frac{s}{\lambda_{\text{mfp}}}\right)} \left(\frac{s}{\lambda_{\text{mfp}}}\right)^n}{n!}. \tag{6.39}$$

The probabilities of electron elastic scatter for $n = 0, 1$ and ≥ 2 scatters are provided in Fig. 6.5 as functions of the ratio of the path length to the elastic scatter mean free path. The probability that no elastic scattering occurs becomes negligible for a thickness of $s \approx 4\lambda_{\text{mfp}}$ and greater, whereas the probability of $n \geq 2$ scatters approaches unity for $s \approx 6\lambda_{\text{mfp}}$ at about the same thickness. Both of these probabilities are monotonic functions of the ratio $\frac{s}{\lambda_{\text{mfp}}}$. To the contrary, the probability of a single scatter occurring is non-monotonic and reaches a maximum of about 0.35 at $s \approx 1.3\,\lambda_{\text{mfp}}$.

Inserting the probability of n collisions given by (6.39) into (6.36) gives

$$G_l = \sum_{n=0}^{\infty} \frac{e^{-\overline{n}}(\overline{n})^n}{n!} \left(\overline{P_l(\cos\theta_1)}\right)^n. \tag{6.40}$$

By recalling the series expansion of the exponential, $e^x = \sum_{n=0}^{\infty} \frac{x^n}{n!}$, we can rewrite (6.40) as

$$\begin{aligned} G_l &= e^{-\overline{n}} \sum_{n=0}^{\infty} \frac{\left(\overline{n}\,\overline{P_l(\cos\theta_1)}\right)^n}{n!} \\ &= e^{-\overline{n}}\, e^{-\overline{n}\,\overline{P_l(\cos\theta_1)}} \\ &= e^{-sQ_l} \end{aligned} \tag{6.41}$$

where we have defined the lth-order transport coefficient as

$$Q_l \equiv \frac{1 - \overline{P_l(\cos\theta_1)}}{\lambda_{\text{mfp}}}. \tag{6.42}$$

The Goudsmit–Saunderson scattering distribution is now of the form

$$f_{GS}(\theta; s) = \frac{1}{4\pi} \sum_{l=0}^{\infty} (2l+1)\, e^{-sQ_l}\, P_l(\cos\theta). \tag{6.43}$$

Let us consider the first-order transport coefficient which from (6.42) is

$$\begin{aligned} Q_1 &= \frac{1 - \overline{P_1(\cos\,\theta_1)}}{\lambda_{\text{mfp}}} \\ &= \frac{2\pi}{\lambda_{\text{mfp}}} \int_{-1}^{1} \mathrm{d}(\cos\,\theta_1) f_1(\theta_1)\,(1 - \cos\,\theta_1) \\ &- 2\pi\rho\left(\frac{N_A}{A}\right)\sigma \int_{-1}^{1} \mathrm{d}(\cos\,\theta_1)\, \frac{\frac{\mathrm{d}\sigma}{\mathrm{d}\Omega}}{\sigma}(1 - \cos\,\theta_1) \\ &= 2\pi\rho\left(\frac{N_A}{A}\right) \int_{-1}^{1} \mathrm{d}(\cos\,\theta_1)\, \frac{\mathrm{d}\sigma}{\mathrm{d}\Omega}(1 - \cos\,\theta_1). \end{aligned} \tag{6.44}$$

By assuming the unscreened Rutherford cross section, we refine further the integral of (6.44) by invoking the small-angle approximations of $\mathrm{d}(\cos\,\theta_1) = -\mathrm{d}\theta_1\theta_1$ and $1 - \cos\,\theta_1 = 2\sin^2\left(\frac{\theta_1}{2}\right) \approx \frac{\theta_1^2}{2}$ and setting the lower limit to be equal to the atomic electron screening angle, χ_0, to yield the expression

$$
\begin{aligned}
Q_1 &\approx \pi\rho\left(\frac{N_\mathrm{A}}{A}\right)\rho_\mathrm{B}^2(E)\int_{\chi_0}^{\pi}\frac{\mathrm{d}\theta_1}{\theta_1} \\
&= \pi\rho\left(\frac{N_\mathrm{A}}{A}\right)\rho_\mathrm{B}^2(E)\ln\left(\frac{\pi}{\chi_0}\right).
\end{aligned}
\tag{6.45}
$$

If we compare the result of (6.45) to (6.20), we see that

$$
\begin{aligned}
Q_1 &= \frac{\rho}{2}\left(\frac{\Im}{\rho}\right) \\
&= \frac{\Im}{2}.
\end{aligned}
\tag{6.46}
$$

That is, the first-order transport coefficient is equal to one-half of the linear scattering power.

We now consider the fundamental single-scattering law that is introduced as input to the Goudsmit–Saunderson theory. We use the exponential-screening form derived in Chap. 5:

$$
\frac{\mathrm{d}\sigma}{\mathrm{d}\Omega} = \left(\frac{\rho_\mathrm{B}(E)}{4}\right)^2 \frac{1}{\left(\left(\frac{\chi_0}{2}\right)^2 + \sin^2\frac{\theta_1}{2}\right)^2}.
$$

We will simplify the following derivations by defining the variable

$$
y = \sin\frac{\theta_1}{2}.
\tag{6.47}
$$

In this case, the total cross section for the single-scatter case can be written as

$$
\begin{aligned}
\sigma &= \int \mathrm{d}\Omega\,\frac{\mathrm{d}\sigma}{\mathrm{d}\Omega} \\
&= 2\pi\left(\frac{\rho_\mathrm{B}(E)}{4}\right)^2\int_0^1 \mathrm{d}y\,\frac{g(y)}{y^3}
\end{aligned}
$$

where $g(y)$ is the multiplicative factor accounting for atomic electron screening which, in our model, is

$$g(y) = \frac{y^4}{\left(y_0'^2 + y^2\right)^2} \tag{6.49}$$

where $y_0' \equiv \frac{\chi_0}{2}$.

We first evaluate the average values of the Legendre polynomials following a single collision and which are used in the transport coefficient. This is performed by the following expansion of the lth-order Legendre polynomial (taken by expanding the Rodrigues' formula of (C.62)):

$$P_l(\cos\theta) = \sum_{k=0}^{l} (-1)^k \frac{(l+k)!}{(l-k)!\,(k!)^2} y^{2k}. \tag{6.50}$$

The average value after a single collision will be

$$\overline{P_l(\cos\theta_1)} = \frac{2\pi}{\sigma}\left(\frac{\rho_{\mathrm{B}}(E)}{4}\right)^2 \sum_{k=0}^{l} (-1)^k \frac{(l+k)\,!}{(l-k)!\,(k!)^2} \int_0^1 \mathrm{d}y\; g(y) y^{2k-3}. \tag{6.51}$$

We isolate the first two terms of the summation of (6.51):

$$\begin{aligned}
\overline{P_l(\cos\;\theta_1)} &= \frac{2\pi}{\sigma}\left(\frac{\rho_{\mathrm{B}}(E)}{4}\right)^2 \left(\int_0^1 \mathrm{d}y\,\frac{g(y)}{y^3} - l(l+1)\int_0^1 \mathrm{d}y\,\frac{g(y)}{y}\right.\\
&\quad \left. + \sum_{k=2}^{l} (-1)^k \frac{(l+k)!}{(l-k)!\,(k!)^2} \int_0^1 \mathrm{d}y\; g(y) y^{2k-3}\right)\\
&= 1 - \frac{2\pi}{\sigma}\left(\frac{\rho_{\mathrm{B}}(E)}{4}\right)^2 \left(l(l+1)\int_0^1 \mathrm{d}y\,\frac{g(y)}{y}\right.\\
&\quad \left. - \sum_{k=2}^{l} (-1)^k \frac{(l+k)!}{(l-k)!\,(k!)^2} \int_0^1 \mathrm{d}y\; g(y) y^{2k-3}\right).
\end{aligned} \tag{6.52}$$

Recalling the definition of the transport coefficient, we have

$$\begin{aligned}
\lambda_{\mathrm{mfp}}\, Q_l &= \frac{2\pi}{\sigma}\left(\frac{\rho_{\mathrm{B}}(E)}{4}\right)^2 \left(l(l+1)\int_0^1 \mathrm{d}y\,\frac{g(y)}{y}\right.\\
&\quad \left. - \sum_{k=2}^{l} (-1)^k \frac{(l+k)!}{(l-k)!\,(k!)^2} \int_0^1 \mathrm{d}y g(y)\, y^{2k-3}\right).
\end{aligned} \tag{6.53}$$

As the major contributions to the integral would be those from scattering events associated with large θ_1 (i.e. small impact parameters and those cases of $y \approx 1 >> y'_0$) where screening is not a factor (i.e. $g(y) = 1$), then

$$\sum_{k=2}^{l} (-1)^k \frac{(l+k)!}{(l-k)!\,(k!)^2} \int_0^1 \mathrm{d}y\, g(y)\, y^{2k-3} \approx \sum_{k=2}^{l} (-1)^k \frac{(l+k)!}{(l-k)!\,(k!)^2\,(2k-2)}$$
$$= l(l+1) \sum_{k=2}^{l} \frac{1}{k}. \tag{6.54}$$

While Goudsmit and Saunderson provided the result of (6.54) without proof, Bethe (1953) provided a derivation of the expression. For convenience and interest, this proof is provided at the end of this chapter in a more transparent manner as part of our derivation of the Molière *pdf.*

However, the condition of $g(y) = 1$ will not be met for very large values of l (corresponding to small χ_0). To demonstrate this, we crudely model the screening function in a discontinuous form

$$g(y) = H(y - y'_0) \tag{6.55}$$

where $H(x)$ is the Heaviside function and y'_0 reflects the effect of the finite radial extent of the atom. Then,

$$\sum_{k=2}^{l} \frac{(l+k)!}{(l-k)!\,(k!)^2} \int_{y'_0}^{1} \mathrm{d}y\, y^{2k-3} = \sum_{k=2}^{l} \frac{(l+k)!}{(l-k)!\,(k!)^2} \frac{1 - y_0'^{(2k-2)}}{(2k-2)}, \tag{6.56}$$

leading to the requirement of $ly'_0 << 1$. Substituting (6.54) into (6.53) gives

$$\lambda_{\mathrm{mfp}} Q_l = \frac{2\pi}{\sigma} \left(\frac{\rho_{\mathrm{B}}(E)}{4} \right)^2 l(l+1) \left(\int_0^1 \mathrm{d}y\, \frac{g(y)}{y} - \sum_{k=2}^{l} \frac{1}{k} \right). \tag{6.57}$$

We now solve for the integral

$$\begin{aligned} \int_0^1 \mathrm{d}y\, \frac{g(y)}{y} &= \int_0^1 \mathrm{d}y\, \frac{y^3}{\left(y_0'^2 + y^2\right)^2} \\ &\approx -\left(\frac{1}{2} + \ln\, y'_0\right) \\ &\approx -\ln\left(1.649\, y'_0\right) \\ &\approx -\ln(0.825\,\chi_0). \end{aligned} \tag{6.58}$$

Substituting this into (6.57) gives

$$\lambda_{\mathrm{mfp}}\, Q_l = \frac{2\pi}{\sigma}\left(\frac{\rho_{\mathrm{B}}(E)}{4}\right)^2 l(l+1)\left(-\sum_{k=2}^{l}\frac{1}{k} - \ln(0.825\,\chi_0)\right). \tag{6.59}$$

From these, the mean value of the Legendre polynomials following multiple scattering after a total path length s,

$$\begin{aligned} G_l &= e^{-\left(\frac{s}{\lambda_{\mathrm{mfp}}}\right)\left(\frac{\pi}{8}\right)\left(\frac{\rho_{\mathrm{B}}^2(E)}{\sigma}\right) l(l+1)\left(-\sum_{k=2}^{l}\frac{1}{k} - \ln(0.825\,\chi_0)\right)} \\ &= e^{\left(\frac{\pi}{8}\right)(N_{\mathrm{S}}s)\rho_{\mathrm{B}}^2(E)\, l(l+1)\left(\sum_{k=2}^{l}\frac{1}{k} + \ln(0.825\,\chi_0)\right)}. \end{aligned} \tag{6.60}$$

Thus, the Goudsmit–Saunderson scattering probability distribution function is

$$f_{GS}(\theta; s) = \frac{1}{4\pi}\sum_{l=0}^{\infty}(2l+1)\, P_l(\cos\,\theta)\, e^{\left(\frac{\pi}{8}\right)(N_{\mathrm{S}}s)\rho_{\mathrm{B}}^2(E)\, l(l+1)\left(\sum_{k=2}^{l}\frac{1}{k} + \ln(0.825\,\chi_0)\right)} \tag{6.61}$$

where $f_{GS}(\theta; s)\sin\theta\, \mathrm{d}\theta$ is the number of charged particles between θ and $\theta + \mathrm{d}\theta$. The path length s means the distance travelled by the charged particle and does not consider the difference between it and the thickness of the medium that the particle traverses. This result is valid for not too large values of l and is also subject to the requirement that

$$\sum_{k=2}^{l}\frac{1}{k} + \ln(0.825\,\chi_0) < 0 \tag{6.62}$$

in order that the exponent of (6.61) remains necessarily negative.

We note here that the general form of the Goudsmit–Saunderson result is

$$f_{GS}(\theta; s) = \frac{1}{2\pi}\sum_{l=0}^{\infty}\left(l + \frac{1}{2}\right) P_l(\cos\theta)\, e^{-N_{\mathrm{S}}s \int \mathrm{d}\theta_1 \sin\,\theta_1\,(1 - P_l(\cos\,\theta_1))\frac{\mathrm{d}\sigma}{\mathrm{d}\theta_1}}. \tag{6.63}$$

If we use the approximation, $P_l(\cos\,\theta) \approx J_0(l\theta)$ for $l >> 1$, the small-angle approximation and replace the summation with an integration, one obtains the form for the distribution function of

$$f(\theta; s) = \int_0^{\infty} \mathrm{d}\eta\, \eta\, J_0(\eta\theta)\, e^{-2\pi N_{\mathrm{S}}s \int_0^{\infty} \mathrm{d}\theta_1 (1 - J_0\eta\theta_1)\frac{\mathrm{d}\sigma}{\mathrm{d}\theta_1}} \tag{6.64}$$

which is of the form obtained by Bothe (1921). Further manipulation of (6.64) is possible if we use the screened scattering cross section. Bielajew (1996) showed that this reduces to

$$f(\theta;\lambda)\theta\,\mathrm{d}\theta = \mathrm{d}\xi\,\xi \int_0^\infty \mathrm{d}z\,z\,J_0(z\xi\theta)\,e^{-\left(\frac{s}{\lambda_{\mathrm{mfp}}}\right)(1-zK_1(z))} \tag{6.65}$$

where, as shown in Appendix C.1, K_1 is the first-order modified Bessel function of the second kind and ξ is the scattering angle normalised to the atomic electron screening angle:

$$\xi = \frac{\theta}{\chi_0}. \tag{6.66}$$

For small angles, the Goudsmit–Saunderson result approximates the Gaussian *pdf*,

$$f_{GS}(\theta;s) = \frac{1}{\sqrt{2\pi\,\overline{\Theta^2}}}\,e^{-\left(\frac{\theta^2}{2\overline{\Theta^2}}\right)}. \tag{6.67}$$

The variance of this distribution was determined by Goudsmit and Saunderson through the numerical evaluation of $f_{GS}(\theta;s)$ for electrons scattered through foils. For aluminium, lead and carbon (the number of terms in the series being as much as 20), they found that the variance could be well described by

$$\overline{\Theta^2} \approx \frac{1}{\sum_l (2l+1)G_l}. \tag{6.68}$$

The Goudsmit–Saunderson theory has been extensively used in Monte Carlo calculations of charged particle transport (e.g. Berger 1963; Kadri et al. 2007).

6.3.6 Williams Theory

Williams (1939) presented a theoretical treatment of the multiple-scattering problem as applied to both fast electrons and high-energy massive particles which, at that time, were restricted to the experimental studies of cosmic rays. He was likely the first to have categorised explicitly the cumulative effect of small-angle deflections as being either ‘multiple or ‘plural’, depending upon whether the number of deflections is sufficient to lead to a Gaussian *pdf*. However, Williams also considered explicitly the contribution of single, large-angle scattering events and evaluated the probability of these and folded these with the Gaussian *pdf* so as to achieve the *pdf* representative of experimental data.

Williams' analysis was that of the scattering angle as projected onto two mutually orthogonal planes with the intersection defined by the particle's incident trajectory. This was due to his interest in explaining observations of scattered projectiles' trajectories in cloud-chamber photographs. In order to avoid this complexity of planar projections, we will not be considering projected scattering angles in our review of Williams' results. However, this will not affect the generality of the conclusions although the resulting formula will differ slightly (mostly through a factor of 2).

Williams considered the scattering problem within the context of atomic electron screening, which establishes a lower limit of integration of the unscreened (Rutherford) differential cross section, and finite nuclear size, which sets the upper limit of this integration. Following our previous discussion, we will neglect this latter effect. We again invoke the small-angle approximation and consider the general problem of a charged particle traversing a small section of medium of thickness s. Let us define a scattering angle, θ_S, such that in traversing s the particle undergoes, on the average, one collision which deflects it through an angle $\theta > \theta_S$. That is,

$$\begin{aligned} Ns \int \mathrm{d}\Omega \, \frac{\mathrm{d}\sigma}{\mathrm{d}\Omega} &= 2\pi N_S s \, \rho_B^2(E) \int_{\theta_S}^{\pi} \frac{\mathrm{d}\theta}{\theta^3} \\ &= \pi \rho_B^2(E) Ns \left(\frac{1}{\theta_S^2} - \frac{1}{\pi^2} \right) \\ &= 1. \end{aligned} \tag{6.69}$$

Because of the rapid drop-off of $\frac{\mathrm{d}\sigma}{\mathrm{d}\Omega}$ for $\theta_S << \pi$, we are led to an expression

$$\theta_S^2 = \pi \rho_B^2(E) \, Ns. \tag{6.70}$$

The angle θ_S could be considered as a limit between the dominances of multiple and single-scattering regimes. Williams was the first to suggest replacing Z^2 in the expression for $\overline{\Theta^2}$ with $Z^2 + Z = Z(Z+1)$ as a means of accounting for the additional effects of the atomic electrons upon the multiple-scattering *pdf*. In our nomenclature, we now include this modification within our expression for the Bohr collision diameter which means scaling its original definition by the factor of $\sqrt{\frac{Z+1}{Z}}$. Note that this subsequent increase in the value of the conventional Bohr diameter is significant for low atomic number media only: e.g. by 8 % for carbon compared to 2.5 % for calcium. However, as it is the square of the Bohr diameter that enters into these calculations, this significance becomes greater: inclusion of the effects of independent scatter from atomic electrons increases $\rho_B^2(E)$ by 16.7 and 5 % for carbon and calcium, respectively, and hence, it is not a negligible factor in the evaluation of multiple scattering.

The derivation of the mean-square scattering angle would require modification so as to set the upper limit of integration equal to θ_S and thus yield

$$\overline{\theta_1^2} = \frac{2\ln\left(\frac{\theta'_S}{\chi_0}\right)}{\frac{1}{\chi_0^2} - \frac{1}{\theta_S^2}}. \tag{6.71}$$

Because of the rapid drop-off in the probability of elastic Coulomb scatter with scattering angle, there will be large numbers of collisions occurring with scattering angles less than θ_S. This number is determined through the calculation of the total cross section by integrating the differential cross section from χ_0 to θ_S yielding

$$\begin{aligned}\overline{n} &= \pi\rho_B^2(E)\,\frac{\rho N_A s}{A}\left(\frac{1}{\chi_0^2} - \frac{1}{\theta_S^2}\right)\\ &= \pi\left(\frac{\theta_S^2}{\chi_0^2} - 1\right).\end{aligned} \tag{6.72}$$

As $\theta_S^2 >> \chi_0^2$ (as to be demonstrated in numerical examples) this simplifies to

$$\overline{n} \approx \pi\left(\frac{\theta_S^2}{\chi_0}\right)^2 \tag{6.73}$$

where it is noted that the value of θ_S is larger than that of χ_0 by at least an order of magnitude. The reader will note that this result follows also from (6.3) and (6.15). By the central-limit theorem, this will result in a Gaussian distribution function in θ with a mean-square scattering angle:

$$\begin{aligned}\overline{\Theta^2} &= \overline{n}\,\overline{\theta_1^2}\\ &= \pi\left(\frac{\theta'_S}{\chi_0}\right)^2\left(\frac{2\ln\left(\frac{\theta_S}{\chi_0}\right)}{\frac{1}{\chi_0^2} - \frac{1}{\theta_S^2}}\right)\\ &= 2\theta_S^2\ln\frac{\theta_S}{\chi_0}.\end{aligned} \tag{6.74}$$

This result demonstrates that $\sqrt{\overline{\Theta^2}} > \theta_S$ through the factor of $2\ln\frac{\theta_S}{\chi_0}$ or, in other words, the effect of the large number of collisions with $\theta < \theta_S$ is to yield a resultant RMS scattering angle greater than θ_S.

Table 6.1 summarises examples of the magnitudes of the angles under consideration here ($\chi_0, \theta'_S, \sqrt{\overline{\Theta^2}}$) as calculated for 100 keV electrons and 250 MeV protons incident to carbon and calcium, our usual elemental representations of soft tissue and bone. As the low-energy electrons have a limited range and the intermediate-energy protons have a much greater one, the thicknesses of the media for the

Table 6.1 Examples of numerical values for derived parameters

Particle and kinetic energy	Scattering medium[a]	ρs (g/cm^2)	$\frac{\hbar c}{pR_{\mathrm{N}}}$	λ_{mfp} (μm)	Mean number of collisions $\overline{n}$	Atomic electron screening angle, χ_0 (radians)	Single/multiple-scattering angle cut-off, θ_S (radians)	RMS scattering angle[a], $\sqrt{\overline{\Theta^2}}$ (radians)	
100 keV electrons	0.1 mm carbon	0.0225	274	0.170	589	2.3×10^{-2}	0.56	1.61	
	0.1 mm calcium	0.0155	183	0.144	695	3.2×10^{-2}	0.80	2.36	
								Equation (6.16)	Equation (6.17)
250 MeV protons	10 cm carbon	22.5	0.126	0.213	470,459	1.1×10^{-5}	7.0×10^{-3}	3.3×10^{-2}	2.8×10^{-2}
	1 cm calcium	1.55	0.084	0.179	55,907	1.5×10^{-5}	3.2×10^{-3}	1.5×10^{-2}	1.3×10^{-2}

[a]Neglecting projectile spin

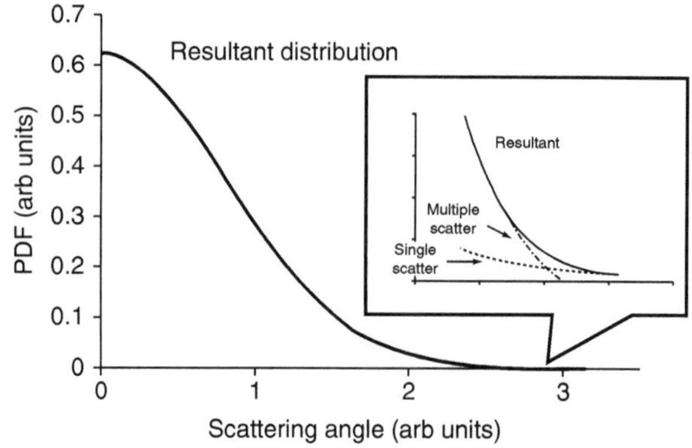

Fig. 6.6 Theoretical scattering distributions (Refer to text for discussion. Based upon a figure by Williams (1939))

electrons are 0.01 cm, and those for the protons are 10 cm of carbon and 1 cm calcium. As $\frac{\hbar c}{pR_N} > \pi$, $\sqrt{\overline{\Theta^2}}$ for the electron examples is calculated from (6.16), whereas, for the proton examples $\frac{\hbar c}{pR_N} < \pi$, $\sqrt{\overline{\Theta^2}}$ is calculated from both (6.16) and (6.17) to investigate the sensitivity of the result to the upper limit of the integral in the numerator of (6.7).

It is assumed that the kinetic energy loss of the particle traversing the thickness s of the medium is to be negligible (explicit consideration of the effect of energy loss upon multiple scattering is given in the Fermi–Eyges theory). However, should there be energy loss, but which is small and occurs at a constant rate over distance, then E can be replaced by the arithmetic average:

$$\overline{E} = \frac{(E_{\text{in}} + E_{\text{out}})}{2}, \tag{6.75}$$

where E_{in} is the particle kinetic energy incident to the medium and E_{out} is that upon its exit. Should the energy loss be extensive, then integration over energy shall be required.

Four features can be noted from the numerical examples given in Table 6.1:

- For a given areal density ρs, the number of elastic Coulomb scatters does not differ markedly between carbon and calcium. This limited Z-dependence of this number will be demonstrated analytically through our derivation of the Molière multiple-scattering theory.
- The atomic electron screening angle, χ_0, is much less than the cut-off angle, θ_S, by one to two orders of magnitude – a result which will prove useful in the mechanics of deriving the Molière theory.
- As expected from our earlier qualitative discussion, the RMS scattering angle, $\sqrt{\overline{\Theta^2}}$, is two to three times larger than the cut-off angle, θ_S.
- For the proton examples shown, there is not a large difference between the values of the RMS scattering angle calculated from (6.16) and (6.17).

Figure 6.6 shows conceptually the theoretical scattering distributions. The main graph shows the resultant distribution a function of scattering angle which is dominated by the Gaussian for small scattering angles. The insert shows the contributions of the multiple-scatter distribution (which is purely Gaussian) and the single-scatter contribution, proportional to $\frac{1}{\theta^3}$, to the resultant distribution at large scattering angles. The resultant and multiple-scatter distributions are coincident until the multiple-scatter distribution becomes less than that of the single-scatter distribution. The resultant distribution then goes over to the single-scatter contribution, thus providing the large-angle scattering tail to the distribution.

6.3.7 Fermi–Eyges Theory

In the Goudsmit–Saunderson and Williams theories, the assumption of a Gaussian *pdf* describing multiple small-angle scatter arose through the central-limit theorem and the large number of multiple Coulomb elastic scattering events that occur. In this, and the following section, we explicitly determine the *pdf* through the derivation of an explicit expression for the function and then by solving this equation. These two sections consider two approaches.

The first approach considered in describing multiple scatter is that beginning with the derivation by Fermi of a calculation of the scattering of cosmic rays within the atmosphere.[8] Because of the high kinetic energies of the cosmic rays for which energy losses would be minimal, these were neglected in Fermi's derivation and the result excluded any kinetic energy losses. This will not be the general case, in particular for charged particles within a condensed medium (e.g. tissue), and hence, energy loss should be recognised explicitly. As the degree of scattering is proportional to $\frac{1}{E^2}$, the scattering power will no longer be constant with distance but will increase as the charged particle loses kinetic energy through its penetration into the medium. Eyges (1948) extended Fermi's derivation so as to include the kinetic energy loss of the particle within the medium. As a result, the Fermi–Eyges theory has had, historically, many applications in therapeutic radiation physics calculations of particle transport for electrons (e.g. Hogstrom et al. 1981; McParland et al. 1988) and protons (Russell et al. 1995).

The Fermi–Eyges theory is derived here in two steps. First, the Fermi diffusion equation for the distribution function is derived. There have been numerous such derivations, but, here, we use as a basis of our derivation the simplified and transparent approach taken by Jette (1988). In the second part, we solve this equation for the distribution function but including Eyges result of a kinetic energy dependence.

Figure 6.7 shows an infinitely narrow beam of mono-energetic charged particles incident along the +z-axis to a semi-infinite medium with the entrance plane defined by the x-y plane. The particle is subsequently scattered into a trajectory described by the polar θ and azimuthal ϕ angles as shown. Due to azimuthal symmetry, ϕ is

[8] This was reported within the review article by Rossi and Greisen (1941).

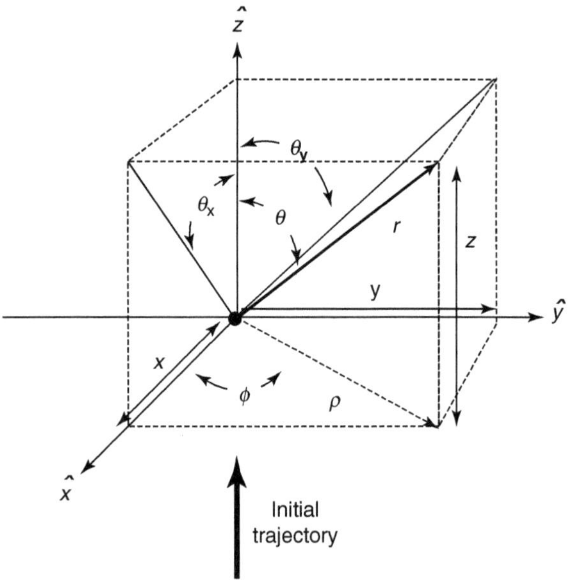

Fig. 6.7 Scattering of a particle with an initial trajectory along the *z*-axis through (θ,ϕ). θ_x and θ_y are the projections of the polar scattering angle onto the *x*-*z* and *y*-*z* planes, respectively

uniformly distributed between 0 and 2π. The angles θ_x and θ_y are the projections of the polar scattering angle onto the *x*-*z* and *y*-*z* planes, respectively, so that

$$\tan\theta_x = \tan\theta\cos\phi \tag{6.76}$$

$$\tan\theta_y = \tan\theta\sin\phi. \tag{6.77}$$

The mean-square values of these projected scattering angles are, in the small scattering angle approximation,

$$\begin{aligned}\overline{\theta_x^2} &= \frac{\int \mathrm{d}\Omega\, \theta_x^2 \frac{\mathrm{d}\sigma}{\mathrm{d}\Omega}}{\int \mathrm{d}\Omega\, \frac{\mathrm{d}\sigma}{\mathrm{d}\Omega}} \\ &= \frac{\int_0^{2\pi} \mathrm{d}\phi \cos^2\phi \int_0^{\pi} \mathrm{d}\theta\, \theta^3 \frac{\mathrm{d}\sigma}{\mathrm{d}\Omega}}{2\pi \int_0^{\pi} \mathrm{d}\theta\, \theta \frac{\mathrm{d}\sigma}{\mathrm{d}\Omega}} \\ &= \frac{\overline{\theta^2}}{2}.\end{aligned} \tag{6.78}$$

Similarly,

$$\overline{\theta_y^2} = \frac{\overline{\theta^2}}{2} \tag{6.79}$$

which gives the sum of the mean-squared projected scattering angles as

$$\overline{\theta^2} = \overline{\theta_x^2} + \overline{\theta_y^2}. \tag{6.80}$$

Recalling the earlier invocation of the central-limit theorem, one conjectures that the scattering *pdf* of the two projected scattering angles are Gaussian. For example,

$$f(\theta_x) = \frac{1}{\sqrt{2\pi\,\overline{\theta_x^2}}}\, e^{-\frac{\theta_x^2}{2\overline{\theta_x^2}}} \tag{6.81}$$

with zero mean and variance $\overline{\theta_x^2}$. An expression for $\overline{\theta_x^2}$ can be obtained by combining the expressions of the linear scattering power given and the projection relationship and using the finite-difference approximation:

$$\begin{aligned} \Im &\approx \frac{\Delta\,\overline{\theta^2}}{\Delta\,s} \\ &\approx \frac{\Delta\,\overline{\theta^2}}{\left(\frac{\Delta\,z}{\cos\ \theta}\right)} \\ &\approx \frac{\Delta\,\overline{\theta^2}}{\Delta\,z}, \end{aligned} \tag{6.82}$$

where the small-angle approximation (i.e. $\cos\ \theta \approx 1$) has again been assumed. The Gaussian *pdf* for θ_x can now be written as

$$f(\theta_x) = \frac{1}{\sqrt{\pi\,\Im\,\Delta z}}\, e^{-\frac{\theta_x^2}{\Im\,\Delta z}}. \tag{6.83}$$

Having taken the *pdf*s for the projected scattering angles to be Gaussian, also it is acceptable to make the same assumption for the *pdf*s describing the change in scattering angle from the mean (e.g. $\Delta\theta_x = \theta_x - \overline{\theta_x}$) over $\Delta\,z$:

$$g(\Delta\theta_x; \Delta z) = \frac{1}{\sqrt{\pi\,\Im\,\Delta\,z}}\, e^{-\frac{(\Delta\theta_x)^2}{\Im\,\Delta z}} \tag{6.84}$$

where, for later purposes, we include $\Delta\,z$, as an explicit argument.

The initial condition for the particle is given by $\delta(x)\,\delta(y)\,\delta(z)\,\delta(\theta)$, where $\delta(x)$ is the Kronecker δ-function. This represents a monodirectional and infinitely narrow incident beam centred at the origin. Then, $f\left(x, \theta_x; y, \theta_y; z\right)\,\Delta x\,\Delta\theta_x\,\Delta y\,\Delta\theta_y$ will equal the probability that at a depth z, the particle will be between x and $x + \Delta x$ and between y and $y + \Delta y$ and have projected scattering angles between θ_x and $\theta_x + \Delta\theta_x$ and between θ_y and $\theta_y + \Delta\theta_y$.

In order to simplify the derivation, the calculation will be restricted initially to the two-dimensional case and the result then extended to three dimensions.

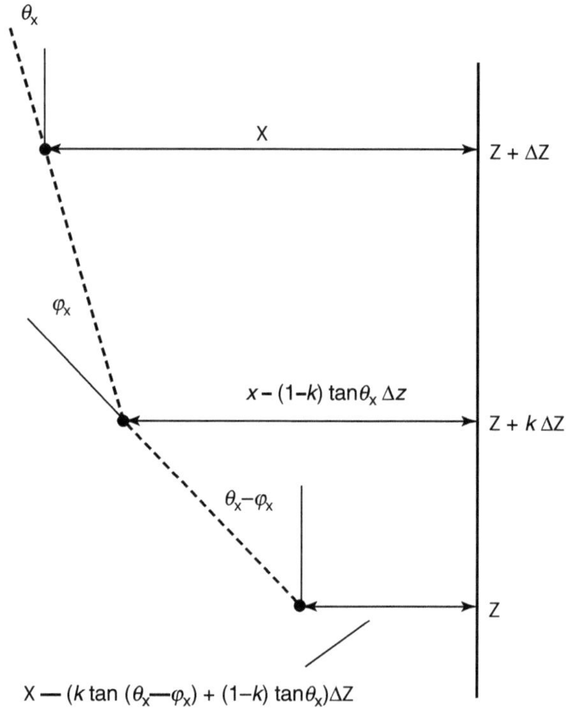

Fig. 6.8 Trajectory of a charged particle twice scattered between depths of z and $z + \Delta z$

The intent is to find, in two dimensions, the variation in $f(x, \theta_x; z)$ with increasing penetration into the medium, z. To do so, consider the trajectory of a particle in two dimensions as shown in Fig. 6.8. A particle is transported from an initial point to a final point through an intermediate scattering event. Working backward, the particle's end point is at a depth $z + \Delta z$ with projected lateral displacement x and projected polar direction θ_x following an intermediate scatter occurring at a depth $z + k\Delta z$, where $0 < 1$. This position is projected back to the particle's original position at a depth z. At the intermediate scattering point, the particle was at a projected lateral displacement

$$x - (1 - k) \tan\, \theta_x\, \Delta z \approx x - (1 - k)\, \theta_x\, \Delta z \tag{6.85}$$

and was scattered through an angle φ_x. Following a back projection of the trajectory from this intermediate point to the depth z, the particle had been scattered through an angle $\theta_x - \varphi_x$ and was at a projected lateral displacement:

$$x - (\tan\, \theta_x + k(\tan(\theta_x - \varphi_x) - \tan \theta_x))\, \Delta z \approx x - (\theta_x - k\varphi_x)\Delta z. \tag{6.86}$$

For any value of k, one obtains the *pdf* $f(x,\theta_x;z + \Delta z)$ by convolving over all φ_x:

$$f(x, \theta_x; z + \Delta z) = \int_{-\infty}^{\infty} d\varphi_x f\,(x - (\theta_x - k\varphi_x)\Delta z,\, \theta_x - v_x; z)\, g(\varphi_x; \Delta z). \tag{6.87}$$

This integral is solved by first expanding $f(x - (\theta_x - k\varphi_x)\Delta z,\, \theta_x - \varphi_x; z)\, g \times (\varphi_x; \Delta z)$ in a Taylor series to second order in θ_x, φ_x and Δz:

$$\begin{aligned} f(x - (\theta_x - k\varphi_x)\Delta z,\, \theta_x - \varphi_x; z)\, g(\varphi_x; \Delta z) \cong f(x, \theta_x; z) \\ - \frac{\partial f(x, \theta_x; z)}{\partial x}(\theta_x - k\varphi_x)\Delta z \\ - \frac{\partial f(x, \theta_x; z)}{\partial \theta_x}\varphi_x \\ + \frac{1}{2}\frac{\partial^2 f(x, \theta_x; z)}{\partial \theta_x^2}\varphi_x^2. \end{aligned} \tag{6.88}$$

Substituting this into the integrand of (6.87) gives

$$\begin{aligned} f(x, \theta_x; z + \Delta z) &= \int_{-\infty}^{\infty} \mathrm{d}\varphi_x f(x, \theta_x; z)\, g(\varphi_x; \Delta z) \\ &\quad - \int_{-\infty}^{\infty} \mathrm{d}\varphi_x \tfrac{\partial f(x,\theta_x;z)}{\partial x}(\theta_x - k\varphi_x)\,\Delta z\, g(\varphi_x; x) \\ &\quad - \int_{-\infty}^{\infty} \mathrm{d}\varphi_x \tfrac{\partial f(x,\theta_x;z)}{\partial \theta_x}\varphi_x\, g(\varphi_x; x) \\ &\quad + \tfrac{1}{2}\int_{-\infty}^{\infty} \mathrm{d}\varphi_x \tfrac{\partial^2 f(x,\theta_x;z)}{\partial \theta_x^2}\varphi_x^2\, g(\varphi_x; x) \\ &= f(x, \theta_x; z) \int_{-\infty}^{\infty} \mathrm{d}\varphi_x\, g(\varphi; \Delta z) \\ &\quad - \tfrac{\partial f(x,\theta_x;z)}{\partial x} \int_{-\infty}^{\infty} \mathrm{d}\varphi_x\,(\theta_x - k\varphi_x)\,\Delta z\, g(\varphi_x; x) \\ &\quad - \tfrac{\partial f(x,\theta_x;z)}{\partial \theta_x} \int_{-\infty}^{\infty} \mathrm{d}\varphi_x\,\varphi_x\, g(\varphi_x; x) \\ &\quad + \tfrac{1}{2}\tfrac{\partial^2 f(x,\theta_x;z)}{\partial \theta_x^2} \int_{-\infty}^{\infty} \mathrm{d}\varphi_x\,\varphi_x^2\, g(\varphi; x). \end{aligned} \tag{6.89}$$

In order to simplify this expression prior to solving, we first define the first three moments of $g(\Delta\theta_x; \Delta z)$:

$$\begin{aligned} \int_{-\infty}^{\infty} \mathrm{d}(\Delta\theta_x)\, g(\Delta\theta_x; \Delta z) &= \frac{1}{\sqrt{\pi\,\Im\,\Delta z}} \int_{-\infty}^{\infty} \mathrm{d}(\Delta\theta_x)\, \exp\left(-\frac{(\Delta\theta_x)^2}{\Im\,\Delta z}\right) \\ &= 1 \end{aligned} \tag{6.90}$$

$$\begin{aligned} \int_{-\infty}^{\infty} \mathrm{d}(\Delta\theta_x)\, \Delta\theta_x\, g(\Delta\theta_x; \Delta z) &= \frac{1}{\sqrt{\pi\,\Im\,\Delta z}} \int_{-\infty}^{\infty} \mathrm{d}(\Delta\theta_x)\, \Delta\theta_x \exp\left(-\frac{(\Delta\theta_x)^2}{\Im\,\Delta z}\right) \\ &= 0 \end{aligned} \tag{6.91}$$

$$\int_{-\infty}^{\infty} \mathrm{d}(\Delta\theta_x)\ (\Delta\theta_x)^2\, g(\Delta\theta_x;\Delta z)x = \frac{1}{\sqrt{\pi\,\Im\,\Delta z}} \int_{-\infty}^{\infty} \mathrm{d}(\Delta\theta_x)\ (\Delta\theta_x)^2 \exp\left(-\frac{(\Delta\theta_x)^2}{\Im\,\Delta z}\right)$$
$$= \frac{\Im\,\Delta z}{2}. \tag{6.92}$$

Using these moments, we write (6.80) in the simplified form

$$f(x,\theta_x;z+\Delta z) = f(x,\theta_x;z) - \theta_x \frac{\partial f(x,\theta_x;z)}{\partial x}\Delta z + \frac{\Im}{4}\frac{\partial^2 f(x,\theta_x;z)}{\partial\theta_x^2}\Delta z. \tag{6.93}$$

Note that in this final expression for the *pdf*, there is no dependence upon the parameter *k*, and as a result, the selection of the intermediate scatter position had been a purely arbitrary one. From this result, we change the finite-difference expression to a differential change in the *pdf* with penetration:

$$\frac{\partial f(x,\theta_x;z)}{\partial z} = \lim_{\Delta z\to 0} \frac{f(x,\theta_x;z+\Delta z) - f(x,\theta_x;z)}{\Delta z}, \tag{6.94}$$

leading to the Fermi diffusion equation in two dimensions follows:

$$\frac{\partial f(x,\theta_x;z)}{\partial z} = -\theta_x \frac{\partial f(x,\theta_x;z)}{\partial x} + \frac{\Im}{4}\frac{\partial^2 f(x,\theta_x;z)}{\partial\theta_x^2}. \tag{6.95}$$

The extension of the Fermi diffusion equation to three dimensions is straightforward:

$$\frac{\partial f}{\partial z} = -\theta_x \frac{\partial f}{\partial x} - \theta_y \frac{\partial f}{\partial y} + \frac{\Im}{4}\left(\frac{\partial^2 f}{\partial\theta_x^2} + \frac{\partial^2 f}{\partial\theta_y^2}\right) \tag{6.96}$$

where, for clarity of expression, the functional dependencies of the *pdf* have been omitted.

We now extend this result by noting that this derivation of the Fermi diffusion equation has ignored the fact that as a particle penetrates within the medium, it loses energy. As the elastic single-scatter cross section has a $\frac{1}{E^2}$ dependence, the scattering power will thus vary with the projectile kinetic energy and, hence, becomes a function of the depth of penetration as the kinetic energy loss will increase with increasing penetration. For the two-dimensional form of the Fermi diffusion equation,

$$\frac{\partial f(x,\theta_x;z)}{\partial z} = -\theta_x \frac{\partial f(x,\theta_x;z)}{\partial x} + \frac{\Im(z)}{4}\frac{\partial^2 f(x,\theta_x;z)}{\partial\theta_x^2} \tag{6.97}$$

where the explicit functional dependency of the scattering power upon z is noted.[9] This expression neglects the fact that once the particle has reached a depth z, it will have actually travelled a distance greater than z due to multiple-scattering events.

Equation (6.97) is solved using the Fourier transform method in Appendix B.2. Here, that method is extended from 1 to 2 dimensions. The Fourier transform pairs for two variables are

$$f(x,\theta_x;z) = \frac{1}{2\pi}\int_{-\infty}^{\infty}\int_{-\infty}^{\infty} \mathrm{d}\xi\,\mathrm{d}\Lambda\, f(\xi,\Lambda;z)\, e^{i(x\xi+\theta_x\Lambda)} \tag{6.98}$$

$$f(\xi,\Lambda;z) = \frac{1}{2\pi}\int_{-\infty}^{\infty}\int_{-\infty}^{\infty} \mathrm{d}x\,\mathrm{d}\theta_x\, f(x,\theta_x;z)\, e^{-i(x\xi+\theta_x\Lambda)}. \tag{6.99}$$

Applying these to the two-dimensional Fermi diffusion equation gives a differential equation in the Fourier transform of the *pdf*:

$$\frac{\partial f}{\partial z} = \xi\frac{\partial f}{\partial \Lambda} - \frac{\Lambda^2\Im(z)}{4}\omega. \tag{6.100}$$

The form of this result can be further simplified by defining the variables:

$$\kappa = z + \frac{\Lambda}{\xi} \tag{6.101}$$

$$z' = z \tag{6.102}$$

to yield

$$\frac{\partial f}{\partial z'} = -\frac{\xi^2(\kappa - z)^2\Im(z)}{4}\omega. \tag{6.103}$$

The solution to this differential equation is

$$f = k(\kappa)\exp\left(-\xi^2\int_{\lambda}^{z'}\mathrm{d}\eta\,\frac{(\kappa-\eta)^2\Im(\eta)}{4}\right). \tag{6.104}$$

[9] In ananisotropic/inhomogeneous medium (such as the body), the scattering power would also be a function of x and y. That level of complexity is not required for this discussion.

The lower limit of the integral has been set to an arbitrary value, λ, and $k(\kappa)$ is specified from the initial condition of the problem. From the previously defined set of initial conditions,

$$f(x,\theta_x;z=0)=\delta(x)\,\delta(\theta_x) \tag{6.105}$$

with the Fourier transform

$$f(\xi,\Lambda;z=0)=\frac{1}{2\pi}. \tag{6.106}$$

Writing out the solution for $\omega(\xi,\Lambda;z)$ in the original variables gives

$$f(\xi,\Lambda;z)=k\left(z+\frac{\Lambda}{\xi}\right)\exp\left(-\xi^2\int\limits_{\lambda}^{z}\mathrm{d}\eta\,\frac{\left(z+\frac{\Lambda}{\xi}-\eta\right)^2\Im(\eta)}{4}\right). \tag{6.107}$$

From the initial conditions, we have

$$k\left(\frac{\Lambda}{\xi}\right)\exp\left(-\xi^2\int\limits_{\lambda}^{0}\mathrm{d}\eta\,\frac{\left(\frac{\Lambda}{\xi}-\eta\right)^2\Im(\eta)}{4}\right)=\frac{1}{2\pi} \tag{6.108}$$

which enables the Fourier transform of the *pdf* to be written as

$$f(\xi,\Lambda;z)=\frac{1}{2\pi}\exp\left(-\xi^2\int\limits_{\lambda}^{z}\mathrm{d}\eta\,\frac{\left(z+\frac{\Lambda}{\xi}-\eta\right)^2\Im(\eta)}{4}\right). \tag{6.109}$$

Prior to taking the inverse Fourier transform in order to recover the *pdf*, this expression is simplified by defining the functions:

$$A_0(z)=\int\limits_{0}^{z}\mathrm{d}\eta\,\frac{\Im(\eta)}{4} \tag{6.110}$$

$$A_1(z)=\int\limits_{0}^{z}\mathrm{d}\eta\,(z-\eta)\frac{\Im(\eta)}{4} \tag{6.111}$$

$$A_2(z)=\int\limits_{0}^{z}\mathrm{d}\eta\,(z-\eta)^2\frac{\Im(\eta)}{4}. \tag{6.112}$$

The inclusion of the linear scattering power within the integrand and assigning it an η dependence allows for the change in scattering power with energy loss as a result of penetration into the medium. In the case where energy loss can be neglected, i.e. $\frac{\Im(\eta)}{4}$ is constant,

$$A_0(z) = \frac{\Im}{4} z \tag{6.113}$$

$$A_1(z) = \frac{\Im}{8} z^2 \tag{6.114}$$

$$A_2(z) = \frac{\Im}{12} z^3 \tag{6.115}$$

from which it follows, for no energy dependence,

$$\sqrt{A_0 A_2 - A_1^2} = \frac{\Im}{8} z. \tag{6.116}$$

These functions allow us to write a simpler form of the Fourier transform of the *pdf*:

$$f(\xi, \Lambda; z) = \frac{1}{2\pi} \exp\left(-A_0 \Lambda^2 + 2A_1 \Lambda \xi + A_2 \xi^2\right) \tag{6.117}$$

where the z-dependencies of the A_i have been omitted for clarity. Taking the inverse Fourier transform of the result will result in the multiple-scattering *pdf*:

$$f(x, \theta_x; z) = \frac{1}{2\pi \sqrt{A_0 A_2 - A_1^2}} \exp\left(-\frac{\theta_x^2 A_2 - 2A_1 x \theta_x + x^2 A_0}{4A_0 \left(A_0 A_2 - A_1^2\right)} \right) \tag{6.118}$$

which is the Fermi–Eyges result for two dimensions. The three-dimensional result is simply given by the product of the 2 two-dimensional *pdfs* in (x,θ_x) and (y,θ_y).

Equation (6.118) provides a Gaussian *pdf*, as the result of the small-angle approximation, in terms of the functions $A_0(z)$, $A_1(z)$ and $A_2(z)$. The depth of penetration, z, is now explicitly included in the *pdf* through these functions. As the linear scattering power has an energy dependence (i.e. a less energetic particle will be more readily scattered than an energetic one), a direct functional dependence upon particle energy is implicit.

6.3.8 *Molière Theory*

6.3.8.1 Introduction

The Fermi–Eyges result just derived is that of a Gaussian *pdf*, i.e. it ignores the contributions of large-angle scatters which arise from single Coulomb scatter.

Williams theory allowed for the contributions of these scatters which will extend the 'tail' of the resultant *pdf*. The Goudsmit–Saunderson result provides the electron multiple-scatter *pdf* as the weighted summation of Legendre polynomials which can allow for a distribution beyond that of a Gaussian. Molière's multiple-scattering theory (Molière 1947, 1948) is the most advanced of these and most used for calculating the effects of the multiple scattering of charged particles. The theory evolved from consideration of consecutive scatters which, in practice, is the solution of the transport equation derived during our discussion of the Fermi–Eyges theory. Molière's theory is both independent of an individual form of the single-scatter cross section and neglects energy loss. On the other hand, the single-scatter cross section is input to the Molière theory through only a single parameter, the Molière screening angle, θ_A. The shape of the multiple-scatter *pdf* is dependent upon a single parameter, b.[10]

As in the Goudsmit–Saunderson theory, the practical evaluation of the Molière multiple-scatter *pdf* is via a series expansion.[11] The first term provides the Gaussian form which reflects complete multiple scatter, and the remaining terms demonstrate the transition through plural to single-scatter regimes, with the latter described by the Rutherford unscreened differential cross section. The original Molière theory does not include the effects of kinetic energy loss. Moreover, the Molière theory does not account for either finite nuclear size or spin except through the single-scattering theory which is used as input.

Because of the importance of the theory, Molière's result is derived in considerable detail here through, primarily, Bethe's (1953) review, which is mathematically more transparent than the original and through incorporating further improvements suggested by Andreo et al. (1993).[12] In our derivation of the Williams theory, we defined thef modified versions of $\rho_{\mathrm{B}}(E)$ and θ_S in order to account for the inclusion of the atomic electrons as coherent scattering centres through the replacement of Z^2 with $Z(Z+1)$. While this approach was not used by Molière in the original derivation, it was included by Bethe in his review of the theory. We will follow Bethe's lead in our investigation of Molière's theory of multiple scattering. After our derivation of Molière's results, we will then examine the refinements suggested by Fano (1954).

Our derivation of a practical means of calculation the Molière *pdf* will be performed in three steps:

- The charged particle transport equation will be presented and then solved using Fourier–Bessel transforms to yield an integral representation of the *pdf*.

---

[10] This symbol of b should not be confused with that of the impact parameter previously introduced. We use this symbol b here as it is used in all of the literature pertaining to the Molière theory and the difference between it, and the impact parameter will be apparent by the context.

[11] It will be shown in the course of the development of the Molière theory that both are closely linked.

[12] Fernández-Varea et al. (1993) have provided a succinct derivation of the Molière theory beginning with the Goudsmit–Saunderson result. The connection between the theories of Molière and Goudsmit–Saunderson is investigated later in this section.

- This representation will be transformed to a considerably simpler integral form which incorporates the Molière parameter b which provides the input of the single-scattering model underlying the multiple-scattering theory. This result is amenable to numerical integration.
- This final integral form is then expanded into a series with the first term representing the Gaussian *pdf* and with additional terms accounting for the 'tails' due to plural and single scatter.

6.3.8.2 Derivation of the Molière Scattering Distribution

The problem to be solved begins by considering a monodirectional and infinitely narrow beam of charged particles incident to a medium of thickness s, physical density ρ, atomic number Z and atomic mass A. By equating this to the diffusion of electrons in the scattering plane, the transport equation is, for the Molière scattering *pdf*,

$$\frac{\partial f_{\mathrm{M}}(\theta; s)}{\partial s} = N\left(\int \mathrm{d}\theta_1 f_{\mathrm{M}}\left(\hat{\theta}'; s\right)\frac{\mathrm{d}\sigma}{\mathrm{d}\theta_1} - f_{\mathrm{M}}(\theta; s)\int \mathrm{d}\theta_1\,\theta_1\,\frac{\mathrm{d}\sigma}{\mathrm{d}\theta_1}\right) \tag{6.119}$$

where $\hat{\theta}' = \hat{\theta} - \hat{\theta}_1$ is the direction vector of the charged particle, prior to the last scatter, in the plane of θ. As before, N is the number density of scattering centres (here, the number of atoms per unit volume). The first term of (6.119) provides the number of particles scattered *into* the angular interval $\mathrm{d}\theta$ about θ, whereas the second terms provides the number scattered *out*.

To solve this differential equation, we invoke the Fourier–Bessel transform derived in Appendix B.3 (and which is the zeroth-order Hankel transform):

$$\begin{aligned} f_{\mathrm{M}}(\theta; s) &= Ha_0[g_{\mathrm{m}}(\eta; s)] \\ &= \int_0^\infty \mathrm{d}\eta\,\eta\,J_0(\eta\theta)\,g_{\mathrm{M}}(\eta; s) \end{aligned} \tag{6.120}$$

and

$$\begin{aligned} f_{\mathrm{M}}(\eta; s) &= Ha_0^{-1}[f_{\mathrm{M}}(\theta; s)] \\ &= \int_0^\infty \mathrm{d}\theta\,\theta\,J_0(\eta\theta)f_{\mathrm{M}}(\theta; s). \end{aligned} \tag{6.121}$$

These transform pairs are now applied to the transport equation to yield

$$\frac{\partial f_{\mathrm{M}}(\eta;s)}{\partial s} = N\left(f_{\mathrm{M}}(\eta;s)\int_{0}^{\infty} \mathrm{d}\theta_1\theta_1 J_0(\eta\theta_1)\frac{\mathrm{d}\sigma}{\mathrm{d}\theta_1} - f_{\mathrm{M}}(\eta;s)\int_{0}^{\infty} \mathrm{d}\theta_1\theta_1\frac{\mathrm{d}\sigma}{\mathrm{d}\theta_1}\right)$$
$$= -f_{\mathrm{M}}(\eta;s)\,N\int_{0}^{\infty} \mathrm{d}\theta_1\,\theta_1\,(1 - J_0(\eta\theta_1))\,\frac{\mathrm{d}\sigma}{\mathrm{d}\theta_1}. \tag{6.122}$$

As the initial condition for the incident charged particle beam is

$$f_{\mathrm{M}}(\theta;0) = \delta(\hat{\theta}) \tag{6.123}$$

due to the assumption of a monodirectional beam, it follows that the corresponding Fourier–Bessel transform of the initial condition is

$$g_{\mathrm{M}}(\eta;0) = \int_{0}^{\infty} \mathrm{d}\theta\,\theta\,J_0(\eta\theta)\,\delta(\hat{\theta})$$
$$= 1. \tag{6.124}$$

The solution to (6.122) is

$$g_{\mathrm{M}}(\eta;s) = e^{(\Omega(\eta;s)-\Omega(0;s))} \tag{6.125}$$

where

$$\Omega(\eta;s) = Ns\int_{0}^{\infty} \mathrm{d}\theta_1\,\theta_1\,\frac{\mathrm{d}\sigma}{\mathrm{d}\theta_1}\,J_0(\eta\theta_1). \tag{6.126}$$

Note that as $J_0(0) = 1$, the parameter for $\eta = 0$ is

$$\Omega(0;s) = N\,s\int_{0}^{\infty} \mathrm{d}\theta_1\,\theta_1\,\frac{\mathrm{d}\sigma}{\mathrm{d}\theta_1} \tag{6.127}$$

and will be equal to the total number of collisions occurring along the path length s.

The Molière multiple-scattering *pdf* in the cumulative scattering angle θ is returned through the inverse Fourier–Bessel transform of (6.125):

$$f_{\mathrm{M}}(\theta;s) = \int_{0}^{\infty} \mathrm{d}\eta\,\eta\,J_0(\eta\theta)\,e^{-Ns\int_{0}^{\infty}\mathrm{d}\theta_1\,\theta_1\,(1-J_0(\eta\theta_1))\frac{\mathrm{d}\sigma}{\mathrm{d}\theta_1}}. \tag{6.128}$$

Because of the explicit inclusion of the scattering differential cross section in this result, it is apparent that as with the Goudsmit–Saunderson theory, the Molière multiple-scattering *pdf* will not be restricted to a particular Coulomb single-scatter theory. In conclusion, the result of (6.128) is *exact*. We will show later how this, in fact, is an analytic approximation to the Goudsmit–Saunderson result.

6.3.8.3 Transformation

Following the approaches of Molière and Bethe, we now proceed to simplify (6.128) through a number of approximations in order to yield a result amenable to practical calculations. We begin by transforming (6.128) to a simpler integral expression and recall that the elastic Coulomb single-scatter differential cross section, with atomic electron screening considered, is given, $\left(\frac{\rho_B(E)}{\chi_0^2+\theta^2}\right)^2$ in the small-angle approximation and under the assumption of exponential atomic electron screening. As we have seen that the atomic electron screening angle χ_0 is small for most of our physical considerations, we can conclude that the differential cross section is largely proportional to θ^{-4} (i.e. it goes over to the unscreened Rutherford form) but becomes more complicated when the magnitude of θ is comparable to that of χ_0 where the effects of screening become evident. We begin this evaluation by considering the ratio of the actual screened cross section to the Rutherford cross section:

$$q(\theta_1) = \frac{\left(\frac{d\sigma_{Scr}}{d\theta_1}\right)}{\left(\frac{d\sigma_{Ruth}}{d\theta_1}\right)}. \tag{6.129}$$

It will be noted from our discussions of Chap. 5 that the ratio of (6.129) represents a multiplicative atomic electron screening factor. We can write this ratio as

$$q(\theta_1) = \frac{\theta_1^4}{\rho_B^2(E)} \frac{d\sigma_{Scr}}{d\theta_1}. \tag{6.130}$$

We can rewrite (6.130) by recalling the unit-probability scattering angle that we derived in the context of the Williams theory from (6.70) and including the approximate coherent contributions of the atomic electrons. Including this in (6.130) gives

$$q(\theta_1) = \pi N s \frac{\theta_1^4}{\theta_S^2(E)} \frac{d\sigma_{Scr}}{d\theta_1}. \tag{6.131}$$

The ratio $q(\theta_1)$ has properties that will prove useful when applied to simplifying the derivation to follow. Clearly, $q(\theta_1) \to 0$ as $\theta_1 \to 0$ due to the θ_1^{-4} dependence of the unscreened cross section. By extension, as the screened and unscreened cross sections will converge for large scattering angles (here, defined by the condition of the scattering angle being much greater than the screening angle, or $\chi_0 << \theta_1$), $q(\theta_1) \to 1$ as $\theta_1 \to \infty$. Further, the most rapid change in $q(\theta_1)$ will necessarily occur for scattering angles comparable to the screening angle, or $\theta_1 \approx \chi_0$. The next step of the derivation is to take the logarithm of (6.125):

$$\begin{aligned} -\ln f_{\mathrm{M}}(\eta; s) &= \Omega(0; s) - \Omega(\eta; s) \\ &= 2\pi N s \int_0^\infty \mathrm{d}\theta_1\, \theta_1\, (1 - J_0(\eta\theta_1)) \frac{\mathrm{d}\sigma_{\mathrm{Scr}}}{\mathrm{d}\theta_1} \\ &= \frac{2\theta_S^2}{\pi} \int_0^\infty \mathrm{d}\theta_1\, \frac{(1 - J_0(\eta\theta_1))}{\theta_1^3} q(\theta_1). \end{aligned} \tag{6.132}$$

In order to solve this integral, we investigate the features of the integrand over the range of θ_1. The limits of this range will be defined by the angles of χ_0 and θ_S. As was shown by Table 6.1, quantitatively, $\chi_0 << \theta_S$ by about one to two orders of magnitude. We now define an angle, θ_k, which is intermediate to these two angles:

$$\chi_0 << \theta_k << \theta_S. \tag{6.133}$$

This definition by inequality is analogous to our earlier definition of the angle $\theta_{(a)}$ of (5.73) in the description of the conditions for atomic electron screening. This splitting of the angular range of the integral will be advantageous for the calculation of the integral of (6.132) when we consider the following attributes of these angles:

- Values of η become important when they are of the order of $\frac{1}{\theta_S}$.
- $q(\theta_1)$ differs appreciably from unity for $\theta_1 \approx \chi_0$.

Having defined θ_k through this angular inequality, we note the following two approximations that will be used in the derivation to follow. First,

$$q(\theta_1) \approx 1 \quad \text{for} \quad \theta_k < \theta_1 < \infty \tag{6.134}$$

which reflects the accuracy of the Rutherford cross section for large scattering angles. Second, from the Taylor series expansion of the zeroth-order Bessel function of the first kind, we have

$$1 - J_0(\eta\theta_1) \approx \left(\frac{\eta\theta_1}{2}\right)^2 \quad \text{for} \quad 0 < \theta_1 < \theta_k. \tag{6.135}$$

(6.134) removes a dependence of the integral upon η over the specified angular range. Using these angular-dependent approximations, the integral of (6.132) can be written as

$$\begin{aligned}\int_0^\infty \mathrm{d}\theta_1 \frac{(1-J_0(\eta\theta_1))}{\theta_1^3} q(\theta_1) &= \int_0^{\theta_k} \mathrm{d}\theta_1 \frac{(1-J_0(\eta\theta_1))}{\theta_1^3} q(\theta_1) \\ &+ \int_{\theta_k}^\infty \mathrm{d}\theta_1 \frac{(1-J_0(\eta\theta_1))}{\theta_1^3} q(\theta_1) \\ &\approx \frac{\eta^2}{4} \int_0^{\theta_k} \mathrm{d}\theta_1 \frac{q(\theta_1)}{\theta_1} + \int_{\theta_k}^\infty \mathrm{d}\theta_1 \frac{(1-J_0(\eta\theta_1))}{\theta_1^3}.\end{aligned} \tag{6.136}$$

To solve this result, we consider the second integral first. Using a change of variables, $x = \eta\theta_1$, the integral is rewritten as

$$\begin{aligned}\int_{\theta_k}^\infty \mathrm{d}\theta_1 \frac{(1-J_0(\eta\theta_1))}{\theta_1^3} &= \eta^2 \int_{\eta\theta_k}^\infty \mathrm{d}x \frac{(1-J_0(x))}{x^3} \\ &= \eta^2 \int_{\eta\theta_k}^\infty \frac{\mathrm{d}x}{x^3} - \eta^2 \int_{\eta\theta_k}^\infty \mathrm{d}x \frac{J_0(x)}{x^3} \\ &= \frac{1}{2\theta_k^2} - \eta^2 \int_{\eta\theta_k}^\infty \mathrm{d}x \frac{J_0(x)}{x^3}.\end{aligned} \tag{6.137}$$

The integral on the right-hand side of (6.137) is solved by multiple integrations by parts. We begin with the first

$$\int_{\eta\theta_k}^\infty \mathrm{d}x \frac{J_0(x)}{x^3} = \int_{\eta\theta_k}^\infty \mathrm{d}r\, w \tag{6.138}$$

where

$$w \equiv J_0(x) \quad \mathrm{d}w = -\mathrm{d}x\, J_1(x)$$

$$\mathrm{d}r \equiv \frac{\mathrm{d}x}{x^3} \quad r \equiv -\frac{1}{2x^2}.$$

Hence,

$$\int_{\eta\theta_k}^{\infty} \mathrm{d}x\, \frac{J_0(x)}{x^3} = \frac{J_0(\eta\theta_k)}{2\,(\eta\theta_k)^2} - \frac{1}{2} \int_{\eta\theta_k}^{\infty} \mathrm{d}x\, \frac{J_1(x)}{x^2}. \tag{6.139}$$

Now, the integral of (6.139) is also solved by parts

$$\int_{\eta\theta_k}^{\infty} \mathrm{d}x\, \frac{J_1(x)}{x^2} = \int_{\eta\theta_k}^{\infty} \mathrm{d}r\, w \tag{6.140}$$

where we have defined

$$w \equiv J_1(x) \quad \mathrm{d}w = \mathrm{d}x\left(J_0(x) - \frac{J_1(x)}{x}\right)$$

$$\mathrm{d}r \equiv \frac{\mathrm{d}x}{x^2} \quad r \equiv -\frac{1}{x}.$$

Then,

$$\begin{aligned}\int_{\eta\theta_k}^{\infty} \mathrm{d}x\, \frac{J_1(x)}{x^2} &= \frac{J_1(\eta\theta_k)}{\eta\theta_k} + \int_{\eta\theta_k}^{\infty} \mathrm{d}x\, \frac{J_0(x)}{x} - \int_{\eta\theta_k}^{\infty} \mathrm{d}x\, \frac{J_1(x)}{x^2} \\ &= \frac{1}{2}\left(\frac{J_1(\eta\theta_k)}{\eta\theta_k} + \int_{\eta\theta_k}^{\infty} \mathrm{d}x\, \frac{J_0(x)}{x}\right).\end{aligned} \tag{6.141}$$

Returning the results of (6.139) and (6.141) to (6.136), we obtain the integration over the upper angular range:

$$\int_{\theta_k}^{\infty} \mathrm{d}\theta_1 \frac{(1 - J_0(\eta\theta_1))}{\theta_1^3} \approx \frac{1 - J_0(\eta\theta_k)}{2\theta_k^2} + \frac{\eta^2}{8} + \frac{\eta^2}{4} \int_{\eta\theta_k}^{\infty} \mathrm{d}x\, \frac{J_0(x)}{x} \tag{6.142}$$

where the Taylor series expansion of the Bessel functions has been used along with the small-argument approximation to give $J_1(\eta\theta_k) \cong \frac{\eta\theta_k}{2}$. This leads to

$$\int_{\theta_k}^{\infty} \mathrm{d}\theta_1\, \frac{1 - J_0(\eta\theta_1)}{\theta_1^3} \approx \left(\frac{\eta}{2}\right)^2 (1 - \gamma_{\mathrm{EM}} + \ln 2 - \ln \eta\theta_k) \tag{6.143}$$

where those additional terms of the order of $(\eta\theta_k)^2$ and higher have been neglected. From these results, we achieve

$$\Omega(0;s) - \Omega(\eta;s) = 2\theta_S^2 \int_0^\infty \mathrm{d}\theta_1 \, \frac{1 - J_0(\eta\theta_1)}{\theta_1^3} \, q(\theta_1)$$
$$= \frac{(\eta\theta_S)^2}{2} \int_0^{\theta_k} \mathrm{d}\theta_1 \, \frac{q(\theta_1)}{\theta_1} + 1 - \gamma_{\mathrm{EM}} + \ln 2 - \ln \eta\theta_k. \quad (6.144)$$

A characteristic scattering angle, known as the Molière screening angle, is now defined through the expression

$$-\ln \theta_A \equiv \lim_{\theta_k \to \infty} \left(\int_0^{\theta_k} \mathrm{d}\theta_1 \, \frac{q(\theta_1)}{\theta_1} + \frac{1}{2} - \ln \theta_k \right). \quad (6.145)$$

It is through (6.145), via $q(\theta_1)$, that the single-scatter Coulomb cross section is input to the Molière theory. Placing (6.145) into (6.144) gives

$$\Omega(0;s) - \Omega(\eta;s) = \frac{(\eta\theta_S)^2}{2} \left(-\ln \eta\theta_A + \frac{1}{2} + \ln 2 - \gamma_{\mathrm{EM}} \right). \quad (6.146)$$

We can simplify the appearance of this result by defining the dimensionless variables $y \equiv \eta\theta_S$ and $\lambda \equiv \frac{\theta}{\theta_S}$. Then, the right-hand side of (6.146) becomes

$$\frac{(\eta\theta_S)^2}{2} \left(-\ln \eta\theta_A + \frac{1}{2} + \ln 2 + \gamma_{\mathrm{EM}} \right) = \frac{y^2}{2} \left(-\ln \left(y \frac{\theta_A}{\theta_S} \right) + \frac{1}{2} + \ln 2 + \gamma_{\mathrm{EM}} \right)$$
$$= \frac{y^2}{2} \left(-\ln \left(\frac{y}{2} \right) + \frac{1}{2} \left(\ln \left(\frac{\theta_S}{\theta_A} \right)^2 + 1 - 2\gamma_{\mathrm{EM}} \right) \right)$$
$$= \frac{y^2}{2} \left(-\ln \frac{y}{2} + \frac{b}{2} \right) \quad (6.147)$$

where the Molière parameter has been defined:

$$b \equiv \ln \left(\frac{\theta_S}{\theta_A} \right)^2 + 1 - 2\gamma_{\mathrm{EM}}$$
$$\equiv \ln \frac{\theta_S}{\theta_A'}. \quad (6.148)$$

Here, the modified characteristic (Molière) screening angle has been defined:

$$\begin{aligned}\theta'_A &= \theta_A\, e^{\gamma_{\mathrm{EM}}-\frac{1}{2}} \\ &\approx 1.08\,\theta_A.\end{aligned} \tag{6.149}$$

This allows the simplification of the expression of (6.148):

$$b = \ln\left(\frac{\left(\frac{\theta_S}{\theta_A}\right)^2}{1.167}\right). \tag{6.150}$$

The elastic Coulomb single-scatter model is provided as input to the Molière theory through b via (6.145). Using these results, the Molière *pdf*, $f_{\mathrm{M}}(\theta, s) = \int_0^\infty \mathrm{d}\eta\, \eta\, J_0(\eta\theta)\, e^{-(\Omega(0;s)-\Omega(\eta;s))}$, can now be written in a simpler form amenable to calculation:

$$f_{\mathrm{M}}(\theta; s)\,\theta\,\mathrm{d}\theta = \lambda\,\mathrm{d}\lambda \int_0^\infty \mathrm{d}y\, y\, J_0(\lambda y)\, e^{\left(\frac{y}{2}\right)^2\left(2\ln\left(\frac{y}{2}\right)-b\right)}. \tag{6.151}$$

This result is, however, dependent upon the angular inequality that provided the definition of the intermediate angle, θ_k, and will thus become invalid should $\eta \approx \frac{1}{\chi_0}$. This condition is the same as that of $y \approx \frac{\theta_S}{\chi_0} \approx e^{-\frac{b}{2}}$. The exponent within the integral of (6.151) is worth considering in some detail as it will help to specify the appropriate limits of integration to use. This exponent has a minimum for $y = 2\, e^{\frac{(b-1)}{2}}$. For values of y exceeding this, the exponential in the integrand of (6.151) becomes infinite. Hence, we set the upper level of the integral to this minimum of the exponent:

$$f_{\mathrm{M}}(\theta; s)\,\theta\,\mathrm{d}\theta = \lambda\,\mathrm{d}\lambda \int_0^{2e^{\frac{(b-1)}{2}}} \mathrm{d}y\, y\, J_0(\lambda y)\, e^{\left(\frac{y}{2}\right)^2\left(2\ln\left(\frac{y}{2}\right)-b\right)}. \tag{6.152}$$

Bethe explained the appropriateness of this replacement of the upper limit of integration by noting that the minimal value of the exponential is equal to $e^{-e^{(b-1)}}$. As, from above, $e^b \approx \left(\frac{\theta'_S}{\theta_A}\right)^2$, this is approximately equal to $\Omega(0; s)$ which is equal to the number of scattering events. This establishes (6.152) to be correct to within an error of the order of $e^{-\frac{\Omega(0;s)}{e}}$, a quantity the magnitude of which is negligible.

6.3.8.4 Series Expansion of Molière's Probability Distribution Function

While the integral form of $f_{\mathrm{M}}(\theta; s)$ can be solved numerically, Molière considered a more practical means of solving for the *pdf* by expanding this integral form. To demonstrate this, we begin by investigating the parameter $\Omega(0; s)$. First, we note that the Molière result was found through the characteristic screening angle θ_A which was derived by Molière from the Thomas–Fermi atomic model discussed in Chap. 5 (we should, however, remind ourselves that the Thomas–Fermi atomic model is a statistical one and, as such, is a poor representation of light atoms, probably for boron and lighter). As noted there, the solution for the Thomas–Fermi potential is obtained numerically, and through this approach, Molière obtained an expression of this characteristic screening angle of the form

$$\theta_A = \chi_0 \sqrt{1.13 + 3.76\zeta^2} \tag{6.153}$$

where the quantity ζ is defined as

$$\zeta \equiv \left(\frac{zZ}{\beta}\right)\alpha. \tag{6.154}$$

From the above, we can write an expression for the total number of elastic scattering events as[13]

$$\begin{aligned}
\Omega(0; s) &= e^b \\
&= \frac{\left(\frac{\theta_S'}{\theta_A}\right)^2}{1.167} \\
&= \pi N s \left(\frac{\alpha \hbar c}{E}\right)^2 \frac{z^2 Z(Z+1)}{1.167\chi_0^2\left(1.13 + 3.76\zeta^2\right)} \\
&= \pi N s \left(a r_\infty\right)^2 \left(\frac{p}{E}\right)^2 \frac{z^2 Z(Z+1)}{1.167\left(z^{\frac{2}{3}} + Z^{\frac{2}{3}}\right)\left(1.13 + 3.76\zeta^2\right)} \\
&= \frac{\pi N s}{1.32}\left(\frac{\beta \hbar c}{m_e}\right)^2 \frac{z^2 Z(Z+1)}{A\left(z^{\frac{2}{3}} + Z^{\frac{2}{3}}\right)\left(1 + 3.33\zeta^2\right)}.
\end{aligned} \tag{6.155}$$

The Z-dependent term of (6.155) is plotted in Fig. 6.9 as a function of atomic number between $Z = 6$ and $Z = 20$, a range of relevance to tissue and over which the Thomas–Fermi atomic model would be considered valid, and for $z = 1$

[13] As per our convention, we will continue to use the Bohr expression of the atomic electron screening parameter in deriving this result rather than the reciprocal of the Thomas–Fermi radius which was used by Bethe.

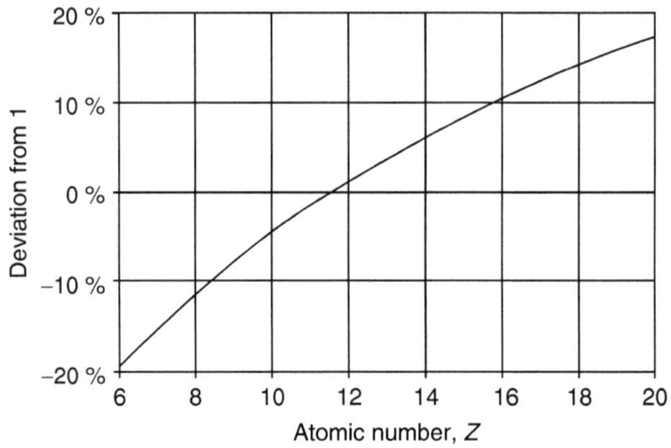

Fig. 6.9 The percentage deviation of the Z-dependent term of Eq. (6.155) as a function of atomic number for elements between carbon and calcium (see text for details)

and $\beta = 1$ and the further approximation of $A = 2Z$. In particular, this term is plotted as a percentage deviation from unity, and it is evident that over this range of atomic number the deviation is within $\pm 20\ \%$ of 1.0. This indicates that the number of elastic Coulomb scatters per unit areal density, ρs, is similar over all elements present in tissue, a phenomenon previously noted in the numerical examples of Table 6.1.

The evaluation of the Molière *pdf* of (6.152) begins with the definition of an expansion parameter related to b through the transcendental equation:

$$B - \ln B = b. \tag{6.156}$$

The transcendental nature of this equation can be shown to follow from the definition of the width of the multiple-scattering distribution. Recall our earlier discussion in which the width is described by the RMS angle, given by the square root of the mean-square angle but now with the limit placed at infinity:

$$\overline{\theta_1^2} = \frac{\int\limits_{\chi_0}^{\infty} \frac{d\theta_1}{\theta_1}}{\int\limits_{\chi_0}^{\infty} \frac{d\theta_1}{\theta_1^3}}. \tag{6.157}$$

However, the upper integration limit being set to infinity clearly leads to a logarithmic divergence in the integral in the numerator of this ratio. One approximation that can be used to resolve this particular problem is to instead set the upper limit of integration to the RMS angle $\sqrt{\overline{\theta_1^2}}$ itself or

$$\overline{\theta_1^2} \equiv \frac{\int_{\chi_0}^{\sqrt{\overline{\theta_1^2}}} \frac{\mathrm{d}\theta_1}{\theta_1}}{\int_{\chi_0}^{\sqrt{\overline{\theta_1^2}}} \frac{\mathrm{d}\theta_1}{\theta_1^3}} = \frac{2\ln\left(\frac{\sqrt{\overline{\theta_1^2}}}{\chi_0}\right)}{\frac{1}{\chi_0^2} - \frac{1}{\overline{\theta_1^2}}} = \frac{\ln\left(\frac{\overline{\theta_1^2}}{\chi_0^2}\right)}{\frac{1}{\chi_0^2} - \frac{1}{\overline{\theta_1^2}}}. \tag{6.158}$$

This circular argument then leads directly to the transcendental equations:

$$\frac{\overline{\theta_1^2}}{\chi_0^2} - \ln\frac{\overline{\theta_1^2}}{\chi_0^2} = 1 \tag{6.159}$$

and

$$\overline{\theta_1^2} - \chi_0^2 \ln\overline{\theta_1^2} = \chi_0^2\left(1 - \ln\chi_0^2\right). \tag{6.160}$$

These two equations indicate that B will be a measure of the width of the Gaussian component of the Molière *pdf*. Numerical approximations to B have since been given by Scott (1963),

$$B = 1.153 + 2.583\ \log_{10}\left(\frac{\theta_S}{\theta_A}\right)^2, \tag{6.161}$$

and Tabata and Ito (1976),

$$B = 2.6 + 2.3863\ \log_{10}\left(\frac{\theta_S}{\theta_A}\right)^2 - \frac{3.234}{\log_{10}\left(\frac{\theta_S}{\theta_A}\right)^2 + 0.994}. \tag{6.162}$$

The value of B will range typically from between about 5 to about 20. Molière stated that $B = 4.5$ as the minimum value required in order to establish enough elastic scattering events in order to establish the multiple-scattering regime.

Continuing the calculation, Molière then defined the reduced scattering angle:

$$\vartheta \equiv \frac{\lambda}{\sqrt{B}} = \frac{\theta}{\theta_S\sqrt{B}} \tag{6.163}$$

and changed the variable of integration y in (6.152) to that of

$$u \equiv y\sqrt{B}. \tag{6.164}$$

The expression of the distribution function is then expanded in a power series in $\frac{1}{B}$:

$$f_{\mathrm{M}}(\theta; s)\,\theta\,\mathrm{d}\theta = \vartheta\,\mathrm{d}\vartheta \sum_{n=0}^{\infty} \frac{f_{\mathrm{M}}^{(n)}(\vartheta)}{B^n} \tag{6.165}$$

where the coefficients of the expansion are given by

$$f_{\mathrm{M}}^{(n)}(\vartheta) = \frac{1}{n!} \int_0^{\infty} \mathrm{d}u\, u\, J_0(\vartheta u) \left(\left(\frac{u}{2}\right)^2 \ln\left(\left(\frac{u}{2}\right)^2 \right) \right)^n e^{-\left(\frac{u}{2}\right)^2}. \tag{6.166}$$

Let us first consider that the zeroth-order coefficient of the power series is

$$f_{\mathrm{M}}^{(0)}(\vartheta) = \int_0^{\infty} \mathrm{d}u\, u\, J_0(\vartheta u)\; e^{-\left(\frac{u}{2}\right)^2}. \tag{6.167}$$

This integral can be calculated using the following property of integer-order Bessel functions:

$$\int_0^{\infty} \mathrm{d}t\, t^{\nu+1}\, e^{-a^2 t^2} J_{\nu}(\varepsilon\, t) = \frac{\varepsilon^{\nu}}{(2a^2)^{\nu+1}}\, e^{-\frac{\varepsilon^2}{4a^2}} \quad \text{where} \quad 0 < \mathrm{Re} \quad a^2. \tag{6.168}$$

This yields the result

$$\begin{aligned} f_{\mathrm{M}}^{(0)}(\vartheta) &= \int_0^{\infty} \mathrm{d}u\, u\, J_0(\vartheta u)\, e^{-\left(\frac{u}{2}\right)^2} = 2e^{-\vartheta^2} \\ &= 2e^{-\vartheta^2} \end{aligned}$$

or, by substituting the expression for the reduced scattering angle,

$$f_{\mathrm{M}}^{(0)}(\theta) = 2e^{-\frac{\theta^2}{\left(\theta_S^2 B\right)}}. \tag{6.169}$$

This result is the Gaussian *pdf* of the multiple-scatter regime. The following two coefficients were provided by Molière in analytical form:

$$f_{\mathrm{M}}^{(1)}(\vartheta) = 2e^{-\vartheta^2}\left(\vartheta^2 - 1\right)\left(\overline{Ei}(\vartheta) - \ln\,\vartheta\right) - 2\left(1 - 2e^{-\vartheta}\right) \tag{6.170}$$

$$\begin{aligned} f_{\mathrm{M}}^{(2)}(\vartheta) &= \tfrac{e^{\vartheta^2}}{4}\left(\Psi^2(2) + \Psi'(2)\right)\left(\vartheta^4 - 4\vartheta^2 + 2\right) \\ &+ \int_0^1 \mathrm{d}x \frac{\left(\ln\left(\frac{x}{(1-x)}\right) - \Psi(2)\right)\left((1-x)^2 e^{x\vartheta^2} - 1 - (\vartheta-2)x - \left(\left(\frac{\vartheta^2}{2}\right) - 2\vartheta + 1\right)x^2\right)}{x^3}. \end{aligned} \tag{6.171}$$

The newly introduced functions in these two expressions are as follows. The function $\overline{Ei}(\vartheta)$ is given by (C.132) and by Jahnke and Emde (1945):

$$\overline{Ei}(\vartheta) = PV \int_{\vartheta}^{\infty} \mathrm{d}x \, \frac{e^{-x}}{x} \quad \text{where} \quad 0 < \vartheta. \tag{6.172}$$

$\Psi(n)$ is a version of the logarithmic derivative of the gamma function (given by (C.172)) and was specified by Molière as

$$\Psi(n) = \frac{\mathrm{d}\ln\,\Gamma(n+1)}{\mathrm{d}n}. \tag{6.173}$$

Clearly, these expressions for the higher-order coefficients of $f_{\mathrm{M}}^{(1)}(\vartheta)$ and $f_{\mathrm{M}}^{(2)}(\vartheta)$ are not easy to solve in these analytical forms. Bethe reported his inability to determine an analytical solution to the integral of (6.172) and the need to resort to a numerical solution. From these results, he provided tabular values of these functions. Fernández-Varea et al. (1993) have summarised cubic-spline fits to Bethe's tabulated data for reduced scattering angles of $\vartheta < 6$ and the following analytical approximations for $\vartheta \geq 6$:

$$f_{\mathrm{M}}^{(1)}(\vartheta) \approx \frac{2}{\vartheta^4}\left(1 - \frac{5}{\vartheta^2}\right)^{-\frac{4}{5}} \tag{6.174}$$

$$f_{\mathrm{M}}^{(2)}(\vartheta) \approx \frac{16}{\vartheta^6}\left(\ln\,\vartheta + \gamma_{\mathrm{EM}} - \frac{3}{2}\right)\left(1 + \frac{9}{\vartheta^2}\right) - \frac{38}{\vartheta^8}. \tag{6.175}$$

As expected, for very large values of ϑ, the function $f_{\mathrm{M}}^{(1)}(\vartheta)$ goes over to the single-scatter law:

$$f_{\mathrm{M}}^{(1)}(\vartheta) \approx \frac{2}{\vartheta^4} \quad \text{large } \vartheta. \tag{6.176}$$

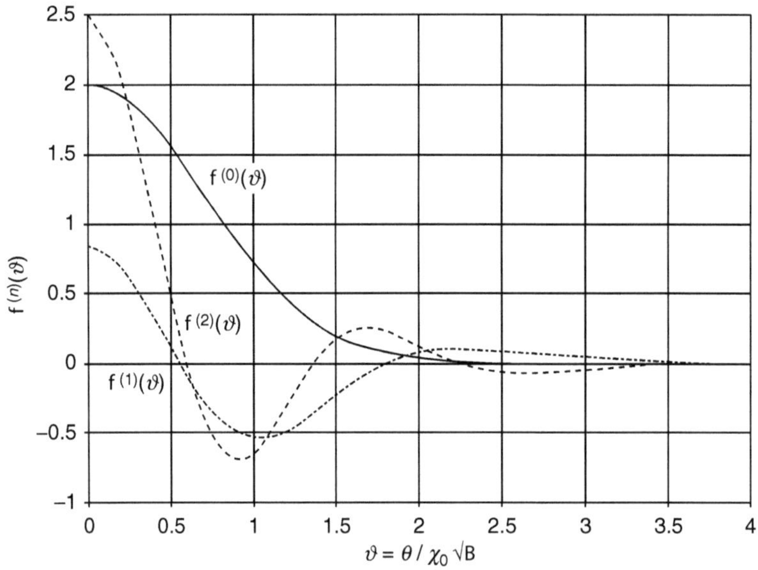

Fig. 6.10 The coefficients $f^{(0)}(\vartheta)$, $f^{(1)}(\vartheta)$ and $f^{(2)}(\vartheta)$ of the Molière expansion of the multiple-scattering probability distribution function as functions of the reduced scattering angle ϑ

Bethe provided expressions for the coefficients of the expansion derived from numerical calculations. These are shown in Fig. 6.10, and it is evident that the coefficients become negligible for values of $\vartheta > 3.5$. While one would expect from the expansion of the Molière *pdf* that the accuracy of the result would increase with the inclusion of an increasing number of terms, Bethe noted that this is not the case due mainly to the omission of electron intrinsic spin and relativity, the effects of which grow at large single-scattering angles. Andreo et al. (1993) have exhaustively studied the limitations to the Molière theory, and the interested reader is referred to that work.

6.3.8.5 Connection Between Molière and Goudsmit–Saunderson Theories

The Goudsmit–Saunderson *pdf* can be rewritten in the form

$$f_{GS}(\theta;t) = \frac{1}{2\pi}\sum_{l=0}^{\infty}\left(l+\frac{1}{2}\right)P_l(\cos\theta)\,e^{-N_{SS}\int \mathrm{d}\theta_1 \sin\theta_1\,(1-P_l(\cos\theta_1))\frac{\mathrm{d}\sigma}{\mathrm{d}\theta_1}}.$$

This result is exact regardless of the magnitude of θ_1. The Molière result may be considered to be an approximation to the Goudsmit–Saunderson result that is achieved through three steps:

- The replacement of the summation over l by an integration over scattering angle
- The use of the small-angle approximation in the integration
- Relating the Legendre polynomials with the zeroth-order Bessel function of the first kind

In Appendix C.2, we show that

$$P_l(\cos\theta) \approx J_0(l\theta).$$

For $l >> 1$ and $\theta << 1$, Bethe proposed another version of this approximation:

$$P_l(\cos\theta) \approx J_0\left(\left(l+\frac{1}{2}\right)\cos\theta\right), \tag{6.177}$$

indicating that it is valid for small θ, regardless of the magnitude of l. Bethe then quoted an Euler summation formula:

$$\sum_{l=0}^{\infty} w\left(l+\frac{1}{2}\right) \approx \int_0^{\infty} \mathrm{d}x\, w(x) + \frac{1}{24}\frac{\mathrm{d}w}{\mathrm{d}x}\bigg|_{x=0} + \cdots. \tag{6.178}$$

Applying (6.178) to the integrand of (6.177), we have

$$w(x) = x J_0(\lambda x)\, e^{\left(\frac{x}{2}\right)^2\left(2\ln\left(\frac{x}{2}\right)-b\right)}$$

and

$$\frac{\mathrm{d}w}{\mathrm{d}x}\bigg|_{x=0} = 1.$$

We are thus left with the relationship between the Goudsmit–Saunderson and Molière theories as

$$f_{GS}(\theta;s) = f_{\mathrm{M}}(\theta;s) + \frac{1}{24} + \cdots. \tag{6.179}$$

Let us consider this particular result. Within the Gaussian region of the *pdf* where $f_{\mathrm{M}}(\theta;s) \approx \frac{1}{\theta_s^2}$, it is evident that the $\frac{1}{24}$ factor can be neglected. However, that would be a condition which is valid only for $\theta < \pi$ or so. But it is clear from (6.178) that the higher-order terms of the Euler formula are required in the single-scattering regime (i.e. high θ_1).

6.3.9 *Contributions of Atomic Electrons II*

6.3.9.1 Fano's Inclusion of Inelastic Interactions with Atomic Electrons

The contribution of the effect of incoherent scattering from the atomic electrons has been so far estimated by the proposal by Williams (1940) to replace Z^2 by $Z(Z+1)$. This replacement implicitly assumes that the Coulomb single-scatter cross sections are the same for nuclear and electronic targets. We know, from our investigations of Chap. 5, that these cross sections will differ at small and large scattering angles. In order to account for this difference and to improve upon the Molière theory, Fano (1954) proposed isolating the nuclear (elastic) and electronic (inelastic) contributions from each other and then calculating for them separately. The prior derivation of the Molière theory has been for elastic scatter only, and in this section, the associated expressions will be denoted by the *el* subscript. Inelastic scatter expressions will be denoted by the *inel* subscript. Thus, (6.131) will be rewritten as

$$q_{\mathrm{el}}(\theta_1) = \pi N_S s \frac{\theta_1^4}{\theta_S^2(E)} \frac{\mathrm{d}\sigma_{\mathrm{Scr,el}}}{\mathrm{d}\theta_1}. \tag{6.180}$$

Clearly, those quantities of (6.180) are the same in (6.131). The corresponding quantity for inelastic scatter is

$$q_{\mathrm{inel}}(\theta_1) = \pi Z N_S s \frac{\theta_1^4}{\theta_S^2(E)} \frac{\mathrm{d}\sigma_{\mathrm{Scr,inel}}}{\mathrm{d}\theta_1} \tag{6.181}$$

where the multiplicative factor of Z is required as there are Z atomic electrons per each scattering centre. The characteristic inelastic screening angle is defined in analogy to (6.145):

$$-\ln\theta_{A,\mathrm{inel}} = \lim_{\theta_k \to \infty} \left(\int_0^{\theta_k} \mathrm{d}\theta_1 \frac{q_{\mathrm{inel}}(\theta_1)}{\theta_1} + \frac{1}{2} - \ln\theta_k \right). \tag{6.182}$$

Then, the Molière *pdf* can be rewritten in the form so as to account for both elastic and inelastic interactions:

$$f_{\mathrm{M}}(\theta; s)\,\theta\,\mathrm{d}\theta = \lambda\,\mathrm{d}\lambda \int_0^{\infty} \mathrm{d}y\, y\, J_0(\lambda y)\, e^{\left(\frac{y}{2}\right)^2 + (Y_{\mathrm{el}} + Y_{\mathrm{inel}})} \tag{6.183}$$

where

$$Y_{\mathrm{el}} = \frac{y^2}{2B} \ln\frac{y}{2} \tag{6.184}$$

and

$$Y_{\text{inel}} = \frac{2y^2}{(Z+1)B} \int_{\theta_k}^{\infty} \mathrm{d}\theta \, \frac{1 - J_0(\theta)}{\theta^3}. \tag{6.185}$$

The expansion parameter is again defined from a transcendental equation, but now with elastic and inelastic components:

$$B - \ln B = b_{\text{el}} + b_{\text{inel}} \tag{6.186}$$

where

$$b_{\text{el}} = \ln\left(\frac{\theta_S}{\theta_A}\right)^2 + 1 - 2\gamma_{\text{EM}} \tag{6.187}$$

and

$$b_{\text{inel}} = \frac{1}{Z+1} \ln\left(\frac{\theta_A}{\theta_{A,\text{inel}}}\right)^2. \tag{6.188}$$

Fano provided an explicit means of the calculation of the refined parameter b:

$$b' = b + \Delta b \tag{6.189}$$

which replaces the right-hand side of (6.186). b is the original Molière expression and where, if the incident projectile is an electron, Z^2 is replaced by $Z(Z+1)$ and

$$\Delta b = \frac{\ln\left(0.16Z^{-\frac{2}{3}}\left(1 + 3.33\left(\frac{\alpha Z}{\beta}\right)\right)\right) - \ln \theta_{A,\text{inel}}^2}{Z+1}. \tag{6.190}$$

and where, if the incident projectile is a charged hadron, the Z^2 term is retained and

$$\Delta b = \frac{\ln\left(1130Z^{-\frac{4}{3}}\left(\frac{\beta^2}{1-\beta^2}\right)\right) - \ln \theta_{A,\text{inel}}^2 - \frac{\beta^2}{2}}{Z}. \tag{6.191}$$

The $\ln \theta_{A,\text{inel}}^2$ is a medium-dependent parameter related to the integration of the incoherent form factor of (5.139); tabulated values were provided by Fano and, then, by Berger and Wang (1988). It is a slowly varying function of Z. Figure 6.11 shows $\ln \theta_{A,\text{inel}}^2$ as a function of the medium's atomic number. From hydrogen to oxygen, its value decreases from −3.6 to −5.0; beyond that, it slowly decreases to a value of −6.3 for lead.

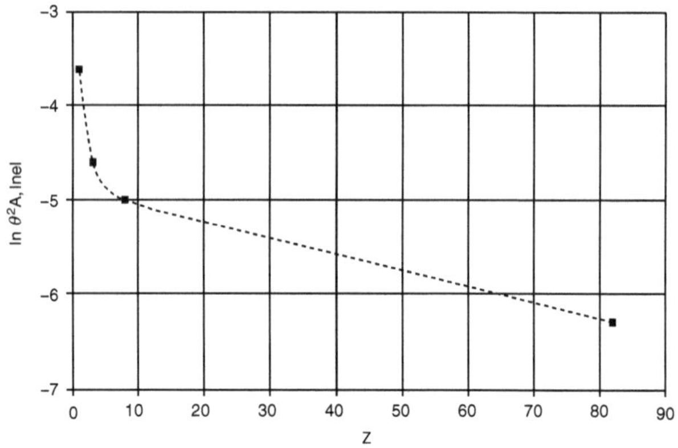

Fig. 6.11 Fano terms to account for inelastic contributions to Molière theory as a function of the medium's atomic number. Line is a guide only

As the terms of Δb are additive quantities, the result will be an increase in the number of scatterings in a given thickness of medium as a function of both the medium's atomic number and of the nature and the velocity of the projectile.

6.3.9.2 Experimental Confirmation of Atomic Electron Inclusion

Molière did not account for the effect of the inelastic contributions of the atomic electrons in the original development of the theory. In his rederivation of the theory, Bethe applied the Williams approximation by replacing Z with $\sqrt{Z(Z+1)}$ in order to account for these contributions. Fano isolated the nuclear and atomic electron contributions and accounted for the different screening angles.

Results from experiment have indicated the validities of these approaches. Shen et al. (1979) determined that Bethe's use of $\sqrt{Z(Z+1)}$ in the Molière theory agreed well with experimental data for protons incident to targets ranging from hydrogen to lead. Gottschalk et al. (1993) measured the multiple scattering of 158.6 MeV protons from 14 different materials (including elements ranging from beryllium to uranium) and compared the results to the Molière theory and concluded that it, with the Fano inelastic scatter contribution correction, was accurate to within 1 %.

6.3.10 Bethe's Derivation of the Goudsmit–Saunderson Formula

As an appendix to his derivation of the Molière theory, Bethe derived the Goudsmit–Saunderson *pdf*. In particular, we show here a modification of Bethe's proof of that result (6.54) by yet again defining an intermediate angle θ_k but which satisfies the inequalities

$$\chi_0 << \theta_k << \frac{1}{l}. \tag{6.192}$$

We then use the exact Rutherford differential cross section for single scatter and neglect screening. In addition, recognising the use of Legendre polynomials, we define the variable $x = \cos\theta_1$ where, from (6.47),

$$y^2 = \frac{1-x}{2}. \tag{6.193}$$

In this case, the exact Rutherford differential cross section can be written in the form

$$\frac{d\sigma}{d\Omega} = \frac{\left(\frac{\rho_B(E)}{4}\right)^2}{y^4} = \frac{\left(\frac{\rho_B(E)}{2}\right)^2}{(1-x)^2}. \tag{6.194}$$

The total cross section is obtained by integrating this expression over all solid angle:

$$\sigma = 2\pi \int_{-1}^{1} d(\cos\theta_1)\,\frac{d\sigma}{d\Omega} = \frac{\pi}{2}\rho_B^2(E) \int_{-1}^{1} \frac{dx}{(1-x)^2}. \tag{6.195}$$

Recalling the definition of lth transport coefficient, we see from (6.195) that we shall need to calculate for an integral of the form

$$K_l = \int_{-1}^{1-\varepsilon} dx\,\frac{1-P_l(x)}{(1-x)^2} \tag{6.196}$$

where a quantity ε has been included within the upper limit of the integral so as to avoid a divergence during the calculation. It will be identified shortly. The integral of the (6.196) is solved by parts:

$$K_l = 2\left.\frac{1-P_l(x)}{1-x}\right|_{-1}^{1-\varepsilon} + 2\int_{-1}^{1-\varepsilon} dx\,\frac{\frac{dP_l(x)}{dx}}{1-x}. \tag{6.197}$$

Let us consider the first term of (6.197):

$$\left.\frac{1-P_l(x)}{1-x}\right|_{-1}^{1-\varepsilon}=\frac{1-P_l(1-\varepsilon)}{\varepsilon}-\frac{1}{2}+\frac{P_l(-1)}{2}. \tag{6.198}$$

We interpret $P_l(1-\varepsilon)$ using the formula for $0<\theta<\theta_k$:

$$1-P_l(\theta)=\frac{l(l+1)}{4}\theta^2. \tag{6.199}$$

By expanding the cosine to second order

$$\cos\theta_k\approx 1-\frac{\theta_k^2}{2}, \tag{6.200}$$

we recognise that

$$\varepsilon\approx\frac{\theta_k^2}{2}. \tag{6.201}$$

Thus,

$$\begin{aligned}1-P_l(1-\varepsilon)&=\frac{l(l+1)}{4}\left(1-(1-\varepsilon)^2\right)\\&=\frac{l(l+1)}{4}\left(1-\left(1+\varepsilon^2-2\varepsilon\right)\right)\\&\approx\frac{l(l+1)}{2}\varepsilon\end{aligned} \tag{6.202}$$

where we ignore the ε^2 term. Substituting (6.202) into (6.198) gives

$$\left.\frac{1-P_l(x)}{1-x}\right|_{-1}^{1-\varepsilon}=\frac{(l(l+1)-1+P_l(-1))}{2}. \tag{6.203}$$

We now consider the integrand of (6.197) by first expanding the denominator:

$$\begin{aligned}\frac{1}{1-x}&=\frac{1}{2}\left(\frac{1+x}{1-x}+1\right)\\&=\frac{1}{2}\left(\frac{1-x^2}{(1-x)^2}+1\right).\end{aligned} \tag{6.204}$$

The reason for this manipulation will be shown when we draw upon the differential equation for the Legendre polynomial given in Appendix C.2. Inserting (6.204) into the integral of (6.197),

$$\int_{-1}^{1-\varepsilon} \mathrm{d}x \, \frac{\frac{\mathrm{d}P_l(x)}{\mathrm{d}x}}{1-x} = \frac{1}{2} \int_{-1}^{1-\varepsilon} \mathrm{d}x \, \frac{\mathrm{d}P_l(x)}{\mathrm{d}x} \left(\frac{1-x^2}{(1-x)^2} + 1 \right)$$
$$= \frac{1}{2} \int_{-1}^{1-\varepsilon} \mathrm{d}x \, \frac{\mathrm{d}P_l(x)}{\mathrm{d}x} \left(\frac{1-x^2}{(1-x)^2} \right) + \frac{1}{2} \int_{-1}^{1-\varepsilon} \mathrm{d}x \, \frac{\mathrm{d}P_l(x)}{\mathrm{d}x}. \tag{6.205}$$

We consider the second, and simpler, integral of (6.205) first:

$$\int_{-1}^{1-\varepsilon} \mathrm{d}x \, \frac{\mathrm{d}P_l(x)}{\mathrm{d}x} = P_l(1-\varepsilon) - P_l(-1)$$
$$= 1 + \frac{l(l+1)}{2}\varepsilon - P_l(-1). \tag{6.206}$$

The first integral of (6.205) is solved by parts:

$$\int_{-1}^{1-\varepsilon} \mathrm{d}x \, \frac{\mathrm{d}P_l(x)}{\mathrm{d}x} \left(\frac{1-x^2}{(1-x)^2} \right) = \left(\frac{\mathrm{d}P_l(x)}{\mathrm{d}x} \right) \left(\frac{1-x^2}{1-x} \right) \Bigg|_{-1}^{1-\varepsilon} - \int_{-1}^{1-\varepsilon} \mathrm{d}x \, \frac{\frac{\mathrm{d}}{\mathrm{d}x}\left((1-x^2) \frac{\mathrm{d}P_l(x)}{\mathrm{d}x} \right)}{(1-x)}. \tag{6.207}$$

As

$$\frac{\mathrm{d}P_l(x)}{\mathrm{d}x} = -\frac{l(l+1)}{2} x, \tag{6.208}$$

we can write the integrated term of (6.207):

$$\left(\frac{\mathrm{d}P_l(x)}{\mathrm{d}x} \right) \left(\frac{1-x^2}{1-x} \right) \Bigg|_{-1}^{1-\varepsilon} = -\frac{l(l+1)}{2} x \left(\frac{1-x^2}{1-x} \right) \Bigg|_{-1}^{1-\varepsilon}$$
$$= -\frac{l(l+1)}{2} (1-\varepsilon) \frac{1-(1-\varepsilon)^2}{\varepsilon}$$
$$= -l(l+1)(1-\varepsilon) \tag{6.209}$$

where the ε^2 term has again been neglected. The integral of (6.207) is determined by noting the numerator of the integrand, and the differential equation for the Legendre polynomial given in Appendix C which is written in the form

$$\frac{\mathrm{d}\left((1-x^2)\frac{\mathrm{d}P_l(x)}{\mathrm{d}x}\right)}{\mathrm{d}x} = -l(l+1)P_l(x). \tag{6.210}$$

Substituting (6.209) and (6.210) into (6.207) gives

$$\int_{-1}^{1-\varepsilon} \mathrm{d}x \frac{\mathrm{d}P_l(x)}{\mathrm{d}x}\left(\frac{1-x^2}{(1-x)^2}\right) = -l(l+1)(1-\varepsilon) + l(l+1)\int_{-1}^{1-\varepsilon} \mathrm{d}x \frac{P_l(x)}{(1-x)}$$
$$= l(l+1)\left(-(1-\varepsilon) + \int_{-1}^{1-\varepsilon} \mathrm{d}x \frac{P_l(x)}{(1-x)}\right). \tag{6.211}$$

Continuing by substituting (6.206) and (6.211) into (6.205) gives

$$\int_{-1}^{1-\varepsilon} \mathrm{d}x \frac{\frac{\mathrm{d}P_l(x)}{\mathrm{d}x}}{1-x} = \frac{l(l+1)}{2}\left(-(1-\varepsilon) + \int_{-1}^{1-\varepsilon} \mathrm{d}x \frac{P_l(x)}{(1-x)}\right) + \frac{1}{2}\left(1 + \frac{l(l+1)}{2}\varepsilon - P_l(-1)\right)$$
$$= \frac{1}{2}\left(1 - P_l(-1) + l(l+1)\left(-1 + \int_{-1}^{1-\varepsilon} \mathrm{d}x \frac{P_l(x)}{1-x}\right)\right) \tag{6.212}$$

where the ε term has at last been discarded. Finally, we achieve the result of

$$K_l = (l(l+1) - 1 + P_l(-1)) + \left(1 - P_l(-1) + l(l+1)\left(-1 + \int_{-1}^{1-\varepsilon} \mathrm{d}x \frac{P_l(x)}{1-x}\right)\right)$$
$$= l(l+1)\int_{-1}^{1-\varepsilon} \mathrm{d}x \frac{\mathrm{d}P_l(x)}{1-x}$$
$$= l(l+1)A_l(x)$$

where the quantity $A_l(x)$ has been defined as the integral. In order to extend the upper limit of the integral from $1-\varepsilon$ to 1, one first takes the difference of

$$A_{l-1}(x) - A_l(x) = \int_{-1}^{1} \mathrm{d}x \frac{P_{l-1}(x) - P_l(x)}{1-x} \tag{6.214}$$

and then uses the relationship

$$l(P_{l-1}(x) - xP_l(x)) = \left(1-x^2\right)\frac{\mathrm{d}P_l(x)}{\mathrm{d}x} \tag{6.215}$$

in order to write (6.214) as

$$A_{l-1}(x) - A_l(x) = \frac{1}{l}\int_{-1}^{1} \mathrm{d}x\,(1+x)\,\frac{\mathrm{d}P_l(x)}{\mathrm{d}x} - \int_{-1}^{1} \mathrm{d}x\,P_l(x). \tag{6.216}$$

The second integral is, for any non-zero l, equal to zero. The first integral is solved by parts:

$$\begin{aligned}\int_{-1}^{1} \mathrm{d}x\,(1+x)\,\frac{\mathrm{d}P_l(x)}{\mathrm{d}x} &= (1+x)P_l(x)\big|_{-1}^{1} - \int_{-1}^{1} \mathrm{d}x\,P_l(x)\\ &= 2.\end{aligned} \tag{6.217}$$

Thus, for non-zero l,

$$A_{l-1}(x) - A_l(x) = \frac{2}{l}. \tag{6.218}$$

Returning this recursive relationship to (6.213), we have

$$\begin{aligned}K_l &= l(l+1)\left(A_1 - 2\sum_{k=2}^{l}\frac{1}{k}\right)\\ &= l(l+1)\left(\frac{K_1}{2} - 2\sum_{k=2}^{l}\frac{1}{k}\right)\end{aligned} \tag{6.219}$$

which we compare to the term in the parentheses of (6.54).

6.3.11 Nuclear and Electronic Contributions

Here, we will compare the relative contributions of the nucleus and atomic electrons to multiple Coulomb scatter. This discussion anticipates the derivation of the soft-collision energy loss in the next part of this book, and the details of that derivation are deferred until then.

Recall the fundamental geometry of an elastic Coulomb scatter as shown in Fig. 4.1. Let the rest masses of the projectile and the target be m_{Proj} and m_{Tgt}, respectively. In the small-angle approximation,

$$\theta_1 \approx \frac{q_\perp}{|\mathbf{p}|} \tag{6.220}$$

where $q_\perp$ is the 3-vector momentum transfer perpendicular to the incident trajectory as a result of the scatter. It will be shown in Part III of this book that this momentum transfer component is given by

$$q_\perp = 2\frac{zZ\,\alpha\hbar c}{\beta b} \tag{6.221}$$

where b is the impact parameter and β is the projectile velocity normalised to that of light. In the nonrelativistic case, it thus follows that the energy transfer to the target is

$$E_{\text{Tgt}} = \frac{2}{m_{\text{Tgt}}}\left(\frac{zZ\,\alpha\hbar c}{\beta b}\right)^2. \tag{6.222}$$

The scatter angle can be written as

$$\theta_1 = \sqrt{\frac{m_{\text{Tgt}}\,E_{\text{Tgt}}}{m_{\text{Proj}}\,E_{\text{Proj}}}} \tag{6.223}$$

and the mean-square single-scattering angle is

$$\begin{aligned}\overline{\theta_1^2} &= \frac{m_{\text{Tgt}}}{m_{\text{Proj}}\,E_{\text{Proj}}}\,Ns\int \mathrm{d}\Omega\,\frac{\mathrm{d}\sigma}{\mathrm{d}\Omega}\,E_{\text{Tgt}} \\ &= \frac{m_{\text{Tgt}}}{m_{\text{Proj}}\,E_{\text{Proj}}}\,\overline{\Delta E_{\text{Tgt}}}.\end{aligned} \tag{6.224}$$

We will ignore the angular limits to the integral as we are interested only at approximations of the nuclear and electronic contributions. These are, respectively,

$$\overline{\theta_{1,\text{nucl}}^2} = \frac{m_{\text{nucl}}}{m_{\text{Proj}}\,E_{\text{Proj}}}\,\overline{\Delta E_{\text{nucl}}} \tag{6.225}$$

and

$$\overline{\theta_{1,\text{elec}}^2} = \frac{m_{\text{elec}}}{m_{\text{Proj}}\,E_{\text{Proj}}}\,\overline{\Delta E_{\text{elec}}} \tag{6.226}$$

where m_{nucl} and m_{elec} are the rest masses of the nucleus and of the electron, respectively. The ratio of these two mean-square scattering angles is

$$\frac{\overline{\theta_{1,\text{nucl}}^2}}{\overline{\theta_{1,\text{elec}}^2}} = \frac{m_{\text{nucl}}}{m_{\text{elec}}}\,\frac{\overline{\Delta E_{\text{nucl}}}}{\overline{\Delta E_{\text{elec}}}}. \tag{6.227}$$

While Part III provides derivations of these mean energy transfers, for the sake of the current argument, it is sufficient to say that (6.227) simplifies to

$$\frac{\overline{\theta^2_{1,\mathrm{nucl}}}}{\overline{\theta^2_{1,\mathrm{elec}}}} \approx Z\frac{L_{\mathrm{nucl}}}{L_{\mathrm{elec}}} \tag{6.228}$$

where the ratio on the right-hand side is that of the nuclear and electronic stopping numbers which are proportional to the logarithm of the maximum and minimum possible energy transfers. It is evident from (6.228) that the nuclear contribution to the scattering will be dominant for high atomic number targets. However, as to be shown that $L_{\mathrm{nucl}} >> L_{\mathrm{elec}}$, the nuclear component will dominate even in low atomic number media.

Bibliography and Further Reading

Abramowitz M, Stegun IA, editors. Handbook of mathematical functions. New York: Dover; 1972.
Bjorken JD, Drell SD. Relativistic quantum mechanics. New York: McGraw-Hill; 1964.
Gottschalk B. On the scattering power of radiotherapy protons. Med Phys. 2010;37:352–67.
McKinley WA, Feshbach H. The Coulomb scattering of relativistic electrons by nuclei. Phys Rev. 1948;74:1759–63.
McParland BJ. A derivation of the electron mass scattering power for electron dose calculations. Nucl Instrum Methods Phys Res. 1989;A274:592–6.
Uehling EA. Penetration of heavy charged particles in matter. Annu Rev Nucl Sci. 1954;4:315–50.

References

Andreo P, Medin J, Bielajew AF. Constraints of the multiple-scattering theory of Molière in Monte Carlo simulations of the transport of charged particles. Med Phys. 1993;20:1315–25.
Berger MJ. Monte Carlo calculation of the penetration and diffusion of fast charged particles. In: Alder B, Ferntech S, Rotesberg M, editors. Methods in computational physics, vol I: Statistical physics. New York: Academic; 1963.
Berger MJ, Wang R. Multiple-scattering angular deflections and energy-loss straggling. In: Jenkins TM, Nelson WR, Rindi A, editors. Monte Carlo transport of electrons and photons. New York: Plenum Press; 1988. p. 21–56.
Bethe HA. Molière's theory of multiple scattering. Phys Rev. 1953;89:1256–66.
Bielajew AF. A hybrid multiple-scattering theory for electron-transport Monte Carlo calculations. Nucl Instrum Methods Phys Res. 1996;B111:195–208.
Bothe W. Das allgemeine Fehlergesetz, die Schwankungen der Feldstärke in einem Dielectrikum und die Zerstreuung der alpha-Strahlen. Z Physik. 1921;5:63–9.
Cooper LN, Rainwater J. Theory of multiple Coulomb scattering from extended nuclei. Phys Rev. 1955;97:492–504.
Eyges L. Multiple scattering with energy loss. Phys Rev. 1948;74:1534–5.
Fano U. Inelastic collisions and the Molière theory of multiple scattering. Phys Rev. 1954;93:117–20.

Fernández-Varea JM, Mayol R, Baro J, Salvat F. On the theory and simulation of multiple elastic scattering of electrons. Nucl Instrum Methods Phys Res. 1993;B73:447–73.

Gottschalk B, Koehler AM, Schneider RJ, Sisterson JM, Wagner MS. Multiple Coulomb scattering of 160 MeV protons. Nucl Instrum Methods Phys Res B. 1993;74:467–90.

Goudsmit S, Saunderson JL. Multiple scattering of electrons. Phys Rev. 1940;57:24–9.

Hogstrom KR, Mills MD, Almond PR. Electron beam dose calculations. Phys Med Biol. 1981;26:445–59.

Jackson JD. Classical electrodynamics. New York: Wiley; 1999.

Jahnke E, Emde F. Tables of functions. New York: Dover Publications; 1945.

Jette D. Electron dose calculation using multiple-scattering theory. A. Gaussian multiple-scattering theory. Med Phys. 1988;15:123–37.

Kadri O, Ivanchenko V, Gharbi F, Trabelsi A. Incorporation of the Goudsmit-Saunderson electron transport theory in the Geant4 Monte Carlo code. Nucl Instrum Methods Phys Res B. 2007;267:3624–32.

Kulchitsky LA, Latyshev GD. The multiple scattering of fast electrons. Phys Rev. 1942;61:254–65.

McParland BJ. Nuclear medicine radiation dosimetry: advanced theoretical principles. London: Springer; 2010.

McParland BJ, Cunningham JR, Woo MK. Electron beam dose calculations for heterogeneous media. Med Phys. 1988;15:489–97.

Molière VG. Theorie der Streuung schneller geladener Teilchen, I. Einzelstreuung am ambgeschirmten Coulomb-Field. Z Naturforschg. 1947;2a:133–45.

Molière VG. Theorie der Streuung schneller geladener Teilchen, II. Mehrfach- und Veilfachstreuung. Z Naturforschg. 1948;3a:78–97.

Rossi B. High-energy particles. New York: Prentice-Hall; 1952.

Rossi B, Greisen K. Cosmic-ray theory. Rev Mod Phys. 1941;13:240–309.

Russell KR, Grusell E, Montelius A. Dose calculations in proton beams: range straggling corrections and energy scaling. Phys Med Biol. 1995;40:1031–43.

Scott WT. The theory of small-angle multiple scattering of fast charged particles. Rev Mod Phys. 1963;35:231–313.

Shen G, Ankenbrandt C, Atac M, Brown R, Ecklund S, Gollon PJ, et al. Measurement of multiple scattering at 50 to 200 GeV/c. Phys Rev D. 1979;20:1584–8.

Tabata T, Ito R. An improved interpolation formula for the parameter B in Molière's theory of multiple scattering. Jpn J Appl Phys. 1976;15:1583–4.

Williams EJ. Concerning the scattering of fast electrons and of cosmic ray particles. Proc R Soc. 1939;A169:531–72.

Williams EJ. Multiple scattering of fast electrons and alpha-particles, and curvature of cloud tracks due to scattering. Phys Rev. 1940;89:293–306.

Part III
Collision Energy Loss

Chapter 7
Introduction to Part III

Contents

> The theory gives a satisfactory description of the influence of impacts for which the minimum distance b between the electron and the passing particle is larger than the atomic dimensions. Quantum mechanical corrections have to be introduced for the very close impacts, when the particle passes through the atom.
>
> –Enrico Fermi (1940)

Abstract This chapter introduces us to Part III, which investigates the mean collision energy loss resulting from the interactions between moving charged particles and matter. Most importantly, it draws the distinction between the definitions of soft and hard collisions which will dominate the first three chapters. It will also present considerations of complex media which will become of interest later in Part III.

B.J. McParland, *Medical Radiation Dosimetry*,
DOI 10.1007/978-1-4471-5403-7_7, © Springer-Verlag London 2014

7.1 Introductory Comments

Excluding the hadron–nuclear interactions based upon the strong nuclear force, charged particles interact with the atoms in the medium they are traversing through via the Lorentz force, $ze(\mathbf{E} + \mathbf{v} \times \mathbf{B})$. In Part II, we investigated those interactions defined as being elastic, i.e. those that are within our definition for which the sum of the projectile and target kinetic energies are equal before and after the scatter. It should again be emphasised that these elastic collisions also include those within which kinetic energy is lost by the projectile and transferred to the medium. This energy transfer is manifested as the recoil energy of the target. As another example, if the energy transfer from the projectile to an atomic electron during a Coulomb scatter is much greater than the electron binding energy, the pre-collision electron can be considered to be quasi-free and stationary with the result being that the collision can be treated as being elastic. Energy throughout both collisions has, by definition, been transferred to an atom within the medium. For the case of the target being an atomic electron, this particle-atomic electron collision is known generally as a *hard collision*[1] and will be the subject of Chap. 9. Such an interaction requires the penetration of the projectile to within the atomic electron 'cloud' followed by a direct and singular collision with one electron. The collision between projectile and electron is thus to be described quantum mechanically. It is inherently obvious that such a case of proximity is unlikely, whereas the interaction of the target with the entire atom at a distance has a greater probability of occurring. This latter condition is referred to as a *soft collision*, the subject of Chap. 8, as the magnitude of the energy transfer is reduced.

In Part III of this book, we will look specifically at those charged particle–matter interactions which transfer kinetic energy to the medium, either through inelastic processes or through, as described in the previous paragraph, what are effectively elastic knock-on collisions between a projectile and an atomic electron target. As noted in Part II, the types of collisions are differentiated into what Bohr (1948) defined as *nuclear collision*, in which the projectile interacts with the atom as a whole, and an *electronic collision* which is that between the projectile and an atomic electron. The former would be largely inelastic (unless the transfer of the projectile's kinetic energy greatly exceeds the binding energy of the electron), resulting in atomic excitation or ionisation with the ejection of low-energy electrons (and the sum of the post-collision kinetic energies being less than that which existed pre-collision), whereas the latter, as previously noted, is an effectively elastic interaction. These categories are analogous to the terms most commonly used in medical radiation physics: *soft collision* and *hard collision*, respectively. We will use the latter categorisations because of their common use in our discipline.

The differentiator between soft and hard collisions in the classical theory is the magnitude of the impact parameter which describes the proximity of the projectile's trajectory to the atom, the atomic electrons and the atomic nucleus. Within each

---

[1] Sometimes referred to as a 'knock-on' collision.

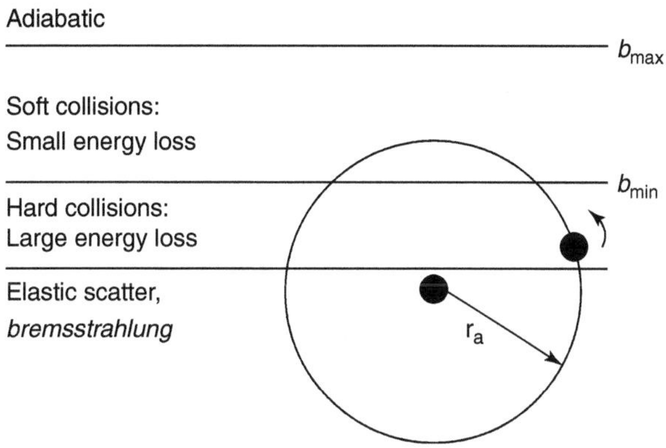

Fig. 7.1 Conceptual differentiation between projectile–atomic target interactions on the basis of impact parameter

category, different physics occurs which we now discuss. In order to set the stage, Fig. 7.1 shows the regions that we shall consider.

At small impact parameters, the projectile interacts directly with the atomic nucleus with the consequence of elastic scatter as described in Chap. 5 or, for an electron or positron projectile, *bremsstrahlung*. At slightly larger values of impact parameter, the projectile interacts directly and singularly with atomic electrons in elastic (or quasi-elastic) scatters (i.e. hard collisions). At even larger values of impact parameter, beyond b_{min}, but less than some value b_{max}, the projectile does not interact with individual electrons but with the atom as a whole via soft collisions where limited amounts of energy are transferred. Beyond a certain impact parameter b_{max}, the projectile does not transfer energy effectively to any component of the atom and the response is regarded as being adiabatic. The above descriptions are peculiar to classical mechanics. In the quantum (Bethe) theory, the metric that differentiates between soft and hard collisions is the momentum transfer (or, equivalently, the energy transfer) resulting from the collision. However, the two metrics of classical and quantum mechanics can be considered to be related through the Heisenberg uncertainty principle, $b \approx \frac{\hbar c}{q}$.

Both conditions are now reviewed in the following two sections.

7.2 Soft Versus Hard Collisions: Impact Parameter Regimes

7.2.1 $b \ll r_a$

Let us begin by considering the case of the impact parameter being much smaller than some atomic radius, r_a, which we use to characterise atomic size. This was, in fact, the spatial regime that we considered in Chap. 5 in the context of elastic

Coulomb screening from a nucleus screened by atomic electrons. For this case of $b \ll r_a$, the projectile can undergo four possible interactions:

- If the projectile is hadronic (and not necessarily charged as in the case of the neutron) and it penetrates into the atom sufficiently that its wavefunction overlaps that of the nucleons within the nucleus, there is the possibility of nuclear excitation or even fragmentation if the projectile energy is comparable to, or sufficiently greater than, the nucleon binding energies. This is, for example, the basis of π^--radiotherapy as discussed in Chap. 1.
- A charged projectile can interact with the nuclear protons which can lead to nuclear excitation and fragmentation.
- A charged projectile can be elastically scattered by the nuclear Coulomb field with negligible recoil, as discussed in Chap. 5.
- If the projectile is an electron or positron, it can interact with the nuclear Coulomb field and create *bremsstrahlung*.

In Part III, none of the above three interactions will be considered.

7.2.2 $b \sim r_a$

Let us consider the case of the impact parameter being of comparable dimensions to that of the atom. In this spatial regime, the projectile can interact directly with single-atomic electrons through the hard collisions previously noted. These interactions are described by explicit projectile–electron interaction cross sections and will be the topic of Chap. 9. We will consider the cases of the projectile being a spinless massive particle (defined as $m \gg m_e$), the electron and the positron. This spatial regime is also characterised by the inequality of $b < b_{\min}$, where $b_{\min}$ is the demarcation between the soft and hard collision regimes and is derived in Chap. 8. The energy lost by the projectile will be relatively high due to the close-range interaction between the projectile and the target atomic electron.

7.2.3 $b_{\min} < b < b_{\max}$

When the magnitude of the impact parameter is far greater than atomic dimensions but still within a distance allowing the particle to interact with the atom, the projectile transfers energy, an ensemble of atomic electrons as a whole. Here, the momentum and energy transfer to the atom is reduced due to the greater distance from the atom and the larger number of electrons potentially influenced.

7.2.4 $b > b_{\max}$

For the case of the impact parameter exceeding some maximum value, the energy exchange becomes negligible and one enters an adiabatic regime. The etymology of this word is of interest: it is derived from the Greek, αδιαβαινετν, meaning 'impassable'. While this word has a clear definition in descriptions of thermodynamical processes, its application to the stopping of energetic charged particles in a medium is somewhat ambiguous in comparison. The momentum/energy transfer is protracted over a long period of time so that the perturbation of the atomic electron is very limited.

7.3 Soft Versus Hard Collisions: Energy and Momentum Transfers, Q and q

7.3.1 *Energy Transfer,* Q

The discussion to this point has been on differentiating between collision regimes, especially those of soft and hard collisions, using the impact parameter. This is a useful approach since, as we shall see throughout our exploration of soft collisions in Chap. 8, the classical theory of Bohr is based upon the impact parameter. On the other hand, the quantum theory of Bethe differentiates between soft and hard collisions on the basis of momentum and/or energy transfer, depending upon the application. In the Bethe description developed in Chap. 8, and which is based upon the Born approximation of Chap. 2, the target atom is in its ground state, and the probability of a transition occurring as a result of the collision is calculated for, using quantum perturbation theory, between this ground state and the available excited states. The differentiator between soft and hard collisions in the Bethe theory is given by a specific value of the energy transfer, Q_C, from the projectile to the atomic system. Within the calculation, this value was somewhat arbitrary in that it was specified so that:

- Q_C does not exceed the atomic electron binding energies.
- The projectile's associated momentum be limited so that its de Broglie wavelength ($\lambda\!\!\!^{-} = \frac{\hbar c}{p}$) is not comparable to nuclear dimensions.

In practice, this combination of requirements requires that the energy transfer limit be within the range of $10\,\text{keV} < Q_c < 100\,\text{keV}$ (Uehling 1954). Of course, one can approximately relate the impact parameter and momentum/energy specifications of Bohr and Bethe theories, respectively, through the Heisenberg principle, $b_{\min} \approx \frac{\hbar c}{\sqrt{2m_e Q_c}}$. It will be found that when the energy losses of the soft (Bethe) collisions and those of hard collisions are combined to yield the total collisional energy loss, this factor of Q_C disappears from the final result.

7.3.2 Momentum Transfer, q

We now expand upon the previous discussion by focussing upon the momentum transfer from the projectile to the target as a result of the collision. Recall from our discussion of two-body collision kinematics that led to (2.86) that the 3-vector momentum transfer of a collision through an angle θ is

$$\begin{aligned} q^2 &= \left|\mathbf{p}_f - \mathbf{p}\right|^2 \\ &= \left|\mathbf{p}_f\right|^2 + |\mathbf{p}|^2 - 2p'p\cos\theta \\ &= (E_f + m)(E + m) - 2m^2 - 2p_f p\cos\theta \end{aligned} \tag{7.1}$$

where the f subscript denotes a post-collision kinematic quantity. The post-collision kinetic energy of the scattered projectile is

$$E_f = E - \Delta E \tag{7.2}$$

where ΔE is the energy lost by the projectile and transferred to the target. We can write the 3-vector momentum of the scattered projectile as

$$\begin{aligned} p_f &= \sqrt{((E - \Delta E) + m)^2 - m^2} \\ &= \sqrt{(E - \Delta E)^2 + 2m(E - \Delta E)} \\ &= \sqrt{p^2 + \Delta E^2 - 2\Delta E(E + m)} \\ &= p\sqrt{1 + \left(\frac{\Delta E}{p}\right)^2 - 2\left(\frac{\Delta E}{p}\right)\frac{(E + m)}{p}} \end{aligned} \tag{7.3}$$

Assuming that the energy loss is small relative to the incident momentum (i.e. $\Delta E \ll p$), we can expand the square root of (7.3) to second order in $\frac{\Delta E}{E}$ and, by including only those terms of up to $\left(\frac{\Delta E}{E}\right)^2$, to give

$$p_f \approx \frac{p^2 - (E + m)\Delta E}{p} - \frac{(m\Delta E)^2}{2p^3}. \tag{7.4}$$

We return (7.4) to (7.1) and, through the small-angle approximation, expand the cosine function to first order ($\cos\theta \approx 1 - \frac{\theta^2}{2}$) to reach

$$\begin{aligned} q &\approx \sqrt{(p\theta)^2 + \left(\frac{m\Delta E}{p}\right)^2} \\ &= \sqrt{(p\theta)^2 + \left(\frac{\Delta E}{\beta\gamma}\right)^2}. \end{aligned} \tag{7.5}$$

We will note that the first term of (7.5) can also arise from (2.86) in the small-angle approximation and will, in general, dominate over the second term as $\Delta E \ll p$. But it is the second term, however, that we are interested in here. This term defines a minimal momentum transfer which occurs, for $\theta = 0$ decreases the more energetic the projectile becomes (i.e. $\beta\gamma$ increasing). Recall that, in Coulomb interactions, the range and spatial resolution of the virtual photons exchanged between projectile and target will be, from the Heisenberg uncertainty principle, of the order of $\frac{\hbar c}{q}$. This increases with decreasing q or, from (7.5), increasing $\beta\gamma$ at a fixed scattering angle. At low values of projectile momentum, where the product $\beta\gamma$ approaches unity, the range is usually comparable to atomic dimensions which limits the number of interactions. However, with increasing projectile momentum, $\beta\gamma$ increases, q decreases and the range of interactions increases along with the number of interactions. It is also interesting to note, independently, from (2.86) and (4.23) that the probability of a Coulomb interaction is proportional to q^{-4} which indicates to us that the probability of the interaction will increase with decreasing q and decreasing resolution. This is in agreement with our classical observations based upon the impact parameter: small values of q are associated with limited spatial resolution through which the projectile interacts with the entire atomic electron ensemble (soft collisions), whereas large values of q are associated with the projectile's ability to resolve individual atomic electrons (hard collisions).

7.4 Summary of Soft and Hard Collisions

The semi-quantitative definitions of soft and hard collisions have been provided with the former being associated with low momentum/energy transfers and the latter associated with high values. By using the impact parameter as a metric, soft collisions are those that occur at extended distances from the atom. As such, the relative probability of a soft collision would be high. Conversely, hard collisions occur when the impact parameter is of the order of atomic dimensions so that the projectile can interact directly with an atomic electron. The relative probability would, then, be much smaller.

Table 7.1 summarises the discussions of the previous subsections.

The net energy transfers of both collision types, i.e. the product of the energy transfer per collision and the probability of the collision, are approximately equal.

Table 7.1 Differentiators between soft and hard collisions

	Probability	Energy transfer
Soft collision	High	Low
Hard collision	Low	High

7.5 Collision Stopping Power Nomenclature

As noted earlier, the linear collision stopping power is the differential rate per unit distance that the kinetic energy of a charged particle is reduced through collisions with atoms in a medium. In this book, this quantity will be denoted by the derivative

$$\left.\frac{\mathrm{d}E}{\mathrm{d}x}\right|_y .$$

It will also be frequently normalised to the physical density of the medium to give the mass collision stopping power:

$$\left.\frac{\mathrm{d}E}{\rho\mathrm{d}x}\right|_y .$$

The subscript y will identify the nature of the calculation used to derive the stopping power. In the medical physics literature, y appears frequently as coll or rad to indicate if the energy loss is due to collisions or to radiation (*bremsstrahlung*). As this book considers collision stopping powers only, this clarification will be unnecessary. Here, the subscript will be either S, H, T or Δ to denote *soft*, *hard*, *total* (the sum of soft and hard) or restricted collision stopping powers, respectively. Occasionally, an additional identifier will be added. For example, the subscript S, *Bohr* will indicate that the collision stopping power is that calculated for soft collisions using the Bohr theory. If a subscript is missing entirely, it is to be implicitly assumed that the stopping power is the total collision stopping power.

The collision stopping power is, in our context, a negative quantity as energy is lost by the projectile through transfers to the medium.[2] However, for convenience and unlike some other authors, we will not include a negative sign in the final expressions of the stopping power, or in the associated figures, as it will be understood to be a negative quantity (or, equivalently, interpreted as the energy gained by the medium per distance travelled by the projectile by which the quantity will be positive).

Having defined our nomenclature of the collision stopping power, the reader should be aware that other authors use different nomenclature for the linear and mass stopping powers. We shall note the uses of various authorities and commonly

[2] The quantity would be positive only if energy were gained from the target such as could be possible in a plasma or if the projectile velocity were much lower than the orbital velocity of an atomic electron that it collides with.

used medical physics textbooks. The International Commission on Radiation Units and Measurements (ICRU) (1984, 1993, 2005) uses two forms:

$$S_{\text{col}}$$

and

$$\left(\frac{\mathrm{d}E}{\mathrm{d}l}\right)_{\text{col}}$$

for the linear collision stopping power (where l is the distance travelled) and

$$\frac{1}{\rho}S_{\text{col}}$$

and

$$\left(\frac{\mathrm{d}E}{\rho\mathrm{d}l}\right)_{\text{col}}$$

for the mass collision stopping power. Attix (1986) uses the expression

$$\left(\frac{\mathrm{d}T}{\rho\mathrm{d}x}\right)_{c}$$

for the mass collision stopping power (where $\mathrm{d}T$ is the differential kinetic energy loss), whereas Johns and Cunningham (1983) use

$$S_{\text{ion}}$$

where the subscript ion is to, as the authors stated, *…emphasize that we are describing energy exchanges that result in ionizations.*

7.6 Media

The concept of the 'medium' through which the charged particle moves and transfers kinetic energy to will require refinement in later developments of Part III.

Our initial investigation of stopping power theory will treat the medium, implicitly, as a cold and dilute gas. Here, *cold* refers to the atoms of the medium being in their ground state prior to any interaction with the projectile. *Dilute gas* specifies that the charged particle interacts with only one atom at a time and that the medium's atoms themselves are sufficiently separated as to be considered independent of each other. The two elemental media we have considered, and will continue to do so, shall be carbon ($Z = 6, A = 12$) and calcium ($Z = 20, A = 40$) as representing the range of biological media between soft tissue and bone. In practice, this selection will have limited effect upon the calculation of the mass collision stopping power which is proportional to the ratio of $\frac{Z}{A}$. The only differentiating effect will lie within the mean excitation energy and the density parameter.

More complex media will be required as we progress our investigations and incorporate modifications to the initial calculations for a cold and dilute gas. We will consider, again, soft tissue and bone. Here, soft tissue will, in general, be represented by skeletal muscle, the composition of which is specified by the International Commission on Radiological Protection (ICRP 1975). Bone is a complex tissue (ICRP 1995), and we consider it here as the skeletal mineral material remaining following the removal of red and yellow bone marrow, cartilage, periarticular tissue and any other soft tissue. Bone is classified into two categories: trabecular (or spongy) bone and cortical (or compact) bone. For our calculations, we will consider cortical (compact) bone and use the composition specified by the ICRU (1964) and the water content and density given by the ICRP (1975).

7.7 Organisation of the Chapters of Part III

Chapter 8 begins this investigation by developing the classical and quantum theories of *soft collisions*, starting with a discussion of the impact parameter more rigorous than that provided previously in this chapter. Our exploration of soft collisions will be approached in two ways. The first is that of classical/semi-classical techniques, based upon perturbation theory. Inevitably, these will lead to the Bohr theory, from which we will derive the soft-collision stopping power, or the differential rate at which energy is lost by the projectile per distance travelled. We then consider the quantum description and follow the approach, based upon the Born approximation, taken by Bethe to describe this energy loss to the atom in nonrelativistic and relativistic conditions. The Bohr and Bethe results are not dissimilar, but we defer the bridging between the two theories, that was eventually performed by Bloch, to Chap. 11, where it has a more natural placement.

In Chap. 9, we investigate *hard collisions* between the projectile and the atomic electron. These will be based solely upon quantum-mechanical descriptions as the nature of the projectile will be important. As noted earlier, we consider those instances of the projectile being massive and without spin, the electron and the positron.

In Chap. 10, we congregate the results of Chaps. 8 and 9 to derive the soft- and hard-collision stopping powers and then combine them to produce the total collision stopping power.

In Chap. 11, we consider the mean excitation energy $\bar{I}$ which is an explicit medium-dependent parameter that appears within the calculation of the stopping power.

Chapter 12 introduces a large variety of considerations we must note in the complete description of charged particle energy loss. Prominent among these is that our consideration of energy loss has been based upon an interaction between a single projectile and a single target. Such an assumption of a binary interaction is not unreasonable in many cases and reflects the non-medical case of the medium being a cold and dilute gas. But tissue represents a condensed and dielectric medium and a charged particle moving within such a medium will interact with many atoms, not just one. Polarisation effects become prominent, and corrections must be made in the stopping power calculation. The Barkas–Andersen effect, which is the difference between the stopping powers in a medium of a particle and its antiparticle, concludes this chapter.

The range, or depth of penetration, of a charged particle is dependent reciprocally upon the stopping power and this is investigated in Chap. 13. We separate the investigations of the ranges of heavy charged particles, such as proton and α-particles, and that of light-charged particles (electrons and positrons) as the latter is dominated by the effects of multiple Coulomb scatter. The practical considerations of charged particle range in electron beam therapy are also reviewed.

Chapter 14 concludes Part III by exploring the means in which stopping power calculations derived for elemental media can be extended to compound media.

References

Attix FH. Introduction to radiological physics and radiation dosimetry. New York: Wiley; 1986.

Bohr N. The penetration of atomic particles through matter. Mat Fys Medd Dan Vid Selsk. 1948;18:1–144.

Fermi E. The ionisation loss of energy in gases and in condensed materials. Phys Rev. 1940;57:485–93.

ICRP. Report of the task group on reference man – International Commission on Radiological Protection Report No 23. Oxford: Pergamon Press; 1975.

ICRP. Basic anatomical and physiological data for use in radiation protection: the skeleton – International Commission on Radiological Protection Report No 70. Oxford: Pergamon Press; 1995.

ICRU. Physical aspects of irradiation – International Commission on Radiological Units and Measurements Report No 10b, published as National Bureau of Standards Handbook 8. Bethesda: ICRU; 1964.

ICRU. Stopping powers and ranges for electrons and positrons – International Commission on Radiological Units and Measurements Report 37. Bethesda: ICRU; 1984.

ICRU. Stopping powers and ranges for protons and alpha particles – International Commission on Radiological Units and Measurements Report 49. Bethesda: ICRU; 1993.

ICRU. Stopping of ions heavier than helium – International Commission on Radiological Units and Measurements Report 73. Oxford: Oxford University Press; 2005.

Johns HE, Cunningham JR. The physics of radiology. 4th ed. Springfield: Charles C Thomas; 1983.

Uehling EA. Penetration of heavy charged particles in matter. Annu Rev Nucl Sci. 1954;4:315–50.

Chapter 8
Soft Collisions

Contents

Abstract This chapter prevents a number of theories associated with soft collisions, that is, those interactions between a charged projectile and the entire atom (which Bohr referred to as *nuclear* collisions). We begin with applying our earlier analysis of projectile momentum and atomic electron screening in elastic scatter to a new analysis of the conditions of soft collisions. Then, the theory of soft-collision energy loss is developed using classical mechanics. First, the Rutherford formula is developed. This is followed by a thorough analysis of the Bohr theory

B.J. McParland, *Medical Radiation Dosimetry*,
DOI 10.1007/978-1-4471-5403-7_8,

from both classical and semi-classical points of view. As an aside, the Fermi model of soft-collision energy loss, which is based upon classical electrodynamic theory, is investigated not only on its own right but also for the foundations that it provides for the derivation of the Bethe quantum-mechanical theory of soft-collision energy loss and the effects of a condensed medium upon collision energy loss as discussed in Chap. 12. Bethe's theory is then developed. Initially, this is for the nonrelativistic regime and is then extended to relativistic energies.

8.1 Introductory Comments

8.1.1 Historical Background

Recall that *soft collisions* are defined as those in which the projectile interacts with the atom as a whole through which a limited amount of energy is transferred and that Bohr characterised these as *nuclear collisions* as the interaction is with the atomic system, the mass of which is dominated by the nucleus. However, we will retain the former description which is more commonly used in medical radiation physics. As explained qualitatively in Chap. 7, soft collisions occur with high probability with the result that the net exchange of energy is comparable to the low-probability hard collisions which transfer much larger amounts of energy per interaction. While a projectile can transfer momentum/energy to atomic electrons or the nucleus, we will see that, on the basis of simple kinematical arguments, energy loss is dominated by projectile–electron interactions.

Descriptions of this projectile–atom interaction have been made through either classical or quantum-mechanical means and both are explored here. It is found that there are, as one would expect, significant similarities between the two under certain conditions, although the explicit bridging between both descriptions is deferred for discussion until later. The classical descriptions of the energy loss of energetic charged particles moving through a medium began with the works of Thomson (1912), Darwin (1912) and, most famously, Bohr (1913, 1915, 1948). Darwin's and Thomson's theories were based upon the assumption that the projectile collided with free electrons. However, both imposed ad hoc maximum impact parameter values in order to avoid the divergence of the Rutherford cross section.

Darwin's theory is of particular interest as it was the impetus for Bohr's investigation of the collision stopping power: Bohr is quoted in correspondence in 1912 to his brother of stating that ... *here C.G. Darwin (grandson of the real Darwin) has just published a theory about this problem, and I felt that it not only wasn't quite right mathematically (however, only slightly wrong but very unsatisfactory in the basic conception, and I have worked out a little theory about it, which, even it isn't much, may perhaps throw some light on certain things connected with the structure of atoms* (Thorsen 1987). On the basis of the ease with which α-particles are capable of ionising atoms, Darwin concluded that the α-particles lost energy by imparting energy to the atomic electrons and assumed that, owing to the high velocity of the α-particles, the electrons could be treated as free. He further assumed that the negative charge of the electrons was uniformly distributed, and he formulated two calculations on the assumption of a spherical distribution or a spherical surface (shell) distribution, each with a radius equal to the atomic radius. This defined b_{max} beyond which the α-particles lost no energy (i.e. energy was lost only when the projectile penetrated into the atom). Darwin's results were in qualitative agreement with the experimental results of Geiger which had yielded the empirical result of the velocity of an α-particle, v, having traversed a thickness x of a medium being $v = \left(v_0^3 - bx\right)^{1/3}$ where v_0 is the initial velocity. Part of Darwin's failure to provide quantitative agreement, in addition to the assumption of the interactions with the atomic electrons as described above, was his neglect of the recoil energy of the nucleus.

Bohr recognised these errors and realised that the binding of the atomic electrons was primary to evaluating the soft-collision stopping power. He determined, as we will show, that the maximum impact parameter should be fixed at a calculable adiabatic limit at which the effective interaction time between the projectile and atom was comparable to the atomic electrons' orbital periods. Critically, for impact parameters less than this maximum impact parameter and greater than a calculated minimum, the atomic electrons can be treated as simple harmonic oscillators set into oscillatory motion by the interaction with the Coulomb field of the projectile. Bohr's 1913 paper considered the ionisations caused by both α- and β-particles. He found good agreement with experimental data for α-particles, but not for β-particles. This discordance was because his calculation was of the *mean* energy loss. It was known from Geiger's work that energy loss is stochastic, a phenomenon that would be expected as energy loss is the result of individual interactions between the projectile and the atoms of the medium it is traversing and is described by a probability distribution function (*pdf*). This statistical description of energy loss is the topic of Part IV of this book. Owing to the large amount of energy loss per unit distance travelled that occurs for an α-particle, the uncertainty of this energy loss and the width of the *pdf* is small relative to the mean energy loss. However, this is not the case for the much lighter β-particle, and this difference was recognised by Bohr, as described in correspondence as … *it proved not only necessary … to determine the energy loss of the β rays …, but also to take into account the probability distribution of the energy loss of the individual β rays* (Thorsen 1987). This stochastic analysis was then presented in his 1915 paper, which we will review in Part IV.

In the 1930s, muons (then referred to as *mesotrons*) were discovered in cosmic rays by Anderson and Neddermeyer (1937). It was noted that the degree of the muon's interaction with air was far greater than with condensed matter having an equal areal density. Fermi hypothesised that part of this difference could be attributable to the higher dielectric constant (relative permittivity) associated with condensed matter: the high dielectric constant would lead to a reduction in the electric field of the rapidly moving muon, resulting in an associated reduction in the ionisation of the medium by the muon. He suggested this in an initial letter (Fermi

1939) which led to a fuller development (Fermi 1940). This reduction in the stopping power is, in the medical radiation physics literature, referred to as the *density effect* and is investigated in detail in Chap. 12. We develop Fermi's theory in this chapter as it lays the foundation not only for our future discussion of the density effect but also for a theoretical derivation of the mean excitation energy provided in Chap. 11 and the relativistic component of the Bethe theory of soft collisions in this chapter.

While Fermi's classical investigations followed the introduction of quantum mechanics by nearly two decades, we will next consider the quantum theory of soft-collision energy loss, although this is not in chronological order. The use of a discrete impact parameter in the Bohr classical theory was necessary as the theory was developed a decade before the introduction of Schrödinger's wave mechanics. With the expanding use of quantum mechanics in the 1920s, initial attempts were made to apply it in the calculation of soft-collision energy loss. An initial attempt was made by Henderson (1922) through restricting the excited electrons to discrete energy levels. Gaunt's theory (1927) was able to reproduce the classical Bohr results through the use of quantum mechanics. But it was through the development of quantum perturbation theory which led to the full and proper development of the quantal derivation of the soft-collision energy loss theory. The quantum-mechanical description of this energy loss will be based exclusively, within this chapter, upon the investigations of Bethe (1930, 1932) who based his derivations upon the Born approximation. The Bethe calculations are separated into the non-relativistic and relativistic regimes.

The striking similarity between the Bohr classical and the Bethe quantal results for the soft-collision energy loss prompted a comparative investigation by Bloch (1933). Although we defer a full examination of Bloch's work until its typical consideration as a perturbative correction factor to the Bethe result, we will derive the initial Bloch result in this chapter. Bloch demonstrated that the Bohr soft-collision energy loss expression was valid in quantum mechanics for a bound system of electrons if the calculated energy loss was to be considered an average over all available atomic electron transitions. This required the use of the dipole approximation which we will derive in this chapter.

8.1.2 Effects of Projectile Momentum and Screening

As a soft collision is that between the projectile and the total atom, we cannot lose sight of the existence of atomic electron screening that will influence this interaction. To this end, let us recall both Fig. 5.8 and the definitions of $\eta(E)$ and $\zeta(E)$ given by (5.69) and (5.72), respectively. The dimensionless factor $\eta(E)$ is the ratio of the Bohr collision diameter to the reduced de Broglie wavelength, $\rho_B(E)/\lambda\!\!\!^{-}$, whereas the dimensionless parameter $\zeta(E)$ is the product of the atomic electron screening parameter and the Bohr collision diameter, $\kappa\rho_B(E)$. In order to characterise the placement of soft collisions within the mapping afforded by Fig. 5.8, we modify the definitions of both $\eta(E)$ and $\zeta(E)$ by including the Bohr velocity v_0, which is equal to αc, so that the Bohr collision diameter is written as

$$\rho_B(E) = \hbar v_0 / E \quad \text{and leads to}$$

$$\eta(E) = 2zZ\frac{v_0}{v} \tag{8.1}$$

and, using the Bohr screening parameter,

$$\zeta(E) = 2\sqrt{z^{2/3} + Z^{2/3}}\, zZ\left(\frac{m_e}{m}\right)\left(\frac{v_0}{v}\right)^2. \tag{8.2}$$

By rearranging these two expressions, we obtain the ratio

$$\frac{\eta^2(E)}{\zeta(E)} = 2\frac{zZ}{\sqrt{z^{2/3} + Z^{2/3}}}\left(\frac{m}{m_e}\right). \tag{8.3}$$

From this ratio, we have the inequality, as we are assuming a heavy projectile, $m > m_e$,

$$\eta(E) > \sqrt{\zeta(E)}. \tag{8.4}$$

This result corresponds to those collisions which are in the region above the line (E) of Fig. 5.8 which represents the case of the atomic electron screening angle being equal to the square root angle of (5.71), $\chi_0 = \sqrt{\theta_{(a)}}$. Recall that $\theta_{(a)}$ is the scattering angle beyond which the Rutherford result is expected to hold. Excluding the case of electron projectiles, unless the projectile velocity is very small (i.e. $v << v_0$) then, from (8.2), $\zeta(E) << 1$. Thus, for fast projectiles, the collisions will fall within the ranges of (A) or (B) in Fig. 5.8 if $\eta(E) > 1$ or $\eta(E) < 1$ respectively. These inequalities are, respectively, equivalent to $\rho_B(E) > \lambda$ and $\rho_B(E) < \lambda$. From (8.1), for slow projectiles, $\eta(E) > 1$ (i.e. Region (A)) always whereas, for rapid projectiles, the case of $\eta(E) > 1$ and the applicability of classical mechanics is dependent upon the product of zZ. In either case, Rutherford scattering holds for angles greater than the $\theta_{(a)}$ for $\eta(E) < 1$. We now consider the case of an electron projectile but exclude quantum-mechanical considerations, such as the Pauli exclusion principle. In this case,

$$\eta(E) = 2Z\left(\frac{v_0}{v}\right) \tag{8.5}$$

and

$$\begin{aligned}\zeta(E) &= 2Z\sqrt{1 + Z^{2/3}}\left(\frac{v_0}{v}\right)^2 \\ &\approx 2Z^{4/3}\left(\frac{v_0}{v}\right)^2.\end{aligned} \tag{8.6}$$

For energetic electrons such that $v >> v_0 Z^{2/3}$, then $\zeta(E) << 1$ with the result that the collisions will be within Range (A) or Range (B) depending upon $\eta(E) > 1$ or $\eta(E) < 1$, respectively. Should the electron projectile velocity be small enough such that the inequality $\eta(E) > 1$, then the collisions will not lie within the Region (F) of Fig. 5.8. Further, it should be noted that for an electron projectile velocity such that $v < Z^{1/3} v_0$, then the reduced de Broglie wavelength of the projectile is such that there is the inequality

$$\begin{aligned} \lambda\!\!\!^{-} &> \frac{\hbar c}{m_e(v_0/c)} Z^{-1/3} \\ &> a_\infty Z^{-1/3}. \end{aligned} \tag{8.7}$$

This immediately leads to the inequality of $\eta(E) < \zeta(E)$ which corresponds to Region G of Fig. 5.8 where quantum-mechanical conditions exist in order to be able to describe the interaction.

The above has described qualitatively the projectile kinematics and atomic electron screening conditions associated with *soft collisions* or *nuclear collisions* as described by Bohr. We now consider the classical and semi-classical descriptions of soft collisions.

8.2 Classical and Semi-classical Theories

8.2.1 Introductory Comments

We shall define classical theory in the context of soft collisions as that which treats the atomic electrons classically. Quantum-mechanical considerations shall play no role in these derivations. We then progress to a semi-classical investigation of the projectile-atom interaction. In the semi-classical treatment, the physics of the nucleus is considered classically, whereas the atomic electrons are described within the realm of quantum mechanics. That is, the Bohr model of the atomic electrons as simple harmonic oscillators capable of absorbing a continuum of energies from Coulomb interactions with a passing projectile is the foundation of the classical theory. In the semi-classical theory, the electrons are capable of occupying only discrete quantum states. In this latter case, the quantum perturbation theory developed in Chap. 2 will be essential.

In either theory, the critical metric will be the impact parameter – the perpendicular distance between the trajectory of the projectile and the target. Here, we will define the origin from which the impact parameter is referenced as being the centre of the nucleus. Clearly, this definition becomes untenable in the quantum theory which we develop afterwards – in fact, energy or momentum transfer is the metric rather than the spatial separation between the projectile and the atom.

8.2.2 *Impact Parameter*

8.2.2.1 Introductory Comments

As noted, the impact parameter is used in differentiating between soft and hard collisions in a classical or semi-classical description. This is important as not only do the magnitudes of the energy transfer differ between both regimes but the fundamental physics of the interactions also differ.

As a preliminary discussion before deriving the Bohr theory, we note that there are three fundamental concepts associated with that theory:

- It is based upon the dipole approximation which neglects all but the leading term of the atomic electron's spatial position within the Fourier transform of the projectile's electric field. This is a long wavelength limit and assumes that the spatial variation of the Coulombic field is small relative to the range of the perturbed electron's motion.
- The energy transfer from the projectile to the target diverges in the dipole approximation for small impact parameter b.
- As noted in Chap. 7, we recognise the need for an adiabatic limit, b_{max}. This arises necessarily when reconciling the binding energies of the atomic electrons and the infinite range of the projectile's Coulomb potential. The adiabatic response is that the electron enters a slow motion but remains bound to the atom.

In this section, we derive a variety of expressions for b_{max} beyond which the atom responds adiabatically to the influence of the projectile and b_{min} below which the projectile undergoes hard collisions with single atomic electrons.

8.2.2.2 Collision Kinematics

Energy and Momentum Transfers

We begin our derivation of the impact parameter by developing a description of the kinematics of the collision. Let us consider the general case of a collision event as demonstrated in Fig. 8.1. As before, we have a charged particle with net electric charge ze, mass m and momentum p incident with an impact parameter b to a bare-charged target. In virtually all of the discussion to follow, the target is considered to be an atomic electron with mass m_e and electric charge e (i.e. $Z = 1$ in our nomenclature). Hence, the Bohr collision diameter is given by $\rho_{\text{B}}(E) = \alpha\hbar cz/E$. It is also taken that the projectile is massive, i.e. $m/(m_e >> 1)$.

The radial vector $r(t)$ between the projectile and target and the associated azimuthal angle $\zeta(t)$ as shown in Fig. 8.1 arc both functions of time t with the limits of $r(t) \rightarrow \pm\infty$ and $\zeta(t) \rightarrow \begin{matrix}\pi - \theta \\ 0\end{matrix}$ as $t \rightarrow \pm\infty$, respectively. The time of $t = 0$ is defined as the time of the closest approach of the projectile to the target.

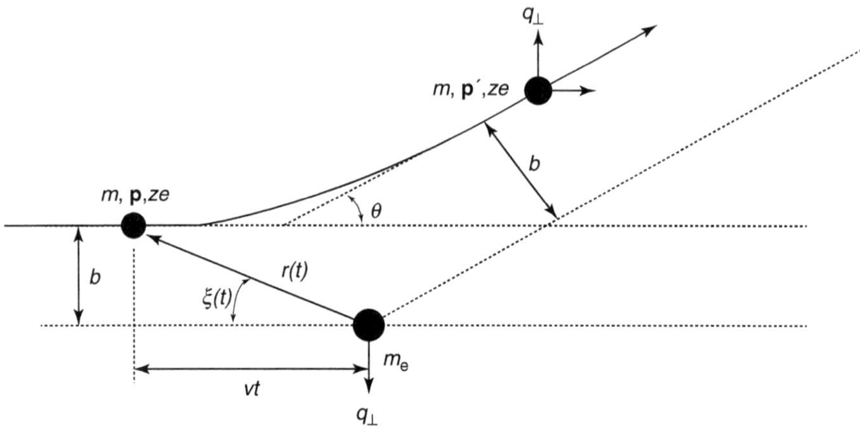

Fig. 8.1 A collision between a projectile and an atomic electron with a net momentum transfer orthogonal to the projectile's incident trajectory

For nonrelativistic kinematics, the magnitude of the electrostatic force between the projectile and the electron as a function of time is

$$|\mathbf{F}(t)| = \frac{\alpha\hbar c\, z}{(r(t))^2} = \frac{\alpha\hbar c\, z}{(vt)^2 + b^2}. \tag{8.8}$$

The vector components of this force parallel to and orthogonal to the original direction of the projectile's motion are

$$F_{||}(t) = |\mathbf{F}(t)| \cos\zeta(t) = \frac{\alpha\hbar c\, z\, vt}{\left((vt)^2 + b^2\right)^{3/2}} \tag{8.9}$$

and

$$F_{\perp}(t) = |\mathbf{F}(t)| \sin\zeta(t) = \frac{\alpha\hbar c\, z\, b}{\left((vt)^2 + b^2\right)^{3/2}}, \tag{8.10}$$

respectively. The two normalised force components are plotted in Fig. 8.2 as a function of time normalised to the ratio, b/v.

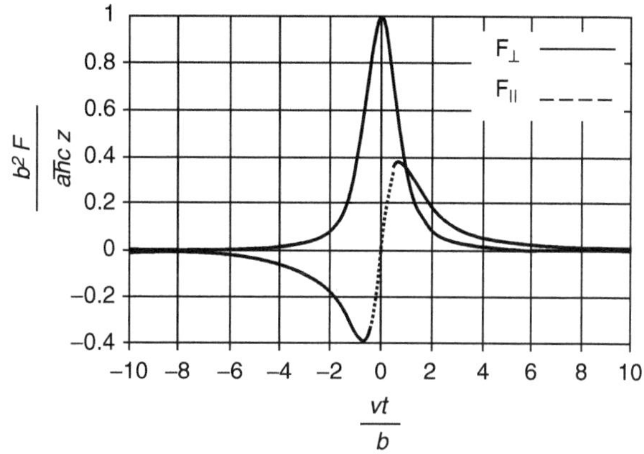

Fig. 8.2 Normalised vector components of the nonrelativistic electrostatic force of Eqs. (8.9) and (8.10)

The component of the force perpendicular to the trajectory is sharply peaked at time $t = 0$, whereas the parallel component is biphasic with negative and positive peaks symmetric about $t = 0$. The relative widths of these distributions decrease as the velocity of the projectile increases (McParland 2010). The corresponding components of the momentum transfer between the projectile and target are determined by integration:

$$q_{\|} = \alpha\hbar c\, z\, v \int_{-\infty}^{\infty} \mathrm{d}t \frac{t}{\left((vt)^2 + b^2\right)^{3/2}} \tag{8.11}$$

and

$$q_{\perp} = \alpha\hbar c\, z\, b \int_{-\infty}^{\infty} \frac{\mathrm{d}t}{\left((vt)^2 + b^2\right)^{3/2}}. \tag{8.12}$$

Completing these integrations, we find that the two net momentum transfer components are

$$q_{\|} = 0 \tag{8.13}$$

and

$$q_{\perp} = -\frac{2\,\alpha\hbar c\, z}{bv}. \tag{8.14}$$

That the net momentum transfer parallel to the projectile's original trajectory over all time is zero is intuitive from Fig. 8.2 due to the equal magnitudes of the peaks in anti-phase. The net energy transfer to the electron resulting from the collision is, thus,

$$\begin{aligned} Q &= \frac{q_\perp^2}{2m_e} \\ &= \frac{2}{m_e}\left(\frac{\alpha\hbar c\, z}{bv}\right)^2 \\ &= 2m_e\left(\frac{r_e z}{b\beta}\right)^2 \end{aligned} \tag{8.15}$$

where r_e is the classical electron radius, $\alpha\hbar c/m_e$.

We now take a temporary diversion to consider the result of (8.15). If we had instead selected the target body to be a nucleus of mass $Am_{\mathrm{N}} \approx 2Zm_{\mathrm{N}}$, where m_{N} is the nucleon mass (≈ 940 MeV), and electric charge Ze, the energy transfer would be

$$Q_{\mathrm{nuc}} = \frac{1}{Am_{\mathrm{N}}}\left(\frac{\alpha\hbar c\, zZ}{bv}\right)^2. \tag{8.16}$$

The ratio of the energy transfer to the nucleus to that of the atomic electron is

$$\begin{aligned} \frac{Q_{\mathrm{nuc}}}{Q} &\approx \frac{Z}{2}\left(\frac{m_e}{m_{\mathrm{N}}}\right) \\ &\approx \frac{Z}{3,675}. \end{aligned} \tag{8.17}$$

This result of (8.17) is rather profound in that it accentuates the difference between the atomic components that contribute to elastic scatter and to energy loss. Whereas, in the context of our discussions of elastic Coulomb scatter in Part II, we found that the *scattering* of a projectile was dominated by its interaction with the atomic nucleus, (8.17) shows that the *stopping* of the projectile is dominated by its interactions with atomic electrons. For the case of carbon and calcium, the energy transferred to the nucleus is about 0.2 % and 0.5 % of that to an atomic electron, respectively. Even for the extreme case of a lead target, this figure is only about 2 %. Hence, in the discussion of soft collisions, energy loss due to the recoil of the nucleus will be neglected.

Resuming our consideration of the target body being a single electron, our discussion of the classical theory will be based upon the case of the projectile having a greater mass than that of the electron. When we do consider the case of an electron projectile, a quantum-mechanical description is of necessity due to the need to account for case of identical fermions and the Pauli exclusion principle. As most of the calculations to follow will be for $m >> m_e$, we recall from Appendix D that the maximum energy transfer to the electron target is

$$Q_{\max} = 2m_e(\gamma\beta)^2 \tag{8.18}$$

where $$\beta = v/c \text{ and } \gamma = 1/\sqrt{1-(v/c)^2}.$$

Thomson's Formula

We now derive a relationship between the energy transfer to the target, the projectile velocity and the impact parameter. This expression will be of use later. We begin with the relationship between the impact parameter and scattering angle given by (4.20),

$$b = \frac{\rho_{\mathrm{B}}(E)}{2} \cot \frac{\theta}{2}, \tag{8.19}$$

and that of the energy transfer to the target and the projectile kinetic energy given by (D.33),

$$Q = \frac{4mm_e}{(m_e+m)^2} E \sin^2\left(\frac{\theta}{2}\right). \tag{8.20}$$

It follows from (8.19) that

$$\sin^2\left(\frac{\theta}{2}\right) = \frac{\left(\frac{\rho_{\mathrm{B}}(E)}{2b}\right)^2}{1+\left(\frac{\rho_{\mathrm{B}}(E)}{2b}\right)^2}. \tag{8.21}$$

Substituting (8.21) into (8.20) gives

$$Q = \frac{4mm_e}{(m_e+m)^2} E \frac{\left(\frac{\rho_{\mathrm{B}}(E)}{2b}\right)^2}{1+\left(\frac{\rho_{\mathrm{B}}(E)}{2b}\right)^2}. \tag{8.22}$$

For a heavy projectile, $m >> m_e$, (8.22) reduces to

$$Q = 4\left(\frac{m_e}{m}\right) E \frac{\left(\frac{\rho_{\mathrm{B}}(E)}{2b}\right)^2}{1+\left(\frac{\rho_{\mathrm{B}}(E)}{2b}\right)^2}.$$

$$= 2m_e r_e^2 \left(\frac{z}{\beta b}\right)^2 \frac{1}{1+\left(\frac{\rho_{\mathrm{B}}(E)}{2b}\right)^2}. \tag{8.23}$$

Effective Interaction Time

In our consideration of the regime corresponding to a soft collision, we will be interested in the duration of the interaction between the projectile and the target as a long protracted collision will allow the atom to respond adiabatically to the influence of the projectile. Following Bohr's argument (1948), we take the collision to be such that the projectile velocity is much greater than the orbital velocity of the sth electron, $v >> v_s$ (recall from the Bohr atomic model that the orbital velocity of an electron in the $n = 1$ orbital of an atom with atomic number Z is equal to $Z\alpha c$)[1] and the impact parameter greater than the Bohr collision diameter, $b >> \rho_B(E)$. Then, the effective duration of the interaction will be of the order of the ratio of the momentum transfer to the maximum force between the projectile and target:

$$t_{\text{eff}} \approx \frac{|q_\perp|}{F_{\max}}. \tag{8.24}$$

$F_{\max}$ will occur for the minimum separation between the projectile and target is b, or $F_{\max} = \alpha\hbar c\, z/b^2$, which, along with (8.14), provides us with an expression for the effective interaction time, which is approximately

$$t_{\text{eff}} \approx \frac{2b}{v}. \tag{8.25}$$

That is, we may consider the target and projectile to interact effectively over a distance of $2b$ along the course of the trajectory.

The above derivation when applied to the relativistic case requires the use of the Lorentz boost of the electric field (Jackson 1999; McParland 2010) to give the components of the momentum transfer as

$$\begin{aligned} q_{||} &= \alpha\hbar c\, zZ\gamma\, v \int_{-\infty}^{\infty} \mathrm{d}t \frac{t}{\left((\gamma v t)^2 + b^2\right)^{3/2}} \\ &= 0 \end{aligned} \tag{8.26}$$

and

$$\begin{aligned} q_\perp &= \alpha\hbar c\, zZ\gamma\, b \int_{-\infty}^{\infty} \frac{\mathrm{d}t}{\left((\gamma v t)^2 + b^2\right)^{3/2}} \\ &= -\frac{2\,\alpha\hbar c\, zZ}{bv}. \end{aligned} \tag{8.27}$$

---

[1] Feynman noted the interesting quandary in orbital velocity that would arise for any elements that exist with $Z > 137$.

As these components are the same as those obtained for the nonrelativistic case despite the fact that the force has been modified by the Lorentz factor, γ, it is evident that the effective interaction time must then be reduced by the same factor:

$$t_{\text{eff}} \approx \frac{2b}{\gamma v}. \tag{8.28}$$

Maximum Impact Parameter: The Adiabatic Limit

Recall Fig. 7.1 which provided the boundaries of the regimes corresponding to the atomic adiabatic response, the soft and hard collisions and elastic scatter/*bremsstrahlung*. Here, we will look at the upper limit of the impact parameter beyond which the atom responds adiabatically to the projectile. To define this position, which we term as b_{max}, let us consider the atomic electrons as, in Bohr's words, *an ensemble of harmonic oscillators, each consisting of an electron bound in a quasielastic field of force* (Bohr 1948). The spatial dimension of this force can be approximated by

$$r_S \approx \frac{v_S}{\omega_S} \tag{8.29}$$

where ω_S is the frequency of the sth harmonic oscillator with a corresponding velocity given by

$$v_S = \sqrt{\frac{2I_S}{m_e}}. \tag{8.30}$$

I_S is the binding energy of the sth atomic electron.

The relationships between the effective interaction time and the impact parameter lead us to the consideration of the adiabatic limit to this soft collision. We expect that for a projectile beyond this limit, it can perturb the atom only slowly with the atom eventually returning to its original ground state and without transfer of energy from the projectile to the atom. In order to evaluate b_{max} with this method, we must first provide a description of the atomic electron as a simple harmonic oscillator.

Simple Harmonic Oscillator Analysis

Following the aforementioned guidance of Bohr, we model the bound atomic electron by a classical damped harmonic oscillator. An abbreviated review of the simple harmonic oscillator is provided in Appendix G. Using this model, when perturbed by a force $F(t)$, the time-dependent displacement of the electron is described by the second-order linear differential equation:

$$\begin{aligned}\frac{\mathrm{d}^2 r(t)}{\mathrm{d}t^2} + \Gamma\frac{\mathrm{d}r(t)}{\mathrm{d}t} + \omega_0^2 r(t) &= \frac{F(t)}{m_e} \\ &\equiv g(t)\end{aligned} \tag{8.31}$$

where Γ is the damping constant accounting for the dissipative energy loss (with the unit of inverse time) associated with the excitation of the electron and ω_0 is the oscillator's undamped resonant frequency. Our intent is to determine semi-quantitatively the momentum transfer to the electron as a function of the duration of the electric force from the projectile. To obtain this, we will first solve for the displacement $r(t)$ and assume that the damping of (8.31) is *noncritical* (i.e. under-damped) as defined in Appendix G. The electron is considered to be initially at rest and at the origin prior to the application of this force, $r(t=0)=0$; $(\mathrm{d}r(t=0))/\mathrm{d}t=0$.

In order to solve for the displacement $r(t)$, we first rewrite (8.31) in the form of an operator equation or

$$\boldsymbol{O}_t\, r(t) = g(t) \tag{8.32}$$

where the second-order linear differential operator is

$$\boldsymbol{O}_t \equiv \frac{\mathrm{d}^2}{\mathrm{d}t^2} + \Gamma\frac{\mathrm{d}}{\mathrm{d}t} + \omega_0^2. \tag{8.33}$$

From our discussion of the Green's function in Appendix C.11, the solution of (8.33) is given in terms of the Green's function:

$$r(t) = \int \mathrm{d}t'\, G(t,t')\, g(t'). \tag{8.34}$$

By definition, the Green's function is the solution to the equation

$$\boldsymbol{O}_t\, G(t,t') = \delta(t-t'), \tag{8.35}$$

where the Dirac δ-function is defined in Appendix C.10. As (8.32) shows, we must first solve for the Green's function for the differential operator in order to calculate for $r(t)$. To do so, we begin by taking the Fourier transform of both sides of (8.35) to give

$$\left(-\omega^2 + i\omega\Gamma + \omega_0^2\right) G(\omega) = \frac{1}{\sqrt{2\pi}}. \tag{8.36}$$

From this, we obtain the following expression for the Green's function in frequency space:

$$G(\omega) = \frac{1}{\sqrt{2\pi}} \frac{1}{\left(-\omega^2 + i\omega\Gamma + \omega_0^2\right)} = -\frac{1}{\sqrt{2\pi}} \frac{1}{(\omega - \omega_{P1})(\omega - \omega_{P2})} \tag{8.37}$$

where the poles of the Green's function are

$$\omega_{P\,{1 \atop 2}} = i\frac{\Gamma}{2} \pm \sqrt{\omega_0^2 - \left(\frac{\Gamma}{2}\right)^2} = i\frac{\Gamma}{2} \pm \omega_{\mathrm{D}}. \tag{8.38}$$

In (8.38), the damped frequency has been defined as

$$\omega_{\mathrm{D}} \equiv \sqrt{\omega_0^2 - \left(\frac{\Gamma}{2}\right)^2}. \tag{8.39}$$

Recall that we have specified the harmonic oscillator to be under-damped which is defined by the inequality of $4\omega_0^2 > \Gamma^2$ with the result that the damped frequency will be a real quantity with a magnitude less than that of the resonant frequency. From Appendix C.10, we note that the solution for the electron displacement is given through the inverse Fourier transform:

$$r(t) = \frac{1}{\sqrt{2\pi}} \int_{-\infty}^{\infty} \mathrm{d}\omega\, e^{i\omega t} G(\omega)\, g(\omega) + r_{\mathrm{hom}}(t) \tag{8.40}$$

where $r_{\mathrm{hom}}(t)$ is the solution to the homogeneous form of (8.31) (i.e. for $g(t) = 0$). To solve for the inhomogeneous solution of (8.40), we insert the definition of the Green's function and the Fourier transform of $g(\omega)$ into it:

$$r(t) = -\frac{1}{\sqrt{2\pi}} \int_{-\infty}^{\infty} \mathrm{d}\omega \frac{e^{i\omega t}}{(\omega - \omega_{P1})(\omega - \omega_{P2})} \frac{1}{\sqrt{2\pi}} \int_{-\infty}^{\infty} \mathrm{d}t' e^{-i\omega t'} g(t') = -\frac{1}{2\pi} \int_{-\infty}^{\infty} \mathrm{d}t' g(t') \int_{-\infty}^{\infty} \mathrm{d}\omega \frac{e^{-i\omega(t'-t)}}{(\omega - \omega_{P1})(\omega - \omega_{P2})} \tag{8.41}$$

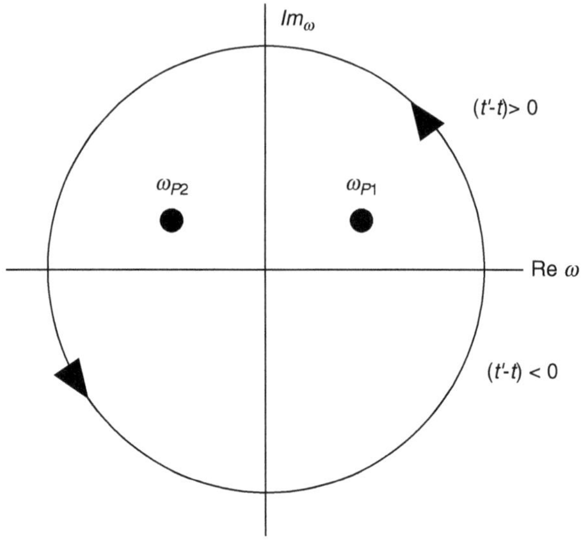

Fig. 8.3 Contour integration to yield the Green's function of Eq. (8.42)

which, from (8.34), returns us the result of

$$G(t,t') = -\frac{1}{2\pi}\int_{-\infty}^{\infty} d\omega \frac{e^{-i\omega(t'-t)}}{(\omega-\omega_{P1})(\omega-\omega_{P2})}. \tag{8.42}$$

This integral is solved using the residue theorem given in Appendix A.4. The two poles are at $i\frac{\Gamma}{2} \pm \omega_D$ in the upper half of the complex ω-plane, as shown in Fig. 8.3.

The contour is along the Re ω axis, and how we manipulate it will be dependent upon the $e^{-i\omega(t'-t)}$ term within the expression for the Green's function. As ω is a complex quantity, we write out its real and imaginary parts explicitly,

$$\omega = \omega_R + i\,\omega_I \tag{8.43}$$

so that

$$e^{-i(\omega_R+i\omega_I)(t'-t)} = e^{\omega_I(t'-t)}\,e^{-i\omega_R(t'-t)}. \tag{8.44}$$

Let us begin with the case of $(t'-t) > 0$ for which the exponential will be

$$e^{-i\omega(t'-t)} = e^{\omega_I|(t'-t)|}\,e^{-i\omega_R(t'-t)}. \tag{8.45}$$

Hence, for the case of Im $\omega = \omega_{\rm I} < 0$, the exponential term of (8.42) will decay to zero with increasing ω (i.e. the contour extending downwards into the lower plane) and the integral in the Green's function goes to zero for $(t - t') < 0$. On the other hand, for $(t - t') > 0$, the exponential term is

$$e^{-i\omega(t'-t)} = e^{-\omega_I|(t'-t)|}\, e^{-i\omega_R(t'-t)}, \tag{8.46}$$

and the behaviour of the contour is such that as it is moved towards $+i\infty$ (i.e. Im $\omega = \omega_{\rm I} > 0$), it diminishes and the remaining non-zero contours are those encircling the two poles as shown in Fig. 8.3. As a result of the above, the integral of (8.42) is equal to the product of $2\pi i$ and the sum of the residues of the integrand:

$$\begin{aligned}
\int_{-\infty}^{\infty} \mathrm{d}\omega \frac{e^{-i\omega(t'-t)}}{(\omega-\omega_{P1})(\omega-\omega_{P2})} &= 2\pi i \sum \mathrm{Res}\, \frac{e^{-i\omega(t'-t)}}{(\omega-\omega_{P1})(\omega-\omega_{P2})} \\
&= 2\pi i \left(\lim_{\omega\to\omega_{P1}} \left(\frac{e^{-i\omega(t'-t)}}{\omega-\omega_{P2}}\right) + \lim_{\omega\to\omega_{P2}} \left(\frac{e^{-i\omega(t'-t)}}{\omega-\omega_{P1}}\right)\right) \\
&= \frac{2\pi i}{\omega_{P1}-\omega_{P2}} \left(e^{-i\omega_{P1}(t'-t)} - e^{-i\omega_{P2}(t'-t)}\right) \\
&= \frac{\pi i}{\omega_{\rm D}} \left(e^{-i\omega_{P1}(t'-t)} - e^{-i\omega_{P2}(t'-t)}\right) \\
&= \frac{\pi i}{\omega_{\rm D}} e^{-\frac{\Gamma(t-t')}{2}} \left(e^{i\omega_{\rm D}(t-t')} - e^{-i\omega_{\rm D}(t-t')}\right) \\
&= -\frac{2\pi}{\omega_{\rm D}} e^{-\frac{\Gamma(t-t')}{2}} \sin(\omega_{\rm D}(t-t')).
\end{aligned} \tag{8.47}$$

Thus, the Green's function in the time domain for the under-damped simple harmonic oscillator is

$$\begin{aligned}
G(t,t') &= \frac{1}{\omega_D} e^{-\frac{\Gamma(t-t')}{2}} \sin(\omega_{\rm D}(t-t')); \quad t' < t \\
&= 0; \quad \text{otherwise}
\end{aligned} \tag{8.48}$$

Putting this result into (8.41) gives

$$\begin{aligned}
r(t) &= \int_{-\infty}^{\infty} \mathrm{d}t'\, G(t,t')\, g(t') \\
&= \frac{1}{\omega_{\rm D}} \int_{-\infty}^{t} \mathrm{d}t'\, e^{\frac{-\Gamma(t-t')}{2}} (\sin \omega_{\rm D}(t-t'))\, g(t').
\end{aligned} \tag{8.49}$$

The upper limit of the integration is set equal to t as only those events in the past can influence the position of the oscillator at that time.

8.2.2.3 Effect of $t_{\rm eff}$ upon Momentum Transfer to an Atomic Electron

Having obtained the integral expression for the electron's displacement of (8.49), we now calculate the momentum transfer to the atomic electron as a function of the time duration of the application of the external force. We model the time behaviour of this force by a finite temporal duration of a constant force:

$$\begin{aligned} g(t) &= \frac{F_0}{m_e} \quad \text{for} \quad 0 < t < t_{\rm eff} \\ &= 0 \quad \text{otherwise.} \end{aligned} \tag{8.50}$$

This is based upon our earlier derivation of the effective interaction time described by (8.25) and (8.28). Thus, for the time variable within the range $0 \le t \le t_{\rm eff}$, and using the substitution of variable $y = t - t'$, we can solve for the displacement:

$$\begin{aligned} r(t) &= \frac{F_0}{m_e \omega_{\rm D}} \int\limits_0^{t \le t_{\rm eff}} {\rm d}t'\, e^{\frac{-\Gamma(t-t')}{2}} (\sin\, \omega_{\rm D}(t-t')) \\ &= \frac{F_0}{m_e \omega_{\rm D}\left((\Gamma/2)^2 + \omega_{\rm D}^2\right)} \left(e^{-\Gamma y/2}\left(\frac{\Gamma}{2}\sin\, \omega_{\rm D} y + \omega_{\rm D} \cos\, \omega_{\rm D} y\right)\right)_{y=0}^{y=t} \\ &= \frac{F_0}{m_e \omega_{\rm D} \omega_0^2}\left(\omega_{\rm D}\left(1 - e^{-\Gamma t/2} \cos\, \omega_{\rm D} t\right) - e^{-\Gamma t/2} \frac{\Gamma}{2} \sin\, \omega_{\rm D} t\right) \qquad 0 \le t \le t_{\rm eff}. \end{aligned} \tag{8.51}$$

We see that (8.51) reduces to the expected form of $(F_0(1 - \cos\, \omega_0 t))/m_e \omega_0^2$ for an undamped oscillator (i.e. the conditions of $\Gamma = 0$ and $\omega_{\rm D} = \omega_0$). The momentum transferred to the electron is

$$\begin{aligned} q &= m_e \int\limits_{-\infty}^{\infty} {\rm d}t' \,\frac{{\rm d}^2 r(t')}{{\rm d}t'} \\ &= \frac{F_0}{\omega_{\rm D}} \int\limits_0^{t_{\rm eff}} {\rm d}t'\, e^{-\Gamma t'/2}\left(\omega_{\rm D} \cos\, \omega_{\rm D} t' - \frac{\Gamma}{2} \sin\, \omega_{\rm D} t'\right) \\ &= \frac{F_0}{\omega_{\rm D}} e^{-\Gamma t_{\rm eff}/2} \sin\, \omega_{\rm D} t_{\rm eff}. \end{aligned} \tag{8.52}$$

Including (8.25), we see that the momentum transfer to the oscillator is limited by the amount:

$$|q| \le \frac{e \alpha \hbar c\, z}{\omega_{\rm D} b} e^{-2\Gamma b/v}. \tag{8.53}$$

Thus, the momentum transfer to the electron diminishes with increasing effective interaction time or, correspondingly, increasing impact parameter. If the interaction time t_{eff} is sufficiently small compared to $2\pi/\omega_0$ so that $e^{-\Gamma\, t_{\text{eff}}/2} \approx 1$ and $\sin \omega_{\text{D}} t_{\text{eff}} \approx \omega_{\text{D}} t_{\text{eff}}$, then the momentum transfer reduces to

$$q \approx F_0 t_{\text{eff}} \quad \text{small} \quad t_{\text{eff}}. \tag{8.54}$$

This is the impulse delivered as if the electron were to be considered as a free particle. But, from (8.52), we see that

$$q \rightarrow 0 \quad \text{as} \quad t_{\text{eff}} \rightarrow \infty. \tag{8.55}$$

Kinematic Definitions of b_{max}

We can use the above semi-quantitative argument to specify a maximum impact parameter beyond which the atomic response the perturbation caused by the projectile's electric field is adiabatic. Let us neglect the influence of damping (which we have already specified as being small). Then, from (8.52), $q = 0$ for $\omega_0 t_{\text{eff}} = \pi$ (neglecting the case of the argument being equal to zero). Thus, using (8.25), we can approximate the maximum impact parameter beyond which the atomic electron response is adiabatic:

$$\begin{aligned} b_{\text{max}} &= \frac{\pi v}{2\omega_0} \\ &\approx \frac{\beta c}{\omega_0} \end{aligned} \tag{8.56}$$

in which this last approximation appears most commonly in the literature. In the relativistic case, this becomes

$$b_{\text{max}} \approx \frac{\gamma \beta c}{\omega_0}. \tag{8.57}$$

We can modify this expression by defining $\bar{I}$ as the mean excitation energy of the electrons or

$$\bar{I} = \hbar\omega_0, \tag{8.58}$$

thus allowing us to write

$$b_{\text{max}} \approx \frac{\hbar c \gamma \beta}{\bar{I}}. \tag{8.59}$$

The approximation of $b_{\max}$ given above is based upon the momentum transfer to a bound atomic electron represented by a simple harmonic oscillator. Here, we consider another approach which is based also upon kinematics but which recognises that for impact parameters greater than $b_{\max}$, the atomic electrons cannot be ionised or excited to higher states. Recall the energy transfer given by (8.15) and relate it through this inequality to the mean excitation energy:

$$2m_e\left(\frac{r_e z}{b\beta}\right)^2 << \bar{I}. \tag{8.60}$$

From this, we can define a maximum limit to the impact parameter for soft collisions:

$$b_{\max} = \frac{r_e z}{\beta}\sqrt{\frac{2m_e}{\bar{I}}}. \tag{8.61}$$

There are fundamental differences between the definitions of $b_{\max}$ given by (8.59) and (8.61). The former lacks any dependence upon the projectile's electric charge and increases with projectile velocity through the $\gamma\beta$ product. On the other hand, the latter has an explicit dependence upon the projectile's electric charge and an inverse dependence upon the projectile velocity.

We shall consider these differences quantitatively following our derivations of expressions for $b_{\min}$ which separates the soft and hard-collision regimes.

8.2.2.4 Minimum Impact Parameter I

Recoil Distance Definition

If we assume the impulse approximation and that the recoil distance of the target electron is not significantly greater than the impact parameter, we can write this recoil distance using the effective time of interaction given by (8.28) as

$$\begin{aligned} l_{rec} &= \frac{|\mathbf{q}|}{m_e} t_{eff} \\ &\approx \frac{2|\mathbf{q}|}{m_e}\frac{b}{\gamma v}. \end{aligned} \tag{8.62}$$

By inserting the relativistic expression of the momentum transfer of (8.27) into (8.62), we have an expression for the minimum impact parameter for soft collisions:

$$b_{\min} = \frac{4z}{\gamma\beta^2} r_e. \tag{8.63}$$

Maximum Energy Transfer Definition

Another means of defining $b_{\min}$ is through the recognition that hard collisions are defined as those that occur for $b < b_{\min}$. One can specify crudely that the energy transfer to the electron approaches the maximum that is kinematically possible in a hard collision. Assuming that the projectile's mass is much greater than that of the electron ($m >> m_e$) and equating the energy transfer to the electron given by (8.15) and the kinematical maximum given by (8.18) and solving for $b_{\min}$, we obtain

$$b_{\min} = \left(\frac{z}{\gamma\beta^2}\right) r_e \tag{8.64}$$

which is ¼ the value defined by (8.63).

8.2.2.5 Equating of Kinetic and Potential Energies Definition

Another means of specifying $b_{\min}$ is by defining it as the impact parameter between the atom and the projectile at which the potential and kinetic energies are equal or

$$\frac{1}{2}mv^2 = \frac{\alpha\hbar c\, z}{b_{\min}}. \tag{8.65}$$

This leads to

$$\begin{aligned} b_{\min} &= \frac{2\alpha\hbar c\, z}{m\beta^2} \\ &= \frac{2z}{\beta^2}\left(\frac{m}{m_e}\right) r_e. \end{aligned} \tag{8.66}$$

Because of the somewhat arbitrary nature of its derivation, we would not expect this result to be meaningful.

Quantum Mechanical Definition

Yet another approach to defining $b_{\min}$ is quantum mechanical in nature as it arises through the Heisenberg uncertainty principle. As the uncertainty of the impact parameter will be of the order of $\hbar c/m_e\gamma\beta$, the value of which we can use to approximate $b_{\min}$,

$$b_{\min} = \frac{\hbar c}{m_e\gamma\beta}. \tag{8.67}$$

Table 8.1 Numerical examples of impact parameter calculations

	Target and projectile			
	Carbon		Calcium	
	$\bar{I} = 86$ eV		$\bar{I} = 229$ eV	
	$r_{atom} = 7 \times 10^5$ fm		$r_{atom} = 2 \times 10^8$ fm	
Model	Proton ($E = 250$ MeV)	α-particle ($E = 10$ MeV)	Proton ($E = 250$ MeV)	α-particle ($E = 10$ MeV)
b_{max} (fm)				
Simple harmonic oscillator (8.59)	3.6×10^6	3.3×10^5	1.3×10^6	1.3×10^5
Kinematic (8.61)	500	8.4×10^3	310	5.2×10^3
b_{min} (fm)				
Recoil distance (8.63)	24	4.2×10^3	24	4.2×10^3
Maximum energy transfer (8.64)	6	1.1×10^3	6	1.1×10^3
Equating kinetic and potential energies (8.66)	3×10^{-3}	0.2	3×10^{-3}	0.2
Quantum mechanical (8.67)	500	5.3×10^3	500	5.3×10^3

Summary of Impact Parameter Models

We have derived two expressions for b_{max} and four expressions for b_{min}, each based upon differing physical models and assumptions. Here, we compare these expressions to assess which are physically sensible to use in calculations of soft-collision energy loss. To assist us, we will use a quantitative comparison for the cases of a 250-MeV proton and a 10-MeV α-particle, each incident to a carbon atom and to a calcium atom, as shown in Table 8.1. The mean excitation (or ionisation) energy, $\bar{I}$, was calculated from a parameterisation given by Segrè (1977) and which is discussed in Chap. 11. An empirical atomic radius, r_{atom}, for each atom was taken from Slater (1964).

In comparing the models used for calculating the maximum impact parameter, we see that that based upon the simple harmonic oscillator, (8.59) yields numerical results which are at least comparable to atomic dimensions and which are greater than those determined by the kinematical argument by 2–4 orders of magnitude. We would thus conclude that simple harmonic oscillator-based result of (8.59) is the appropriate model to use when calculating b_{max}.

The four models used for calculating the minimum impact parameter offer a considerable range of values. Clearly that based upon the maximum energy transfer to an electron or the equating of the projectile's kinetic and potential energies are unrealistic and are subsequently dismissed. The remaining two calculations provide comparable results. In practice, one uses the maximum value of b_{min} in a calculation (Rossi 1952; McParland 2010). Thus, we specify

$$b_{min} = \tfrac{4z}{\gamma\beta^2} r_e \quad \beta < 4\alpha z \tag{8.68}$$

$$b_{min} = \tfrac{r_e}{\alpha\gamma\beta} \quad \beta > 4\alpha z \tag{8.69}$$

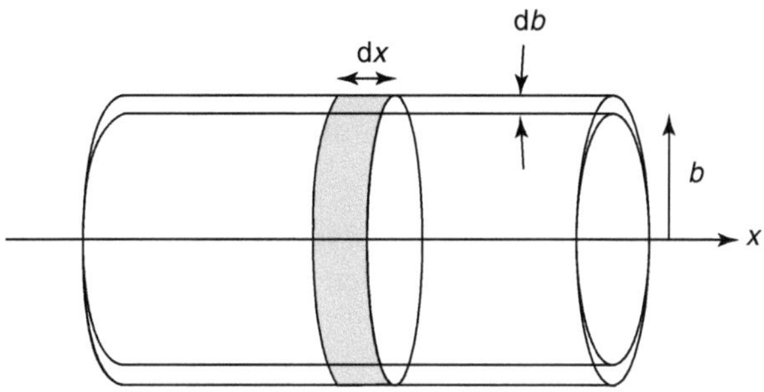

Fig. 8.4 Geometry used for calculation of the Rutherford energy loss formula

8.2.3 Rutherford Energy Loss Formula

8.2.3.1 Differential Cross Section in Energy Transfer

The above discussions have been in the context of a projectile interacting with a single atomic electron or a single atom. We now consider the more realistic case of the projectile interacting with an ensemble of atoms (Fig. 8.4) in order to calculate for the energy lost through collisions as a function of the distance travelled.

The projectile's trajectory is along the x-axis, and we isolate those electrons within a cylinder, the axis of which is coincident with the trajectory. We use the arguments surrounding (4.1) so as to write the probability of a collision between a projectile and an atom for an impact parameter between b and $b + db$ and a differential extent of travel dx in the form of

$$\Pr(b)\,\mathrm{d}b\,\mathrm{d}x = N\,\mathrm{d}\sigma\,\mathrm{d}x \tag{8.70}$$

where, from (4.1), the differential cross section is

$$\mathrm{d}\sigma = |2\pi\, b\,\mathrm{d}b| \tag{8.71}$$

and, as before, N is the scattering centre density, which, in this case, will be the electron density given by

$$N = \rho N_{\mathrm{A}}\left(\frac{Z}{A}\right). \tag{8.72}$$

We solve for the differential of (8.71) by first rearranging the relationship between the impact parameter and energy transfer given by (8.15)

$$b = \frac{r_e z}{\beta}\sqrt{\frac{2m_e}{\mathrm{Q}}} \tag{8.73}$$

and then by differentiating to obtain

$$\mathrm{d}b = -\frac{r_e z}{2\beta}\frac{\sqrt{2m_e}}{Q^{3/2}}\,\mathrm{d}Q. \tag{8.74}$$

Substituting (8.73) and (8.74) into (8.71) gives the Coulomb collision differential cross section in energy transfer:

$$\frac{\mathrm{d}\sigma}{\mathrm{d}Q} = 2\pi m_e \left(\frac{r_e z}{\beta Q}\right)^2. \tag{8.75}$$

It is interesting to consider an alternative method demonstrated by Jackson (1999) to derive (8.75). This begins with the Rutherford differential cross section in solid angle as given by (4.23)

$$\frac{\mathrm{d}\sigma}{\mathrm{d}\Omega} = \left(\frac{\rho_{\mathrm{B}}(E)}{4}\right)^2 \frac{1}{\sin^4\theta/2}$$

and the 4-vector momentum transferred

$$\begin{aligned} \mathrm{Q}^2 &= -(\mathbf{p}-\mathbf{p}') \\ &= 4p^2 \sin^2\frac{\theta}{2} \end{aligned} \tag{8.76}$$

for elastic scatter so as to create a Lorentz-invariant form of the cross section. Note that the momentum is $p = \gamma\beta m$ in our use of the units of energy for momentum. The Rutherford differential cross section in this 4-vector momentum transferred is

$$\frac{\mathrm{d}\sigma}{\mathrm{d}\mathrm{Q}^2} = 2\pi \sin\theta \frac{\mathrm{d}\sigma}{\mathrm{d}\Omega}\left(\frac{\mathrm{d}\mathrm{Q}^2}{\mathrm{d}\theta}\right)^{-1} \tag{8.77}$$

where

$$\frac{\mathrm{d}\mathrm{Q}^2}{d\theta} = 4p^2 \sin\frac{\theta}{2}\cos\frac{\theta}{2}. \tag{8.78}$$

Returning this to (8.77), expanding the expression for the Bohr collision diameter and using the trigonometric relationship

$$\sin\theta = 2\sin\frac{\theta}{2}\cos\frac{\theta}{2}, \tag{8.79}$$

we reach

$$\frac{\mathrm{d}\sigma}{\mathrm{d}\mathrm{Q}^2} = \pi\left(\frac{\alpha\hbar c}{4Ep\sin^2\frac{\theta}{2}}\right)^2 \tag{8.80}$$

Substituting $E = p^2/2m$ and recalling the definition of Q^2, we have the differential cross section in 4-momentum vector transferred:

$$\frac{\mathrm{d}\sigma}{\mathrm{d}\mathrm{Q}^2} = 4\pi\left(\frac{\alpha\hbar c\, z}{\beta \mathrm{Q}^2}\right)^2. \tag{8.81}$$

In the electron's rest frame, $\mathrm{Q}^2 = 2m_e Q^2$, so

$$\begin{aligned}\frac{\mathrm{d}\sigma}{\mathrm{d}Q} &= 2m_e\frac{\mathrm{d}\sigma}{\mathrm{d}\mathrm{Q}^2}\\ &= 2\pi m_e\left(\frac{r_e z}{\beta Q}\right)^2\end{aligned} \tag{8.82}$$

where we have used the definition of the electron classical radius to obtain the same result as (8.75).

Having obtained this differential cross section in energy transfer, we can now write the probability of an energy transfer of between Q and Q + dQ occurring within a differential path length dx in the form of

$$\begin{aligned}\Pr(b)\,\mathrm{d}b\,\mathrm{d}x &= 2\pi m_e\rho N_\mathrm{A}\left(\frac{Z}{A}\right)\left(\frac{r_e z}{\beta Q}\right)^2 \mathrm{d}x\,\mathrm{d}Q\\ &= 2\pi m_e N_\mathrm{A} r_e^2\rho\left(\frac{Z}{A}\right)\left(\frac{z}{\beta Q}\right)^2 \mathrm{d}x\,\mathrm{d}Q\\ &= C\rho\left(\frac{Z}{A}\right)\left(\frac{z}{\beta Q}\right)^2 \mathrm{d}x\,\mathrm{d}Q\end{aligned} \tag{8.83}$$

where we have defined the constant (and one which will appear throughout this book):

$$\begin{aligned}C &\equiv 2\pi m_e N_\mathrm{A} r_e^2\\ &= 0.1535\, MeV\, atom\ \mathrm{cm}^2\, /\, mol.\end{aligned} \tag{8.84}$$

Equation (8.83) is the Rutherford energy loss formula for charged particle energy loss, and it demonstrates succinctly the functional dependence of Coulomb-based energy loss with atomic electrons[2]:

- As this result varies as $1/Q^2$, the probability of a soft collision (small energy transfer Q) is much more likely than that of a hard collision (in agreement with the qualitative arguments of Chap. 7).

[2] And not to be confused with the Rutherford differential cross section we derived in Chap. 4.

- As the probability varies as $1/\beta^2$, slow moving particles (small β) will spend a protracted interaction time with the atom, thus leading to a greater probability of an interaction.
- As expected, the probability is proportional to the electron density of the medium as a consequence of the collision energy loss being the result of interactions with the atomic electrons.
- The probability increases with the square of the projectile electric charge.
- The collision probability is independent of the projectile mass.

8.2.3.2 Rutherford Soft-Collision Stopping Power

Recall that the stopping power is the kinetic energy lost by a moving projectile per unit distance travelled as it decelerates through the medium that it is travelling through. We derive these from the Rutherford result just reached. From (8.75), we have the definition of the differential cross section in energy transfer. The stopping cross section is defined as the integral of this differential cross section weighted by energy loss over the range of available energy losses, or

$$S \equiv \int_{Q_{\min}}^{Q_{\max}} \mathrm{d}\sigma\, Q. \tag{8.85}$$

Multiplying this stopping cross section by the scattering centre density, N, yields the linear collision stopping power (appropriate selection of the nature of N can provide the mass collision stopping power as to be shown). The result of (8.85) indicates an important and implicit aspect of this calculation in that the integration represents the incoherent summation of energy losses. This is allowable if the energy losses are uncorrelated, i.e. the charged particle transfers energy to a single atom at a time. In other words, the medium is implicitly a dilute gas with no atom–atom interactions. The extension of this simple model of the medium to the more realistic case of condensed matter is the topic of Chap. 12. The stopping power cross section will figure significantly in our discussions of the stochastic nature of collision energy loss in Part IV.

In the classical theory, there is a one-to-one relationship between the energy transfer and the impact parameter. This is not the case in semi-classical theory in which the atomic electrons are treated quantum mechanically. It is clear that a minimum energy transfer must be specified as the integrand of (8.85) is divergent in Q otherwise. Evaluating $\mathrm{d}\sigma$ from (8.75) and inserting this into (8.85) gives

$$\begin{aligned} S &\equiv 2\pi m_e \left(\frac{r_e z}{\beta}\right)^2 \int_{Q_{\min}}^{Q_{\max}} \frac{\mathrm{d}Q}{Q} \\ &= 2\pi m_e \left(\frac{r_e z}{\beta}\right)^2 \ln \frac{Q_{\max}}{Q_{\min}}. \end{aligned} \tag{8.86}$$

Both finite and non-zero values of the energy transfer are necessary in order to avoid logarithmic divergences. We now define the ranges of the energy transfer using our recent detailed discussion of the minimum and maximum values of the impact parameter. Clearly, $Q_{\max}$ will be set by the kinematics of the collision and is given by (8.18) for a projectile more massive than the electron target. On the other hand, $Q_{\min}$ for a soft collision can be defined through a maximum impact parameter using (8.15):

$$Q_{\min} = 2m_e \left(\frac{r_e z}{b_{\max} \beta} \right)^2. \tag{8.87}$$

From these specifications of the minimum and maximum energy transfers, the Rutherford stopping cross section is

$$S = 4\pi m_e \left(\frac{r_e z}{\beta} \right)^2 \ln \left(\frac{\gamma \beta^2 b_{\max}}{r_e z} \right). \tag{8.88}$$

The Rutherford mass soft-collision stopping power is the product of S and the number of collision centres (electrons) per unit volume:

$$\begin{aligned} \left. \frac{\mathrm{d}E}{\rho\, \mathrm{d}x} \right|_{S,\mathrm{Ruth}} &= \left(N_{\mathrm{A}} \frac{Z}{A} \right) S \\ &= 2C \frac{Z}{A} \left(\frac{z}{\beta} \right)^2 \ln \left(\frac{\gamma \beta^2 b_{\max}}{r_e z} \right), \end{aligned} \tag{8.89}$$

where Z refers to the atomic number of the medium and not the charge of the target. Note that the letter S has been included in the subscript to note that the result is for the soft-collision regime. As $b_{\max}$ resides within the logarithm of (8.89), the sensitivity of the calculation to the model used to derive the impact parameter is reduced. Our discussion pertaining to Table 8.1 led us to the conclusion that the expression for $b_{\max}$ given by (8.59) was the most appropriate to use in the classical collision stopping power calculation. Substituting this expression into (8.89) gives the Rutherford soft-collision mass collision stopping power of

$$\left. \frac{\mathrm{d}E}{\rho\, \mathrm{d}x} \right|_{S,\mathrm{Ruth}} = 2C \frac{Z}{A} \left(\frac{z}{\beta} \right)^2 \left(\ln \left(\frac{\beta^3 c}{r_e \omega_0 z} \right) - \ln(1 - \beta^2) \right). \tag{8.90}$$

8.2.4 *Soft-Collision Energy Transfer*

8.2.4.1 Introductory Comments

In this section, we explore in detail the Bohr classical theory of soft collisions between a charged particle and an atom. Bohr (1913, 1915) provided the definitive classical theory of soft-collision energy loss. As in our earlier derivation of the adiabatic limit of the impact parameter for soft collisions, we will use his enlightened model of the atomic electron system as being an ensemble of electrons each bound so as to be considered under-damped harmonic oscillators. The goal of our derivation of the Bohr theory will be the expression of the soft-collision stopping power which is the energy transferred from the projectile to atoms per unit distance travelled for those cases of the impact parameter between b_{min} and b_{max}. Hence, the first step will be to again calculate the energy transfer, Q, from the projectile to a simple harmonic oscillator representing an atomic electron, but now as an explicit function of the projectile's electric field at the position of the oscillator. This result will be shown as being proportional to the modulus squared of the Fourier transform of the projectile's electric field, $|E(\omega)|^2$ at the position of the electron. This Fourier transform will be evaluated in two independent approaches – one based upon the kinematics of the interaction and the other from a dipole approximation to the interaction. The latter is done as this also develops an interim result that will be used later in the calculation of oscillator strengths for multi-electron atoms. Having evaluated an expression for Q, we then investigate these oscillator strengths before then providing an expression for the Bohr mass soft-collision stopping power,

$$\frac{\mathrm{dE}}{\mathrm{d}x}\Big|_{S,\mathrm{Bohr}}.$$

8.2.4.2 Dipole Approximation

We now calculate for an expression of the energy transfer, Q, from the projectile to the simple harmonic oscillator target due to a soft collision. As the charged projectile passes by the atom, the electron is subject to the force $-e\mathbf{E}(\mathbf{r},t)$ due to the particle's electric field. The resulting equation of motion is given by (8.31) where

$$g(\mathbf{r},t) = -\frac{e}{m_e}\mathbf{E}(\mathbf{r},t). \tag{8.91}$$

The obvious difference between (8.31) and (8.91) is that the driving force now contains a spatial dependence. This is problematic in that it will lead to a non-linear result when the equation of motion is solved. In order to avoid this undesirable outcome, we recall that soft collisions occur at large impact parameters. As a result, the spatial variation of the electric field at the position of the electron will be limited through the $1/r$ dependence of the electric field, and we can simplify the problem by thus ignoring the spatial dependence of (8.91) and write

$$g(t) = -\frac{e}{m_e}\mathbf{E}(t). \tag{8.92}$$

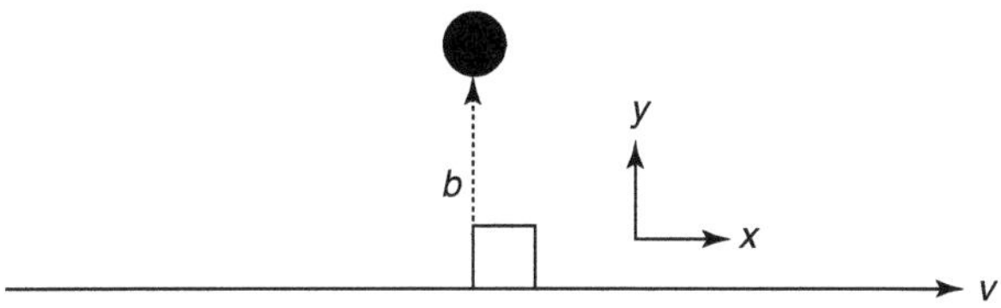

Fig. 8.5 Definition of the minimum vectorial impact parameter, b_{min} between the projectile trajectory and the target atom. The co-ordinate system shown is to be used in the calculation of the oscillator strengths

While the above brief argument is intuitive, we can obtain the same result through an argument based upon a more rigorous fashion by applying the dipole approximation (Sigmund 2006). As this analysis will prove useful later in our discussion of oscillator strengths, we explore it further here. The electric field experienced by the electron is given by the gradient of the scalar potential:

$$\mathbf{E}(\mathbf{r},t) = -\nabla V(\mathbf{r},t) \tag{8.93}$$

where the electric potential for the point-charge projectile is

$$V(\mathbf{r},t) = \frac{\alpha\hbar c\, z}{e\,|\mathbf{r} - (\mathbf{b} + \mathbf{v}t)|}. \tag{8.94}$$

Here, **b** is the vectorial impact parameter, defined in Fig. 8.5, and **v** is the projectile's velocity. From Fig. 8.5, we see that $\mathbf{b} \bullet \mathbf{v} = 0$.

To solve for the electric field given by (8.93), we recall our solution of the Bethe integral and recognise that we can write the Fourier transform of the function $1/r$ as

$$\begin{aligned} F\left[\frac{1}{r}\right] &= \frac{1}{(2\pi)^{3/2}} \int d^3\mathbf{r}\, \frac{e^{-i\left(\frac{\mathbf{p}\bullet\mathbf{r}}{\hbar c}\right)}}{r} \\ &= \frac{1}{(2\pi)^{3/2}} \int d^3\mathbf{r}\, \frac{e^{-i\mathbf{k}\bullet\mathbf{r}}}{r} \\ &= \sqrt{\frac{2}{\pi}}\frac{1}{k^2} \end{aligned} \tag{8.95}$$

where we have used the definition of the wavevector, $\mathbf{k} = \mathbf{p}/\hbar c$. Hence, using the inverse Fourier transform, we can write $1/r$ as

$$\begin{aligned} \frac{1}{r} &= \sqrt{\frac{2}{\pi}} F^{-1}\left[\frac{1}{k^2}\right] \\ &= \frac{1}{2\pi^2} \int_{-\infty}^{\infty} d^3\mathbf{k}\, \frac{e^{i\mathbf{k}\bullet\mathbf{r}}}{k^2} \end{aligned} \tag{8.96}$$

so as to allow us to perform the calculation in $\boldsymbol{k}$-space.

We can now write the electric field as

$$
\begin{aligned}
\mathbf{E}(\mathbf{r},t) &= -\left(\frac{\alpha\hbar c}{e}\right) z\nabla \frac{1}{|\mathbf{r}-(\mathbf{b}+\mathbf{v}t)|} \\
&= -\frac{1}{2\pi^2}\left(\frac{\alpha\hbar c}{e}\right) z\nabla \int_{-\infty}^{\infty} \mathrm{d}^3\mathbf{k}\, \frac{e^{i\mathbf{k}\bullet(\mathbf{r}-(\mathbf{b}+\mathbf{v}t))}}{k^2} \\
&= -i\frac{1}{2\pi^2}\left(\frac{\alpha\hbar c}{e}\right) z \int_{-\infty}^{\infty} \mathrm{d}^3\mathbf{k}\,\mathbf{k}\, \frac{e^{i\mathbf{k}\bullet(\mathbf{r}-(\mathbf{b}+\mathbf{v}t))}}{k^2}.
\end{aligned} \tag{8.97}
$$

Let us now consider the exponential within the integrand of (8.97)

$$
e^{i\mathbf{k}\bullet(\mathbf{r}-(\mathbf{b}+\mathbf{v}t))} = e^{i\mathbf{k}\bullet\mathbf{r}} e^{-i\mathbf{k}\bullet(\mathbf{b}+\mathbf{v}t)}. \tag{8.98}
$$

Assuming small displacements of the electron from its equilibrium position due to the electric field relative to $\frac{1}{k}$ (which is the long wavelength limit expected in a soft collision following our earlier discussion), the exponent in $\mathbf{r}$ in (8.98) can be approximated by

$$
e^{i\mathbf{k}\bullet\mathbf{r}} \approx 1. \tag{8.99}
$$

This removes the spatial dependence that is present within (8.97)

$$
\mathbf{E}(t) = -i\frac{1}{2\pi^2}\left(\frac{\alpha\hbar c}{e}\right) z \int_{-\infty}^{\infty} \mathrm{d}^3\mathbf{k}\,\mathbf{k}\, \frac{e^{-i\mathbf{k}\bullet(\mathbf{b}+\mathbf{v}t)}}{k^2}. \tag{8.100}
$$

We will recall this expression when we later calculate both its Fourier transform and the atomic electron harmonic oscillator strength. The truncation of

$$
e^{i\mathbf{k}\bullet\mathbf{r}} = \sum_{n=0}^{\infty} \frac{1}{n!} (\mathbf{k}\bullet\mathbf{r})^n \tag{8.101}
$$

at zeroth order as given by (8.99) known as the dipole approximation.

8.2.4.3 Energy Transfer and the Electric Field

The Fourier transform of the electron displacement follows from our previous discussion which was centred around the equation of motion. Substituting (8.92) for $g(t)$ into the equation of motion of (8.31) and taking the inverse Fourier transforms of both sides, we have

$$\int_{-\infty}^{\infty} d\omega \left(-\omega^2 + i\Gamma\omega + \omega_0^2\right) r(\omega)\, e^{i\omega t} = -\left(\frac{e}{m_e}\right) \int_{-\infty}^{\infty} d\omega\, E(\omega)\, e^{i\omega t}. \tag{8.102}$$

We multiply both sides of (8.102) by $e^{-i\omega t'}$ and then integrate over t':

$$\begin{aligned}\int_{-\infty}^{\infty} d\omega \left(-\omega^2 + i\Gamma\omega + \omega_0^2\right) r(\omega) \int_{-\infty}^{\infty} dt'\, e^{i\omega(t-t')}\\ = -\left(\frac{e}{m_e}\right) \int_{-\infty}^{\infty} d\omega\, E(\omega) \int_{-\infty}^{\infty} dt'\, e^{i\omega t(t-t')}.\end{aligned} \tag{8.103}$$

Recalling the integral definition of the Dirac δ-function derived in Appendix C.9, this result leads us to an expression for the Fourier transform of the electron displacement or

$$\mathbf{r}(\omega) = -\left(\frac{e}{m_e}\right)\left(\frac{\mathbf{E}(\omega)}{\left(\omega_0^2 - \omega^2\right) + i\Gamma\omega}\right). \tag{8.104}$$

Having obtained this result, we now focus on the energy loss of the projectile which is equal to the energy transferred to the target:

$$Q = -e \int_{-\infty}^{\infty} dt\, \frac{d\mathbf{r}(t)}{dt} \mathbf{E}(t). \tag{8.105}$$

We solve the integral of (8.105) by first writing the integrand in terms of the product of the two inverse Fourier transforms:

$$\begin{aligned}Q &= -e \int_{-\infty}^{\infty} dt \left(\frac{1}{\sqrt{2\pi}} \frac{d}{dt} \int_{-\infty}^{\infty} d\omega\, e^{i\omega t} \mathbf{r}(\omega)\right)\left(\frac{1}{\sqrt{2\pi}} \int_{-\infty}^{\infty} d\omega'\, e^{i\omega' t}\, \mathbf{E}(\omega')\right)\\ &= -i\left(\frac{e}{2\pi}\right) \int_{-\infty}^{\infty} d\omega' \int_{-\infty}^{\infty} d\omega \left(\int_{-\infty}^{\infty} dt\, e^{i(\omega+\omega')t}\right) \omega\mathbf{r}(\omega)\, \mathbf{E}(\omega')\\ &= -ie \int_{-\infty}^{\infty} d\omega' \int_{-\infty}^{\infty} d\omega\, \delta(\omega+\omega')\, \omega\mathbf{r}(\omega)\, \mathbf{E}(\omega')\\ &= -ie \int_{-\infty}^{\infty} d\omega\; \omega\, \mathbf{r}(\omega)\, \mathbf{E}(-\omega)\end{aligned} \tag{8.106}$$

where the sifting property of the Dirac δ-function, $\int_{-\infty}^{\infty} dx' f(x')\, \delta(x'+x) = f(-x)$, has been used.

We shall need to consider this result and the functional dependence of the electric field. As the electric field is a real quantity, i.e. $\mathbf{E}(-\omega) = \mathbf{E}^*(\omega)$, (8.106) can be written as

$$Q = ie \int_{-\infty}^{\infty} d\omega\ \omega\, \mathbf{r}(\omega)\, \mathbf{E}^*(\omega)$$
$$= 2e\,\mathrm{Re}\left(i \int_{0}^{\infty} d\omega\ \omega\, \mathbf{r}(\omega)\, \mathbf{E}^*(\omega) \right). \tag{8.107}$$

We now insert the expression for $\mathbf{r}(\omega)$ of (8.104) into the integrand of (8.107):

$$Q = \left(\frac{2e^2}{m_e}\right) \mathrm{Re}\left(i \int_0^{\infty} d\omega \frac{\omega\, |\mathbf{E}(\omega)|^2}{(\omega_0^2 - \omega^2) - i\Gamma\omega} \right)$$
$$= \left(\frac{2e^2}{m_e}\right) \mathrm{Re}\left(i \int_0^{\infty} d\omega \frac{\omega\, |\mathbf{E}(\omega)|^2 \left((\omega_0^2 - \omega^2) - i\Gamma\omega\right)}{(\omega_0^2 - \omega^2)^2 + \Gamma^2\omega^2} \right)$$
$$= \left(\frac{2e^2}{m_e}\right) \mathrm{Re}\left(\int_0^{\infty} d\omega \frac{\omega\, |\mathbf{E}(\omega)|^2 \left(\Gamma\omega + i(\omega_0^2 - \omega^2)\right)}{(\omega_0^2 - \omega^2)^2 + \Gamma^2\omega^2} \right)$$
$$= \left(\frac{2e^2}{m_e}\right) \left(\int_0^{\infty} d\omega \frac{\Gamma\,\omega^2\, |\mathbf{E}(\omega)|^2}{(\omega_0^2 - \omega^2)^2 + \Gamma^2\omega^2} \right). \tag{8.108}$$

Recall that we have specified previously that the harmonically bound atomic electron is considered as an under-damped system or $\Gamma << \omega_0$. As a result, we can simplify (8.108) through noting that the response will be highly peaked about the resonant frequency ω_0. As a result, we can use the approximation of $\mathbf{E}(\omega) \approx \mathbf{E}(\omega_0)$ to enable us to extract $|\mathbf{E}(\omega)|^2$ from the integrand and evaluate it at ω_0. This leads us to the expression for the energy transfer:

$$Q = \left(\frac{2e^2}{m_e}\right) |\mathbf{E}(\omega_0)|^2 \int_0^{\infty} d\omega \frac{\Gamma\omega^2}{(\omega_0^2 - \omega^2)^2 + \Gamma^2\omega^2}. \tag{8.109}$$

The integral is solved using the substitution of variable, $u = \frac{\omega}{\Gamma}$,

$$\int_0^{\infty} d\omega \frac{\Gamma\omega^2}{(\omega_0^2 - \omega^2)^2 + \Gamma^2\omega^2} = \int_0^{\infty} du \frac{u^2}{(p^2 - u^2)^2 + u^2} \tag{8.110}$$

where we have defined the dimensionless constant $p \equiv \omega_0/\Gamma$. This will be a large quantity due to our assumption of the harmonically bound atomic electron being an under-damped system. The integrand is then rearranged so as to take advantage of the fact that $p >> 1$:

$$\int_0^\infty du \frac{u^2}{\left(p^2-u^2\right)^2+u^2} = \int_0^\infty \frac{du}{\left(\frac{(u-p)(u+p)}{u}\right)^2+1} = \int_{-p}^\infty \frac{dx}{\left(\frac{x\,(x+2p)}{x+p}\right)^2+1} \tag{8.111}$$

where an additional change of variable $x = u - p$ has been used. As p is large, the integrand can be simplified considerably:

$$\int_{-p}^\infty \frac{dx}{\left(\frac{x(x+2p)}{x+p}\right)^2+1} \approx \int_{-p}^\infty \frac{dx}{\left(\frac{1}{p^2}\right)\left(x^2+2px\right)^2+1} \approx \int_{-p}^\infty \frac{dx}{1+4x^2}. \tag{8.112}$$

Because of the assumption of an under-damped system and large p, the lower integration limit can be set to $-\infty$. Hence, the integral remaining to be solved is

$$\int_0^\infty d\omega \frac{\Gamma\omega^2}{\left(\omega_0^2-\omega^2\right)^2+\Gamma^2\omega^2} \approx \int_{-\infty}^\infty \frac{dx}{1+4x^2}. \tag{8.113}$$

An integral of this form is solved using residue theory as was demonstrated as an explicit example in Appendix A.4. The result is

$$\int_{-\infty}^\infty \frac{dx}{1+4x^2} = \frac{\pi}{2}. \tag{8.114}$$

Returning this result to (8.108), we find a succinct expression of the energy transfer from the projectile to the harmonically bound atomic electron[3]:

$$Q = \left(\pi\frac{e^2}{m_e}\right)|\mathbf{E}(\omega_0)|^2. \tag{8.115}$$

[3] That the units of the right-hand side of (8.115) should be equal to that of energy may perhaps seem peculiar to the reader, but may be verified using dimensional analysis. To confirm that this is so, first, using SI units, note that the unit for the Fourier transform of the electric field is V•s/m and that that for $e^2|\mathbf{E}(\omega_0)|^2$ is equal to kg^2•m^2•s^{-2}. Hence, the unit for the right-hand side of (8.99) is kg•m^2•s^{-2} = N•m = J.

Let us write the Fourier transform in (8.115) out in full so as to investigate the dependence of the energy transfer upon the harmonic oscillator's natural frequency:

$$
\begin{aligned}
Q &= \left(\pi \frac{e^2}{m_e}\right) \left| \frac{1}{\sqrt{2\pi}} \int_{-\infty}^{\infty} \mathrm{d}t\, \mathbf{E}(t)\, e^{-i\omega_0 t} \right|^2 \\
&= \left(\frac{e^2}{2m_e}\right) \left| \int_{-\infty}^{\infty} \mathrm{d}t\, \mathbf{E}(t)\, e^{-i\omega_0 t} \right|^2 .
\end{aligned} \tag{8.116}
$$

It is seen immediately that for the case of $\omega_0 = 0$, we have the impulse approximation of

$$
Q = \frac{q^2}{2m_e}. \tag{8.117}
$$

However, as ω_0 increases, the calculation approaches the adiabatic response of the atom. For example, let the magnitude of the electric field $|\mathbf{E}(t)|$ be approximated as a constant, $|\mathbf{E}|^2$. Then,

$$
= \frac{e^2}{2m_e} \frac{|\mathbf{E}|^2}{\omega_0^2} \tag{8.118}
$$

where it is clear that the energy transfer $Q \to 0$ as the resonant frequency $\omega_0 \to \infty$.

8.2.4.4 The Electric Field at the Position of the Electron

We have derived an expression for the energy transfer as being proportional to the squared modulus of the Fourier transform of the projectile's electric field at the position of the electron and evaluated at the resonant frequency of the harmonic oscillator. Hence, this Fourier transform is now to be determined, and we shall perform this evaluation via two different paths. The first considers the Lorentz-boosted electric fields that are readily found and the second extends our earlier use of the dipole approximation. Although both approaches yield the same result, the course of this second calculation will yield a result to be useful later in our derivation of the harmonic oscillator strengths required in the calculation of the Bohr collision stopping power as determined by Bloch.

8.2.4.5 Kinematics-Based Derivation

Beginning our first derivation of $|\mathbf{E}(\omega_0)|^2$, consider the relativistic expressions of the orthogonal and parallel components of the force given by (8.9) and (8.10). The corresponding electric field components as functions of time are

$$E_{\perp}(t) = \left(\frac{\alpha\hbar c}{e}\right)\frac{z}{b^2}\frac{\gamma}{\left[1+\left(\frac{\gamma v t}{b}\right)^2\right]^{\frac{3}{2}}} \tag{8.119}$$

$$E_{||}(t) = -\left(\frac{\alpha\hbar c}{e}\right)\frac{z}{b^3}\frac{\gamma v t}{\left[1+\left(\frac{\gamma v t}{b}\right)^2\right]^{\frac{3}{2}}}. \tag{8.120}$$

We now calculate the Fourier transforms of these components into frequency space. That of the orthogonal component of the electric field is

$$\begin{aligned} E_{\perp}(\omega) &= \frac{1}{\sqrt{2\pi}}\int_{-\infty}^{\infty} \mathrm{d}t\, e^{-i\omega t} E_{\perp}(t) \\ &= \frac{1}{\sqrt{2\pi}}\left(\frac{\alpha\hbar c}{e}\right)\frac{\gamma z}{b^2}\int_{-\infty}^{\infty} \mathrm{d}t\,\frac{e^{-i\omega t}}{\left[1+\left(\frac{\gamma v t}{b}\right)^2\right]^{\frac{3}{2}}}. \end{aligned} \tag{8.121}$$

The integral in (8.121) is simplified by using the substitution of variable, $x = \gamma v t/b$,

$$\begin{aligned} E_{\perp}(\omega) &= \frac{1}{\sqrt{2\pi}}\left(\frac{\alpha\hbar c}{e}\right)\frac{z}{bv}\int_{-\infty}^{\infty} \mathrm{d}x\,\frac{e^{-i\left(\frac{\omega b}{\gamma v}\right)x}}{[1+x^2]^{\frac{3}{2}}} \\ &= \sqrt{\frac{2}{\pi}}\left(\frac{\alpha\hbar c}{e}\right)\frac{z}{bv}\int_{0}^{\infty} \mathrm{d}x\,\frac{\cos\left(\frac{\omega b}{\gamma v}x\right)}{[1+x^2]^{\frac{3}{2}}} \end{aligned} \tag{8.122}$$

with the last step following from $\cos\left(\frac{\omega b}{\gamma v}x\right)$ and $\sin\left(\frac{\omega b}{\gamma v}x\right)$ being even and odd functions, respectively. The integral is a form of the modified Bessel function, as shown in Appendix C.1,

$$K_{\nu}(y) = \frac{2^{\nu}\Gamma\left(\nu+\frac{1}{2}\right)}{\sqrt{\pi}y^{\nu}}\int_{0}^{\infty} \mathrm{d}x\,\frac{\cos(xy)}{[1+x^2]^{\nu+\frac{1}{2}}} \tag{8.123}$$

where ν is either 0 or an integer. For example,

$$K_0(y) = \int_{0}^{\infty} \mathrm{d}x\,\frac{\cos(xy)}{\sqrt{1+x^2}} \tag{8.124}$$

and

$$K_1(y) = \frac{1}{y} \int_0^{\infty} \mathrm{d}x \, \frac{\cos(xy)}{[1+x^2]^{\frac{3}{2}}}. \tag{8.125}$$

The expression of the Fourier transform of the electric field orthogonal to the projectile trajectory is now

$$\begin{aligned} E_{\perp}(\omega) &= \sqrt{\frac{2}{\pi}} \frac{z\,\alpha\hbar c}{ebv} \int_0^{\infty} \mathrm{d}x \frac{\cos\left(\frac{\omega b}{\gamma v} x\right)}{[1+x^2]^{\frac{3}{2}}} \\ &= \sqrt{\frac{2}{\pi}} \frac{z\,\alpha\hbar c}{ebv} \left(\frac{\omega b}{\gamma v}\right) K_1\left(\frac{\omega b}{\gamma v}\right). \end{aligned} \tag{8.126}$$

We can now derive an expression for the Fourier transform of the electric field component parallel to the particle's trajectory:

$$E_{||}(\omega) = -\frac{1}{\sqrt{2\pi}} \left(\frac{\alpha\hbar c}{e}\right) \frac{z\gamma v}{b^3} \int_{-\infty}^{\infty} \mathrm{d}t \, \frac{t e^{-i\omega t}}{\left[1+\left(\frac{\gamma v t}{b}\right)^2\right]^{\frac{3}{2}}}. \tag{8.127}$$

Using, again, the substitution of variable, $x = \gamma v t/b$,

$$E_{||}(\omega) = -\frac{1}{\sqrt{2\pi}\, v} \left(\frac{\alpha\hbar c}{e}\right) \left(\frac{z}{b\gamma v}\right) \int_{-\infty}^{\infty} \mathrm{d}x \, \frac{x e^{-i\left(\frac{\omega b}{\gamma v}\right)x}}{[1+x^2]^{\frac{3}{2}}}. \tag{8.128}$$

This integral is solved by parts (for clarity, we will define the variable $y \equiv \omega b/\gamma v$):

$$\begin{aligned} \int_{-\infty}^{\infty} \mathrm{d}x \, \frac{x e^{-ixy}}{[1+x^2]^{\frac{3}{2}}} &= \int_{-\infty}^{\infty} \mathrm{d}r\, s \\ &= s\,r\Big|_{-\infty}^{\infty} - \int_{-\infty}^{\infty} \mathrm{d}s\, r \end{aligned} \tag{8.129}$$

Where

$$s \equiv e^{-ixy} \qquad \mathrm{d}s = -\mathrm{d}x \; i\, y\, e^{ixy}$$
$$\mathrm{d}r \equiv \mathrm{d}x \frac{x}{[1+x^2]^{\frac{3}{2}}} \qquad r = -\frac{1}{\sqrt{1+x^2}}$$

to give

$$\int_{-\infty}^{\infty} \mathrm{d}x \, \frac{x e^{-i\left(\frac{\omega b}{\gamma v}\right)x}}{[1+x^2]^{\frac{3}{2}}} = -2i\left(\frac{\omega b}{\gamma v}\right)\int_{0}^{\infty} \mathrm{d}x \, \frac{\cos\left(\frac{\omega b}{\gamma v}x\right)}{\sqrt{1+x^2}}$$
$$= -2i\left(\frac{\omega\, b}{\gamma v}\right)K_0\left(\frac{\omega\, b}{\gamma v}\right). \tag{8.130}$$

The Fourier transform of the electric field component parallel to the particle's trajectory is now

$$E_{||}(\omega) = -i\sqrt{\frac{2}{\pi}}\left(\frac{\alpha\hbar c}{e}\right)\frac{z\omega}{\gamma^2 v^2}K_0\left(\frac{\omega b}{\gamma v}\right). \tag{8.131}$$

The squared modulus of the electric field in frequency space and evaluated at the harmonic oscillator's resonant frequency is thus

$$|\mathbf{E}(\omega_0)|^2 = E_{\perp}^2(\omega_0) + E_{||}^2(\omega_0)$$
$$= \frac{2}{\pi}\left(\frac{\alpha\hbar c}{e}\right)^2\left(\frac{z\omega_0}{\gamma v^2}\right)^2\left[\frac{K_0^2\left(\frac{\omega_0 b}{\gamma v}\right)}{\gamma^2} + K_1^2\left(\frac{\omega_0 b}{\gamma v}\right)\right]. \tag{8.132}$$

In its nonrelativistic form, this result appears as

$$|\mathbf{E}(\omega_0)|^2 = \frac{2}{\pi}\left(\frac{\alpha\hbar c}{e}\right)^2\left(\frac{z\omega_0}{v^2}\right)^2\left[K_0^2\left(\frac{\omega_0 b}{v}\right) + K_1^2\left(\frac{\omega_0 b}{v}\right)\right]. \tag{8.133}$$

8.2.4.6 Dipole Approximation-Based Calculation

We now repeat a derivation of the solution for $|\mathbf{E}(\omega_0)|^2$ that resulted in (8.133) but through the use of the dipole approximation. There are two reasons for doing so. First, it provides completeness of the derivation. Second, it provides an interim result which will be of later use.

Recall our earlier derivation of the electric field that was based upon the dipole approximation of (8.99). The Fourier transform of that result is

$$\begin{aligned}
\mathbf{E}(\omega) &= \frac{1}{\sqrt{2\pi}} \int_{-\infty}^{\infty} \mathrm{d}t \, e^{i\omega t} \, \mathbf{E}(t) \\
&= -\frac{i}{2^{3/2}\,\pi^{5/2}} \left(\frac{\alpha\hbar c}{e}\right) z \int_{-\infty}^{\infty} \mathrm{d}t \, e^{i\omega t} \int_{-\infty}^{\infty} \mathrm{d}^3\mathbf{k} \, \mathbf{k} \frac{e^{-i\mathbf{k}\bullet(\mathbf{b}+\mathbf{v}t)}}{k^2} \\
&= -\frac{i}{2^{3/2}\,\pi^{5/2}} \left(\frac{\alpha\hbar c}{e}\right) z \int_{-\infty}^{\infty} \mathrm{d}^3\mathbf{k} \, \mathbf{k} \frac{e^{-i\mathbf{k}\bullet\mathbf{b}}}{k^2} \int_{-\infty}^{\infty} \mathrm{d}t \, e^{i(\omega-\mathbf{k}\bullet\mathbf{v})t} \\
&= -\frac{i}{\pi\sqrt{2\pi}} \left(\frac{\alpha\hbar c}{e}\right) z \int_{-\infty}^{\infty} \mathrm{d}^3\mathbf{k} \, \mathbf{k} \frac{e^{-i\mathbf{k}\bullet\mathbf{b}}}{k^2} \, \delta(\omega - \mathbf{k}\bullet\mathbf{v})
\end{aligned} \tag{8.134}$$

where we have used the integral definition of the Dirac δ-function. The integral of (8.134) will arise in the next section in the context of evaluating the oscillator strengths. Recalling the geometry of the interaction as described by Fig. 8.5, we can state that

$$\begin{aligned}
\mathbf{k} &= \frac{\omega}{v}\hat{\mathbf{x}} + k_\perp \hat{\mathbf{y}} \\
\mathbf{v} &= v\hat{\mathbf{x}}.
\end{aligned} \tag{8.135}$$

Thus,

$$\mathbf{k}\bullet\mathbf{v} = \omega \tag{8.136}$$

and (8.134) becomes

$$\begin{aligned}
\mathbf{E}(\omega) &= -\frac{i}{\pi\sqrt{2\pi}} \left(\frac{\alpha\hbar c}{e}\right) \left(\frac{z}{v}\right) \int \mathrm{d}^2\mathbf{k}_\perp \frac{\left(\frac{\omega}{v}\right)\hat{\mathbf{x}} + k_\perp \hat{\mathbf{y}}}{\left(\frac{\omega}{v}\right)^2 + k_\perp^2} e^{-i\,\mathbf{k}_\perp\bullet\mathbf{b}} \\
&= \mathbf{E}_x(\omega) + \mathbf{E}_\perp(\omega)
\end{aligned} \tag{8.137}$$

where

$$\mathbf{E}_x(\omega) = -\frac{i}{\pi\sqrt{2\pi}} \left(\frac{\alpha\hbar c}{e}\right) \left(\frac{z\omega}{v^2}\right) \int \mathrm{d}^2\mathbf{k}_\perp \frac{e^{-i\,\mathbf{k}_\perp\bullet\mathbf{b}}}{\left(\frac{\omega}{v}\right)^2 + k_\perp^2} \hat{\mathbf{x}} \tag{8.138}$$

and

$$\begin{aligned}
\mathbf{E}_\perp(\omega) &= -\frac{i}{\pi\sqrt{2\pi}} \left(\frac{\alpha\hbar c}{e}\right) \left(\frac{z}{v}\right) \int \mathrm{d}^2\mathbf{k}_\perp \frac{k_\perp}{\left(\frac{\omega}{v}\right)^2 + k_\perp^2} e^{-i\,\mathbf{k}_\perp\bullet\mathbf{b}} \, \hat{\mathbf{y}} \\
&= -\frac{1}{\pi\sqrt{2\pi}} \left(\frac{\alpha\hbar c}{e}\right) \left(\frac{z}{v}\right) \frac{\partial}{\partial b} \int \mathrm{d}^2\mathbf{k}_\perp \frac{e^{-i\,\mathbf{k}_\perp\bullet\mathbf{b}}}{\left(\frac{\omega}{v}\right)^2 + k_\perp^2} \hat{\mathbf{y}}.
\end{aligned} \tag{8.139}$$

We now write the Fourier transform of the electric field as

$$\mathbf{E}(\omega) = \frac{1}{\pi\sqrt{2\pi}}\left(\frac{\alpha\hbar c}{e}\right)\left(\frac{z}{v}\right)\left(-i\frac{\omega}{v}\hat{\mathbf{x}} + \frac{\partial}{\partial b}\hat{\mathbf{y}}\right)\int \mathrm{d}^2\mathbf{k}_\perp \frac{e^{-i\,\mathbf{k}_\perp \bullet \mathbf{b}}}{\left(\frac{\omega}{v}\right)^2 + k_\perp^2}. \tag{8.140}$$

As noted before, there are only two co-ordinates, we evaluate the integral of (8.140) in cylindrical co-ordinates and by separating the integral into these, we obtain

$$\int \mathrm{d}^2\mathbf{k}_\perp \frac{e^{-i\,\mathbf{k}\bullet\mathbf{b}}}{(\omega/v)^2 + k_\perp^2} = \int_0^\infty \mathrm{d}k_\perp \frac{k_\perp}{(\omega/v)^2 + k_\perp^2}\int_0^{2\pi} \mathrm{d}\varphi\, e^{-ik_\perp b\cos\varphi} \tag{8.141}$$

From Appendix C.1, we can write the angular integral as, $\int_0^{2\pi} d\varphi\, e^{-ik_\perp b\cos\varphi} = 2\pi J_0(k_\perp b)$ which reduces (8.141) to

$$\int \mathrm{d}^3\mathbf{k}_\perp \frac{e^{-i\,\mathbf{k}_\perp\bullet\mathbf{b}}}{(\omega/v)^2 + k_\perp^2} = 2\pi\int_0^\infty \mathrm{d}k_\perp \frac{k_\perp}{(\omega/v)^2 + k_\perp^2} J_0(k_\perp b). \tag{8.142}$$

The integral of (8.142) is of the Hankel–Nicholson type with the solution given by Abramowitz and Stegun (1972):

$$\int_0^\infty \mathrm{d}x \frac{x^{\nu+1}}{(x^2+z^2)^{\mu+1}} J_\nu(ax) = \frac{a^\mu z^{\nu-\mu}}{2^\mu \Gamma(\mu+1)} K_{\nu-\mu}(az) \tag{8.143}$$

subject to the requirements of

$$a > 0;\ \operatorname{Re} z > 0; -1 < \operatorname{Re}\nu < 2\operatorname{Re}\mu + \frac{3}{2}.$$

This leads us to the general solution of the integral of (8.140) as

$$\int_0^\infty \mathrm{d}x \left(\frac{x}{x^2+z^2}\right) J_0(ax) = K_0(az). \tag{8.144}$$

Hence,

$$2\pi\int_0^\infty \mathrm{d}k_\perp \left(\frac{k_\perp}{(\omega/v)^2 + k_\perp^2}\right) J_0(bk_\perp) = K_0(bk_\perp). \tag{8.145}$$

Returning this result to (8.139),

$$\mathbf{E}(\omega) = \sqrt{\frac{2}{\pi}}\left(\frac{\alpha\hbar c}{e}\right)\left(\frac{z}{v}\right)\left(-i\frac{\omega}{v}\hat{\mathbf{x}} + \frac{\partial}{\partial b}\hat{\mathbf{y}}\right)K_0\left(\frac{\omega b}{v}\right). \tag{8.146}$$

From Appendix C.1, $\frac{dK_0(y)}{y} = -K_1(y)$ which leads us to

$$\mathbf{E}(\omega) = -\sqrt{\frac{2}{\pi}}\left(\frac{\alpha\hbar c}{e}\right)\left(\frac{z\omega}{v^2}\right)\left(iK_0\left(\frac{\omega b}{v}\right)\hat{\mathbf{x}} + K_1\left(\frac{\omega b}{v}\right)\hat{\mathbf{y}}\right). \tag{8.147}$$

And the amplitude evaluated at the oscillator's resonant frequency is

$$|\mathbf{E}(\omega_0)|^2 = \frac{2}{\pi}\left(\frac{\alpha\hbar c}{e}\right)^2\left(\frac{z\omega_0}{v}\right)^2\left[K_0^2\left(\frac{\omega_0 b}{v}\right) + K_1^2\left(\frac{\omega_0 b}{v}\right)\right] \tag{8.148}$$

which is the same result previously derived from the collision kinematics leading to (8.133).

8.2.4.7 Energy Transfer as a Function of Impact Parameter

Although originally derived by Bohr, we will combine the two contemporary approaches provided by Sigmund (2006) and by McParland (2010) as these then lead naturally to the Bohr soft-collision stopping power.

Using the definition of the classical radius of the electron and the relationship between the resonant frequency and the maximum impact parameter, $b_{\max}$, the results of (8.133) and (8.148) lead to an expression of the energy transfer to a harmonically bound electron as a function of impact parameter:

$$\begin{aligned} Q(b) &= \left(\pi\frac{e^2}{m_e}\right)|\mathbf{E}(b)|^2 \\ &= 2m_e\left(\frac{r_e}{b_{\max}}\right)^2\left(\frac{z}{\beta}\right)^2\left[\frac{K_0^2\left(\frac{b}{b_{\max}}\right)}{\gamma^2} + K_1^2\left(\frac{b}{b_{\max}}\right)\right]. \end{aligned} \tag{8.149}$$

It is evident that the first and second terms within the square brackets of (8.149) reflect the effects of the momentum transfer to the electron parallel to and orthogonal to, respectively, the direction of the projectile.

We note here the small- and large-argument limits of the modified Bessel functions which are summarised in Appendix C.1:

$$\begin{aligned} K_0(y) &\approx \ln\left(\tfrac{2}{y}\right) - \gamma_{\mathrm{EM}} && \text{as} \quad y \to 0 \\ K_1(y) &\approx \tfrac{1}{y} && \text{as} \quad y \to 0 \\ K_{0,1}(y) &\approx \sqrt{\frac{\pi}{2y}}e^{-y} && \text{as} \quad y \to \infty. \end{aligned} \tag{8.150}$$

In particular, if we apply the large-argument limit to (8.149), corresponding to large impact parameters, we are presented with the result of

$$Q(b) \approx \pi m_e \left(\frac{z}{\beta}\right)^2 \frac{r_e^2}{b_{\max} b} \left(\frac{1}{\gamma^2}+1\right) e^{-2b/b_{\max}} \quad \text{(large } b\text{)}. \tag{8.151}$$

The $e^{-2b/b_{\max}}/b$ factor of (8.151) introduces the rapid cut-off in energy transfer for large impact parameters as required in the adiabatic region.

We next evaluate the stopping cross section. As the stopping cross section involves the integration of the energy transfer of (8.149) over impact parameter, (8.151) demonstrates the presence of a logarithmic divergence as $b \to 0$. McParland (2010) avoided this problem by setting the lower limit of integration equal to $b_{\min}$. Sigmund (2006) provided an alternative, but more elegant, solution following that of Bohr and which we summarise here. This separates the integration over impact parameter into two integrations: one for large impact parameters for which (8.149) and another for small impact parameters for which the Thomson result of (8.23) is used. The demarcation between the two ranges of integration will be shown to be arbitrary as it disappears when the results are combined. However, we will demonstrate that it can be defined on the basis of the differences between the physics of the two processes.

We begin by considering (8.149) as being valid only for impact parameters beyond an arbitrary value of b_{mid} and write it as

$$Q_{\text{dist}}(b) = 2m_e r_e^2 \left(\frac{z}{\beta b}\right)^2 f_{\text{dist}}(b) \tag{8.152}$$

where the modification factor is

$$f_{\text{dist}}(b) = \left[\left(\left(\frac{b}{b_{\max}}\right)\frac{K_0\left(\frac{b}{b_{\max}}\right)}{\gamma}\right)^2 + \left(\left(\frac{b}{b_{\max}}\right)K_1\left(\frac{b}{b_{\max}}\right)\right)^2\right]. \tag{8.153}$$

The *dist* subscript reflects that this calculation is for collisions for which the projectile is at an extended distance from the atom. From our derivation of the Thomson formula of (8.23), we can write the corresponding result for close collisions:

$$Q_{\text{close}}(b) = 2m_e r_e^2 \left(\frac{z}{\beta b}\right)^2 f_{\text{close}}(b) \tag{8.154}$$

where

$$\begin{aligned} f_{\text{close}}(b) &= \frac{1}{1+\left(\frac{\rho B(E)}{2b}\right)^2} \\ &= \frac{1}{1+\upsilon/\left(\frac{b}{b_{\max}}\right)^2} \end{aligned} \tag{8.155}$$

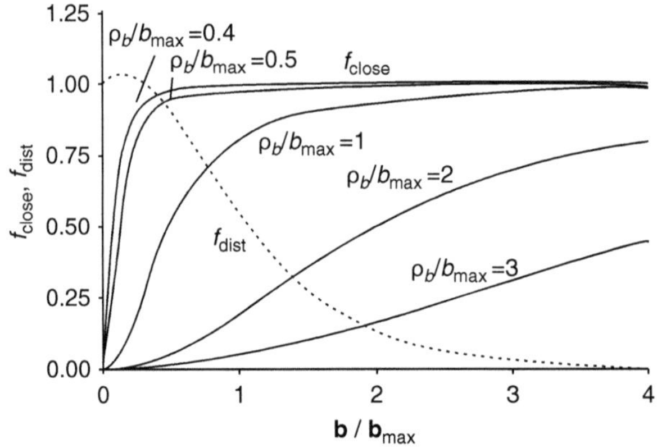

Fig. 8.6 Energy loss modification factors for distance and close collisions as functions of the normalised impact parameter. Note that the energy loss modification associated with close interactions has a dependence upon the Bohr collision diameter which that of the distant interaction does not. This figure is developed after similar productions by Neufeld (1954) and Sigmund (1996)

where, for convenience, we have defined the ratio

$$\upsilon \equiv \left(\frac{\rho_{B(E)}}{2b_{max}}\right)^2. \tag{8.156}$$

Why we have introduced $f_{close}(b)$ shall become apparent soon.

We now consider the two modification factors for the nonrelativistic case of $\gamma = 1$. Note that only $f_{close}(b)$ has an energy dependence through its incorporation of the Bohr collision diameter. These modification factors of $f_{close}(b)$ and $f_{dist}(b)$ are plotted in Fig. 8.6. From the small- and large-argument behaviours of these two functions, we can explain the bases of the shapes of these curves. Clearly, from (8.155), we have

$$\begin{aligned} f_{close}(b) &\to 0 \quad as \quad \frac{b}{b_{max}} \to 0 \\ f_{close}(b) &\to 1 \quad as \quad \frac{b}{b_{max}} \to \infty. \end{aligned} \tag{8.157}$$

The behaviour of $f_{dist}(b)$ requires further reflection as this shall be important in our calculation of the Bohr stopping power. From (8.150), we see that

$$\begin{aligned} f_{dist}(b) &\approx 1 + \left(\frac{b}{b_{max}} \ln\left(\frac{b}{b_{max}}\right)\right)^2 \quad as \quad \frac{b}{b_{max}} \to 0 \\ f_{dist}(b) &\approx \pi\left(\frac{b}{b_{max}}\right) e^{-2(b/b_{max})} \quad as \quad \frac{b}{b_{max}} \to \infty. \end{aligned} \tag{8.158}$$

Now having developed the energy transfers for the two categories of impact parameter, we now proceed to evaluating the Bohr soft-collision stopping power.

8.2.5 Bohr Soft-Collision Stopping Cross Section

Equation (8.158) reminds of the need to remove the logarithmic divergence associated with small impact parameters. As a result, in order to calculate the collision stopping cross section, we must split the integration of the stopping cross section given by (8.85) into two regions of impact parameter (the energy transfer being directly related to impact parameter). The first region is that from zero to the arbitrarily defined b_{mid} over which we shall use $f_{\text{close}}(b)$ and the second being that ranging from b_{mid} to infinity over which we shall use $f_{\text{dist}}(b)$. Thus,

$$\begin{aligned} S &= S_{\text{close}} + S_{\text{dist}} \\ &= 2\pi \left(\int_0^{b_{\text{mid}}} \mathrm{d}b\, b\, Q_{\text{close}}(b) + \int_{b_{\text{mid}}}^{\infty} \mathrm{d}b\, b\, Q_{\text{dist}}(b) \right). \end{aligned} \tag{8.159}$$

Essentially, we have considered b_{mid} as corresponding to that impact parameter for which $f_{\text{close}}(b_{\text{mid}})$ is the corresponding intersection between the $f_{\text{close}}(b_{\text{mid}})$ and $f_{\text{dist}}(b_{\text{mid}})$ shown in Fig. 8.6 for the relevant ratio of $\rho_{\text{B}}(E)/b_{\text{max}}$.

Let us begin with the lower range of impact parameter:

$$\begin{aligned} S_{\text{close}} &= 2\pi \int_0^{b_{\text{mid}}} \mathrm{d}b\, b\, Q_{\text{close}}(b) \\ &= 4\pi\, m_e r_e^2 \left(\frac{z}{\beta}\right)^2 \int_0^{b_{\text{mid}}} \mathrm{d}b\, \frac{f_{\text{close}}(b)}{b}. \end{aligned} \tag{8.160}$$

The integral is solved for by using the change of variable $y = b/b_{\text{max}}$,

$$\begin{aligned} \int_0^{b_{\text{mid}}} \mathrm{d}b\, \frac{f_{\text{close}}(b)}{b} &= \int_0^{b_{\text{mid}}/b_{\text{max}}} \frac{\mathrm{d}y}{y\left(1+\frac{\upsilon}{y^2}\right)} \\ &= \frac{1}{2} \ln \left(1 + \frac{\left(\frac{b_{\text{mid}}}{b_{\text{max}}}\right)^2}{\upsilon} \right) \\ &= \frac{1}{2} \ln \left(1 + \left(\frac{2 b_{\text{mid}}}{\rho_{\text{B}}(E)} \right)^2 \right). \end{aligned} \tag{8.161}$$

Thus,

$$S_{\text{close}} = 2\pi\, m_e\, r_e^2 \left(\frac{z}{\beta}\right)^2 \ln \left(1 + \left(\frac{2\, b_{\text{mid}}}{\rho_{\text{B}}(E)} \right)^2 \right). \tag{8.162}$$

In analogy to the development leading to the result of (8.162):

$$S_{\text{dist}} = 4\pi\, m_e r_e^2 \left(\frac{z}{\beta}\right)^2 \int_{b_{\text{mid}}}^{\infty} db\, \frac{f_{\text{dist}}(b)}{b}. \tag{8.163}$$

We evaluate the integral using (8.153) and the change of variable, $y = b/b_{\max}$,

$$\int_{b_{\text{mid}}}^{\infty} db\, \frac{f_{\text{dist}}(b)}{b} = \int_{y_{\text{mid}}}^{\infty} dy\, y \left(\left(\frac{K_0(y)}{\gamma}\right)^2 + yK_1^2(y) \right) \tag{8.164}$$

where $y_{\text{mid}} = b_{\text{mid}}/b_{\max}$. We now expand the integral and note that $1/\gamma^2 = 1 - \beta^2$ to give

$$\int_{y_{\text{mid}}}^{\infty} dy\, y \left(\left(\frac{K_0(y)}{\gamma}\right)^2 + yK_1^2(y) \right) = \int_{y_{\text{mid}}}^{\infty} dy\, y \left(K_0^2(y) - K_1^2(y)\right) - \beta_2 \int_{y_{\text{mid}}}^{\infty} dy\, yK_0^2(y). \tag{8.165}$$

To solve for these two integrals, we recall the derivatives of $dK_0(y)/dy = -K_1(y)$ and $dK_1(y)/dy = -(K_0(y) + (K_1(y)/y))$. Then,

$$\frac{d}{dy}\left(yK_0(y)\, K_1(y)\right) = -y\left(K_0^2(y) + K_1^2(y)\right) \tag{8.166}$$

and

$$\frac{d}{dy}\left(y\left(K_1^2(y) - K_0^2(y)\right)\right) = -2yK_0^2(y). \tag{8.167}$$

Thus,

$$\int_{y_{\text{mid}}}^{\infty} dy\, y\left(K_0^2(y) - K_1^2(y)\right) = y_{\text{mid}}\, K_0(y_{\text{mid}})\, K_1(y_{\text{mid}}) \tag{8.168}$$

and

$$\int_{y_{\text{mid}}}^{\infty} dy\, yK_0^2(y) = \frac{y_{\text{mid}}^2}{2}\left(K_1^2(y_{\text{mid}}) - K_0^2(y_{\text{mid}})\right). \tag{8.169}$$

These integrations give

$$\int_{b_{\mathrm{mid}}}^{\infty} \mathrm{d}b \, \frac{f_{\mathrm{dist}}(b)}{b} = y_{\mathrm{mid}} \, K_0(y_{\mathrm{mid}}) \, K_1(y_{\mathrm{mid}}) - \frac{\beta^2 y_{\mathrm{mid}}^2}{2} \left(K_1^2(y_{\mathrm{mid}}) - K_0^2(y_{\mathrm{mid}}) \right), \quad (8.170)$$

leading to

$$S_{\mathrm{dist}} = 4\pi \, m_e r_e^2 \left(\frac{z}{\beta}\right)^2 \left(\left(\frac{b_{\mathrm{mid}}}{b_{\mathrm{max}}}\right) K_0\left(\frac{b_{\mathrm{mid}}}{b_{\mathrm{max}}}\right) K_1\left(\frac{b_{\mathrm{mid}}}{b_{\mathrm{max}}}\right) \right.$$
$$\left. - \frac{\beta^2}{2} \left(\frac{b_{\mathrm{mid}}}{b_{\mathrm{max}}}\right)^2 K_1^2\left(\frac{b_{\mathrm{mid}}}{b_{\mathrm{max}}}\right) - K_0^2\left(\frac{b_{\mathrm{mid}}}{b_{\mathrm{max}}}\right) \right). \quad (8.171)$$

Let us now reflect upon the results of (8.162) and (8.171). S_{close} diverges as $b_{\mathrm{mid}} \to \infty$ and S_{dist} diverges as $b_{\mathrm{max}} \to 0$. Consequently, we select a value of b_{mid} such that the inequality

$$\frac{\rho_{\mathrm{B}}(E)}{b_{\mathrm{max}}} << \frac{b_{\mathrm{mid}}}{b_{\mathrm{max}}} << 1 \quad (8.172)$$

holds. (8.172) corresponds to the far left section of Fig. 8.6. In this case, (8.162) becomes

$$S_{\mathrm{close}} \approx 4\pi \, m_e \, r_e^2 \left(\frac{z}{\beta}\right)^2 \ln\left(\frac{2 \, b_{\mathrm{mid}}}{\rho_{\mathrm{B}}(E)}\right). \quad (8.173)$$

Then, using the small-argument approximations of the modified Bessel functions, we have

$$K_0\left(\frac{b_{\mathrm{mid}}}{b_{\mathrm{max}}}\right) \approx \ln\left(\frac{2 \, b_{\mathrm{max}}}{b_{\mathrm{mid}}}\right) - \gamma_{\mathrm{EM}}$$
$$= \ln\left(\frac{2 \, b_{\mathrm{max}}}{b_{\mathrm{mid}}} \, e^{-\gamma_{\mathrm{EM}}}\right) \quad (8.174)$$

and

$$K_1\left(\frac{b_{\mathrm{mid}}}{b_{\mathrm{max}}}\right) \approx \frac{2 \, b_{\mathrm{max}}}{b_{\mathrm{mid}}}. \quad (8.175)$$

We can then write the distance collision stopping cross section as

$$S_{\mathrm{dist}} \approx 4\pi \, m_e r_e^2 \left(\frac{z}{\beta}\right)^2 \left(\ln\left(\frac{2e^{-\gamma_{\mathrm{EM}}} b_{\mathrm{max}}}{b_{\mathrm{mid}}}\right) - \frac{\beta^2}{2} \right.$$
$$\left. + \frac{\beta^2}{2} \left(\frac{b_{\mathrm{mid}}}{b_{\mathrm{max}}}\right)^2 \left(\ln\left(\frac{2e^{-\gamma_{\mathrm{EM}}} b_{\mathrm{max}}}{b_{\mathrm{mid}}}\right)\right)^2 \right)$$
$$= 4\pi \, m_e r_e^2 \left(\frac{z}{\beta}\right)^2 \left(\ln\left(\frac{2e^{-\gamma_{EM}} b_{\mathrm{max}}}{b_{\mathrm{mid}}}\right) - \frac{\beta^2}{2} \right) \quad (8.176)$$

where the term proportional to $(b_{\mathrm{mid}}/b_{\mathrm{max}})^2$ is neglected as a result of the inequality of (8.172). The complete stopping cross section is obtained from the sum of (8.173) and (8.176):

$$S = 4\pi\, m_e r_e^2 \left(\frac{z}{\beta}\right)^2 \ln\left(\frac{4e^{-\gamma_{\mathrm{EM}}} b_{\mathrm{max}}}{\rho_{\mathrm{B}}(E)}\right) - \frac{\beta^2}{2}. \tag{8.177}$$

As b_{mid} cancels in the summation of the logarithms, it is clear that its exact value is not critical as long as the inequality of (8.172), which allowed the small-argument approximations of the modified Bessel functions of (8.174) and (8.175) to be used. To continue our derivation of the Bohr stopping cross section, we use the adiabatic impact parameter of the relativistic form of (8.57) and the relativistic form of the Bohr impact parameter. Then, the logarithm of (8.177) becomes, after manipulation,

$$\begin{aligned}\ln\left(\frac{4e^{-\gamma_{\mathrm{EM}}} b_{\mathrm{max}}}{\rho_{\mathrm{B}}(E)}\right) &= \ln\left(\frac{2e^{-\gamma_{\mathrm{EM}}}\gamma^2\beta^3 c}{\omega_0\, r_e\, z}\right)\\ &= \ln\left(\frac{2e^{-\gamma_{\mathrm{EM}}}\beta^3 c}{\omega_0\, r_e\, z}\right) - \ln\left(1-\beta^2\right).\end{aligned} \tag{8.178}$$

Returning this result to (8.177) and recalling our definition of the constant C given by (8.84), we can write

$$S = 2\frac{C}{N_{\mathrm{A}}}\left(\frac{z}{\beta}\right)^2 \left(\ln\left(\frac{2e^{-\gamma_{\mathrm{EM}}}\beta^3 c}{\omega_0\, r_e\, z}\right) - \ln\left(1-\beta^2\right) - \frac{\beta^2}{2}\right). \tag{8.179}$$

8.2.6 Bohr Soft-Collision Mass Stopping Power

The Bohr soft-collision mass stopping power is then obtained by multiplying the collision cross section of (8.179) by the density of scattering centres to yield

$$\left.\frac{\mathrm{d}E}{\rho\,\mathrm{d}x}\right|_{S,\mathrm{Bohr}} = 2C\left(\frac{Z}{A}\right)\left(\frac{z}{\beta}\right)^2 \left(\ln\left(\frac{2e^{-\gamma_{\mathrm{EM}}}\beta^3 c}{\omega_0\, r_e\, z}\right) - \ln\left(1-\beta^2\right) - \frac{\beta^2}{2}\right). \tag{8.180}$$

It is of interest to compare this Bohr result with the Rutherford formula:

$$\begin{aligned}\left.\frac{\mathrm{d}E}{\rho\,\mathrm{d}x}\right|_{S,\mathrm{Ruth}} &= 2C\left(\frac{Z}{A}\right)\left(\frac{z}{\beta}\right)^2 \left(\ln\left(\frac{\beta^3 c}{\omega_0\, r_e\, z}\right) - \ln\left(1-\beta^2\right)\right)\\ \left.\frac{\mathrm{d}E}{\rho\,\mathrm{d}x}\right|_{S,\mathrm{Bohr}} &= \left.\frac{\mathrm{d}E}{\rho\,\mathrm{d}x}\right|_{S,\mathrm{Ruth}} + \Delta\left(\left.\frac{\mathrm{d}E}{\rho\,\mathrm{d}x}\right|_{S}\right)\end{aligned} \tag{8.181}$$

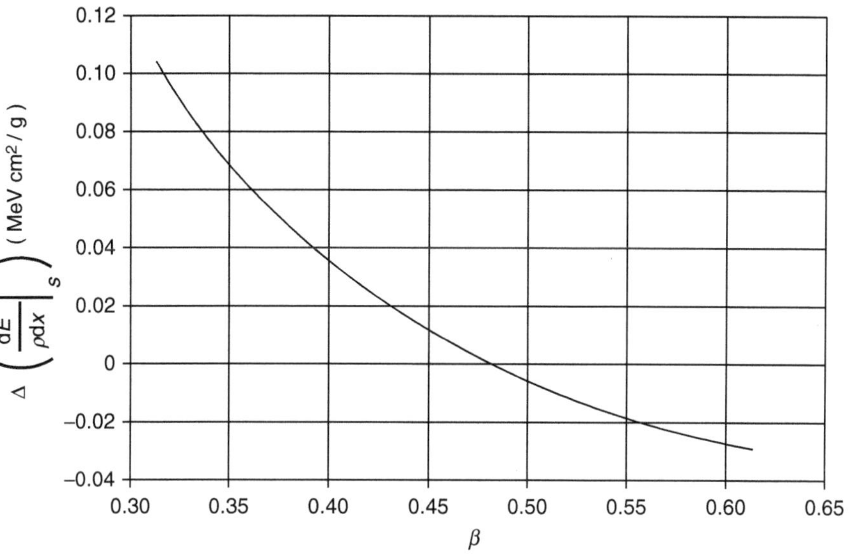

Fig. 8.7 The difference between the Bohr and Rutherford mass soft-collision stopping powers, given by Eq. (8.182), for protons in a medium for which $A = 2Z$ as a function of proton normalised velocity

where

$$\Delta\left(\frac{dE}{\rho\,dx}\bigg|_S\right) = 2C\left(\frac{Z}{A}\right)\left(\frac{z}{\beta}\right)^2\left(\ln 2 - \gamma_{EM} - \frac{\beta^2}{2}\right). \tag{8.182}$$

The difference between the Bohr and Rutherford mass soft-collision stopping powers given by (8.182) is shown in Fig. 8.7 for protons in a medium where $Z/A = 1/2$, which is the case for both ^{12}C and ^{40}Ca. The range of β shown corresponds to proton kinetic energies of between 50 and 250 MeV.

As to be shown when we compare the Rutherford, Bohr and Bethe results later in this chapter, we will note that this difference is only a few percent or so of the mass soft-collision stopping power. In particular, we note that the Rutherford and Bohr results are equal for $\beta = \sqrt{2\,(\ln 2 - \gamma_{EM})} \approx 0.48$. Below this normalised velocity, the Bohr mass soft-collision stopping power exceeds that of the Rutherford calculation. On inspection of (8.182), it should be noted that the $(\ln 2 - \gamma_{EM})$ contribution arises from the small-argument approximation of $K_0(b/b_{max})$ in the Bohr derivation of the energy transfer result of (8.149). Similarly, the $-\beta^2/2$ contribution arises from the small-argument limit of $K_1(b/b_{max})$ in the same equation. These indicate, through (8.138), (8.139) and (8.140), the roles of longitudinal and transverse components of the electric fields, respectively. On the other hand, the Rutherford result follows from the energy transfer that is calculated for the transverse case as shown in (8.15).

8.2.7 Excitation

We transit now from the classical theory of Bohr derived above to a semi-classical theory in which the atomic electrons are described quantum mechanically while the nucleus, when explicitly considered, is described classically. The fundamental difference for our consideration is that the atomic electron is no longer treated as a classical simple harmonic oscillator but, following excitation, can occupy only discrete energy levels.

In the Bohr classical theory, the energy gained from the projectile by the atomic electron was continuous and is related to the impact parameter. The introduction of quantum mechanics means that instead of this continuous transition, the electron has available a discrete number of energy levels to move to, each with a given transition probability. That is, for a given impact parameter b, the mean energy gained by the electron is given by the summation over all available final states of

$$\bar{Q}(b) = \sum_f \lambda_{f0}\left(E_f - E_0\right) \tag{8.183}$$

where λ_{f0} is the transition probability from state $|0\rangle$ to state $|f\rangle$ with the associated energies of E_0 and E_f, respectively. We implicitly assume that the electron is in the ground state before the collision, although the derivation to follow is applicable to any initial state. Clearly, our initial task is to derive this transition probability. The reader will recall the development of the quantum perturbation theory in Sect. (2.4) that will be relevant to our discussion. The time-dependent atomic wavefunction is given by (2.24) which we rewrite using the Dirac *bra–ket* nomenclature summarised in Appendix E;

$$\Psi(\mathbf{r}, t) = \sum_f a_f(t)\, e^{-i\frac{E_f t}{\hbar}}|f\rangle. \tag{8.184}$$

Recall from (2.33) that the coefficients themselves can be expanded into a perturbative series:

$$a_f(t) = \sum_{m=0}^{\infty} \lambda^m a_{f,m}(t) \tag{8.185}$$

where the perturbation coefficient λ is taken to be small. As a consequence of this, we can terminate the series at $m = 1$ allowing us to write

$$a_{f,1}(t) = -\frac{i}{\hbar}\int_0^t \mathrm{d}t'\, e^{i\omega_{f0}t'}\,\langle f|U(\mathbf{r}, t')|0\rangle \tag{8.186}$$

and where the perturbing potential is

$$U(\mathbf{r},t) = -\alpha\hbar cz\sum_{j=1}^{Z}\frac{1}{\left|\mathbf{r_j}-(\mathbf{b}+\mathbf{v}t)\right|}. \tag{8.187}$$

The summation is over all atomic electrons, and the associated Bohr frequency is

$$\omega_{f0} = \frac{E_f - E_0}{\hbar}. \tag{8.188}$$

The derivation of this result may be found within the derivation of (2.31); however, it should be noted that the perturbing Coulomb potential here has a spatial dependence unlike that considered in Sect. 2.4. We now obtain an expression for the perturbation coefficient relevant to the Coulomb interaction:

$$a_{f,1}(t) = i\,\alpha cz\int_0^t \mathrm{d}t'\,e^{i\omega_{f0}t'}\,\langle f|\sum_{j=1}^{Z}\frac{1}{\left|\mathbf{r_j}-(\mathbf{b}+\mathbf{v}t')\right|}|0\rangle. \tag{8.189}$$

As has been noted numerous times throughout our discussions of scattering theory in Chaps. 2 and 4, the observer can detect only the asymptotic final state of the scatter, i.e. that as $t \longrightarrow +\infty$. We will denote the coefficient for this asymptotic case as $a_{f,1}(t \rightarrow +\infty)$ and which is solved for by writing the matrix element of (8.189) as the inverse Fourier transform of its Fourier transform into $\boldsymbol{k}$-space:

$$\begin{aligned} a_{f,1}(t \rightarrow +\infty) &= \frac{i}{2\pi^2}\,\alpha cz\int_0^t \mathrm{d}t'\,e^{i\omega_{f0}t'}\int_{-\infty}^{\infty}\mathrm{d}^3\mathbf{k}\,\frac{1}{k^2}\langle f|\sum_{j=1}^{Z}e^{i\mathbf{k}\bullet\left(\mathbf{r}_j-(\mathbf{b}+\mathbf{v}t')\right)}|0\rangle \\ &= \frac{i}{2\pi^2}\,\alpha cz\int_{-\infty}^{\infty}\mathrm{d}^3\mathbf{k}\,\frac{e^{-i\mathbf{k}\bullet\mathbf{b}}}{k^2}\langle f|\sum_{j=1}^{Z}e^{i\mathbf{k}\bullet\mathbf{r}_j}|0\rangle\int_0^t \mathrm{d}t'\,e^{i\left(\omega_{f0}-\mathbf{k}\bullet\mathbf{v}\right)t'} \\ &= -\frac{1}{\pi}\,\alpha cz\int_{-\infty}^{\infty}\mathrm{d}^3\mathbf{k}\,\frac{e^{-i\mathbf{k}\bullet\mathbf{b}}}{k^2}F_{f0}(\mathbf{k})\delta\left(\omega_{f0}-\mathbf{k}\bullet\mathbf{v}\right). \end{aligned} \tag{8.190}$$

Here, we have defined the matrix element as

$$F_{f0}(\mathbf{k}) \equiv \langle f|\sum_{j=1}^{Z}e^{i\mathbf{k}\bullet\mathbf{r}_j}|0\rangle. \tag{8.191}$$

The reader may recognise this as being the atomic (or inelastic) form factor. It is the sum over all atomic electron co-ordinates of the exponential operation in momentum transfer space (through the relationship of $\mathbf{q} = \hbar c\,\mathbf{k}$) between the ground state and the *f*th excited state. We now expand the exponent of (8.191)

$$e^{i\mathbf{k}\bullet\mathbf{r}_j} = 1 + \sum_{n=1}^{\infty} \frac{1}{n!}\left(i\,\mathbf{k}\bullet\mathbf{r}_j\right)^n \tag{8.192}$$

and truncate at first order as in the case of the dipole approximation

$$e^{i\mathbf{k}\bullet\mathbf{r}_j} \approx 1 + i\,\mathbf{k}\bullet\mathbf{r}_j. \tag{8.193}$$

Returning this result to the expression for the matrix element gives

$$F_{f0}(\mathbf{k}) \approx \langle f||0\rangle + i\langle f|\mathbf{k}\bullet\mathbf{r}_j|0\rangle. \tag{8.194}$$

Due to the orthonormality of states, this leads to

$$F_{f0}(\mathbf{k}) = i\mathbf{k}\bullet\langle f|\sum_{j=1}^{Z}\mathbf{r}_j|0\rangle. \tag{8.195}$$

Inserting this expression for the form factor into (8.190) gives

$$a_{f,1}(t\to+\infty) = -\frac{1}{\pi}\,\alpha cz\langle f|\sum_{j=1}^{Z}\mathbf{r}_j|0\rangle\bullet\int_{-\infty}^{\infty}\mathrm{d}^3\mathbf{k}\,\frac{\mathbf{k}}{k^2}e^{-i\mathbf{k}\bullet\mathbf{b}}\delta(\omega_{f0}-\mathbf{k}\bullet\mathbf{v}) \tag{8.196}$$

It will be noted that the integral of (8.196) is the same as that of (8.134) and which had been solved:

$$\begin{aligned}\int_{-\infty}^{\infty}\mathrm{d}^3\mathbf{k}\,\frac{\mathbf{k}}{k^2}e^{-i\mathbf{k}\bullet\mathbf{b}}\delta(\omega_{f0}-\mathbf{k}\bullet\mathbf{v}) &= \frac{1}{v}\int_{-\infty}^{\infty}\mathrm{d}^3\mathbf{k}\,\frac{\left(\frac{\omega_{f0}}{v}\right)\hat{\mathbf{x}}+k_\perp\mathbf{y}}{\left(\frac{\omega_{f0}}{v}\right)^2+k_\perp^2}e^{-i\mathbf{k}\bullet\mathbf{b}}\\ &= \left(\frac{1}{v}\right)\left(-i\left(\frac{\omega_{f0}}{v}\right)\hat{\mathbf{x}}+\frac{\partial}{\partial b}\hat{\mathbf{y}}\right)K_0\left(\frac{\omega_{f0}\,b}{v}\right)\\ &= -\left(\frac{\omega_{f0}}{v^2}\right)\left(iK_0\left(\frac{\omega_{f0}\,b}{v}\right)\hat{\mathbf{x}}+K_1\left(\frac{\omega_{f0}\,b}{v}\right)\hat{\mathbf{y}}\right).\end{aligned} \tag{8.197}$$

Thus,

$$\begin{aligned}a_{f,1}(t\to+\infty) &= \frac{1}{\pi}\alpha c\left(\frac{\omega_{f0}\,z}{v^2}\right)\langle f|\sum_{j=1}^{Z}\mathbf{r}_j|0\rangle\\ &\bullet\left(iK_0\left(\frac{\omega_{f0}\,b}{v}\right)\hat{\mathbf{x}}+K_1\left(\frac{\omega_{f0}\,b}{v}\right)\hat{\mathbf{y}}\right).\end{aligned} \tag{8.198}$$

The expectation value of the harmonic oscillator's energy after the collision and as $t \to +\infty$ is

$$\langle \Psi(\mathbf{r}, t \to +\infty) | H | \Psi(\mathbf{r}, t \to +\infty) \rangle = \sum_f |a_f(t \to +\infty)|^2 E_f. \tag{8.199}$$

where the summation is over all final states. Then, as the atom was in its ground state prior to the collision, the mean energy acquired by the atom as a result of the soft collision is

$$\begin{aligned}\overline{Q} &= \langle \Psi(\mathbf{r}, t \to +\infty) | H | \Psi(\mathbf{r}, t \to +\infty) \rangle - \langle \Psi(\mathbf{r}, t \to -\infty) | H | \Psi(\mathbf{r}, t \to -\infty) \rangle \\ &= \sum_f \left| a_{f,1}(t \to +\infty) \right|^2 E_{f0}\end{aligned} \tag{8.200}$$

where, for convenience, we have defined $E_{f0} \equiv E_f - E_0$. From our discussion of quantum perturbation theory of Chap. 2, we recognise that the $\left| a_{f,1}(t \to +\infty) \right|^2$ is the transition probability from the ground state $|0\rangle$ to the final state $|f\rangle$:

$$\begin{aligned}\lambda_{f0}(b) &= \left| a_{f,1}(t \to +\infty) \right|^2 \\ &= \frac{1}{\pi^2} \left(\frac{\alpha}{\hbar c} \right)^2 \left(\frac{E_{f0} z}{\beta} \right)^2 \left| \langle f | \sum_{j=1}^{Z} \mathbf{r}_j | 0 \rangle \right|^2 \\ &\times \left(K_0^2 \left(\frac{E_{f0}\, b}{\beta \hbar c} \right) + K_1^2 \left(\frac{E_{f0}\, b}{\beta \hbar c} \right) \right).\end{aligned} \tag{8.201}$$

We return this result to (8.183); we obtain an expression for the mean energy transfer:

$$\begin{aligned}\bar{Q}(b) =& \frac{1}{\pi^2} \left(\frac{\alpha}{\hbar c} \right)^2 \left(\frac{z}{\beta} \right)^2 \sum_f E_{f0}^3 \left| \langle f | \sum_{j=1}^{Z} \mathbf{r}_j | 0 \rangle \right|^2 \\ &\times \left(K_0^2 \left(\frac{E_{f0}\, b}{\beta \hbar c} \right) + K_1^2 \left(\frac{E_{f0}\, b}{\beta \hbar c} \right) \right)\end{aligned} \tag{8.202}$$

The excitation cross section will thus be

$$\begin{aligned}S &= \sum_f E_{f0} \int \mathrm{d}^3\mathbf{b}\, \lambda_{f0}(b) \\ &= 2\pi \sum_f E_{f0} \int_0^\infty \mathrm{d}b\, b\, \lambda_{f0}(b).\end{aligned} \tag{8.203}$$

Inserting (8.201) into (8.203) would demonstrate the need to again address the logarithmic divergence of the modified Bessel function for small impact parameters. This would be done using the method described previously, but is not repeated here. We do, however, consider the component associated with large impact parameters in the next section.

We can simplify the above results by incorporating the dipole oscillator strength derived in Appendix G:

$$f_{f0} = \frac{2}{3}\frac{m_e}{(\hbar c)^2 Z} E_{f0} \left| \langle f | \sum_{j=1}^{Z} \mathbf{r}_j | 0 \rangle \right|^2 .$$

The transition probability can be rewritten as

$$\lambda_{f0}(b) = \frac{3}{2\pi^2}\left(\frac{\alpha z}{\beta}\right)^2 \frac{ZE_{f0}}{m_e} f_{f0}\left(K_0^2\left(\frac{E_{f0}\,b}{\beta\hbar c}\right) + K_1^2\left(\frac{E_{f0}\,b}{\beta\hbar c}\right)\right) \tag{8.204}$$

and the mean energy transferred as

$$\bar{Q}(b) = \frac{3}{2\pi^2}\left(\frac{\alpha z}{\beta}\right)^2 \frac{Z}{m_e} \sum_f E_{f0}^2 f_{f0}\left(K_0^2\left(\frac{E_{f0}\,b}{\beta\hbar c}\right) + K_1^2\left(\frac{E_{f0}\,b}{\beta\hbar c}\right)\right). \tag{8.205}$$

If we compare the semi-classical expression for the mean energy transfer of (8.205) with the classical result of (8.149), the latter including the expression of (8.59) for the adiabatic limit to the impact parameter, we see that the former uses averaging over energies of the continuous electron transitions, whereas the latter uses the weighted summation over discrete transitions.

8.2.8 Mean Resonant Frequency

The soft-collision stopping power contains a term of $-\ln\omega_0$ per a single resonant frequency. In practice, we must contend with the fact that the charged particle is interacting with multi-electron atoms also. As a result, it will be necessary to evaluate a mean resonant frequency relevant to the stopping power. This will be the weighted sum of the logarithms of the transitional frequencies, the weights being given by the dipole oscillator strengths:

$$\ln\bar{\omega} = \sum_f f_{f0} \ln\omega_{f0}. \tag{8.206}$$

This result is, essentially, the weighted geometric mean of the transitional frequencies.

8.2.9 Fermi Theory of Soft-Collision Energy Loss

> The great theoretical importance of this conclusion justifies a careful investigation of possible alternative explanations of the observed difference in absorption. I have therefore considered the following effect which seems to explain the observations, at least to some extent, without assuming a decay of the mesotron.
>
> Enrico Fermi (1939)

8.2.9.1 Introductory Comments

As discussed in the Introduction to this chapter, the experimental results of the difference between the measured muon energy losses in air and condensed matter led Fermi to the consideration of the effect of the relative permittivity between the two as being the cause. This will lead, as to be shown in Chap. 12, to the introduction of the density effect correction to the stopping power. Here, we will show how this consideration also leads to an expression of the stopping power for heavy particles and presents a precursor for both our later discussion of the density effect and the relativistic component of the Bethe soft-collision energy loss theory.

The derivation of the stopping power for a decelerating charged particle in a medium will require the outcome of the calculation of the electric field of that particle. This, in turn, requires us to determine the associated scalar and vector potentials that generate this field the ohmic response of the medium. As usual, we shall do this calculation in the two-dimensional Fourier space of wavevector and frequency $(\mathbf{k}, \omega)$ rather than that of space and time $(\mathbf{r}, t)$. We begin this derivation through evaluating the current density, divided into longitudinal and transverse components (i.e. parallel and orthogonal to $\mathbf{k}$) using Maxwell's equations.

Note that some of the constituents of these derivations parallel those given in Section (13.3) of Jackson (1999), Section (5.2) of Sigmund (2006) and Section (7.3.7.6) of McParland (2010).

8.2.9.2 Classical Electrodynamics In Vacuo

Maxwell's equations for the electric ($\mathbf{E}$) and magnetic ($\mathbf{B}$) fields in vacuo are, in SI units:

$$\begin{aligned}
\nabla \bullet \mathbf{E} &= \frac{\rho_e}{\varepsilon_0} \\
\nabla \bullet \mathbf{B} &= 0 \\
\nabla \times \mathbf{E} &= -\frac{\partial \mathbf{B}}{\partial t} \\
\nabla \times \mathbf{B} &= \mu_0 \left(\mathbf{J} + \varepsilon_0 \frac{\partial \mathbf{E}}{\partial t} \right)
\end{aligned} \tag{8.207}$$

where, in SI units,

- ρ_e is the charge density ($\mathrm{C/m^3}$)
- $\mathbf{J}$ is the current density ($\mathrm{C/s} \bullet \mathrm{m}^2$)
- ε_0 is the permittivity of free space $\left(8.854 \times 10^{-12}\,\mathrm{C^2\,s^2/kg} \bullet \mathrm{m^3}\right)$
- μ_0 is the permeability of free space $\left(4\pi \times 10^7\,\mathrm{m\,kg/C^2}\right)$.

Recall that

$$c = \frac{1}{\sqrt{\varepsilon_0 \mu_0}}. \tag{8.208}$$

The electric and magnetic fields are also interrelated through the expression for the Lorentz force:

$$\mathbf{F} = q(\mathbf{E} + \mathbf{v} \times \mathbf{B}) \tag{8.209}$$

which describes the force upon a charge q moving with velocity $\mathbf{v}$ within electric $\mathbf{E}$ and magnetic $\mathbf{B}$ fields.

Of the four Maxwell's equations of (8.207), the first is Gauss' law, the second reflects the observed absence of magnetic monopoles, the third is Faraday's law and the fourth is a version of Ampère's law. All of the electromagnetic quantities in $(\mathbf{r}, t)$-space are of the usual phasor form of

$$f(\mathbf{r}, t) = f_0\, e^{i(\mathbf{k} \bullet \mathbf{r} - \omega t)}. \tag{8.210}$$

The Fourier transforms of the various derivatives, etc. that we shall encounter in this section are provided in Appendix B.1.

In the derivation to follow, unless necessary so as to avoid ambiguity, we remove the explicit $(\mathbf{r}, t)$ functional dependence so as to lend clarity to the expressions.

From the condition of the zero divergence of the magnetic field as expressed by $\nabla \bullet \mathbf{B} = 0$, we can define a vector potential $\mathbf{A}$ through the vector algebra requirement that the *divergence* of the *curl* of a vector function be equal to zero, e.g. $\nabla \bullet (\nabla \times \mathbf{A}) = 0$. This leads to the definition of the vector potential through

$$\mathbf{B} = \nabla \times \mathbf{A} \tag{8.211}$$

which further leads us to, through a rearrangement of Faraday's law,

$$\begin{aligned} \nabla \times \mathbf{E} + \frac{\partial \mathbf{B}}{\partial t} &= \mathbf{0} \\ \nabla \times \mathbf{E} + \frac{\partial}{\partial t}(\nabla \times \mathbf{A}) &= \mathbf{0} \\ \nabla \times \left(\mathbf{E} + \frac{\partial \mathbf{A}}{\partial t}\right) &= \mathbf{0}. \end{aligned} \tag{8.212}$$

From vector algebra, a vector which has zero *curl* can be written as the gradient of a scalar quantity. In our case, we identify this scalar as the electric potential, or

$$\mathbf{E} = -\left(\nabla V + \frac{\partial \mathbf{A}}{\partial t}\right). \tag{8.213}$$

Ultimately, we wish to calculate the electric field within a medium due to the moving charged particle. We do this by using, first, the in vacuo form of Maxwell's equations to generate explicit expressions for the scalar and vector potentials as functions of the electric charge and current densities. To initiate this, we consider the relationship connecting the scalar and vector potentials and the electric field of (8.213) in Gauss' law:

$$-\nabla \bullet \left(\nabla V + \frac{\partial \mathbf{A}}{\partial t}\right) = \frac{\rho_e}{\varepsilon_0},$$

or

$$\nabla^2 V + \frac{\partial}{\partial t}(\nabla \bullet \mathbf{A}) = -\frac{\rho_e}{\varepsilon_0}. \tag{8.214}$$

We then write Ampère's law in the form

$$\begin{aligned}\nabla \times \mathbf{B} &= \mu_0 \mathbf{J} + \mu_0 \varepsilon_0 \frac{\partial \mathbf{E}}{\partial t} \\ &= \mu_0 \mathbf{J} + \frac{1}{c^2}\frac{\partial \mathbf{E}}{\partial t}\end{aligned} \tag{8.215}$$

where (8.208) has been used. Then, using the vector identity of the double *curl*, $\nabla \times (\nabla \times \mathbf{A}) = \nabla(\nabla \bullet \mathbf{A}) - \nabla^2 \mathbf{A}$, the definition of the vector potential of (8.199) and incorporating (8.215), we obtain the inhomogeneous coupled equation:

$$\nabla^2 \mathbf{A} - \frac{1}{c^2}\frac{\partial^2 \mathbf{A}}{\partial t^2} - \nabla\left(\nabla \bullet \mathbf{A} + \frac{1}{c^2}\frac{\partial V}{\partial t}\right) = -\mu_0 \mathbf{J}. \tag{8.216}$$

Before initiating further manipulations of the above two results, we now consider the concept of the gauge transformation. A gauge is a particular choice of scalar and vector potentials that satisfy (8.211) and (8.213) simultaneously. For example, consider the transformation of the vector potential in which the gradient of a function $\Xi(\mathbf{r}, t)$ is added to it:

$$\mathbf{A} \to \mathbf{A}' = \mathbf{A} + \nabla \Xi. \tag{8.217}$$

Then, from (8.211) and the vector identity that the *curl* of a *divergence* is equal to zero, the magnetic field **B** remains unchanged. But, from (8.213), this transformation of the vector potential alters the electric field **E**:

$$\mathbf{E} = -\left(\nabla\left(V + \frac{\partial \Xi}{\partial t}\right) + \frac{\partial \mathbf{A}}{\partial t}\right). \tag{8.218}$$

However, we can return the original expression for the electric field if we transform the scalar potential:

$$V \to V' = V - \frac{\partial \Xi}{\partial t}. \tag{8.219}$$

Thus, the simultaneous transformations of the vector and scalar potentials given by (8.217) and (8.218) leave the electric and magnetic fields unchanged. $\Xi(\mathbf{r}, t)$ is known as a *gauge function*, and a most useful gauge for the potentials is given by the *Coulomb gauge*, which is given by the zero divergence of the vector potential:

$$\nabla \bullet \mathbf{A} = 0. \tag{8.220}$$

This leads (8.214) to be written as

$$\nabla^2 V = -\frac{\rho_e}{\varepsilon_0}, \tag{8.221}$$

that is, the scalar potential satisfies Poisson's equation in the Coulomb gauge and has the integral solution

$$V(\mathbf{r}, t) = \frac{1}{4\pi\varepsilon_0} \int \mathrm{d}^3\mathbf{r}' \, \frac{\rho_e(\mathbf{r}', t)}{|\mathbf{r} - \mathbf{r}'|}. \tag{8.222}$$

As this is an unretarded potential, the scalar potential is referred to as the instantaneous Coulomb potential. It is because of this that the zero divergence of the vector potential is defined as the Coulomb gauge. Similarly, from (8.216),

$$\nabla^2 \mathbf{A} - \frac{1}{c^2}\left(\frac{\partial^2 \mathbf{A}}{\partial t^2} - \nabla \frac{\partial V}{\partial t}\right) = -\mu_0 \mathbf{J}. \tag{8.223}$$

As the *curl* of the vector potential is equal to zero, the vector potential is termed to be *irrotational*. We can thus separate the current density into orthogonal components,

$$\mathbf{J} = \mathbf{J}_{||} + \mathbf{J}_{\perp}, \tag{8.224}$$

where the first term is referred to as the *longitudinal* current as it is irrotational,

$$\nabla \times \mathbf{J}_{||} = 0, \tag{8.225}$$

and the second term is referred to as the *transverse* current as

$$\nabla \bullet \mathbf{J}_{\perp} = 0. \tag{8.226}$$

Substituting (8.224) into (8.223), we obtain

$$\nabla^2 \mathbf{A} - \frac{1}{c^2}\left(\frac{\partial^2 \mathbf{A}}{\partial t^2} - \nabla \frac{\partial V}{\partial t}\right) = -\mu_0 \left(\mathbf{J}_{||} + \mathbf{J}_{\perp}\right). \tag{8.227}$$

This separates the coupling between potentials and leads to the longitudinal and transverse expressions:

$$\varepsilon_0 \nabla \frac{\partial V}{\partial t} = -\mathbf{J}_{||} \tag{8.228}$$

$$\nabla^2 \mathbf{A} - \frac{1}{c^2}\frac{\partial^2 \mathbf{A}}{\partial t^2} = -\mu_0 \mathbf{J}_{\perp}. \tag{8.229}$$

We see that the solution of the wave equation for the vector potential is in terms of the transverse current. As a result, the *Coulomb gauge* is also referred to as the *transverse gauge*. Further, as the scalar potential (instantaneous Coulomb potential) of (8.222) goes to zero at distances far from the electric charge, this gauge is also referred to as the *radiation gauge* owing to the transverse radiation fields resulting from the vector potential **A**.

It will now prove necessary to transform these results into $(\mathbf{k}, \omega)$ -space, respectively,

$$\mathbf{k} \bullet \mathbf{J}(\mathbf{k}, \omega) = \omega \rho_e(\mathbf{k}, \omega) \tag{8.230}$$

$$\varepsilon_0 k^2 V(\mathbf{k}, \omega) = \rho_e(\mathbf{k}, \omega) \tag{8.231}$$

$$\varepsilon_0 \omega \mathbf{k} V(\mathbf{k}, \omega) = -\mathbf{J}_{||}(\mathbf{k}, \omega) \tag{8.232}$$

$$\left(k^2 - \left(\frac{\omega}{c}\right)^2\right) \mathbf{A}(\mathbf{k}, \omega) = \mu_0 \mathbf{J}_{\perp}(\mathbf{k}, \omega). \tag{8.233}$$

As the ultimate goal of this discussion is to determine the energy loss of the charged particle as it moves through a medium, our prior experiences suggest that this can be achieved through the derivation of an expression for the electric field.

We now begin this derivation by first investigating the response of the medium to the energetic charged particle.

8.2.9.3 Response of the Medium

Having now transformed the field equations in vacuo on Fourier space, we now consider the response of the medium to the electric field of the charged particle on the basis of these results. From this consideration, we are required to derive the net scalar and vector potentials that contribute to the energy loss of that particle.

This is done by first partitioning both the charge and current densities into two components: the first is a direct consequence of the projectile, and the second is that induced into the medium by the projectile. We thus write the electric charge density as the sum given by

$$\rho_e = \rho_{e,\text{proj}} + \rho_{e,\text{med}} \tag{8.234}$$

and the current density as

$$\mathbf{J} = \mathbf{J}_{\text{proj}} + \mathbf{J}_{\text{med}} \tag{8.235}$$

where the *proj* and *med* subscripts refer to the component due to the projectile and that due to the response of the medium, respectively. In order to evaluate the longitudinal and transverse components of the current density induced into the medium, we introduce Ohm's law. We begin by writing the longitudinal and transverse components of the electric field in $(\mathbf{k}, \omega)$-space:

$$\mathbf{E}_{||}(\mathbf{k}, \omega) = -i\mathbf{k}V(\mathbf{k}, \omega) \tag{8.236}$$

$$\mathbf{E}_{||}(\mathbf{k}, \omega) = i\omega\mathbf{A}(\mathbf{k}, \omega). \tag{8.237}$$

Implicitly assuming a linear response of the medium to the projectile's electric field, we have the longitudinal and transverse components of the current densities:

$$\begin{aligned}\mathbf{J}_{\text{med},||}(\mathbf{k}, \omega) &= \sigma_{||}(k, \omega)\mathbf{E}_{||}(\mathbf{k}, \omega) \\ &= -i\,\mathbf{k}\sigma_{||}(k, \omega)V(\mathbf{k}, \omega)\end{aligned} \tag{8.238}$$

and

$$\begin{aligned}\mathbf{J}_{\text{med},\perp}(\mathbf{k}, \omega) &= \sigma_{\perp}(k, \omega)\mathbf{E}_{\perp}(\mathbf{k}, \omega) \\ &= i\,\omega\sigma_{\perp}(k, \omega)\mathbf{A}(\mathbf{k}, \omega).\end{aligned} \tag{8.239}$$

The medium's conductivity has been assumed to have longitudinal and transverse components. Inserting (8.238) and (8.239) into (8.232) yields

$$\varepsilon_0\omega\mathbf{k}V(\mathbf{k},\omega) = \mathbf{J}_{\text{med},||}(\mathbf{k},\omega) + \mathbf{J}_{\text{proj},||}(\mathbf{k},\omega). \tag{8.240}$$

Then, by inserting (8.238) into (8.240) and solving for $\mathbf{J}_{\text{proj},||}(\mathbf{k},\omega)$, we achieve the result:

$$\mathbf{J}_{\text{proj},||}(\mathbf{k},\omega) = \varepsilon_0\omega\mathbf{k}\left(1 + i\frac{\sigma_{||}(\mathbf{k},\omega)}{\varepsilon_0\omega}\right)V(\mathbf{k},\omega). \tag{8.241}$$

We now take the scalar product of (8.241) with $\mathbf{k}$ and recall (8.230) to provide the result of

$$\begin{aligned}\mathbf{k} \bullet \mathbf{J}_{\text{proj},||}(\mathbf{k},\omega) &= \varepsilon_0\omega k^2\left(1 + i\frac{\sigma_{||}(\mathbf{k},\omega)}{\varepsilon_0\omega}\right)V(\mathbf{k},\omega) \\ &= \omega\rho_{e,\text{proj}}(\mathbf{k},\omega).\end{aligned} \tag{8.242}$$

We can then define the longitudinal dielectric constant (or relative permittivity) for this result as

$$\varepsilon_{R,||}(\mathbf{k},\omega) \equiv \left(1 + i\frac{\sigma_{||}(\mathbf{k},\omega)}{\varepsilon_0\omega}\right) \tag{8.243}$$

so as to rewrite (8.242) as

$$\varepsilon_0\,\varepsilon_{R,||}(\mathbf{k},\omega)k^2V(\mathbf{k},\omega) = \rho_{e,\text{proj}}(\mathbf{k},\omega). \tag{8.244}$$

By adding $\varepsilon_0\,k^2V(\mathbf{k},\omega)$ to both sides and rearranging, we obtain the result of

$$\varepsilon_0\,k^2V(\mathbf{k},\omega) = \rho_{e,\text{proj}}(\mathbf{k},\omega) - \varepsilon_0\,k^2\big(\varepsilon_{R,||}(\mathbf{k},\omega) - 1\big)V(\mathbf{k},\omega). \tag{8.245}$$

We compare this result with Poisson's equation in $(\mathbf{k},\omega)$-space given by (8.231) and the decomposition of the charge density given by (8.234). We can now identify the charge density due to the medium's response to the charged particle:

$$\rho_{e,\text{med}}(\mathbf{k},\omega) = -\varepsilon_0\,k^2\big(\varepsilon_{R,||}(\mathbf{k},\omega) - 1\big)V(\mathbf{k},\omega). \tag{8.246}$$

Thus, (8.244) and (8.246) define both of the charge density components through which we can evaluate the scalar potential that exists within the medium.

Let us now consider the transverse component through which we can evaluate the vector potential within the medium. We begin with (8.233) and (8.239) to give

$$\left(k^2 - \left(\frac{\omega}{c}\right)^2\right)\mathbf{A}(\mathbf{k},\omega) = \mu_0\mathbf{J}_{\text{proj},\hat{}}(\mathbf{k},\omega) + i\omega\mu_0\sigma_\perp(\mathbf{k},\omega)\mathbf{A}(\mathbf{k},\omega). \tag{8.247}$$

Rearranging this result gives

$$\left(k^2 - \left(\frac{\omega}{c}\right)^2 \varepsilon_{R,\perp}\right) \mathbf{A}(\mathbf{k},\omega) = \mu_0 \mathbf{J}_{\text{proj},\perp}(\mathbf{k},\omega) \tag{8.248}$$

where we have now defined the transverse dielectric function as

$$\varepsilon_{R,\perp}(\mathbf{k},\omega) = \left(1 + i\frac{c^2\mu_0\sigma_\perp(\mathbf{k},\omega)}{\omega}\right). \tag{8.249}$$

8.2.9.4 Evaluation of the Electric Field

We have now evaluated the responses of the medium to the movement of the charged particle via the scalar and vector potentials and the charge and current density. From these we can now calculate the electric field.

Again, we consider a charged particle of charge ze with a constant velocity $\mathbf{v}$. The charge density corresponding to the particle (which is implicitly assumed to have no spatial dimension) is[4]

$$\rho_{e,\text{proj}}(\mathbf{r},t) = ze\,\delta(\mathbf{r} - \mathbf{v}t). \tag{8.250}$$

As our calculation will be performed in $(\mathbf{k},\omega)$-space, the Fourier transform of this density is

$$\rho_{e,\text{proj}}(\mathbf{k},\omega) = \frac{ze}{(2\pi)^{3/2}}\,\delta(\omega - \mathbf{k}\bullet\mathbf{v}). \tag{8.251}$$

Using (8.244), we obtain the scalar potential:

$$V(\mathbf{k},\omega) = \frac{ze}{(2\pi)^{3/2}\varepsilon_0\,\varepsilon_{R,||}(\mathbf{k},\omega)k^2}\,\delta(\omega - \mathbf{k}\bullet\mathbf{v}). \tag{8.252}$$

We next consider the current density due to the projectile which is

$$\mathbf{J}_{\text{proj}}(\mathbf{r},t) = ze\,\mathbf{v}\delta(\mathbf{r} - \mathbf{v}t). \tag{8.253}$$

In $(\mathbf{k},\omega)$-space, this is

$$\mathbf{J}_{\text{proj}}(\mathbf{k},\omega) = \frac{ze}{(2\pi)^{3/2}}\mathbf{v}\,\delta(\omega - \mathbf{k}\bullet\mathbf{v}). \tag{8.254}$$

---

[4] For the astute reader applying dimensional analysis mentally throughout these derivations, the SI units of this result in $(\mathbf{r},t)$ space are indeed C/m^3 as the units of the Dirac δ-function are the reciprocal of its argument.

As we shall need the transverse component for the vector potential, we can extract it through (8.254) to lead to

$$\mathbf{J}_{\text{proj},\perp}(\mathbf{k},\omega) = \frac{ze}{(2\pi)^{3/2}}\left(\mathbf{v} - \frac{\mathbf{k}}{k^2}\mathbf{k}\bullet\mathbf{v}\right)\delta(\omega - \mathbf{k}\bullet\mathbf{v}). \tag{8.255}$$

By inserting this result into (8.248), we obtain the vector potential in Fourier space:

$$\begin{aligned}\mathbf{A}(\mathbf{k},\omega) &= \frac{\mu_0}{\left(k^2 - \left(\frac{\omega}{c}\right)^2 \varepsilon_{R,\perp}(\mathbf{k},\omega)\right)}\mathbf{J}_{\text{proj},\perp}(\mathbf{k},\omega)\\ &= \frac{\mu_0\, ze}{(2\pi)^{3/2}\left(k^2 - \left(\frac{\omega}{c}\right)^2 \varepsilon_{R,\perp}(\mathbf{k},\omega)\right)}\left(\mathbf{v} - \frac{\mathbf{k}}{k^2}\mathbf{k}\bullet\mathbf{v}\right)\delta(\omega - \mathbf{k}\bullet\mathbf{v}).\end{aligned} \tag{8.256}$$

These results now allow us to evaluate the electric field in $(\mathbf{k},\omega)$-space, $\mathbf{E}(\mathbf{k},\omega)$, through the Fourier transform of (8.213):

$$\begin{aligned}\mathbf{E}(\mathbf{k},\omega) &= -i(\mathbf{k}V(\mathbf{k},\omega) - \omega\mathbf{A}(\mathbf{k},\omega)) = -i\frac{ze}{(2\pi)^{\frac{3}{2}}k^2}\,\delta(\omega - \mathbf{k}\bullet\mathbf{v})\\ &\times\left(\frac{\mathbf{k}}{\varepsilon_0\,\varepsilon_{R,||}(\mathbf{k},\omega)} - \frac{\omega\mu_0}{\left(1 - \left(\frac{\omega}{kc}\right)^2 \varepsilon_{R,\perp}(\mathbf{k},\omega)\right)}\left(\mathbf{v} - \frac{\mathbf{k}}{k^2}\mathbf{k}\bullet\mathbf{v}\right)\right).\end{aligned} \tag{8.257}$$

The electric field is thus given by the inverse Fourier transform of (8.257):

$$\begin{aligned}\mathbf{E}(\mathbf{r},t) = &-i\frac{ze}{(2\pi)^3}\int\frac{\mathrm{d}^3\mathbf{k}}{k^2}\\ &\int \mathrm{d}\omega\left(\frac{\mathbf{k}}{\varepsilon_0\,\varepsilon_{R,||}(\mathbf{k},\omega)} - \frac{\omega\mu_0}{\left(1 - \left(\frac{\omega}{kc}\right)^2 \varepsilon_{R,\perp}(\mathbf{k},\omega)\right)}\left(\mathbf{v} - \frac{\mathbf{k}}{k^2}\mathbf{k}\bullet\mathbf{v}\right)\right)\\ &\times e^{-i(\mathbf{k}\bullet\mathbf{r} - \omega t)}\delta(\omega - \mathbf{k}\bullet\mathbf{v}).\end{aligned} \tag{8.258}$$

We note that $\mathbf{v} = \mathbf{r}/t$ and that $\mu_0 = 1/\varepsilon_0 c^2$. The integration over ω is trivial owing to the presence of the Dirac δ-function:

$$\begin{aligned}\mathbf{E}(\mathbf{r},t) = &-i\frac{ze}{(2\pi)^3\varepsilon_0}\int\frac{\mathrm{d}^3\mathbf{k}}{k^2}\\ &\left(\frac{\mathbf{k}}{\varepsilon_{R,||}(\mathbf{k},\mathbf{k}\bullet\mathbf{v})} - \frac{\mathbf{k}\bullet\mathbf{v}}{c^2\left(1 - \left(\frac{\mathbf{k}\bullet\mathbf{v}}{kc}\right)^2 \varepsilon_{R,\perp}(\mathbf{k},\mathbf{k}\bullet\mathbf{v})\right)}\left(\mathbf{v} - \frac{\mathbf{k}}{k^2}\mathbf{k}\bullet\mathbf{v}\right)\right).\end{aligned} \tag{8.259}$$

By using $\mathbf{k} = k\hat{\mathbf{k}}$ and noting that $\mathbf{k} \bullet \mathbf{r} = kr\cos\theta$ (as we are evaluating the integral in spherical co-ordinates), (8.259) reduces to

$$\mathbf{E}(\mathbf{r},t) = -i\frac{\alpha\hbar cz}{2\pi^2 e}\int \mathrm{d}k\, k \int_{-1}^{1} \mathrm{d}(\cos\theta)\left(\frac{\hat{\mathbf{k}}}{\varepsilon_{R,||}(\mathbf{k},\mathbf{k}\bullet\mathbf{v})} - \beta^2\frac{(\hat{\mathbf{r}} - \hat{\mathbf{k}}\cos\theta)\cos\theta}{1-\beta^2\cos^2\theta\,\varepsilon_{R,\perp}(\mathbf{k},\mathbf{k}\bullet\mathbf{v})}\right). \tag{8.260}$$

We recall the Dirac δ-function of (8.258) and recognise that we can replace $\mathbf{k}\bullet\mathbf{v} = kv\cos\theta = k\beta c\cos\theta = \omega$. We thus rewrite the integration over the polar angle and the corresponding integration variable as

$$\mathrm{d}(\cos\theta) = \frac{\mathrm{d}\omega}{k\beta c} \tag{8.261}$$

and

$$\cos\theta = \frac{\omega}{k\beta c}. \tag{8.262}$$

We can then rewrite the expression of the electric field given by (8.260) as

$$\mathbf{E}(\mathbf{r},t) = -i\frac{\alpha\hbar cz}{2\pi^2 e\beta c}\int \mathrm{d}k \int_{-k\beta c}^{k\beta c} \mathrm{d}\omega\left(\frac{\hat{\mathbf{k}}}{\varepsilon_{R,||}(\mathbf{k},\omega)} - \beta^2\frac{\left(\hat{\mathbf{r}} - \hat{\mathbf{k}}\left(\frac{\omega}{k\beta c}\right)\right)\left(\frac{\omega}{k\beta c}\right)}{1-\left(\frac{\omega}{kc}\right)^2\varepsilon_{R,\perp}(\mathbf{k},\omega)}\right). \tag{8.263}$$

We also recall, from the Dirac δ-function seen earlier, that $\omega = \mathbf{k}\bullet\mathbf{v} = (kr/t)\cos\theta$.

8.2.9.5 Evaluation of the Stopping Power

Having calculated an expression for the electric field, we calculate for the energy loss per unit distance (the stopping power) which is equivalent to the force acting upon the projectile:

$$\begin{aligned}\mathbf{F}(\mathbf{r},t) &= ze\,\mathbf{E}(\mathbf{r},t)\\ &= -i\frac{\alpha\hbar cz^2}{2\pi^2\beta c}\int \mathrm{d}k \int_{-k\beta c}^{k\beta c} \mathrm{d}\omega\left(\frac{\hat{\mathbf{k}}}{\varepsilon_{R,||}(\mathbf{k},\omega)} - \beta^2\frac{\left(\hat{\mathbf{r}} - \hat{\mathbf{k}}\left(\frac{\omega}{k\beta c}\right)\right)\left(\frac{\omega}{k\beta c}\right)}{1-\left(\frac{\omega}{kc}\right)^2\varepsilon_{R,\perp}(\mathbf{k},\omega)}\right)\end{aligned} \tag{8.264}$$

and the linear collision stopping power is[5]

$$\frac{\mathrm{d}E}{\mathrm{d}x} = \frac{\mathbf{v} \bullet \mathbf{F}(\mathbf{r}, t)}{v}. \tag{8.265}$$

As $\mathbf{v} = r/t\,\hat{\mathbf{r}}$ and $\hat{\mathbf{r}} \bullet \hat{\mathbf{k}} = \omega/k\beta c$, and by writing $r/t = \beta c$, we have

$$\frac{\mathrm{d}E}{\mathrm{d}x} = -i\frac{\alpha\hbar c z^2}{2\pi^2(\beta c)^2}\int\frac{\mathrm{d}k}{k}\int\limits_{-k\beta c}^{k\beta c}\mathrm{d}\omega\,\omega\left(\frac{1}{\varepsilon_{R,||}(\mathbf{k},\omega)} - \beta^2\frac{1-\left(\dfrac{\omega}{k\beta c}\right)^2}{1-\left(\dfrac{\omega}{kc}\right)^2\varepsilon_{R,\perp}(\mathbf{k},\omega)}\right). \tag{8.266}$$

Let us now reflect upon this result. The first (longitudinal) term of the integrand is the consequence of the scalar potential, whereas the second (transverse) term is due to the vector potential. Most importantly, the contribution of the transverse term becomes significant only at relativistic velocities where $\beta \to 1$.

Excluding the relativistic component of (8.266), including the expression for $\varepsilon_{R,||}(\mathbf{k},\omega)$ given by (8.243), we have for small β an expression for the longitudinal component of the linear soft-collision stopping power:

$$\left.\frac{\mathrm{d}E}{\mathrm{d}x}\right|_{S,||} = \frac{\alpha\hbar c z^2}{2\pi^2(\beta c)^2}\int\frac{\mathrm{d}k}{k}\int\limits_{-k\beta c}^{k\beta c}\mathrm{d}\omega\,\omega\frac{-i+\sigma_{||}\dfrac{(k,\omega)}{\varepsilon_0\omega}}{1+\left(\sigma_{||}\dfrac{(k,\omega)}{\varepsilon_0\omega}\right)^2}. \tag{8.267}$$

We recognise that the stopping power is a real quantity, so (8.267) becomes

$$\left.\frac{\mathrm{d}E}{\mathrm{d}x}\right|_{S,||} = \frac{\alpha\hbar}{2\pi^2\varepsilon_0 c}\left(\frac{z}{\beta}\right)^2\int\frac{\mathrm{d}k}{k}\int\limits_{-k\beta c}^{k\beta c}\mathrm{d}\omega\frac{\sigma_{||}(k,\omega)}{1+\left(\sigma_{||}\dfrac{(k,\omega)}{\varepsilon_0\omega}\right)^2}. \tag{8.268}$$

At this point, we will depart from our consideration of the energy loss of a moving charged particle in a condensed medium and, again, return to the concept of the medium being a cold, monatomic and dilute gas. We will return to our exploration of above when we consider the relativistic features of the Bethe quantum theory of energy loss in the following section and the density effect in Chap. 12.

8.3 Quantum Theory

8.3.1 Introductory Comments

In this section, we follow Bethe's approach in deriving a quantum theory of energy loss which is based upon the Born expansion. As has been studied thoroughly, the

[5] This derivation emphasises the fact that the unit of the linear stopping power is that of force.

foundation of the classical derivation is the impact parameter. This spatial description is not applicable to the quantum derivation which is, instead, based upon the momentum transfer q to the atom. The approaches could be considered, crudely, to be related reciprocally through the Heisenberg uncertainty principle via $q \approx \hbar c/b$. Hence, in the quantum theory, a soft collision is considered to be one in which the energy loss from the projectile is less than a prespecified value. This may suggest arbitrariness that will influence the result. But we should recall a similar approach in defining an arbitrary impact parameter to differentiate between close and distant collisions in the classical theory. That arbitrariness was limited by the inequality of (8.172), although this separation between regions of impact parameter does not appear in the final classical result. A similar result will occur here. However, there is some physics justification in the selection of the threshold energy value, and as we are ultimately interested in the total collision stopping power, we will see that when we combine the Bethe soft collision result with the hard-collision results of the following chapter, this threshold energy disappears from the resulting formula.

8.3.2 Collision Kinematics

As before, we define the kinematics of the collision between the projectile and an atomic electron as shown in Fig. 8.1. However, as the quantum theory is based upon the magnitude of the energy transfer rather than the impact parameter, we further consider the momentum transfer resulting from the collision using Fig. 8.8. The energy transfer is calculated from that quantity as to be shown. A projectile with momentum $\boldsymbol{p}$ and kinetic energy E collides with an atomic electron and scatters through the angle θ with momentum $\boldsymbol{p}'$ and kinetic energy E'. The momentum transfer, now allowing for energy loss, (i.e. the pre- and post-collision momenta are not equal, $p \neq p'$), is

$$q = \sqrt{p^2 + p'^2 - 2pp' \cos\theta}. \tag{8.269}$$

In this derivation, the target atomic electron will be considered as being unbound and initially at rest, which is a reasonable assumption for cases of sufficiently high incident kinetic energies such that the orbital velocity of the atomic electron can be ignored. However, if the projectile speed is comparable to the orbital speeds of the atomic electrons, this assumption becomes untenable and corrections must be made.[6] The kinetic energy of the recoil electron is equal to the energy transfer as the atomic electron's binding energy is neglected,

$$Q = E - E', \tag{8.270}$$

[6] These 'shell' correction factors are discussed in Chap. 12.

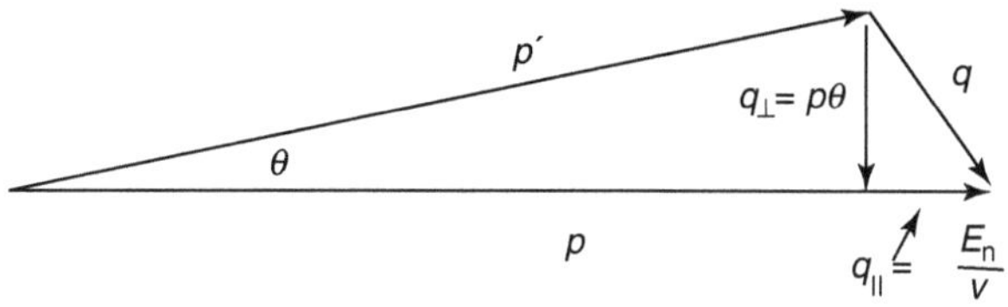

Fig. 8.8 Momentum transfer in the small-angle approximation

which is related to the magnitude of the momentum transfer through the relativistic relationship,

$$(Q + m_e)^2 = q^2 + m_e^2. \tag{8.271}$$

Upon rearrangement of (8.271), we have

$$Q\left(1 + \frac{Q}{2m_e}\right) = \frac{q^2}{2m_e} \tag{8.272}$$

which, for the condition of small energy transfer (i.e. $Q << 2m_e$), reduces to the familiar nonrelativistic form:

$$Q = \frac{q^2}{2m_e}. \tag{8.273}$$

As considered in our discussion of the semi-classical theory, the atomic electron will be excited from its ground state $|0\rangle$ to the state $|f\rangle$ as a result of the collision. The energy transfer Q need not equal the energy of the excited state, E_f, as the recoil electron cannot be considered in isolation as a consequence of it being part of an ensemble of atomic electrons. As a result, the energy transferred to it can be shared among the other atomic electrons.[7]

It is straightforward to calculate the component of the momentum transfer **q** that is parallel to the incident momentum **p** from the excitation energy,

$$\begin{aligned} E_f &\approx \frac{p^2}{2m_e} - \frac{|\mathbf{p} - \mathbf{q}|^2}{2m_e} \\ &\approx \frac{2\mathbf{p} \bullet \mathbf{q} - q^2}{2m_e} \\ &\approx \frac{\mathbf{q} \bullet \mathbf{p}}{m_e}. \end{aligned} \tag{8.274}$$

[7] The energy of the ground state is considered to be equal to zero.

Terms of the order of q^2 are neglected as we are considering soft collisions. This leads to the longitudinal component of the momentum transfer:

$$q_{||} \equiv \frac{E_f}{\beta}. \tag{8.275}$$

As this is a soft collision and q is small, the small-angle approximation can be used and the component of $\mathbf{q}$ perpendicular to $\mathbf{p}$ is

$$q_{\perp} = p\theta. \tag{8.276}$$

The squared magnitude of the momentum transfer is

$$\begin{aligned} q^2 &= q_{||}^2 + q_{\perp}^2 \\ &= \left(\frac{E_f}{\beta}\right)^2 + (p\theta)^2 \end{aligned} \tag{8.277}$$

and the energy transfer to the electron is, consequently,

$$\begin{aligned} Q &= \frac{q^2}{2m_e} \\ &= \frac{\left((E_f/\beta)^2 + (p\theta)^2\right)}{2m_e}. \end{aligned} \tag{8.278}$$

8.3.3 *Bethe Nonrelativistic Soft-Collision Cross Section*

8.3.3.1 Introductory Comments

We now begin our derivation of the quantum theory of soft collisions in earnest. We separate Bethe's derivation into nonrelativistic and relativistic parts. This is necessary as the overall interaction between the projectile and the atomic electrons is handled in the Coulomb gauge in which there are two types of electromagnetic interactions.

The first is the static unretarded potential between the projectile and electrons with a direct coupling between the initial and final states. From our discussion of quantum perturbation theory in Chap. 2, this allows the first-order use of this theory, and the transition rate is given by Fermi's Golden Rule. The foundations of the mechanics of this derivation were laid down when we evaluated the differential cross section of the Coulomb interaction between a charged particle and an atom in an elastic scatter. Here, we repeat the evaluation, but for the inelastic case of $\mathbf{p}' \neq \mathbf{p}$.

The second type of interaction between the projectile and the atomic electrons becomes significant in the relativistic regime, and this is through the emission and absorption of virtual photons. Because this is the result of the interaction between the particle currents with the quantised transverse vector potential, this is the transverse excitation and will be discussed in the following section. We have already derived its foundations in the previous section of this chapter.

8.3.3.2 Nonrelativistic Soft-Collision Cross Section

The unretarded potential leads to the longitudinal calculation, as demonstrated earlier. From (8.266), this is seen to be the 'nonrelativistic' component. This is a misnomer as the component is always present: it is the transverse component that becomes appreciable only as $\beta \to 1$. Using Fermi's Golden Rule, the transition rate from the initial state to the final state is, for longitudinal excitation, given by

$$\lambda_{f0,\|} = \frac{2\pi}{\hbar} |\langle \mathbf{p}', f | U | \mathbf{p}, 0 \rangle|^2 \, \rho_f. \tag{8.279}$$

The matrix element includes all initial and final states of the projectile and atomic system. Again, assuming a plane-wave representation of the projectile, the initial state is the same as that for the elastic scatter calculation:

$$\langle \mathbf{r} \mid \mathbf{p}, 0 \rangle = \frac{1}{\sqrt{L^3}} e^{i\mathbf{p} \bullet \mathbf{r}/\hbar c} \, |0\rangle.$$

The initial atomic state is in our usual form:

$$|0\rangle = \psi_0(\mathbf{r}_1 \ldots \mathbf{r}_Z). \tag{8.280}$$

However, the final state differs now from the elastic calculation derived previously due to the inelasticity of the interaction in which the projectile's momentum and the atomic state following the interaction both differ from those prior to the interaction,

$$\langle \mathbf{r} \mid \mathbf{p}', f \rangle = \frac{1}{\sqrt{L^3}} e^{i\mathbf{p}' \bullet \mathbf{r}/\hbar c} \, |f\rangle, \tag{8.281}$$

and the atomic wavefunction in its final state is

$$|f\rangle = \psi_f(\mathbf{r}_1 \ldots \mathbf{r}_Z). \tag{8.282}$$

Instead of defining the phase-space factor on the basis of the elastic scatter form and which was based upon momentum transfer, we rewrite it in terms of the energy transfer. Beginning with

$$\rho_f = \frac{1}{4\pi^2\beta'}\left(\frac{L}{\hbar c}\right)^3 p'^2\,\mathrm{d}p'\;\mathrm{d}(\cos\theta). \tag{8.283}$$

We use, from our discussion of the collision kinematics,

$$\mathrm{d}(\cos\theta) = \left(\frac{q}{pp'}\right)\mathrm{d}q \tag{8.284}$$

and

$$q = \sqrt{Q(Q+2m_e)} \tag{8.285}$$

to lead us to

$$q\,\mathrm{d}q = m_e(1+Q/m_e)\,\mathrm{d}Q. \tag{8.286}$$

From this, we can write the phase-space factor as

$$\rho_f = \frac{m_e}{4\pi^2\beta'}\left(\frac{L}{\hbar c}\right)^3\left(\frac{p'}{p}\right)\left(1+\frac{Q}{m_e}\right)\mathrm{d}Q'. \tag{8.287}$$

We can simplify this result by recognising that this calculation is for a soft collision for which the energy transfer is much less than the projectile's kinetic energy (i.e. $Q << E$), and hence, we can use the quasi-elastic approximations of $p'/p \approx \beta'/\beta \approx 1$ to give

$$\rho_f \approx \frac{m_e}{4\pi^2\beta}\left(\frac{L}{\hbar c}\right)^3\left(1+\frac{Q}{m_e}\right)\mathrm{d}Q. \tag{8.288}$$

This phase-space term will be common to both longitudinal and transverse excitations.

We now consider the calculation of the matrix element due to the unretarded Coulomb interactions between the projectile and the atomic nucleus and atomic electrons. We can demonstrate that the matrix element due to the Coulomb potential between the projectile and the nucleus is equal to zero, unlike the case for elastic scatter as demonstrated by (5.134). The potential between the projectile and a point nucleus is extracted from the total atomic potential given by

$$U_{\text{nuc}} = \frac{\alpha\hbar c\, zZ}{R}. \tag{8.289}$$

The corresponding matrix element for the nuclear-projectile Coulomb interaction is described by the usual expansion:

$$\begin{aligned}\langle \mathbf{p}',f|U_{\text{nuc}}|\mathbf{p},0\rangle &= \int \mathrm{d}^3\mathbf{r}' \int \mathrm{d}^3\mathbf{r}\, \langle \mathbf{p}',f \mid \mathbf{r}\rangle\langle \mathbf{r}|U_{\text{nuc}}|\mathbf{r}'\rangle\langle \mathbf{r}' \mid \mathbf{p},0\rangle \\ &= \frac{1}{L^3}\int \mathrm{d}^3\mathbf{R}\, e^{i\frac{\mathbf{q}\bullet\mathbf{R}}{\hbar c}} \prod_{j=1}^{Z} \mathrm{d}^3\mathbf{r}_j\, \psi_f^*(\mathbf{r}_1,\ldots\mathbf{r}_Z)U(\mathbf{R})\psi_0(\mathbf{r}_1,\ldots\mathbf{r}_Z) \\ &= -zZ\frac{\alpha\hbar c}{L^3}\int \mathrm{d}^3\mathbf{R}\,\frac{e^{i\frac{\mathbf{q}\bullet\mathbf{R}}{\hbar c}}}{R}\prod_{j=1}^{Z} d^3\mathbf{r}_j\,\psi_f^*(\mathbf{r}_1,\ldots\mathbf{r}_Z)\psi_0(\mathbf{r}_1,\ldots\mathbf{r}_Z) \\ &= 0. \end{aligned} \tag{8.290}$$

The fact that the result is equal to zero is due to the orthonormality of the initial and final states and the fact that there is no functional dependence within the Coulomb potential with the nucleus that is dependent upon the spatial co-ordinates of the atomic electrons.

Having now dismissed the need to consider the role of the nucleus in our calculation of the inelastic atomic scatter, we next proceed to evaluate the matrix element due to the unretarded Coulomb interactions between the projectile and the atomic electrons. This is calculated in the same fashion as in the evaluation of the elastic scatter cross section beginning with extracting the Coulomb potential between the projectile and the atomic electrons:

$$U_{\text{elec}}(\mathbf{r}) = -(z\,\alpha\hbar c)\sum_{j=1}^{Z}\frac{1}{\left|\mathbf{R}-\mathbf{r}_j\right|}. \tag{8.291}$$

The matrix element will be

$$\langle \mathbf{p}',f|U_{\text{elec}}|\mathbf{p},0\rangle = \int \mathrm{d}^3\mathbf{r}\,\mathrm{d}^3\mathbf{r}'\,\langle \mathbf{p}',f|\,\mathbf{r}\rangle\,\langle \mathbf{r}|\,U_{\text{elec}}\,|\mathbf{r}'\rangle\,\langle \mathbf{r}'|\,\mathbf{p},0\rangle. \tag{8.292}$$

Solving this matrix element is identical to that performed for the elastic scatter:

$$\begin{aligned}\langle \mathbf{p}',f|U_{\text{elec}}|\mathbf{p},0\rangle &= -4\pi\frac{z\,\alpha\hbar c}{L^3}\left(\frac{\hbar c}{q}\right)^2\langle f|\sum_{j=1}^{Z}e^{i\frac{\mathbf{q}\bullet\mathbf{r}_j}{\hbar c}}|0\rangle \\ &= -4\pi\frac{z\,\alpha\hbar c}{L^3}\left(\frac{\hbar c}{q}\right)^2 F_{f0}(\mathbf{q},Z). \end{aligned} \tag{8.293}$$

The transition rate can be considered to be independent of azimuthal angle if the target atom is in the s-state (i.e. spherically symmetric) or if the target atoms in the medium are randomly oriented (as is the case in medical irradiation). This permits one to replace the vector momentum transfer $\mathbf{q}$ in the argument of the inelastic form factor with its scalar value, q, to give $F_{f0}(q, Z)$. The squared amplitude of the matrix element is, in terms of the energy transfer,

$$
\begin{aligned}
|\langle \mathbf{p}', n|U_{\text{elec}}|\mathbf{p}, 0\rangle|^2 &= 16\pi^2 \frac{(z\,\alpha\hbar c)^2}{L^6} \left(\frac{\hbar c}{q}\right)^4 |F_{f0}(q, Z)|^2 \\
&= 4\pi^2 \frac{(z\,\alpha\hbar c)^2}{L^6} \frac{(\hbar c)^4}{m_e^2 Q^2 \left(1 + \dfrac{Q}{2m_e}\right)^2} |F_{f0}(q, Z)|^2 .
\end{aligned}
\tag{8.294}
$$

The inelastic Coulomb scatter transition rate for longitudinal excitation is, thus,

$$
\begin{aligned}
\lambda_{f0,||} &= \frac{2\pi}{\hbar} |\langle \mathbf{p}', n|U|\mathbf{p}, 0\rangle|^2 \rho_f \\
&= 2\pi \frac{(z\,\alpha\hbar c)^2 c}{\beta L^3 m_e} \frac{|F_{f0}(q, Z)|^2}{Q^2 \left(1 + \dfrac{Q}{2m_e}\right)^2} \left(1 + \frac{Q}{m_e}\right) \mathrm{d}Q .
\end{aligned}
\tag{8.295}
$$

The differential cross section is the transition rate normalised to the incident particle flux, v/V, or

$$
\begin{aligned}
\mathrm{d}\sigma_{||} &= \frac{L^3}{\beta c} \lambda_{f0,||} \\
&= 2\pi \frac{z^2 m_e r_0^2}{\beta^2} \frac{|F_{f0}(q, Z)|^2}{Q^2 \left(1 + \dfrac{Q}{2m_e}\right)^2} \left(1 + \frac{Q}{m_e}\right) \mathrm{d}Q .
\end{aligned}
\tag{8.296}
$$

The inelastic form factor $F_{f0}(q, Z)$ is related to the generalised (dipole) oscillator strength, GOS (which has already been seen in the derivation of the Bohr energy loss), which is a generalisation of the oscillator strength (see, e.g. Fernández-Vaera 1998) given, in the nomenclature used here, by

$$
f_{f0}(q, Z) = \frac{E_f}{Q} \left|F_{f0}(q, Z)\right|^2 . \tag{8.297}
$$

Analytical representations for the GOS are available for atomic hydrogen, and the free electron gas for which the initial and final states are calculable analytically. However, in the more practical cases of heavier atoms and molecules representative of tissues, these wavefunctions are calculated numerically. This discussion will be limited to the simple details of the GOS required for obtaining an expression for the stopping power. The inelastic form factor can be related to the GOS per unit energy transfer:

$$
\left.\frac{df_{f0}(q, Z)}{\mathrm{d}E_f}\right|_{Q=0} = Z f_{f0}(q, Z). \tag{8.298}
$$

The energy-weighted sum of the oscillator strengths equals the total number of electrons in the atom which, in integral form, is

$$\int \mathrm{d}E_f \frac{\mathrm{d}f_{f0}(q,Z)}{\mathrm{d}E_f} = Z. \tag{8.299}$$

The GOS can also be used to describe the mean ionisation energy of the atomic system as the first energy moment of the oscillator strength distribution:

$$\begin{aligned}\ln \bar{I} &= \frac{\int \mathrm{d}E_f \dfrac{\mathrm{d}f_{f0}(q,Z)}{\mathrm{d}E_f} \ln E_f}{\int \mathrm{d}E_f \dfrac{\mathrm{d}f_{f0}(q,Z)}{\mathrm{d}E_f}} \\ &= \frac{1}{Z} \int \mathrm{d}E_f \frac{\mathrm{d}f_{f0}(q,Z)}{\mathrm{d}E_f} \ln E_f. \end{aligned} \tag{8.300}$$

The GOS per unit energy transfer has a useful interpretation at low momentum transfers. We can write the GOS per unit excitation energy in the Dirac *bra–ket* form:

$$\begin{aligned}\frac{\mathrm{d}f_{f0}(q,Z)}{\mathrm{d}E_f} &= \frac{1}{Q} \left|F_{f0}(q,Z)\right|^2 \\ &= \frac{1}{Q} \left| \langle f | \sum_{j=1}^{Z} e^{i\frac{\mathbf{q}\bullet\mathbf{r}_j}{\hbar c}} |0\rangle \right|^2 \\ &= \frac{2m_e}{q^2} \left| \langle f | \sum_{j=1}^{Z} e^{i\frac{\mathbf{q}\bullet\mathbf{r}_j}{\hbar c}} |0\rangle \right|^2 \end{aligned} \tag{8.301}$$

where, for low momentum transfer, the nonrelativistic relationship between momentum and energy transfer has been used. Following our earlier exploration of the dipole approximation, we now expand the exponential to first order for small q:

$$\begin{aligned}\frac{\mathrm{d}f_{f0}(q,Z)}{\mathrm{d}E_f} &\approx \frac{2m_e}{q^2} \left| Z\langle f|0\rangle + \frac{i}{\hbar c} \langle f| \sum_{j=1}^{Z} \mathbf{q}\bullet\mathbf{r}_j |0\rangle \right|^2 \\ &\approx -\frac{2m_e}{3(\hbar c)^2} \left| \langle f| \sum_{j=1}^{Z} r_j |0\rangle \right|^2 \quad \text{small } q \end{aligned} \tag{8.302}$$

where the orthogonality relationship has been used. Note that $\mathrm{d}f_{f0}(q,Z)/\mathrm{d}E_f \rightarrow \mathrm{d}f_{f0}(Z)/\mathrm{d}E_f$ for small q, where $\mathrm{d}f_{f0}(Z)/\mathrm{d}E_f$ is the optical oscillator strength per unit excitation energy and which is proportional to the square of the dipole-matrix element.

Having introduced the GOS and its properties at low q, the expression of the differential cross section for longitudinal excitation is then simplified:

$$\begin{aligned} d\sigma_{f,\|} &= 2\pi \frac{z^2 m_e r_e^2}{\beta^2} \frac{|F_{f0}(q,Z)|^2}{Q^2\left(1+\frac{Q}{2m_e}\right)^2}\left(1+\frac{Q}{m_e}\right) dQ \\ &= 2\pi \frac{z^2 m_e r_e^2}{\beta^2} \frac{\left(1+\frac{Q}{m_e}\right)}{Q^2\left(1+\frac{Q}{2m_e}\right)^2} \frac{df_{f0}(q,Z)}{dE_f} dQ \\ &= \frac{C}{N_A}\left(\frac{z}{\beta}\right)^2 \frac{\left(1+\frac{Q}{m_e}\right)}{Q^2\left(1+\frac{Q}{2m_e}\right)^2} \frac{df_{f0}(q,Z)}{dE_f} dQ. \end{aligned} \tag{8.303}$$

For small energy transfers (i.e. $Q << m_e$), the cross section simplifies to

$$d\sigma_{f,\|} = \frac{C}{N_A}\left(\frac{z}{\beta}\right)^2 \frac{dQ}{Q} \frac{df_{f0}(q,Z)}{dE_f}. \tag{8.304}$$

For calculational convenience, the kinetic variable is temporarily changed from the energy transfer to the momentum transfer:

$$d\sigma_{f,\|} = \frac{2C}{N_A}\left(\frac{z}{\beta}\right)^2 \frac{dq}{q} \frac{df_{f0}(q,Z)}{dE_f}. \tag{8.305}$$

Integrating over the momentum transfer yields the total cross section:

$$\sigma_{f,\|} = \frac{2C}{N_A}\left(\frac{z}{\beta}\right)^2 \int_{q_{\min}}^{q_{\max}} dq \, \frac{1}{q} \frac{df_{f0}(q,Z)}{dE_f}. \tag{8.306}$$

As the magnitude momentum transfer q is small for soft collisions, then $df_{f0}(q,Z)/dE_f$ can be removed from the integrand to obtain an expression for the total cross section:

$$\begin{aligned} \sigma_{f,\|} &= \frac{2C}{N_A}\left(\frac{z}{\beta}\right)^2 \frac{df_{f0}(q,Z)}{dE_f} \int_{q_{\min}}^{q_{\max}} \frac{dq}{q} \\ &= \frac{2C}{N_A}\left(\frac{z}{\beta}\right)^2 \frac{df_{f0}(q,Z)}{dE_f} \ln\left(\frac{q_{\max}}{q_{\min}}\right). \end{aligned} \tag{8.307}$$

For later convenience when expressions for soft- and hard-collision stopping powers are merged to give the complete collision stopping power, instead of using the momentum transfer, the use of the energy transfer will be more convenient as a means of defining soft and hard collisions. The collision is said to be 'soft' if $Q < Q_C$ and to be 'hard' if $Q > Q_C$, where Q_C is the transition energy transfer. The exact value of Q_C is insignificant as this quantity will eventually cancel out when the expressions for the soft- and hard-collision stopping powers are summed, as to be shown in Chap. 9. Even so, realistic limits should be applied to Q_C in order to ensure that the necessary approximations used in the derivations remain valid. Q_C must exceed atomic binding energies, but it must also not be sufficiently great that the projectile's de Broglie wavelength becomes comparable to nuclear dimensions. A value of between 10 and 100 keV for Q_C would allow both conditions to be simultaneously met (Uehling 1954).

The limits of the momentum transfers which define a soft collision are now calculated. The lower limit, $q_{\min}$, is given by the longitudinal momentum transfer, $q_{||} = E_f/\beta$, and $q_{\max}$ will be set by the predefined energy transfer separation between soft and hard collisions:

$$q_{\max} = \sqrt{2m_e Q_C}. \tag{8.308}$$

Applying these limits to the ratio of the momentum transfer gives the cross section for longitudinal excitation:

$$\begin{aligned} \sigma_{f,||} &= \frac{2C}{N_A}\left(\frac{z}{\beta}\right)^2 \frac{\mathrm{d}f_{f0}(q,Z)}{\mathrm{d}E_f} \ln\left(\frac{\sqrt{2m_e Q_C}}{E_f}\beta\right) \\ &= \frac{C}{N_A}\left(\frac{z}{\beta}\right)^2 \frac{\mathrm{d}f_{f0}(q,Z)}{\mathrm{d}E_f} \ln\left(\frac{2m_e Q_C}{E_f^2}\beta^2\right) \end{aligned} \tag{8.309}$$

8.3.4 Bethe Nonrelativistic Mass Soft-Collision Stopping Power

The mean energy transfer per unit particle fluence due to the longitudinal component is given by

$$\begin{aligned} \bar{Q}_{||} &= \int \mathrm{d}E_f\, \sigma_{f,||} \\ &= \frac{C}{N_A}\left(\frac{z}{\beta}\right)^2 \int \mathrm{d}E_f \frac{\mathrm{d}f_{f0}(q,Z)}{\mathrm{d}E_f} \ln\left(\frac{2m_e Q_C}{E_f^2}\beta^2\right). \end{aligned} \tag{8.310}$$

The integral is solved by splitting the logarithm under the assumption that all energy variables have the same unit:

$$\int \mathrm{d}E_f \frac{\mathrm{d}f_{f0}(q,Z)}{\mathrm{d}E_f} \ln\left(\frac{2m_e Q_C}{E_f^2}\beta^2\right) = \int \mathrm{d}E_f \frac{\mathrm{d}f_{f0}(q,Z)}{\mathrm{d}E_f} \left(-2 \ln E_f + \ln\left(2m_e Q_C \beta^2\right)\right)$$
$$= -2\int \mathrm{d}E_f \frac{\mathrm{d}f_{f0}(q,Z)}{\mathrm{d}E_f} \ln E_f + \ln\left(2m_e Q_C \beta^2\right) \int \mathrm{d}E_f \frac{\mathrm{d}f_{f0}(q,Z)}{\mathrm{d}E_f}. \tag{8.311}$$

We now consider the definition of the GOS, allowing us to write the two terms of (8.311) as

$$-2\int \mathrm{d}E_f \frac{\mathrm{d}f_{f0}(q,Z)}{\mathrm{d}E_f} \ln E_f = Z \ln \frac{1}{\bar{I}^2} \tag{8.312}$$

and

$$\ln\left(2m_e Q_C \beta^2\right) \int \mathrm{d}E_f \frac{\mathrm{d}f_{f0}(q,Z)}{\mathrm{d}E_f} = Z \ln\left(2m_e Q_C \beta^2\right). \tag{8.313}$$

Putting these both into (8.311) gives

$$\int \mathrm{d}E_f \frac{\mathrm{d}f_{f0}(q,Z)}{\mathrm{d}E_f} \ln\left(\frac{2m_e Q_C}{E_f^2}\beta^2\right) = Z \ln\left(\frac{2m_e Q_C \beta^2}{\bar{I}^2}\right). \tag{8.314}$$

Returning this to (8.310) gives

$$\bar{Q}_{||} = \frac{CZ}{N_A}\left(\frac{z}{\beta}\right)^2 \ln\left(\frac{2m_e Q_C \beta^2}{\bar{I}^2}\right). \tag{8.315}$$

Then, the Bethe soft-collision stopping power for nonrelativistic projectiles (i.e. longitudinal excitations only) is

$$\left.\frac{\mathrm{d}E}{\rho\,\mathrm{d}x}\right|_{S,\mathrm{Bethe},||} = 2C\left(\frac{Z}{A}\right)\left(\frac{z}{\beta}\right)^2 \ln\left(\frac{\sqrt{2m_e Q_C}\beta}{\bar{I}}\right). \tag{8.316}$$

(8.316) is the nonrelativistic form of the Bethe mass soft-collision stopping power. We now consider its extension to the relativistic regime.

8.3.5 Bethe Relativistic Mass Soft-Collision Stopping Power

The derivation of the Bethe nonrelativistic result considered only that due to the static Coulomb potential only (i.e. longitudinal excitation). The exchange process of transverse photons (from the vector potential) and which is relevant only at relativistic energies between the projectile and an electron has two intermediate states (assume that the electron can be treated as at being rest) and is, thus, a second-order process. In the first, the projectile emits a photon of momentum $\mathbf{q} = \mathbf{p} - \mathbf{p}'$ and which is absorbed by the electron to give it a momentum $\mathbf{q}$. In the second, the electron emits a photon with momentum $-\mathbf{q}$, to give it a momentum $\mathbf{q}$. The photon is absorbed by the projectile to give it a momentum $\mathbf{p}' = \mathbf{p} - \mathbf{q}$. In this case, as there is no direct coupling between the initial and final states, the transition rate is taken up to second order,

$$\lambda_{f0} = \frac{2\pi}{\hbar}\left|\langle \mathbf{p}',f|U|\mathbf{p},0\rangle + \sum_j \frac{\langle \mathbf{p}',f|U|\mathbf{k},j\rangle\langle \mathbf{k},j|U|\mathbf{p},0\rangle}{E_0 - E_j}\right|^2 \rho_f, \tag{8.317}$$

and where the summation is over the intermediate states of the interaction. One can use the second-order process of Compton scatter as a model of the calculation. The reader is referred to the derivation of the Klein–Nishina cross section in Chap. 6 of McParland (2010), the discussion provided in the review articles by Fano (1963) and Ahlen (1980) for details of this calculation. The probability of the photon emission by the projectile is proportional to the matrix element of $z(\alpha\hbar c/e)(\alpha \bullet \hat{\varepsilon}_m)$ $e^{-i\mathbf{q}\bullet\mathbf{r}/\hbar c}$ where α is the Dirac velocity operator of the projectile, $\hat{\varepsilon}_m$ is the photon's unit polarisation vector and where $m = 1, 2$. The absorption of this photon by the jth atomic electron is proportional to a matrix element of $z(\alpha\hbar c/e)(\alpha \bullet \hat{\varepsilon}_m)\, e^{i\mathbf{q}\bullet\mathbf{r}/\hbar c}$.

We can derive the transverse term through our earlier consideration of the longitudinal and transverse dielectric constants. Recall that in our earlier derivation of Fermi's theory of the stopping power in a dielectric medium, the transverse component become significant only when $\beta \to 1$. Here, we now derive this component fully and lead ourselves to the Bethe relativistic soft-collision stopping power.

8.3.5.1 Longitudinal Dielectric Function

The longitudinal dielectric function can be obtained from the expression of (8.246):

$$\varepsilon_{R,||}(\mathbf{k},\omega) = 1 - \frac{\rho_{e,\text{med}}(\mathbf{k},\omega)}{\varepsilon_0\, k^2\, V(\mathbf{k},\omega)}. \tag{8.318}$$

This requires us to first evaluate $\rho_{e,\text{med}}(\mathbf{k},\omega)$ which we will do by evaluating the electron density for a single atom using first-order perturbation theory. We use the co ordinate system in which the nucleus is set at the position $\mathbf{R}$. The matrix element for the potential energy of an electron resulting from a scalar potential $V(\mathbf{k},\omega)$ is given by the inverse Fourier transform,

$$\langle f|U(\mathbf{r},t)\mid 0\rangle = -e\int d^3\mathbf{k}\int d\omega\, V(\mathbf{k},\omega)\, e^{i(\mathbf{k}\bullet\mathbf{R}-\omega t)}F_{f0}(\mathbf{k}), \tag{8.319}$$

where $F_{f0}(\mathbf{k})$ is now for a *single* electron only rather than an ensemble:

$$F_{f0}(\mathbf{k}) = \int d^3\mathbf{r}\psi_f^*(\mathbf{r})\, e^{i\mathbf{k}\bullet\mathbf{r}}\psi_0(\mathbf{r}) \tag{8.320}$$

and

$$F_{0f}(\mathbf{k}) = \int d^3\mathbf{r}\psi_0^*(\mathbf{r})\, e^{i\mathbf{k}\bullet\mathbf{r}}\psi_f(\mathbf{r}). \tag{8.321}$$

Recall that the time-dependent wavefunction can be expanded,

$$\Psi(\mathbf{r},t) = \sum_f a_f(t)\, e^{-iE_f t/\hbar}\,|f\rangle, \tag{8.322}$$

where the coefficients are themselves expanded, $a_f(t) = \delta_{f0} + \sum_{j=1}^{\infty} a_{f,j}(t)$. As in our previous use of first-order perturbation theory, we truncate this expansion of the coefficients to $a_{f,1}(t)$. Then, the perturbation coefficient (to first order) is, from (8.186),

$$\begin{aligned} a_{f,1}(t) &= \frac{ie}{\hbar}\int_{-\infty}^{t} dt'\, e^{i\omega_{f0}t'}\int d^3\mathbf{k}\int d\omega\, V(\mathbf{k},\omega)\, e^{i(\mathbf{k}\bullet\mathbf{R}-\omega t')}F_{f0}(\mathbf{k}) \\ &= \frac{ie}{\hbar}\int_{-\infty}^{t} dt'\int d^3\mathbf{k}e^{i\mathbf{k}\bullet\mathbf{R}}\int d\omega\, V(\mathbf{k},\omega)\, e^{i(\omega_{f0}-\omega)t'}F_{f0}(\mathbf{k}). \end{aligned} \tag{8.323}$$

The integration over the dummy time variable is

$$\int_{-\infty}^{t} dt'\, e^{i(\omega_{f0}-\omega)t'} = \frac{1}{i}\frac{e^{i(\omega_{f0}-\omega)t}}{\omega_{f0}-\omega-i\Omega}. \tag{8.324}$$

An imaginary coefficient $i\Omega$ has been included in the denominator of the integrand for managing the singularity that arises when $\omega = \omega_{f0}$. This inclusion of the coefficient leads to a distortion of the contour along the $\mathrm{Re}\,\omega$ axis so as to yield an infinitesimal semicircle below the point $\omega = \omega_{f0}$ and avoid that singularity. The perturbation coefficient is, thus,

$$a_{f,1}(t) = \frac{e}{\hbar} \int \mathrm{d}^3\mathbf{k} e^{i\mathbf{k}\bullet\mathbf{R}} \int \mathrm{d}\omega\, V(\mathbf{k},\omega)\, F_{f0}(\mathbf{k}) \frac{e^{i\left(\omega_{fo}-\omega\right)t}}{\omega_{f0} - \omega - i\Omega}. \tag{8.325}$$

The atomic electron density is given by the squared modulus of (8.322), weighted by the electric charge prior to squaring. To first order in the potential, the induced electric charge density is, from (8.322) and (8.325),

$$\rho_{e,\mathrm{med}}(\mathbf{r},t) = -\frac{e^2}{\hbar} \sum_{j=0}^{f} \left(\left(\psi_0^*(\mathbf{r}-\mathbf{R}) \psi_j(\mathbf{r}-\mathbf{R}) \int d^3\mathbf{k} \int d\omega\, V(\mathbf{k},\omega)\, e^{i\mathbf{k}\bullet\mathbf{R}} \right.\right.$$
$$\left.\left. \times F_{jo}(\mathbf{k}) \frac{e^{-i\omega t}}{\omega_{j0} - \omega - i\Omega} \right) + \mathrm{c.c} \right). \tag{8.326}$$

Let us take the medium to be a dilute monatomic gas with each atom having a single electron. The net polarisation charge is the integral of (8.326) integrated over all space and scaled by the number density of atoms in the medium,

$$\rho_{e,\mathrm{pol}}(\mathbf{r},t) = \frac{N_{\mathrm{A}}\rho}{A} \int \mathrm{d}^3\mathbf{R} \rho_{e,\mathrm{med}}(\mathbf{r},t), \tag{8.327}$$

where ρ is the physical density of the medium. As

$$\int \mathrm{d}^3\mathbf{R}\, \psi_0^*(\mathbf{r}-\mathbf{R})\, e^{i\mathbf{k}\bullet\mathbf{R}} \psi_f(\mathbf{r}-\mathbf{R}) = F_{f0}(-\mathbf{k})\, e^{i\mathbf{k}\bullet\mathbf{r}} \tag{8.328}$$

and the complex conjugate of the potential in Fourier space is

$$V^*(\mathbf{k},\omega) = V(-\mathbf{k},-\omega), \tag{8.329}$$

then (8.327) simplifies to

$$\rho_{e,\mathrm{pol}}(\mathbf{k},\omega) = -\frac{N_{\mathrm{A}}\rho}{A} \left(\frac{e}{\hbar}\right)^2 V(\mathbf{k},\omega)$$
$$\times \sum_{j=0}^{f} \left(\frac{F_{oj}(-\mathbf{k})\, F_{j0}(\mathbf{k})}{\omega_{j0} - \omega - i\Omega} - \frac{F_{oj}(\mathbf{k})\, F_{j0}(-\mathbf{k})}{\omega_{j0} + \omega + i\Omega} \right). \tag{8.330}$$

We thus achieve an expression for the longitudinal dielectric function:

$$\varepsilon_{R,||}(\mathbf{k},\omega) = 1 + \frac{N_{\mathrm{A}}A}{\rho\varepsilon_0} \left(\frac{e}{\hbar k}\right)^2 \sum_{j=0}^{f} \left|F_{jo}(\mathbf{k})\right|^2 \left(\frac{1}{\omega_{j0} - \omega - i\Omega} + \frac{1}{\omega_{j0} + \omega + i\Omega} \right). \tag{8.331}$$

Incorporating the definition of the plasma frequency ω_P of the medium given in Appendix H and the conversion factor $\alpha\hbar c$, (8.331) can be written as

$$\varepsilon_{R,||}(\mathbf{k},\omega) = 1 + \frac{m_e}{\hbar}\left(\frac{\omega_P}{kc}\right)^2 \sum_{j=0}^{f} \left|F_{jo}(\mathbf{k})\right|^2 \left(\frac{1}{\omega_{j0} - \omega - i\Omega} + \frac{1}{\omega_{j0} + \omega + i\Omega}\right). \tag{8.332}$$

8.3.5.2 Longitudinal Soft-Collision Stopping Power

The nonrelativistic (longitudinal) component of the stopping power is, from (8.266),

$$\left.\frac{\mathrm{d}E}{\mathrm{d}x}\right|_{S,||} = -i\frac{\alpha\hbar c z^2}{2\pi^2(\beta c)^2}\int\frac{\mathrm{d}k}{k}\int_{-k\beta c}^{k\beta c}\mathrm{d}\omega\frac{\omega}{\varepsilon_{R,||}(\mathbf{k},\omega)}. \tag{8.333}$$

To evaluate an expression for the linear stopping power, we first evaluate the integral over ω. We begin by using the binomial expansion to first order of $1/1 + x \approx 1 - x$ to give

$$\begin{aligned}\frac{1}{\varepsilon_{R,||}(\mathbf{k},\omega)} &= \frac{1}{1 + \frac{m_e}{\hbar}\left(\frac{\omega_P}{kc}\right)^2 \sum_{j=0}^{f} \left|F_{jo}(\mathbf{k})\right|^2 \left(\frac{1}{\omega_{j0} - \omega - i\Omega} + \frac{1}{\omega_{j0} + \omega + i\Omega}\right)}\\ &\approx 1 - \frac{m_e}{\hbar}\left(\frac{\omega_P}{kc}\right)^2 \sum_{j=0}^{f} \left|F_{jo}(\mathbf{k})\right|^2 \left(\frac{1}{\omega_{j0} - \omega - i\Omega} + \frac{1}{\omega_{j0} + \omega + i\Omega}\right).\end{aligned} \tag{8.334}$$

We next use the Cauchy principal value and the Sokhotski–Plemelj theorem described in Appendix A.5 to rewrite the quantity in the parentheses:

$$\begin{aligned}\lim_{\Omega\to 0}\left(\frac{1}{\omega_{j0} - \omega - i\Omega} + \frac{1}{\omega_{j0} + \omega + i\Omega}\right) &= \mathrm{PV}\left(\frac{1}{\omega_{j0} - \omega}\right)\\ &+ \mathrm{PV}\left(\frac{1}{\omega_{j0} + \omega}\right) + i\pi\big(\delta(\omega_{j0} - \omega) - \delta(\omega_{j0} + \omega)\big).\end{aligned} \tag{8.335}$$

Inserting this result into (8.334) gives

$$\begin{aligned}\frac{1}{\varepsilon_{R,||}(\mathbf{k},\omega)} =& 1 - \frac{m_e}{(2\pi)^{3/2}\hbar}\left(\frac{\omega_P}{kc}\right)^2 \sum_{j=0}^{f} \left|F_{jo}(\mathbf{k})\right|^2\\ &\times\left(\mathrm{PV}\left(\frac{1}{\omega_{j0} - \omega}\right) + \mathrm{PV}\left(\frac{1}{\omega_{j0} + \omega}\right) + i\pi\big(\delta(\omega_{j0} - \omega) - \delta(\omega_{j0} + \omega)\big)\right).\end{aligned} \tag{8.336}$$

Putting this result into the integral of (8.333), we achieve

$$\left.\frac{\mathrm{d}E}{\mathrm{d}x}\right|_{S,||} = -i\frac{\alpha\hbar c z^2}{2\pi^2(\beta c)^2}\left(\int\frac{\mathrm{d}k}{k}\int_{-k\beta c}^{k\beta c}\mathrm{d}\omega\,\omega - \right.$$

$$\left(1-\frac{m_e}{\hbar}\left(\frac{\omega_P}{kc}\right)^2\sum_{j=0}^{f}\left|F_{jo}(\mathbf{k})\right|^2\int_{-k\beta c}^{k\beta c}\mathrm{d}\omega\,\omega\left(PV\left(\frac{1}{\omega_{j0}-\omega}\right)+PV\left(\frac{1}{\omega_{j0}+\omega}\right)\right.\right.$$

$$\left.\left.\left. + i\pi\big(\delta(\omega_{j0}-\omega)-\delta(\omega_{j0}+\omega)\big)\right)\right)\right). \tag{8.337}$$

We recall that the stopping power is a real quantity and, as such, (8.316) reduces to

$$\left.\frac{\mathrm{d}E}{\mathrm{d}x}\right|_{S,||} = \frac{\alpha\hbar c z^2 m_e}{2\pi^2\,\hbar(\beta c)^2}\left(\frac{\omega_P}{c}\right)^2 \times\left(\int\frac{\mathrm{d}k}{k^3}\sum_{j=0}^{f}\left|F_{jo}(\mathbf{k})\right|^2\int_{-k\beta c}^{k\beta c}\mathrm{d}\omega\,\omega\,\big(\delta(\omega_{j0}+\omega)-\delta(\omega_{j0}-\omega)\big)\right). \tag{8.338}$$

The integration over ω uses the sifting property of the Dirac δ-function:

$$\begin{aligned}\int_{-k\beta c}^{k\beta c}\mathrm{d}\omega\,\omega\big(\delta(\omega_{j0}+\omega)-\delta(\omega_{j0}-\omega)\big) &= 2\int_{\omega_{jo}-k\beta c}^{\omega_{jo}+k\beta c}\mathrm{d}x\,x\,\delta(x)\\ &= 2\omega_{jo}.\end{aligned} \tag{8.339}$$

$$\left.\frac{\mathrm{d}E}{\mathrm{d}x}\right|_{S,||} = \frac{\alpha\hbar c z^2}{\pi\hbar}\left(\frac{z\omega_P}{\beta c^2}\right)^2\sum_{j=0}^{f}\omega_{j0}\int_{k\beta c>\omega_{jo}}\frac{\mathrm{d}k}{k^3}\left|F_{jo}(\mathbf{k})\right|^2. \tag{8.340}$$

Incorporating the definition of the plasma frequency and using the energy transfer $Q = (\hbar c k)^2/2m_e$, we reduce (8.340) to

$$\left.\frac{\mathrm{d}E}{\mathrm{d}x}\right|_{S,||} = C\left(\frac{Z}{A}\right)\left(\frac{z}{\beta}\right)^2\sum_{j=0}^{f}\hbar\omega_{j0}\int_{\Re}\frac{\mathrm{d}Q}{Q^2}\left|\frac{F_{jo}(\mathbf{k})}{\sqrt{2\pi}}\right|^2 \tag{8.341}$$

where the range $\Re$ of integration is for $Q > \left(\hbar\omega_{j0}\right)^2/2m_e\beta^2$.

The extension of (8.341) to the relativistic regime requires the inclusion of the transverse term. Ahlen (1980) and Sigmund (2006) derived this extension in different manners. Ahlen followed the approach of Fano (1963) which uses the quantized form of the vector potential, whereas Sigmund uses the transverse component of the relative permittivity. In either case, the final expression of the combined (nonrelativistic and relativistic) Bethe mass soft-collision stopping power can be written as

$$\left.\frac{\mathrm{d}E}{\mathrm{d}x}\right|_{S,\mathrm{Bethe}} = C\left(\frac{Z}{A}\right)\left(\frac{z}{\beta}\right)^2 \sum_{j=0}^{f} \hbar\omega_{j0} \int_{\Re} \frac{\mathrm{d}Q}{Q^2}\left|\frac{F_{jo}(\mathbf{k})}{\sqrt{2\pi}}\right|^2 \left(1+\frac{\dfrac{2m_e\beta^2 Q}{(\hbar\omega_{jo})^2}-1}{\left(\dfrac{2m_e Q}{(\hbar\omega_{jo})^2}-1\right)^2}\right). \tag{8.342}$$

Fano considered (8.342) in three regions of the energy transfer, Q. The first region is for sufficiently small Q such that the dipole approximation holds. The second region is that in which the longitudinal term of (8.342) alone contributes. The maximum energy transfer from the projectile to the atom within this region is less than the electron's rest mass. The third region is that for which the atomic electrons can be considered to be free. In our result, we evaluate (8.342) up to the energy transfer Q_C which separates Fano's second and third regions. In this case, (8.342) becomes

$$\left.\frac{\mathrm{d}E}{\rho\,\mathrm{d}x}\right|_{S,\mathrm{Bethe}} = 2C\left(\frac{Z}{A}\right)\left(\frac{z}{\beta}\right)^2 \sum_{j} f_{jo}\left(\ln\left(\frac{\sqrt{2m_e Q_\mathrm{C}}\beta}{\hbar\omega_{j0}}\right) - \ln(1-\beta^2) - \beta^2\right). \tag{8.343}$$

When we compare the Bethe mass soft-collision stopping power to that of Bohr's given by (8.180), we see that (excluding the argument of the logarithm) the only difference is that of the inclusion of a factor of an additional $-\beta^2/2$ term which arises from close collisions. Finally, we can rewrite (8.343) as

$$\left.\frac{\mathrm{d}E}{\rho\,\mathrm{d}x}\right|_{S,\mathrm{Bethe}} = C\left(\frac{Z}{A}\right)\left(\frac{z}{\beta}\right)^2 \left(\ln\left(\frac{2m_e Q_\mathrm{C}\beta}{\bar{I}^2}\right) - \ln(1-\beta^2) - \beta^2\right). \tag{8.344}$$

8.4 Quantitative Comparison of Rutherford, Bohr and Bethe Results for a Massive Projectile

In this section, we compare quantitatively the mass soft-collision stopping powers yielded by the three mass soft-collision stopping powers we have derived from the Rutherford, Bohr and Bethe results.

The medium selected is carbon as an elemental representation of soft tissue. As with light elements, we take $A = 2Z$ and we have selected arbitrarily $Q_\mathrm{c} = 50\,\mathrm{keV}$. These mass soft-collision stopping powers are shown in Fig. 8.9 as a function of proton energy ranging from 0.1 to 10^6 MeV.

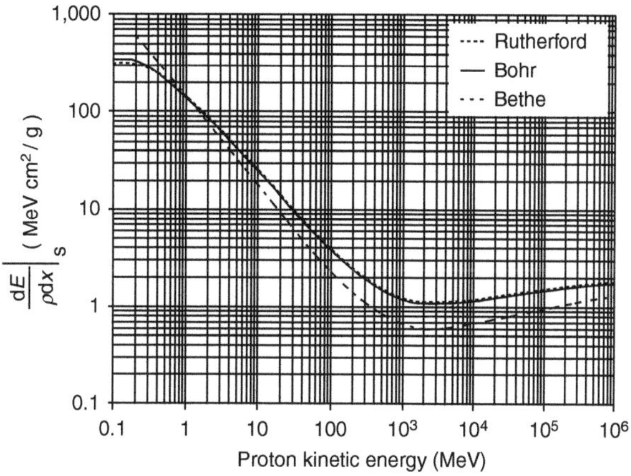

Fig. 8.9 Mass soft-collision stopping powers for protons in carbon as calculated by the Rutherford (Eq. 8.90), Bohr (Eq. 8.180) and Bethe (Eq. 8.344) formula. A value of 50 keV is assumed for Q_C for the Bethe calculation

There is very little difference between the Rutherford and Bohr results, so we shall exclude the former from discussion. Although the Bohr and Bethe results are not in agreement quantitatively, they both exhibit the same general qualitative features: an initial decrease with kinetic energy due to the dominant $1/\beta^2$ factor to a broad minimum at a kinetic energy of about 2–3 GeV. This is then followed by a slow increase. The minimum, which occurs at kinetic energies of about two to three times the rest mass energy of the massive particle, is known colloquially as the minimum-ionising region.

8.5 Summary

In this chapter, we reviewed the soft-collision stopping powers using classical, semi-classical and quantum theories. We have seen that there are similarities among these results, which leads to the consideration of linking the classical and quantum-mechanical results. Discussion of the bridging is deferred to Chap. 12 following our introduction of the hard-collision stopping powers.

Bibliography

Attix FH. Introduction to radiological physics and radiation dosimetry. New York: Wiley; 1986.

Berger MJ, Seltzer SM. Stopping powers and ranges of electrons and positrons. NBSIR 82-2550-A. Washington, DC: National Bureau of Standards; 1983.

Bjorken JD, Drell SD. Relativistic quantum mechanics. New York: McGraw Hill; 1964.

Bloch F. Zur Bremsung rasch bewegter Teilchen beim Durchgang durch Materie. Ann Phys. 1933a;16:285–320.

Brice DK, Sigmund P. Secondary electron spectrum from dielectric theory. Matt Fys Medd Dan Vid Selsk. 1980;40:1–34.

Enz PC, editor. Pauli lectures on physics: vol 6. Selected topics on field quantization. Cambridge: Wolfgang Pauli. Massachusetts Institute of Technology; 1973.
Gould RJ. Electromagnetic processes. Princeton: Princeton University Press; 2006.
Green G. An essay on the application of mathematical analysis to the theories of electricity and magnetism; 1828.
ICRU. Stopping powers and ranges for electrons and positrons. ICRU Report 37. Bethesda: International Commission on Radiation Units and Measurements; 1984.
ICRU. Stopping powers and ranges for protons and alpha particles. ICRU Report 49. Bethesda: International Commission on Radiation Units and Measurements; 1993.
Jordan EG, Balmain KG. Electromagnetic waves and radiating systems. Englewood Cliffs: Prentice Hall Inc; 1968.
Ziegler JF. The stopping of energetic light ions in elemental matter. J Appl Phys. 1999;85:1249–72.

References

Abramowitz M, Stegun IA, editors. Handbook of mathematical functions. New York: Dover Publications; 1972.
Ahlen SP. Theoretical and experimental aspects of the energy loss of relativistic heavy ionizing particles. Rev Mod Phys. 1980;52:121–73 (erratum Rev Mod Phys 1980; 52: 653).
Anderson C, Neddermeyer S. New evidence for the existence of a particle intermediate between the proton and electron. Phys Rev. 1937;52:1003–4.
Bethe HA. Zur Theorie des Durchgangs schneller Korpuskularstrahlen durch Materie. Ann Phys. 1930;5:324–400.
Bethe HA. Bremsformel für Elektronen relativisticher Geschwindigkeit. Z Physik. 1932;76:293–9.
Bloch F. Zur Bremsun rasch bewegter Telichen beim Durchgang durch Materie. Ann Physik. 1933;16:285–320.
Bohr N. On the theory of the decrease of velocity of moving electrified particles on passing through matter. Philos Mag. 1913;25:10–31.
Bohr N. On the decrease of velocity of swiftly moving electrified particles in passing through matter. Philos Mag. 1915;30:581–612.
Bohr N. The penetration of atomic particles through matter. Matt Fys Medd Dan Vild Selsk. 1948;18(8):1–144.
Darwin CG. A theory of the absorption and scattering of the α rays. Philos Mag. 1912;23:901–21.
Fano U. Penetration of protons, alpha particles and mesons. Annu Rev Nucl Sci. 1963;13:1–66.
Fermi E. The absorption of mesotrons in air and in condensed materials. Phys Rev. 1939;56:1242.
Fermi E. The ionization loss of energy in gases and in condensed materials. Phys Rev. 1940;47:485–92.
Fernández-Vaera JM. Monte Carlo simulation of the inelastic scattering of electrons and positrons using optical-data models. Rad Phys Chem. 1998;53:235–45.
Gaunt JA. The stopping power of hydrogen atoms of α-particles according to the new quantum theory. Proc Camb Philos Soc. 1927;23:732–54.
Henderson GH. The decrease in energy of α particles passing through matter. Philos Mag. 1922;44:680–8.
Jackson JD. Classical electrodynamics. New York: Wiley; 1999.
McParland BJ. Nuclear medicine radiation dosimetry: advanced theoretical principles. London: Springer; 2010.
Neufeld J. Electron capture and loss by moving ions in dense media. Phys Rev. 1954;96:1470–8.
Rossi B. High-energy particles. New York: Prentice-Hall; 1952.
Segrè E. Nuclei and particles. 2nd ed. Reading: Benjamin/Cummings; 1977.

Sigmund P. Low-velocity limit of Bohr's stopping power formula. Phys Rev A. 1996;54:3113–17.
Sigmund P. Particle penetration and radiation effects. Berlin: Springer; 2006.
Slater JC. Atomic radii in crystals. J Chem Phys. 1964;41:3199–204.
Thomson JJ. Ionization by moving electrified particles. Philos Mag. 1912;6–23:449.
Thorsen J (ed). The penetration of charged particles through matter. In: Niels Bohr: collected works. Amsterdam: Elsevier NV; 1987.
Uehling EA. Penetration of heavy charged particles in matter. Annu Rev Nucl Sci. 1954;4:315–50.

Chapter 9
Hard Collisions

Contents

Abstract This chapter presents the differential cross sections in energy transfer for hard collisions between a target electron and massive spin-0 and spin-½ and electron and positron projectiles. Using the collision stopping power, these are then used to derive the corresponding hard-collision stopping powers.

9.1 Introductory Comments

In Chap. 7, we defined soft and hard collisions, and in Chap. 8, we derived the soft-collision stopping power through classical, semi-classical and quantum-mechanical means. The differentiation between the soft- and hard-collision stopping powers is characterised, in the classical consideration, by the impact parameter and the quantum-mechanical view by energy/momentum transfers to the atomic system.

The hard-collision stopping power is defined as the direct interactions between the projectile and the atomic electron. In practical medical radiation calculations, the initial assumption is that that atomic electron is well approximated as being at rest as the velocity of the projectile is, in general, much greater generally than the electron's orbital velocity. However, this is not the case universally. Let us consider this quantitatively. The maximum orbital velocity for an electron in calcium is

B.J. McParland, *Medical Radiation Dosimetry*,
DOI 10.1007/978-1-4471-5403-7_9, © Springer-Verlag London 2014

equal to about $20\alpha c$ or $\beta \approx 0.15$. This velocity corresponds to an electron with a kinetic energy of about 5 keV or a proton with a kinetic energy of about 15 MeV. Hence, a correction factor to account for the non-zero orbital velocity will be required as the projectile decelerates and is discussed in Chap. 12.

As the hard collision is considered as a two-body interaction, the physical description is inherently quantum mechanical. As a result, the nature, and the intrinsic spin, of the projectile must be taken into account. This is especially true for electron projectiles where exchange effects must be recognised. In medical radiation applications, we will consider here the following projectiles:

- Electron, spin ½
- Positron, spin ½
- Proton, spin ½
- Massive projectiles, spin 0 or spin 1

In this chapter, we derive the hard-collision stopping power beginning with the calculations of the differential cross section in energy transfer from a projectile to an atomic electron, $\frac{\mathrm{d}\sigma}{\mathrm{d}Q}$. It will be assumed that the scatter is elastic in that the binding energy of the atomic electron is negligible in comparison to the projectile's kinetic energy and, that in the case of an ion projectile, electron capture from the target atom does not occur. From this differential cross section, we then calculate the hard mass collision stopping power, $\left.\frac{\mathrm{d}E}{\rho\,\mathrm{d}x}\right|_H$, through the stopping cross section of (8.85).

9.2 Differential Cross Sections in Energy Transfer

9.2.1 Massive Projectiles (m >> m_e)

9.2.1.1 Introductory Comments

We consider here those cases of a projectile with a mass much greater than that of an electron. Two considerations will be made to the intrinsic spin of the projectile: spin-0 and spin-½ and spin-1. Derivations beyond those of Chap. 8 are not provided, but the original publications are cited.

9.2.1.2 Spin-0 Projectiles

Consider a massive spin-zero projectile of charge ze and with velocity β. The differential cross section for the energy transfer between Q and $Q + \mathrm{d}Q$ to an electron at rest was derived by Bhabha (1938). This would also be the calculation for which spin is neglected and only relativistic effects are noted:

$$\begin{aligned}\frac{d\sigma}{dQ} &= 2\pi r_e^2 m_e \left(\frac{z}{\beta}\right)^2 \frac{1}{Q^2}\left(1-\beta^2\frac{Q}{Q_{\max}}\right)\\ &= \frac{C}{N_A}\left(\frac{z}{\beta}\right)^2 \frac{1}{Q^2}\left(1-\beta^2\frac{Q}{Q_{\max}}\right) \quad \text{Spin-0} \end{aligned} \tag{9.1}$$

where $Q_{\max}$ is the maximum energy transfer to the electron, an expression for this being given by (D.15). For the case of a low projectile momentum $\left(\text{i.e. } p << \frac{m^2}{m_e}\right)$, $Q_{\max} = 2m_e\left(\frac{p}{m}\right)^2$. If we compare this result to the Rutherford formula of (8.82), we see that the result (9.1) introduces a $\left(-\beta^2\frac{Q}{Q_{\max}}\right)$ term to reduce the cross section from the Rutherford value. The Rutherford expression, which was derived classically, represents the nonrelativistic form of the Bhabha massive spin-0 projectile differential cross section.

9.2.1.3 Spin-½ Projectiles (Protons)

For the case of the projectile having intrinsic spin-½, (9.1) is modified (Bhabha 1938; Massey and Corben 1939):

$$\frac{d\sigma}{dQ} = \frac{C}{N_A}\left(\frac{z}{\beta}\right)^2 \frac{1}{Q^2}\left(1-\beta^2\frac{Q}{Q_{\max}}+\frac{1}{2}\left(\frac{Q}{E+m}\right)^2\right) \quad \text{Spin-}1/2. \tag{9.2}$$

Comparing (9.2) with (9.1), we see that accounting for the projectile's spin-½ has added an additional factor of ½ $\left(\frac{Q}{E+m}\right)^2$ to the spin-0 calculation.

9.2.2 Projectiles of Electron Mass ($m = m_e$)

9.2.2.1 Introductory Comments

This section considers the cases of electron and positron projectiles. Electron–electron (Møller) scatter was studied in Sect. 4.3.7 and the differential cross section in solid angle obtained which accounted for the indistinguishability between projectile and target and exchange effects. The discussion here on Møller scatter may be considered as an addendum to that given earlier.

We have not yet considered positron–electron (Bhabha) scatter.

9.2.2.2 Møller Scatter

The full relativistic Møller differential cross section in energy transfer is, for a relativistic incident electron of kinetic energy E (where $E >> m_e$) transferring an energy between Q and $Q + \mathrm{d}Q$ to another electron (Møller 1932; Rohrlich and Carlson 1954),

$$\begin{aligned}\frac{\mathrm{d}\sigma}{\mathrm{d}Q} &= \frac{\pi}{E}\left(\frac{\alpha\hbar c}{E}\right)^2\left[\left(\frac{E}{Q}\right)^2 + \left(\frac{Q}{E-Q}\right) + \left(\frac{\gamma-1}{\gamma}\right)^2 - \left(\frac{2\gamma-1}{\gamma^2}\right)\left(\frac{E}{Q}\right)\left(\frac{Q}{E-Q}\right)\right]\\ &= \frac{C}{2N_\mathrm{A}}\frac{m_e}{E^3}\left(\left(\frac{E}{Q}\right)^2 + \left(\frac{Q}{E-Q}\right) + \left(\frac{\gamma-1}{\gamma}\right)^2 - \left(\frac{2\gamma-1}{\gamma^2}\right)\left(\frac{E}{Q}\right)\left(\frac{Q}{E-Q}\right)\right).\end{aligned} \tag{9.3}$$

This describes the probability that, following Møller scatter, the final state has an electron with a kinetic energy of Q and the other has a kinetic energy of $E - Q$. Thus, all possible outcomes of the scatter are obtained for Q ranging in value from 0 to $\frac{E}{2}$. By convention, the exiting electron with the highest kinetic is considered to have been the projectile electron.

9.2.2.3 Bhabha Scatter

We now consider the case of a positron projectile with kinetic energy E. The Feynman diagrams for positron–electron (Bhabha) scatter are given in Fig. 9.1. While the first graph is similar to that of Møller scatter given in Fig. 4.9, the fact that it is possible for the electron and positron to annihilate requires the provision of an additional graph accounting for the production of a virtual photon from which the exiting electron–positron pair is created. The differential cross section that an incident positron with kinetic energy $E >> m_e$ will suffer a kinetic energy loss between Q and $Q + dQ$ which is transferred to the target electron is (Bhabha 1936)

$$\begin{aligned}\frac{\mathrm{d}\sigma}{\mathrm{d}Q} = \frac{C}{N_\mathrm{A}}\frac{1}{E^2}&\left[\left(\frac{E}{Q}\right)^2 - \left(\frac{\gamma^2-1}{\gamma^2}\right)\left(\frac{E}{Q}\right) + \frac{1}{2}\left(\frac{\gamma-1}{\gamma}\right)^2\right.\\ &- \left(\frac{\gamma-1}{\gamma+1}\right)\left\{\left(\frac{\gamma+2}{\gamma}\right)\left(\frac{E}{Q}\right) - 2\left(\frac{\gamma^2-1}{\gamma^2}\right) + \left(\frac{\gamma-1}{\gamma}\right)^2\left(\frac{E}{T}\right)\right\}\\ &\left.+\left(\frac{\gamma-1}{\gamma+1}\right)^2\left\{\frac{1}{2} + \frac{1}{\gamma} + \frac{3}{2\gamma^2} - \left(\frac{\gamma-1}{\gamma}\right)^2\left(\frac{Q}{E}\right)\left(1 - \left(\frac{Q}{E}\right)\right)\right\}\right].\end{aligned} \tag{9.4}$$

This is taken from the sum of the probabilities that, in the final state, either the electron or the positron have an energy between Q and $Q + \mathrm{d}Q$.

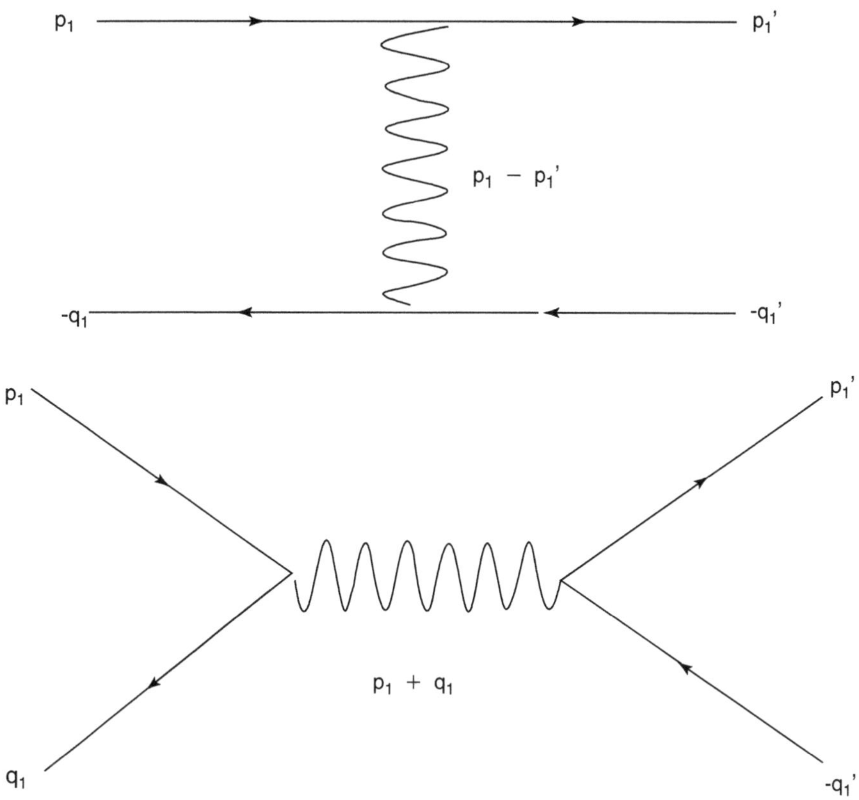

Fig. 9.1 Feynman diagrams for positron–electron Coulomb scatter. *Top*: direct interaction; *bottom*: annihilation intermediate-state interaction

9.3 Hard-Collision Stopping Powers

9.3.1 Introductory Comments

Having now obtained the differential cross sections in energy transfer to atomic electrons at rest for massive projectiles and electron and positron projectiles, we now evaluate the hard-collision stopping powers for these different categories. As the calculations for the hard collisions have been based upon quantum mechanics and use the energy transfer Q between the projectile and target as the metric, we logically combine these hard-collision stopping power results with those obtained previously for the Bethe soft collision in order to yield the total collision stopping power.

The electron projectile will pose a condition in terms of the indistinguishability between projectile and recoil electrons in the final state.

9.3.2 Massive Projectiles ($m >> m_e$)

As the differential cross sections in energy transfer for a massive spin-½ projectile reduces to that for a massive spin-0 projectile at low kinetic energies, the spin-0 case is considered first. The general expression for the mass hard-collision stopping power is, from (8.85),

$$\left.\frac{\mathrm{d}E}{\rho \mathrm{d}x}\right|_H = N_{\mathrm{A}}\left(\frac{Z}{A}\right)\int_{Q_C}^{Q_{\max}} \mathrm{d}Q\, Q\, \frac{\mathrm{d}\sigma}{\mathrm{d}Q} \tag{9.5}$$

where the integral limits are Q_C, which we recall as separating the regimes of soft and hard collisions, and the maximum energy transferred to the target electron, $Q_{\max}$, which is set by the relevant kinematics. The mass hard-collision stopping power for a massive spin-0 particle is obtained by substituting (9.1) into (9.5):

$$\begin{aligned}\left.\frac{\mathrm{d}E}{\rho \mathrm{d}x}\right|_H &= N_{\mathrm{A}}\left(\frac{Z}{A}\right)\int_{Q_C}^{Q_{\max}} \mathrm{d}Q\, Q\, \frac{\mathrm{d}\sigma}{\mathrm{d}Q} \\ &= C\left(\frac{Z}{A}\right)\left(\frac{z}{\beta}\right)^2\int_{Q_C}^{Q_{\max}} \frac{\mathrm{d}Q}{Q}\left[1-\beta^2\frac{Q}{Q_{\max}}\right] \\ &= C\left(\frac{Z}{A}\right)\left(\frac{z}{\beta}\right)^2\left[\ln\left(\frac{Q_{\max}}{Q_C}\right)-\beta^2\left(\frac{Q_{\max}-Q_C}{Q_{\max}}\right)\right].\end{aligned} \tag{9.6}$$

In a hard collision, the energy transfer is assumed to be sufficiently high ($Q_{\max} >> Q_C$) to allow this to be simplified to the form

$$\left.\frac{\mathrm{d}E}{\rho \mathrm{d}x}\right|_H = C\left(\frac{Z}{A}\right)\left(\frac{z}{\beta}\right)^2\left[\ln\left(\frac{Q_{\max}}{Q_C}\right)-\beta^2\right] \quad \text{Spin-0.} \tag{9.7}$$

We now consider the mass hard-collision stopping power for a massive spin-½ projectile by substituting (9.2) into (9.5):

$$\begin{aligned}\left.\frac{\mathrm{d}E}{\rho \mathrm{d}x}\right|_H &= C\left(\frac{Z}{A}\right)\left(\frac{z}{\beta}\right)^2\int_{Q_C}^{Q_{\max}} \frac{\mathrm{d}Q}{Q}\left[1-\beta^2\frac{Q}{Q_{\max}}+\frac{1}{2}\left(\frac{Q}{E+m}\right)^2\right] \\ &= C\left(\frac{Z}{A}\right)\left(\frac{z}{\beta}\right)^2\left[\ln\left(\frac{Q_{\max}}{Q_C}\right)-\beta^2\left(\frac{Q_{\max}-Q_C}{Q_{\max}}\right)+\frac{1}{4}\frac{Q_{\max}^2-Q_C^2}{(E+m)^2}\right] \\ &= C\left(\frac{Z}{A}\right)\left(\frac{z}{\beta}\right)^2\left[\ln\left(\frac{Q_{\max}}{Q_C}\right)-\beta^2+\left(\frac{Q_{\max}}{2(E+m)}\right)^2\right] \quad \text{Spin-}^1\!/_2\end{aligned} \tag{9.8}$$

where $Q_{\max} > Q_C$. Note that for a massive spin-½ projectile with a rest mass much greater than the maximum energy transfer, the last squared term in the square brackets can be neglected, and this result reduces to the simpler spin-0 expression.

9.3.3 *Electron Projectiles*

The restricted mass hard-collision stopping power for an electron projectile is

$$\left.\frac{\mathrm{d}E}{\rho\,\mathrm{d}x}\right|_{H,\Delta} = N_A\left(\frac{Z}{A}\right)\int_{Q_C}^{\Delta}\mathrm{d}Q\,Q\,\frac{\mathrm{d}\sigma}{\mathrm{d}Q} \tag{9.9}$$

where the Møller differential cross section is to be used in the integral. Unlike the derivations of the massive particle hard-collision stopping powers, an upper limit of integration $\Delta \le Q_{\max} = \frac{T}{2}$ has been specified so as to ignore those energy transfers greater than Δ. Although the restricted mass collision stopping power can be defined for any projectile, the discussion here will be limited to the electron projectile for dosimetric interest, as described in Chap. 10.

The mass collision stopping power for the Møller cross section is

$$\begin{aligned}\left.\frac{\mathrm{d}E}{\rho\,\mathrm{d}x}\right|_{H,\Delta} &= C\left(\frac{Z}{A}\right)\frac{1}{\beta^2E^2}\int_{Q_C}^{\Delta}\mathrm{d}Q\,Q\left[\left(\frac{E}{Q}\right)^2+\left(\frac{Q}{E-Q}\right)+\left(\frac{\gamma-1}{\gamma}\right)^2\right.\\ &\quad\left.-\left(\frac{2\gamma-1}{\gamma^2}\right)\left(\frac{E}{Q}\right)\left(\frac{Q}{T-Q}\right)\right]\\ &= C\left(\frac{Z}{A}\right)\frac{1}{\beta^2}\left[\ln\frac{\Delta(E-\Delta)}{Q_C(E-Q)}+\left(\frac{E}{E-\Delta}-\frac{E}{E-Q}\right)\right.\\ &\quad\left.+\left(\frac{\gamma-1}{\gamma}\right)^2\left(\frac{\Delta^2-Q_C^2}{2E^2}\right)-\left(\frac{2\gamma-1}{\gamma^2}\right)\ln\left(\frac{E-Q_C}{E-\Delta}\right)\right].\end{aligned} \tag{9.10}$$

This expression can be simplified by normalising Δ and Q_C to the projectile electron's kinetic energy:

$$\delta \equiv \frac{\Delta}{E} \tag{9.11}$$

$$\tau_C \equiv \frac{Q_C}{E} \tag{9.12}$$

to allow a simpler form of (9.10)

$$\left.\frac{\mathrm{d}E}{\rho\,\mathrm{d}x}\right|_{H,\Delta} = C\left(\frac{Z}{A}\right)\frac{1}{\beta^2}\left[\ln\frac{\delta(1-\delta)}{\tau_C(1-\tau_C)} + \left(\frac{1}{1-\delta} - \frac{1}{1-\tau_C}\right) + \left(\frac{\gamma-1}{\gamma}\right)^2\left(\frac{\delta^2-\tau_C^2}{2}\right) - \left(\frac{2\gamma-1}{\gamma^2}\right)\ln\left(\frac{1-\tau_C}{1-\delta}\right)\right]. \tag{9.13}$$

In Chap. 8, it was noted that the selection of Q_C is somewhat arbitrary. Hence, we can ensure a value of $\tau_C << 1$ which allows us to use $1-\tau_C \approx 1$ and $\delta^2 - \tau_C^2 \approx \delta^2$. Then this expression for the electron restricted mass hard-collision stopping power to:

$$\left.\frac{\mathrm{d}E}{\rho\,\mathrm{d}x}\right|_{H,\Delta} = C\left(\frac{Z}{A}\right)\frac{1}{\beta^2}\left[\ln\frac{\delta(1-\delta)}{\tau_C} + \left(\frac{1}{1-\delta} - 1\right) + \left(\frac{\gamma-1}{\gamma}\right)^2\left(\frac{\delta^2}{2}\right) + \left(\frac{2\gamma-1}{\gamma^2}\right)\ln(1-\delta)\right] \quad \text{–Electron (restricted).} \tag{9.14}$$

The unrestricted mass hard-collision stopping power is that for $\Delta = Q_{\max} = \frac{E}{2}$ or $\delta = \frac{1}{2}$, and

$$\left.\frac{\mathrm{d}E}{\rho\,\mathrm{d}x}\right|_{H,\Delta=\frac{E}{2}} = C\left(\frac{Z}{A}\right)\frac{1}{\beta^2}\left[\ln\frac{1}{4\tau_C} + 1 + \frac{1}{8}\left(\frac{\gamma-1}{\gamma}\right)^2 - \left(\frac{2\gamma-1}{\gamma^2}\right)\ln 2\right] \quad \text{–Electron (unrestricted).} \tag{9.15}$$

which corresponds to the result originally given by Rohrlich and Carlson (1954).

The unrestricted mass hard-collision stopping power for a positron projectile is calculated as for an electron project but with the Bhabha differential cross section.

Bibliography and Further Reading

Ahlen SP. Theoretical and experimental aspects of the energy loss of relativistic heavy ionizing particles. Rev Mod Phys. 1980;52:121–73 (erratum Rev Mod Phys 1980; 52: 653).

Attix FH. Introduction to radiological physics and radiation dosimetry; 1986.

Berger MJ, Seltzer SM. Stopping powers and ranges of electrons and positrons. NBSIR 82-2550-A. Washington, DC: National Bureau of Standards; 1983.

Fano U. Studies in penetration of charged particles in matter, Nuclear Science Series Report No 39. Washington, DC: National Academy of Sciences; 1964. p. 1–338.
ICRU. Stopping powers and ranges for electrons and positrons. ICRU Report 37. Bethesda: International Commission on Radiation Units and Measurements; 1984.
ICRU. Stopping powers and ranges for protons and alpha particles. ICRU Report 49. Bethesda: International Commission on Radiation Units and Measurements; 1993.
Uehling EA. Penetration of heavy charged particles in matter. Annu Rev Nucl Sci. 1954;4:315–50.
Ziegler JF. The stopping of energetic light ions in elemental matter. J Appl Phys. 1999;85:1249–72.

References

Bhabha HJ. The scattering of positrons by electrons with exchange on Dirac's theory of the positron. Proc R Soc A. 1936;154:195–206.
Bhabha JH. On the penetrating component of cosmic radiation. Proc R Soc A. 1938;164:257–94.
Massey HSW, Corben HS. Elastic collisions of mesons with electrons and protons. Proc Camb Philos Soc. 1939;35:463–72.
Møller C. Zur theorie des durchgangs schneller electronen durch materie. Ann Physik. 1932;14:531–85.
Rohrlich F, Carlson BC. Positron-electron differences in energy loss and multiple scattering. Phys Rev. 1954;93:38–44.

Chapter 10
Total and Restricted Collision Stopping Powers and Theory of the Mean Energy Expended to Create an Ion Pair

Contents

Abstract This chapter combines the soft- and hard-collision stopping powers derived in the two previous chapters to yield the total collision stopping power (but still not accounting for the higher-order effects to be studied in Chap. 12). For electron projectiles, we consider the restricted hard-collision stopping power derived in Chap. 9 to provide the restricted stopping power for electrons, a quantity of importance in practical radiation dosimetry evaluations. The main part of this chapter is, however, assigned to the theoretical investigation of the mean energy required to create an ion pair. This is a fundamental quantity not only to practical radiation dosimetry but also to our understanding of the ionisation resulting from the interaction of radiation with matter. We develop the generalised Fowler equation, which is essentially a transport equation, to yield the number of ion pairs created per unit energy transfer.

B.J. McParland, *Medical Radiation Dosimetry*,
DOI 10.1007/978-1-4471-5403-7_10,

10.1 Introductory Comments

10.1.1 Organisation of Chapter

In this chapter, we cover three endpoints of much of our discussions of the soft- and hard-collision stopping powers in Chaps. 8 and 9. Further refinements remain and these are discussed in Chaps. 11, 12 and 14. The first is the combination of these to yield expressions for the complete collision stopping power. We then further discuss the practical matters of the stopping powers which are restricted in terms of kinetic energy transfer – a consideration of essential necessity in practical medical dosimetry. Finally, we consider the required energy expended to generate ionisation and how theory can guide us in its exploration.

Each component of this chapter is now introduced individually and in detail.

10.1.2 Total Collision Stopping Power

The total collision stopping power is, as expected, the summation of the soft-collision stopping power, which reflects energy transfer to the atom as a whole and investigated in Chap. 8, and the hard-collision stopping power, which is a result of the interaction between the projectile and an individual atomic electron. In Chap. 8, we explored the soft-collision stopping power in both classical (Bohr) and quantum (Bethe) theories. The former was based upon the use of the spatial impact parameter as the metric dividing between the regimes of soft and hard collisions. Contrary to this, the latter based the differentiation between soft and hard collisions upon momentum or energy transfer (i.e. Q_C) from the projectile to the atom. The latter is compatible with our approach of Chap. 9 in which hard collisions were considered solely through quantum-mechanical models. Hence, in this chapter, we will combine the foundation of the total collision stopping power through the summation of the Bethe soft-collision stopping power and the quantum-mechanical representations of the hard-collision stopping powers of Chap. 9.

10.1.3 Restricted Collision Stopping Power

The absorbed dose is the ratio of the energy deposited by a charged particle within a volume within the medium to that volume (ICRU 1980). That energy frequently leads to ionisation, as will be discussed quantitatively in this chapter. The ionisation results in the generation of secondary electrons, i.e. atomic electrons ejected with kinetic energy and referred to as δ-rays. As these carry energy *away* from the interaction site, these do not contribute to the absorbed dose at the interaction site. Hence, for radiation dosimetry calculations, it is necessary to use the restricted collision stopping power which excludes those interactions that result in energy being transported at a distance from the interaction site. In this chapter, we will consider the restricted collision stopping power for electrons only.

10.1.4 Mean Energy Expended to Create an Ion Pair, W

The transfer of kinetic energy from a decelerating charged particle to atoms in the medium results in excitation and ionisation of the medium. Ionisation within tissue leads to disruptions in the cellular DNA which culminates in biological damage. Hence, the ability to calculate the ionisation yield for a given irradiation condition in order to estimate such effects is desirable. A metric for calculating this yield on the basis of energy lost by a charged particle is the mean energy expended to create an ion pair and which is denoted by W.

An important application of W results from the fact that absorbed dose calibrations in the clinical radiation setting are generally derived from measurements of the amount of ions (either in the form of collected charge or electric current) produced within a gas cavity within a medium. Consider, for example, a small volume of gas contained within an irradiated medium. Electrons traversing the medium create an absorbed dose, and the electron fluence traversing the cavity will create ionisations within the gas. Bragg–Gray cavity theory enables the determination of the absorbed dose from a measurement the collected electric charge in the gas,

$$D_{\text{med}} = \left(\frac{\mathrm{d}Q}{\mathrm{d}m_{\text{gas}}}\right) W_{\text{gas}} \left.\frac{\mathrm{d}E}{\rho\,\mathrm{d}x}\right|_{\text{gas}}^{\text{med}}, \tag{10.1}$$

where $\mathrm{d}Q/\mathrm{d}m_{\text{gas}}$ is the electric charge released through ionisations per unit mass within the cavity,[1] W_{gas} is the mean energy expended that for the gas and $\mathrm{d}E/\rho\,\mathrm{d}x|_{\text{gas}}^{\text{med}}$ is the ratio of the mass collision stopping power of the medium for electrons to that of the gas.

10.2 Total Collision Stopping Power

10.2.1 Massive Projectiles $(m \gg m_e)$

Neglecting spin, the Bethe total mass collision stopping power is given by the sum of the Bethe soft mass collision power expression of (8.344) and the hard mass collision stopping power of a massive spin-0 projectile from (9.7):

[1] The symbol for electric charge used here, Q, is not to be confused with that for energy transfer.

$$
\begin{aligned}
\frac{\mathrm{d}E}{\rho \mathrm{d}x} &= \left.\frac{\mathrm{d}E}{\rho \mathrm{d}x}\right|_S + \left.\frac{\mathrm{d}E}{\rho \mathrm{d}x}\right|_H \\
&= C\left(\frac{Z}{A}\right)\left(\frac{z}{\beta}\right)^2\left[\ln\left(\frac{2m_\mathrm{e}Q_\mathrm{C}}{\bar{I}^2}\beta^2\right) - \ln(1-\beta^2) - \beta^2 + \ln\left(\frac{Q_\mathrm{max}}{Q_C}\right) - \beta^2\right] \\
&= C\left(\frac{Z}{A}\right)\left(\frac{z}{\beta}\right)^2\left[\ln\left(\frac{2m_\mathrm{e}Q_\mathrm{max}}{\bar{I}^2}\beta^2\right) - \ln(1-\beta^2) - 2\beta^2\right].
\end{aligned}
\tag{10.2}
$$

As desired, the Q_C factors in the logarithmic terms of the soft- and hard-collision stopping powers cancel. As noted before, the maximum energy transfer to an electron for the case of a heavy projectile (i.e. one with mass $m \gg m_\mathrm{e}$) is $Q_\mathrm{max} = 2\, m_\mathrm{e}\gamma^2\beta^2$. Inserting this gives the complete mass collision stopping power for a massive spin-0 projectile:

$$
\begin{aligned}
\frac{dE}{\rho dx} &= C\left(\frac{Z}{A}\right)\left(\frac{z}{\beta}\right)^2\left[\ln\left(\frac{2m_\mathrm{e}Q_\mathrm{max}}{\bar{I}^2}\beta^2\right) - \ln(1-\beta^2) - 2\beta^2\right] \\
&= C\left(\frac{Z}{A}\right)\left(\frac{z}{\beta}\right)^2\left[\ln\left(\frac{4m_\mathrm{e}^2}{\bar{I}^2}\beta^2\right) - 2\ln(1-\beta^2) - 2\beta^2\right] \\
&= 2C\left(\frac{Z}{A}\right)\left(\frac{z}{\beta}\right)^2\left[\ln\left(\frac{2m_\mathrm{e}}{\bar{I}}\beta\right) - \ln(1-\beta^2) - \beta^2\right]
\end{aligned}
\qquad \text{Spin-0} \tag{10.3}
$$

Similarly, the total mass collision stopping power or a massive spin-½ projectile is given by the sum of (8.344) and (9.2):

$$
\begin{aligned}
\frac{dE}{\rho dx} &= C\left(\frac{Z}{A}\right)\left(\frac{z}{\beta}\right)^2\left[\ln\left(\frac{2m_\mathrm{e}Q_\mathrm{C}}{\bar{I}^2}\beta^2\right) - \ln(1-\beta^2) - \beta^2\right. \\
&\quad \left. + \ln\left(\frac{Q_\mathrm{max}}{Q_\mathrm{C}}\right) - \beta^2 + \left(\frac{Q_\mathrm{max}}{2(E+m)}\right)^2\right] \\
&= C\left(\frac{Z}{A}\right)\left(\frac{z}{\beta}\right)^2\left(\ln\left(\frac{4m_\mathrm{e}^2}{\bar{I}^2}\beta^2\right) - 2\ln(1-\beta^2) - 2\beta^2 + \left(\frac{Q_\mathrm{max}}{2(E+m)}\right)^2\right) \\
&= 2C\left(\frac{Z}{A}\right)\left(\frac{z}{\beta}\right)^2\left(\ln\left(\frac{2m_\mathrm{e}}{\bar{I}}\beta\right) - \ln(1-\beta^2) - \beta^2 + \frac{1}{8}\left(\frac{Q_\mathrm{max}}{(E+m)}\right)^2\right) \\
&= 2C\left(\frac{Z}{A}\right)\left(\frac{z}{\beta}\right)^2\left(\ln\left(\frac{2m_\mathrm{e}}{\bar{I}}\beta\right) - \ln(1-\beta^2) - \beta^2 + \frac{1}{2}\left(\frac{m_\mathrm{e}}{(E+m)}\frac{\beta^2}{(1-\beta^2)}\right)^2\right).
\end{aligned}
\qquad \text{Spin-}1/2 \tag{10.4}
$$

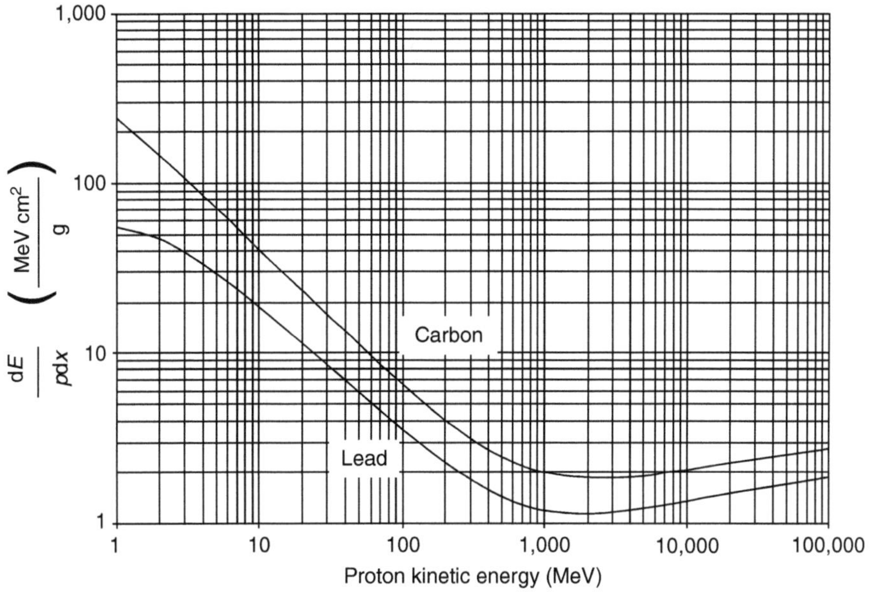

Fig. 10.1 Mass collision stopping powers for protons in carbon and lead as calculated from (10.4). These results exclude the density and shell correction factors that are derived in Chap. 12

As $m \gg m_e$, it is clear that the mass collision stopping power for a massive spin-½ projectile rapidly reduces to that for a spin-0 projectile at low kinetic energies (i.e. $(\beta^2/1-\beta^2)^2 \to 0$).

For the following quantitative consideration of the total collision stopping power, we will replace our usual high-Z elemental medium of calcium with that of lead. The reason for doing so shall become apparent. Figure 10.1 shows the mass collision stopping power calculated for a proton in carbon and lead as a function of the proton's kinetic energy. The two extremes in terms of atomic number are deliberately selected for a reason to be demonstrated shortly. Both curves of $\mathrm{d}E/\rho\,\mathrm{d}x$ exhibit the same characteristic behaviour of a decrease with increasing kinetic energy, due to the $1/\beta^2$ factor, to a broad minimum of about 1–2 MeV cm^2/g which occurs at a particle kinetic energy equal to about three times its rest mass. This is referred to as the minimally ionising region. With increasing kinetic energy, the mass collision stopping power increases logarithmically at large β due to the $\ln(\gamma\beta)$ term. This increase could be quenched, however, in condensed media due to polarisation, as discussed in Chap. 12. It will be noted from Fig. 10.1 that for a given projectile kinetic energy, the mass collision stopping power for carbon ($Z = 6$) is greater than that for lead ($Z = 82$). This is a result of the combined facts of the energy loss being dominated by interactions with atomic electrons over those with the nucleus and the electron density being proportional to Z/A. As a result, the ratio of the carbon to lead mass collision stopping powers is (excluding the effect of the mean excitation energy which is limited due to its placement within the logarithm)

$$\frac{\left(\frac{Z}{A}\right)_C}{\left(\frac{Z}{A}\right)_{Pb}} = \frac{\left(\frac{6}{12}\right)}{\left(\frac{82}{208}\right)} = 1.268.$$

In other words, the mass collision stopping power for carbon is greater than that for lead due to the greater number of electrons per unit mass.

10.2.2 Electron and Positron Projectiles

To obtain the unrestricted mass collision stopping power of an electron projectile, we sum the Bethe mass soft-collision stopping power (8.344) and the Møller mass hard-collision stopping power (9.10):

$$\begin{aligned}\left.\frac{\mathrm{d}E}{\rho dx}\right|_{e^-} &= C\left(\frac{Z}{A}\right)\left(\frac{1}{\beta}\right)^2\left[\ln\left(\frac{2m_\mathrm{e}Q_\mathrm{C}}{\bar{I}^2}\gamma^2\beta^2\right) - \beta^2 + \ln\left(\frac{E}{4Q_\mathrm{C}}\right)\right. \\ &\quad \left. +1 + \frac{1}{8}\left(\frac{\gamma-1}{\gamma}\right)^2 - \left(\frac{2\gamma-1}{\gamma^2}\right)\ln 2\right] \\ &= C\left(\frac{Z}{A}\right)\left(\frac{1}{\beta}\right)^2\left[\ln\left(\frac{m_\mathrm{e}E}{2\bar{I}^2}\gamma^2\beta^2\right) + f^-(\gamma)\right] \end{aligned} \tag{10.5}$$

where E is the electron kinetic energy and the parameter

$$f^-(\gamma) \equiv 1 - \beta^2 + \frac{1}{8}\left(\frac{\gamma-1}{\gamma}\right)^2 - \left(\frac{2\gamma-1}{\gamma^2}\right)\ln 2. \tag{10.6}$$

By using the relativistic relationship,

$$\begin{aligned} m_\mathrm{e}\gamma^2\beta^2 &= m_\mathrm{e}\left(\gamma^2 - 1\right), \\ &= E(\gamma+1) \end{aligned}$$

the (unrestricted) electron mass hard-collision stopping power is simplified to

$$\left.\frac{\mathrm{d}E}{\rho dx}\right|_{e^-} = C\left(\frac{Z}{A}\right)\left(\frac{1}{\beta}\right)^2\left[\ln\left(\frac{E^2(\gamma+1)}{2\bar{I}^2}\right) + f^-(\gamma)\right]. \tag{10.7}$$

Similarly, to obtain the positron complete collision stopping power, we add the Bethe mass soft- and the Bhabha mass hard-collision stopping powers:

$$\left.\frac{\mathrm{d}E}{\rho dx}\right|_{e^+} = C\left(\frac{Z}{A}\right)\frac{1}{\beta^2}\left[\ln\left(\frac{E^2(\gamma+1)}{2\bar{I}^2}\right) + f^+(\gamma)\right] \tag{10.8}$$

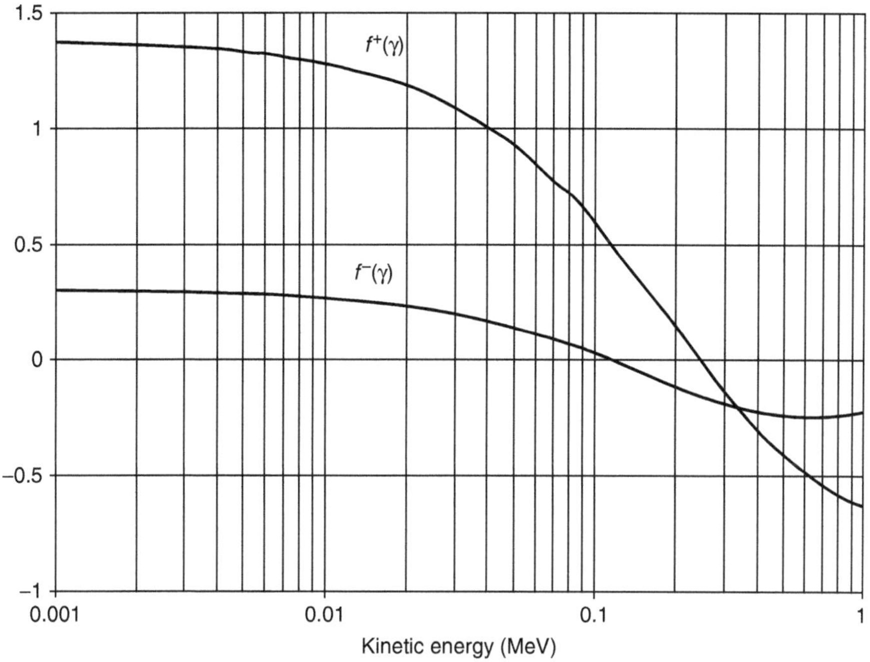

Fig. 10.2 The functions $f^-(\gamma)$ and $f^+(\gamma)$ of the mass collision stopping powers of electrons and positrons, respectively, as a function of the kinetic energy of the electron or positron

where

$$f^+(\gamma) = 2\ln 2 - \frac{\beta^2}{12}\left[23 + \frac{14}{(\gamma+1)} + \frac{10}{(\gamma+1)^2} + \frac{4}{(\gamma+1)^3}\right]. \tag{10.9}$$

A comparison of these electron and positron mass collision stopping powers shows that the difference between the electron and positron mass collisions stopping powers resides within the difference between $f^-(\gamma)$ and $f^+(\gamma)$. These two terms are plotted in Fig. 10.2 as functions of the kinetic energy of the incident electron/positron (only the unrestricted version of $\Delta = E/2$ is used for the electron term). At low energies, $f^+(\gamma)$ exceeds $f^-(\gamma)$ by about a factor of 3 and decreases with positron kinetic energy to equal $f^-(\gamma)$ at about 0.32 MeV. Note that both $f^-(\gamma)$ and $f^+(\gamma)$ become negative with increasing kinetic energy. However, when we compare the electron and positron unrestricted mass collision stopping powers for carbon and lead shown in Fig. 10.3, we see that there is little difference between the two which reflects the dominance of the logarithmic term in the collision stopping power expression over the magnitudes of $f^-(\gamma)$ and $f^+(\gamma)$. The mass collision stopping power curves exhibit the same characteristics as those derived earlier for the proton. A clear comparison of the curves shows that while the mass collision stopping

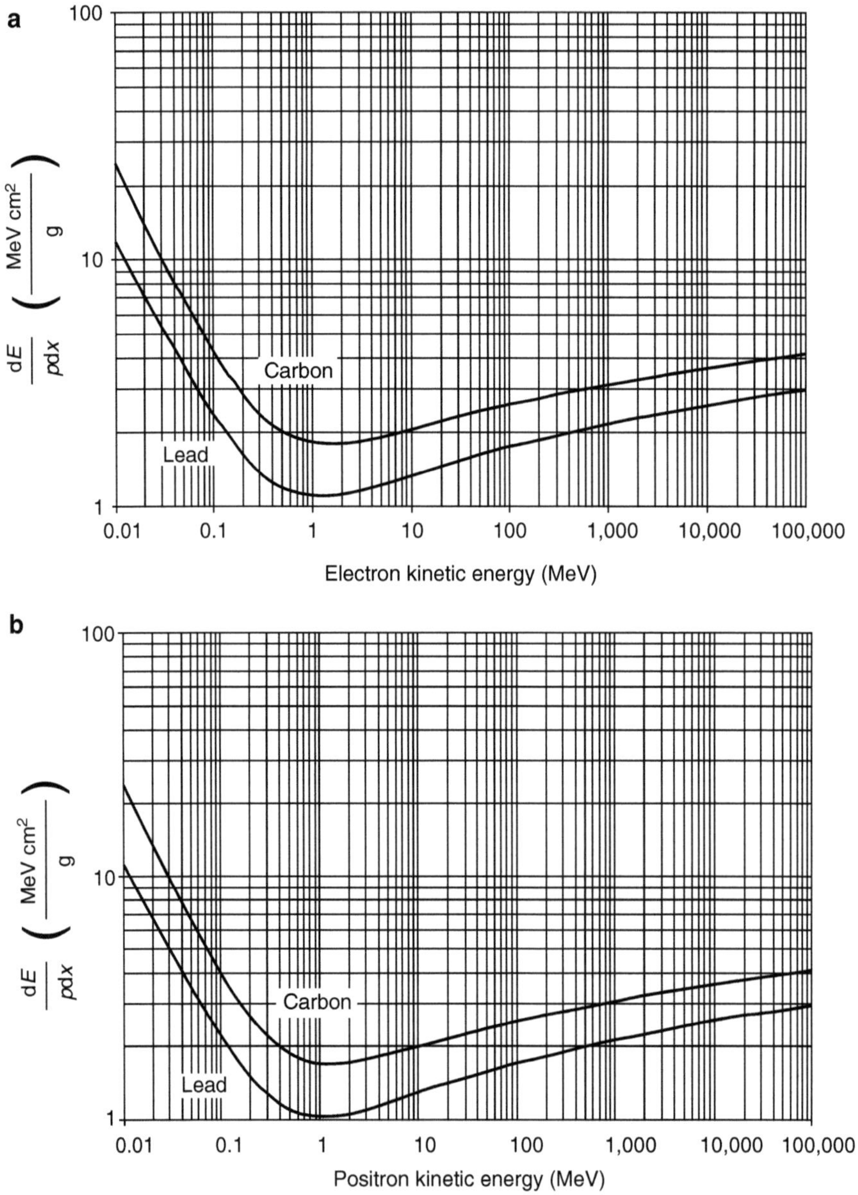

Fig. 10.3 Mass collision stopping powers for electrons and positrons in carbon and lead calculated from Eqs. (10.5) and (10.8), respectively. These results exclude the density and shell correction factors that are derived in Chap. 12

power of protons, electrons and positrons are roughly equal in the minimally ionising region, the kinetic energy at which this region occurs is equal to roughly equal to three times the particle rest mass (3 GeV for protons and 1.5 MeV for electrons and positrons) or for a speed of the order of $\beta \approx 0.8$.

10.3 Restricted Collision Stopping Power

An important refinement in the definition of the collision stopping power is the introduction of the *restricted* collision stopping power. Recall that the collision stopping power describes the mean rate of energy transfer from the charged projectile to the medium through both soft and hard collisions. This energy transfer can ionise an atom with the result that the ejected electron can carry considerable kinetic energy away a significant distance from the interaction site. The calculation of the energy deposition in a small volume using the mass collision stopping power will thus *overestimate* the amount of energy deposited within that. To correct for this, the restricted stopping power contains the sum of the soft-collision energy transfers and the energy transferred in hard collisions which result in δ-rays with kinetic energies which are less than a specified cut-off energy, Δ.

For example, the electron-restricted mass collision stopping power is

$$\begin{aligned}\left.\frac{dE}{\rho dx}\right|_{\Delta,e^-} &= \left.\frac{dE}{\rho dx}\right|_{S,e^-} + \left.\frac{dE}{\rho dx}\right|_{H,\Delta,e^-}\\ &= C\left(\frac{Z}{A}\right)\left(\frac{1}{\beta}\right)^2\left[\ln\left(\frac{2m_e Q_C}{I^2}\gamma^2\beta^2\right) - \beta^2 + \ln\left(\frac{\Delta\left(1-\frac{\Delta}{E}\right)}{Q_C}\right)\right.\\ &\quad \left. + \left(\frac{1}{1-\left(\frac{\Delta}{E}\right)} - 1\right) + \left(\frac{\gamma-1}{\gamma}\right)^2\frac{\Delta^2}{2E^2} + \left(\frac{2\gamma-1}{\gamma}\right)\ln\left(1-\frac{\Delta}{E}\right)\right]\\ &= C\left(\frac{Z}{A}\right)\left(\frac{1}{\beta}\right)^2\left(\ln\left(\frac{2\Delta(E-\Delta)}{I^2}(\gamma+1)\right) + f^-(\gamma,\Delta)\right)\end{aligned} \tag{10.10}$$

where

$$f^-(\gamma,\Delta) = \left(\frac{\Delta}{T-\Delta}\right) - \beta^2 + \left(\frac{2\gamma-1}{\gamma}\right)\ln\left(\frac{E-\Delta}{E}\right) + \left(\frac{\gamma-1}{\gamma}\right)^2\frac{\Delta^2}{2E^2}. \tag{10.11}$$

Note that if Δ is set equal to the maximum energy transfer specified by the kinematics (i.e. $E/2$ for electron and E for positrons), the restricted collision stopping power is equal to the collision stopping power.

Figure 10.4 shows the restricted mass collision stopping power for electrons in carbon as a function of the secondary electron kinetic energy.

10.4 Sources of Stopping Power Data

Although we have not yet concluded our derivation of the final collision stopping power where higher-order corrections are required, it is useful to cite sources of collision stopping power data that are available to the user.

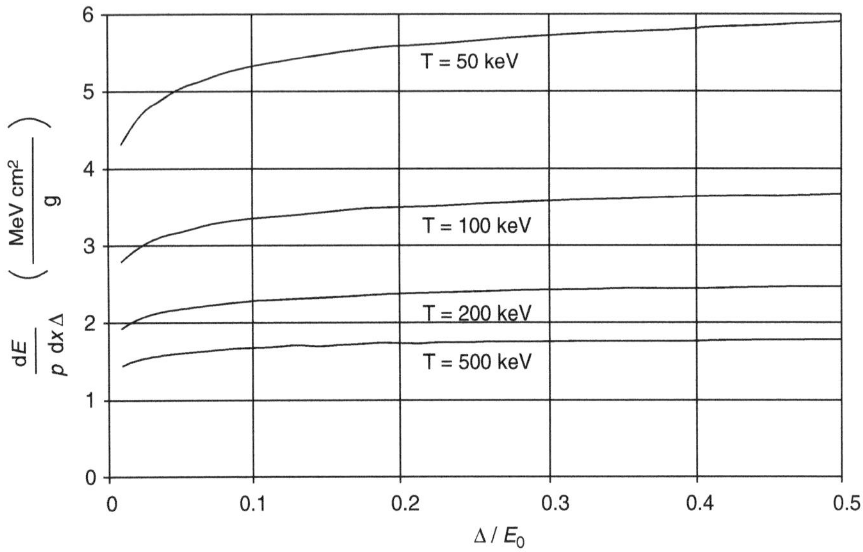

Fig. 10.4 Restricted mass collision stopping power for electrons in carbon as a function of the secondary electron kinetic energy normalised to the incident kinetic energy

The ICRU has produced three comprehensive reports containing tabular compilations of stopping powers. ICRU Report No 37 (1984) provides tables of both collision and radiative stopping powers for electrons and positrons in a variety of media for kinetic energies. ICRU Report No 49 (1993) provides tables of both electronic (hard) and nuclear (soft) collision stopping powers for protons and α-particles. ICRU Report No 73 (2005) provides tables of collision stopping powers for 16 different types of ions ranging from lithium to argon.

The National Institute of Science and Technology in the United States provides web-based access to a database containing the stopping powers of electrons, protons and α-particles, based upon the data contained within ICRU Reports No 37 and No 49, at http://www.nist.gov/pml/data/star/index.cfm.

10.5 Mean Energy Required to Create an Ion Pair, *W*

10.5.1 Introductory Comments

The immediate consequence of ionisation in a liquid or gaseous medium is the creation of an electron–ion pair. In some solids, ionisation can elevate electrons into the conduction band, thus forming an electron–hole pair (following the convention used by ICRU Report No 31 (1979), the term 'ion pair' will also be used to describe both electron–ion and electron–hole pairs). All ionising radiations create energetic electrons within the irradiated medium. Any one of these electrons can cause further ionisations provided that its kinetic energy exceeds an ionisation potential of an atom or molecule in the medium. Thus, the foundation of any theory to calculate the mean energy required to create an ion pair, W, will be based upon knowledge of the inelastic cross sections of electron–atom or electron–molecule collisions. Should the primary radiation not be electrons, then additional interaction knowledge is required. In the case of photon irradiation, photon interaction cross sections (e.g. photoelectric effect and Compton scatter) need be known and incorporated within the calculation. If the primary radiation is composed of hadrons or heavy ions, additional complications arise as the secondary ionising particles could include complex entities such as protons or nuclear fragments. As a consequence, calculations of W are largely limited to electron irradiation.

Inevitably, any theory of calculating W must address two fundamental empirically noted questions:

- Why is the magnitude of W what it is?
- Why is W relatively insensitive to the kinetic energy, mass and charge of the projectile (subject to the projectile's velocity being greater than those of the valence electrons)?

The first attempt to calculate W was by Bethe (1930); however, the result presented a correlation between the ionisation potential and W whereas experimental data had demonstrated a near independence. In 1946, Fano (1946) adopted a 'bookkeeping' approach, essentially a model of electron transport, to improve the estimate of W. This bookkeeping mechanism was introduced by Fowler (1923) through an integral equation of electron transport and which is derived in Sect. 10.5.3. Inherent to such a calculation is the need for accurate cross sections for inelastic electron–atom/ electron–molecule collisions. For example, Erskine (1954) used cross sections calculated using the first-order Born approximation to estimate W in helium, and Miller (1956) compiled a set of cross sections for helium from experimental and theoretical sources.

Prior to approaching the transport-equation-based methods of estimating W, we first consider the systematics of the ratio of W to the ionisation potential and their dependence upon underlying physics.

10.5.2 Ionisation Energy Balance

Platzman (1961) presented an equation of energy balance for a charged particle with kinetic energy E_0 which is stopped in a gaseous medium. Then, we may write

$$E_0 = N_e(\bar{E}_{\text{Ion}} + \bar{\varepsilon}) + N_{\text{Exc}}\bar{E}_{\text{Exc}} \tag{10.12}$$

where

- N_e is the total number of electrons generated through ionisation
- N_{Exc} is the total number of excited atomic states created
- $\bar{E}_{Ion}$ is the mean energy required to produce an ion
- $\bar{\varepsilon}$ is the mean energy of the secondary electrons which are insufficiently energetic to cause further ionizations
- $\bar{E}_{Exc}$ is the mean energy of the discrete excited atomic states

By definition, the mean energy expended to produce an ion pair is given by

$$W = \frac{E_0}{N_e} = (\bar{E}_{Ion} + \bar{\varepsilon}) + \left(\frac{N_{Exc}}{N_e}\right)\bar{E}_{Exc}. \tag{10.13}$$

and the ratio of W to the first ionisation energy[2] I is

$$\frac{W}{I} = \frac{\bar{E}_{Ion}}{I} + \frac{\bar{\varepsilon}}{I} + \left(\frac{N_{Exc}}{N_e}\right)\frac{\bar{E}_{Exc}}{I}. \tag{10.14}$$

It is possible to describe qualitatively the magnitude of the three terms on the right-hand side of (10.14) from first principles. First, one would expect

$$\frac{\bar{E}_{Ion}}{I} > 1 \tag{10.15}$$

due to the fact that excitation and other non-ionisation channels exist, especially for molecules where there are additional rotational and vibrational modes available. Similarly, one would also expect that

$$\frac{\bar{E}_{Exc}}{I} < 1 \tag{10.16}$$

although this ratio is not that much less than unity as the energy levels of most discrete excited states are relatively near I. The ratio $\bar{\varepsilon}/I$ will be small with values typically about 0.3 for noble gases and smaller for molecular gases. Finally, the ratio of the number of discrete excited states to the number of electrons generated, N_{Exc}/N_e is estimated in ICRU Report Number 31 (1979) to be about 0.3 for closed-shell atoms and close to 1 for closed-shell molecules. Overall, the magnitude of W/I is greater than unity with values ranging from about 1.7 for noble gases to up to 3.2 for alkaline earths.

[2] This is not the same as the mean excitation energy discussed in earlier chapters.

10.5.3 Fowler Equation for W

10.5.3.1 Introduction

As described in the previous section, a critical piece of knowledge required in understanding *W* within a material is the number of ion pairs that are generated. This will require two linked processes (Inokuti 1975):

- An understanding of the magnitudes and energy dependencies of inelastic cross sections (the sum of excitation/molecular disassociation and ionisation cross sections) for electron–matter interactions
- Tracking (or bookkeeping) of the secondary electrons generated through ionizations

The above could be addressed by the Monte Carlo method, but analytical means are also possible as first developed by Fowler (1923). The method was expanded upon by Inokuti (1975). The Fowler equation provides a reasonable qualitative description of the number of ion pairs produced by an energetic electron within a medium. Importantly, it also allows a mean of calculating *W* using cross sections of electron–atom/electron–molecule collisions.

We begin with an initial derivation of the Fowler equation by assuming the medium to be an ideal gas with a single ionisation energy. This simplification is adequate for the derivation. However, practical considerations are required when a real gas is considered as the medium. These include the effects of dissociation occurring in competition with autoionisation, the presence of multiple ionisation thresholds associated with different states and the possibility of multiple ionisations occurring as a result of a single interaction. Following this initial derivation, we then extend the result to a more general expression for a real gas and yield a generalised form of the Fowler equation.

10.5.3.2 Simple Fowler Equation

In our initial derivation of the Fowler equation, we begin with the consideration of an electron with kinetic energy E_0 in a medium. In interacting with the atomic electrons of an atom or molecule, one of two processes can occur, as shown in Fig. 10.5.

The first process is one in which the incident electron excites an atomic electron (or disassociates the molecule) from the ground state to an excited state through the transfer of an amount of energy ΔE_j, where j denotes a discrete state of excitation. The incident electron then continues travelling within the medium with a kinetic energy $E_0 - \Delta E_j$. The second process is that of ionisation in which the final state consists of a pair of electrons – the scattered incident electron with a kinetic energy equal to $E_0 - \Delta E$, where ΔE is the kinetic energy transferred, and the ejected atomic electron with a kinetic energy equal to $\Delta E - I$, where I is the single ionisation energy. Obviously, ionisation cannot occur unless $\Delta E > I$.

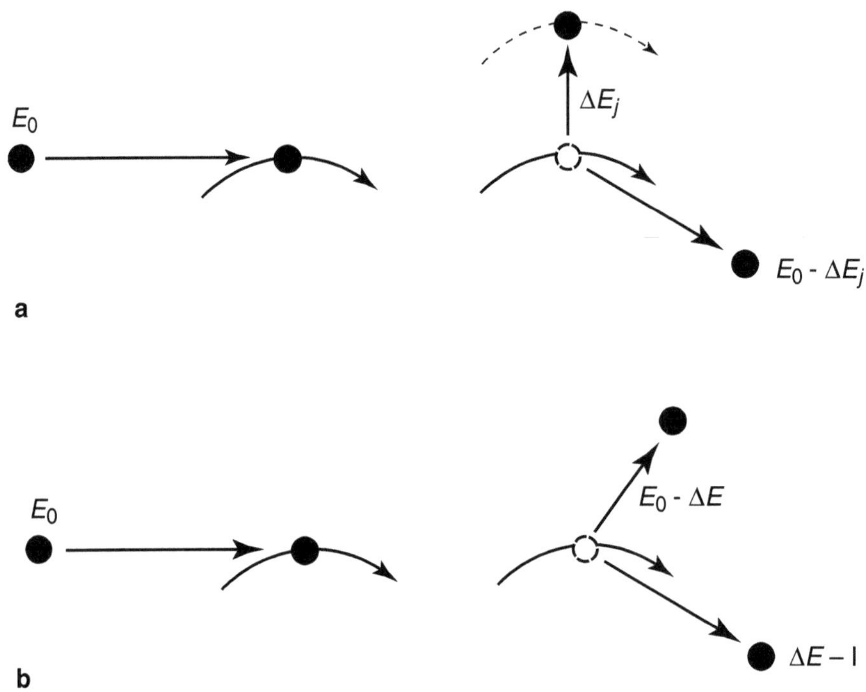

Fig. 10.5 The two processes considered in the Fowler equation describing electron–atom/molecule interactions: (**a**) excitation or molecular disassociation and (**b**) ionisation

The total inelastic cross section is the sum of the ionisation cross section and the excitation cross sections summed over all available final states:

$$\sigma_{\mathrm{Inel}}(E_0) = \sigma_{\mathrm{Ion}}(E_0) + \sum_j \sigma_j(E_0). \tag{10.17}$$

In order to simplify the presentations of the derivation, we define the following probabilities on the basis of these total cross sections:

- The probability of an excitation of an atomic electron to the jth-atomic state:

$$p_j(E_0) = \frac{\sigma_j(E_0)}{\sigma_{\mathrm{Inel}}(E_0)}. \tag{10.18}$$

- The probability of an ionisation occurring:

$$p_{\mathrm{Ion}}(E_0) = \frac{\sigma_{\mathrm{Ion}}(E_0)}{\sigma_{\mathrm{Inel}}(E_0)}. \tag{10.19}$$

The differential of the ionisation probability with energy is

$$\frac{dp_{\text{Ion}}(E_0, \Delta E)}{d\Delta E} = \frac{\frac{d\sigma_{\text{Ion}}(E_0)}{d\Delta E}}{\sigma_{\text{Inel}}(E_0)}. \tag{10.20}$$

From (10.17), (10.18) and (10.19), inclusive,

$$p_{\text{Ion}}(E_0) + \sum_j p_j(\Delta E_j) = 1. \tag{10.21}$$

We define $N(E_0)$ as the mean number of ion pairs that are produced ultimately by the interactions of an electron with an initial kinetic energy E_0 with the atoms and molecules of the medium until the kinetic energy falls below the ionisation energy, I. The mean number of ion pairs produced through the initial ionisation is

$$N_{\text{Ion},1}(E_0) = p_{\text{Ion}}(E_0). \tag{10.22}$$

Note that this describes only the initial ionisation and does not account for subsequent ionisations from the secondary electrons. The ionisation product is a pair of electrons with kinetic energies equal to $E_0 - \Delta E$ and $\Delta E - I$. The subsequent mean number of ion pairs produced by these two secondary electrons is

$$\begin{aligned} N_{\text{Ion},2}(E_0) &= \int_I^{\frac{1}{2}(E_0+I)} d(\Delta E) \frac{dp_{Ion}(E_0, \Delta E)}{d\Delta E} N(E_0 - \Delta E) \\ &+ \int_I^{\frac{1}{2}(E_0+I)} d(\Delta E) \frac{dp_{\text{Ion}}(E_0, \Delta E)}{d\Delta E} N(\Delta E - I) \\ &= \int_I^{\frac{1}{2}(E_0+I)} d(\Delta E) \frac{dp_{Ion}(E_0, \Delta E)}{d\Delta E} (N(E_0 - \Delta E) + N(\Delta E - I)). \end{aligned} \tag{10.23}$$

We now consider the excitation/disassociation process. Here, an ion pair has not been formed, but a secondary electron with kinetic energy $E_0 - \Delta E_j$ continues into the medium following the interaction. The mean number of ion pairs that are eventually generated as a result of this initial interaction is given by

$$N_{\text{exc}}(E_0) = \sum_j p_j(E_0) N(E_0 - \Delta E_j). \tag{10.24}$$

Hence, the total number of ion pairs that are generated from a single electron with kinetic energy E_0 is given by the summation of (10.21), (10.22) and (10.23) or

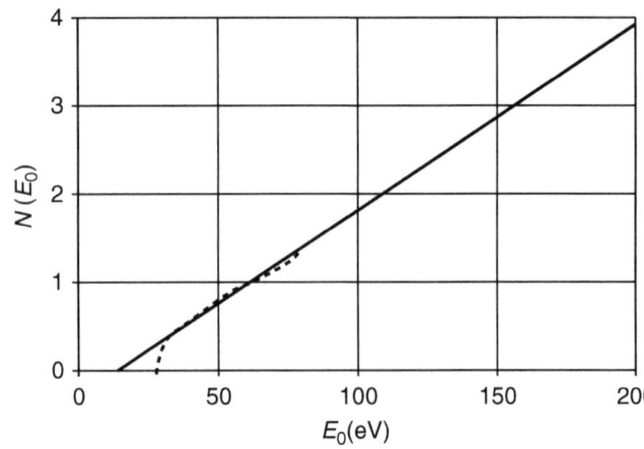

Fig. 10.6 The number of ion pairs generated by electrons in helium gas as a function of electron kinetic energy as calculated from the Fowler equation. The *solid line* is an extrapolation to from high-to-low energies whereas the *dotted line* indicates the result of the solution to the Fowler equation at low energies (Redrawn from Inokuti (1975) and Miller (1956))

$$N(E_0) = p_{\text{Ion}}(E_0) + \int_I^{\frac{1}{2}(E_0+I)} \text{d}(\Delta E)\frac{\text{d}p_{\text{Ion}}(E_0,\Delta E)}{d\Delta E}(N(E_0-\Delta E)+N(\Delta E-I)) + \sum_j p_j(E_0)N(E_0-\Delta E_j). \tag{10.25}$$

This is a simple form of the Fowler equation, which is a Volterra integral equation of the second kind, for the case of only a single secondary electron produced through the ionisation. The right-hand side enumerates the different contributions to$N(E_0)$ that occur as the result of an initial inelastic interaction by an electron with kinetic energy E_0. An example of the numerical solution for gaseous helium is shown in Fig. 10.6 in which the number of ion pairs produced as a function of initial electron kinetic energy are shown. There is very nearly a linear behaviour in the extrapolation, with an abscissa intercept equal to about 14 eV, but with a calculated cut-off at about 25 eV.

Solutions of Fowler's equation reveal many common characteristics demonstrated by Fig. 10.6. Except for the region of $E_0 \approx I$, $N(E_0)$ is a linear function of E_0 with a slope equal to $1/W$. For electron kinetic energies of the order of the ionisation potential, the behaviour of $N(E_0)$ becomes erratic as a reflection of uncertainties in the cross sections $\sigma_{\text{Ion}}(E_0)$ and $\sigma_j(E_0)$ at low energies.

10.5.3.3 Calculation of *W* from the Simple Fowler Equation

Due to the linearity of the $N(E_0)$ versus E_0 plot at large values of E_0 and as the extrapolation back to $N(E_0) = 0$ yields an intercept of $E_0 = U > 0$, one can approximate $N(E_0) = 0$ by a linear function for $E_0 \gg I$:

$$N(E_0) \cong \frac{E_0 - U}{W}. \tag{10.26}$$

Substituting (10.26) into (10.25) provides us with

$$\frac{E_0 - U}{W} = p_{\text{Ion}}(E_0) + \int_I^{\frac{1}{2}(E_0+I)} \mathrm{d}(\Delta E) \frac{\mathrm{d}p_{\text{Ion}}(E_0, \Delta E)}{\mathrm{d}\Delta E} \left(\frac{E_0 - \Delta E - I - 2U}{W} \right) + \sum_j p_j(E_0) \frac{E_0 - \Delta E_j - U}{W}. \tag{10.27}$$

We can then solve for W by rearranging (10.27) and recalling (10.18), (10.19) and (10.21) to give

$$W \cong \sum_j \Delta E_j \frac{\sigma_j(E_0)}{\sigma_{\text{Ion}}(E_0)} + I + U. \tag{10.28}$$

The reader considering (10.28) will note that it cannot be vigorously correct as the right-hand side contains E_0 as a variable, whereas the left-hand side is a constant. However, as empirical data demonstrate that W is a linear function of E_0 for high electron kinetic energies, it would be reasonable to evaluate (10.28) in that energy regime. Thus, a solution of (10.28) requires knowledge of the intercept, U.

Inokuti (1975) described a method of determining this intercept. At high kinetic energies (i.e. $E_0 \gg I$, from which it follows that $E_0 \gg \Delta E_j$), one can use (10.26) to substitute for $N(E_0)$, $N(E_0 - \Delta E_j)$ and $N(E_0 - \Delta E)$ in (10.25). However, as the magnitude of $(\Delta E - I)$ can be comparable to I, (10.26) cannot be used to substitute for $N(\Delta E - I)$. Thus, our use of (10.26) to yield (10.28) will have introduced an error through the $\int_I^{1/2(E_0+I)} \mathrm{d}(\Delta E)(\mathrm{d}p_{\text{Ion}}(E_0, \Delta E)/\mathrm{d}\,\Delta E)N(\Delta E - I)$ term. To resolve this, consider the error caused by the use of (10.26) in that term,

$$\varepsilon = \int_I^{\frac{1}{2}(E_0+I)} \mathrm{d}(\Delta E) \frac{\mathrm{d}p_{\text{Ion}}(E_0, \Delta E)}{\mathrm{d}\Delta E} N(\Delta E - I) - \int_I^{\frac{1}{2}(E_0+I)} \mathrm{d}(\Delta E) \frac{\mathrm{d}p_{\text{Ion}}(E_0, \Delta E)}{\mathrm{d}\Delta E} \frac{(\Delta E - I - U)}{W}, \tag{10.29}$$

and the fact that,

$$N(E_0) = 0 \quad \text{for} \quad E_0 < I. \tag{10.30}$$

The lower limit of the first integral of (10.29) is actually equal to $2I$, and hence, it would be reasonable to replace $N(\Delta E - I)$ with (10.25) in that integral to yield an estimate of the error as

$$\varepsilon = -\int_{I}^{2I} \mathrm{d}\,(\Delta E) \frac{\mathrm{d}p_{Ion}(E_0, \Delta E)}{\mathrm{d}\,\Delta E} \frac{(\Delta E - I - U)}{W} \tag{10.31}$$

provided that $(\Delta E - I)$ is larger than I. Setting the error given by (10.31) equal to zero, we obtain an estimate of the intercept:

$$U = \frac{\int_{I}^{2I} \mathrm{d}(\Delta E)\Delta E\left(\frac{\mathrm{d}p_{\mathrm{Ion}}(E_0, \Delta E)}{\mathrm{d}\Delta E}\right)}{\int_{I}^{2I} \mathrm{d}(\Delta E)\left(\frac{\mathrm{d}p_{\mathrm{Ion}}(E_0, \Delta E)}{\mathrm{d}\Delta E}\right)} - I$$
$$\equiv \Delta E_1 - I \tag{10.32}$$

where ΔE_1 is the average energy transfer for ionisations occurring for $I < \Delta E < 2I$. The ejected electron (δ-ray) is incapable of further ionisations and is referred to as a subionisation electron. Substituting (10.32) into (10.28) gives us an improved estimation of the mean energy expended to create an ion pair:

$$W \cong \sum_j \Delta E_j \frac{\sigma_j(E_0)}{\sigma_{\mathrm{Ion}}(E_0)} + \Delta E_1 \tag{10.33}$$

The first term of (10.33) is the mean energy per ion pair that is diverted into excitations, whereas the second term is the mean energy required in an ionisation.

10.5.3.4 Generalised Form of the Fowler Equation

Recall that the above derivation was for an ideal gas medium with a single ionisation energy. To extend this result to more realistic media, we must account for the effects of dissociation, multiple ionisation threshold energies and multiple ionisations in order to yield a generalised form of the Fowler equation.

Dissociation Competing with Autoionisation

A non-ionising interaction will be grouped with the discrete ionisation term as the final state has only one electron. As a result, we can include this within the third term of (10.25), $\sum_j p_j(E_0) N(E_0 - \Delta E_j)$. However, there will be a difference in that

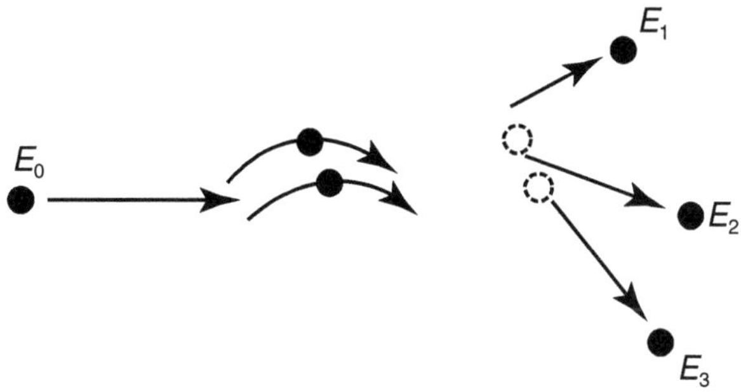

Fig. 10.7 Multiple ionisation case in which two atomic electrons are ejected during a single inelastic collision

ΔE_j in that it may exceed the first ionisation energy and this third term may include an integration over a continuous spectrum of ΔE_j to account for the continuum.

Multiple Ionisation Threshold Energies

Instead of a single ionisation threshold, for a real gas, we may allow for multiple energies, I_1, I_2, I_3 .... This would replace the single integral of the second term of (10.25) to be replaced by a summation of integrals, with each term of the sum corresponding to a single threshold energy.

Multiple Ionisations

This describes interactions with the final state following an inelastic collision having three or more electrons, where two or more electrons are δ-rays. Before deriving the full form of the Fowler equation for m-fold ionizations, we first consider the simpler case of double ionisation in which there are three electrons in the final state, as shown in Fig. 10.7, so as to explain the calculation principles.

The probability of a double-ionisation result (that in which three electrons are present the final state with the kinetic energies of E_1, E_2 and E_3 per inelastic collision with a projectile electron of kinetic energy E_0) is defined as, $\mathrm{Pr}^{(2)}(E_1, E_2, E_3; E_0)$, where the (2) superscript denotes the double-ionisation nature of this interaction. The total double-ionisation probability is the integral of this probability over all possible final-state kinetic energies:

$$p_{\mathrm{Ion}}^{(2)} = \int \mathrm{d}^3 E\, \mathrm{Pr}^{(2)}(E_1, E_2, E_3; E_0) \tag{10.34}$$

where

$$\mathrm{d}^3 E \equiv \mathrm{d}E_1\, \mathrm{d}E_2\, \mathrm{d}E_3. \tag{10.35}$$

The kinematical constraint upon the limits of integration of (10.34) are dictated by

$$E_0 - I^{(2)} = E_1 + E_2 + E_3. \tag{10.36}$$

Let us now consider the application of this double-ionisation process to the original Fowler's equation of (10.25). With the addition of the two extra electrons, we now add an additional term to the right-hand side of that equation or $2p_{\mathrm{Ion}}^{(2)} + \int d^3E\, Pr^{(2)}(E_1, E_2, E_3; E_0)\,(N(E_1) + N(E_2) + N(E_3))$.

We now embark upon deriving the generalised form of the Fowler equation for an m-fold ionisation which will incorporate the effects of dissociation and multiple ionization threshold energies. The final state will have $m+1$ electrons with kinetic energies of $E_1, E_2, \ldots, E_{m+1}$. The remnant ion is left in a final quantum state, k, with an ionisation threshold, $I_k^{(m)}$. Extending from our discussion of double ionisation, the net probability per inelastic collision of an m-fold ionisation with the ion in a final-state k is

$$p_{\mathrm{Ion},k}^{(m)}(E_0) = \int \mathrm{d}^{m+1}E\, Pr_k^{(m)}(E_1, E_2, \ldots, E_{m+1}; E_0) \tag{10.37}$$

where

$$\mathrm{d}^{m+1}E \equiv \mathrm{d}E_1 \mathrm{d}E_2 \ldots \mathrm{d}E_{m+1}. \tag{10.38}$$

The net probability of the m-fold ionisation occurring regardless of the final ion state is

$$p_{\mathrm{Ion}}^{(m)}(E_0) = \sum_k p_{\mathrm{Ion},k}^{(m)}(E_0). \tag{10.39}$$

The total probability of an ionisation event per elastic collision is

$$\sum_m p_{\mathrm{Ion},k}^{(m)}(E_0) = p_{\mathrm{Ion}}(E_0). \tag{10.40}$$

and the expectation value of the number of ejected electrons is

$$\overline{m(E_0)} = \sum_m m p_{\mathrm{Ion}}^{(m)}(E_0). \tag{10.41}$$

Then, the generalised Fowler equation for m-fold ionisations is

$$N(E_0) = \sum_m m p_{\mathrm{Ion}}^{(m)}(E_0) + \sum_m \sum_k \int \mathrm{d}^3 E_{m+1} Pr_k^{(m)}(E_1, E_2, \ldots, E_{m+1}; E_0) \sum_{\mu=1}^{m+1} N(E_\mu)$$
$$+ \sum_j p_j(E_0) N(E_0 - \Delta E_j). \tag{10.42}$$

The kinematical constraint on the limits of the $(m+1)$-fold integral is

$$E_0 - I_k^{(m)} = \sum_\mu^{m+1} T_\mu. \tag{10.43}$$

We can repeat the earlier process of estimating the energy expended in the production of an ion pair. We begin by first substituting (10.26) into (10.42), and then incorporate (10.21) and (10.37), (10.38), (10.39), (10.40) and (10.41), inclusive, to obtain an estimate of W as derived from the generalised Fowler equation

$$W = \sum_j \Delta E_j \frac{\overline{p_j(E_0)}}{\overline{m(E_0)}} + \bar{I} + U \tag{10.44}$$

where a mean ionisation energy has been defined:

$$\bar{I} = \frac{\sum_m \sum_k I_k^{(m)} p_{\mathrm{Ion},k}(E_0)}{\overline{m(E_0)}}. \tag{10.45}$$

The result of (10.44) is a generalization of (10.28) for m-fold ionisation. U is the mean energy of a subionisation electron and is obtained in the same manner of the optimization process used in deriving (10.32).

10.5.4 Spencer–Fano Equation for W

Another transport-equation approach to calculating for the number of ion pairs was proposed by Spencer and Fano (1954; Douthat 1975) and uses the concept of the degradation spectrum. Imagine a mono-energetic beam of electrons with kinetic energy E_0 incident to a gas. Within this medium, these electrons will, through deceleration and the production of secondary electrons, yield a net flux of electrons with a continuous energy spectrum. The path length of all electrons with kinetic energies between E and $E + \mathrm{d}E$ is $y(E_0, E)\,\mathrm{d}E$, which is also a descriptor of the electron spectrum. This spectrum will not be derived here; one can note that in the extreme case of the continuous-slowing-down approximation (discussed in Chap. 13) for in which the projectile loses only a small amount of energy, the number of ion pairs produced is

$$N(E_0) = \rho_{\text{Molec}} \int_{I}^{E_0} \mathrm{d}E \, \sigma_{\text{Ion}}(E) \, y(E_0, E) \tag{10.46}$$

where ρ_{Molec} is the number of molecules per unit volume in the medium (i.e. the molecular number density) and I is the first ionisation energy. A scaling property of $y(E_0, E) \, \mathrm{d}E$ was found by Douthat (1975) who showed that the quantity ρ_{Molec} $(E/E_0) \ln(E_0/I) \, \sigma_{\text{Ion}}(E) \, y(E_0, E)$ plotted as a function of the variable,

$$\xi = \frac{\ln\left(\dfrac{E}{I}\right)}{\ln\left(\dfrac{E_0}{I}\right)}$$

was virtually independent of the projectile electron's kinetic energy, E. Fano and Spencer (1975) subsequently defined the quantity

$$z(\xi) = \rho_{\text{Molec}} \, \sigma_{\text{St}}(E) \left(\frac{\ln\left(\dfrac{E}{I}\right)}{\xi} \right) \left(\frac{E}{E_0}\right) y(E, E_0) \tag{10.47}$$

where $\sigma_{\text{St}}(\xi)$ is the stopping cross section. $z(\xi)$ describes the amount of energy transferred to the secondary electrons and which is used by some to cause further ionisations. Fano and Spencer referred to $z(\xi)$ as the *recycling parameter* and has a limited range of values, e.g. $\int_0^1 \mathrm{d}\xi \, z(\xi)$ is reported to lie within the range of 1.3–1.6 in helium for electron kinetic energies between 2 and 100 keV.

Using the above, the expression for $N(E_0)$ can now be rewritten as

$$N(E_0) = E_0 \int_0^1 \mathrm{d}\xi \, z(\xi) \, \frac{\sigma_{\text{Ion}}(\xi)}{\sigma_{\text{St}}(\xi)} \tag{10.48}$$

which leads to an expression of the mean energy required to produce an ion pair:

$$\begin{aligned} W(E_0) &= \frac{E_0}{N(E_0)} \\ &= \frac{1}{\int_0^1 \mathrm{d}\xi z(\xi) \dfrac{\sigma_{\text{Ion}}(\xi)}{\sigma_{\text{St}}(\xi)}}. \end{aligned} \tag{10.49}$$

It should be noted that $\Delta E \, \sigma_{\text{Ion}}(\xi)/\sigma_{\text{St}}(\xi)$ is the number of ion pairs produced directly by the projectile electron within the energy interval $E - \Delta E$ to E. The integrand of (10.49) is insensitive to E.

Bibliography and Further Reading

Attix FH. Introduction to radiological physics and radiation dosimetry. New York: Wiley; 1986.

Berger SM, Bergstrom Jr PM. Changes in the U.S. primary standard for the air kerma from gamma ray beams. J Res Natl Inst Stand Technol. 2003;108:359–81.

Inokuti M, Douthat DA, Rau ARP. Degradation spectra and ionisation yields of electrons in gases. In: Booz J, Ebert HG, Smith BGR, editors. Proceedings of the fifth symposium on microdosimetry, Report No EUR-5452. Luxembourg: Commission of the European Communities; 1976.

Johns HE, Cunningham JR. The physics of radiology. 4th ed. Springfield: Charles C Thomas; 1983.

McGinnies RT. Energy spectrum resulting from electrons slowing down, NBS Circular No 597. Washington, DC: National Bureau of Standards; 1959.

Podgoršak EB. Radiation physics for medical physicists. Berlin: Springer; 2010.

Rohrlich F, Carlson BC. Positron-electron differences in energy loss and multiple scattering. Phys Rev. 1954;3:38–44.

Thomson JJ, Thomson GP. Conduction of electricity through gases, vol. II. 3rd ed. Cambridge: Cambridge University Press; 1933.

References

Bethe H. Zur theorie des durchgangs schneller korpuskularstrahlen durch materie. Ann Phys. 1930;5:324–400.

Douthat DA. Energy deposition by electrons and degradation spectra. Radiat Res. 1975;64:141–52.

Erskine GA. Calculation of the energy per ion pair for alpha-particles in helium. Proc R Soc Lond A. 1954;224:362–73.

Fano U. On the theory of ionization yield of radiations in different substances. Phys Rev. 1946;70:44–52.

Fano U, Spencer LV. Quasi-scaling of electron degradation spectra. Int J Radiat Phys Chem. 1975;7:63–76.

Fowler RH. Contributions to the theory of the motion of α-particles through matter. Part II. Ionizations. Proc Camb Philos Soc. 1923;21:521–40.

ICRU. Average energy required to produce an ion pair. Report No 31. Bethesda: International Commission on Radiation Units and Measurements; 1979.

ICRU. Radiation quantities and units. Report No 33. Bethesda: International Commission on Radiation Units and Measurements; 1980.

ICRU. Stopping powers for electrons and positrons. Report No 37. Bethesda: International Commission on Radiation Units and Measurements; 1984.

ICRU. Stopping powers and ranges for protons and alpha particles. Report No 49. Bethesda: International Commission on Radiation Units and Measurements; 1993.

ICRU. Stopping of ions heavier than helium. Report No 73. Bethesda: International Commission on Radiation Units and Measurements; 2005.

Inokuti M. Ionization yields in gases under electron irradiation. Radiat Res. 1975;64:6–22.

Miller WF. Mean energy per ion pair for electrons in helium. Bull Am Phys Soc. 1956;1:202.

Platzman RL. Total ionization in gases by high-energy particles: an appraisal of our understanding. Int J Appl Radiat Isot. 1961;10:116–27.

Spencer LV, Fano U. Energy spectrum resulting from electron slowing down. Phys Rev. 1954;93:1172–81.

Chapter 11
Mean Excitation Energy

Contents

Abstract The mean excitation energy, $\bar{I}$, appears in the logarithmic term of the soft collision stopping power. In practice, it is determined through extraction from measurements of the energy loss of a charged particle traversing a medium. However, there remains significant interest in understanding the theoretical aspects of the parameter, including deriving it ab initio and discovering that quite simplistic models can recreate the fundamental physical properties of $\bar{I}$. In this chapter, theoretical models of increasing complexity are considered in the calculation of the mean excitation energy. The chapter is concluded with a review of the mechanisms of experimental determination of $\bar{I}$ and parameterisations of the quantity. Although the latter are of limited practical use since tabulated values of $\bar{I}$ exist, it is of historical interest to consider them.

B.J. McParland, *Medical Radiation Dosimetry*,
DOI 10.1007/978-1-4471-5403-7_11,

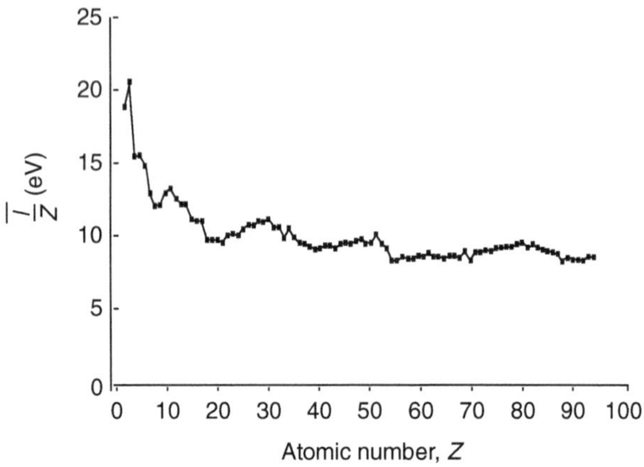

Fig. 11.1 Mean excitation energy normalised to the atomic number for elements as a function of atomic number (Data are from ICRU Report 37 (1984))

11.1 Introductory Comments

The mean logarithmic excitation energy is defined by, following our discussion of Chap. 8,

$$\ln \bar{I} = \sum_{j=0}^{f} f_{j0} \ln\left(E_j - E_0\right). \tag{11.1}$$

$\ln \bar{I}$ is a property of the medium alone and is independent of the nature of the charged projectile. Further, as $\bar{I}$ resides within the argument of a logarithm, changes and uncertainties in its value can have limited effects upon the value of the collision stopping power.

Values of $\bar{I}$ can be determined through both theoretical or experimental means. Because of the complexities inherent to the calculation and the need of knowledge of the atomic wavefunction, ab initio calculations of $\bar{I}$ have tended to be limited to monatomic gases of low atomic number. As such, the number of elements for which $\bar{I}$ has been calculated is small with the result that $\bar{I}$ is obtained largely through measurement. Both theoretical and experimental methods of evaluating $\bar{I}$ are investigated in this chapter.

We begin our discussion of the mean excitation energy with the theoretical means of determining the quantity. This is done, at first, by a very simple (and crude) estimate based upon the ratio of the Bohr velocity and atomic radius. Despite its crudeness, the estimate yields a surprisingly reasonable result. Then, following the work of Bloch (1933), we derive a more elegant means of calculating $\bar{I}$ using the Thomas–Fermi model of the atom that we derived in Chap. 5. We also examine the optical dipole oscillator strength and the local plasma approximation as approaches to calculating $\bar{I}$.

Following this, we look at the considerations that must be accounted for in the experimental determination of $\bar{I}$ and the means of how it is measured. We then consider parameterisations of $\bar{I}$ as a function of Z which can be useful in practical medical radiation physics calculations instead of using tabular data.

Additional discussions of the mean excitation energy can be found in those reviews by Uehling (1954) and Fano (1964) and in ICRU Reports 37 (1984) and 49 (1993).

The general characteristics of $\bar{I}$ can be seen in Fig. 11.1 in which the mean excitation energy normalised to the atomic number for elements is shown as a function of atomic number. These data are from ICRU Report 37 and are largely based upon the $\bar{I}$ values obtained from energy loss transmission measurements of high-energy protons, as described in Sect. 11.3. It can be seen that this ratio decreases rapidly to a constant value of about 10 eV for Z greater than about 15. However, it is also seen that superimposed upon this general trend are discontinuities which can be attributed to the effects of atomic shell structure.

11.2 Theoretical Evaluations of $\bar{I}$

11.2.1 Introductory Comments

A theoretical evaluation of $\bar{I}$ will likely have limited application to medical radiation physics calculations. The desired accuracy in the numerical value of the quantity will be best provided by experiment, as discussed in Sect. 11.3. Even so, a theoretical understanding of $\bar{I}$ can be revelatory by showing facets of the underlying physics that plays a role in the features of the mean excitation energy. In this section, we review a number of theoretical means of calculating $\bar{I}$ ab initio.

11.2.2 Ratio of Bohr Velocity to Bohr Radius

We can estimate crudely the dependence of the excitation energy upon the atomic number of the atom by treating the atomic electrons as oscillators and by approximating the atomic electron orbital frequency with the ratio of the Bohr velocity to the Bohr radius:

$$\begin{aligned}\omega &\approx \frac{v_0}{r_\infty} \\ &\approx \frac{Z\alpha c}{\left(\frac{\hbar c}{\alpha m_e}\right)} \\ &\approx Z\frac{\alpha^2 m_e}{\hbar}.\end{aligned} \tag{11.2}$$

Then, the mean excitation energy is

$$\begin{aligned}\bar{I} &\approx \hbar\omega \\ &\approx Z\alpha^2 m_e \\ &\approx 27Z\,\text{eV}.\end{aligned} \tag{11.3}$$

There should be no doubt that this is a crude model. However, the result suggests that the ratio of the mean excitation energy of a monatomic medium to its atomic number, $\frac{\bar{I}}{Z}$, is a constant, which was shown to be reasonably true for $Z > 15$ in Fig. 11.1. The constant of proportionality in (11.3) is to be expected to be an overestimate as we have used the Bohr velocity for the ground state, $v_0 = Z\alpha c$, which is the fastest of the atomic electrons. If we were to instead use the arithmetic mean and approximate a mean Bohr velocity, $\bar{v}_0 \approx \frac{Z\alpha c}{2}$, we obtain

$$\bar{I} \approx 13.5\,Z\,\text{eV}. \tag{11.4}$$

Quantitatively, this new result is closer to the results obtained from experiment as shown in Fig. 11.1.

We now attempt to develop further this 'back-of-the-envelope' calculation further by returning to the Thomas–Fermi statistical model of the atom that we developed in Chap. 5.[1]

11.2.3 Thomas–Fermi Model

Another means of estimating $\bar{I}$ is through the statistical model of Thomas and Fermi in describing the atom. This model was derived in Sect. 5.3.5 in the context of describing the atomic electron screening of the nuclear Coulomb potential. Recall that the radius of the Thomas–Fermi atom is defined as

$$\begin{aligned}r_{TF} \equiv& \frac{(3\pi)^{\frac{2}{3}}}{2^{\frac{7}{3}}} \frac{r_\infty}{Z^{\frac{1}{3}}} \\ \approx& 0.885 \frac{r_\infty}{Z^{\frac{1}{3}}}.\end{aligned}$$

We will crudely model the atom as being a homogeneous sphere of negative charge $-Ze$ surrounding a point charge $+Ze$. The sphere, or electron 'cloud', will be subject to two sources of potential energy, one due to the self-energy of the electrons, which we refer to as U_e, and the other due to that of the Coulomb interaction with the nucleus, which is referred to as U_n. In addition, there will be a kinetic energy contribution. Let us now evaluate these three sources.

[1] However, we must still recall the *caveats* articulated in Chap. 5 of the approximate nature of the Thomas–Fermi model, particular with respect to the minimum number of atomic electrons required to provide a statistically valid result.

First, we evaluate the self-energy, U_e. Consider the electron sphere to have a radius r_{TF} for a net charge of $-Ze$. Because of the assumption of the negative charge homogeneity, the negative charge density within the sphere is

$$\rho_e = \frac{3Ze}{4\pi r_{TF}^3}. \tag{11.5}$$

Then, the self-energy term is

$$U_e = \frac{\rho_e}{3\varepsilon_0} \int_0^{r_{TF}} d^3\mathbf{r}\, r^2. \tag{11.6}$$

The integration of (11.6) results in

$$\begin{aligned} U_e &= \frac{3}{5} \frac{(Ze)^2}{4\pi\varepsilon_0 r_{TF}} \\ &= \frac{3}{5} \alpha\hbar c \frac{Z^2}{r_{TF}} \end{aligned} \tag{11.7}$$

where the definition of the fine-structure constant has been introduced in order to simplify the form of the expression.

The second term consists of the energy that is required to bring the nucleus from infinity to the edge of the electron cloud and then to the centre. We label these two individual terms as U_{n1} and U_{n2}, respectively. The first is simply

$$\begin{aligned} U_{n1} &= -\frac{1}{4\pi\varepsilon_0} \frac{(Ze)^2}{r_{TF}} \\ &= -\alpha\hbar c \frac{Z^2}{r_{TF}}. \end{aligned} \tag{11.8}$$

To calculate U_{n2}, we note that the electric field within the sphere is

$$\begin{aligned} |\mathbf{E}(r)| &= \frac{Ze}{4\pi\varepsilon_0 r^2} \left(\rho_e \frac{4\pi}{3} r^3 \right) \\ &= \frac{Ze\rho_e}{3\varepsilon_0} r. \end{aligned} \tag{11.9}$$

Then, the energy required to bring the nucleus to the centre of the electron cloud is

$$\begin{aligned} U_{n2} &= -\frac{1}{2} |\mathbf{E}(r)| r \\ &= -\frac{1}{2} \alpha\hbar c \frac{Z^2}{r_{TF}}. \end{aligned} \tag{11.10}$$

Thus, the potential energy of the electron cloud containing the nucleus is the sum of U_{e}, U_{n1} and U_{n2}, or

$$U_{\mathrm{e}} = -\frac{9}{10}\alpha\hbar c\frac{Z^2}{r_{\mathrm{TF}}}. \tag{11.11}$$

We now calculate the kinetic energy using the Fermi gas model (see Section 3.3 of McParland (2010) for a derivation of that model). The mean energy of the electron states is given by

$$E_{\mathrm{e}} = \frac{3}{5}Z\frac{p_{\mathrm{F}}^2}{2m_{\mathrm{e}}} \tag{11.12}$$

where p_{F} is the Fermi momentum given by

$$p_{\mathrm{F}} = \hbar c k_{\mathrm{F}} \tag{11.13}$$

and k_{F} is the Fermi wave number. This latter quantity is related to the electron number density, ρ_{e}, through

$$k_{\mathrm{F}}^3 = 3\pi^2\rho_{\mathrm{e}}. \tag{11.14}$$

As ρ_{e} is a number density rather than the charge density used earlier, then

$$\rho_{\mathrm{e}} = \frac{Z}{\left(\frac{4\pi}{3}\right)r_{\mathrm{TF}}^3}. \tag{11.15}$$

By including (11.13), (11.14) and (11.15) in (11.12), we at last achieve an expression for the kinetic energy term:

$$E_{\mathrm{e}} = \frac{9}{10}\left(\frac{\pi}{4}\right)^{\frac{2}{3}}\frac{(\hbar c)^2}{m_{\mathrm{e}}}\frac{Z^{\frac{5}{3}}}{r_{\mathrm{TF}}^2}. \tag{11.16}$$

Summing the terms, the system's total energy is

$$\begin{aligned} E_{\mathrm{tot}} &= U_{\mathrm{e}} + E_{\mathrm{e}} \\ &= \frac{9}{10}\left(-\alpha\hbar c\frac{Z^2}{r_{\mathrm{TF}}} + \left(\frac{\pi}{4}\right)^{\frac{2}{3}}\frac{(\hbar c)^2}{m_{\mathrm{e}}}\frac{Z^{\frac{5}{3}}}{r_{\mathrm{TF}}^2}\right). \end{aligned} \tag{11.17}$$

This energy is minimised when the quantity within the parentheses is equal to zero. Or,

$$\alpha\hbar c\frac{Z^2}{r_{\mathrm{TF}}} = \left(\frac{\pi}{4}\right)^{\frac{2}{3}}\frac{(\hbar c)^2}{m_{\mathrm{e}}}\frac{Z^{\frac{5}{3}}}{r_{\mathrm{TF}}^2}. \tag{11.18}$$

Simple algebra applied to (11.18) leads to

$$r_{\mathrm{TF}} = \left(\frac{\pi}{4}\right)^{\frac{2}{3}} \frac{r_{\infty}}{Z^{\frac{1}{3}}} \tag{11.19}$$

which, numerically, is the same result that we obtained in (5.101). From the results of (5.101) and (11.18), we can write the potential energy term of (11.7) as

$$\begin{aligned} U_{\mathrm{e}} &= -\frac{9}{10}\alpha\hbar c \frac{Z^2}{r_{\mathrm{TF}}} \\ &= -\frac{9}{10}\left(\frac{4}{\pi}\right)^{\frac{2}{3}} \frac{\alpha\hbar c}{r_{\infty}} Z^{\frac{7}{3}} \end{aligned} \tag{11.20}$$

and the kinetic term as

$$E_{\mathrm{e}} = \frac{9}{10}\left(\frac{4}{\pi}\right)^{\frac{2}{3}} \frac{\alpha\hbar c}{r_{\infty}} Z^{\frac{7}{3}}. \tag{11.21}$$

Hence, we see that the magnitudes of both terms are equal to

$$\frac{9}{10}\left(\frac{4}{\pi}\right)^{\frac{2}{3}} \frac{\alpha\hbar c}{a_{\infty}} Z^{\frac{7}{3}}. \tag{11.22}$$

The energy per electron is, thus, (11.22) divided by Z, and the Thomas–Fermi velocity is straightforward to obtain through

$$\frac{m_{\mathrm{e}} v_{\mathrm{TF}}^2}{2} = \frac{9}{10}\left(\frac{4}{\pi}\right)^{\frac{2}{3}} \frac{\alpha\hbar c}{a_{\infty}} Z^{\frac{4}{3}}, \tag{11.23}$$

leading to

$$\begin{aligned} v_{\mathrm{TF}} &= \frac{3}{5}\left(\frac{4}{\pi}\right)^{\frac{1}{3}} \alpha c Z^{\frac{2}{3}} \\ &= \frac{3}{5}\left(\frac{4}{\pi}\right)^{\frac{1}{3}} v_0 Z^{\frac{2}{3}} \\ &\approx 0.65 v_0 Z^{\frac{2}{3}}. \end{aligned} \tag{11.24}$$

Echoing our earlier derivation, the atomic electron orbital frequency can be approximated by the ratio of the Thomas–Fermi velocity to the Thomas–Fermi atomic radius:

$$\begin{aligned}\omega &= \frac{v_{\mathrm{TF}}}{r_{\mathrm{TF}}} \\ &= \frac{0.65 v_0\, Z^{\frac{2}{3}}}{0.885\left(\frac{r_\infty}{Z^{\frac{1}{3}}}\right)} \\ &= 0.734 Z \frac{v_0}{r_\infty}.\end{aligned} \tag{11.25}$$

Substituting the values for the Bohr velocity and radius, and again using $\bar{I} = \hbar\omega$, we obtain

$$\bar{I} = 20\, Z\, \mathrm{eV}. \tag{11.26}$$

As before, we have a linear dependence upon the atomic number. But we have calculated for a K-shell velocity and if we then crudely halve it to obtain a mean, we have the result

$$\bar{I} \approx 10\, Z\, \mathrm{eV} \tag{11.27}$$

in reasonable agreement with our earlier estimate for $Z > 15$.

11.2.4 Optical Dipole Oscillator Strength

Recall our earlier definition of the mean excitation energy from Chap. 8:

$$\ln \bar{I} = \frac{1}{Z} \int dE_f \frac{df_{f0}(q,Z)}{dE_f} \ln E_f \tag{11.28}$$

with the normalisation of

$$\int dE_f \frac{df_{f0}(q,Z)}{dE_f} = Z. \tag{11.29}$$

Note that the numerical values of $\bar{I}$ and E_f must be in a consistently used set of units (e.g. eV) so that the argument of the logarithm is a pure number (11.28) is the first energy moment of the optical (generalised) oscillator strength distribution, or GOS. This was given in Chap. 8 as

$$\frac{df_{f0}(q,Z)}{dE_f} = \frac{1}{Q} \left|F_{f0}(q,Z)\right|^2. \tag{11.30}$$

Recall from (8.301) that

$$F_{f0}(q,Z) = \langle f|\sum_{j=1}^{Z} \mathrm{e}^{i\frac{\mathbf{q}\bullet\mathbf{r}_j}{\hbar c}}|0\rangle. \tag{11.31}$$

In Appendix C.5.2, we demonstrated how the Hermite polynomial can be used to evaluate the above matrix element. As an example, let us consider an atom with a single electron, $Z = 1$. Then,

$$F_{f0}(q,1) = \langle f|\mathrm{e}^{i\frac{qr}{\hbar c}}|0\rangle \tag{11.32}$$

where

$$\langle f|\mathrm{e}^{iqx}|0\rangle = \frac{1}{\sqrt{2^f f!}}\left(i\frac{q}{\hbar c\xi}\right)^f \mathrm{e}^{-\left(\frac{q}{2\hbar c\xi}\right)^2} \tag{11.33}$$

and where

$$\xi = \sqrt{\frac{m_e\omega_0}{\hbar c^2}}. \tag{11.34}$$

Then,

$$\left|\langle f|\mathrm{e}^{iqx}|0\rangle\right|^2 = \frac{1}{2^f f!}\left(i\frac{q}{\hbar c\xi}\right)^{2f} \mathrm{e}^{-2\left(\frac{q}{2\hbar c\xi}\right)^2}. \tag{11.35}$$

This then gives

$$f_{f0}(q,1) = \frac{1}{(f-1)!}\left(\frac{Q}{\hbar\omega_0}\right)^{f-1} \mathrm{e}^{-\frac{Q}{\hbar\omega_0}} \tag{11.36}$$

where $f = 1, 2, 3\ldots$ and $Q = \frac{q^2}{2m_e}$.

If we recall the Poisson probability function, $\Pr(x) = \frac{\mu^x \mathrm{e}^{-\mu}}{x!}$, we see that (11.36) is a Poisson distribution for $x = f - 1$ and $\mu = \frac{Q}{\hbar\omega_0} = \frac{q^2}{2m_e\hbar\omega_0}$. Figure 11.2 shows $f_{f0}(q,1)$ as a function of the normalised energy transfer for the final states ranging from $f = 1$ to $f = 5$. The increase in $f_{f0}(q,1)$ with f at large $\frac{Q}{\hbar\omega_0}$ can be readily interpreted. We can see from Fig. 11.2 that the maximum in $f_f(q,1)$ increases in increasing Q with increasing final state number, f. The maximum in the GOS can be demonstrated through a trivial differentiation of (11.36) to occur for the condition of the final state energy being $f\hbar\omega_0 = Q + \hbar\omega_0$. Thus for the condition of the energy transfer exceeding $\hbar\omega_0$ ($Q >> \hbar\omega_0$), then the final state energy is $f\hbar\omega_0 \approx Q$.

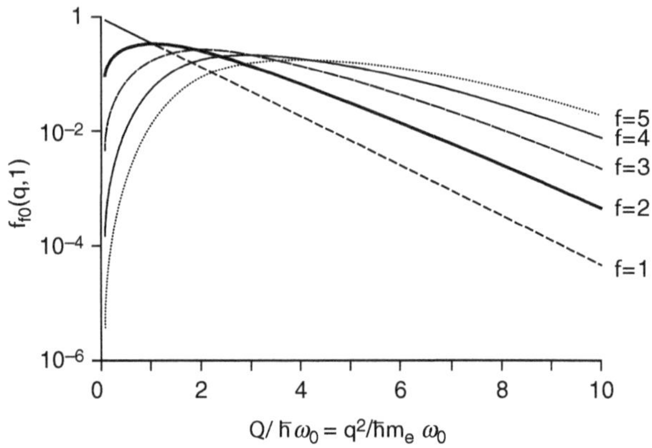

Fig. 11.2 Generalised oscillator strength for a one-electron atom as a function of the normalised energy transfer for final states ranging from 1 to 5 as calculated from Eq. (11.36)

That is, in the hard collision, the energy transfer approaches that of the free electron value and the atomic electron binding energy can be ignored.

Another approach to calculating the mean excitation energy from (11.28) is to use the moments of this distribution as defined by Dalgarno (1960):

$$M(m) = \int \mathrm{d}E_f \frac{\mathrm{d}f_{f0}(Z)}{\mathrm{d}E_f} E_f^m \tag{11.37}$$

where, from the normalisation requirement, the zeroth moment is

$$M(0) = Z. \tag{11.38}$$

The logarithm of the mean ionisation potential can be written by using $\frac{\mathrm{d}E_f^m}{\mathrm{d}m} = E_f^m \ln E_f$ so that

$$\ln \bar{I} = \frac{\left.\frac{\mathrm{d}M(m)}{\mathrm{d}m}\right|_{m=0}}{M(0)}. \tag{11.39}$$

In ICRU Report 37 (1984), it is noted that the moments for $m = -1, 1, 2$ can be determined theoretically and that for $m = -2$ can be extracted from measured data. An analytical fit to these four moments and can be made, from which the ratio in (11.39) can be determined.

11.2.5 Free Electron/Local Plasma Model

In Appendix H, we derive the plasma frequency, ω_P,

$$\omega_{\mathrm{P}} = c\sqrt{\frac{4\pi\eta\alpha\hbar c}{m_{\mathrm{e}}}}. \tag{11.40}$$

where η is the number density of electric charges. By inserting all of the physical constants, we can rewrite (11.40) as

$$\omega_{\mathrm{P}} = \left(1.59 \times 10^{4}\right)\sqrt{\eta} \quad \mathrm{s}^{-1}. \tag{11.41}$$

Lindhard and Scharff (1953) proposed the model of a free electron gas to estimate the excitation energy. In our nomenclature, they provided

$$\ln I = \frac{1}{Z}\int \mathrm{d}^3\mathbf{r}\eta(\mathbf{r}) \ln\left(\chi \frac{\hbar\omega_{\mathrm{p}}(\mathbf{r})}{m_{\mathrm{e}}}\right) \tag{11.42}$$

Where χ is a numerical constant to account for atomic electron binding (i.e. $\chi > 1$).

11.2.6 Condensed Media

In Chap. 8, we used Fermi's macroscopic theory of electromagnetic interactions with matter to develop a stopping theory in which the relative dielectric permittivity is prominent. Fano (1956) developed a related theory for condensed media based upon treating the atomic electrons as quantum oscillators and yielded an expression for the mean excitation energy:

$$\ln \bar{I} = \frac{2}{\pi\omega_{\mathrm{P}}^2}\int \mathrm{d}\omega \ \mathrm{Im}\left(-\frac{1}{\varepsilon_R(\omega)}\right)\omega \ln(\hbar\omega). \tag{11.43}$$

11.3 Experimental Determination of $\bar{I}$

Recall the expressions for the total mass collision stopping power that were derived in Chap. 10 for a variety of charged particle projectiles. This is not yet a complete expression as we must account for when the projectile velocity is comparable to those of the orbiting atomic electrons (shell correction), the Barkas–Andersen effect, the Bloch correction factor and of polarisation of the medium (density effect). These additional considerations are to be investigated in the following chapter. We will neglect these effects in the following discussion in order to demonstrate the principle behind how empirical energy loss data can be used to estimate the mean excitation energy. The experimental conditions that allow us to neglect these effects will be examined subsequently.

The total mass collision stopping power, where all of the above effects are to grouped within the factor Ξ, is given by

$$\frac{\mathrm{d}E}{\rho \mathrm{d}x} = 2C\left(\frac{Z}{A}\right)\left(\frac{z}{\beta}\right)^2\left(\ln\left(\frac{2m_e\beta^2}{\bar{I}}\right) - \ln\left(1-\beta^2\right) - \beta^2 - \Xi\right). \tag{11.44}$$

Consider an experimental arrangement in which a charged particle beam of known and well-defined energy E_0 (and 'good' geometry so as to avoid the influence of scattering) is incident to a medium of area thickness $\rho\,\Delta x$. The energy of the exiting particles, E_{ex}, is measured, and the measured energy loss is

$$\Delta E_{\mathrm{meas}} = E_0 - E_{\mathrm{ex}} \tag{11.45}$$

where we have used our convention of describing the stopping power as a positive quantity. One can then approximate (11.44) with

$$\frac{\Delta E_{\mathrm{meas}}}{\rho \Delta x} = 2C\left(\frac{Z}{A}\right)\left(\frac{z}{\beta}\right)^2\left(\ln\left(\frac{2m_e\beta^2}{\bar{I}}\right) - \ln\left(1-\beta^2\right) - \beta^2 - \Xi\right). \tag{11.46}$$

As all of the quantities except for the mean excitation energy are known, one can invert (11.46) to yield an expression for $\ln \bar{I}$,

$$\ln \bar{I} = \ln\left(2m_e\beta^2\right) - \ln\left(1-\beta^2\right) - \beta^2 - \Xi - \frac{\Delta E_{\mathrm{meas}}}{\rho \Delta x}\left(\frac{Z}{2CA}\right)\left(\frac{\beta}{z}\right)^2. \tag{11.47}$$

Clearly, $\bar{I}$ and m_e must be in the same units of energy. As the kinematics of the projectile and the atomic properties and thickness of the medium is considered known, then a measurement of the energy loss is sufficient to yield a value of $\bar{I}$, providing that adequate accounting of the factor Ξ has been made. The resulting uncertainty in the value of $\bar{I}$ is of interest. It is straightforward to derive this from (11.47) from the quadrature sum (ICRU 1984):

$$\Delta \bar{I} = \bar{I}\sqrt{\left(\Delta L_{\mathrm{meas}}\right)^2 + \left(\Delta \Xi\right)^2} \tag{11.48}$$

where

$$L_{\mathrm{meas}} = \frac{\Delta E_{\mathrm{meas}}}{\rho \Delta x}\left(\frac{Z}{2CA}\right)\left(\frac{\beta}{z}\right)^2. \tag{11.49}$$

It is obvious from the above that an accurate determination of the mean excitation potential from measurement requires accurate measurements of the energy loss traversing the medium and the projectile's incident velocity and accurate knowledge of the various correction factors contained within Ξ.

Figure 11.3 presents schematically the two principle means of measuring the stopping power and, hence, determining the mean ionisation potential. The incident velocity of the charged particle beam will be accurately known as this is selected through electromagnetic momentum filters to yield an almost mono-energetic beam.

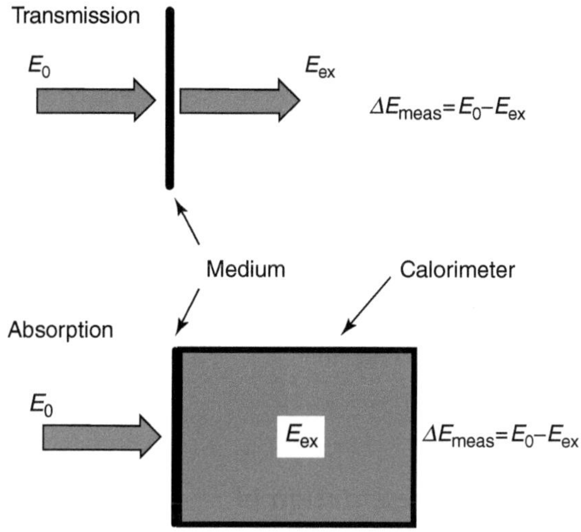

Fig. 11.3 Two experimental approaches for measuring the energy loss within a known thickness of medium from which a value of the mean ionisation potential can be determined. In the transmission measurement, the exiting energy of the charged particle beam is measured through, for example, electromagnetic means to determine its momentum. In the absorption method, the detector is an absorber sufficiently thick to stop the exiting particle beam and the energy of the exiting beam determined from the rise in temperature of the calorimeter

Knowing this velocity and the type of projectile, the incident kinetic energy is also known. The medium that the beam traverses and loses energy within must be thin so that ΔE_{meas} is small and that any variation around it is also small (this 'energy straggling' is the topic of Part IV). As the calculated collision stopping power is the mean, but the measured energy loss is the most probable, one must also be aware of the effects of this energy straggling. For a sufficiently thin section of medium, the energy loss probability distribution function is Gaussian, and the mean and the most probable quantities are the same. Should the medium be insufficiently thin, the energy loss probability distribution function becomes asymmetric and the most probable (measured) and mean (calculated) energy loss values will differ. For condensed materials, the medium is a thin foil, the thickness and constituency of which must be precisely known. One must then measure the exiting kinetic energy in order to determine E_{ex}. This can, again, be done using a magnetic spectrometer for momentum analysis using a position-sensitive detector as in the case of a transmission measurement. Another means of measuring the kinetic energy of the exiting charged particles is through 'time of flight'. The particle traverses two exit detectors of known separation. By measuring the time difference between the two detections, the velocity and, hence, the kinetic energy are determined.

An approach differing from this transmission measurement is to use complete absorption of the beam. In this case, the exiting beam is incident to an absorber sufficiently thick to be opaque to the exiting beam at its energy. Calorimetry is then performed: the temperature rise is measured and related to the total energy deposited (i.e. the time-integrated exit energy).

The absorption method appears to be simpler: calorimetry measures the rise in temperature in the absorber as a result of the energy deposition within it integrated over time. Knowing this temperature gain and the specific heat of the absorber, one can calculate the energy deposition. However, corrections must be made for nonthermal channels which are available. These include molecular morphology changes, chemical reactions and the emission of photons or other charged particles. Once these corrections are applied, this method does avoid the issue of the difference between the mean and the most probable energy loss. Andersen et al. (1966) describe the use of a liquid helium calorimeter in this method.

11.4 Parametric Representation of $\bar{I}$

Tabulated values of the mean excitation potential for the elements are available in ICRU Reports 37 and 49 (ICRU 1984, 1993). The genesis of these data is through primarily stopping measurements of high-energy protons. The ICRU has reported that it is currently re-evaluating these data (ICRU 2005).

Parametric forms of $\bar{I}$ can prove useful in calculation algorithms rather than relying upon look-up tables of values of $\bar{I}$. The simplest form is that due originally to Bloch (1933) and which we evaluated in Sects. 11.2.1 and 11.2.2:

$$\bar{I} = kZ \tag{11.50}$$

where k is a constant equal to about 10 eV for the Thomas–Fermi model derived in Sect. 11.2.2. This is demonstrated in Fig. 11.4. It is clearly a crude parametrisation with reasonable validity only for $Z > 20$. Hence, this is not an appropriate formula to use.

A commonly used parametrisations of the mean excitation energy as a function of Z is that presented by Segrè (1977):

$$\bar{I} = 9.1\left(1 + \frac{1.9}{Z^{\frac{1}{3}}}\right)\text{eV} \quad \text{for} \quad Z \geq 4. \tag{11.51}$$

This parametrisation is shown in Fig. 11.4. It is evident that it overestimates the mean excitation energy for $Z > 20$.

As tissues have low atomic numbers (the greatest of practical concern being 20 for calcium), Fig. 11.4 would suggest that in medical radiation physics applications, tabulated values of $\bar{I}$ are the most sensible to use.

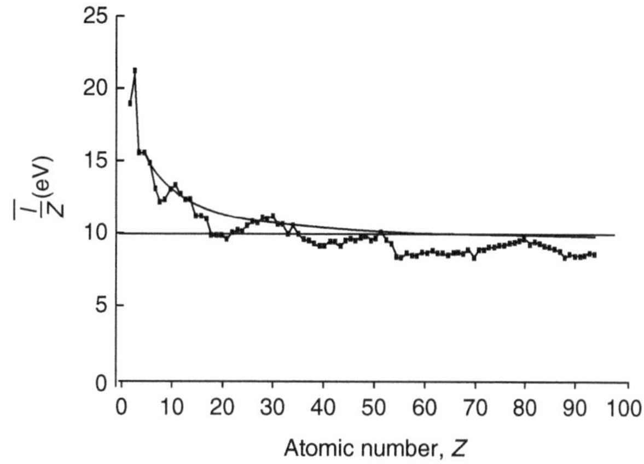

Fig. 11.4 Mean excitation energy normalised to atomic number for elements as a function of atomic number (Data points are from ICRU Report 37 (1984)). The *horizontal line* at 10 eV is that estimated from the Thomas–Fermi calculation of Sect. 11.2.2. The *curve* is the parameterisation from Segrè (1977) given by Eq. (11.51)

Bibliography and Further Reading

Jackson JD. Classical electrodynamics. 3rd ed. New York: Wiley; 1999.
Rossi B. High-energy particles. New York: Prentice-Hall; 1952.

References

Andersen HH, Garfinkel AF, Hanke CC, Sørensen H. Stopping power of aluminium for 1–12 MeV protons and deuterons. Mat FysMedd Dan Vid Selsk. 1966;35:1–24.
Bloch F. Bremsvermögen von Atomen mit mehreren Elektronen. Z Phys. 1933;81:363–76.
Dalgarno A. The stopping powers of atoms. Proc Phys Soc (Lond). 1960;76:422.
Fano U. Atomic theory of electromagnetic interactions in dense materials. Phys Rev. 1956;103:1212–8.
Fano U. Studies in penetration of charged particles in matter. Nuclear Science Series Report No. 39. Washington, DC: National Academy of Sciences; 1964. p. 1–338.
ICRU. Stopping powers for electrons and positrons. Report No 37. Bethesda: International Commission on Radiation Units and Measurements; 1984.
ICRU. Stopping powers and ranges for protons and alpha particles. Report No 49. Bethesda: International Commission on Radiation Units and Measurements; 1993.
ICRU. Stopping of ions heavier than helium. Report No 73. International Commission on Radiation Units and Measurements. Oxford: Oxford University Press; 2005.
Lindhard J, Scharff M. Energy loss in matter by fast particles of low charge. Mat Fys Medd Dan Vid Selsk. 1953;27:1–31.
McParland BJ. Nuclear medicine radiation dosimetry: advanced theoretical concepts. London: Springer; 2010.
Segrè E. Nuclei and particles. 2nd ed. Reading: Benjamin/Cummings; 1977.
Uehling EA. Penetration of heavy charged particles in matter. Ann Rev Nucl Sci. 1954;4:315–50.

Chapter 12
Higher-Order Corrections to the Collision Stopping Power

Contents

Abstract The development of the collision stopping power to this point has been based upon the use of the first term of the Born series, or the *first Born approximation*. Associated with this have been a number of simplifications and approximations: in particular, the collision stopping power is proportional to the square of the projectile's charge, the atomic electrons are effectively stationary in the reference frame of the

B.J. McParland, *Medical Radiation Dosimetry*,
DOI 10.1007/978-1-4471-5403-7_12, © Springer-Verlag London 2014

projectile and the projectile does not polarise the medium. In this chapter, we derive a variety of correction factors to be applied to our results obtained so far. We do so in the context of the stopping number which is considered as an expansion of the collision stopping. The zeroth-order (or kinematic) term is that of Chap. 10, and additional corrections are treated as higher-order perturbations. Corrections for nonnegligible atomic electron motion and medium polarisation are independent of the projectile's charge. Additional perturbations are in powers of this charge and are the Barkas–Andersen effects and the Bloch correction.

12.1 Introductory Comments

We have derived the soft- and hard-collision stopping powers, combined them to yield the total collision stopping power and have derived the expenditure of energetic charged particle energy required to yield ionisation. These results have been built upon the foundation of the first term of the Born series applied to a simple model of the medium composed of a cold and dilute monoatomic gas. To extend this to realistic cases, we must consider the inclusion of additional terms, or correction factors, to account for changes in the calculated stopping power caused by realistic situations. We may thus consider the collision stopping power expression derived in Chap. 10 as being a zeroth-order result.

This chapter presents these additional correction factors which contribute to the stopping power expression so as to correct for the simplifying assumptions used in its derivation. In effect, these additional terms may be considered as being perturbative in that they are small in magnitude corrections proportional to powers of the projectile's electric charge (up to second order).[1] These factors account for the physics neglected in our derivation so far: atomic electron motion, a condensed medium and the Bloch correction. In order to place the derivations of these higher order into an organised fashion, we introduce the concept of the *stopping number*, a dimensionless entity of which we have already derived its zeroth-order term in Chap. 10.

12.2 Stopping Number

12.2.1 Definition

A useful means of combining the additional correction terms with the expression of the collision stopping power is through the definition of the stopping number.

[1] Recall that there is a z^2 in the multiplier of the stopping power calculation. Thus a perturbative term of order z^2 describes a contribution to the stopping power that is proportional to z^4.

Consider the expression for the mass collision stopping power, disregarding projectile spin, given by (10.3). We can write it in the following general form:

$$\frac{\mathrm{d}E}{\rho \mathrm{d}x} = 2C\left(\frac{Z}{A}\right)\left(\frac{z}{\beta}\right)^2 L(\beta). \tag{12.1}$$

$L(\beta)$ is known as the *stopping number* because it is a dimensionless modifier of $2C(Z/A)(z/\beta)^2$. We shall write it as the summation of three further dimensionless terms, each weighted by powers of the projectile electric charge (normalised to the unit charge, e), and which may be considered as a perturbative expression. That is,

$$\begin{aligned} L(\beta) &= \sum_{j=0}^{2} z^j L_j(\beta) \\ &= L_0(\beta) + \sum_{j=1}^{2} z^j L_j(\beta) \\ &= L_0(\beta) + L_z(\beta) \end{aligned} \tag{12.2}$$

We can see from (10.3) that the zeroth-order term of this expansion (i.e. with no explicit dependence), $L_0(\beta)$, can itself be written as a series

$$L_0(\beta) = \sum_{k=0}^{2} L_{0k}(\beta). \tag{12.3}$$

We now identify the individual terms within these summations.

12.2.2 Zeroth-Order Stopping Number, $L_0(\beta)$

This summation contains three terms. The first is the kinematical term which was derived in our application of the first-order term of the Born expansion to the theoretical development of the collision stopping power. The second accounts for the effects of atomic electron motion. The third includes the perturbative effects of the surrounding medium upon the projectile's electric field and the resulting reduction in its electric field leading to a subsequent decrease in the stopping power.

12.2.2.1 Kinematical Term $L_{00}(\beta)$

The first term in $L_0(\beta)$ is that of (10.3),

$$L_{00}(\beta) = \left[\ln\left(\frac{2m_{\mathrm{e}}}{\bar{I}}\beta^2\right) - \ln\left(1-\beta^2\right) - \beta^2\right]. \tag{12.4}$$

This term is dependent solely upon the kinematics of the projectile and the mean excitation energy of the medium, which was the topic of Chap. 11.

This is the only term in the stopping power expression developed on the basis of the assumptions outlined in Sect. 12.1.

12.2.2.2 Atomic Electron Motion Term $L_{01}(\beta)$

The next term in $L_0(\beta)$ corrects for the fact that the derivation of the Bethe soft-collision stopping power had assumed that the projectile velocity is much greater than those of the orbiting atomic electrons. This led to the kinematical approximation that the target electron was at rest. But, as the projectile decelerates, its velocity eventually becomes comparable to those of the atomic electrons, and the validity of this assumption becomes increasingly untenable. Consequently, an additional factor must be added to correct for this:

$$L_{01}(\beta) = -\frac{C_e(\beta, Z)}{Z}. \tag{12.5}$$

We will refer to $L_{01}(\beta)$ as the atomic electron motion term and to $C_{\mathrm{e}}(\beta, Z)$ as the atomic shell correction term so as to be consistent with use in the literature. $L_{01}(\beta)$ will be derived in Sect. 12.3.

12.2.2.3 Condensed Medium Term $L_{02}(\beta)$

The next term in $L_0(\beta)$ allows the stopping power calculation to be applied to a condensed medium by correcting for the assumption that the medium is a cold dilute gas in which case, the energetic charged projectile interacts with a single atom at a time. On the other hand, in a condensed medium, such as tissue, the Coulomb field of the charged projectile will interact with multiple atoms leading to the polarisation of the medium resulting in a reduction of the penetration of the electric field. With this reduction of the electric field, the collision stopping power will be diminished. The form of this factor is conventionally given in the form of

$$L_{02}(\beta) = -\frac{\delta}{2}. \tag{12.6}$$

We will refer to $L_{02}(\beta)$ as the condensed medium term and to δ as the *density effect term* so as to maintain consistency with the literature $L_{02}(\beta)$, and the physics of the polarisability of a medium by an energetic charged particle is the topic of Sect. 12.4.

12.2.3 *z-Dependent Stopping Number, $L_z(\beta)$*

12.2.3.1 Barkas–Andersen Term $zL_1(\beta)$

As to be discussed in Sect. 12.5, it was discovered experimentally during the 1950s that the range (i.e. the integration of the reciprocal of the stopping power over energy and the topic of Chap. 13) of a particle differed minutely from that of its antiparticle. This is known as the Barkas–Andersen effect in which the stopping power for a positively charged particle is slightly greater than that for a negatively charged particle with the same mass and velocity (i.e. its antiparticle with the same kinetic energy).

This discovery led to the realisation that the collision stopping power had to have a dependence upon the sign of the projectile's electric charge, whereas the stopping power derivations yielded a z^2 dependence only. That is, it must include a perturbative effect proportional to z^n, where n is an odd integer. Hence, the first term in $L_z(\beta)$ is $zL_1(\beta)$. The perturbative term is dimensionless, and the multiplier z which, when combined with the $(z/\beta)^2$ term in the multiplier of (12.1), produces a contribution to the collision stopping power that is proportional to z^3 and, thus, a dependence upon the sign of the projectile's charge.

12.2.3.2 Bloch Correction Term $z^2L_2(\beta)$

As described in Chaps. 1 and 8, the first third of the twentieth century presented two theories of the soft-collision stopping power – that derived by Bohr from classical mechanics and that derived by Bethe from quantum mechanics. Despite the vastly differing starting points, as exemplified, for example, by the use of impact parameter by Bohr and the use of the Born approximation and momentum/energy transfer by Bethe, the results of both theories were so similar that the consideration of 'bridging' between the two had to have been had. This was done by Bloch, and this resulted in a term within the associated correction term proportional to z^2 to lead to an overall z^4 dependence in the collision stopping power. We extend our earlier discussion of Chap. 8 of the Bloch correction term in Sect. 12.6.

12.3 Atomic Electron Motion Term, $L_{01}(\beta)$

12.3.1 *Introductory Comments*

Our derivations of the collision stopping power have ignored the orbital velocities of the atomic electrons within the medium by assuming implicitly that the projectile velocity far exceeds that of the atomic electron. That is, in terms of the kinematics of the collision, the electron is considered to be at rest. This assumption is usually valid for a fast projectile and a low-Z atom. However, it begins to break down as the projectile decelerates to velocities approaching those of the orbital electrons or for the case of high-Z atoms where the orbital velocities can approach significant fractions of the speed of light. Thus, a kinematical correction is required, through $L_{01}(\beta)$, to account for the change in the collision stopping power induced by the nonnegligible orbital velocities.

Clearly, the correction factor, as a function of projectile velocity, will change as the projectile decelerates. The K-shell electrons will be affected first as these are the most tightly bound and have the highest orbital velocities. These will then be followed by contributions related to the L- and M-shell electrons. $L_{01}(\beta)$ leads to a reduction in the stopping power as will be shown although the magnitude of the correction is rather small (a few percent) for low atomic number media, but increases considerably for high-Z media.

12.3.2 Effects of a Medium of Moving Electrons upon Stopping

To demonstrate the effects of moving target electrons upon the stopping power of an energetic charged particle, we first consider a simple case of a heavy charged particle traversing with velocity **v** a medium which we shall consider a case of free electrons uniformly distributed and with a common velocity $\mathbf{v}_\mathrm{e}$ (Sigmund 2006). This is a first attempt at modelling the interaction between an energetic charged particle and atomic electrons in which the latter are in motion; the effects of binding are neglected. Here, $m \gg m_\mathrm{e}$ and, in the reference frame of the projectile which exists largely in the centre-of-mass, *i* the projectile would perceive itself as being bombarded by electrons with an incident velocity equal to

$$\mathbf{v}_{\mathbf{e,Ref}} = \mathbf{v}_\mathbf{e} - \mathbf{v}. \tag{12.7}$$

If the number density of the electrons is equal to ρ_e, then, with the assumptions and definitions above, the current density is uniform and equal to

$$\mathbf{J}_\mathrm{e} = \rho_\mathrm{e}\mathbf{v}_{\mathbf{e,Ref}} \tag{12.8}$$

in units of *electron*/cm^2 • s. Then the number of electrons scattered per unit time interval is equal to

$$|\mathbf{J}_\mathrm{e}|\mathrm{d}t\,\mathrm{d}\sigma\big(v_{\mathrm{e,Ref}}, \theta_{\mathrm{CM}}\big) = \rho_\mathrm{e}\,\mathbf{v}_{\mathbf{e,Ref}}\mathrm{d}t\,\mathrm{d}\sigma\big(v_{\mathrm{e,Ref}}, \theta_{\mathrm{CM}}\big) \tag{12.9}$$

where the scattering angle θ_{CM} is measured within the scattering system's centre-of-mass. Recall that we are operating within the projectile's reference frame in which the electrons are considered to be the projectiles. Those electrons that are scattered by the ion through an angle θ_{CM} within the centre-of-mass will transfer a mean momentum to the projectile equal to

$$\Delta \mathbf{p} = m_e \mathbf{v}_{\mathbf{e,Ref}} (1 - \cos\,\theta_{CM}). \tag{12.10}$$

It is implicitly noted here that as a consequence of the uniformly moving and isotropic flux of electrons, the only non-zero component of the momentum transfer is that which is parallel to $\mathbf{v}_{\mathbf{e,Ref}}$. The net force upon the ion as a consequence of these momentum transfers is simply the derivative of the momentum transfer or

$$\begin{aligned} \mathbf{F} &= \frac{\mathrm{d}\mathbf{p}}{\mathrm{d}t} \\ &= \rho_e\, m_e\, v_{e,Ref}\, \mathbf{v}_{\mathbf{e,Ref}} \int_{-1}^{1} \mathrm{d}(\cos\theta_{CM}) \frac{\mathrm{d}\sigma(v_{e,Ref}, \theta_{CM})}{\mathrm{d}\Omega} (1 - \cos\theta_{CM}). \end{aligned} \tag{12.11}$$

Now, in order to make the above description more analogous to that of atomic electrons, we extend from the case of incident electrons (in the centre-of-mass) with uniform velocities to that where the electrons exhibit a distribution of velocities, $f(\mathbf{v}_{e,})$, as will be the case for the projection being incident to atomic electrons. In this case, the force of (12.11) is modified to the integral

$$\mathbf{F} = \rho_e\, m_e \int \mathrm{d}^3 v_e f(\mathbf{v}_e) v_{e,Ref} \mathbf{v}_{\mathbf{e,Ref}} \int_{-1}^{1} \mathrm{d}(\cos\theta_{CM}) \frac{\mathrm{d}\sigma(v_{e,Ref}, \theta_{CM})}{\mathrm{d}\Omega} (1 - \cos\theta_{CM}). \tag{12.12}$$

From (8.265), the energy lost per unit distance, or the linear stopping power, is

$$\frac{\mathrm{d}E}{\mathrm{d}x} = \hat{\mathbf{v}} \bullet \mathbf{F} \tag{12.13}$$

where $\hat{\mathbf{v}}$ is the unit vector in the direction of the ion's velocity. Then,

$$\begin{aligned} \frac{\mathrm{d}E}{\mathrm{d}x} &= \rho_e\, m_e \int \mathrm{d}^3 v_e f(\mathbf{v}_e) v_{e,Ref} \hat{\mathbf{v}} \\ &\bullet \mathbf{v}_{\mathbf{e,Ref}} \int_{-1}^{1} \mathrm{d}(\cos\theta_{CM}) \frac{\mathrm{d}\sigma(v_{e,Ref}, \theta_{CM})}{\mathrm{d}\Omega} (1 - \cos\theta_{CM}). \end{aligned} \tag{12.14}$$

This can be rewritten in a simpler form through the application of the stopping cross section S of (8.85):

$$\frac{\mathrm{d}E}{\mathrm{d}x} = \rho_e S. \tag{12.15}$$

The stopping cross section per electron is

$$S = m_e \int \mathrm{d}^3 v_e f(\mathbf{v}_e)|\mathbf{v}_e - \mathbf{v}|\hat{\mathbf{v}} \bullet (\mathbf{v} - \mathbf{v}_e)\sigma^{(1)}(|\mathbf{v}_e - \mathbf{v}|) \tag{12.16}$$

where the first-order transport cross section per electron is

$$\sigma^{(1)}(v) = \int_{-1}^{1} \mathrm{d}(\cos\theta_{CM}) \frac{\mathrm{d}\sigma(v_{e,Ref}, \theta_{CM})}{\mathrm{d}\Omega}(1 - \cos\theta_{CM}). \tag{12.17}$$

It can be seen how (12.17) is related to the first-order transport coefficient of the Goudsmit–Saunderson multiple Coulomb scatter theory given by (6.44).

If we consider the case of the ion velocity much greater than that of the electron, $v_e << v$, then (12.16) reduces, and we can define

$$S_0(v) \equiv m_e v^2 \sigma^{(1)}(v) \tag{12.18}$$

from which we can obtain the transport cross section per electron,

$$\sigma^{(1)}(v) = \frac{S_0(v)}{m_e v^2}. \tag{12.19}$$

Substituting (12.19) into (12.16) gives

$$S(v) = \int \mathrm{d}^3 v_e f(\mathbf{v}_e) \frac{(\mathbf{v} - \mathbf{v}_e)}{|\mathbf{v}_e - \mathbf{v}|} \hat{\mathbf{v}} S_0(|\mathbf{v} - \mathbf{v}_e|). \tag{12.20}$$

In order to evaluate this expression for the stopping cross section, we assume isotropicity in the velocity distribution (i.e. $f(\mathbf{v}_e) = f(v_e)$) and then perform a Taylor expansion of $|\mathbf{v} - \mathbf{v}_e|$ to second order in v_e (whilst recalling that $\hat{\mathbf{v}} = \mathbf{v}/v$) to give

$$|\mathbf{v} - \mathbf{v}_e| = v\left(1 - \frac{\mathbf{v} \bullet \mathbf{v}_e}{v^2} + \frac{(v\, v_e)^2 - (\mathbf{v} \bullet \mathbf{v}_e)^2}{2\, v^4} \cdots\right). \tag{12.21}$$

Applying this to (12.20) gives an expansion of the stopping cross section (Sigmund 1982) in the form

$$S(v) \approx S_0(v) + \left(\frac{\overline{v_e^2}}{v^2}\right)\left(1 - \frac{1}{3}S_0(v) + \frac{1}{3}v\frac{dS_0(v)}{dv} + \frac{1}{6}v^2\frac{d^2S_0(v)}{dv^2}\cdots\right) \tag{12.22}$$

where the mean-square electron velocity is

$$\overline{v_e^2} = \int d^3v_e f(v_e)v_e^2. \tag{12.23}$$

The stopping cross section result of (12.22) demonstrates that the orbital motion of the atomic electrons will introduce terms that will modify the stopping power which are of the order of $\overline{v_e^2}/v^2$. That is, it predicts that the effect of atomic motion upon the stopping power is to reduce it and that this effect will increase with the atomic number of the medium (and, thus, increasing orbital velocity) and decrease with increasing projectile velocity. Binding has, however, been neglected. This effect is now considered within the context of the Bethe theory.

12.3.3 Quantum-Mechanical Theory

The derivation of the Bethe soft-collision theory was based upon the first-order of the Born expansion. Thus, reflecting upon the results of Sect. 8.3.3 and assuming that the contributions of other perturbative effects (Bloch, Barkas–Andersen and density, as to be discussed later) can be neglected, we can write the correction to the stopping number due to shell corrections as

$$\begin{aligned}\Delta L &= L_{\text{Born}} - L_{00}(\beta) \\ &= L_{\text{Born}} - \left[\ln\left(\frac{2m_e}{I}\beta^2\right) - \ln\left(1-\beta^2\right) - \beta^2\right]\end{aligned} \tag{12.24}$$

(Walske 1952; Sigmund 2006). In the nonrelativistic regime,

$$\begin{aligned}-\ln\left(1-\beta^2\right) - \beta^2 &= -\left(\ln\left(1-\beta^2\right) + \beta^2\right) \\ &\approx -\left(-\beta^2 + \beta^2\right) \\ &= 0\end{aligned} \tag{12.25}$$

leaving us with a more commonly seen form

$$\Delta L = L_{\text{Born}} - \ln\left(\frac{2m_e}{I}\beta^2\right). \tag{12.26}$$

The Born term is, following from our discussions in Sect. 8.3.3,

$$L_{\text{Born}} = \frac{1}{2} \sum_j \int_{E_j^2 2m_e/\beta^2}^{\infty} dQ \frac{f_{j0}(Q)}{Q}. \tag{12.27}$$

Recalling the definition of the logarithm of the mean excitation energy of (8.300) and splitting the integral of (12.27) at $Q = 2m_e\beta^2$, we can rewrite (12.24) as

$$\Delta L = \frac{1}{2} \sum_j \left(\int_{2m_e\beta^2}^{\infty} dQ \frac{f_{j0}(Q)}{Q} - \int_0^{E_j^2/2m_e\beta^2} dQ \frac{(f_{j0}(Q) - f_{j0}(0))}{Q} \right). \tag{12.28}$$

This result follows as a consequence of the Thomas–Reiche–Kuhn rule (Appendix G.2.3) which gives

$$\sum_j \int_0^{2m_e\beta^2} dQ \frac{(f_{j0}(Q) - f_{j0}(0))}{Q} = 0. \tag{12.29}$$

The result of (12.28) shows that the atomic electron motion correction $\Delta L \to 0$ as the projectile velocity $\beta \to \infty$. While (12.28) has been evaluated numerically (Walske 1952, 1956; Khandelwal and Merzbacher 1966), an analytical form was derived by Fano (1963) for a one-electron atom:

$$\Delta L = -\left(\frac{\overline{v_e^2}}{v^2}\right) - \frac{1}{2}\left(\frac{\overline{v_e^4}}{v^4}\right) - \frac{5\pi}{3}\left(\frac{\alpha c}{v}\right) r_e^3 |\psi(\mathbf{r})|^2 \tag{12.30}$$

where $|\psi(\mathbf{r})|^2$ is the modulus squared of the electronic wavefunction.

12.3.4 *The Atomic Shell Correction Factor,* $C_e(\beta, Z)$

Consider a multi-electron atom with atomic number Z. We could write the change in stopping number due to orbital electron motion in the form

$$Z\,\Delta L = \sum_\mu Z_\mu\, \Delta L_\mu \tag{12.31}$$

where the summation is over the atomic shells (e.g. $\mu = K, L, M \ldots$), Z_μ is the number of electrons in the μth shell and ΔL_μ is the contribution of the electrons in that shell to the overall correction. As $L_{01}(\beta)$ is a term which is added to the stopping number and reflects a negative quantity (reduction in collision stopping power), then, from (12.31), we write

$$\begin{aligned} L_{01}(\beta) &= \frac{\sum_{\mu} Z_{\mu}\,\Delta L_{\mu}}{Z} \\ &\equiv -\frac{C_{\rm e}(\beta, Z)}{Z} \end{aligned} \tag{12.32}$$

where $C_{\rm e}(\beta, Z)/Z$ is a positive quantity known as the *shell correction factor*. From the above, we can also write the summation of the contributions over the atomic electron shells:

$$C_{\rm e}(\beta, Z) = \sum_{\mu} C_{\mu}(\beta, Z). \tag{12.33}$$

Contributions from the K- (Walske 1952; Khandelwal 1968), L- (Walske 1956; Khandelwal 1968) and M-shell (Khandelwal and Merzbacher 1966) electrons have been calculated. Bichsel (1992) has presented calculations for higher-order shells.

12.3.5 Parameterisation of $C_{\rm e}(\beta, Z)$

In the Preface, we noted how the required accuracy of radiation absorbed dose assessment in therapeutic applications is demanding. As a consequence, even the small-magnitude effects of the atomic shell correction in low-Z media, such as tissue, must be accounted for accurately. In many numerical calculations (e.g. Monte Carlo) non-tabulated data are preferable to look-up tables. One useful application of this approach is the atomic shell correction factor. Barkas (1962) provided a useful semiempirical parameterisation of $C_{\rm e}(\beta, Z)$:

$$\begin{aligned} C_{\rm e}(\beta, Z) = \bar{I}^2 \Bigg\{ &\left(\frac{0.42237}{(\gamma\beta)^2} + \frac{0.0304}{(\gamma\beta)^4} - \frac{0.00038}{(\gamma\beta)^6} \right) \times 10^{-6} \\ &+ \bar{I} \left(\frac{3.858}{(\gamma\beta)^2} - \frac{0.1668}{(\gamma\beta)^4} + \frac{0.00158}{(\gamma\beta)^6} \right) \times 10^{-9} \Bigg\} \end{aligned} \tag{12.34}$$

where the units of the mean excitation energy $\bar{I}$ are eV. $\bar{I}$ also contains the implicit Z-dependence of $C_{\rm e}(\beta, Z)$. The validity of this parameterisation of $C_{\rm e}(\beta, Z)$ is limited to the product $\gamma\beta > 0.13$. Figure 12.1 shows the shell correction term $C_{\rm e}(\beta, Z)/Z$ for protons in carbon and lead as functions of proton kinetic energy. The differences between the magnitudes of the correction factors with Z (atomic electron orbital velocity) and energy are clearly evident.

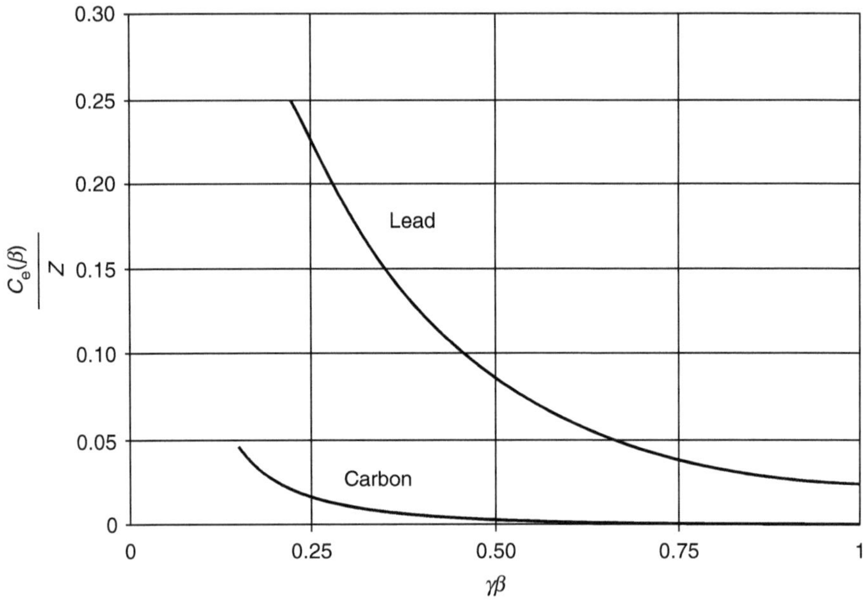

Fig. 12.1 Shell correction terms for carbon and lead as functions of $\gamma\beta$

12.4 Condensed Medium Term, $L_{02}(\beta)$

12.4.1 Introductory Comments

As noted earlier, the projectile's electric field has been implicitly assumed to be in vacuo, and the medium through which it is travelling assumed to be a cold and dilute monoatomic gas (Halpern and Hall 1948; Fano 1956). Such simplifications ignore any response of a condensed medium to this moving electric field. Should the particle be travelling in a dense dielectric medium (such as tissue), atoms will be polarised, creating an array of electric dipoles which generate a secondary electric field limiting the particle's electric field at a distance and reducing the stopping power. This effect is expected to be significant at high particle speeds, as shown by Fig. 8.2, where, as $\beta \rightarrow 1$, the electric field parallel to the trajectory flattens and the orthogonal component extends such that the dielectric response of the medium will limit the relativistic rise in stopping power. Because the magnitude of this dielectric response is directly related to the number of secondary electric dipoles, it will clearly depend upon the medium's physical density.[2] This reduction in stopping power is

[2] Because the interatomic spacings in a gas are greater than those in a solid or liquid, the wider dispersion of atoms in the gaseous phase will limit the dielectric response such that the mass collision stopping power for a given medium will be greater for the condensed phases than for the vapour phase and, hence, this interest in this phenomenon in tissue.

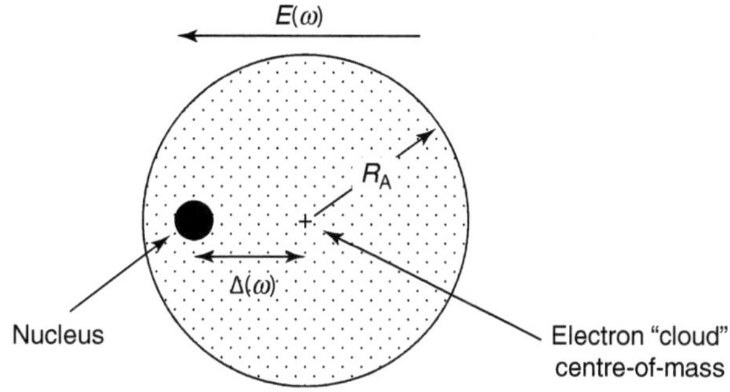

Fig. 12.2 Electronic polarisation due to an external electric field

characterised by the condensed medium term $L_{02}(\beta)$ of the stopping number. It was Fermi who first recognised the need of a dielectric correction as the $L_{00}(\beta)$ term of the stopping number goes to infinity due to the $-\ln(1-\beta^2)$ term as $\beta \to 1$.

We have already examined much of the effect of a condensed medium upon the stopping power of an energetic charged particle in our investigation of the Fermi theory of soft-collision energy loss provided earlier in Sect. 8.2.9. We will extend our results of that section and use the semi-classical work of Fermi (1940) to guide this derivation.

12.4.2 Electronic Polarisation

We begin first with a medium not much different from that we have considered to this point. We shall treat it is an isotropic monoatomic gas with interatomic distances sufficiently large that an atom can be treated in isolation.

The atom is considered to consist of a negatively charged and mobile spherical electron cloud of charge $-Ze$ and mass Zm_e which, in its unperturbed state, is centred on a fixed nucleus of charge Ze. As a charged particle in motion passes by the atom, it will perturb the position of this electron cloud, leading to the polarisation of the atom, as demonstrated in Fig. 12.2. At high particle speeds, the shortness of the time duration of the electric field 'pulse' experienced by the atom, as shown in Fig. 8.2, will be dominated by high-frequency components, and in frequency space, one can consider the atom as being exposed to a high-frequency electric field. The displacement of the electron cloud, Δ, will be limited by the short duration of the pulse, and one can thus effectively ignore the spatial variation of the field over the atom as a result (in analogy to the result of (8.92)). The electron cloud will also experience a restorative force due to the mutual attraction between it and the nucleus and thus go into oscillatory motion. The attractive force can be calculated by modelling the electron cloud as a uniformly charged sphere of radius R_A with an electric charge density, $(3/4\pi)Ze/R_A^3$. The radial electric field due to this charge at a distance Δ from the centre of the atom is found by applying Gauss' law to the sphere of radius Δ:

$$\varepsilon_0 \left(4\pi\Delta^2\right)E = \frac{4\pi}{3}\Delta^3\left(\frac{3}{4\pi}\frac{Ze}{R_{\mathrm{A}}^3}\right)$$

from which one obtains an expression for the magnitude of the electric field in terms of the displacement of the atomic electron cloud,

$$E = \frac{Ze}{4\pi\varepsilon_0}\frac{\Delta}{R_{\mathrm{A}}^3}. \tag{12.35}$$

The restorative force acting upon the displaced electron cloud is $F = ZeE$ or

$$\begin{aligned} F &= \frac{(Ze)^2}{4\pi\,\varepsilon_0 R_{\mathrm{A}}^3}\Delta \\ &= \eta\Delta \end{aligned} \tag{12.36}$$

where we have defined the quantity $\eta = (Ze)^2/4\pi\varepsilon_0 R_{\mathrm{A}}^3$ for later ease in display. This leads to

$$ZeE = \frac{(Ze)^2}{4\pi\varepsilon_0}\frac{\Delta}{R_{\mathrm{A}}^3}$$

from which the magnitude of the product of the charge and its displacement, electric dipole moment, is

$$\begin{aligned} p &= Ze\,\Delta \\ &= 4\pi\varepsilon_0 R_{\mathrm{A}}^3 E. \end{aligned} \tag{12.37}$$

By describing the electric dipole moment as a vector, we can write

$$\begin{aligned} \mathbf{p} &= 4\pi\varepsilon_0 R_{\mathrm{A}}^3 \mathbf{E} \\ &\equiv \alpha_{\mathrm{e}}\mathbf{E} \end{aligned} \tag{12.38}$$

where α_{e} is defined as the electronic polarisability. It can be noted that the magnitude of the ratio of the electronic polarisability to the permittivity of free space, $\alpha_{\mathrm{e}}/\varepsilon_0 = 4\pi R_{\mathrm{A}}^3$, is equal to the atomic volume.

Much like the previous analysis of the simple harmonic oscillator model of a single electron in our derivation of the dipole approximation in Sect. 8.2.4, the resulting equation of motion of the displaced electron cloud is

$$Zm_{\mathrm{e}}\frac{\mathrm{d}^2\Delta(t)}{\mathrm{d}t^2} = Ze\mathbf{E}(t) - \eta\Delta(t) - \Gamma Zm_{\mathrm{e}}\frac{\mathrm{d}\Delta(t)}{\mathrm{d}t} \tag{12.39}$$

where a damping force $\Gamma Z m_e \mathrm{d}\Delta(t)/\mathrm{d}t$ with a positive damping constant has been allowed for. This differential equation is readily solvable as demonstrated in our earlier work and, in frequency space, is given by (8.104)

$$\Delta(\omega) = \frac{(e/m_e)\mathbf{E}(\omega)}{(\omega_0^2 - \omega^2) + i\Gamma\omega}$$

with a resonant frequency given by

$$\omega_0 = c\sqrt{Z\frac{r_0}{R_A^3}}. \tag{12.40}$$

As the atomic radius R_A of the order of 10^{-10} m, the resonant frequency will consequently be of the order of 10^{16}–10^{17} s^{-1} with a corresponding wavelength in the ultraviolet range of the electromagnetic spectrum. The electric dipole moment in frequency space is

$$\begin{aligned}\mathbf{p}(\omega) &= Ze\,\Delta(\omega)\\ &= Z\frac{\left(\frac{e^2}{m_e}\right)E(\omega)}{(\omega_0^2 - \omega^2) + i\Gamma\omega}.\end{aligned} \tag{12.41}$$

Next, the scale is expanded from that of the individual atom to the macroscopic medium which is taken to be linear and isotropic.[3] The polarisation is the number of electric dipoles per unit volume, or

$$\mathbf{P}(\omega) = \rho_e\frac{\left(\frac{e^2}{m_e}\right)\mathbf{E}(\omega)}{(\omega_0^2 - \omega^2) + i\Gamma\omega}. \tag{12.42}$$

The electric flux density vector is related to the electric field and the polarisation in frequency space by

$$\begin{aligned}\mathbf{D}(\omega) &= \varepsilon_0\mathbf{E}(\omega) + \mathbf{P}(\omega)\\ &\equiv \varepsilon_0\mathbf{E}(\omega) + \chi_e(\omega)\varepsilon_0\mathbf{E}(\omega)\\ &= \varepsilon_0(1 + \chi_e(\omega))\mathbf{E}(\omega)\\ &= \varepsilon_0\varepsilon_R(\omega)\,\mathbf{E}(\omega)\end{aligned} \tag{12.43}$$

[3] Otherwise, the electric susceptibility and dielectric permittivity scalars would be replaced by tensors.

where $\chi_e(\omega)$ is the electric susceptibility and is defined through (8.243) and the associated discussion. (12.43) can be algebraically manipulated to relate the polarisation to the electric field by

$$\mathbf{P}(\omega) = \varepsilon_0(\varepsilon_R(\omega) - 1)\mathbf{E}(\omega). \tag{12.44}$$

By equating these two expressions for the polarisation, the relative dielectric constant can be written as

$$\begin{aligned}\varepsilon_R(\omega) &= 1 + \frac{(\rho_e/\varepsilon_0)(e^2/m_e)}{(\omega_0^2 - \omega^2) + i\Gamma\omega} \\ &= 1 + \frac{\omega_P^2}{(\omega_0^2 - \omega^2) + i\Gamma\omega}\end{aligned} \tag{12.45}$$

where we have used the definition of the plasma frequency from Appendix H. The dielectric dispersion formulae follow from (12.45) when the real and imaginary components are isolated:

$$\begin{aligned}\varepsilon_R(\omega) = &\left(1 + \omega_P^2\left(\frac{\omega_0^2 - \omega^2}{(\omega_0^2 - \omega^2)^2 + \Gamma^2\omega^2}\right)\right) - i\,\omega_P^2\left(\frac{\Gamma\omega}{(\omega_0^2 - \omega^2)^2 + \Gamma^2\omega^2}\right) \\ &\equiv \varepsilon_R{}'(\omega) - i\varepsilon_R{}''(\omega)\end{aligned} \tag{12.46}$$

As the imaginary component of the relative dielectric permittivity is proportional to the damping constant, it will describe the power loss within the dielectric.

12.4.3 Electromagnetic Fields in a Dielectric Medium

The properties of the electromagnetic fields associated with an energetic charged particle in a dielectric medium have been largely derived already in Sect. 8.2.9 in the context of the Fermi theory of the soft-collision stopping power. From those, we can write the electric and magnetic fields in $(\mathbf{k}, \omega)$-space as

$$\mathbf{E}(\mathbf{k}, \omega) = i\frac{ze\delta(\omega - \beta c k_{||})}{2\pi\varepsilon_0\varepsilon_R(\omega)\left(k^2 - \frac{\omega^2\varepsilon_R(\omega)}{c^2}\right)}\left(\omega\varepsilon_R(\omega)\frac{\beta}{c}\hat{\mathbf{v}}_{||} - \mathbf{k}\right) \tag{12.47}$$

and

$$\mathbf{H}(\mathbf{k}, \omega) = i\frac{ze\beta c}{2\pi\left(k^2 - \frac{\omega^2\varepsilon_R(\omega)}{c^2}\right)}\delta(\omega - \beta c k_{||})\mathbf{k} \times \hat{\mathbf{v}}_{||}. \tag{12.48}$$

12.4.4 Energy Loss in a Dielectric Medium

Using (8.265) we developed an expression for the linear collision stopping power in a medium. We repeat a similar calculation by first deriving the energy loss in a dielectric medium now that we have derived the electromagnetic fields $\mathbf{E}(\mathbf{k},\omega)$ and $\mathbf{H}(\mathbf{k},\omega)$. The energy loss to a single electron at the position associated with the impact parameter, b, is calculated at the vectorial position,

$$\mathbf{x} = b\,\hat{\mathbf{v}}_{\perp}, \tag{12.49}$$

and is equal to the electromagnetic energy flow described by the magnitude of the Poynting vector. We now calculate this vector.

First, we apply the inverse Fourier transform to (12.47) so as to yield an expression for the electric field in the spatial domain (but still retain the frequency dependence):

$$\begin{aligned}\mathbf{E}(b\hat{\mathbf{v}}_{\perp},\omega) &= \frac{1}{(2\pi)^{3/2}}\int_{-\infty}^{\infty} \mathrm{d}^3\mathbf{k}\,\mathbf{E}(\mathbf{k},\omega)\,\mathrm{e}^{ib\mathbf{k}\bullet\hat{\mathbf{v}}_{\perp}} \\ &= i\frac{ze}{(2\pi)^{5/2}\varepsilon_0\varepsilon_R(\omega)}\int_{-\infty}^{\infty}\mathrm{d}^3\mathbf{k}\frac{\left(\omega\varepsilon_R(\omega)\frac{\beta}{c}\hat{\mathbf{v}}_{||}-\mathbf{k}\right)}{\left(k^2-\frac{\omega^2\varepsilon_R(\omega)}{c^2}\right)}\mathrm{e}^{ib\mathbf{k}\bullet\hat{\mathbf{v}}_{\perp}}\delta\left(\omega-\beta ck_{||}\right) \\ &= i\frac{ze}{(2\pi)^{5/2}\varepsilon_0\varepsilon_R(\omega)}\int_{-\infty}^{\infty}\mathrm{d}^3\mathbf{k}\frac{\left(-k_x\hat{\mathbf{v}}_x+\left(\omega\varepsilon_R(\omega)\frac{\beta}{c}-k_{||}\right)\hat{\mathbf{v}}_{||}-k_{\perp}\hat{\mathbf{v}}_{\perp}\right)}{\left(k_x^2+k_{||}^2++k_{\perp}^2-\frac{\omega^2\varepsilon_R(\omega)}{c^2}\right)}\mathrm{e}^{ibk_{\perp}}\delta\left(\omega-\beta ck_{||}\right).\end{aligned} \tag{12.50}$$

We have confronted integrals of this form throughout Sect. 8.2.4. In order to evaluate the integral of (12.50), we apply the substitution of variable $\xi = \beta ck_{||}$ to take to the form

$$\begin{aligned}\mathbf{E}(b\hat{\mathbf{v}}_{\perp},\omega) = i&\frac{ze}{(2\pi)^{5/2}\varepsilon_0\varepsilon_R(\omega)\beta c} \\ &\times\int_{-\infty}^{\infty}\mathrm{d}k_x\int_{-\infty}^{\infty}\mathrm{d}\xi\int_{-\infty}^{\infty}\mathrm{d}k_{\perp}\frac{-k_x\hat{\mathbf{v}}_x+\left(\omega\varepsilon_R(\omega)\frac{\beta}{c}-\frac{\xi}{\beta c}\right)\hat{\mathbf{v}}_{||}-k_{\perp}\hat{\mathbf{v}}_{\perp}}{\left(k_x^2+\left(\frac{\xi}{\beta c}\right)^2+k_{\perp}^2-\frac{\omega^2\varepsilon_R(\omega)}{c^2}\right)} \\ &\times\mathrm{e}^{ibk_{\perp}}\delta(\omega-\xi).\end{aligned} \tag{12.51}$$

The integration over ξ is trivial due to the δ-function and results in

$$\mathbf{E}(b\hat{\mathbf{v}}_\perp,\omega) = i\frac{ze}{(2\pi)^{5/2}\varepsilon_0\varepsilon_R(\omega)\beta c} \times \int_{-\infty}^{\infty} \mathrm{d}k_x \int_{-\infty}^{\infty} \mathrm{d}k_\perp \frac{-k_x\hat{\mathbf{v}}_x + \frac{\omega}{\beta c}\left(\varepsilon_R(\omega)\beta^2 - 1\right)\hat{\mathbf{v}}_{||} - k_\perp\hat{\mathbf{v}}_\perp}{\left(k_x^2 + k_\perp^2 + \lambda(\omega)^2\right)} \mathrm{e}^{ibk_\perp} \tag{12.52}$$

where the quantity

$$\lambda^2(\omega) = \left(\frac{\omega}{\beta c}\right)^2 \left(1 - \beta^2\varepsilon_R(\omega)\right) \tag{12.53}$$

has been defined. It is of interest to note the similarity between the expressions of the Fourier transform of the electric field given by (12.52) with that of (8.140). To solve the integral of (12.52), we split it up into the integral expression for the Fourier transform of the electric field of three integrals along each orthogonal direction. Each are then solved separately:

$$\mathbf{E}(b\hat{\mathbf{v}}_\perp,\omega) = i\frac{ze}{(2\pi)^{5/2}\varepsilon_0\varepsilon_R(\omega)\beta c}\left(I_x\hat{\mathbf{v}}_x + I_{||}\hat{\mathbf{v}}_{||} + I_\perp\hat{\mathbf{v}}_\perp\right) \tag{12.54}$$

where the integrals are

$$\begin{aligned} Ix &= -\int_{-\infty}^{\infty} \mathrm{d}k_\perp \mathrm{e}^{ibk_\perp} \int_{-\infty}^{\infty} \mathrm{d}k_x \frac{k_x}{\left(k_x^2 + k_\perp^2 + \lambda(\omega)^2\right)}. \\ &= 0 \end{aligned} \tag{12.55}$$

and

$$\begin{aligned} I_{||} &= \frac{\omega}{\beta c}\left(\varepsilon_R(\omega)\beta^2 - 1\right) \int_{-\infty}^{\infty} \mathrm{d}k_\perp \mathrm{e}^{ibk_\perp} \int_{-\infty}^{\infty} \frac{\mathrm{d}k_x}{\left(k_x^2 + k_\perp^2 + \lambda(\omega)^2\right)} \\ &= \frac{\pi\omega}{\beta c}\left(\varepsilon_R(\omega)\beta^2 - 1\right) \int_{-\infty}^{\infty} \mathrm{d}k_\perp \frac{\mathrm{e}^{ibk_\perp}}{\sqrt{k_\perp^2 + \lambda(\omega)^2}} \\ &= \frac{2\pi\omega}{\beta c}\left(\varepsilon_R(\omega)\beta^2 - 1\right) \int_{-\infty}^{\infty} \mathrm{d}k_\perp \frac{\cos bk_\perp}{\sqrt{k_\perp^2 + \lambda(\omega)^2}} \\ &= \frac{2\pi\omega}{\beta c}\left(\varepsilon_R(\omega)\beta^2 - 1\right) \int_{0}^{\infty} \frac{\mathrm{d}k_\perp}{\lambda(\omega)} \frac{\cos\left(\lambda(\omega)b\left(\frac{k_\perp}{\lambda(\omega)}\right)\right)}{\sqrt{1 + \left(\frac{k_\perp}{\lambda(\omega)}\right)^2}} \\ &= \frac{2\pi\omega}{\beta c}\left(\varepsilon_R(\omega)\beta^2 - 1\right)K_0(\lambda(\omega)b) \end{aligned} \tag{12.56}$$

and

$$\begin{aligned} I_\perp &= -\int_{-\infty}^{\infty} \mathrm{d}k_\perp k_\perp \mathrm{e}^{ibk_\perp} \int_{-\infty}^{\infty} \frac{\mathrm{d}k_x}{\left(k_x^2 + k_\perp^2 + \lambda(\omega)^2\right)} \\ &= -\pi \int_{-\infty}^{\infty} \mathrm{d}k_\perp \frac{k_\perp \mathrm{e}^{ibk_\perp}}{\sqrt{k_\perp^2 + \lambda(\omega)^2}} \\ &= -i\pi \frac{\mathrm{d}}{\mathrm{d}b} \int_{-\infty}^{\infty} \mathrm{d}k_\perp \frac{\mathrm{e}^{ibk_\perp}}{\sqrt{k_\perp^2 + \lambda(\omega)^2}} \\ &= -i2\pi \frac{\mathrm{d}}{\mathrm{d}b} \int_{-\infty}^{\infty} \mathrm{d}k_\perp \frac{\cos bk_\perp}{\sqrt{k_\perp^2 + \lambda(\omega)^2}} \\ &= -i2\pi \frac{\mathrm{d}}{\mathrm{d}b} K_0(\lambda(\omega)b) \\ &= i2\pi\lambda(\omega)K_1(\lambda(\omega)b). \end{aligned} \tag{12.57}$$

The electric field at the position of the electron $b\hat{\mathbf{v}}_\perp$ is, then,

$$\begin{aligned} \mathbf{E}(b\hat{\mathbf{v}}_\perp, \omega) &= \frac{ze}{(2\pi)^{3/2}\varepsilon_0\varepsilon_R(\omega)\beta c} \\ &\quad \times \left(\lambda(\omega)K_1(\lambda(\omega)b)\hat{\mathbf{v}}_\perp - i\frac{\omega}{\beta c}\left(1 - \varepsilon_R(\omega)\beta^2\right)K_0(\lambda(\omega)b)\hat{\mathbf{v}}_{||}\right). \end{aligned} \tag{12.58}$$

One can see from the definition of λ^2 that, for a real relative dielectric constant,

$$\lambda^2 > 0 \quad \text{for} \quad \beta < \frac{1}{\sqrt{\varepsilon_R(\omega)}} \tag{12.59}$$

$$\lambda^2 < 0 \quad \text{for} \quad \beta > \frac{1}{\sqrt{\varepsilon_R(\omega)}}. \tag{12.60}$$

In other words, λ can only be real if the particle speed is less than the phase velocity of the medium. It should be noted that if polarisation is neglected, i.e. $\varepsilon_R(\omega) = 1$, then $\lambda^2(\omega) = \left(\frac{\omega}{\gamma\beta c}\right)^2$ and the expression for the electric field reduces to

$$
\begin{aligned}
\mathbf{E}(b\hat{\mathbf{v}}_\perp,\omega) &= \frac{ze}{(2\pi)^{3/2}\varepsilon_0\beta c}\left(\frac{\omega}{\gamma\beta c}K_1\left(\frac{\omega}{\gamma\beta c}b\right)\hat{\mathbf{v}}_\perp - i\frac{\omega}{\gamma^2\beta c}K_0\left(\frac{\omega}{\gamma\beta c}b\right)\hat{\mathbf{v}}_{||}\right)\\
&= \frac{ze}{(2\pi)^{3/2}\varepsilon_0}\frac{\omega}{\gamma(\beta c)^2}\left(K_1\left(\frac{\omega}{\gamma\beta c}b\right)\hat{\mathbf{v}}_\perp - i\frac{1}{\gamma}K_0\left(\frac{\omega}{\gamma\beta c}b\right)\hat{\mathbf{v}}_{||}\right) \quad \text{for} \quad \varepsilon_R(\omega)=1
\end{aligned} \tag{12.61}
$$

which is the result which was derived previously for a heavy charged particle interacting with a harmonically bound electron.

In a fashion similar to the above calculation for the electric field, one can next calculate the magnetic field strength at the position of the target electron:

$$
\begin{aligned}
\mathbf{H}(b\hat{\mathbf{v}},\omega) &= -i\frac{ze\beta c}{(2\pi)^{5/2}}\int_{-\infty}^{\infty} \mathrm{d}^3\mathbf{k}\frac{\mathbf{k}\times\hat{\mathbf{v}}_{||}}{\left(k^2-\frac{\omega^2\varepsilon_R(\omega)}{c^2}\right)}\mathrm{e}^{ib\mathbf{k}\bullet\hat{\mathbf{v}}_\perp}\delta(\omega-\beta c k_{||})\\
&= -i\frac{ze\beta c}{(2\pi)^{5/2}}\int_{-\infty}^{\infty}\mathrm{d}k_x\int_{-\infty}^{\infty}\mathrm{d}k_{||}\int_{-\infty}^{\infty}\mathrm{d}k_\perp\frac{k_x\hat{\mathbf{v}}_\perp - k_\perp\hat{\mathbf{v}}_x}{\left(k_x^2+k_{||}^2+k_\perp^2-\frac{\omega^2\varepsilon_R(\omega)}{c^2}\right)}\mathrm{e}^{ibk_\perp}\delta(\omega-\beta c k_{||})\\
&= -i\frac{ze}{(2\pi)^{5/2}}\int_{-\infty}^{\infty}\mathrm{d}k_\perp \mathrm{e}^{ibk_\perp}\int_{-\infty}^{\infty}\mathrm{d}k_x\frac{k_x\hat{\mathbf{v}}_\perp - k_\perp\hat{\mathbf{v}}_x}{\left(k_x^2+k_{||}^2+\lambda^2(\omega)\right)}\\
&= -i\frac{ze}{(2\pi)^{5/2}}(I_x\hat{\mathbf{v}}_x + I_\perp\hat{\mathbf{v}}_\perp),
\end{aligned} \tag{12.62}
$$

where the integral components are

$$
\begin{aligned}
I_x &= -\int_{-\infty}^{\infty}\mathrm{d}k_\perp k_\perp \mathrm{e}^{ibk_\perp}\int_{-\infty}^{\infty}\frac{\mathrm{d}k_x}{\left(k_x^2+k_\perp^2+\lambda(\omega)\right)}\\
&= i\,2\pi\lambda(\omega)K_1(\lambda(\omega)b)
\end{aligned} \tag{12.63}
$$

and

$$
\begin{aligned}
I_\perp &= \int_{-\infty}^{\infty}\mathrm{d}k_\perp \mathrm{e}^{ibk_\perp}\int_{-\infty}^{\infty}\mathrm{d}k_x\frac{k_x}{\left(k_x^2+k_\perp^2+\lambda^2(\omega)\right)}\\
&= 0
\end{aligned} \tag{12.64}
$$

to give

$$
\mathbf{H}(b\hat{\mathbf{v}},\omega) = \frac{ze}{(2\pi)^{3/2}}\lambda(\omega)K_1(\lambda(\omega)b)\hat{\mathbf{v}}_x. \tag{12.65}
$$

Now that the electric and magnetic fields have been derived in Fourier space at the position of the electron at $b\,\hat{\mathbf{v}}_\perp$, the energy loss due to collisions at an impact parameter of $b > b_{\min}$ can be readily calculated from the energy flow through a cylinder of radius $b_{\min}$ centred on the particle's trajectory through the inverse Fourier transform. The energy flow through this cylinder is equal to the power loss of the particle:

$$\left.\frac{\mathrm{d}E}{\mathrm{d}x}\right|_{b\geq b_{\min}} = \frac{1}{\beta c}\frac{\mathrm{d}E}{\mathrm{d}t}. \tag{12.66}$$

The power flow is given by the outgoing component of the Poynting vector, $\mathbf{P} = \mathbf{E}\times\mathbf{H}$,

$$\begin{aligned}\mathbf{P} &= \left(E_{||}\hat{\mathbf{v}}_{||} + E_\perp\hat{\mathbf{v}}_\perp\right)\times H_x\hat{\mathbf{v}}_x \\ &= E_\perp H_x\hat{\mathbf{v}}_{||} - E_{||}H_x\hat{\mathbf{v}}_\perp.\end{aligned} \tag{12.67}$$

The outgoing component of the vector is $-\,E_{||}H_x$ and

$$\begin{aligned}\left.\frac{\mathrm{d}E}{\mathrm{d}x}\right|_{b\geq b_{\min}} &= -\frac{2\pi\,b_{\min}}{\beta c}\int_{-\infty}^{\infty}\mathrm{d}x E_{||}H_x \\ &= -2\pi\,b_{\min}\int_{-\infty}^{\infty}\mathrm{d}t E_{||}H_x \\ &= -2\pi\,b_{\min}\left(\frac{1}{2\pi}\int_{-\infty}^{\infty}\mathrm{d}t\int_{-\infty}^{\infty}\mathrm{d}\omega\int_{-\infty}^{\infty}\mathrm{d}\omega' E_{||}(\omega)H_x(\omega)\mathrm{e}^{-i(\omega+\omega')t}\right) \\ &= -2\pi\,b_{\min}\int_{-\infty}^{\infty}\mathrm{d}\omega\,E_{||}(\omega)H_x^*(\omega) \\ &= -4\pi\,b_{\min}\mathrm{Re}\left(\int_0^{\infty}\mathrm{d}\omega\,E_{||}(\omega)\,H_x^*(\omega)\right) \\ &= \frac{(ze)^2\,b_{\min}}{2\pi^2\varepsilon_0(\beta c)^2}\mathrm{Re}\left(\int_0^{\infty}\mathrm{d}\omega\,(i\omega\lambda^*(\omega))\left(\frac{1}{\varepsilon_R(\omega)}-\beta^2\right)K_0(\lambda(\omega)b_{\min})\,K_1(\lambda^*(\omega)b_{\min})\right).\end{aligned} \tag{12.68}$$

Although i can be removed from the integrand through the use of $\mathrm{Re}\,iz = -\mathrm{Im}\;z$, it is retained in anticipation of a future complex integration. In order to have the result of a non-zero stopping power, this integral must have a real component which requires that either λ, and hence, or $\varepsilon_R(\omega)$ be complex (recall, e.g. the result of (8.243)). Even if the relative dielectric permittivity were to be real, it is possible for the stopping power to be non-zero when λ is complex which is a result of the particle speed exceeding the phase velocity of the medium. This energy loss is manifested as the Čerenkov radiation which is to be discussed. On the other hand, if the particle speed were to be less than the phase velocity, λ would be complex only if $\varepsilon_R(\omega)$ was. Because of the $\beta^2\varepsilon_R(\omega)$ term in the expression for λ^2, this effect becomes significant at high projectile speeds. One can simplify the expression for the stopping power in a dielectric medium by taking advantage of this and limiting the derivation to the extreme relativistic case of $\beta \approx 1$ for which

$$\lambda^2 \approx \left(\frac{\omega}{c}\right)^2 (1 - \varepsilon_R(\omega)) \quad \text{for} \quad \beta \approx 1. \tag{12.69}$$

From the earlier discussion of electronic polarisation, the integral will be significant for ω in the ultraviolet region (10^{16}–10^{17} s^{-1}). As $b_{\min}$ is of the order of the atomic radius, R_{A}, the argument of the modified Bessel function will thus be of the order of

$$|\lambda b| < \left(\frac{\omega_0 R_{\mathrm{A}}}{c}\right) \approx 0.003.$$

As a result, one can use our now familiar small-argument limits:

$$K_0(y) \approx \ln\frac{2\,\mathrm{e}^{-\gamma_{\mathrm{EM}}}}{y} \quad \text{and} \quad K_1(y) \approx \frac{1}{y}.$$

Substituting these limits into the stopping power expression gives

$$\begin{aligned}
\frac{\mathrm{d}E}{\mathrm{d}x} &= \frac{(ze)^2}{2\pi^2\varepsilon_0(\beta c)^2}\mathrm{Re}\left(\int_0^\infty \mathrm{d}\omega(i\omega)\left(\frac{1}{\varepsilon_R(\omega)} - \beta^2\right)\ln\frac{2\mathrm{e}^{-\gamma_{\mathrm{EM}}}}{\lambda(\omega)b_{\min}}\right)\\
&= \frac{(ze)^2}{2\pi^2\varepsilon_0(\beta c)^2}\mathrm{Re}\left(\int_0^\infty \mathrm{d}\omega(i\omega)\left(\frac{1}{\varepsilon_R(\omega)} - \beta^2\right)\ln\frac{2\mathrm{e}^{-\gamma_{\mathrm{EM}}}}{\left(\frac{\omega b_{\min}}{\beta c}\right)\sqrt{1-\beta^2\varepsilon_{R(\omega)}}}\right)\\
&= \frac{(ze)^2}{2\pi^2\varepsilon_0(\beta c)^2}\mathrm{Re}\left(\int_0^\infty \mathrm{d}\omega(i\omega)\left(\frac{1}{\varepsilon_R(\omega)} - \beta^2\right)\left(\ln\left(\frac{2e^{-\gamma_{\mathrm{EM}}}\beta c}{\omega b_{\min}}\right) - \frac{1}{2}\ln(1-\beta^2\varepsilon_R(\omega))\right)\right).
\end{aligned} \tag{12.70}$$

(Note that, in the limit of $\varepsilon_R(\omega) \to 1$, $-\frac{1}{2}\ln\left(1 - \beta^2\varepsilon_R(\omega)\right) \to \ln\gamma$.) In making the extreme relativistic case even more explicit, set $\beta = 1$,

$$\frac{\mathrm{d}E}{\mathrm{d}x} = \frac{(ze)^2}{2\pi^2\,\varepsilon_0\,c^2}\mathrm{Re}\left(\int_0^\infty \mathrm{d}\omega\,(i\omega)\left(\frac{1}{\varepsilon_R(\omega)} - 1\right)\left(\ln\left(\frac{2\mathrm{e}^{-\gamma_{EM}}\,c}{\omega b_{\min}}\right) - \frac{1}{2}\ln(1-\varepsilon_R(\omega))\right)\right). \tag{12.71}$$

For a small damping constant Γ, the imaginary part of the relative dielectric permittivity can be neglected, and the expression for the stopping power becomes

$$\begin{aligned}\frac{\mathrm{d}E}{\mathrm{d}x} &= \frac{(ze)^2}{2\pi^2\varepsilon_0 c^2}\mathrm{Re}\left(\int_0^\infty \mathrm{d}\omega(i\omega)\left(\frac{-\frac{\omega_P^2}{(\omega_0^2-\omega^2)+i\Gamma\omega}}{1+\frac{\omega_P^2}{(\omega_0^2-\omega^2)+i\Gamma\omega}}\right)\times\left(\ln\left(\frac{2\mathrm{e}^{-\gamma_{\mathrm{EM}}}c}{\omega b_{\min}}\right)-\frac{1}{2}\ln\left(-\frac{\omega_P^2}{(\omega_0^2-\omega^2)+i\Gamma\omega}\right)\right)\right)\\ &= \frac{(ze)^2}{2\pi^2\varepsilon_0 c^2}\mathrm{Re}\left(i\int_0^\infty \mathrm{d}\omega\omega\left(\frac{\omega_P^2}{(\omega_0^2-\omega^2+\omega_P^2)+i\Gamma\omega}\right)\left(\ln\left(\frac{2\mathrm{e}^{-\gamma_{\mathrm{EM}}}}{\omega b_{\min}}\right)+\frac{1}{2}\ln\left(\frac{(\omega^2-\omega_0^2)+i\Gamma\omega}{\omega_P^2}\right)\right)\right).\end{aligned} \tag{12.72}$$

The integral of (12.72)

$$\begin{aligned}&i\int_0^\infty \mathrm{d}\omega\omega\left(\frac{\omega_P^2}{(\omega_0^2-\omega^2+\omega_P^2)+i\Gamma\omega}\right)\left(\ln\left(\frac{2\mathrm{e}^{-\gamma_{\mathrm{EM}}}}{\omega b_{\min}}\right)+\frac{1}{2}\ln\left(\frac{(\omega^2-\omega_0^2)+i\Gamma\omega}{\omega^2}\right)\right)\\ &= i\int_0^\infty \mathrm{d}\omega\omega\left(\frac{\omega_P^2}{(\omega_0^2-\omega^2+\omega_P^2)+i\Gamma\omega}\right)\left(\ln\left(\frac{2\mathrm{e}^{-\gamma_{\mathrm{EM}}}}{\omega_P b_{\min}}\right)+\frac{1}{2}\ln\left(\frac{(\omega^2-\omega_0^2)+i\Gamma\omega}{\omega_P^2}\right)\right)\end{aligned} \tag{12.73}$$

is determined using the Cauchy–Goursat theorem (Appendix A.3) by changing the integration over positive real ω (i.e. $0 \leq \omega \leq \infty$) to that over positive imaginary ω minus the integration over the quarter circle to infinity, as shown in Fig. 12.3, with the intent of isolating the integration I_X. It can be seen from the integrand of (12.73) that poles occur for

$$\left(\omega_0^2-\omega^2+\omega_P^2\right)+i\Gamma\omega=0$$

and

$$\frac{(\omega^2-\omega_0^2)+i\Gamma\omega}{\omega^2}=0.$$

In both cases, ω would be imaginary, and the poles of the integrand exist in the lower half plane. As there are no poles within the quarter circle of the contour, the total integral is equal to zero, or

$$\mathrm{Re}(I_X+I_Y+I_{SC})=0.$$

The integral down the imaginary axis is evaluated first using $\omega = i\Omega$:

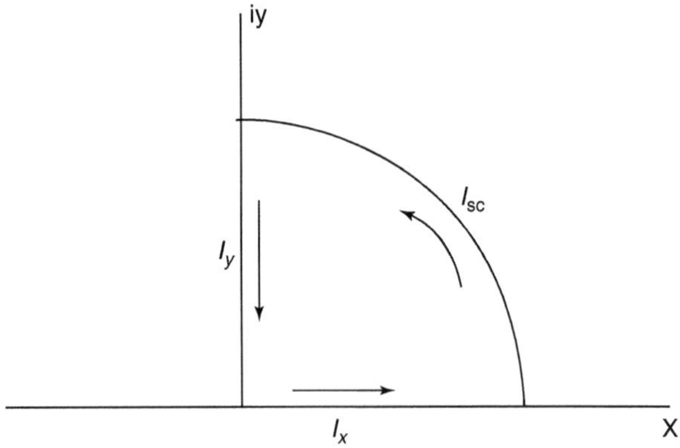

Fig. 12.3 Integration contour in the complex plane $z = x + iy$ used to evaluate the integral of Eq. (12.73)

$$I_Y = -i\int_0^{\infty} d\Omega\,\Omega\left(\frac{\omega_P^2}{(\omega_0^2+\Omega^2+\omega_P^2)-i\Gamma\Omega}\right)$$
$$\times\left(\ln\left(\frac{2e^{-\gamma_{EM}}c}{\omega_P b_{min}}\right)+\frac{1}{2}\ln\left(\frac{-(\Omega^2+\omega_0^2+\Gamma\Omega)}{(i\Omega)}\right)\right). \quad (12.74)$$

It is clear that, as this integral is pure imaginary (i.e. $\mathrm{Re}\, I_Y = 0$), its contribution to the stopping power will be zero. As a result,

$$\mathrm{Re}\, I_X = -\mathrm{Re}\, I_{SC}. \quad (12.75)$$

In order to calculate the integral over the quarter circle, I_{SC}, we write ω in terms of complex polar co-ordinates, or

$$\omega = re^{i\theta}, \quad (12.76)$$

where r is a constant. Thus,

$$\mathrm{Re} I_X = -\mathrm{Re}\, I_{SC}$$
$$= -\mathrm{Re}\int_0^{\pi/2} d\theta r^2 e^{i2\theta}\left(\frac{\omega_P^2}{\left(\omega_0^2 - r^2e^{i2\theta}+\omega_p^2\right)+i\Gamma re^{i\theta}}\right)\left(\ln\left(\frac{2e^{-\gamma_{EM}}c}{\omega_{P\min}^b}\right)\right. \quad (12.77)$$
$$\left.\left.+\frac{1}{2}\ln\left(\frac{\left(r^2e^{i2\theta}-\omega_0^2\right)+i\Gamma re^{i\theta}}{r^2}e^{-i2\theta}\right)\right)\right).$$

Maintaining the assumption that the damping constant Γ is small and constant,

$$
\begin{aligned}
&- \mathrm{Re}\int_0^{\pi/2} \mathrm{d}\theta r^2 \mathrm{e}^{i2\theta}\left(\frac{\omega_P^2}{\left(\omega_0^2 - r^2\mathrm{e}^{i2\theta} + \omega_p^2\right) + i\Gamma r\mathrm{e}^{i\theta}}\right)\left(\ln\left(\frac{2\mathrm{e}^{-\gamma_{\mathrm{EM}}}c}{\omega_{P\min}^b}\right)\right.\\
&\left.+\frac{1}{2}\ln\left(\frac{\left(r^2\mathrm{e}^{i2\theta} - \omega_0^2\right) + i\Gamma r\mathrm{e}^{i\theta}}{r^2}\mathrm{e}^{-i2\theta}\right)\right)\\
&= \mathrm{Re}\, I_1 + \mathrm{Re}\, I_2
\end{aligned}
$$

where, by rearranging the variables within the integrand,

$$
\begin{aligned}
I_1 =& \omega_P^2 \ln\left(\frac{2e^{-\gamma_{\mathrm{EM}}}c}{\omega_P b_{\min}}\right)\int_0^{\pi/2} \mathrm{d}\theta \frac{\mathrm{e}^{i2\theta}}{\left(\frac{\omega_0^2+\omega_P^2}{r^2}\right) - \mathrm{e}^{i2\theta} + i\left(\frac{\Gamma}{r}\right)\mathrm{e}^{i\theta}}\\
&\approx -\frac{\pi\omega_P^2}{2}\ln\frac{2\mathrm{e}^{-\gamma_{\mathrm{EM}}}}{b\omega_P}.
\end{aligned}
\tag{12.78}
$$

The facts that the ratio $\Gamma/|\omega|$ is negligible and that $\omega^2 >> \left|\omega_0^2 + \omega_P^2\right|$ have been used. The second integral is

$$
\begin{aligned}
I_2 =& \frac{1}{2}\int_0^{\pi/2} \mathrm{d}\theta r^2\mathrm{e}^{i2\theta}\left(\frac{\omega_P^2}{\left(\omega_0^2 - r^2\mathrm{e}^{i2\theta} + \omega_p^2\right) + i\Gamma r\mathrm{e}^{i\theta}}\right)\left(\ln\left(\frac{\left(r^2\mathrm{e}^{i2\theta} - \omega_0^2\right) + i\Gamma r\mathrm{e}^{i\theta}}{r^2}\mathrm{e}^{-i2\theta}\right)\right)\\
=& \frac{1}{2}\int_0^{\pi/2} \mathrm{d}\theta \frac{\mathrm{e}^{i2\theta}}{\left(\frac{\omega_0^2+\omega_P^2}{r^2}\right) - \mathrm{e}^{i2\theta} + i\left(\frac{\Gamma}{r}\right)\mathrm{e}^{i\theta}}\left(\ln\left(\frac{\left(r^2\mathrm{e}^{i2\theta} - \omega_0^2\right) + i\Gamma r\mathrm{e}^{i\theta}}{r^2}\mathrm{e}^{-i2\theta}\right)\right)\\
\approx& -\frac{1}{2}\int_0^{\pi/2} \mathrm{d}\theta\left(\ln\left(\frac{\left(r^2\mathrm{e}^{i2\theta} - \omega_0^2\right) + i\Gamma r\mathrm{e}^{i\theta}}{r^2}\mathrm{e}^{-i2\theta}\right)\right)\\
\approx& -\frac{1}{2}\int_0^{\pi/2} \mathrm{d}\theta \ln\left(\left(\mathrm{e}^{i2\theta} - \left(\frac{\omega_0}{r}\right)^2 + i\frac{\Gamma}{r}\mathrm{e}^{i\theta r}\right)\mathrm{e}^{-i2\theta}\right)\\
\approx& -\frac{1}{2}\int_0^{\pi/2} \mathrm{d}\theta \ln\left(1 - \left(\frac{\omega_0}{r}\right)^2\mathrm{e}^{-i2\theta} + i\frac{\Gamma}{r}\mathrm{e}^{-i\theta r}\right)\\
\approx& \; 0.
\end{aligned}
$$

Thus,

$$
\begin{aligned}
&\mathrm{Re}\left(i\int_0^{\infty} \mathrm{d}\omega\omega\left(\frac{\omega_P^2}{\left(\omega_0^2 - \omega^2 + \omega_P^2\right) + i\Gamma\omega}\right)\ln\left(\frac{2\mathrm{e}^{-\gamma_{\mathrm{EM}}}c}{\omega_p b_{\min}}\right) + \frac{1}{2}\ln\left(\frac{\left(\omega^2 - \omega_0^2\right) + i\Gamma\omega}{\omega^2}\right)\right)\\
&\approx \frac{\pi\omega_P^2}{2}\ln\frac{2\mathrm{e}^{-\gamma_{\mathrm{EM}}}}{b\omega_P}.
\end{aligned}
\tag{12.79}
$$

One now writes the linear stopping power in a dielectric medium in the extreme relativistic limit as

$$\left.\frac{\mathrm{d}E}{\mathrm{d}x}\right|_{b>b_{\min}} = \frac{(ze)^2\omega_P^2}{4\pi\varepsilon_0 c^2}\ln\frac{2\mathrm{e}^{-\gamma_{\mathrm{EM}}}\,c}{b_{\min}\omega_P}. \tag{12.80}$$

One can recast this as the mass collision stopping power for $\beta \approx 1$,

$$\frac{\mathrm{d}E}{\rho\,\mathrm{d}x} = 2C\left(\frac{Z}{A}\right)z^2\ln\frac{2\mathrm{e}^{-\gamma_{\mathrm{EM}}}\,c}{b_{\min}\omega_P}, \tag{12.81}$$

and where the identifier of $b > b_{\min}$ has been removed. A significant difference between this expression, which accounts for the medium's dielectric response to the charged particle, and the relativistic forms of those that do not is that the former is no longer a function of the medium's atomic structure, which appears in the form of $\bar{I}$ or $\bar{\omega}$ within the logarithmic term, but rather of the medium's electron density which appears through the electron plasma frequency. That is, in the extreme relativistic regime, the mass collision stopping powers of two media are equal provided that the electron densities and the Z/A ratios are the same.

In the absence of any polarisation, the corresponding relativistic mass collision stopping power is, where for clarity the Bohr result is used with $\beta \approx 1$,

$$\left.\frac{\mathrm{d}E}{\mathrm{d}x}\right|_{b>b_{\min}} = 2C\left(\frac{Z}{A}\right)z^2\ln\frac{2\mathrm{e}^{-\gamma_{\mathrm{EM}}}\,\gamma\,c}{b_{\min}\bar{\omega}}. \tag{12.82}$$

An analytical form of the density correction term given by the difference between these two mass collision stopping powers is

$$\delta = 2C\left(\frac{Z}{A}\right)z^2\ln\frac{\gamma\omega_P}{\bar{\omega}}. \tag{12.83}$$

As the density effect results in the reduction of the electric field and, hence, the collision stopping power, it is a negative quantity. We then write its contribution to the stopping number summation as

$$L_{02}(\beta) = -\frac{\delta}{2}. \tag{12.84}$$

12.4.5 Sternheimer–Peierls Parameterisation

While the above investigation has provided a theoretical expression for the polarisation/density effect (with considerable simplifications and approximations), practical calculations benefit from a parameterisation of the effect that is frequently required in dosimetry calculations. A frequently used parameterisation which can accelerate the calculation of δ is that of Sternheimer and Peierls (1971) who presented an expression for δ applicable to both condensed media and gases

$$\delta = (2\ln 10)x + \Phi \quad x \geq x_1 \tag{12.85}$$

$$\delta = (2\ \ln 10)x + \Phi + a(x_1 - x)^n \quad x_0 \leq x \leq x_1 \tag{12.86}$$

$$\delta = 0 \quad x < x_0 \tag{12.87}$$

where the kinematical variable is, for a particle of mass m and momentum p,

$$x = \log_{10}\left(\frac{p}{m}\right) \tag{12.88}$$

where the parameters of x_0 and x_1 are defined below. This reproduces the logarithmic increase in the density effect with $p/m \approx E/m \approx \gamma$ at high energies. The remaining quantities in this parameterisation are

$$\Phi = -2\ln\left(\frac{\bar{I}}{\hbar\omega_P}\right) - 1 \tag{12.89}$$

and

$$a = \frac{-\Phi - (2\ln 10)x_0}{(x_1 - x_0)^n}. \tag{12.90}$$

The values of x_0 and x_1 also depend upon the values of the mean excitation energy, $\bar{I}$, and the material phase of the medium. For liquid and solid media, these are given in Table 12.1.

In all cases, n is equal to 3. The relative magnitude of δ and its growth with particle speed are shown in Fig. 12.4 which plots fitted values for δ for carbon and lead as a function of p/m; for example, the value of δ for carbon reaches 2 for $p/m = 9.8$, corresponding to a kinetic energy of about 5 MeV for electrons.

Table 12.1 Sternheimer–Peierls parameter values

$\bar{I}$ (eV)	$-\Phi$	x_0	x_1
<100	≤3.681	0.2	2.0
<100	>3.681	$-0.326\,\Phi - 1.0$	2.0
≥100	≤5.215	0.2	3.0
≥100	>5.215	$-0.326\,\Phi - 1.5$	3.0

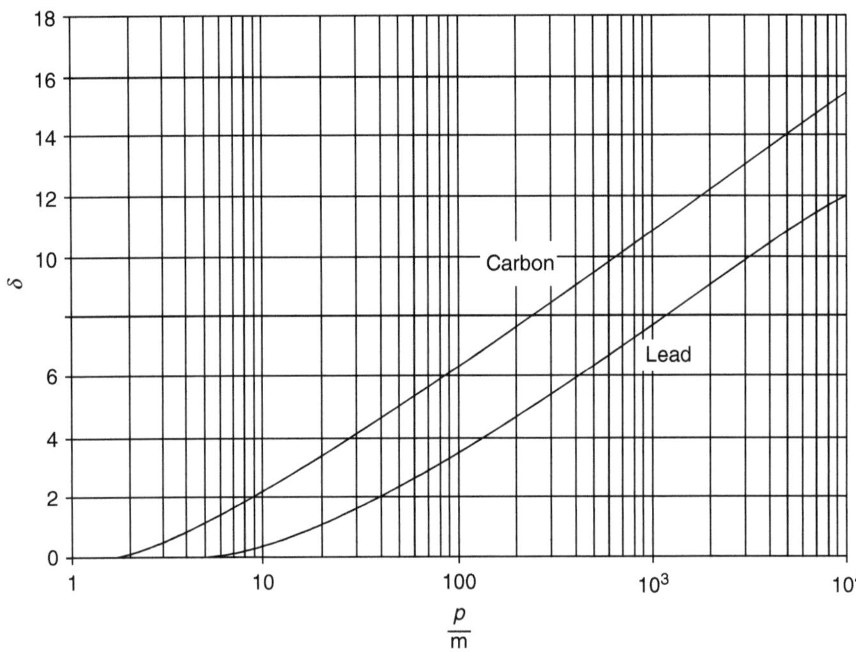

Fig. 12.4 Polarisation/density correction terms for carbon and lead as functions of the ratio of the particle momentum normalised to particle mass

12.4.6 Čerenkov Radiation

In 1934, Čerenkov (1934) reported his observations of the eponymous radiation following the irradiation of various liquids to γ-rays. Shortly afterwards, Tamm and Frank (1937) published a theoretical explanation of this observation. Although the magnitude of the energy loss associated with Čerenkov radiation is negligible for medical radiation dosimetry considerations, the radiation itself must be accounted for in some dosimetry devices based upon the use of scintillator fibres (e.g. Beddar et al. 1992).

The energy loss per unit distance travelled to regions within the medium with impact parameters greater than a value, $b_{\min}$, is given by (12.82). In that derivation, it was assumed that $b_{\min}$ is of the order of atomic dimensions such that the product

$|\lambda\, b_{\min}| << 1$. This assumption allowed the use of the small-argument limits of the modified Bessel functions. Consequently, the final result represents the rate of local energy deposition with distance. On the other hand, by permitting $|\lambda\, b_{\min}| << 1$, one obtains the rate of energy deposited at great distances per distance travelled from the projectile's trajectory. Recalling the large-argument expression for the modified Bessel function,

$$\left.\frac{\mathrm{d}E}{\mathrm{d}x}\right|_{b>b_{\min}} = \frac{(ze)^2}{4\pi\varepsilon_0(\beta c)^2}\mathrm{Re}\left(\int_0^{\infty}\mathrm{d}\omega\left(i\omega\sqrt{\frac{\lambda^*(\omega)}{\lambda(\omega)}}\right)\left(\frac{1}{\varepsilon_R(\omega)}-\beta^2\right)\mathrm{e}^{-(\lambda(\omega)+\lambda^*(\omega))b_{\min}}\right), \tag{12.91}$$

we consider the effects of

$$\lambda(\omega) = \left(\frac{\omega}{\beta c}\right)\sqrt{1-\beta^2\varepsilon_R(\omega)}. \tag{12.92}$$

In general, $\mathrm{Re}\,\lambda(\omega) > 0$ and the exponential term in the integrand will, as a consequence, rapidly attenuate the energy loss with distance from the trajectory as expected. However, $\lambda(\omega)$ will be purely imaginary if the damping constant Γ is negligible, which allows $\varepsilon_R(\omega)$ to be purely real resulting in $\beta^2\,\varepsilon_R(\omega) >> 1$. For purely imaginary $\lambda(\omega)$,

$$\sqrt{\frac{\lambda^*(\omega)}{\lambda(\omega)}} = i$$

and

$$\mathrm{e}^{-(\lambda(\omega)+\lambda^*(\omega))b_{\min}} = 1.$$

Thus, for the conditions of purely real $\varepsilon_R(\omega)$ and $\beta^2\,\varepsilon_R(\omega) \gg 1$, the stopping power expression simplifies to

$$\left(\frac{\mathrm{d}E}{\mathrm{d}x}\right) = z^2\hbar\int_{\omega_1}^{\infty}\mathrm{d}\omega\,\omega\left(1-\frac{1}{\beta^2\varepsilon_R(\omega)}\right) \tag{12.93}$$

where the lower-frequency limit is specified by the requirement of

$$\varepsilon_R(\omega > \omega_1) > \frac{1}{\beta^2} \tag{12.94}$$

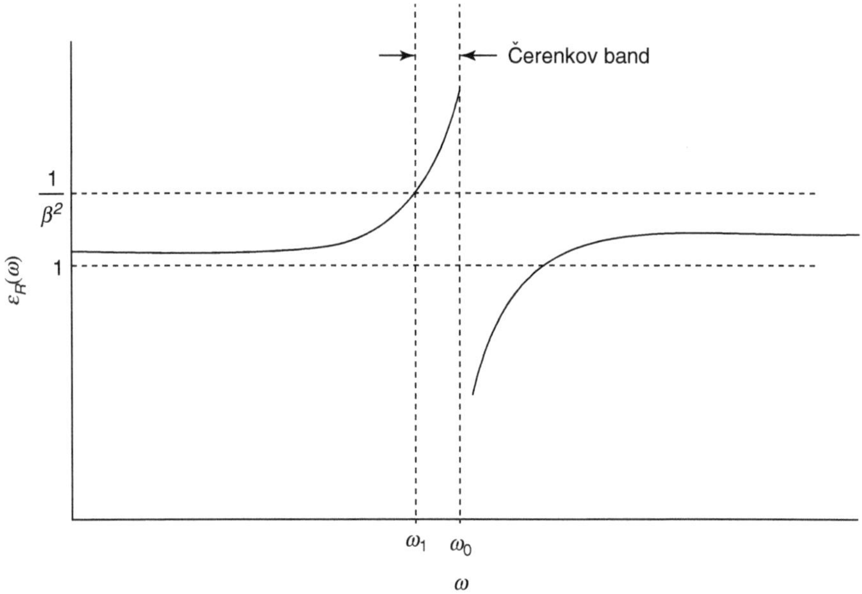

Fig. 12.5 The relative dielectric permittivity (assumed to be purely real) of a medium as a function of frequency. The Čerenkov band, within which Cerenkov radiation can occur, is shown

and there is no longer a dependence of the energy transfer upon $b_{\min}$. This result describes the rate per distance travelled by the projectile at which energy is radiated. The projectile's speed must exceed the phase velocity of the medium for the given frequency in order for Čerenkov radiation to occur.

In Fig. 12.5, the real component of $\varepsilon_R(\omega)$ is plotted as a function of ω for negligible energy absorption, i.e. $\Gamma = 0$. There is a discontinuity at $\omega = \omega_0$, and the *Čerenkov band* is shown for which $\beta^2\,\varepsilon_R(\omega) >> 1$ and which identifies the lower-frequency integration limit.

Before investigating the spectrum of Čerenkov radiation, the frequently used geometrical description of a charged particle moving through a dielectric medium at a speed βc exceeding the phase velocity is investigated. The index of refraction of the medium is $n(\omega) = \sqrt{\varepsilon_R(\omega)}$; however, for conciseness, the frequency dependence is ignored. The electric field will thus propagate with speed c/n, and for the condition of $\beta c > c/n$, Huygens' constructions are used for the electromagnetic waves emitted by the particle as shown in Fig. 12.6. Let the particle be at point A at time $t = 0$. In the time that it takes the particle to travel the distance AB, which is equal to

$$T = \frac{AB}{\beta c}, \tag{12.95}$$

a wavefront emitted at $t = 0$ (i.e. at point A) will have reached point C, where

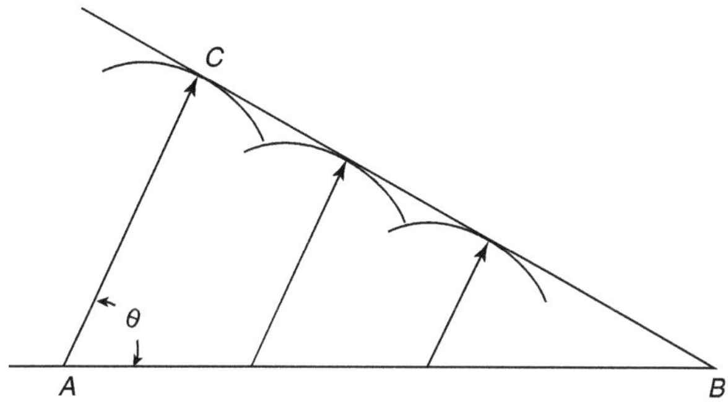

Fig. 12.6 Huygens' reconstruction of the coherent wavefront of Čerenkov radiation

$$AC = T\frac{c}{n}. \tag{12.96}$$

Thus,

$$\begin{aligned}\theta &= \cos^{-1}\frac{AC}{AB}\\ &= \cos^{-1}\frac{1}{\beta n}.\end{aligned} \tag{12.97}$$

Note that Fig. 12.6 shows a single plane only and that due to axial symmetry about AB, the Čerenkov wavefronts form a cone about the charged particle's trajectory. As Čerenkov radiation can only occur if the particle speed $\beta > 1/n$, this effect will be limited to high energies in most media of dosimetric interest (e.g. as $n = 1.5$ for Perspex, the threshold value for β is 0.67). The threshold energy for which Čerenkov radiation will occur follows as $\beta_{\mathrm{Thr}} = 1/n$ where

$$\beta_{\mathrm{Thr}} = \sqrt{1 - \left(\frac{m}{E_{\mathrm{Thr}} + m}\right)^2} \tag{12.98}$$

where E_{Thr} is the threshold kinetic energy of the charged particle and m is its mass.

Figure 12.7 shows the variation of the threshold energy as a function of refractive index for electrons and for those electrons set in motion by backscattered photons following incoherent scatter. It can be seen that within the energy range of 0.1–0.3 MeV, Čerenkov radiation presents a means of detecting γ-rays in low-Z media where the probability of a photoelectric absorption is much less than that of incoherent scatter. The differential spectrum in frequency of Čerenkov radiation is provided by the integrand of (12.93). The number of photons emitted per cm of projectile path length with a frequency between ω and $\omega + \delta\omega$ is

$$\mathrm{d}N = \frac{\alpha}{c}\left(1 - \frac{1}{(\beta n)^2}\right)\mathrm{d}\omega. \tag{12.99}$$

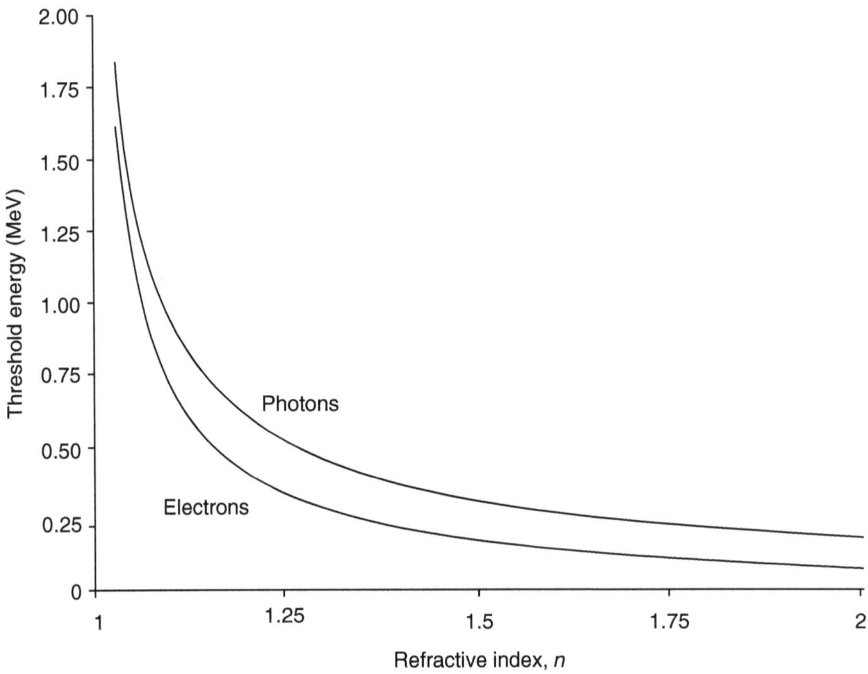

Fig. 12.7 Čerenkov radiation energy threshold for electrons and for Compton-scattered electrons set in motion by backscattered photons (labelled as 'photons') as functions of the refractive index

As $d\omega = -2\pi c \, \frac{d\lambda}{\lambda^2}$, we can rewrite this more conveniently in terms of the photon wavelength:

$$\frac{dN}{d\lambda} = -\frac{2\pi\alpha}{\lambda^2}\left(\beta^2 n^2 - 1\right). \tag{12.100}$$

This spectrum is plotted as a function of photon wavelength for Perspex ($n = 1.5$) for values of β of 0.67 (threshold), 0.75 and 1 in Fig. 12.8. For protons, this threshold value corresponds to a kinetic energy of about 375 MeV. The $1/\lambda^2$ dependence shows that the Čerenkov spectrum is dominated by short wavelength photons. Integrating the Čerenkov spectrum yields the number of Čerenkov photons emitted per cm between the spectrum limits of λ_{min} and λ_{max}:

$$N = 2\pi\alpha\left(1 - \frac{1}{(\beta n)^2}\right)\left(\frac{1}{\lambda_{min}} - \frac{1}{\lambda_{max}}\right). \tag{12.101}$$

Using $\cos\theta = 1/\beta n$, one can also write the spectrum as

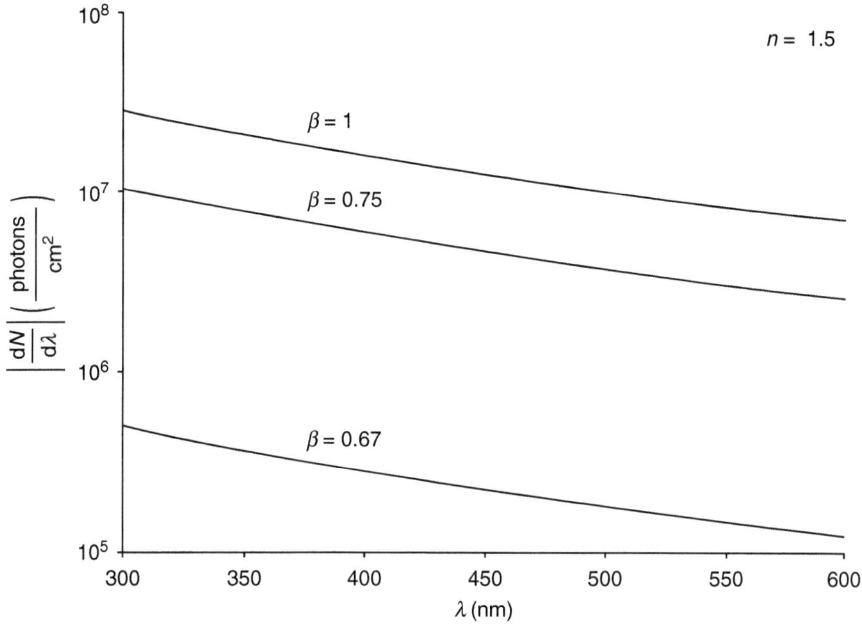

Fig. 12.8 Number of Čerenkov photons emitted per unit path length per unit wavelength in Perspex ($n = 1.5$) for three values of β

$$\frac{\mathrm{d}N}{\mathrm{d}\lambda} = 2\pi\,\alpha\,\sin^2\theta\left(\frac{1}{\lambda_{\max}} - \frac{1}{\lambda_{\min}}\right). \tag{12.102}$$

Within the spectrum limits of 300 and 600 nm, a total of $764\sin^2\theta$ Čerenkov photons are emitted per cm of particle path length, which clearly demonstrates the low level of light output. As for minimally ionising particles in a low-Z medium this corresponds to a conversion efficiency of about 0.2 %; the contribution of Čerenkov radiation to the overall energy loss of a charged particle traversing a medium is small and can be neglected in energy deposition calculations. Thin plastic scintillator fibres have proven useful for high spatial-resolution measurements of absorbed dose distributions of charged particle beams. A background signal due to Čerenkov radiation can be generated within the optical transmission fibres and must be accounted for.

12.5 Barkas–Andersen Term, $L_1(\beta)$

12.5.1 History

The derivations of the complete stopping power of a charged particle incorporating both soft and hard components led to, as shown in Chap. 10, a result proportional to the square of the projectile's electric charge, z^2. That is, the collision stopping powers of a charged particle and its antiparticle for the same medium and kinetic energy would be identical under this result. The first suggestion that this may not be true was discovered experimentally on the basis of charged particle range.

One of the simplest means of assessing the stopping power of a charged particle is to measure its range into a medium: high stopping powers lead to a short range and *vice versa*. The range of a charged particle in a medium is the topic of Chap. 13, and the range of a charged particle is equal to the integral over its kinetic energy of the reciprocal of the collision stopping power.[4] Clearly, high stopping power results in short ranges, and low stopping power results in long ranges.

The Barkas group first noted the effects of the difference in collision stopping powers between particles and their antiparticles through measuring the minute differences between the ranges of π^- and π^+ mesons stopped in a medium (Smith et al. 1953; Barkas et al. 1956). While this could have been ascribable to a difference in mass between the particle and the antiparticle, this hypothesis was dismissed in a further study and the range difference was concluded to be reflective of higher-order perturbations to the first-order Born series used in the Bethe derivation of the soft-collision stopping power (Barkas et al. 1963). It was noted that this observation of the possible contribution of an energy loss effect proportional to z^n, where $n > 2$, would also lead to the result that the ratio of the stopping power of an α-particle ($z = 2$) would not be exactly four times that of a proton ($z = 1$) with the same velocity that would have been expected through the $(z/\beta)^2$ term of the collision stopping power expression. That such an effect existed was confirmed by Andersen et al. (1969) who measured at the differences between the stopping of protons and deuterons with kinetic energies between 5 and 13.5 MeV and that of α-particles with kinetic energies between 8 and 20 MeV in aluminium and tantalum.

An example of the Barkas–Andersen effect is shown in Fig. 12.9 which shows the mass collision stopping powers of protons and antiprotons in silicon. That for protons is taken from ICRU Report No 39, and that for antiprotons are taken from measurements performed on the Low Energy Antiproton Ring (LEAR) at CERN by Medenwaldt et al. (1991) and Møller et al. (1997). At low kinetic energies, the mass collision stopping power of protons exceeds that of antiprotons, but it eventually equalises.

One can surmise that the effect exemplified by Fig. 12.9 is indicative of polarisation (an effect already examined in considerable detail). That is, a negatively charged projectile (i.e. the antiproton) passing by the electron cloud surrounding an atom repels this cloud so as to increase the effective impact parameter and reduce the collision stopping power. On the other hand, a positively charged projectile (i.e. the proton) attracts the cloud so as to decrease the effective impact parameter and, subsequently, increase the collision stopping power.

---

[4] Recall that we are not considering the infinitely small contribution of radiative energy loss for heavy charged particles in medical applications.

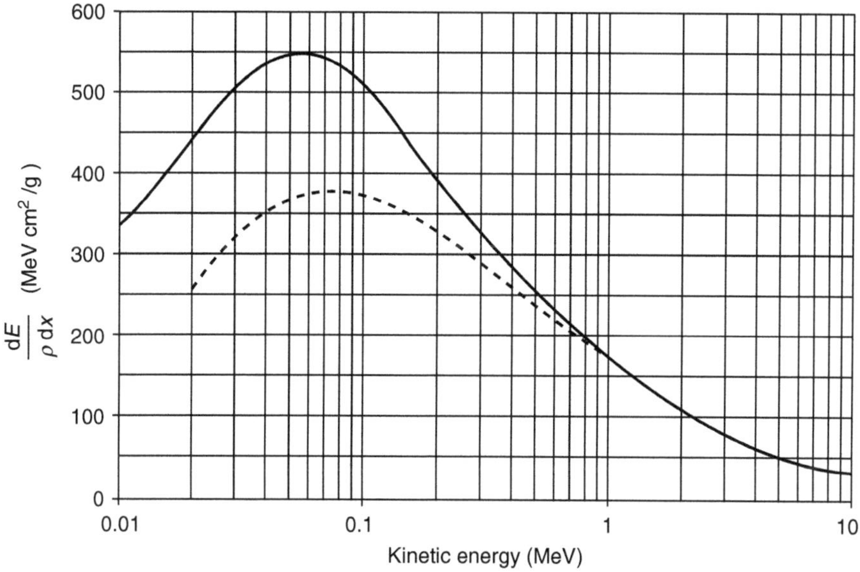

Fig. 12.9 Mass collision stopping powers of protons (*solid line*) and antiprotons (*dashed line*) in silicon as functions of kinetic energy. The proton curve is derived from data provided by ICRU Report No 39 (1993). The antiproton curve is derived from data provided by Medenwaldt et al. (1991) and Møller et al. (1997)

We will consider the Barkas–Andersen effect, but shall do so only to a limited degree as it is prominent only at low kinetic energies of the projectile and, hence, is of a reduced concern to medical radiation physics. A full examination of the effect is summarised by Sigmund (2006).

12.5.2 Derivation of $L_1(\beta)$ from Atomic Electron Screening

Lindhard (1976) proposed an estimate of the Barkas–Andersen effect based upon the Bohr classical theory of soft-collision energy loss. The potential between the projectile and an atomic electron is considered to be screened. Assuming a Yukawa type of screening potential (as used in Chap. 5), we can write the Coulomb potential energy as

$$U(r) = -\frac{z\,\alpha\hbar c}{r}\,\mathrm{e}^{-\omega_0 r/\beta c} \tag{12.103}$$

where $\beta c/\omega_0$ is a measure of the adiabatic radius. Expanding the exponential to first order, we have

$$V(r) \approx -\frac{z\,\alpha\hbar c}{re} + \frac{z\alpha\hbar c\,\omega_0}{\beta c}. \tag{12.104}$$

The Coulomb potential has been shifted by an amount which is independent of the distance between the projectile and the electron. Hence, the collision energy has also been shifted:

$$\begin{aligned} E &= \frac{m_e\beta^2}{2} - \frac{z\,\alpha\hbar c\,\omega_0}{\beta c} \\ &= \frac{m_e\beta^2}{2}\left(1 - \frac{2z\,\alpha\hbar c\,\omega_0}{m_e\beta^3 c}\right). \end{aligned} \tag{12.105}$$

As it is the reciprocal of this expression that appears within the collision stopping power, we use the binomial expansion to first order;

$$\frac{1}{E} \approx \frac{2}{m_e\beta^2}\left(1 + \frac{2z\,\alpha\hbar c\,\omega_0}{m_e\beta^3 c}\right). \tag{12.106}$$

Thus, we can write a crude form of the nonrelativistic form of the soft-collision stopping power:

$$\left.\frac{\mathrm{d}E}{\rho\,\mathrm{d}x}\right|_S = 2C\left(\frac{Z}{A}\right)\left(\frac{z}{\beta}\right)^2\left(1 + \frac{2z\,\alpha\hbar c\,\omega_0}{m_e\beta^3 c}\right)\ln\left(\frac{2m_e\beta^2}{\bar{I}}\right). \tag{12.107}$$

We can thus identify from (12.107)

$$L_1(\beta) = \frac{2\,\alpha\hbar c\,\omega_0}{m_e\beta^3 c}\ln\left(\frac{2m_e\,\beta^2}{\bar{I}}\right) \tag{12.108}$$

in our nomenclature of the stopping number.

While accounting for screening has introduced the necessary additive Barkas–Andersen term with a z^3-dependence, we see that the result diverges at low velocities. The dimensionless parameter $\left(z\,\alpha\hbar c\,\omega_0/m_e\,\beta^3 c\right)$ was also derived by Ashley et al. (1972a) by extending the Bohr soft-collision theory by an additional order in z.

12.5.3 Additional Comments

Ashley et al. (1972b, 1974) provided a thorough theoretical evaluation of the Barkas–Andersen effect. From this work, the Barkas–Andersen correction term can be written in the form

$$L_1(\beta) = k\frac{F_{\mathrm{ARB}}(b/\sqrt{x})}{\sqrt{Z x^3}} \tag{12.109}$$

where

$$x = \frac{(\beta/\alpha)^2}{Z}. \tag{12.110}$$

k is a constant approximately equal to $\sqrt{2}$ and b is a scaled impact parameter with a value between 1.3 and 1.8 for elements heavier than lithium. ICRU Report No 49 (1993) provides tabulated values of b. For example, for Z between 3 and 10 $b = 1.8$ and for Z between 19 and 25, $b = 1.4$. $F_{\mathrm{ARB}}(b/\sqrt{x})$ is a numerically evaluated function and which is tabulated in Ashley et al. (1974), and b is related to the minimum impact parameter. Values for this correction term are provided in ICRU Report No 49 (1993).

12.6 Bloch Correction Term, $L_2(\beta)$

12.6.1 Introduction

Recall the soft-collision stopping power expressions derived from the Bohr classical theory (8.180),

$$\left.\frac{\mathrm{d}E}{\rho\,\mathrm{d}x}\right|_{S,\mathrm{Bohr}} = 2C\left(\frac{Z}{A}\right)\left(\frac{z}{\beta}\right)^2\left(\ln\left(\frac{2\mathrm{e}^{-\gamma_{\mathrm{EM}}} m_{\mathrm{e}}\beta^3\gamma^2}{\omega_0\,\alpha\hbar\, z}\right) - \frac{\beta^2}{2}\right),$$

and the Bethe quantal theory (8.344),

$$\left.\frac{\mathrm{d}E}{\rho\,\mathrm{d}x}\right|_{S,\mathrm{Bethe}} = 2C\left(\frac{Z}{A}\right)\left(\frac{z}{\beta}\right)^2\left(\ln\left(\frac{\sqrt{2} m_{\mathrm{e}} Q c\beta\,\gamma}{\bar{I}}\right) - \frac{\beta^2}{2}\right),$$

which have been slightly rearranged and the definition for the classical electron radius used explicitly. The similarity between the two results, with the difference residing within the logarithmic term, is curious.

In the early 1930s, Bloch reconciled the Bohr classical and Bethe quantum-mechanical calculations of the soft-collision stopping power by demonstrating that the Bohr result was valid quantum mechanically if the Bohr energy loss were to be interpreted as a mean value over all possible atomic electron transitions (Bloch 1933; Ahlen 1980). We now demonstrate this following the approach shown by Lindhard and Sørensen (1996).

12.6.2 Derivation

In Sect. 4.3.2, we demonstrated that the Rutherford differential cross section for Coulomb scatter from an unscreened charge could be correctly derived using the first term of the Born expansion. However, as subsequently demonstrated in Sect. 4. 3.3, incorporating the next term of the Born expansion led to a divergence in the result. This was resolved through the consideration of bound electrons and incorporating the effects of screening.

The approach of Lindhard and Sørensen is to evaluate the difference between the stopping cross sections of free Coulomb scattering and the Born expansion applied to the free Coulomb scattering, or

$$\Delta S = S_{\mathrm{FC}} - S_{\mathrm{FC},B}. \tag{12.111}$$

It is assumed that a heavy projectile is incident to the electron. We further assume that the effect of bound atomic electrons is accurately contained within $S_{\mathrm{FC},B}$, thus allowing us to write an improved stopping cross section

$$S = S_{\mathrm{Bethe}} + \Delta S \tag{12.112}$$

where S_{Bethe}. The stopping cross section is evaluated using a method of phase shifts (*cf* Sect. 2.5):

$$S = m_e \beta^2 \sigma \tag{12.113}$$

where

$$\sigma = \frac{4\pi}{k^2} \sum_{l=0}^{\infty} (l+1)\sin^2(\delta_l - \delta_{l+1}) \tag{12.114}$$

following from (2.142). As developed in our discussion of the phase shift analysis of free Coulomb scatter of Sect. 4.3.5, we have, from (4.109),

$$\delta_l = \arg \Gamma\left(l + 1 + i\frac{k\rho_B(E)}{2}\right). \tag{12.115}$$

Using the identity (Abramowitz and Stegun 1972) for $\zeta = x + iy$,

$$\arg \Gamma(1+\zeta) = \arg \Gamma(\zeta) + \tan^{-1}\frac{y}{x}, \tag{12.116}$$

we have

$$\begin{aligned}\delta_l - \delta_{l+1} &= \arg\Gamma\left(1 + l + i\frac{k\rho_B(E)}{2}\right) - \arg\Gamma\left(1 + (l+1) + i\frac{k\rho_B(E)}{2}\right)\\ &= \arg\Gamma\left(1 + l + i\frac{k\rho_B(E)}{2}\right) - \arg\Gamma\left(1 + l + i\frac{k\rho_B(E)}{2}\right)\\ &\quad + \tan^{-1}\left(\frac{k\rho_B(E)}{2l}\right) - \tan^{-1}\left(\frac{k\rho_B(E)}{2l}\right) - \tan^{-1}\left(\frac{k\rho_B(E)}{2(l+1)}\right)\\ &= -\tan^{-1}\left(\frac{k\rho_B(E)}{2(l+1)}\right).\end{aligned} \tag{12.117}$$

This results in

$$\sin^2(\delta_l - \delta_{l+1}) = \frac{(k\rho_B(E)/2)^2}{(k\rho_B(E)/2)^2 + (l+1)^2}. \tag{12.118}$$

Substituting (12.118) into (12.114) gives the free Coulomb scatter total cross section as

$$\sigma = \pi\rho_B^2(E)\sum_{l=0}^{\infty}\frac{(l+1)}{(k\rho_B(E)/2)^2 + (l+1)^2}. \tag{12.119}$$

The free Coulomb stopping cross section is, recalling the definition of the Bohr collision diameter,

$$S_{\text{FC}} = \frac{4\pi(\alpha\hbar c)^2 z^2}{m_e\beta^2}\sum_{l=0}^{\infty}\frac{(l+1)}{(k\rho_B(E)/2)^2 + (l+1)^2}. \tag{12.120}$$

We should reflect briefly upon this result. The summation of (12.120) exhibits a divergence for large l. This is perhaps to be expected through the recognition that l is a measure of the angular momentum and, hence, a measure of the impact parameter – the divergence is a reflection of the logarithmic divergence of the collision stopping power at large impact parameters.

The first-term Born expansion approximation of (12.120) requires an expansion in z. We first include explicitly the expression for the Bohr collision diameter in the denominator of (12.120):

$$S_{\text{FC}} = \frac{4\pi(\alpha\hbar c)^2 z^2}{m_e\beta^2}\sum_{l=0}^{\infty}\frac{(l+1)}{(k\,\alpha\hbar c\,z/2E)^2 + (l+1)^2}. \tag{12.121}$$

Sigmund (2006) argued that in anticipation of this expansion, the Bohr collision diameter term in the denominator of (12.120) can be removed so as to leave

$$S_{\mathrm{FC},B} = \frac{4\pi(\alpha\hbar c)^2 z^2}{m_e\beta^2}\sum_{l=0}^{\infty}\frac{1}{(l+1)}. \tag{12.122}$$

Substituting (12.121) and (12.122) into (12.111) leaves us with

$$\begin{aligned}\Delta S = S_{\mathrm{FC}} - S_{\mathrm{FC},B} &= \frac{4\pi(\alpha\hbar c)^2 z^2}{m_e\beta^2}\sum_{l=0}^{\infty}\frac{(l+1)}{(k\rho_B(E)/2)^2+(l+1)^2} - \frac{1}{(l+1)}\\ &= -\frac{4\pi(\alpha\hbar c)^2 z^2}{m_e\beta^2}\left(z\left(\frac{\alpha}{\beta}\right)\right)^2\sum_{l=0}^{\infty}\frac{1}{(l+1)\left((z(\alpha/\beta))^2+(l+1)^2\right)}\\ &= -\frac{4\pi(\alpha\hbar c)^2 z^2}{m_e\beta^2}\left(z\left(\frac{\alpha}{\beta}\right)\right)^2\sum_{n=1}^{\infty}\frac{1}{n\left((z(\alpha/\beta))^2+n^2\right)}.\end{aligned} \tag{12.123}$$

Recalling the series expansion of the digamma function from Appendix C.13,

$$\mathrm{Re}\,(\psi(1+iy)) = -\psi(1) + y^2\sum_{n=0}^{\infty}\frac{1}{n(y^2+n^2)}. \tag{12.124}$$

We can then write the Bloch correction term as, recalling that $\psi(1) = -\gamma_{EM}$,

$$L_2(\beta) = -\gamma_{\mathrm{EM}} - \mathrm{Re}\left(\psi\left(1+i\frac{z\alpha}{\beta}\right)\right). \tag{12.125}$$

This result is shown in Fig. 12.10, along with the corresponding results for the Bethe and Bohr differences.

Now consider the asymptotic behaviours of the Bloch correction term due to the projectile charge and speed. To do so, use the asymptotic formula for the digamma function:

$$\mathrm{Re}\psi(1+iy) = \ln y + \frac{1}{12y^2} + \frac{1}{120y^4} + \frac{1}{252y^6} + \ldots \quad \text{for} \quad y \to \infty.$$

Thus, for a slow heavy charged particle (i.e. $\alpha z/\beta >> 1$), the Bloch correction term is

$$z^2 L_2(\beta) \approx -\gamma_{\mathrm{EM}} - \ln\left(\alpha\frac{z}{\beta}\right). \tag{12.126}$$

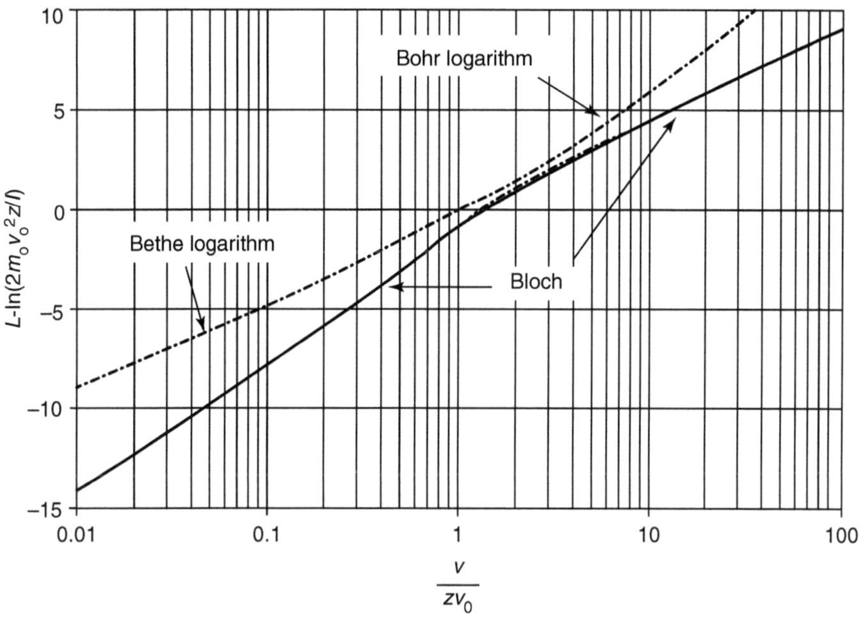

Fig. 12.10 Bloch result along with the Bohr and Bethe logarithms. Redrawn from Sigmund (1997)

In this limit, the Bloch correction leads to the Bohr classical form of the collision stopping power. On the other hand, one does not achieve the Bethe result in the relativistic case of $\alpha z/\beta \rightarrow 0$ due to an error in the original derivation and which is discussed by Ahlen (1980). This has a negligible calculational consequence; however, as the Bloch term becomes insignificant in such a case, it is not problematic.

Bibliography and Further Reading

Attix FH. Introduction to radiological physics and radiation dosimetry. New York: Wiley; 1986.
Jackson JD. Classical electrodynamics. 3rd ed. New York: Wiley; 1999.
Johns HE, Cunningham JR. The physics of radiology. 4th ed. Springfield: Charles C Thomas; 1983.
Nye JF. Physical properties of crystals. Oxford: Clarendon; 1957.
Podgoršak EB. Radiation physics for medical physicists. Berlin: Springer; 2010.
Ziegler JF. The stopping of energetic light ions in elemental matter. J Appl Phys. 1999;85:1249–72.

References

Abramowitz M, Stegun IA, editors. Handbook of mathematical functions. New York: Dover Publications; 1972.
Ahlen SP. Theoretical and experimental aspects of the energy loss of relativistic heavily ionizing particles. Rev Mod Phys. 1980;52:121–73 (*erratum* Rev Mod Phys 1980; 52: 653).
Andersen HH, Simonsen H, Sørensen H. An experimental investigation of charge-dependent deviations from the Bethe stopping power formula. Nucl Phys A. 1969;125:171–5.
Ashley JC, Ritchie RH, Brandt W. Z_1^3 effect in the stopping power of matter for charged particles. Phys Rev B. 1972a;5:2393–7.
Ashley JC, Ritchie RH, Brandt W. Z_1^3-dependent stopping power and range contributions. Phys Rev B. 1972b;5:2402–8.
Ashley JC, Anderson VE, Ritchie RH, Brandt W. Z_1^3 effects on the stopping power of matter for charged particle: Tables of functions. Document no 021195. New York: National Auxiliary Publication Service; 1974
Barkas WH. Technical report UCRL-10292. University of California Lawrence Radiation Laboratory; 1962 Aug.
Barkas WH, Birnbaum W, Smith FM. Mass ratio method applied to the measurement of L-meson masses and the energy balance in pion decay. Phys Rev. 1956;101:778–95.
Barkas WH, Dyer JN, Heckman HH. Resolution of the Σ^- mass anomaly. Phys Rev Let. 1963;11:26–8 (*erratum* Phys Rev Let 1963; 11: 138).
Beddar AS, Mackie TR, Attix FH. Cerenkov light generated in optical fibres and other light pipes irradiated by electron beams. Phys Med Biol. 1992;37:925–35.
Bichsel H. Stopping power and ranges of fast ions in heavy elements. Phys Rev. 1992; A46:5761–73.
Bloch F. Zur Bremsung rasch bewegter Teilchen beim Durchgang durch Materie. Ann Physik. 1933;16:285–320.
Čerenkov PA. Visible emission of clean liquids by action of γ radiation. Doklady Akad Nauk SSSR. 1934;2:451.
Fano U. Atomic theory of electromagnetic interactions in dense materials. Phys Rev. 1956;103:1202–18.
Fano U. Penetration of protons, alpha particles, and mesons. Ann Rev Nucl Sci. 1963;13:1–66.
Fermi E. The ionization loss of energy in gases and in condensed materials. Phys Rev. 1940;57:485–93.
Halpern O, Hall H. The ionization loss of energy of fast charged particles in gases and condensed bodies. Phys Rev. 1948;73:477–86.
ICRU. Stopping powers and ranges for protons and alpha particles. Report no 49. Bethesda: International Commission on Radiation Units and Measurements; 1993
Khandelwal GS, Merzbacher E. Stopping power of M electrons. Phys Rev. 1966;144:349–52.
Khandelwal GS. Shell corrections for K- and L-electrons. Nucl Phys A. 1968;116:97–111.
Lindhard J. The Barkas effect – or Z31, Z41-corrections to stopping of swift charged particles. Nucl Instrum Method. 1976;132:1–5.
Lindhard J, Sørensen AH. On the relativistic theory of stopping of heavy ions. Phys Rev A. 1996;53:2443–56.
Medenwaldt R, Møller SP, Uggerhøj E, Worm T, Hvelpund P, Knudsen H, Elsener K, Morenzoni E. Measurement of the stopping power of silicon for antiprotons between 0.2 and 3 MeV. Nucl Instrum Method Phys Res B. 1991;58:1–5.
Møller SP, Uggerhøj E, Bluhme H, Knudsen H, Mikkelsen U, Paludan K, Morenzoni E. Direct measurement of the stopping power for antiprotons of light and heavy targets. Phys Rev A. 1997;56:2930–9.
Sigmund P. Kinetic theory of particle stopping in a medium with internal motion. Phys Rev A. 1982;26:2497–517.

Sigmund P. Charge-dependent electronic stopping of swift nonrelativistic heavy ions. Phys Rev A. 1997;56:3781–93.

Sigmund P. Particle penetration and radiation effects. Berlin: Springer; 2006.

Smith FM, Birnbaum W, Barkas WH. Measurements of meson masses and related quantities. Phys Rev. 1953;91:765–6.

Sternheimer RM, Peierls RF. General expression for the density effect for the ionization loss of charged particles. Phys Rev B. 1971;3:3681–92.

Tamm IE, Frank IM. Coherent radiation of fast electrons in a medium. Doklady Akad Nauk SSSR. 1937;14:107.

Walske MC. The stopping power of K-electrons. Phys Rev. 1952;88:1283–9.

Walske MC. Stopping power of L-electrons. Phys Rev. 1956;101:940–4.

Chapter 13
Charged Particle Range

Contents

Abstract This chapter considers one of the main manifestations of charged particle energy loss that of the penetration into matter. It begins with the derivation of the continuous slowing down approximation (CSDA) range for heavy charged particles and presents the Bragg peak. As electrons remain currently the most commonly used charged particles in radiation oncology, we review the depth–dose curves of electrons in soft tissue. Unlike energetic heavy charged particles, multiple Coulomb scatter is a

B.J. McParland, *Medical Radiation Dosimetry*,
DOI 10.1007/978-1-4471-5403-7_13, © Springer-Verlag London 2014

significant effect upon the radiation transport of electrons and cannot be excluded in the determination of the range, as has already been examined in the context of the Fermi–Eyges theory of Sect. 6.3.7. We conclude the chapter with an examination of the ranges of positrons in matter and the probabilities of in-flight annihilation.

13.1 Introductory Comments

It follows immediately that as an energetic charged particle transfers its kinetic energy to the medium that it is traversing through, it will lose all of this energy and eventually come to rest (or, more properly, come to thermal energies). The distance between from where the particle entered the medium (either at an interface as in the case of an external beam of particles or within the medium itself as in the case of a radioactive source of charged particles) to where it has stopped is referred to as the *range* of the charged particle. More formally, as to be shown in Part IV of this book, as the energy loss process is a stochastic one, the range is the expectation value of the path length that the particle follows until it has stopped. In this chapter, we introduce the basic concepts and definitions of range in the context of charged particles.

Much of this chapter should be considered in parallel with the discussion of multiple Coulomb scatter presented in Chap. 6.

13.2 Geometry

Figure 13.1 shows the geometry specific to a charged particle which is incident to a medium at an angle χ to the norm. The particle penetrates the medium and is assumed to follow a tortuous path length due to multiple Coulomb scatter until it comes to rest at the points with the co-ordinates of (x_S, y_S, z_S).

We now consider the description of this geometry in some detail. As the particle penetrates the medium, it becomes both subject to elastic multiple Coulomb scatter from the atomic nuclei and the transfers of kinetic energy to the medium via collisions with the atomic electrons. The former leads to deviations of the particle's trajectory from its original. The latter leads the deceleration of the particle until it reaches its terminus at (x_S, y_S, z_S). The *path length* of the projectile is the total distance travelled by the particle, and the *projected range* (R_P) is the distance along the direction of the original trajectory to the point projected along the perpendicular from (x_S, y_S, z_S). The length of this perpendicular is known as the *transverse projected range*. The *penetration* or *projected* range is the depth within the medium at which the particle stops (i.e. x_S). Finally, the *spread* is defined as the distance along the surface (i.e. the yz-plane) from the entrance point to the projection of the stopping point (x_S, y_S, z_S) at the surface, $(0, y_S, z_S)$.

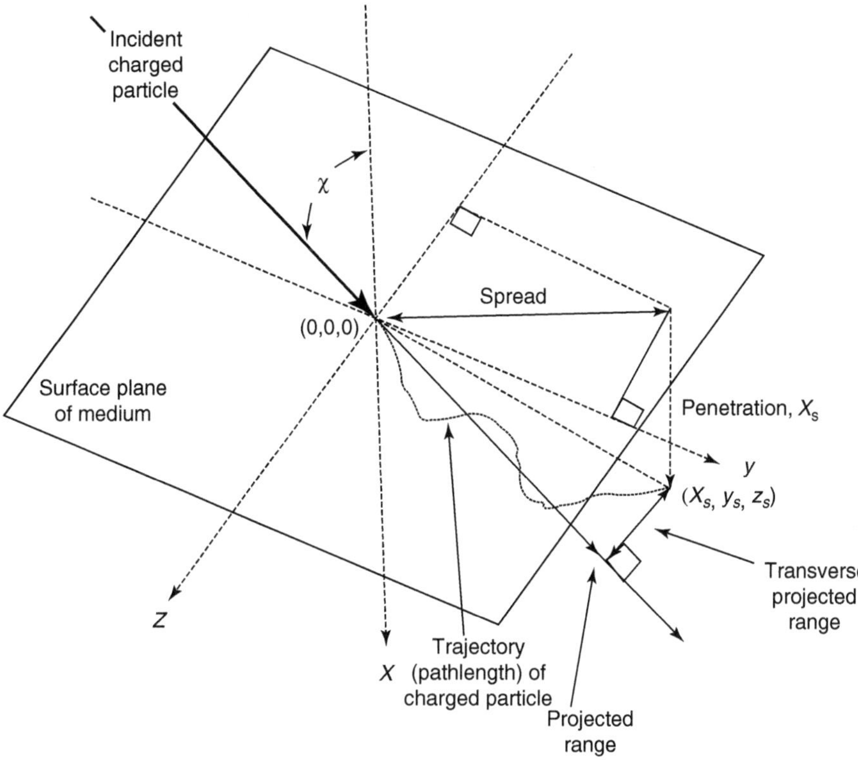

Fig. 13.1 A charged particle is incident to the surface of a medium at an angle of χ to the normal. It follows a non-linear trajectory until it is stopped at (x_s, y_s, z_s). Shown are the definitions of the projected range, spread, penetration and the transverse projected range

13.3 Continuous Slowing Down Approximation

13.3.1 Definition

As the name suggests, in the continuous slowing down approximation (CSDA), it is assumed that the energy loss is continuous with distance as the particle decelerates (ICRU 1984). It neglects any influence of the effects of multiple Coulomb scatter described in Chap. 6 with the result that a straight-line trajectory is assumed. Thus, for a particle with an initial kinetic energy E_0, the CSDA range is given by

$$\Re_{\mathrm{CSDA}}(E_0) = \int_0^{E_0} \frac{\mathrm{d}E}{\frac{\mathrm{d}E}{\mathrm{d}x}}. \tag{13.1}$$

Note that we have used our convention of considering the stopping power as a positive quantity. The full evaluation of $\Re_{\mathrm{CSDA}}$ would include the radiative energy loss contribution. However, as noted in the Preface to this book, the consideration of this contribution as being negligible is justifiable for the tissues and energy ranges of interest to medical radiation dosimetry. As $\frac{\mathrm{d}E}{\mathrm{d}x}$ is the expectation value of the rate of energy loss per unit distance through collisions, $\Re_{\mathrm{CSDA}}$ represents, correspondingly, the expectation value of the particle's range. The use of the linear collision stopping power in (13.1) yields the units of $\Re_{\mathrm{CSDA}}$ as distance, whereas the use of the mass collision stopping power yields the units of mass per distance squared.

13.3.2 Calculation of $\Re_{\mathrm{CSDA}}$ for Heavy Charged Particles

Recall the mass collision stopping power for a massive spin-½ particle of (10.4):

$$\frac{\mathrm{d}E}{\rho \mathrm{d}x} = 2C\left(\frac{Z}{A}\right)\left(\frac{z}{\beta}\right)^2\left[\ln\left(\frac{2m_\mathrm{e}\beta^2}{I}\right) - \ln(1-\beta^2) - \beta^2 + \left(\frac{1}{2}\frac{m_\mathrm{e}}{(E+m)}\frac{\beta^2}{1-\beta^2}\right)^2\right].$$

Let us assume that the particle is sufficiently nonrelativistic, allowing us to neglect the last three terms within the brackets to remain with

$$\frac{\mathrm{d}E}{\rho \mathrm{d}x} = 2C\left(\frac{Z}{A}\right)\left(\frac{z}{\beta}\right)^2 \ln\left(\frac{2m_\mathrm{e}\beta^2}{I}\right). \tag{13.2}$$

Inserting this form into (13.1),

$$\begin{aligned}\rho\Re_{\mathrm{CSDA}}(E_0) &= \int_0^{E_0} \frac{\mathrm{d}E}{\frac{\mathrm{d}E}{\rho \mathrm{d}x}} \\ &= \frac{1}{2C}\left(\frac{A}{Z}\right)\frac{1}{z^2}\int_0^{E_0} \mathrm{d}E \frac{\beta^2}{\ln\left(\frac{2m_\mathrm{e}\beta^2}{I}\right)} \\ &= \frac{1}{C}\left(\frac{A}{Z}\right)\frac{1}{mz^2}\int_0^{E_0} \mathrm{d}E \frac{E}{\ln\left(\left(\frac{4E}{I}\right)\left(\frac{m_\mathrm{e}}{m}\right)\right)}\end{aligned} \tag{13.3}$$

where $\beta^2 = \frac{2E}{m}$ has been used.

13.3.3 Scaling of $\Re_{\mathrm{CSDA}}$

The expression of the CSDA range given by (13.3) allows us to evaluate the CSDA range of an arbitrary heavy charged particle from that known for a reference particle. To demonstrate this, let us consider that reference particle to be a proton, in which case we will rewrite (13.3) as

$$\begin{aligned}\rho\Re_{\mathrm{CSDA,p}}(E_0) &= \frac{1}{C}\left(\frac{A}{Z}\right)\frac{1}{m_\mathrm{p}}\int_0^{E_0} \mathrm{d}E_\mathrm{p}\frac{E_\mathrm{p}}{\ln\left(\left(\frac{4E_\mathrm{p}}{I}\right)\left(\frac{m_\mathrm{e}}{m_\mathrm{p}}\right)\right)}\\ &= \frac{1}{C}\left(\frac{A}{Z}\right)m_\mathrm{p}\int_0^{E_0} \mathrm{d}\left(\frac{E_\mathrm{p}}{m_\mathrm{p}}\right)\frac{\left(\frac{E_\mathrm{p}}{m_\mathrm{p}}\right)}{\ln\left(4\left(\frac{E_\mathrm{p}}{m_\mathrm{p}}\right)\left(\frac{m_\mathrm{e}}{I}\right)\right)}\end{aligned} \tag{13.4}$$

where the p subscript refers to the proton. Now consider another heavy charged particle, for which we will assign the subscript, x. Then, from (13.4),

$$\begin{aligned}\rho\Re_{\mathrm{CSDA,x}}(E'_0) &= \frac{1}{C}\left(\frac{A}{Z}\right)\frac{m_\mathrm{x}}{z_\mathrm{x}^2}\int_0^{E'_0} \mathrm{d}\left(\frac{E_\mathrm{x}}{m_\mathrm{x}}\right)\frac{\left(\frac{E_\mathrm{x}}{m_\mathrm{x}}\right)}{\ln\left(4\left(\frac{E_\mathrm{x}}{m_\mathrm{x}}\right)\left(\frac{m_\mathrm{e}}{I}\right)\right)}\\ &= \frac{1}{C}\left(\frac{A}{Z}\right)m_\mathrm{p}\left(\frac{m_\mathrm{x}}{m_\mathrm{p}}\right)\frac{1}{z_\mathrm{x}^2}\int_0^{E'_0} \mathrm{d}\left(\frac{E_\mathrm{x}}{m_\mathrm{p}}\frac{m_\mathrm{p}}{m_\mathrm{x}}\right)\frac{\left(\frac{E_\mathrm{x}}{m_\mathrm{p}}\right)\left(\frac{m_\mathrm{p}}{m_\mathrm{x}}\right)}{\ln\left(4\left(\frac{E_\mathrm{x}}{m_\mathrm{p}}\right)\left(\frac{m_\mathrm{p}}{m_\mathrm{x}}\right)\left(\frac{m_\mathrm{e}}{I}\right)\right)}\\ &= \left(\frac{m_\mathrm{x}}{m_\mathrm{p}}\right)\frac{1}{z_\mathrm{x}^2}\frac{1}{C}\left(\frac{A}{Z}\right)m_\mathrm{p}\int_0^{E'_0} \mathrm{d}\left(\frac{E'_\mathrm{x}}{m_\mathrm{p}}\right)\frac{\left(\frac{E'_\mathrm{x}}{m_\mathrm{p}}\right)}{\ln\left(4\left(\frac{E'_\mathrm{x}}{m_\mathrm{p}}\right)\left(\frac{m_\mathrm{e}}{I}\right)\right)}\\ &= \left(\frac{m_\mathrm{x}}{m_\mathrm{p}}\right)\frac{1}{z_\mathrm{x}^2}\rho\Re_{\mathrm{CSDA,p}}(E'_0)\end{aligned} \tag{13.5}$$

where

$$E'_0 = E_0\frac{m_\mathrm{p}}{m_\mathrm{x}}. \tag{13.6}$$

Thus, the CSDA range of the arbitrary charged particle can be approximated by a scaled CSDA range of a proton. For example, consider a 10-MeV α-particle where $\frac{m_\mathrm{p}}{m_\alpha} = 3.97$ and, from (13.6), $E_0 = 2.52$ MeV. Then, from (13.5), we can estimate the CSDA range of the α-particle on the basis of that determined for the proton by $\rho\Re_{\mathrm{CSDA},\alpha}(10\,\mathrm{MeV}) = 0.248\,\rho\,\Re_{\mathrm{CSDA,p}}(2.52\,\mathrm{MeV})$.

13.3.4 Examples of $\Re_{\mathrm{CSDA}}$ for Heavy Charged Particles

The CSDA ranges of protons and α-particles for kinetic energy ranges typical of therapeutic applications are shown in Figs. 13.2 and 13.3, respectively.

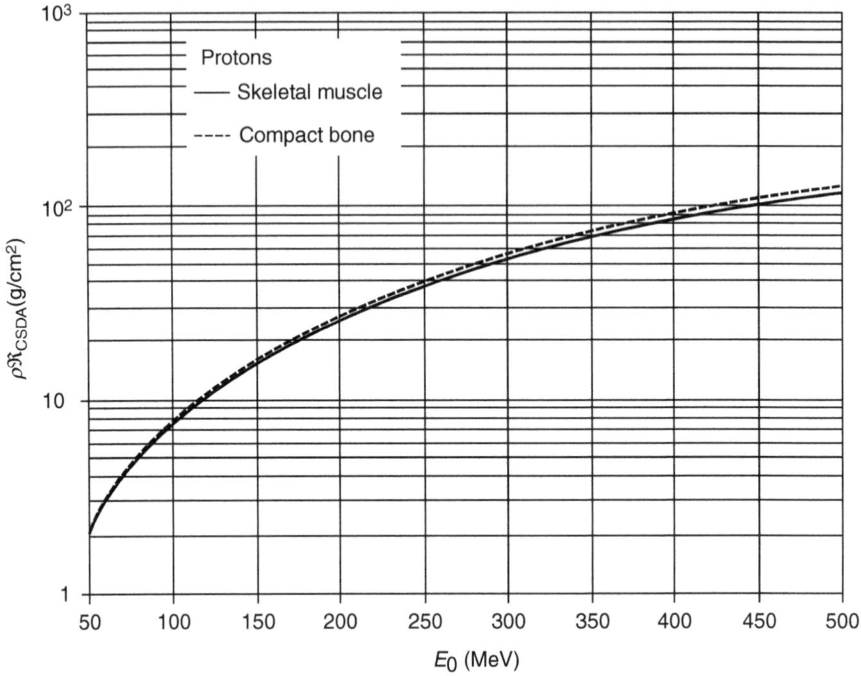

Fig. 13.2 The CSDA range of protons in skeletal muscle and compact (cortical) bone as a function of kinetic energy representative of therapeutic applications (*Curves* are generated from date provided in ICRU Report 49 (ICRU 1993))

It is clear from a comparison of the ordinates of Figs. 13.2 and 13.3 for the therapeutic applications of protons and α-particles that the ranges are markedly different by a few orders of magnitude.

13.4 Projected Range

13.4.1 Definition

As shown in Fig. 13.1, the projected range x_S is the maximum perpendicular penetration of the charged particle into the medium. It is defined here as an expectation value

$$\overline{x_S} = \int_0^\infty \mathrm{d}x_S\, x_S \left| \frac{\mathrm{d}N}{\mathrm{d}x_S}(x_S) \right| \tag{13.7}$$

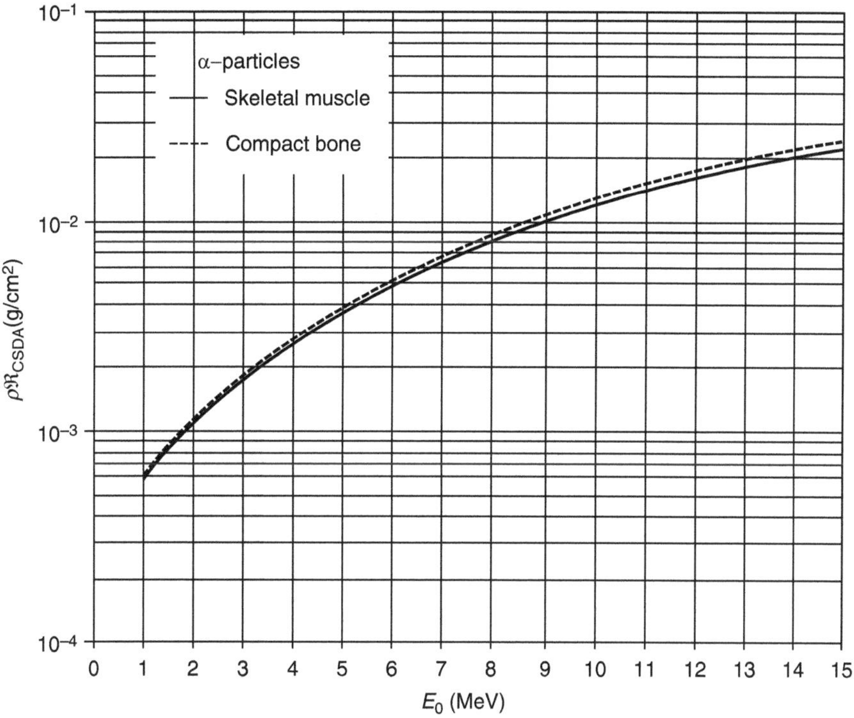

Fig. 13.3 The CSDA range of α-particles in skeletal muscle and compact (cortical) bone as a function of kinetic energy representative of therapeutic applications (*Curves* are generated from date provided in ICRU Report 49 (ICRU 1993))

where $\frac{dN}{dx_S}(x_S)$ is the rate at which particles are stopped per unit depth and which is normalised to

$$\int_0^\infty dx_S \left|\frac{dN}{dx_S}(x_S)\right| = 1. \tag{13.8}$$

The absolute value sign is a formality as the rate is negative due to the loss of particles as they stop.

13.4.2 Detour Factor

Consider the obliquity angle of χ in Fig. 13.1 to be equal to zero, i.e. the projectile is incident normally to the medium. The co-ordinate x_s is that at which the particle has finally decelerated to rest with an expectation value denoted by the projected range

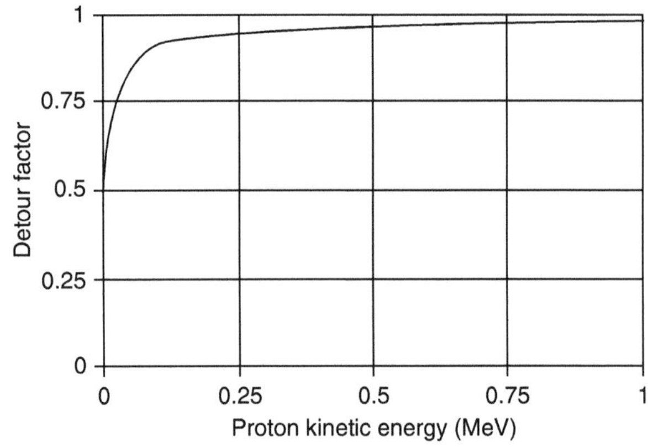

Fig. 13.4 Detour factor for protons in carbon in the form of amorphous graphite with a physical density of 2 g/cm^3 as a function of proton kinetic energy

of (13.7). The *detour factor* is defined as the ratio of the average of this penetration depth to the CSDA range:

$$\text{Detour factor} = \frac{\overline{x_S}}{\Re_{\text{CSDA}}}. \tag{13.9}$$

Hence, this is a numerical quantity which will be less than unity due to the effects of multiple scattering resulting in $\overline{x}_S < \Re_{\text{CSDA}}$ as the CSDA approximation neglects the perturbations in the particle trajectory caused by multiple scattering. For heavy charged particles with kinetic energies of interest in medical radiation physics, such as protons and α-particles, the detour factor does not vary significantly from unity as demonstrated by Fig. 13.4.

13.5 Bragg Peak

As the stopping power is relatively constant in the minimally ionising region and has a β^{-2} dependence at lower energies, a charged particle penetrating a medium will lose energy at a near-constant rate with depth until, as it slows down beyond the minimally ionising region, it loses its energy at an ever-increasing rate until it has thermalised. Figure 13.5 shows the curve of relative energy deposition as a function of depth for a heavy charged particle, such as an α-particle, where multiple scattering can be neglected and the particle travels very nearly along a straight line. Only near the end of its range does the relative energy loss increase from its near-continuous value; this rapid increase is known as the 'Bragg' peak (Bragg and Kleeman 1905).

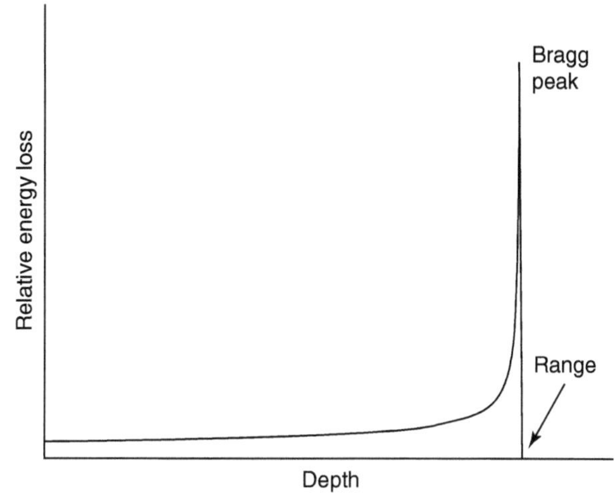

Fig. 13.5 Energy loss (deposition) as a function of depth for a heavy charged particle (e.g. proton or α-particle)

13.6 Electron and Positron Ranges

13.6.1 Introduction

When a charged particle penetrates a medium, it is subject to multiple elastic scatter through the Coulomb interaction with the nucleus and to kinetic energy loss through the Coulomb interactions with atomic electrons. In Chap. 6, it was demonstrated that elastic scatter is inversely proportional to the projectile mass. On the other hand, it was shown in Chaps. 8, 9 and 10 that energy loss is independent of projectile mass. As a result, we observe the Bragg peak for heavy charged particles slowing down in a medium, where the influence of multiple scatter is minimal. But this will not be the case for light particles such as electrons and positrons where the influence of multiple scatter becomes predominant.

In this section, we consider, first, the range of electrons stopping in a medium. In particular, we will tend to focus on external beams of electrons which are used to treat superficial lesions while sparing normal tissue at depth. Here, the combined effects of multiple scatter and collision energy loss are of great importance within the clinic.

While positrons have the same mass as electrons, the physics of their interactions with matter affects their range and energy loss. The first immediately obvious difference between electrons and positrons that we must consider in evaluating the range of a positron is that the positron is subject to in-flight annihilation with an electron. We will see, however, that at the kinetic energies and media typical of medical applications of positrons (i.e. positron emission tomography), this effect is very limited and can, in practice, be neglected. More importantly, though, is the fact that the Coulomb interaction of a positron with an atomic electron is described by the Bhabha cross section whereas the electron–electron interaction is described

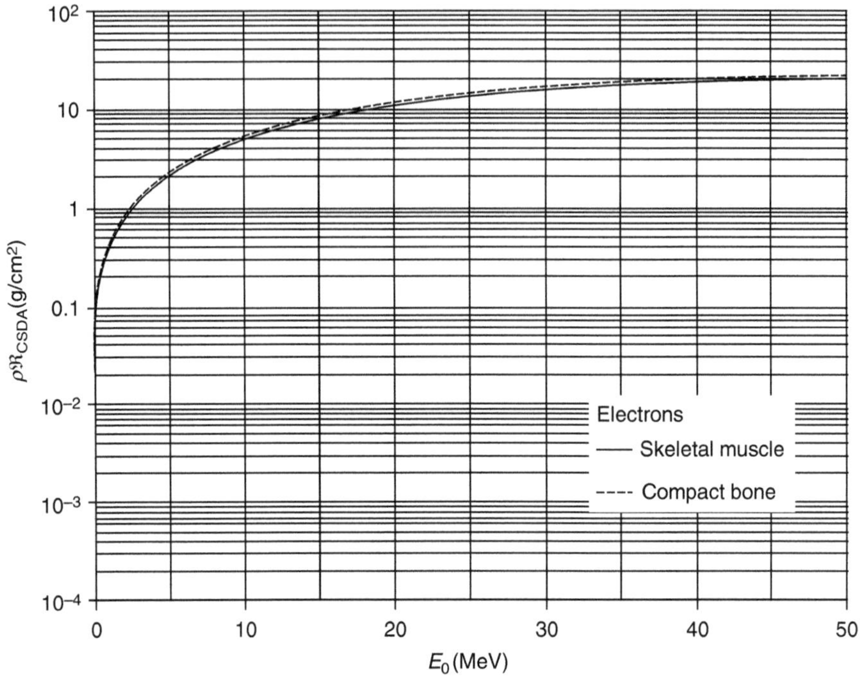

Fig. 13.6 CSDA range of electrons in skeletal muscle and compact (cortical) bone as a function of initial kinetic energy

by the Møller cross section: the latter allows for particle exchange. In addition, the positron can lose all of its kinetic energy in such an interaction whereas, by definition, the electron can lose, at most, half of its kinetic energy in an interaction with another electron. Hence, there will be differences in the ranges between positrons and electrons and these are reviewed here.

13.6.2 Electron CSDA Range

Although multiple scatter will affect the electron range, as to be shown later, it remains of interest to consider the CSDA range which neglects the influence of multiple scatter. Figure 13.6 shows the CSDA range for electrons as a function of kinetic energy typical of external beam therapy.

A superficial consideration of Fig. 13.6 would indicate that $\rho\Re_{\text{CSDA}} \approx \frac{E_0}{2}$ where $\rho\Re_{\text{CSDA}}$ is in units of $\frac{\text{g}}{\text{cm}^2}$ and E_0 is in units of MeV. This reflects the minimally ionising condition for low atomic number media.

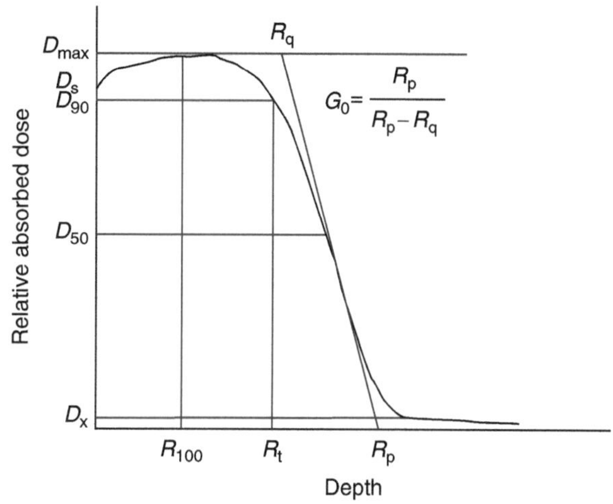

Fig. 13.7 A representative central-axis depth–dose curve for a therapeutic electron beam. The parameters are discussed in the text. The parameters characterising the depth–dose curve are described in the text

13.6.3 *Electron Beam Central-Axis Depth–Dose Curves*

13.6.3.1 Introductory Comments

In external beam radiotherapy, the central-axis depth–dose curve is the absorbed dose (usually measured in water or some other low-Z medium) along the central axis of the radiation beam as a function of depth. As shown earlier, this will be in the form of the Bragg peak for a heavy charged particle, but the shape is markedly different for electrons due to multiple scatter, as shown in Fig. 13.7. This is a typical depth–dose curve for a therapeutic electron beam in which the beam is incident from the left (in air) to the medium (typically water) with the boundary given by the ordinate.[1] The width of the beam (i.e. the dimension orthogonal to the central axis) is sufficiently large that lateral electron equilibrium exists (i.e. the number of electrons scattered out of a small volume of medium on the central axis is exactly equal to that scattered in). In such a case, the central-axis depth–dose curve is independent of beam size.

In external electron beam therapy, the typical electron beam kinetic energies range from 5 to 25 MeV, although same facilities have used higher-energy beams.[2] The general features are an increase in absorbed dose from the surface up to a

[1] In general practice, this curve is derived from a measured depth–ionisation curve for which the ordinate is the ionisation measured in an ionisation chamber. This is converted to absorbed dose through the application of a number of factors, including the ratio of the collision stopping power of the medium to that of the gas within the chamber (usually air). This ratio has an energy dependence which is equivalent to a depth dependence.

[2] In radionuclide therapy in which the radiation is emitted internally, the typical energies of the β-particles are of the order of 1 MeV or less. This is due to the vector delivering the β-emitting radionuclide to the target and long ranges are usually not required.

maximum (the breadth of which will be dependent upon the electron's kinetic energy as to be shown) and then a drop-off with depth due to the finite range of the electrons within the medium. Finally, the depth–dose curve diminishes to a non-zero absorbed dose tail which is due to *bremsstrahlung* generated in the electron accelerator and beam-shaping devices and, to a much lesser degree, within the medium.

The depth–dose curve is characterised by a number of dosimetry and spatial parameters. D_S is the absorbed dose at the surface of the medium. As it is difficult to measure the absorbed dose at the interface between air and the medium and because the radiation-sensitive layers of the epidermis are at an approximate depth of 0.5 mm, this depth is that at which D_S is measured. D_{90} is the absorbed dose which is 90 % of the maximum, D_{max}. The range in the medium between the surface and the distal depth at which the depth dose is D_{90} is considered to be the therapeutic region, and this distal depth is denoted by R_t. The depth at which the depth–dose curve is at a maximum is denoted by R_{100}. Beyond this depth, the absorbed dose drops precipitously as the electrons reach their range to eventually reach the *bremsstrahlung* tail.

The extrapolation of the tangent to the inflection point to zero absorbed dose defines the practical range, R_P. A useful rule of thumb is that R_P in cm of water is equal to the incident electron beam energy, in MeV, divided by 2, as was suggested earlier in discussion of the CSDA range of electrons. This evolved from the parametric fit by Markus (1961, 1964) to measured depth–dose data:

$$E_M(0) = 0.48 + 1.95 R_P \tag{13.10}$$

where $E_M(z)$ is the modal kinetic energy of the electron beam at the depth z in the medium (13.10) has been shown to be valid to be within about 2 % for energies up to about 25 MeV. A further refinement proposed by Brahme and Svensson (1976) is useful to much higher kinetic energies (50 MeV) and accounts for the effects of radiative energy loss:

$$E_M(0) = 0.22 + 1.98 R_P + 0.0025 R_P^2. \tag{13.11}$$

Clearly, the combined effects of stopping and multiple scatter will lead to a reduction in the kinetic energies of the electrons. Harder (1966) demonstrated that the modal energy at depth within the phantom can be well modelled by a linear dependence:

$$E_M(z) = E_M(0)\left(1 - \frac{z}{R_p}\right). \tag{13.12}$$

The extrapolation of the tangent in the opposing direction leads to an intersection with the horizontal line of D_{max} to the depth R_q and the gradient given by

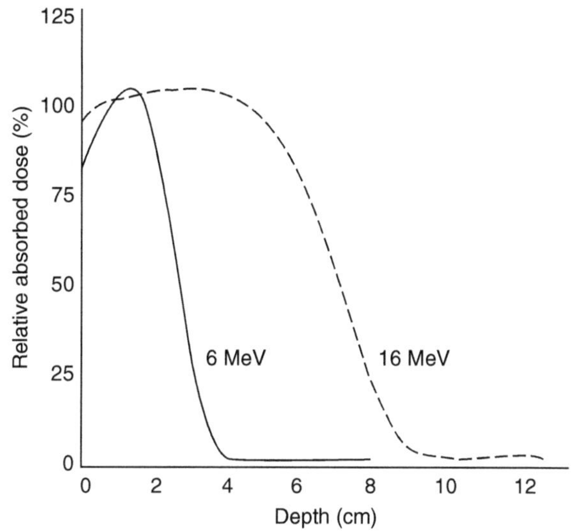

Fig. 13.8 Percentage depth–dose curves for 6- and 16-MeV electron beams incident to water

$$G_0 = \frac{R_P}{R_P - R_q} \qquad (13.13)$$

is a useful measure of how rapidly with depth the depth–dose curve drops off.

The relationships between energy and depth given by (13.10), (13.11) and (13.12) are important as they provide measures of the electron beam energy at depth within the phantom. In many practical clinical settings, the measurement is performed using a calibrated ionisation chamber in which the ionisation within a gas volume is collected, and this yields a depth–ionisation curve. This is necessarily corrected to the depth–dose curve (where the absorbed dose is to the medium) through the use of collision stopping powers which, as they are energy dependent, require knowledge of the projectiles' energies.

Why electrons and heavy charged particles exhibit such differing depth–dose characteristics is now explained within the context of the energy dependence of the electron beam depth–dose curve.

13.6.3.2 Role of Multiple Scatter

Figure 13.8 presents the depth–dose curves in water for electrons of 6- and 16-MeV energy. Both curves are normalised to the maximum absorbed dose for each.

Both curves exhibit differences in magnitude of penetration and in shape: the low-energy beam demonstrates a reduced surface absorbed dose and a sharp gradient, G_0, whereas the high-energy beam depth–dose curve is relatively flat with depth and then decreases less sharply. The *bremsstrahlung* tail for the 6-MeV beam is less than that of the 16-MeV beam. The four physical factors that influence these results are:

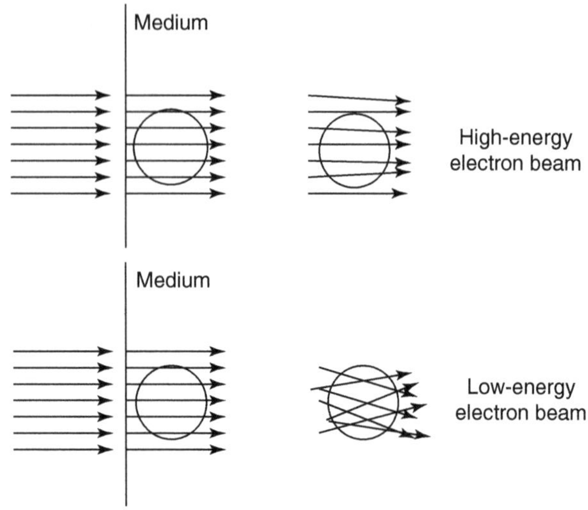

Fig. 13.9 Qualitative description of the shape and range variations in the electron beam central-axis depth–dose curve due to multiple scattering of the electrons. The *arrows* represent electron tracks and the *circles* represent volumes of the medium (Refer to text for details)

- The linear scattering power is approximately inversely proportional to the electron kinetic energy as shown by (6.29).
- The linear stopping power for electrons is approximately inversely proportional to the square of the electron velocity – hence, the range will be approximately proportional to the electron kinetic energy (and which led to the empirical relationship between R_P and beam energy noted in the previous section).
- Range straggling (to be discussed in Part IV) caused by the stochastic nature of energy loss leads to a drop-off slower than the Bragg peak.
- *Bremsstrahlung* yield increases with electron kinetic energy (*cf* Section 7.6.6 of McParland (2010)).

The second and third points explain the increased penetration with energy and the greater amount of x-ray photons in the 'tail' of the depth–dose curve with energy. The first point is the cause for the shape of the curves.

Figure 13.9 provides a qualitative explanation for the role that Coulomb scattering of electrons in the medium plays in varying the shape of the electron beam depth–dose curve.

The two cases of a fluence of high- and low-energy electrons incident from the left from air to a medium (water or soft tissue) are shown. The arrows represent the tracks of electrons across volumes, represented by circles, within the medium. Recall that the absorbed dose is the energy absorbed per unit mass (or the product of unit volume and density) and that the energy deposition is proportional to the length of the track of the charged per unit mass. Hence, the greater the length and number of tracks crossing a volume, the greater the absorbed dose.

Elastic scattering of the electrons begins once the particles enter the medium. At the surface, this scattering is minimal whereas, at a depth, the differences in multiple scattering with energy becomes apparent. Hence, the absolute absorbed doses for low- and high-energy electron beams at the surface will be very nearly the same as the number of and length of tracks crossing the volume. As the degree of multiple scatter is inversely proportional to the electron kinetic energy, at depth, the number and length of track lengths across the volume for high-energy electrons will not be altered much from that at the surface. At depth, the increased scattering of the low-energy electrons leading to more oblique tracks across the volume will increase the overall length of the traversing tracks and the amount of absorbed dose at depth. As the central-axis depth–dose curve is normalised to the maximum absorbed dose, the result is that the surface dose of a low-energy electron beam is lower than that of a high-energy electron beam.

13.6.4 Positron In-Flight Annihilation

In Chap. 9, we considered the slight differences between the collision stopping powers of electrons and positrons. Because of these differences, we will expect that there will be a slight difference in the ranges. However, in addition to the finite range defined by Coulomb interactions with atomic electrons, the positrons are subject to in-flight annihilation with the electrons. We will review this effect here; positron annihilation at rest will not be considered (this is discussed in medical applications in Section 7.9 of McParland (2010)).

Bethe (1935) examined this problem using Dirac's theory. Kinematically, positron in-flight annihilation can yield one or two photons due to the simultaneous conservation of momentum and energy and the fact that in the initial state, the positron is not at rest. As a single-photon final state requires a nucleus to take up the required recoil momentum/energy and, hence, the unlikely probability of a positron interacting with a tightly bound K-shell electron, he limited his analysis to the two-photon final state. Using Dirac's result of the probability of positron annihilation and his own nonrelativistic expression of collision energy loss explored in Chaps. 8 and 9, Bethe summarised the probability of positron in-flight annihilation for a positron with kinetic energy E in a medium with atomic number Z in the form

$$\frac{P_{\mathrm{ann}}}{\left(\frac{\mathrm{d}E}{\mathrm{d}x}\right)}\,\mathrm{d}E = s\left(\frac{E}{m_{\mathrm{e}}}\right)\mathrm{d}\left(\frac{E}{m_{\mathrm{e}}}\right) \tag{13.14}$$

where the function

$$s\left(\frac{E}{m_{\mathrm{e}}}\right) = \frac{f_P\left(\frac{E}{m_{\mathrm{e}}}\right)}{4\ln\left(\frac{m_{\mathrm{e}}}{I_{\mathrm{H}}Z}\right)} \tag{13.15}$$

and where he defined I_H as an energy approximately equal to that of the ionisation of hydrogen. The function $f_P\left(\frac{E}{m_e}\right)$ is

$$f_P\left(\frac{E}{m_e}\right)=\left(\frac{1}{\left(\frac{E}{m_e}\right)^2\left(1+\left(\frac{E}{m_e}\right)\right)}\right)\left(\left(\left(1+4\left(\frac{E}{m_e}\right)+\left(\frac{E}{m_e}\right)^2\right)\times\right.\right.$$
$$\left.\ln\left(\frac{E}{m_e}\right)\sqrt{\left(\frac{E}{m_e}\right)^2-1}\right)-\left(\left(\frac{E}{m_e}\right)+3\right)\sqrt{\left(\frac{E}{m_e}\right)^2-1}\right). \tag{13.16}$$

From these results, the total probability of the positron in-flight annihilation leading to a two-photon final state is given by

$$\mathrm{Pr}_{2\gamma}(E_0)=\frac{F(E_0)}{4\,\ln\left(\frac{m_e}{I_H Z}\right)} \tag{13.17}$$

where

$$F(E_0)=\int_0^{E_0}\mathrm{d}\left(\frac{E}{m_e}\right)f_P\left(\frac{E}{m_e}\right) \tag{13.18}$$

and where E_0 is the initial energy of the positron. The quantity $F(E_0)$ is shown as a function of E_0 in Fig. 13.10. For example, let us consider the 1.523-MeV energy of one of the positrons emitted by ^{82}Ru which gives $F(E_0)\approx 1.85$. Then the probability of a two-photon in-flight annihilation is about 1.2 % in soft tissue. We can conclude that the contribution of positron in-flight annihilation to reducing the range of positrons in low atomic number media representative of tissue can be neglected.

13.6.5 Positron Range

As noted earlier, the stopping powers of positrons and electrons differ due to the use of the Bhabha and Møller cross sections, respectively, for hard collisions and, whereas the positron can lose all of its kinetic energy in a Coulomb interaction with an electron, by the definition that the maximum energy transfer between electrons is one-half of the incident kinetic energy. We thus expect that the range of positrons in a medium will be different from that of electrons.

Figure 13.11 shows the ratio of CSDA ranges for positrons to electrons in liquid water as a function of initial kinetic energy. Note that, by its definition, the CSDA range is that for which the positron comes to rest and does not include the effects of in-flight annihilation.

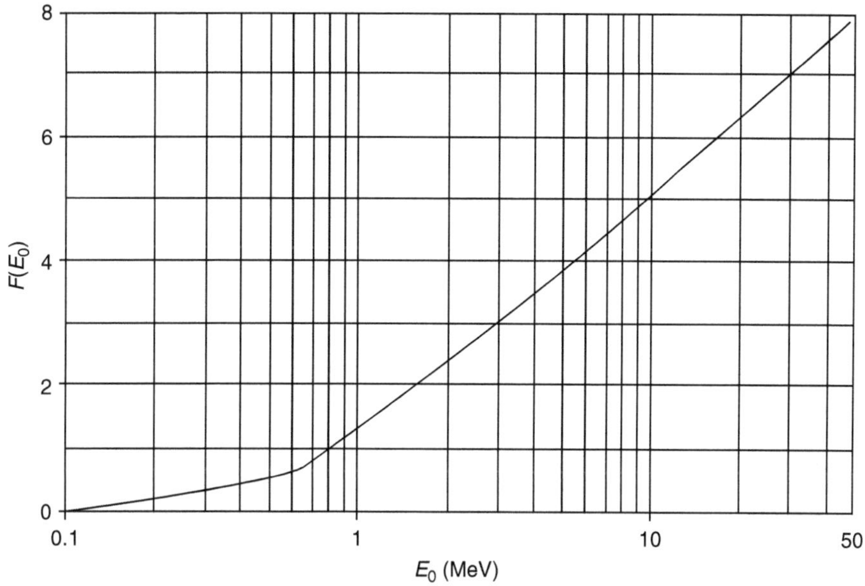

Fig. 13.10 The factor of Eq. (13.18) for calculating the probability of positron in-flight annihilation as a function of positron initial kinetic energy (Redrawn from Bethe 1935)

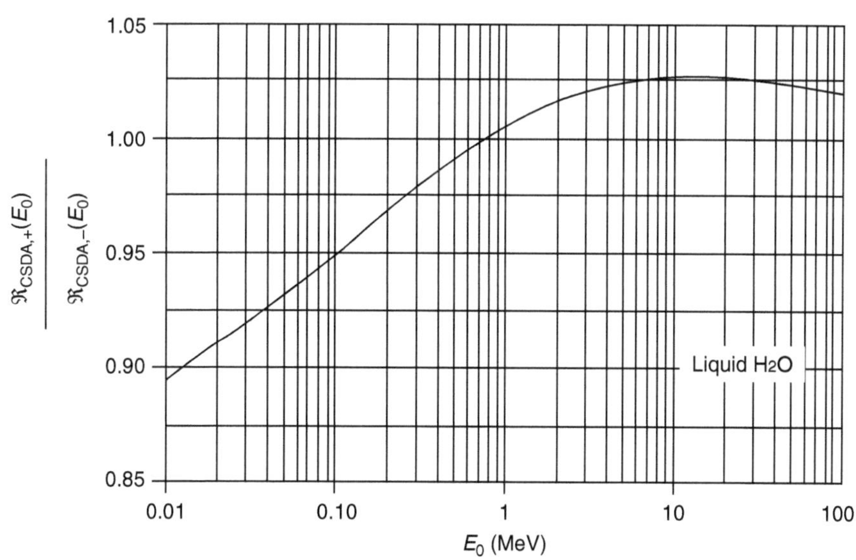

Fig. 13.11 Ratio of the CSDA ranges of positrons to electrons in liquid water as function of initial kinetic energy (Drawn using numerical data from ICRU 1984)

Bibliography

Bjorken JD, Drell SD. Relativistic quantum mechanics. New York: McGraw-Hill; 1964.
Dutreix A, Bridier A. Dosimetry for external beams of photon and electron radiation. In: Kase KR, Bjärngard BE, Attix FE, editors. The dosimetry of ionizing radiation. Orlando: Academic; 1985.
Podgoršak EB. Radiation physics for medical physicists. Berlin: Springer; 2010.

References

Bethe HA. On the annihilation radiation of positrons. Proc R Soc Lond. 1935;A150:129–41.
Bragg W, Kleeman M. On the alpha particles of radium. Phil Mag. 1905;10:600–602.
Brahme A, Svensson H. Specification of electron beam quality from the central-axis depth absorbed-dose distribution. Med Phys. 1976;3:95–102.
Harder D. Spectra of primary and secondary electrons in material irradiated by fast electrons. In, 'Biophysical Aspects of Radiation Quality', Tech Rep Ser – IAEA No. 1966;58:140–146.
ICRU. Stopping powers for electrons and positrons – International Commission on Radiological Units and Measurements Report 37. Bethesda: ICRU; 1984.
ICRU. Stopping powers and ranges for protons and alpha particles – International Commission on Radiological Units and Measurements Report 49. Bethesda: ICRU; 1993.
Markus B. Energiebestimmung schneller Elektronen aus Tiefendosiskurven. Strahlentherapie. 1961;116:280–9.
Markus B. Beiträge zur entwicklung der dosimetry schneller elektronen. Strahlentherapie. 1964;142:33–8.
McParland BJ. Nuclear medicine radiation dosimetry: advanced theoretical principles. London: Springer; 2010.

Chapter 14
Collision Energy Loss in Compound Media

Contents

Abstract Our calculations have, to this point, been for elemental media, in particular carbon, calcium and lead (some data have been shown graphically for striated muscle, compact bone and liquid water, but these are derived from data presented elsewhere). It is necessary to take the foundation of the elemental media calculations and develop the calculations for realistic compound media. This is almost invariably done within the context of the Bragg additivity rule in which energy loss, and other physical parameters, for a compound medium is evaluated as the weighted sum of the energy losses incurred by the individual elements, with the weights being the relative proportions of these elements. There are deficiencies in such an approach as it assumes that there is no change in electronic structure as a result of bonding or with the different material phases of the medium. In this chapter, the Bragg additivity rule is derived ab initio and the practicalities of its deviations are examined and quantified.

B.J. McParland, *Medical Radiation Dosimetry*,
DOI 10.1007/978-1-4471-5403-7_14, © Springer-Verlag London 2014

14.1 Introductory Comments

Throughout our numerical and graphical examples given in Parts II and III, we have restricted our derivations to elemental (i.e. monatomic) media. This is deliberate as we wished to defer our exploration of compound media until we had attained a suitable foundation in understanding collision energy loss. In particular, we have used carbon and calcium as crude elemental representations of soft tissue and bone. Of course, both are, in fact, compounds and Table 14.1 summarises the fractions by weight of the atomic constituents for various tissues. Hydrogen, carbon and oxygen are the dominant constituents of soft tissues, with the addition of calcium for bone. Thus, a dosimetry calculation for tissue must be extended to account for the aggregate effects within a compound medium.

The most commonly used method of calculating the stopping power of a compound medium is to approximate it by a weighted linear sum of the stopping powers of the elemental constituents. This approximation is known as the Bragg additivity rule. It is reasonably accurate in that it has been reported capable of reproducing measured stopping power data in compound media to within about 20 % (although literature reviews, cited below, indicate that the uncertainties can approach 50 % in some cases). As expected by the values quoted in the previous sentence, the validity of the rule has been studied in numerous cases, but the relative discrepancies between the actual stopping power in a compound medium and that calculated by the rule from elemental data can vary widely, leading frequently to confusion in some applications.

The accuracy of the Bragg additivity rule will be limited as the collision stopping power in a medium is dependent upon the atomic electrons' orbital structures and excitation channels. As these will vary, not only with the nature of chemical bonding, but also with the phase of the medium, it is not at all surprising that the assumption inherent to the rule – that the electronic structures of the elements remain unaltered when a compound is formed – leads to inaccuracies. Thwaites (1983, 1992) has surveyed the relevant literature up to those dates and reviewed over 100 studies of the Bragg additivity rule and concluded that inaccuracies in the rule were most prominent for low atomic number media (hydrocarbons) at low charged particle kinetic energies. For such combinations, deviations as great as 50 % can occur. However, at the same time, the accuracies of stopping power measurements at these energies are also limited, thus making assessments of the failure of the Bragg additivity rule somewhat ambiguous.

In this chapter, we will derive the Bragg additivity rule from first principles and then investigate its applications in calculating the mean excitation potential, the effective ratio of atomic number to atomic mass number and the density correction parameter for a compound medium. In the specific context of the mean excitation potential, we will consider explicit means of assigning values for hydrocarbons on the basis of the structures of the chemical bonds.

Table 14.1 Constituents of tissue: examples (as fraction by weight exceeding 0.001)

$Z =$		Hydrogen	Carbon	Nitrogen	Oxygen	Sodium	Magnesium	Phosphorus	Sulphur	Chlorine	Potassium	Calcium
Tissue	ρ (g/cm^3)	1	6	7	8	11	12	15	16	17	19	20
Adipose tissue (ICRP)	0.92	0.119	0.637	0.008	0.232	–	–	–	–	0.001	–	–
Bone, cortical (ICRP)	1.85	0.047	0.144	0.042	0.446	–	0.002	0.105	0.003	–	–	0.210
Muscle, skeletal (ICRP)	1.04	0.101	0.108	0.028	0.754	0.001	–	0.002	0.002	0.001	0.003	–
Muscle, striated (ICRP)	1.04	0.102	0.123	0.035	0.729	0.001	–	0.002	0.005	–	0.003	–

Adapted from ICRU (1993)

14.2 The Bragg Additivity Rule

14.2.1 Derivation

Bragg and Kleeman (1905) proposed a means of calculating for the effects of a compound medium upon an energetic charged particle based upon the individual constituents. This proposal evolved out of their experimental studies of the stopping of α-particles in the hydrocarbon–halogen gases of CH_3Br and CH_3I in which they investigated this stopping as a function of atomic number. They isolated the stopping in iodine and bromine by subtracting the stopping contributions of carbon and hydrogen in these compounds using the stopping powers evaluated separately in pure carbon and hydrogen targets. They concluded that the stopping power in a compound can be estimated by the mass-weighted linear sum of the stopping powers of the individual atomic constituents.

We can derive the Bragg additivity rule for collision energy losses using the approach suggested by Sigmund (2006). Consider a medium composed of a mixture of n noble gases held at low pressure. In these cases, there are no bonds between the constituents and we can treat the medium as a mixture of independent elemental constituents. Then, the mean energy loss of a charged particle traversing a thickness Δx of this medium is

$$\overline{\Delta E} = \Delta x \sum_{j=1}^{n} N_j \sum_k Q_k \sigma_{jk} \tag{14.1}$$

where N_j is the number of atoms of the jth noble gas per unit volume, Q_k is the kinetic energy transfer from the particle to the atom for the kth type of transfer and σ_{jk} is the total cross section for the energy transfer Q_k for the jth noble gas. The effective stopping cross section for this medium is, where N is the total number of atoms per unit volume,

$$\begin{aligned} S_{\text{eff}} &= \frac{\overline{\Delta E}}{N \Delta x} \\ &= \sum_{j=1}^{n} \frac{N_j}{N} \sum_k Q_k \sigma_{jk} \\ &= \sum_{j=1}^{n} f_j S_j \end{aligned} \tag{14.2}$$

where f_j is the relative abundance of the jth constituent and $S_j = \sum_k Q_k \sigma_{jk}$ is the corresponding stopping cross section. Thus, the stopping in a compound made up of noninteracting elements is the linear combination of the stopping for each of the compound's constituents.

This result is valid as the selection of elemental noble gas constituents at low pressure allowed the treatment of the medium as independent constituents and which ensured the linearity of the outcome. More realistic cases of compounds and material phases rather than an aggregate of noble gases present conditions for which the Bragg additivity rule can only provide an approximation.

14.2.2 Applications I: Mass Collision Stopping Power

From the above, the mass collision stopping power of a compound of n elemental constituents can be written in terms of the Bragg additivity rule as

$$\left.\frac{\mathrm{d}E}{\rho\,\mathrm{d}x}\right|_{\mathrm{Compound}} = \sum_{j=1}^{n} w_j \left.\frac{\mathrm{d}E}{\rho\,\mathrm{d}x}\right|_j \tag{14.3}$$

where w_j is the fraction by weight of the jth elemental constituent of the compound and $\left.\frac{\mathrm{d}E}{\rho\,\mathrm{d}x}\right|_j$ is the mass collision stopping power for that constituent. (14.3) is based upon the implicit assumption that the chemical and bonding changes affect the mass collision stopping power to a negligible extent.

It is also possible to approximate the mass collision stopping power of a compound medium by using the stopping number and replacing the medium-dependent quantities of $\bar{I}$, $\frac{Z}{A}$ and δ with appropriate quantities calculated using the Bragg additivity rule. This is demonstrated in the next section.

14.2.3 Applications II: Mean Excitation Energy

14.2.3.1 Introduction

Recall from its definition that the mean excitation energy has a dependence of the collision stopping power upon the atomic electron structure within the medium. This can be seen through its definition from our discussions of Chap. 8 of $ln\,\bar{I} = \sum_{j=0}^{f} f_{j0} \ln\left(\varepsilon_j - \varepsilon_0\right)$ where the dependence enters through the dipole oscillator strength, the number of available final states and their associated energies. The uncertainty in the estimation of $ln\,\bar{I}$ for a compound medium has been reported to be of the order of 10–15 %. This uncertainty can have surprisingly profound effects in therapeutic calculations, for example, Besemer et al. (2013) pointed out that uncertainties in $\bar{I}$ limit the accuracy with which the ranges of protons in tissue can be evaluated. The consequence is that a more conservative proton therapy planning procedure is followed in order to reduce the normal tissue complication probability associated with the uninvolved rectal wall. This leads to a limitation of the absorbed dose to the prostate gland and, potentially, a loss in tumour control probability.

14.2.3.2 Calculation Using the Bragg Additivity Rule

In order to estimate the $\bar{I}$ of a compound, we first define the mean ratio of atomic number to atomic mass number for a compound,

$$(Z/A)|_{\text{Compound}} = \sum_{j=1}^{n} w_j (Z/A)_j, \tag{14.4}$$

where w_j is, again, the fraction by weight of the jth constituent of the compound medium. The mean ratio of (14.4) is the number of electrons per unit mass of the molecule. The mean excitation potential is, as we have seen Fig. 11.1, nearly a linear function of atomic number for Z greater than about 15. Because of this and because it is the logarithm of the mean excitation potential that influences the collision stopping power, we evaluate this logarithm from the weighted sum of the associated logarithms for the elemental constituents. The weighting factor of each term in the sum is proportional to the fraction by weight and the relative number of electrons per constituent:

$$\ln \bar{I}|_{\text{Compound}} = \frac{\sum_{j=1}^{n} w_j \left(\frac{Z}{A}\right)_j \ln \bar{I}_j}{\left(\frac{Z}{A}\right)|_{\text{Compound}}}. \tag{14.5}$$

The expression of (14.5) is crude in that it assumes that $\ln \bar{I}$ will vary only with the relative number of electrons. As noted in our earlier discussion of the mean excitation energy, this is an insufficient model in that it neglects the effects of atomic electron structure upon this potential and which, in turn, varies with the nature of the chemical bonding of the atomic constituents and the material phase (gas, liquid or solid) of the medium. This can be addressed in a variety of manners.

14.2.3.3 Measurement

As explained in Sect. 11.3, the mean excitation energy is determined usually by the energy loss measurements with protons with kinetic energies such that the effects of atomic electron orbital velocities (the shell correction) and the Barkas–Andersen phenomena can be neglected. Tabulated values of $\bar{I}$ (e.g. ICRU Reports 37 (1984) and 49 (1993)) for compounds are derived from such empirical determinations.

Table 14.2 Mean excitation energies for atomic constituents in the Bragg additivity rule depending upon compound phase

	$\bar{I}$(eV)	
Constituent	Gas	Liquid/solid
Hydrogen	19.2	19.2
Carbon	70	81
Nitrogen	82	82
Oxygen	97	106

ICRU (1984)

Table 14.3 Mean excitation energies for atomic constituents in the Bragg additivity rule for condensed organic compounds

Constituent	Type of binding	$\bar{I}$(eV)
Hydrogen	Saturated	19.0 ± 0.8
	Unsaturated	16.0 ± 0.8
Carbon	Saturated	81.1 ± 2.5
	Unsaturated	79.8 ± 2.3
	Highly chlorinated	69.0 ± 3.7
Nitrogen	Amines, nitrates, etc.	105.7 ± 10.6
	In ring	81.9 ± 7.0
Oxygen	–O–	105.7 ± 10.6
	O=	81.9 ± 7.0

From ICRU (1984) based upon data from Thompson (1952)

14.2.3.4 Discrete Assignments

A means of accounting for the phase of the medium is to assign explicit values of $\bar{I}$ on the basis of values derived for compounds, but not to the extent of abandoning the Bragg additivity rule. These are provided in Table 14.2 on the basis of the constituent being either a gas or in condensed form.

In addition to the material phase of the medium, the nature of the molecular bond will affect $\bar{I}$ and a similar means of discrete assignment of values to the quantity can be applied. It is found that the accuracy of the Bragg additivity rule in the calculation of the mean excitation energy improves with increasing atomic number. This is attributable to the increased dependence of $\bar{I}$ on the more tightly bound inner atomic electrons which are insensitive to chemical bonding effects.

Thompson (1952) compared the measured stopping powers of 270-MeV protons in liquid H_2, N_2 and O_2 and in graphite with those of condensed organic media containing these elements. He determined that the greatest deviation from the Bragg additivity rule occurred for hydrocarbons and were of the order of a few percent. He also evaluated values of $\bar{I}$ to use in the Bragg additivity rule as based upon the nature of the chemical bond for condensed organic compounds. These are summarised in Table 14.3. ICRU Report 37 concluded that the variations in mean excitation energy with chemical bond as proposed by Thompson for use within the Bragg additivity rule were plausible but that further experimental work was required in order to validate the results. The report concluded, however, that Thompson's approach was, overall, complicated and did not yield any significant improvement over the simpler approach of Table 14.2 being used where experimental data do not exist. The reader is referred to Chap. 5 of that report for tabulations of numerical values of the mean excitation potential for compounds including tissue.

14.2.4 Applications III: Density Effect

In an argument similar to that just discussed, it could be possible to define an effective density effect parameter of the compound as

$$\delta|_{\text{Compound}} = \frac{\sum_{j=1}^{n} w_j (Z/A)_j \delta_j}{(Z/A)|_{\text{Compound}}}. \tag{14.6}$$

However, ICRU Report 37 reported that the value of the density effect parameter evaluated from (14.6) is usually inaccurate and notes that it is usually necessary to calculate this parameter directly for the compound of interest. It is noted that the theory of the density effect was derived under the assumption that the medium is a homogeneous monoatomic medium and that this assumption leads to the error in determining the density effect using (14.6).

14.3 Considerations

The assumption inherent to the derivation of (14.2) leading to (14.3) of the energy loss in a medium of an unbound mixture of elements being equivalent for that in a medium with the same fraction by weight of this mixture but within a bound state is, of course, subject to refinements. The assumption implies that the valence electron structures are unaffected which, of course, is not true.

Let us look briefly at two considerations in the application of the Bragg additivity rule to collision energy loss (Sigmund et al. 2005).

14.3.1 High Projectile Velocity

At high projectile velocities, the stopping number will be dominated by the zeroth-order term of the zeroth-order term of the stopping number expansion given by (12.4):

$$L_{00}(\beta) = ln\left(\frac{2m_e\beta^2}{\bar{I}}\right) - ln(1-\beta^2) - \beta^2.$$

The effects of changes in the valence electron structure enter through $\ln \bar{I}$. As these effects would be diminished for heavy atoms, such as calcium, the effects at high projectile velocities will be dominated by hydrogen, carbon and oxygen in soft tissue.

14.3.2 Shell and Barkas–Andersen Corrections

With decreasing projectile velocity, the shell correction closes the excitation channels made available for inner-orbit electrons sequentially. Thus, it is evident that the valence effects become significant for the projectile velocity such that the stopping power is dominated by valence electrons. Both the shell and Barkas–Andersen corrections increase in magnitude with reduced projectile velocity and are influenced by valence effects, although the direction of the effect will differ between negative and positive projectiles.

14.4 Summary

The Bragg additivity rule is widely used in the calculation of collision stopping powers in compound media using data obtained from, or calculated for, elemental media. We have seen that it can be used in the manner of calculating the collision stopping power for each atomic constituent of the compound individually and then using their mass-weighted sums to yield that for the compound. We have also demonstrated that one can calculate effective values of Z/A and $\ln \bar{I}$ using the Bragg additivity rule. This is not possible for δ which must be calculated separately for the compound. These are then input into the expression for the mass collision stopping power to yield the result for the compound.

However, the Bragg additivity rule is an approximation: the differences in atomic electron orbital structure that occur when atomic constituents are combined to form compounds will affect the stopping power beyond the assumption of the simple linearity that is inherent to the Bragg additivity rule.

The ICRU has recommended that tabulated data for $\ln \bar{I}$ be used wherever possible rather than a Bragg rule-based calculation. Tabulated values of the mean excitation energy for compounds are available from ICRU Reports 37 (1984) and 49 (1993) and some websites, including that of "The Stopping and Range of Ions in Matter" provided by Ziegler at http://www.srim.org/SRIM/Compounds.htm.

Bibliography and Additional Reading

Ahlen SP. Theoretical and experimental aspects of the energy loss of relativistic heavily ionizing particles. Rev Mod Phys. 1980;52:121–73.

Attix FH. Introduction to radiological physics and radiation dosimetry. New York: Wiley; 1986.

References

Besemer A, Paganetti H, Bednarz B. The clinical impact of uncertainties in the mean excitation energy of human tissues during proton therapy. Phys Med Biol. 2013;58:887–902.

Bragg WH, Kleeman R. On the alpha particles of radium and their loss of range in passing through various atoms and molecules. Philos Mag. 1905;10:318–40.

ICRU. Stopping powers for electrons and positrons – International Commission on Radiological Units and Measurements Report 37. Bethesda: ICRU; 1984.

ICRU. Stopping powers and ranges for protons and alpha particles – International Commission on Radiological Units and Measurements Report 49. Bethesda: ICRU; 1993.

Sigmund P. Particle penetration and radiation effects. Berlin: Springer; 2006.

Sigmund P, Sharma A, Schinner A, Fettouhi A. Valence structure effects in the stopping of swift ions. Nucl Instrum Methods B. 2005;230:1–6.

Thompson TJ. Effect of chemical structure on stopping powers for high-energy protons. Report UCRL-1910. University of California Radiation Laboratory; 1952

Thwaites DI. Bragg's rule of stopping power additivity: a compilation and summary of results. Radiat Res. 1983;95:495–518.

Thwaites DI. Departures from Bragg's rule of stopping power additivity for ions in dosimetric and related materials. Nucl Instrum Methods Phys Res. 1992;B69:53–63.

Part IV
Stochastic Collision Energy Loss

Chapter 15
Introduction to Part IV

Contents

Abstract This chapter introduces us to Part IV, the subject of which is the stochastic nature of the energy loss of charged particles through collisions.

15.1 Introductory Comments

In Part III, we evaluated the energy loss and stopping power of a charged particle decelerating in a medium. The change in the particle's kinetic energy was due to the combination of energy loss to the atom as a whole (soft collisions) and through individual collisions with atomic electrons (hard collisions). Classical and quantum theories were both considered, as were various effects that result due to the interaction of the particle and the medium. Perturbative effects reflecting the real cases of medical exposures to charged particles were then developed.

However, in all of those calculations, it has been assumed that energy loss with distance travelled is a continuous function. That is, for a charged particle that has traversed a distance t through a medium, the energy loss is calculated by integrating the linear collision stopping power over distance traversed:

$$\Delta E = \int_0^t \mathrm{d}x \, \frac{\mathrm{d}E(x)}{\mathrm{d}x}. \tag{15.1}$$

This expression is, of course, an approximation which can be considered valid for specific conditions only. We recognise that the energy transferred from the

B.J. McParland, *Medical Radiation Dosimetry*,
DOI 10.1007/978-1-4471-5403-7_15,

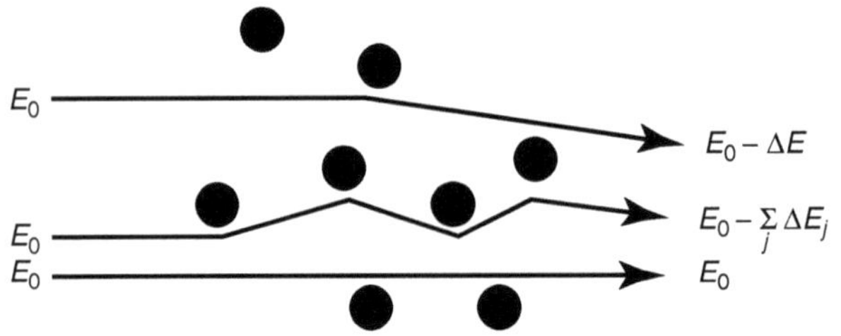

Fig. 15.1 Principle of energy straggling. A mono-energetic beam of charged particles is incident to a medium from the left. Each particle is incident to a different number of atoms at different impact parameters. The result is that the exiting beam exhibits a spectrum of kinetic energies reflecting this (Redrawn from Sigmund (2006))

charged particle to the medium occurs through discrete interactions with atoms. Hence, the energy loss process is not a continuous one, as described by (15.1), but is, instead, stochastic, as demonstrated by Fig. 15.1.

Let us imagine a mono-energetic beam of charged particles incident to a medium from the left. The medium is amorphous, or a liquid, and hence channelling is not possible. As a result, the number of atoms along the trajectories of the charged particles will differ and each particle will undergo a different degree of energy loss (unless the thickness of the medium exceeds the range of the particles leading to all of the particles having stopped). However, after traversing a given thickness of the medium which is less than the particle range, the kinetic energies of the exiting beam of particles, which was initially mono-energetic beam of particles, will display a spectrum. This phenomenon is referred to as *energy straggling* and is a consequence of a number of factors.

As can be seen from Fig. 15.1, the particles will undergo different numbers of interactions at different impact parameters and, thus, experience different amounts of energy transfers. Quantum mechanical considerations also result in variations in energy loss. The energy losses of this beam of particles will be characterised by a probability distribution function with a mean given by (15.1) and with the exiting particle beam having an energy spectrum created by this *pdf*.

The *pdf* describing the energy losses within the medium need not necessarily be symmetric, i.e. the mean and most probable energy losses being the same as in a Gaussian distribution. This will, in general, follow from the central-limit theorem for large numbers of interactions. There are the asymmetric distributions of Vavilov and Landau that arise for thin thicknesses of media, and the transitions between the three *pdf*s are functions of this thickness. An important point to note with respect to an asymmetric collision energy loss *pdf* is that, in general, it is the most probable energy loss that is measured, and not the mean. In our discussion here, we will treat the Vavilov *pdf* as the general case where the Landau and Gaussian *pdf*s are limiting cases for thin and thick absorbers.

15.2 Organisation of Part IV

Part IV consists of three chapters.

In Chap. 16, we review the descriptive statistics of the collision energy losses of moving charged particles. The genesis of this discussion was that given by Bohr in his work of 1915 and which he summarised articulately in his monograph of 1948 (Bohr 1915, 1948). The mean energy loss has already been introduced through the collision stopping power, but now we consider the distribution of energy losses which are described fundamentally by the variance of the energy loss. This variance is known as the straggling parameter which, when multiplied by the areal density of the scattering centres, yields the variance of the energy loss *pdf*. The limiting straggling parameter is derived and we examine the calculation of the straggling parameter in both classical (Bohr) and quantum (Bethe/Born) approaches, calling upon our experiences of earlier chapters.

We then follow this discussion by deriving variations of the Chapman–Kolmogorov and Bothe–Landau formulae in Chap. 17. The Chapman–Kolmogorov equation is not unique to its description of the stochastical nature of charged particle energy loss, but it is a foundation of that description. It is a one-dimensional continuity equation in the *pdf* of the energy loss of a charged particle decelerating within a medium. A solution to the Chapman–Kolmogorov equation is given by the Bothe–Landau equation which provides the mathematical foundations for the derivations of the energy loss *pdf*s.

These are provided in Chap. 18. There, we examine these derivations in two ways. First, we solve the Chapman–Kolmogorov equation for the specific case of a Gaussian *pdf* and find that we can derive an expression for the variance of the *pdf* related to the Bohr straggling parameter (Rossi 1952; Kase and Nelson 1978). Second, we delve into a more complete exploration of the energy straggling *pdf* by using the Bothe–Landau solution to yield the general solution of Vavilov (1957) and demonstrate its ability to yield the Landau (1944) and Gaussian (Kase and Nelson 1978) results as limiting cases. We conclude the chapter by considering numerical means of estimating these distributions.

References

Bohr N. On the decrease of velocity of swiftly moving electrified particles in passing through matter. Philos Mag. 1915;30:581–612.

Bohr N. The penetration of atomic particles through matter. Mat Fys Medd Dan Vid Selsk. 1948;18:1–144.

Kase KR, Nelson WR. Concepts of radiation dosimetry. New York: Pergamon; 1978.

Landau LD. On the energy loss of fast particles by ionization. J Phys USSR. 1944;8:201.

Rossi B. High-energy particles. New York: Prentice-Hall; 1952.

Sigmund P. Particle penetration and radiation effects. Berlin: Springer; 2006.

Vavilov PV. Ionization losses of high-energy heavy particles. Sov Phys JETP. 1957;5:749–51.

Chapter 16
Collision Statistics

Contents

Abstract Here, the statistical metrics used in the description of stochastic energy loss (or *energy straggling*) are derived. The stopping cross section has been previously derived and is the first moment of collision energy loss/transfer. Higher-order moments will be required in descriptions of the statistical nature of collision energy loss. Clearly, the first to be derived will be the second moment as this is necessary for the derivation of the variance of the probability distribution function. The derivation of the third moment will be deferred to Chap 18 when it is properly developed in the context of asymmetric collision energy loss probability distribution functions. The straggling parameter, a contributor to the second moment, is derived as is that due to Bohr. Finally, the effects of a condensed medium upon energy straggling are considered.

16.1 Introductory Notes

While in the subsequent chapters we will deal with specific statistical models of energy lost, this chapter is concerned with the general statistical description of energy loss, or energy straggling. This will lead to a focus on the width of the energy loss *pdf*, characterised by either its variance or the straggling parameter.

B.J. McParland, *Medical Radiation Dosimetry*,
DOI 10.1007/978-1-4471-5403-7_16,

In the discussion presented in this chapter, it will be assumed initially that the collision events follow the Poisson distribution with the interactions being statistically independent, as this allows a considerable simplification in the calculation of the width of the *pdf*. This simplification is based upon the assumption that projectile–atom collisions resulting in different energy transfers are statistically independent due to a wide separation between the atoms of a medium. Such an assumption is appropriate for the dilute medium, such as a gas, that we have used frequently in earlier derivations. However, the validity of this assumption comes into question should the energetic charged particle be traversing a medium for which the atomic separations are much reduced.

The chapter begins with the definition of the mean energy loss on the basis of a statistical description founded on the stopping cross section, a quantity we have made much use of in the past. It then considers the mean-squared energy loss and the variance of the energy loss *pdf* and the definition of the straggling parameter. This latter quantity is then explored further through classical (Bohr) and quantum mechanical (Bethe) means. The chapter concludes with an examination of the effects of the density of the medium upon the straggling parameter where the correlations between projectile–atom collisions must be accounted for.

16.2 Stochastic Descriptions of Energy Loss

16.2.1 Stopping Cross Section and Straggling Parameter

Consider a charged particle with kinetic energy E traversing a medium. Over a given path length of the medium, considered to be shorter than the complete range of the particle in that medium, the net energy loss is given by the summation of the energy losses that occur during the discrete interactions between that particle and the atoms of the medium:

$$\Delta E = \sum_j n_j \Delta E_j \tag{16.1}$$

where ΔE_j is the energy transferred from the projectile to an atom during a collision of the type j and n_j is the number of interactions of that type. Note that as ΔE_j is the energy transfer resulting from a single collision, $\Delta E_j = Q$. The mean collision energy loss over this path length is averaged over this number of interactions:

$$\overline{\Delta E} = \sum_j \overline{n_j} \Delta E_j \tag{16.2}$$

Obviously, the mean number of interactions is

$$\overline{n_j} = N\Delta x\,\sigma_j \tag{16.3}$$

where N is again the number density of scattering centres in the medium, Δx is the thickness of the medium in which the interactions occur and σ_j is the total cross section for the jth interaction type. From the definition of the linear collision stopping power (in a finite consideration), we have

$$\begin{aligned}\frac{\overline{\Delta E}}{\Delta x} &= N\sum_j \sigma_j\,\Delta E_j \\ &= NS \end{aligned}\tag{16.4}$$

where the stopping cross section, S, of (8.85) has been written in the discrete sense. The variance of the energy loss is, by definition,

$$\Omega^2 = \overline{\left(\Delta E - \overline{\Delta E}\right)^2} \tag{16.5}$$

Using (16.2), the variance can be written as

$$\begin{aligned}\Omega^2 &= \sum_{j,k} \overline{\left(n_j - \overline{n_j}\right)\left(n_k - \overline{n_k}\right)\Delta E_j \Delta E_k} \\ &= \sum_j \overline{\left(n_j - \overline{n_j}\right)^2}\Delta E_j^2 + \sum_{j\neq k} \overline{\left(n_j - \overline{n_j}\right)\left(n_k - \overline{n_k}\right)\Delta E_j \Delta E_k} \\ &= \sum_j \overline{\left(n_j - \overline{n_j}\right)^2}\Delta E_j^2 + \sum_{j\neq k} \overline{\left(n_j - \overline{n_j}\right)\left(n_k - \overline{n_k}\right)}\Delta E_j \Delta E_k \\ &= \sum_j \overline{\left(n_j - \overline{n_j}\right)^2}\Delta E_j^2. \end{aligned}\tag{16.6}$$

The fact that the summation for $j \neq k$ has been set equal to zero is a consequence of the assumed statistical independence of the interactions. However, from our discussions in Chaps. 8 and 12 of the interactions of charged particles in condensed media, special attention needs to be paid to the degree of validity of this assumption that atomic interactions are independent. As we shall see, energy straggling is dependent upon the material state of the medium. Accepting that this assumption is acceptable, we will further assume that Poisson statistics is valid and write the variance as

$$\begin{aligned}\Omega^2 &= \sum_j \overline{n_j}\Delta E_j^2 \\ &= N\Delta x \sum_j \sigma_j\,\Delta E_j^2 \\ &= N\Delta x W \end{aligned}\tag{16.7}$$

where the straggling parameter, W, has been introduced. We now rewrite the discrete representations of S of (16.4) and W of (16.7) into the continuous regime and in terms of the energy transfer in single collisions:

$$S = \int_0^{Q_{\max}} \mathrm{d}Q \frac{\mathrm{d}\sigma(Q)}{\mathrm{d}Q}\, Q \tag{16.8}$$

and

$$W = \int_0^{Q_{\max}} \mathrm{d}Q \frac{\mathrm{d}\sigma(Q)}{\mathrm{d}Q}\, Q^2. \tag{16.9}$$

16.2.2 *Partial Wave Derivation of the Straggling Parameter*

We now call upon our earlier developments of the partial wave and phase shift analyses of Chaps. 2 and 4 and the results of Appendix C.2 to expand the expressions of (16.8) and (16.9) for the stopping cross section and straggling parameter in terms of the transport cross section introduced in Sect. 6.3.5 of Chap. 6 which describes the Goudsmit–Saunderson theory of multiple elastic Coulomb scatter. That is, we define the first- and second-order transport equations as

$$\begin{aligned}\sigma^{(1)} &= \int \mathrm{d}\theta \frac{\mathrm{d}\sigma(\theta)}{\mathrm{d}\theta}(1-\cos\theta)\\ &= \frac{4\pi}{k^2}\sum_{l=0}^{\infty}(l+1)\sin^2(\delta_l-\delta_{l+1})\end{aligned} \tag{16.10}$$

and

$$\begin{aligned}\sigma^{(2)} &= \int \mathrm{d}\theta \frac{\mathrm{d}\sigma(\theta)}{\mathrm{d}\theta}(1-\cos\,\theta)^2\\ &= \frac{4\pi}{k^2}\sum_{l=0}^{\infty}(l+1)2\,\sin^2\left((\delta_l-\delta_{l+1})-\left(\frac{l+2}{2l+3}\right)\sin^2(\delta_l-\delta_{l+2})\right).\end{aligned} \tag{16.11}$$

An evaluation of the straggling parameter of (16.9) will thus draw upon (16.11) to provide

$$
\begin{aligned}
W &= \left(\frac{E}{2}\right)^2 \sigma^{(2)} \\
&= \frac{4\pi}{k^2}\left(\frac{m_e(\beta c)^2}{2}\right)^2 (k\rho_B(E))^2 \sum_{l=0}^{\infty}(l+1)\left(2\sin^2(\delta_l-\delta_{l+1})-\left(\frac{l+2}{2l+3}\right)\sin^2(\delta_l-\delta_{l+2})\right)
\end{aligned}
\tag{16.12}
$$

To evaluate the result of (16.12), let us recall Chap. 2 and (12.118) and write

$$
W = \pi\left(m_e(\beta c)^2\right)^2 \rho_B^2(E) \sum_{l=0}^{\infty} \frac{(l+1)\left(l+2+\left((k\rho_B(E))^2/2\right)\right)}{\left((l+1)^2+(k\rho_B(E)/2)^2\right)\left((l+2)^2+(k\rho_B(E)/2)^2\right)}
\tag{16.13}
$$

The fraction within the summation can be subsequently split:

$$
\begin{aligned}
&\sum_{l=0}^{\infty} \frac{(l+1)\left(l+2+\left((k\rho_B(E))^2/2\right)\right)}{\left((l+1)^2+((k\rho_B(E))/2)^2\right)\left((l+2)^2+((k\rho_B(E))/2)^2\right)} \\
&= \sum_{l=0}^{\infty}\left(\frac{(l+1)+\left((k\rho_B(E))^2/4\right)}{(l+1)^2+\left((k\rho_B(E))^2/4\right)} - \frac{(l+2)+\left((k\rho_B(E))^2/4\right)}{(l+2)^2+\left((k\rho_B(E))^2/4\right)}\right).
\end{aligned}
\tag{16.14}
$$

Having done so, we can see that all of the terms within the summation, except for that of the first for $l = 0$, cancel, leaving the summation equal to unity. This result, and the definitions of k and $\rho_B(E)$, provides us with a much reduced expression for the straggling parameter per target electron:

$$
W_S = 4\pi z^2 (\alpha\hbar c)^2
\tag{16.15}
$$

or $W_S \approx 26.1\, z^2\ \text{MeV}^2\,\text{fm}^2$. The S subscript indicates that this result is for a single-electron target atom.

16.2.3 Bohr Straggling Parameter

We now use the Bohr theory as a straight-forward means of evaluating the straggling parameter (Sigmund 2006). Recall our earlier definition of the quantity given by (16.9). In the simple and classical model in which the cross section is defined in terms of the impact parameter, we may write the straggling parameter as

$$W = 2\pi \int_{0}^{\infty} \mathrm{d}b\, b \Delta E^2(b) \tag{16.16}$$

Following from our derivation of the Bohr soft-collision energy transfer resulting in (8.153), we can consider the following. Recall our earlier definition of $f_{\mathrm{dist}}(b)$ of (8.153) and $f_{\mathrm{close}}(b)$ of (8.155). Replicating this allows us to recast (16.16) as, for close and distant collisions, respectively,

$$W_{\mathrm{dist}} = \frac{W_S}{\left(\frac{b_{\max}}{\rho_B(E)}\right)^2} \int_{\left(\frac{\omega_0 b_{\mathrm{mid}}}{\beta c}\right)}^{\infty} \mathrm{d}x\, x \left(|K_0(x)|^2 + |K_1(x)|^2 \right)^2 \tag{16.17}$$

where b_{mid} was defined in Sect. 8.2.5. (16.17) is calculated through numerical integration. For close interactions we have

$$W_{\mathrm{close}} = \frac{W_S}{1 + \left(\rho_B(E)/2b_{\mathrm{mid}}\right)^2}. \tag{16.18}$$

The derivation to this point has been for an atom with a single electron. To extend this to an atom with Z electrons leads to the Bohr straggling parameter (Bohr 1915, 1948)

$$W_{\mathrm{Bohr}} = 4\pi z^2 Z (\alpha \hbar c)^2 \tag{16.19}$$

and the Bohr straggling variance is the product of the Bohr straggling parameter and the areal density of the medium that the particle beam is impinging:

$$\Omega^2_{\mathrm{Bohr}} = 4\pi z^2 Z (\alpha \hbar c)^2 N \Delta x \tag{16.20}$$

Figure 16.1 presents the ratio of W/W_{Bohr} as a function of $b_{\max}/\rho_B(E)$ for distant and close interactions and their sum. Recall that $\rho_B(E)$ is dependent reciprocally upon the projectile energy, so increasing values of $b_{\max}/\rho_B(E)$ represent increasing energy.

We see that the Bohr straggling parameter is the limiting quantity in this application of the Bohr model. It is only for energies such as $(b_{\max}/\rho_B(E)) > 10$ that the Bohr straggling parameter accurately reproduces the straggling parameter. However, there are additional factors that have not been accounted for and which will affect the magnitude of the straggling. These include the perturbative effects explored in Chap. 12 – those of atomic electron motion and the Barkas–Andersen effect.

We now consider our investigation of straggling in the realm of quantum mechanics.

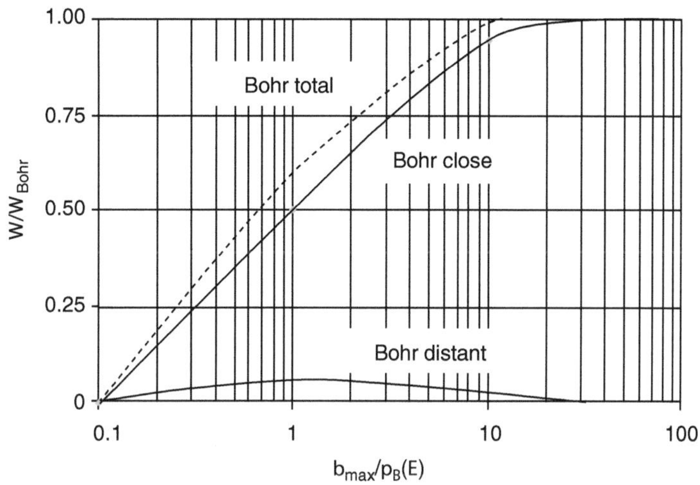

Fig. 16.1 Straggling parameter normalised to that of the Bohr straggling parameter for close and distant interactions and their sum

16.2.4 Straggling Parameter in the Born Approximation

In deriving the straggling parameter in the Born approximation (i.e. the Bethe model), we replicate our previous work of Sects. 8.3.3 and 12.3.3. As we can write the stopping cross section as

$$S = \int \mathrm{d}\sigma\,(Q)\,Q \sum_j f_{j0}(Q) \tag{16.21}$$

where $\mathrm{d}\sigma(Q)$ is the free Coulomb cross section for energy transfer Q, it follows that the straggling parameter, normalised to the Bohr straggling parameter, is given by the integration over energy transfer:

$$\frac{W}{W_{\mathrm{Bohr}}} = \frac{1}{2m_e\beta^2} \sum_j E_j \int\limits_{E_j^2/2m_e\beta^2}^{\infty} \mathrm{d}Q\, \frac{f_{j0}(Q)}{Q}. \tag{16.22}$$

As before, to solve this integral, we split it into low (long distance)- and high (short distance)-energy transfer components with an arbitrary energy transfer Q_0 separating the two regions. Then,

$$\left.\frac{W}{W_{\text{Bohr}}}\right|_{\text{dist}} = \frac{1}{2m_e\beta^2}\sum_j E_j \int\limits_{E_j^2/2m_e\beta^2}^{Q_0} dQ\frac{f_{j0}(Q)}{Q}$$

$$= \frac{1}{2m_e\beta^2}\sum_j \varepsilon_j f_{jo} \int\limits_{E_j^2/2m_e\beta^2}^{Q_0} dQ\frac{f_{j0}(Q)}{Q}$$

$$= \frac{1}{2m_e\beta^2}\sum_j \varepsilon_j f_{jo} ln\frac{2m_e\beta^2 Q_0}{I_S^2} \tag{16.23}$$

where I_S is defined through

$$\sum_j f_{j0}E_j \ln I_S = \sum_j f_{j0}E_j\, ln\, \varepsilon_j \tag{16.24}$$

and I_S and $E_j$are in the same units. We should now reflect upon the definition of this quantity given by (16.24) and that of the mean excitation potential which exists within the Bethe stopping power logarithm. As demonstrated by Sigmund (2006), these have very different magnitudes, different by factors of 10 or greater for media with atomic numbers in excess of 10.

Recalling our discussion of the sum rule of Appendix G, we can write for a single-electron atom

$$\sum_j E_j f_{j0}(Q) = Q + \frac{2}{3m_e}\langle 0|p^2|0\rangle$$

$$= Q + \frac{2}{3}\, m_e\overline{v_e^2} \tag{16.25}$$

where $\overline{v_e^2}$ is the mean-square velocity of an atomic electron in the atom's ground state. We return the result of (16.25) into (16.23):

$$\left.\frac{W}{W_{\text{Bohr}}}\right|_{\text{dist}} = \frac{1}{2m_e\beta^2}\left(Q + \frac{2}{3}m_e\overline{v_e^2}\right) ln\frac{2m_e\beta^2 Q_0}{I_S^2}. \tag{16.26}$$

We continue the derivation for the close interaction

$$\left.\frac{W}{W_{\text{Bohr}}}\right|_{\text{close}} = \frac{1}{2m_e\beta^2}\int\limits_{Q_0}^{2m_e\beta^2} \frac{dQ}{Q}\sum_j \varepsilon_j f_{j0}(Q)$$

$$= \frac{1}{2m_e\beta^2}\int\limits_{Q_0}^{2m_e\beta^2} \frac{dQ}{Q}\left(Q + \frac{2}{3}m_e\overline{v_e^2}\right) \tag{16.27}$$

where (16.25) has been used. Solving for this integral and summing it with (16.26) give the complete straggling parameter (Livingston and Bethe 1937)

$$\frac{W}{W_{\text{Bohr}}} \approx 1 + \frac{2}{3}\frac{\overline{v_e^2}}{(\beta c)^2} ln \frac{2m_e\beta^2}{I_S} \tag{16.28}$$

where we have noted that $Q_0 \ll 2m_e\beta^2$. As with our observation of the classical result shown in Fig. 16.1 for $W \to W_{\text{Bohr}}$ at high projectile velocities, the quantum mechanical result of (16.28) demonstrates the same phenomenon. Note that the result of (16.28) has been derived for a single-electron atom. Should it be otherwise, the result is

$$\frac{W}{W_{\text{Bohr}}} \approx 1 + \frac{2}{3Z}\frac{\overline{v_e^2}}{(\beta c)^2} ln \frac{2m_e\beta^2}{I_S} \tag{16.29}$$

but the contribution of Z in the denominator would go to unity in the event of the atomic electrons of the target atom being uncorrelated (Sigmund 2004). We can see from (16.29) that the term of $\ln \beta^2/\beta^2$ will yield a maximum in W/W_{Bohr} as a function of energy. This is the Bethe–Livingston correction which permits $W/W_{\text{Bohr}} > 1$, a consequence of the orbital velocities of the atomic electrons.

16.3 Effects of a Condensed Medium on Energy Straggling

Recall that our derivation of the straggling parameter was based upon the assumption of the statistical independence of successive collisions which allowed the calculation of (16.6). We have highlighted previously that this assumption is acceptable for a dilute gas in which the separations are large between atoms. However, it becomes questionable in the case of an amorphous condensed medium within which atoms would be in close proximity to each other.

Sigmund (1978) addressed this particular problem of correlated scattering in dense media. His approach to the problem begins by considering a beam of charged particles incident to a medium of thickness Δx containing η collision centres (atoms). The jth centre is positioned at the co-ordinate $\mathbf{r}_j$. Consider the beam to be collinear with a straight-line trajectory entering the medium at the point $\mathbf{r}_0$ and the vectorial impact parameter $\mathbf{b}_{\perp j}$ between the trajectory and the jth scattering centre. The total energy lost will be the summation of the energies transferred to the atoms:

$$\Delta E = \sum_{j=1}^{\eta} \Delta E_j(r_0 - \mathbf{b}_{\perp j}). \tag{16.30}$$

The mean energy is obtained by integrating (16.30) over the cross-sectional area of the beam of particles, Λ:

$$\overline{\Delta E} = \frac{1}{\Lambda}\int_{\Lambda} \mathrm{d}^2\mathbf{r}_0 \Delta E = \frac{1}{\Lambda}\int_{0}^{\infty} \mathrm{d}^2\mathbf{r}_0 \Delta E \tag{16.31}$$

as, in the absence of multiple scatter, the energy loss is considered to be equal to zero outside the area of the beam. Recalling the definition of the stopping cross section given in (16.4) and assuming a monoatomic medium, we have $\overline{\Delta E} = N\,\Delta x\,S$. Hence, (16.4) and (16.31) are equal. Here, N is the number density of atoms, or $N = \rho N_A/A$.

Similarly, the mean-square energy loss is

$$\begin{aligned}\overline{\Delta E^2} &= \frac{1}{\Lambda}\int_{0}^{\infty} \mathrm{d}^2\mathbf{r}_0 \sum_{j=1}^{\eta}\sum_{k=1}^{\eta} \Delta E(\mathbf{r}_0 - \mathbf{r}_j)\Delta E(\mathbf{r}_0 - \mathbf{r}_k) \\ &= \frac{1}{\Lambda}\left(\int_{0}^{\infty} \mathrm{d}^2\mathbf{r}_0 \Delta E^2(\mathbf{r}_0 - \mathbf{r}_j) + \int_{0}^{\infty} \mathrm{d}^2\mathbf{r}_0 \sum_{j\neq k}^{\eta} \Delta E(\mathbf{r}_0 - \mathbf{r}_j)\Delta E(\mathbf{r}_0 - \mathbf{r}_k)\right) \\ &= \frac{\eta}{\Lambda}\int_{0}^{\infty} \mathrm{d}^2\mathbf{r}_0 \Delta E^2(\mathbf{r}_0 - \mathbf{r}_j) + \frac{\eta}{\Lambda}\sum_{j\neq k}^{\eta}\int_{0}^{\infty} \mathrm{d}^2\mathbf{r}_0 \Delta E(\mathbf{r}_0 - \mathbf{r}_j)\Delta E(\mathbf{r}_0 - \mathbf{r}_k) \\ &= \frac{\eta}{\Lambda}\left(W + \sum_{j=1, j\neq k}^{\eta}\int_{0}^{\infty} \mathrm{d}^2\mathbf{r}_0 \Delta E(\mathbf{r}_0)\Delta E(\mathbf{r}_0 + \delta\mathbf{r}_{j-k})\right).\end{aligned} \tag{16.32}$$

Here, we have isolated the summations into those of $j = k$ and that of $j \neq k$ and replaced the variable $\mathbf{r}_0$ with $\mathbf{r}_0 + \mathbf{r}_j$. W is the atomic straggling parameter of (16.7) and the quantity $\delta\mathbf{r}_{j-k}$ is a component of the interatomic distance vector between the jth and kth centres. The $j \neq k$ requirement within the summation of (16.32) specifies that this interatomic distance vector component cannot point from one collision centre to itself. We can replace the summation with an integration over $\delta\mathbf{r}$ which is the separation over all atoms:

$$\begin{aligned}\overline{\Delta E^2} &= \frac{\eta}{\Lambda}\left(W + \frac{\eta}{\Lambda}S^2\right) \\ &= \frac{\eta}{\Lambda}W + \left(\frac{\eta}{\Lambda}S\right)^2 \\ &= N\Delta x\,W + (N\Delta x\,S)^2 \\ &= N\Delta x\,W + \left(\overline{\Delta E}\right)^2\end{aligned} \tag{16.33}$$

where (16.7) has been used.

We now consider the role that the atomic proximities play in establishing the correlations between collisions and the subsequent effect upon energy straggling. Let the probability that an atom is found at a distance $\mathbf{r}$ from another atom and is within the differential volume element $d^3\mathbf{r}$ be given by $N\int d^3\mathbf{r}\vartheta_2(\mathbf{r})$ where $\vartheta_2(\mathbf{r})$ is a pair correlation function.

The second term of (16.32) produces the square mean of the energy loss. We next replace the summation with an integration

$$\sum_{j=1, j\neq k}^{\eta}\int_0^\infty d^2\mathbf{r}_0\Delta E(\mathbf{r}_0)\Delta E(\mathbf{r}_0+\delta\mathbf{r}_{j-k}) = N\int_0^\infty d^2\mathbf{r}_0\int d^3\mathbf{r}\vartheta_2(\mathbf{r})\Delta E(\mathbf{r}_0)\Delta E(\mathbf{r}_0+\delta\mathbf{r}) \tag{16.34}$$

We manipulate this integral by using $\vartheta_2(\mathbf{r}) = (\vartheta_2(\mathbf{r})-1)+1$ to give

$$\begin{aligned} N\int_0^\infty d^2\mathbf{r}_0\int d^3\mathbf{r}\vartheta_2(\mathbf{r})\Delta E(\mathbf{r}_0)\Delta E(\mathbf{r}_0+\delta\mathbf{r}) &= N\int_0^\infty d^2\mathbf{r}_0\int d^3\mathbf{r}(\vartheta_2(\mathbf{r})-1)\Delta E(\mathbf{r}_0)\Delta E(\mathbf{r}_0+\delta\mathbf{r}) \\ &+ N\int_0^\infty d^2\mathbf{r}_0\int d^3\mathbf{r}\Delta E(\mathbf{r}_0)\Delta E(\mathbf{r}_0+\delta\mathbf{r}) \end{aligned} \tag{16.35}$$

We can then write

$$\begin{aligned} \overline{\Delta E^2} = N\Delta x W &+ N^2\Delta x\int d^3\mathbf{r}\int d^2\mathbf{r}_0\Delta E(\mathbf{r}_0)\,\Delta E(\mathbf{r}_0+\delta\mathbf{r})(\vartheta_2(\mathbf{r})-1) \\ &+ N^2\Delta x\int d^3\mathbf{r}\int d^2\mathbf{r}_0\Delta E(\mathbf{r}_0)\Delta E(\mathbf{r}_0+\delta\mathbf{r}). \end{aligned} \tag{16.36}$$

This second integral of (16.36) is

$$\begin{aligned} N^2\Delta x\int d^3\mathbf{r}\int d^2\mathbf{r}_0\Delta E(\mathbf{r}_0)\Delta E(\mathbf{r}_0+\delta\mathbf{r}) &= (N\Delta x S)^2 \\ &= \left(\overline{\Delta E}\right)^2. \end{aligned} \tag{16.37}$$

Then, from (16.5) and defining the variable $\varsigma = \mathbf{r}_0 + \delta\mathbf{r}$, we have

$$\Omega^2 = N\,\Delta x\,W + N^2\,\Delta x \int \mathrm{d}^3\mathbf{r}_0 \Delta E(\mathbf{r}_0) \int \mathrm{d}^2\varsigma\,\Delta E(\varsigma) \int \mathrm{d}^3\mathbf{r}\,(\vartheta_2(\mathbf{r}) - 1)\delta(\varsigma - \mathbf{r}_0 - \delta\mathbf{r}) \tag{16.38}$$

If we should assume an isotropic medium, then $\Delta E(\mathbf{r}_0) = \Delta E(r_0)$ and $\vartheta_2(\mathbf{r}) = \vartheta_2(r)$. Hence, integration of (16.38) over the angular components of the variables leads to the result that the straggling parameter W in a dilute gas is altered to $W + \Delta W$ in a dense amorphous medium due to the correlations between projectile–atom collisions where, using our full expression for N,

$$\Delta W = 2\left(\frac{\rho N_\mathrm{A}}{A}\right) \int \mathrm{d}^3\mathbf{r}_0 \Delta E(r_0) \int \mathrm{d}^2\varsigma\,\Delta E(\varsigma) \int \mathrm{d}r\,\frac{(\vartheta_2(r) - 1)}{\sqrt{1 - \left(\frac{|\mathbf{r}_0 - \varsigma|^2}{r^2}\right)}} \tag{16.39}$$

While the atomic pair correlation function $\vartheta_2(r)$ can be measured experimentally, we will consider it intuitively at the expectation that $\vartheta_2(r)$ would be equal to zero for at small distances r separating the atoms within the medium. For this case, the ΔW of (16.39) would be negative due to the $\vartheta_2(r) - 1$ factor within the integrand. We may conclude, as a consequence, that the straggling parameter will decrease with the increasing density of the medium.

Bibliography and Further Reading

James F. Statistical methods in experimental physics. 2nd ed. Hackensack: World Scientific Publishing; 2006.

Knoll GF. Radiation detection and measurement. 4th ed. New York: John Wiley; 2010.

Rossi B. High-energy particles. New York: Prentice-Hall; 1952.

References

Bohr N. On the decrease of velocity of swiftly moving electrified particles in passing through matter. Philos Mag. 1915;30:581–612.

Bohr N. The penetration of atomic particles through matter. Mat Fys Medd Dan Vid Selsk. 1948;18:1–144.

Livingston MS, Bethe HA. Nuclear physics. C. Nuclear dynamics, experimental. Rev Mod Phys. 1937;9:245–390.

Sigmund P. Statistics of particle penetration. Mat Fys Medd Dan Vid Selsk. 1978;40:1–50.

Sigmund P. Stopping of heavy ions, Springer traces of modern physics, vol. 204. Berlin: Springer; 2004.

Sigmund P. Particle penetration and radiation effects. Berlin: Springer; 2006.

Chapter 17
The Chapman–Kolmogorov and Bothe–Landau Equations

Contents

Abstract The Gaussian and asymmetric collision energy loss *pdfs* that are derived in the following chapter arise from solutions to the two equations derived here. The form of the Chapman–Kolmogorov equation derived here is an integro-differential equation in the collision energy loss *pdf* $f(x, \Delta E)$ where $f(x, \Delta E)\, d(\Delta E)$ is the probability that a charged particle has lost kinetic energy between ΔE and $\Delta E + d(\Delta E)$ after penetrating to a depth x in a medium. The Bothe–Landau equation, also derived in this chapter, is a solution of the Chapman–Kolmogorov equation, and we will see in the following chapter how it can yield a Gaussian *pdf* of collision energy loss, provided a number of important assumptions are held. The asymmetric collision energy loss *pdfs* of Vavilov and Landau can be derived directly from the Chapman–Kolmogorov equation.

17.1 Introductory Comments

In the last chapter, we began our study of energy loss straggling by deriving a simple descriptive numerical quantity, the straggling parameter, which is proportional to the variance of the energy loss *pdf* and the second moment of energy loss. Prior to that, in

B.J. McParland, *Medical Radiation Dosimetry*,
DOI 10.1007/978-1-4471-5403-7_17, © Springer-Verlag London 2014

Part III, we derived the mean energy loss, through the collision stopping power, and which has been recast in the form of the stopping cross section throughout Part III and is also the first moment of energy loss.[1] These two metrics are insufficient to fully characterise the *pdf*. In particular, the *pdf* does not necessarily have to be symmetric; in which case, the mean and most probable energy losses will not be equal. Hence, we must be capable of deriving the full *pdf* of energy loss which not only provides the mean and most probable energy losses and the variance of the distribution but also provides a complete description of the morphology of the *pdf* in a graphical representation as a function of ΔE.

Calculations of various energy loss *pdf*s describing the stochastic behaviour of collision energy loss are provided in the next chapter. The purpose of this chapter is to develop two tools that will prove useful, as will their variants, in the following chapter to derive charged particle energy loss *pdf*s. The loss of kinetic energy as a beam of charged particles penetrates into a medium is readily considered in terms of a continuity equation. The development of this provides us with the Chapman–Kolmogorov equation which is an integro-differential equation in the energy loss *pdf* $f(x, \Delta E)$, where $f(x, \Delta E)\, d(\Delta E)$ is the probability that a charged particle has lost kinetic energy between ΔE and $\Delta E + d(\Delta E)$ after penetrating to a depth x in a medium, and is based upon stochastic energy loss during penetration being a Markov process. This equation is derived here.

Solutions to the Chapman–Kolmogorov equation are provided by the Bothe–Landau equation and its variants. Employing a number of simplifications, one converts the integro-differential form of the Chapman–Kolmogorov to a partial differential equation which can be solved using the Fourier or Laplace transform approaches. In this chapter, we derive a form of the Bothe–Landau equation (Bothe 1921; Landau 1944) using the Fourier transform (which will be shown in the next chapter to yield a Gaussian energy loss *pdf* provided certain assumptions are satisfied). Landau (1944) and Vavilov (1957) provided solutions to the Chapman–Kolmogorov equation, but discussion of these is deferred to that chapter.

17.2 Markov Process

A Markov process is a sequence of n statistically independent indexed events occurring such that the conditional probability of the random variable x_{n+1} being equal to the value j is dependent only upon the nth event. That is, the conditional probability of a future state given the current state is independent of any knowledge of all states prior to the nth event:

[1] The third moment of energy loss and its use is a topic of Chap. 18.

$$\begin{aligned}&\Pr\left(x_{n+1}=j|x_n=i_n,x_{n-1}=i_{n-1},\ldots,x_0=i_0\right)\\&\quad=\Pr\left((x_n-1)+y_n=j|x_n=i_n,x_{n-1}=i_{n-1},\ldots,x_0=i_0\right)\\&\quad=\Pr\left(x_{n+1}=j|x_n=i_n\right).\end{aligned} \tag{17.1}$$

The process is a representation of the sequential energy losses that occur through uncorrelated projectile–atom collisions.

A practical example of a Markov process is that of standing in the rank at the airport awaiting a taxi. Assume that a taxi arrives at the stand once every a specified time interval. Let x_n be the number of people awaiting a taxi at the start of the nth time interval and let y_n be the number of people who join the queue at the taxi rank during the interval. Further, let us assume that the arrival of these customers occurs randomly with time. Hence,

$$\begin{aligned}x_1&=y\\x_{n+1}&=(x_n-1)+y_n\quad\text{for }\ 1\le x_n.\end{aligned} \tag{17.2}$$

In addition, the number of people in the queue during the nth time interval is taken to be $i_n \geq 1$. Then, the conditional probability for the number of j people in the queue during the $(n+1)$th time interval is

$$\begin{aligned}&\Pr\left(x_{n+1}=j\,|\,x_n=i_n,x_{n-1}=i_{n-1},\ldots,x_0=i_0\right)\\&\quad=\Pr\left((x_n-1)+y_n=j\,|\,x_n=i_n,x_{n-1}=i_{n-1},\ldots,x_0=i_0\right)\\&\quad=\Pr\left(y_n=j+1-i_n\,|\,x_n=i_n,x_{n-1}=i_{n-1},\ldots,x_0=i_0\right)\\&\quad=\Pr\left(y_n=j+1-i_n\right).\end{aligned} \tag{17.3}$$

That is, the probability of y_n being the number of people in the queue during the nth time interval is independent of the probabilities of the numbers of people being in the $(n-1)$th time intervals and earlier. Hence, there is no dependence upon time or prior sequence. The Markov series in this context is considered as being *time homogeneous.* A compilation of a Markov series can be written in the form of an $i \times j$ transition probability matrix:

$$\mathbf{P}=\begin{pmatrix}\mathrm{Pr}_{00} & \mathrm{Pr}_{01} & \ldots & \mathrm{Pr}_{0j}\\ \mathrm{Pr}_{10} & \mathrm{Pr}_{11} & \ldots & \mathrm{Pr}_{1j}\\ \vdots & \vdots & \vdots & \vdots\\ \mathrm{Pr}_{i0} & \mathrm{Pr}_{i1} & \ldots & \mathrm{Pr}_{ij}\end{pmatrix} \tag{17.4}$$

where the conditional probabilities making up the matrix entries are

$$\mathrm{Pr}_{ij}=\mathrm{Pr}(\mathrm{x_{n+1}}=\mathrm{j}\,|\,\mathrm{x_n}=\mathrm{i}). \tag{17.5}$$

It subsequently follows that it is required that the sum of the matrix elements along each row be equal to the sum of total probabilities or

$$\sum_{j} \mathrm{Pr}_{ij} = 1. \tag{17.6}$$

Now consider a chain of n statistically independent events which gives

$$\left(\mathrm{Pr}_{ij}\right)^{n} = \mathrm{Pr}(\mathrm{x_{k+n}} = \mathrm{j} \mid \mathrm{x_k} = \mathrm{i}). \tag{17.7}$$

17.3 Chapman–Kolmogorov Equation

17.3.1 Definition

The Chapman–Kolmogorov equation relates joint probability distributions. On the basis of the above discussion of the Markov process, this equation follows in the discrete form:

$$\begin{aligned}
\left(\mathrm{Pr}_{ij}\right)^{m+n} &= \mathrm{Pr}(\mathrm{x_{m+n}} = \mathrm{j} \mid \mathrm{x_0} = \mathrm{i}) \\
&= \sum_{k} \mathrm{Pr}(x_{m+n} = j, x_n = k \mid x_0 = i) \\
&= \sum_{k} \mathrm{Pr}(x_{m+n} = j \mid x_n = k, x_0 = i)\, \mathrm{Pr}(x_n = k \mid x_0 = i) \\
&= \sum_{k} \left(\mathrm{Pr_{ik}}\right)^{m} \left(\mathrm{Pr_{kj}}\right)^{n}.
\end{aligned} \tag{17.8}$$

In a continuous form, if we define $\mathrm{Pr}(x_1, x_2, \ldots, x_n)$ as the joint probability distribution of a collection of random variables $\{x_1, x_2, \ldots, x_n\}$, the Chapman–-Kolmogorov equation is

$$\mathrm{Pr}(\mathrm{x_1}, \mathrm{x_2}, \ldots, \mathrm{x_{n-1}}) = \int_{-\infty}^{\infty} \mathrm{dx_n}\, \mathrm{Pr}(\mathrm{x_1}, \mathrm{x_2}, \ldots, \mathrm{x_n}) \tag{17.9}$$

For example, from the Markov process of Sect. 17.2, the Chapman–Kolmogorov equation for a total of three events is

$$\mathrm{Pr}(\mathrm{x_3} \mid \mathrm{x_1}) = \int_{-\infty}^{\infty} \mathrm{dx_2}\, \mathrm{Pr}(\mathrm{x_3} \mid \mathrm{x_2})\, \mathrm{Pr}(\mathrm{x_2} \mid \mathrm{x_1}). \tag{17.10}$$

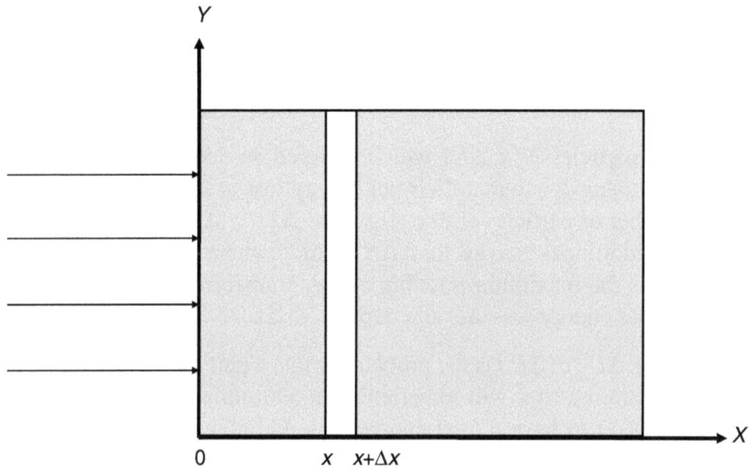

Fig. 17.1 Semi-infinite slab to which a beam of charged particles is incident from the $-x$ side

That is, the probability of being in the final state 3 beginning with the initial state 1 through an intermediate state 2 is the product of the probabilities of going from $1 \rightarrow 2$ and $2 \rightarrow 3$, integrated over all available intermediate states.

Before we proceed to derive the Chapman–Kolmogorov equation for charged particle energy loss, it will be useful to remind ourselves of the definitions of energy loss and energy transfer. The energy loss incurred by a charged particle in traversing a medium or foil, ΔE, is the result of multiple collisions during which numerous energy transfers (each described by the quantity Q) have occurred. Thus, Q refers to energy losses resulting from single events. In those conditions where the foil is sufficiently thin that only a single collision can occur, then $\Delta E = Q$; otherwise, $\Delta E > Q$.

17.3.2 One-Dimensional Continuity Equation

Consider a slab of material of physical density ρ and atomic mass number A. The slab is of semi-infinite extent (i.e. the widths in the y- and z-directions extend between $\pm\infty$ and from 0 to $+\infty$ in the x-direction). Incident to this slab is a mono-energetic beam of charged particles with kinetic energy E_0, as shown in Fig. 17.1. The entrance face is at $x = 0$. We wish to calculate the *pdf* describing the kinetic energy spectrum of the charged particles at depth in the medium. Our model is simplified by restraining it to being one dimensional and by ignoring multiple Coulomb scatter, with the latter assumption meaning that the particles travel in straight lines parallel to the x-axis.

Our first task is to derive the number of particles that have reached a penetration depth of x plus an additional small depth Δx with a net energy loss ΔE, $N(x + \Delta x, \Delta E)$. Similarly, the number of particles that have penetrated to a depth

x (and not yet penetrated Δx) with a net energy loss ΔE is given by $N(x, \Delta E)$. $N(x+\Delta x, \Delta E)$ will consist of:

- Those particles $N(x, \Delta E)$ and which did not lose any more kinetic energy by crossing Δx
- Minus those particles $N(x, \Delta E)$ which suffered an additional energy loss as a result of traversing Δx so that their net energy loss at $x+\Delta x$ will exceed ΔE
- Plus the number of particles with energy loss $\Delta E' < \Delta E$ at a depth x and which undergo an additional energy loss $\Delta E - \Delta E'$, where $\Delta E - \Delta E' \leq Q_{\max}$ and where $Q_{\max}$ is the maximum possible energy transferred, traversing Δx so that they have a net energy loss ΔE at a depth $x + \Delta x$

$\Pr(\Delta x, \Delta E \to \Delta E')\, d(\Delta E')$ is the probability that a particle having had an energy loss ΔE in penetrating to x will experience an additional energy loss $\Delta E' < \Delta E$ through crossing Δx to have a total energy loss $\Delta E'$ at a depth of $x+\Delta x$. Hence, considering the particle number accounting as described above, the number of particles with energy loss ΔE at the depth $x+\Delta x$ is

$$N(x+\Delta x, \Delta E) = N(x, \Delta E) - \int_{\Delta E}^{\infty} \mathrm{d}(\Delta E')\ \Pr(\Delta x, \Delta E \to \Delta E')\, N(x, \Delta E) + \int_{0}^{\Delta E} \mathrm{d}(\Delta E')\ \Pr(\Delta x, \Delta E' \to \Delta E)\, N(x, \Delta E') \tag{17.11}$$

The first integral describes the reduction in $N(x+\Delta x, \Delta E)$ due to particles with energy loss ΔE at x experiencing an energy loss while passing through Δx. The second integral yields the increase in $N(x+\Delta x, \Delta E)$ due to particles with energy loss $\Delta E' < \Delta E$ undergoing that energy loss within the thickness Δx to bring the net energy loss at $x + \Delta x$ to ΔE. From (17.11), we can calculate the differential change with distance in particle number with energy loss ΔE at depth x:

$$\begin{aligned}\frac{\partial N(x, \Delta E)}{\partial x} &\equiv \lim_{\Delta x \to 0} \frac{N(x+\Delta x, \Delta E) - N(x, \Delta E)}{\Delta x} \\ &= \lim_{\Delta x \to 0} \left[- \int_{\Delta E}^{\infty} \mathrm{d}(\Delta E') \frac{\Pr(\Delta x, \Delta E \to \Delta E')}{\Delta x} N(x, \Delta E) \right. \\ &\quad \left. + \int_{0}^{\Delta E} \mathrm{d}(\Delta E') \frac{\Pr(\Delta x, \Delta E' \to \Delta E)}{\Delta x} N(x, \Delta E') \right] \\ &= - \int_{\Delta E}^{\infty} \mathrm{d}(\Delta E')\, \omega(\Delta E' - \Delta E)\, N(x, \Delta E) + \int_{0}^{\Delta E} \mathrm{d}(\Delta E')\, \omega(\Delta E' - \Delta E)\, N(x, \Delta E')\end{aligned} \tag{17.12}$$

where we have defined the differential probability of an energy loss δE occurring over a path length Δx

$$\omega(\delta E) \equiv \lim_{\Delta x \to 0} \frac{\Pr(\Delta x, \delta E)}{\Delta x}. \tag{17.13}$$

We now recognise that $\omega(\delta E)\, d(\delta E)$ is the probability distribution per unit thickness of medium for the energy loss in δE. Due to our original assumption that Δx is only thin enough to allow a single collision such that the energy loss is less than the kinematical maximum of $Q_{\max}$, this differential probability must go to zero for energy losses exceeding $Q_{\max}$ or

$$\omega(\delta E) = 0; \quad \delta E > Q_{\max}. \tag{17.14}$$

Using (17.13) and (17.14) along with a change of variables, the integro-differential equation of (17.12) can be rewritten as

$$\frac{\partial N(x, \Delta E)}{\partial x} = -\int_0^{Q_{\max}} \mathrm{d}Q\, \omega(Q)\, N(x, \Delta E) + \int_0^{\min(Q_{\max}, \Delta E)} \mathrm{d}Q\, \omega(Q) N(x, \Delta E - Q). \tag{17.15}$$

The *pdf* describing the collision energy loss is

$$f(x, \Delta E) = \frac{N(x, \Delta E)}{\int_0^{E_0} \mathrm{d}(\Delta E')\, N(x, \Delta E')}. \tag{17.16}$$

That is, $f(x, \Delta E)\, \mathrm{d}(\Delta E)$ is the probability that the particle has undergone an energy loss ΔE by depth x. Hence, we can modify (17.15) so as to write a continuity equation for the *pdf* in the integro-differential form:

$$\frac{\partial f(x, \Delta E)}{\partial x} = \int_0^{\min(Q_{\max}, \Delta E)} \mathrm{d}Q\, \omega(Q) f(x, \Delta E - Q) - f(x, \Delta E) \int_0^{Q_{\max}} \mathrm{d}Q\, \omega(Q). \tag{17.17}$$

This is our variant of the Chapman–Kolmogorov equation for the *pdf* describing charged particle energy loss in a medium.

17.4 Bothe–Landau Equation

17.4.1 Introductory Comments

The result of (17.17) provides us with a means of evaluating the *pdf* of (17.16). Rossi (1952) and Kase and Nelson (1978) have presented the methodology that we will use here to derive the Bothe–Landau equation. The procedure is to first convert the integro-differential equation of (17.17) to a second-order linear partial differential equation. This requires a number of simplifying assumptions in order to achieve this result and these are discussed in the next section. We then solve this partial differential equation by using our well-exercised method of the Fourier transform in order to yield the Bothe–Landau equation.

17.4.2 Assumptions

Three assumptions are required in order to justify our conversion of (17.17) to a second-order linear partial differential equation.

The first is that the mean energy loss of the particle having penetrated to the depth x is small relative to the particle's kinetic energy:

$$\overline{\Delta E} \ll E_0. \tag{17.18}$$

The need for this particular assumption will become apparent in Chap. 18.

The second assumption is that over the penetration depth x, the collision stopping power is constant so that the mean energy loss can be extracted from the integral and is

$$\begin{aligned}\overline{\Delta E} &= \int_0^x \mathrm{d}x' \,\frac{\mathrm{d}E(x')}{\mathrm{d}x'} \\ &= x\,\frac{\mathrm{d}E}{\mathrm{d}x}.\end{aligned} \tag{17.19}$$

The third assumption is that the *pdf* is a slowly varying function of ΔE. This will allow us to expand the *pdf* in the first integrand of (17.17) into a second-order Taylor's series:

$$f(x,\Delta E - Q) \cong f(x,\Delta E) - \frac{\partial f(x,\Delta E)}{\partial(\Delta E)}Q + \frac{1}{2}\frac{\partial^2 f(x,\Delta E)}{\partial(\Delta E)^2}Q^2. \tag{17.20}$$

17.4.3 Derivation of the Bothe–Landau Equation

Recall from (17.14) that the energy loss is limited by the collision kinematics. This allows us to replace the upper limit of the second integral of (17.17) with ∞ and to rewrite (17.17) in terms of a single integral:

$$\frac{\partial f(x,\Delta E)}{\partial x}=\int_0^\infty \mathrm{d}Q\,\omega(Q)\,(f(x,\Delta E-Q)-f(x,\Delta E)) \tag{17.21}$$

Inserting the Taylor expansion of (17.20) into (17.21) gives us

$$\frac{\partial f(x,\Delta E)}{\partial x}=-\frac{\partial f(x,\Delta E)}{\partial(\Delta E)}\int_0^\infty \mathrm{d}Q\,\omega(Q)Q+\frac{1}{2}\frac{\partial^2 f(x,\Delta E)}{\partial(\Delta E)^2}\int_0^\infty \mathrm{d}Q\,\omega(Q)Q^2. \tag{17.22}$$

The first integral is related to the stopping cross section, S,

$$\begin{aligned}\int_0^\infty \mathrm{d}Q\,\omega(Q)\,Q &= \left(\frac{\rho N_A}{A}\right)S\\ &=\frac{\overline{\Delta E}}{x}\\ &\equiv k_S.\end{aligned} \tag{17.23}$$

The second integral is related to the straggling parameter W of (16.9):

$$\int_0^\infty \mathrm{d}Q\,\omega(Q)Q^2\equiv k_\Omega^2. \tag{17.24}$$

To demonstrate this from the definition of the straggling parameter and noting that

$$\omega(Q)=\left(\frac{\rho N_A}{A}\right)\frac{\mathrm{d}\sigma(Q)}{\mathrm{d}Q}, \tag{17.25}$$

it can be seen that

$$k_\Omega^2=\left(\frac{\rho N_A}{A}\right)W. \tag{17.26}$$

We must now reflect upon the significance of the powers of Q in these two integrands. As k_Ω^2 contains a factor of Q^2 as shown in (17.24), the effects of soft collisions will be much less upon it than upon k_S which contains only a single power of Q as shown in (17.23). Thus, the total collision cross section is used in evaluating

$\omega(Q)$ for k_S (which we have implicitly done through the use of $\overline{\Delta E}$) and only the hard collision component is used for k_Ω^2.

Equation (17.22) is now to be solved for $f(x, \Delta E)$ through our frequent use of the Fourier transform. The associated transform pairs required for solving (17.22) are

$$f(x, \Delta E) = \frac{1}{\sqrt{2\pi}\,\hbar} \int_{-\infty}^{\infty} \mathrm{d}\tau\, \mathrm{e}^{i\left(\frac{\tau \Delta E}{\hbar}\right)} f(x, \tau)$$

$$f(x, \tau) = \frac{1}{\sqrt{2\pi}} \int_{-\infty}^{\infty} \mathrm{d}(\Delta E)\, \mathrm{e}^{-i\left(\frac{\tau \Delta E}{\hbar}\right)} f(x, \Delta E) \tag{17.27}$$

Thus, (17.22) in Fourier space is

$$\frac{\partial f(x, \tau)}{\partial x} = -\left(ik_S \frac{\tau}{\hbar} + \frac{1}{2} \left(k_\Omega \frac{\tau}{\hbar} \right)^2 \right) f(x, \tau) \tag{17.28}$$

The solution of (17.28) is

$$f(x, \tau) = f(0, \tau)\, \mathrm{e}^{-\left(i\left(\frac{k_S \tau}{\hbar}\right) + \frac{1}{2}\left(\frac{k_\Omega \tau}{\hbar}\right)^2 \right) x}. \tag{17.29}$$

The initial condition is determined by noting that the charged particle will not have undergone any energy loss at $x = 0$ (i.e. $\Delta E = 0$) or

$$f(0, \Delta E) = \delta(\Delta E) \tag{17.30}$$

which, from prior derivations, is known to have the Fourier transform of $\frac{1}{\sqrt{2\pi}}$. Thus, the solution of (17.28) is

$$f(x, \tau) = \frac{1}{\sqrt{2\pi}}\, \mathrm{e}^{-\left(i\left(\frac{k_S \tau}{\hbar}\right) + \frac{1}{2}\left(\frac{k_\Omega \tau}{\hbar}\right)^2 \right) x}. \tag{17.31}$$

Taking the inverse Fourier transform to yield the *pdf* for ΔE, we obtain the expression:

$$\begin{aligned} f(x, \Delta E) &= \frac{1}{\sqrt{2\pi}\,\hbar} \int_{-\infty}^{\infty} \mathrm{d}\tau\, \mathrm{e}^{i\left(\frac{\tau \Delta E}{\hbar}\right)} \left(\frac{1}{\sqrt{2\pi}}\, \mathrm{e}^{-\left(i\left(\frac{k_S \tau}{\hbar}\right) + \frac{1}{2}\left(\frac{k_\Omega \tau}{\hbar}\right)^2 \right) x} \right) \\ &= \frac{1}{2\pi\hbar} \int_{-\infty}^{\infty} \mathrm{d}\tau\, \mathrm{e}^{-i\left(\frac{\left(\overline{\Delta E} - \Delta E\right)}{\hbar}\right)\tau - \frac{1}{2}\left(\frac{k_\Omega \tau}{\hbar}\right)^2 x} \end{aligned} \tag{17.32}$$

where (17.23) and (17.24) have been used. The result of (17.32) is a version of the Bothe–Landau equation for our problem.

We must not, however, forget the assumptions of (17.18), (17.19), and (17.20), inclusive, used in the derivations above. We will demonstrate in the next chapter how (17.32) can be solved and yield a Gaussian *pdf*. The kinematical restrictions on k_Ω^2 that yield this Gaussian result will also be derived.

Bibliography and Further Reading

McParland BJ. Nuclear medicine dosimetry: Advanced theoretical principles. Berlin: Springer; 2010.

Sigmund P. Particle penetration and radiation effects. Berlin: Springer; 2006.

References

Bothe W. Die Gültigkeitsgrenzen des Gaußschen Fehlergesetzes für unabhängige Elementarfehlerquellen. Z Physik. 1921;4:161–77.

Kase K, Nelson WR. Concepts of radiation dosimetry. New York: Pergamon Press; 1978.

Landau L. On the theory of energy loss of fast particles by ionization. J Phys USSR. 1944;8:201–6.

Rossi B. High-energy particles. New York: Prentice Hall; 1952.

Vavilov PV. Ionization losses of high-energy heavy particles. So Phys JETP. 1957;5:749–51.

Chapter 18
Probability Distribution Functions for Collision Energy Loss

Contents

Abstract In this chapter, the *pdfs* describing charged particle collision energy losses of three different conditions are derived. These results are of $f(x, \Delta E)$ where $f(x, \Delta E)\,\mathrm{d}(\Delta E)$ represents the probability that a charged particle has lost kinetic energy between ΔE and $\Delta E + \mathrm{d}(\Delta E)$ while penetrating into a depth x within a medium.

The differentiator between these *pdfs* will be the depth of penetration x or, equivalently, the thickness of the medium x. Large thicknesses result in a symmetric Gaussian *pdf* (under the requirement that the assumptions specified in the

B.J. McParland, *Medical Radiation Dosimetry*,
DOI 10.1007/978-1-4471-5403-7_18, © Springer-Verlag London 2014

previous chapter are met). Reduced thicknesses result in the asymmetric *pdf*s of the Vavilov and Landau theories. All three conditions are derived in this chapter.

While the asymmetric results noted above provide analytical results that can be solved readily through numerical means, practical calculations, particularly for the rapid requirements of Monte Carlo simulations, are of interest. Such methods are reviewed.

18.1 Introductory Comments

We now derive *pdf*s for collision energy loss using our variation of the Bothe–Landau equation as our starting tool. However, in so doing, we must not lose sight of the assumptions used to simplify its derivation and which will limit inevitably its applications. Following from the works of Bohr (1915, 1948), Landau (1944) and Vavilov (1957), and the descriptions given by Rossi (1952) and Kase and Nelson (1978), these *pdf*s are categorised here in terms of the distance within the medium that the charged particles traverse. Equivalently, we can consider the energy losses resulting from the traversal of sections of media (foils) by a beam of charged particles.

18.1.1 Thin Foils

A charged particle traversing a thin section of medium will undergo relatively few interactions with the atoms of that medium. As a result, the energy loss spectrum will closely resemble that of the Rutherford formula of (8.82). Due to the $1/Q^2$ dependence of the collision cross section, the peak of the spectrum will be at lower energy transfers and a tail will extend to higher energy transfers. There will be one critical difference between the *pdf* and the expression for the spectrum in that the *pdf* must be normalisable or $\int_0^\infty \mathrm{d}(\Delta E) f(x, \Delta E) = 1$.

18.1.2 Moderately Thick Foils

Now consider a foil sufficiently thick such that the charged particle traversing it undergoes a larger number of collisions that transfer energy. As such, we can invoke the central-limit theorem and consider the energy loss *pdf* to approach that of a Gaussian in this case. We will shortly demonstrate that this is obtainable from the Bothe–Landau formula derived in Chap. 17. However, we must temper this consideration by recalling the simplifying assumptions used in the derivation of our version of the Bothe–Landau formula as given in Sect. 17.4.2. In particular, the limit for the foil thickness in this calculation will be defined by (17.19). As you will recall, that assumption is that of the collision energy loss being constant over the thickness of medium being considered. Or, in the same consideration, that the energy dependence of the collision cross section is negligible. With increasing thickness and greater energy loss, that assumption becomes no longer tenable.

18.1.3 Thick Foils

For a thick foil where the assumption of (17.19) is no longer valid, we must consider explicitly the change in the rate of energy loss with decreasing charged particle energy. The simple Gaussian result for the moderately thick foil will have broken down, and different considerations, similar to those used in the development of the Fermi–Eyges theory of multiple Coulomb scatter with energy loss, must be considered.

18.1.4 Development of the Chapter

Having already developed the Bothe–Landau formula, we will use it immediately to calculate the Gaussian *pdf* of the energy loss for charged particles traversing a moderately thick foil. Here, the central-limit theorem, and our implicit assumption of Poisson statistics in its derivation, yields a Gaussian result provided we remain within the constraints of Sect. 17.4.2.

We will then embark upon the more general solution of Vavilov (1957). His solution of the Chapman–Kolmogorov equation was based upon the application of the Laplace transform rather than our use of the Fourier transform. The result that he achieved is remarkable: it returns the Landau result of energy loss in thin foils and the Gaussian result of energy loss in moderately thick foils.

Following this, we then consider the problem of very thick foils for which there are many collisions, and the change in the rate of energy loss with distance as the particle decelerates can no longer be neglected. In modern medical radiation physics practice, this problem is attacked with Monte Carlo-based solutions. However, to provide a clearer understanding of the problem, we analyse it analytically.

18.2 Derivation of the Gaussian *pdf* from the Bothe–Landau Equation

In Chap. 17, we derived an integral expression for the energy loss *pdf* in the form of the Bothe–Landau formula. Adopting the simplification assumptions of Sect. 17.4.2, we proceed to solve the Bothe–Landau equation by first manipulating algebraically the exponent within it:

$$i\tau\frac{\left(\overline{\Delta E}-\Delta E\right)}{\hbar}+\frac{1}{2}\left(\frac{k_\Omega\tau}{\hbar}\right)^2 x=\left(\frac{k_\Omega\tau}{\hbar}\sqrt{\frac{x}{2}}+i\frac{\overline{\Delta E}-\Delta E}{k_\Omega\sqrt{x}}\right)^2+\frac{\left(\overline{\Delta E}-\Delta E\right)^2}{2k_\Omega^2 x}. \tag{18.1}$$

This manipulation allows part of the integrand to be extracted from the integral so as to allow the *pdf* to be rewritten as

$$f(x,\Delta E)=\frac{1}{2\pi\hbar}\,\mathrm{e}^{-\frac{\left(\overline{\Delta E}-\Delta E\right)^2}{2k_\Omega^2 x}}\int\limits_{-\infty}^{\infty}\mathrm{d}\tau\,\mathrm{e}^{-\left(\frac{k_\Omega\tau}{\hbar}\sqrt{\frac{x}{2}}+i\frac{\overline{\Delta E}-\Delta E}{k_\Omega\sqrt{2x}}\right)^2}. \tag{18.2}$$

In order to simplify the integral prior to beginning to solve it, we perform a change of variables in (18.2) and create the complex variable

$$z=\frac{k_\Omega\tau}{\hbar}\sqrt{\frac{x}{2}}+i\frac{\overline{\Delta E}-\Delta E}{k_\Omega\sqrt{2x}}. \tag{18.3}$$

This enables us to rewrite the integral of (18.2) in the much simpler complex form

$$\int\limits_{-\infty}^{\infty}\mathrm{d}\tau\,\mathrm{e}^{-\left(\frac{k_\Omega\tau}{\hbar}\sqrt{\frac{x}{2}}+i\frac{\overline{\Delta E}-\Delta E}{k_\Omega\sqrt{2x}}\right)^2}=\frac{\hbar}{k_\Omega}\sqrt{\frac{2}{x}}\int\limits_{-\infty+iy_0}^{\infty+iy_0}\mathrm{d}z\,\mathrm{e}^{-z^2} \tag{18.4}$$

where

$$y_0=\frac{\overline{\Delta E}-\Delta E}{k_\Omega\sqrt{2x}}. \tag{18.5}$$

The integral is solved by writing it in another complex form $\int\limits_{-K+iy_0}^{K+iy_0}\mathrm{d}z\,\mathrm{e}^{-z^2}$ where K is a real constant to be allowed later to go to ∞. As the integrand e^{-z^2} is holomorphic on and within the contour and there are no singularities within the contour, the Cauchy–Goursat theorem states that

$$\int\limits_C \mathrm{d}z\,\mathrm{e}^{-z^2}=0. \tag{18.6}$$

This integral was solved in Appendix A.3 and the result is

$$\int_{-\infty+iy_0}^{\infty+iy_0} \mathrm{d}z\, \mathrm{e}^{-z^2} = \sqrt{\pi}. \tag{18.7}$$

Then (18.4) becomes

$$\int_{-\infty}^{\infty} \mathrm{d}\tau\, \mathrm{e}^{-\left(\frac{k_\Omega \tau}{\hbar}\sqrt{\frac{x}{2}}+i\frac{\overline{\Delta E}-\Delta E}{k_\Omega\sqrt{2x}}\right)^2} = \frac{\hbar}{k_\Omega}\sqrt{\frac{2\pi}{x}}. \tag{18.8}$$

The above result provides the Gaussian probability distribution function for energy loss ΔE at a depth x:

$$f(x,\Delta E) = \frac{1}{\sqrt{2\pi\, k_\Omega^2 x}}\, \mathrm{e}^{-\frac{\left(\overline{\Delta E}-\Delta E\right)^2}{2k_\Omega^2 x}}, \tag{18.9}$$

with a mean $\overline{\Delta E}$ and a variance $k_\Omega^2 x$.

To demonstrate that this result is a *pdf*, it is necessary to require that the normalisation

$$\int_0^\infty \mathrm{d}(\Delta E) f(x,\Delta E) = 1 \tag{18.10}$$

holds true. As the maximum energy loss cannot exceed the incident kinetic energy, (8.10) can be written as

$$\int_0^{E_0} \mathrm{d}(\Delta E) f(x,\Delta E) = 1. \tag{18.11}$$

Inserting the expression for the *pdf* of (18.9) into this integral

$$\frac{1}{\sqrt{2\pi\, k_\Omega^2 x}}\, \mathrm{e}^{-\frac{\left(\overline{\Delta E}-\Delta E\right)^2}{2k_\Omega^2 x}} = \frac{1}{\sqrt{\pi}} \int_{y_1}^{y_2} \mathrm{d}y\, \mathrm{e}^{-y^2} \tag{18.12}$$

where the change of variables

$$y = \frac{\overline{\Delta E} - \Delta E}{\sqrt{2k_\Omega^2 x}} \tag{18.13}$$

has been used. The limits of integration used in (18.12) are

$$y_1 = \frac{\overline{\Delta E} - E_0}{\sqrt{2k_\Omega^2 x}} \tag{18.14}$$

and

$$y_2 = \frac{\overline{\Delta E}}{\sqrt{2k_\Omega^2 x}}. \tag{18.15}$$

Recall the simplifying assumptions of Sect. 17.4.2 used in the derivation of the Bothe–Landau equation. As these lead to $E_0 - \overline{\Delta E} \approx E_0$, then the lower limit of integration of (18.14) can be written as

$$y_1 \approx \frac{-E_0}{\sqrt{2k_\Omega^2 x}}. \tag{18.16}$$

Allowing a small standard deviation, i.e.

$$\sqrt{2k_\Omega^2 x} \ll \overline{\Delta E} \tag{18.17}$$

and

$$\sqrt{2k_\Omega^2 x} \ll E_0, \tag{18.18}$$

then the limits of integration of (18.12) can be replaced by $\pm\infty$. As

$$\frac{1}{\sqrt{\pi}} \int_{-\infty}^{\infty} \mathrm{d}y\, \mathrm{e}^{-y^2} = 1, \tag{18.19}$$

we see that $f(x, \Delta E)$ is the *pdf* sought and is a Gaussian distribution with mean and most probable energy loss

$$\overline{\Delta E} = x\frac{\mathrm{d}E}{\mathrm{d}x} \tag{18.20}$$

and variance

$$\sigma^2 = k_\Omega^2 x. \tag{18.21}$$

In Chap. 10, we derived expressions for $\mathrm{d}E/\mathrm{d}x$ which yield the mean of (18.20). Now, let us calculate an explicit expression of the variance given by (18.21). Recall from our discussion of Sect. 17.4.3 that only hard collisions need be considered in the calculation of the straggling parameter. Hence, in our example, we will evaluate k_Ω^2 using the Bhabha hard-collision cross section of (9.1) for a massive projectile of mass m, where spin is neglected. Then, from (17.24), (17.25) and (D.15),

$$\begin{aligned}\sigma^2 &= \left(\frac{\rho N_\mathrm{A} x}{A}\right)\int_0^\infty \mathrm{d}Q\,\frac{\mathrm{d}\sigma\,(Q)}{\mathrm{d}Q}Q^2\\ &= \left(\frac{\rho C x}{A}\right)\left(\frac{z}{\beta}\right)^2\int_0^{Q_\mathrm{max}} \mathrm{d}Q\left(1-\beta^2\frac{Q}{Q_\mathrm{max}}\right)\\ &= \left(\frac{\rho C x}{A}\right)\left(\frac{z}{\beta}\right)^2 Q_\mathrm{max}\left(1-\frac{\beta^2}{2}\right)\\ &= \left(\frac{2\rho C m_\mathrm{e} x}{A}\right)\left(\frac{z}{\beta}\right)^2\left(\left(1+\frac{E_0}{m}\right)^2-1\right)\left(1-\frac{\beta^2}{2}\right).\end{aligned} \tag{18.22}$$

The validity of the result of (18.9) is dependent upon those constraints as noted in Sect. 17.4.2. We thus specify some necessary inequalities. First, from the requirements of the central-limit theorem, the width of this Gaussian *pdf* (2σ) must be greater than the maximum energy transferred in a single collision, or

$$Q_\mathrm{max}^2 \ll k_\Omega^2 x \tag{18.23}$$

The second is from (8.18):

$$k_\Omega^2 x \ll E_0^2. \tag{18.24}$$

Third, and finally, from (8.17) we have

$$k_\Omega^2 x \ll \left(\overline{\Delta E}\right)^2. \tag{18.25}$$

Combining these results, we have the following constraints:

$$Q_\mathrm{max}^2 \ll k_\Omega^2 x \ll E_0^2 \tag{18.26}$$

with

$$\left(\overline{\Delta E}\right)^2 \ll E_0^2\,. \tag{18.27}$$

In addition, note that

$$Q_{\max} \ll \left(\frac{\rho C x}{A}\right)\left(\frac{z}{\beta}\right)^2\left(1-\frac{\beta^2}{2}\right) \tag{18.28}$$

is also a required constraint.

These constraints enable us to define a dimensionless parameter with which to use to test the validity of the Gaussian result:

$$\kappa = \frac{C\rho(Z/A)(z/\beta)^2 x}{Q_{\max}}. \tag{18.29}$$

Clearly, from the above inequalities, the Gaussian result is valid for large κ. This parameter will prove useful in our subsequent derivation of the more general Vavilov *pdf* and permit us to differentiate between those regions of absorber thickness for which the Landau, Vavilov and Gaussian *pdf*s are valid as to be summarised.

18.3 Vavilov Theory

18.3.1 Introductory Comments

We have demonstrated that a symmetric Gaussian *pdf* for the energy loss is associated with a large value of κ, which corresponds to both a thick path length x and, through the $1/\beta^2$ factor, to nonrelativistic particles. For small κ, which corresponds to a thin absorber and/or increasingly relativistic charged particles, the mean energy loss over that absorber will also decrease. However, because the probability of a hard collision remains constant, its relative contribution to the statistical distribution of energy losses will increase as the value of the parameter κ decreases. This leads to a high-energy tail in the energy loss spectrum and an increased asymmetry in the *pdf* for ΔE. This introduction of asymmetry for small κ will inevitably require us to consider the third-order moment, $\int_0^{Q_{\max}} \mathrm{d}Q\omega(Q)\,Q^3$, in addition to our use of k_Ω^2 in the description of the energy loss *pdf*.

Historically, it was Landau (1944) who first derived a solution to the general Bothe–Landau equation leading to an asymmetric energy loss *pdf*. He achieved this through the use of Laplace transforms. Landau's result, however, was derived for the Rutherford differential cross section (which is equivalent to neglecting the β^2 $(Q/Q_{\max})$ term in the Bhabha cross section of (9.1)) and was dependent upon a variety of approximations.

A more general and exact solution to Bothe–Landau equation was provided by Vavilov (1957), again using the Laplace transform as a means of solution. The Vavilov *pdf* reproduces the Gaussian *pdf* as a limit for moderately large[1] κ and the Landau result for $\kappa \to 0$.

18.3.2 *Derivation of the Vavilov* pdf

This derivation follows Vavilov's approach and solves the integro-differential continuity equation for $f(x, \Delta E)$ by using the Laplace transform with the following transform pairs (Appendix B.2):

$$f(x,\tau) = \int_0^\infty \mathrm{d}(\Delta E)\, \mathrm{e}^{-\tau \Delta E/\hbar} f(x, \Delta E) \tag{18.30}$$

$$f(x, \Delta E) = \frac{1}{2\pi\hbar i} \int_{K-i\infty}^{K+i\infty} \mathrm{d}\tau\, \mathrm{e}^{\tau \Delta E/\hbar} f(x,\tau) \tag{18.31}$$

$$\omega(\tau) = \int_0^\infty \mathrm{d}Q\, \mathrm{e}^{-\tau Q/\hbar}\, \omega(Q) \tag{18.32}$$

$$\omega(Q) = \frac{1}{2\pi\hbar i} \int_{K-i\infty}^{K+i\infty} \mathrm{d}q\, \mathrm{e}^{\tau Q/\hbar}\, \omega(\tau) \tag{18.33}$$

where K is an arbitrary real constant. We begin by rewriting the recasting the continuity equation of (17.17) as

$$\frac{\partial f(x, \Delta E)}{\partial x} = \int_0^\infty \mathrm{d}Q\, \omega(Q) f(x, \Delta E - Q) - f(x, \Delta E) \int_0^\infty \mathrm{d}Q\, \omega(Q)\,. \tag{18.34}$$

The upper limit of integration is set to infinity as a result of (17.14). Note that the first integral of (18.34) is a convolution such that when applying the Laplace

[1] This condition has been frequently described as $\kappa \to \infty$. However, this would imply that the collision stopping power remains constant with increasing depth which, at great amounts of penetration, is an assumption that no longer holds. We will refer to the Gaussian *pdf* as being valid for *moderately thick targets* or *foils*.

transform to both sides of the equation, we use the convolution theorem of Appendix B.4. This gives

$$\begin{aligned}\frac{\partial f(x,\tau)}{\partial x} &= \omega(\tau) f(x,\tau) - f(x,\tau) \int_0^\infty \mathrm{d}Q\,\omega(Q) \\ &= f(x,\tau)\left(\omega(\tau) - \int_0^\infty \mathrm{d}Q\,\omega(Q)\right)\end{aligned} \tag{18.35}$$

which has the solution

$$f(x,\tau) = f(0,\tau) \exp\left(\left(\omega(\tau) - \int_0^\infty \mathrm{d}Q\,\omega(Q)\right) x\right). \tag{18.36}$$

For no energy loss at the point of penetrating the foil, the obvious result of the initial condition is

$$\begin{aligned}f(0,\tau) &= \int_0^\infty \mathrm{d}Q\,\mathrm{e}^{-\tau Q/\hbar}\,\delta(Q) \\ &= 1.\end{aligned} \tag{18.37}$$

Then, the complete form of the Laplace transform of the *pdf* is obtained:

$$\begin{aligned}f(x,\tau) &= \exp\left(\left(\omega(\tau) - \int_0^\infty \mathrm{d}Q\,\omega(Q)\right) x\right) \\ &= \exp\left(\left(\int_0^\infty \mathrm{d}Q\,\mathrm{e}^{-\tau Q/\hbar}\,\omega(Q) - \int_0^\infty \mathrm{d}Q\,\omega(Q)\right) x\right) \\ &= \exp\left(x \int_0^\infty \mathrm{d}Q\omega(Q)\left(\mathrm{e}^{-\tau Q/\hbar} - 1\right)\right)\end{aligned} \tag{18.38}$$

as $\omega(Q) = 0$ for $Q > Q_{\max}$. The *pdf* is recovered from this result by calculating its inverse Laplace transform[2]:

[2] Integrating over all possible energy transfers Q gives the total energy loss ΔE of the charged particles traversing x.

$$f(x, \Delta E) = \frac{1}{2\pi\hbar i} \int_{K-i\infty}^{K+i\infty} d\tau\, e^{\tau \Delta E/\hbar} f(x, \tau)$$
$$= \frac{1}{2\pi\hbar i} \int_{K-i\infty}^{K+i\infty} d\tau \exp\left(\frac{\tau \Delta E}{\hbar} + x \int_0^\infty dQ\omega(Q) \left(e^{-\tau Q/\hbar} - 1 \right) \right). \quad (18.39)$$

We now solve for the integral in (18.39) by integrating over τ. The exponent within the integrand is first reduced in order to perform this:

$$\frac{\tau \Delta E}{\hbar} + x \int_0^\infty dQ\omega(Q) \left(e^{-\frac{\tau Q}{\hbar}} - 1 \right) = \frac{\tau \Delta E}{\hbar} + x \int_0^\infty dQ\omega(Q) e^{-\tau Q/\hbar} - x \int_0^\infty dQ\omega(Q)$$
$$= \frac{\tau \left(\Delta E - \overline{\Delta E} \right)}{\hbar} - x \int_0^{Q_{max}} dQ\, \omega(Q) \left(1 - e^{-\tau Q/\hbar} - \frac{\tau Q}{\hbar} \right) \quad (18.40)$$

where (17.23) has been used to include an explicit representation of the mean energy loss over the foil thickness, x.

As per our tendency, $\omega(Q)$ is calculated for the Bhabha cross section of a massive spin-0 projectile. Combining (9.1), the definition of $\omega(Q)$ from (17.25) and the definition of the κ parameter of (18.29), we have

$$\omega(Q) = \left(\frac{\kappa Q_{max}}{x} \right) \frac{1}{Q^2} \left(1 - \beta^2 \frac{Q}{Q_{max}} \right)$$
$$= \frac{\kappa}{xQ} \left(\frac{Q_{max}}{Q} - \beta^2 \right). \quad (18.41)$$

Putting this into the result of (18.40) gives

$$\frac{\tau \left(\Delta E - \overline{\Delta E} \right)}{\hbar} - x \int_0^{Q_{max}} dQ\, \omega(Q) \left(1 - e^{-\frac{\tau Q}{\hbar}} - \frac{\tau Q}{\hbar} \right)$$
$$= \frac{\tau \left(\Delta E - \overline{\Delta E} \right)}{\hbar} - \kappa \int_0^{Q_{max}} dQ \frac{1}{Q} \left(\frac{Q_{max}}{Q} - \beta^2 \right) \left(\left(1 - e^{-\frac{\tau Q}{\hbar}} \right) - \frac{\tau Q}{\hbar} \right)$$
$$= \frac{\tau \left(\Delta E - \overline{\Delta E} \right)}{\hbar} - \kappa Q_{max} \int_0^{Q_{max}} dQ \left(\frac{1 - e^{-\frac{\tau Q}{\hbar}}}{Q^2} \right) + \kappa\beta^2 \int_0^{Q_{max}} dQ \left(\frac{1 - e^{-\frac{\tau Q}{\hbar}}}{Q} \right)$$
$$+ \frac{\tau \kappa Q_{max}}{\hbar} \int_0^{Q_{max}} \frac{dQ}{Q} - \frac{\tau \kappa \beta^2 Q_{max}}{\hbar}. \quad (18.42)$$

The three integrals of (18.42) are now be solved.

18.3.2.1 $\int_0^{Q_{\max}} \mathrm{d}Q \left(\frac{1-\mathrm{e}^{-\frac{\tau Q}{\hbar}}}{Q^2}\right)$

This first integral is simplified following the change of variables: $u = \tau Q/\hbar$

$$\int_0^{Q_{\max}} \mathrm{d}Q \left(\frac{1-\mathrm{e}^{-\tau Q/\hbar}}{Q^2}\right) = \frac{\tau}{\hbar} \int_0^{\tau Q_{\max}/\hbar} \mathrm{d}u \, \frac{1-\mathrm{e}^{-u}}{u^2}. \tag{18.43}$$

This is readily solved through integration by parts:

$$\begin{aligned} f(u) &= -\frac{1}{u} & \frac{\mathrm{d}f}{\mathrm{d}u} &= \frac{1}{u^2} \\ \frac{\mathrm{d}g}{\mathrm{d}u} &= \mathrm{e}^{-u} & g(u) &= 1-\mathrm{e}^{-u} \end{aligned} \tag{18.44}$$

to give

$$\frac{\tau}{\hbar} \int_0^{\tau Q_{\max}/\hbar} \mathrm{d}u \, \frac{1-\mathrm{e}^{-u}}{u^2} = \frac{\tau}{\hbar} \left(-\left(\frac{1-\mathrm{e}^{-u}}{u}\right)\Bigg|_0^{\tau Q_{\max}/\hbar} + \int_0^{\tau Q_{\max}/\hbar} \mathrm{d}u \frac{\mathrm{e}^{-u}}{u} \right). \tag{18.45}$$

The first term requires special consideration and is solved through l'Hôpital's rule

$$\lim_{u\to 0} \frac{1-\mathrm{e}^{-u}}{u} = 1. \tag{18.46}$$

Thus, the first integral of (18.42) can be written as

$$\int_0^{Q_{\max}} \mathrm{d}Q \left(\frac{1-\mathrm{e}^{-\tau Q/\hbar}}{Q^2}\right) = \frac{\tau}{\hbar} - \frac{1-\mathrm{e}^{-\tau Q_{\max}/\hbar}}{Q_{\max}} + \frac{\tau}{\hbar} \int_0^{\tau Q_{\max}/\hbar} \mathrm{d}u \frac{\mathrm{e}^{-u}}{u}. \tag{18.47}$$

18.3.2.2 $\int_0^{Q_{\max}} \mathrm{d}Q \left(\frac{1-\mathrm{e}^{-\tau Q/\hbar}}{Q}\right)$

The second integral of (18.42) is described through the definition of the exponential integral in Appendix C.7:

$$\int_0^{Q_{\max}} \mathrm{d}Q \left(\frac{1-\mathrm{e}^{-\tau Q/\hbar}}{Q}\right) = -\mathrm{Ei}\left(-\frac{\tau Q_{\max}}{\hbar}\right) + \ln\frac{\tau Q_{\max}}{\hbar} + \gamma_{\mathrm{EM}} \tag{18.48}$$

where Ei(x) is the exponential integral, $\mathrm{Ei}(x) = \int_{-x}^{\infty} \mathrm{d}t\, \frac{\mathrm{e}^{-t}}{t}$.

18.3.2.3 $\int_0^{Q_{\max}} \frac{\mathrm{d}Q}{Q}$

In order to be consistent with the result of the first integral given by (18.47), the third integral of (18.42) is rewritten using a change of variable, $u = \tau Q/\hbar$, or

$$\int_0^{Q_{\max}} \frac{\mathrm{d}Q}{Q} = \int_0^{\frac{\tau Q_{\max}}{\hbar}} \frac{\mathrm{d}u}{u}. \tag{18.49}$$

The exponent of the integrand of the expression for the energy loss *pdf* given by (18.39) can now be reduced considerably to

$$\begin{aligned}
&\frac{\tau\left(\Delta E - \overline{\Delta E}\right)}{\hbar} - \kappa Q_{\max}\int_0^{Q_{\max}} \mathrm{d}Q\left(\frac{1-\mathrm{e}^{-\tau Q/\hbar}}{Q^2}\right) + \kappa\beta^2\int_0^{Q_{\max}} \mathrm{d}Q\left(\frac{1-\mathrm{e}^{-\tau Q/\hbar}}{Q}\right) \\
&+\frac{\tau\kappa Q_{\max}}{\hbar}\int_0^{Q_{\max}}\frac{\mathrm{d}Q}{Q} - \frac{\tau\kappa\beta^2 Q_{\max}}{\hbar} \\
&= \frac{\tau\left(\Delta E - \overline{\Delta E}\right)}{\hbar} - \kappa Q_{\max}\left(\frac{\tau}{\hbar} - \frac{1-e^{-\tau Q_{\max}/\hbar}}{Q_{\max}} + \frac{\tau}{\hbar}\int_0^{\tau Q_{\max}/\hbar} \mathrm{d}u\frac{e^{-u}}{u}\right) \\
&\quad + \kappa\beta^2\left(-\mathrm{Ei}\left(-\frac{\tau Q_{\max}}{\hbar}\right) + \ln\frac{\tau Q_{\max}}{\hbar} + \gamma_{\mathrm{EM}}\right) + \frac{\tau\kappa Q_{\max}}{\hbar}\int_0^{\tau Q_{\max}/\hbar}\frac{\mathrm{d}u}{u} - \frac{\tau\kappa\beta^2 Q_{\max}}{\hbar} \\
&= \frac{\tau\left(\Delta E - \overline{\Delta E}\right)}{\hbar} - \frac{\kappa\tau Q_{\max}}{\hbar}\left(1+\beta^2\right) + \kappa\left(1-\mathrm{e}^{-\zeta}\right) + \kappa\left(\beta^2+\zeta\right)\left(-\mathrm{Ei}(\zeta) + \ln\zeta + \gamma_{\mathrm{EM}}\right)
\end{aligned} \tag{18.50}$$

where we have defined the variable

$$\zeta \equiv \frac{\tau Q_{\max}}{\hbar}. \tag{18.51}$$

Returning this expression for the exponent to the inverse Laplace transform for the energy loss *pdf* yields the form attributable to Vavilov:

$$f(x,\Delta E) = \frac{1}{2\pi i\, Q_{\max}}\, e^{\kappa\left(1+\beta^2\gamma_{\mathrm{EM}}\right)} \int_{K-i\infty}^{K+i\infty} dz\, \exp\left[\zeta\lambda_V + \kappa\left(\left(\zeta+\beta^2\right)\left(-\mathrm{Ei}(-\zeta) + \ln\zeta\right) - e^{-z}\zeta\right)\right]. \tag{18.52}$$

The dimensionless Vavilov parameter has been defined as

$$\lambda_{\mathrm{V}} = \frac{\Delta E - \overline{\Delta E}}{Q_{\max}} - \kappa\left(1+\beta^2-\gamma_{\mathrm{EM}}\right). \tag{18.53}$$

To solve (18.52), we integrate it over the imaginary axis (i.e. we set $K = 0$ and let $z = iy$):

$$f(x,\Delta E) = \frac{1}{2\pi\, Q_{\max}}\, e^{\kappa\left(1+\beta^2\gamma_{\mathrm{EM}}\right)} \int_{-\infty}^{\infty} dy \exp\left[iy\lambda_V + \kappa\left(\left(iy+\beta^2\right)\left(-\mathrm{Ei}(-iy) + \ln(iy)\right) - e^{-iy}\right)\right]. \tag{18.54}$$

The logarithm with complex argument requires further evaluation. From complex variable theory (Churchill et al. 1974), we have two results depending upon the sign of the argument:

$$\begin{aligned} \ln iy &= i\frac{\pi}{2} + \ln|y| \qquad y > 0 \\ &= -i\frac{\pi}{2} + \ln|y| \qquad y < 0. \end{aligned} \tag{18.55}$$

We next draw our attention to the exponential integral with complex argument. From Appendix C.7, this is

$$\begin{aligned} \mathrm{Ei}(-iy) &= \mathrm{Ci}(|y|) - i\,\mathrm{Si}(y) + i\frac{\pi}{2} \qquad y > 0 \\ &= \mathrm{Ci}(|y|) - i\,\mathrm{Si}(y) - i\frac{\pi}{2} \qquad y < 0 \end{aligned} \tag{18.56}$$

where $\mathrm{Ci}(x)$ and $\mathrm{Si}(x)$ are the cosine and sine integrals. By inserting the expressions of (18.55) and (18.56) into (18.54) and defining the two functions,

$$g_1(y) = \beta^2(\ln|y| - \mathrm{Ci}(y)) - \cos y - y\,\mathrm{Si}(y) \tag{18.57}$$

$$g_2(y) = y(\ln|y| - \mathrm{Ci}(y)) + \sin y + \beta^2\,\mathrm{Si}(y), \tag{18.58}$$

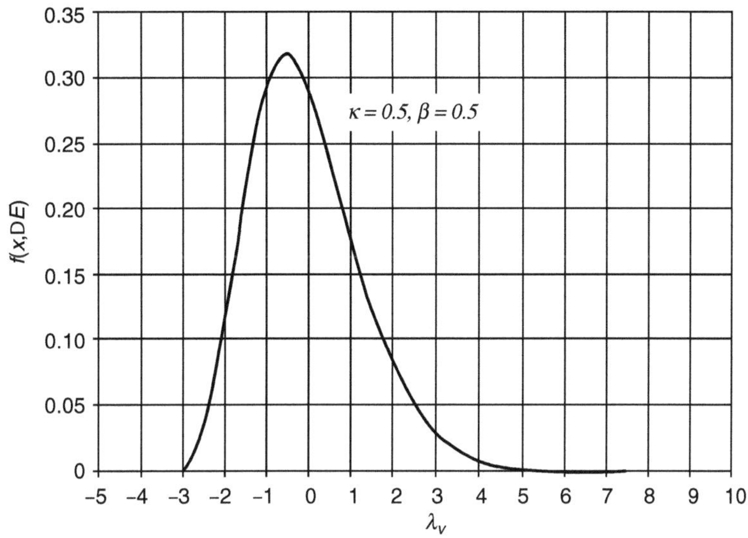

Fig. 18.1 Vavilov energy loss *pdf* for $\kappa = 0.5$, $\beta = 0.5$ calculated using the Edgeworth polynomial expansion described in Sect. 18.4.3

a much simpler integral form of the Vavilov *pdf* is obtained:

$$f(x, \Delta E) = \left(\frac{1}{\pi \kappa Q_{\max}}\right) e^{\kappa(1+\beta^2 \gamma_{\mathrm{EM}})} \int_0^\infty \mathrm{d}y \, \mathrm{e}^{\kappa g_1(y)} \cos(\lambda_V y + \kappa g_2(y)). \quad (18.59)$$

This result can be solved numerically and through the use of approximations for the cosine and sine integrals. A variety of other analytical and numerical approximations also exist, and these will be discussed shortly. Figure 18.1 shows the Vavilov *pdf* for $\kappa = 0.5$ and $\beta = 0.5$ as a function of the Vavilov parameter, λ_V. This curve shows the asymmetry resulting from reduced mean energy loss due to reduced foil thickness and the increased tail due to the constant high-energy transfer component.

18.3.3 *Landau Limit of Vavilov* pdf *for* $\kappa \to 0$

The Vavilov results of (18.52), (18.54) and (18.59) are general in form. The κ parameter is prominent in both the multiplier of the integral and the Vavilov parameter. We now consider the first of the two limiting conditions of the Vavilov *pdf*. The first is that of thin absorbers or relativistic particles for which $\kappa \to 0$. This will yield the result of Landau (1944). We begin with the full result of the Vavilov *pdf* given by (18.52) and applying the change of variables

$$p \equiv \kappa\zeta \tag{18.60}$$

from which we achieve the expression

$$\begin{aligned} f(x,\Delta E) &= \frac{1}{2\pi\, i\,\kappa\, Q_{\max}} \mathrm{e}^{\kappa\left(1+\beta^2\gamma_{\mathrm{EM}}\right)} \\ &\times \int\limits_{K-i\infty}^{K+i\infty} \mathrm{d}p \, \exp\left[p\frac{\lambda_V}{\kappa} + \kappa\left(\left(\frac{p}{\kappa}+\beta^2\right)\left(-\mathrm{Ei}\left(-\frac{p}{\kappa}\right) + \ln\, p - \ln\, \kappa\right) - \mathrm{e}^{-\frac{p}{\kappa}}\right)\right] \\ &= \frac{1}{2\pi\, i\,\kappa\, Q_{\max}} \mathrm{e}^{\kappa\left(1+\beta^2\gamma_{\mathrm{EM}}\right)} \int\limits_{K-i\infty}^{K+i\infty} \mathrm{d}p \, \exp\left[p\left(\frac{\lambda_V}{\kappa} - \ln\kappa\right) + p\left(-\mathrm{Ei}\left(-\frac{p}{\kappa}\right) + \ln p\right)\right. \\ &\left. +\kappa\beta^2\left(-\mathrm{Ei}\left(-\frac{p}{\kappa}\right) + \ln p - \ln\kappa\right) - \kappa\,\mathrm{e}^{-p/\kappa}\right]. \end{aligned} \tag{18.61}$$

By recognising that $\mathrm{Ei}(-p/\kappa) \to 0$ as $\kappa \to 0$, we can see that the exponent goes to $(p(\lambda_V/\kappa - \ln\kappa) + p\ln p)$ for small κ. Thus, we can write the *pdf* in this limit and, which is the Landau *pdf*,

$$f(x,\Delta E) = \frac{1}{2\pi i\kappa Q_{\max}} \int\limits_{K-i\infty}^{K+i\infty} \mathrm{d}p\, \mathrm{e}^{p(\lambda_{\mathrm{L}}+\ln p)}. \tag{18.62}$$

λ_{L} is the Landau parameter and is related to the Vavilov parameter of (18.53) by

$$\begin{aligned} \lambda_{\mathrm{L}} &= \frac{\lambda_{\mathrm{V}}}{\kappa} - \ln\kappa \\ &= \frac{\Delta E - \overline{\Delta E}}{\kappa Q_{\max}} - \left(1+\beta^2\right) + \gamma_{\mathrm{EM}} - \ln\,\kappa. \end{aligned} \tag{18.63}$$

The Landau *pdf* can be written in a form more readily amenable to numerical integration by first changing variables and defining

$$y \equiv -ip \tag{18.64}$$

so as to obtain the form

$$f(x,\Delta E) = \frac{1}{\pi\,\kappa Q_{\max}} \int\limits_{0}^{\infty} \mathrm{d}y\, \mathrm{e}^{-\left(\frac{\pi}{2}\right)y} \cos(y\ln y + \lambda_{\mathrm{L}} y). \tag{18.65}$$

Figure 18.2 shows the product $\kappa Q_{\max} f(x,\Delta E)$ as a function of the Landau parameter λ_{L}.

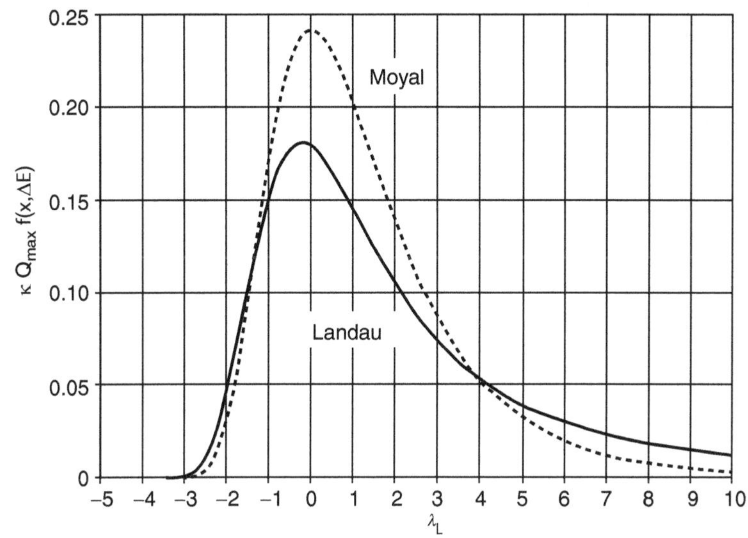

Fig. 18.2 Weighted Landau and Moyal probability distribution functions as functions of the Landau parameter. The Moyal *pdf* is discussed in Sect. 18.4.2

The maximum of the weighted energy loss *pdf* $\kappa Q_{\max} f(x, \Delta E)$ occurs for a Landau parameter of $\lambda_{L,\max} = 0.8 - \gamma_{EM} = -0.22278$ (Kölbig and Schorr 1984) from which the most probable energy loss can be calculated by noting that the Landau parameter can be written as

$$\lambda_L = \frac{\Delta E - (\Delta E)_{MP}}{\kappa Q_{\max}} + \lambda_{L,\max} \tag{18.66}$$

where $(\Delta E)_{MP}$ is the most probable energy loss. This quantity can be solved for by using (18.63) and (18.66):

$$\begin{aligned}(\Delta E)_{MP} &= \overline{\Delta E} + \kappa Q_{\max}\left(\lambda_{L,\max} + 1 + \beta^2 + \ln\kappa - \gamma_{EM}\right) \\ &= \overline{\Delta E} + \kappa Q_{\max}\left(0.2 + \beta^2 + \ln\kappa\right).\end{aligned} \tag{18.67}$$

18.3.4 Gaussian Limit of Vavilov pdf *for Large* κ

We now consider the next limiting condition of the Vavilov *pdf* for large κ, corresponding to thicker absorbers or slower particles. However, we again recall two points:

- The thickness of the absorber must not be so great such that the assumption of a constant collision stopping power with depth (17.19) becomes invalid (i.e. we do not go to the extent of taking the limit of $\kappa \rightarrow \infty$).

- As the Gaussian and Vavilov *pdfs* are symmetric and asymmetric in energy loss, respectively, the latter must include the effects of higher-order moments of energy transfer.

To yield the *pdf* for large κ, we begin with the Vavilov *pdf* in the form of its inverse Laplace transform of (18.39) and expand the exponential term in energy transfer, τQ, in a Taylor series. We allow the expansion to the order of $(\tau Q)^3$ so as to allow for the necessary presence of the asymmetry in the Vavilov result:

$$\begin{aligned} f(x,\Delta E) &= \frac{1}{2\pi\hbar i}\int\limits_{K-i\infty}^{K+i\infty} \mathrm{d}\tau \exp\left(\frac{\tau\Delta E}{\hbar} - x\int\limits_0^\infty \mathrm{d}Q\omega(Q)\left(1-\mathrm{e}^{-\tau Q/\hbar}\right)\right) \\ &= \frac{1}{2\pi\hbar i}\int\limits_{K-i\infty}^{K+i\infty} \mathrm{d}\tau \exp\left(\frac{\tau\Delta E}{\hbar} - x\int\limits_0^\infty \mathrm{d}Q\omega(Q)\left(\frac{\tau Q}{\hbar} - \frac{(\tau Q/\hbar)^2}{2} + \frac{(\tau Q/\hbar)^3}{6}\right)\right) \\ &\approx \frac{1}{2\pi\hbar i}\int\limits_{K-i\infty}^{K+i\infty} \mathrm{d}\tau \exp\left(\frac{\tau\left(\Delta E - \overline{\Delta E}\right)}{\hbar} + \left(\frac{\tau}{\hbar}\right)^2 \frac{x}{2}k_\Omega^2 - \left(\frac{\tau}{\hbar}\right)^3 \frac{x}{6}\varphi\right). \end{aligned} \tag{18.68}$$

The third moment of the expansion is

$$\varphi = \int\limits_0^{Q_{\max}} \mathrm{d}Q\omega(Q)\,Q^3. \tag{18.69}$$

As a Gaussian result for the *pdf* will be obtained if the integral is evaluated only up to k_Ω^2, the third-moment contribution through the $\exp(-\tau^3 x/6\varphi)$ term provides the necessary asymmetry of the Vavilov *pdf* which, as it is proportional to e^{-x}, decreases with increasing absorber thickness as required. Again, we consider the Bhabha cross section for a massive spin-0 projectile, for which

$$k_\Omega^2 = \frac{\kappa Q_{\max}^2}{x}\left(1 - \frac{\beta^2}{2}\right) \tag{18.70}$$

and

$$\varphi = \frac{\kappa Q_{\max}^3}{x}\left(\frac{1}{2} - \frac{\beta^2}{3}\right). \tag{18.71}$$

The *pdf* of (18.68) is solved for by the substitution of variables. First, we define

$$u = \left(\frac{x\varphi}{2}\right)^{1/3}\left(\tau - \frac{k_\Omega^2}{\varphi}\right) \tag{18.72}$$

from which

$$\mathrm{d}\tau = \left(\frac{2}{x\varphi}\right)^{1/3} \mathrm{d}u. \tag{18.73}$$

For simplicity in the derivation, we further define the variable

$$\eta \equiv \left(\frac{x\varphi}{2}\right)^{1/3}. \tag{18.74}$$

These lead to a simplified expression of the Taylor series-expanded version of the Vavilov *pdf*:

$$f(x, \Delta E) = \frac{\mathrm{e}^{at-\frac{a^3}{3}}}{2\pi i\,\eta} \int_{K-i\infty}^{K+i\infty} \mathrm{d}u\, \mathrm{e}^{ut-\frac{u^3}{3}} \tag{18.75}$$

where we have defined the quantities

$$a \equiv \eta\frac{k_\Omega^2}{\varphi} = \left(1 - \frac{\beta^2}{2}\right)\left(\frac{2\kappa}{\left(1-\frac{2}{3}\beta^2\right)^2}\right)^{\frac{1}{3}} \tag{18.76}$$

and

$$t \equiv \frac{\Delta E - \overline{\Delta E}}{\eta} + a^2. \tag{18.77}$$

We now integrate the integral of (18.75) over the imaginary axis:

$$\begin{aligned}\frac{1}{2\pi i}\int_{K-i\infty}^{K+i\infty} \mathrm{d}u\, \mathrm{e}^{ut-\frac{u^3}{3}} &= \frac{1}{2\pi}\int_{-\infty}^{\infty} \mathrm{d}y\, \mathrm{e}^{\frac{y^3}{3}+iyt} \\ &= \frac{1}{\pi}\int_0^{\infty} \mathrm{d}y\, \cos\left(yt + \frac{y^3}{3}\right).\end{aligned} \tag{18.78}$$

This result will be recognised as the Airy function, Ai(t), which is defined in Appendix C.3. As a result, the *pdf* of (18.75) is now written as

$$f(x, \Delta E) = \frac{e^{at - a^3/3}}{\eta} \mathrm{Ai}(t). \tag{18.79}$$

Let us now consider the effect of large values of κ upon the expression of the energy loss *pdf* given by (18.79).

- From (18.76), large κ results in $a \to (2\kappa)^{1/3}$ (i.e. large a).
- From (18.77), large a results in large t.

As t is the argument of the Airy function, we now consider the limiting form of this function resulting from large κ. This form is given by (C.89)

$$\mathrm{Ai}(t) \to \frac{e^{-2/3\, t^{3/2}}}{2\sqrt{\pi}\, t^{1/4}} \text{ for large } t. \tag{18.80}$$

Thus, for large t, the *pdf* of (18.79) can now be written in the form of

$$f(x, \Delta E) = \frac{e^{\left(at - a^3/3 - 2/3t^{3/2}\right)}}{2\eta\sqrt{\pi}\, t^{1/4}} \text{ for large } t. \tag{18.81}$$

This result can be further manipulated by noting from (18.77) that

$$t^{1/4} \to \sqrt{a} \text{ for large } \kappa. \tag{18.82}$$

This allows us to modify part of the exponent of (18.81)

$$at - 2/3t^{3/2} \to \frac{a^2}{3} - \frac{\upsilon^2}{4a} \text{ for large } \kappa. \tag{18.83}$$

where

$$\upsilon = \frac{\Delta E - \overline{\Delta E}}{\eta} \ll a. \tag{18.84}$$

We are then led to a further reduced expression of the energy loss *pdf*:

$$f(x, \Delta E) = \frac{e^{-\upsilon^2/4a}}{2\eta\sqrt{\pi a}}. \tag{18.85}$$

Substituting the previously defined expressions for η and a, we achieve finally the Gaussian result arising from the Vavilov *pdf* for large κ:

Table 18.1 Selection criteria for collision energy loss probability distribution function

Range of κ	Appropriate *pdf*	Equations
$\kappa \leq 0.01$	Landau	(18.62) and (18.65)
$\kappa \leq 10$	Vavilov	(18.54) and (18.59)
$10 \leq \kappa$[a]	Gaussian	(18.9) and (18.86)

[a]Upper limit set by the requirement that the collision stopping power remains constant with penetration

$$f(x, \Delta E) = \frac{1}{\sqrt{2\pi k_{\Omega}^2 x}} \mathrm{e}^{-\left(\frac{\left(\Delta E - \overline{\Delta E}\right)^2}{2 k_{\Omega}^2 x}\right)} \tag{18.86}$$

which is the same result of (18.9).

18.3.5 Separation Between Landau, Vavilov and Gaussian pdf*s*

We now summarise the limiting conditions of the Vavilov *pdf* for which its form, or its limiting Landau and Vavilov forms, should be used. These are provided in Table 18.1.

18.3.6 Electrons

A calculation of the energy straggling of electrons will be confounded by the effects of the multiple Coulomb scatter discussed in Chap. 6. Of course, for the heavy charged particles considered within our derivations in this chapter, the effects of scatter have been, justifiably, neglected. However, in principle, the effect of energy straggling can be accounted for by modifying the Landau or Vavilov distributions with the appropriate collision cross section. In particular, the Vavilov *pdf* is calculated using the Møller cross section for electron–electron collisions. Chibani (2002) simplified this calculation by approximating the Møller cross section by a fourth-order polynomial in order to allow the Vavilov *pdf* to be calculated analytically (this is also done for positron projectiles by doing the same for the Bhabha positron–electron collision cross section).

However, as we explored in our derivation of the Fermi–Eyges theory in Chap. 6, it is not possible for a practical analytical calculation of the stochastic energy losses for electrons to neglect the effects of multiple Coulomb scatter. Hence, it is often more practicable to use Monte Carlo simulations to evaluate the energy loss of the light electrons and positrons in traversing a medium.

18.3.7 Atomic Electron Binding Effects

In the case of a high-Z thin absorber, energy losses through resonant transfers to atomic electrons can become important. Such effects are manageable through the convolution of the Landau distribution with a Gaussian function. In this application, because it is easy to convolve two Gaussian functions, the Landau *pdf* is represented by a weighted sum of four Gaussian pdfs (Blunck and Leisegang 1951). Even though there are difficulties with this approach that have been stated to require addressing (Findlay and Dusautoy 1980), they have relatively little immediate application to medical radiation dosimetry.

18.4 Approximations of the Asymmetric Energy Loss *pdfs*

18.4.1 Introductory Comments

The Vavilov energy loss *pdf* of (18.59) and the Landau energy loss *pdf* of (18.65) are solvable through numerical integration. This does not make these forms readily applicable to rapid calculation as required by, for example, Monte Carlo modelling of charged particle transport. To address this, a variety of analytical approximations, or numerical evaluations, of these *pdfs* have been suggested. This Section provides a brief summary of some of the more long-standing methods.

18.4.2 Moyal Analytical Representation

Moyal (1955) suggested a closed analytical form of an asymmetric energy loss *pdf* as an approximation to the Landau *pdf*:

$$f(x, \Delta E) = \frac{1}{\sqrt{2\pi}\,\kappa\, Q_{\max}} \mathrm{e}^{-\frac{1}{2}(\lambda + \exp(-\lambda))} \tag{18.87}$$

where the coefficient is

$$\lambda = \frac{\Delta E - (\Delta E)_{\mathrm{MP}}}{\kappa\, Q_{\max}}. \tag{18.88}$$

The most probable energy loss $(\Delta E)_{\mathrm{MP}}$ is given by the expression of (18.67). The Moyal result $\kappa\, Q_{\max} f(x, \Delta E)$ was plotted earlier in Fig. 8.2 for comparison with the corresponding Landau quantity. As can be seen, the Moyal expression is not an entirely accurate reproduction of the Landau result due to both the differing positions of the most probable energy loss (the Moyal curve peaks at $\lambda_{\mathrm{L}} = 0$, whereas the Landau curve peaks at $\lambda_{\mathrm{L}} = -0.22278$) and the difference between the magnitudes of the maximum values of the two *pdfs*. Important in our consideration of the asymmetry in energy loss *pdfs* is the observation that the Moyal approximation underestimates the high-energy loss tail of the Landau *pdf*. Even so, the Moyal approximation clearly has calculational advantages over the Landau integral result and is used as its approximation.

An improvement upon the Moyal approximation was proposed by Rotondi and Montagna (1990) and is of the form

$$f_{\kappa,\beta}(x,\Delta E) = \left(\frac{a_1}{\kappa Q_{\max}}\right)\exp\left[-a_2\left(\lambda + a_5\lambda^2\right) - a_3\exp\left(-a_4\left(\lambda + a_6\lambda^2\right)\right)\right] \quad (18.89)$$

where the six parameters, a_i, are functions of κ and β and were determined by fitting (18.89) to the numerical solution for the Vavilov *pdf* for $\kappa \leq 3$. These are provided as the weighted sum of the products of two Chebyshev polynomials, and the reader is referred to that publication for further details.

18.4.3 Edgeworth Series

For values of κ that are not too small, the Vavilov distribution is Gaussian-like. The deviation from the Gaussian *pdf* can be described in terms of an Edgeworth series described in Appendix C.5.4 (Symon 1948; Rotondi and Montagna 1990; Van Ginneken 2000). Following the nomenclature of Rotondi and Montagna, the Vavilov *pdf* is approximated by the weighted series:

$$\begin{aligned} f(x,\Delta E) = \frac{\mathrm{e}^{-\frac{\Delta E^2}{2\sigma^2}}}{\sqrt{2\pi}\sigma}\Bigg[1 &+ \frac{1}{3!}\frac{\mu_3}{\sigma^3}H_3\left(\frac{\Delta E}{\sigma}\right) + \frac{1}{4!}\left(\frac{\mu_4}{\sigma^4} - 3\right)H_4\left(\frac{\Delta E}{\sigma}\right) \\ &+ \frac{1}{5!}\left(\frac{\mu_5}{\sigma^5} - 10\frac{\mu_3}{\sigma^3}\right)H_5\left(\frac{\Delta E}{\sigma}\right) + \frac{10}{6!}\left(\frac{\mu_3}{\sigma^3}\right)^2 H_6\left(\frac{\Delta E}{\sigma}\right) \\ &+ \frac{35}{7!}\frac{\mu_3}{\sigma^3}\left(\frac{\mu_4}{\sigma^4} - 3\right)H_7\left(\frac{\Delta E}{\sigma}\right) + \frac{280}{9!}\left(\frac{\mu_3}{\sigma^3}\right)^3 H_9\left(\frac{\Delta E}{\sigma}\right)\Bigg] \end{aligned} \quad (18.90)$$

where H_i is the ith Hermite polynomial (derived in Appendix C.5.1) and the coefficients of μ_i and σ are given by the moments of the Vavilov distribution

$$\mu_n = x\int_0^{Q_{\max}} \mathrm{d}Q\,\omega(Q)Q^n \qquad n = 2,3 \quad (18.91)$$

$$\mu_4 = 3\mu_2^2 + \left(x\int_0^{Q_{\max}} \mathrm{d}Q\,\omega(Q)Q^4\right)^2 \quad (18.92)$$

$$\mu_5 = 10\mu_2\mu_3 + x \int_0^{Q_{\max}} \mathrm{d}Q\omega(Q)\, Q^5 \tag{18.93}$$

$$\sigma^2 = \mu_2. \tag{18.94}$$

This Edgeworth series result was reported to be valid for $0.29 \leq \kappa$ and for $\lambda_{\mathrm{Lo}} \leq \lambda \leq \lambda_{\mathrm{Hi}}$ where the limits λ_{Lo} and λ_{Hi} define the limits of 0 and 1 in the cumulative distribution function. These latter quantities are determined from empirical fits.

18.4.4 Distorted Log-Normal Distribution

Chibani (1998, 2002) has described two algorithms for calculating the Vavilov *pdf* in the interval $\kappa \in [0.01,1.0]$. The first algorithm is valid for $\kappa \in [0.01,0.3]$ and describes the Vavilov *pdf* as the convolution of log-normal and Poisson distributions. The second method, for $\kappa \in [0.3,1.0]$, takes advantage of the similarity of the shape of the log-normal distribution to that of the Vavilov *pdf*. The reader is referred to these two papers for further details.

18.5 Very Thick Foils

We will not consider the statistical fluctuations in kinetic energy loss within thick foils to any great degree. The conditions of such foils mean that the assumptions of Sect. 17.4.2 are no longer valid as the rate of energy loss will vary with depth. As such, the Chapman–Kolmogorov equation of (17.17) adopts an explicit dependence upon the initial projectile energy:

$$\begin{aligned}\frac{\partial f(E_0,x,\Delta E)}{\partial x} = & \int_0^{\min(Q_{\max},\Delta E)} \mathrm{d}Q\, \omega(Q) f(E_0,x,\Delta E - Q) - f(E_0,x,\Delta E) \\ & \times \int_0^{Q_{\max}} \mathrm{d}Q\, \omega(Q).\end{aligned} \tag{18.95}$$

This does not allow the analytical application of the convolution theorem as used in the derivation of (18.35), although a numerical solution can be had. Further discussion may be found in Sigmund (2006).

Bibliography and Further Reading

Abramowitz M, Stegun IA, editors. Handbook of mathematical functions. New York: Dover; 1972.
Ahlen SP (1980) Theoretical and experimental aspects of the energy loss of relativistic heavily ionizing particles. Rev Mod Phys 52: 121–173 (*erratum* Rev Mod Phys 1980; 52: 653)
Bjorken JD, Drell SD. Relativistic quantum mechanics. New York: McGraw-Hill; 1964.
Inokuti M. Inelastic collisions of fast charged particles with atoms and molecules – the Bethe theory revisited. Rev Mod Phys. 1971;43:297–347.
James F. Statistical methods in experimental physics. 2nd ed. Singapore: World Scientific; 2006.
Schorr B. Programs for the Landau and the Vavilov distributions and the corresponding random numbers. Comp Phys Comm. 1974;7:215–24.
Schorr B. Numerical inversion of a class of characteristic functions. BIT Num Math. 1975;15:94–102.
Uehling EA. Penetration of heavy charged particles in matter. Ann Rev Nucl Sci. 1954;4:315–50.

References

Blunck O, Leisegang S. Zum Energieverlust energiereicher Elektronen in dunnen Schichten. Z Phys. 1951;130:641–9.
Bohr N. On the decrease of velocity of swiftly moving electrified particles in passing through matter. Phils Mag. 1915;30:581–612.
Bohr N. The penetration of atomic particles through matter. Mat Fys Medd Dan Vid Selsk. 1948;18:1–144.
Chibani O. New algorithms for the Vavilov distribution calculation and the corresponding energy loss sampling. IEEE Trans Nucl Sci. 1998;45:2288–92.
Chibani O. Energy-loss straggling algorithms for Monte Carlo electron transport. Med Phys. 2002;29:2374–83.
Churchill RV, Brown JW, Verhey RF. Complex variables and applications. 3rd ed. New York: McGraw-Hill; 1974.
Findlay DJS, Dusautoy AR. Improvements to the Blunck-Leisegang energy loss straggling distribution. Nucl Instr Meth. 1980;174:531–3.
Kase KR, Nelson WR. Concepts of radiation dosimetry. New York: Pergamon Press; 1978.
Kölbig KS, Schorr B. Asymptotic expansions for the Landau density and distribution functions. Comput Phys Commun. 1984;32:121–31.
Landau LD. On the energy loss of fast particles by ionization. Jour Phys USSR. 1944;8:201.
Moyal JE. Theory of ionization fluctuations. Phil Mag. 1955;46:263–80.
Rossi B. High-energy particles. New York: Prentice-Hall; 1952.
Rotondi A, Montagna P. Fast calculation of Vavilov distribution. Nucl Instr Meth in Phys Res. 1990;B47:215–23.
Sigmund P. Particle penetration and radiation effects. Berlin: Springer; 2006.
Symon KR (1948) Fluctuations in energy lost by high energy charged particles in passing through matter. Thesis, Harvard University (unpublished)
Van Ginneken A. Edgeworth series for collision energy loss and multiple scattering. Nucl Instr Meth. 2000;B160:460–70.
Vavilov PV. Ionization losses of high-energy heavy particles. Sov Phys JETP. 1957;5:749–51.

Appendices

Contents

B.J. McParland, *Medical Radiation Dosimetry*,
DOI 10.1007/978-1-4471-5403-7, © Springer-Verlag London 2014

Abstract These eight appendices provide background derivations and reviews of mathematical concepts used in the book.

A Integration of Complex Variables

A.1 Introduction

The integration of a function in complex variables is a powerful tool with innumerable applications in applied physics, engineering and mathematics. In this Appendix, summaries of complex integration methods as applied to the derivations provided in this book are provided.

A.2 Holomorphicity

A function f of the complex variable $z = x + iy$ is defined as being holomorphic, or analytic, at the point z_0, if its derivative exists at z_0 and at each point z in the neighbourhood of z_0. The first of two necessary (but not sufficient) conditions for holomorphicity is that within a domain D that $f(z)$ be continuous throughout the domain. The second is that the Cauchy–Riemann equations are satisfied; i.e. for the function defined as $f(z) = u(x,y) + iv(x,y)$ in the neighbourhood of the point $z_0 = x_0 + iy_0$, then the existence of the derivative $\frac{\mathrm{d}f}{\mathrm{d}z}$ at z_0 requires that the Cauchy–Riemann equations

$$\frac{\mathrm{d}u}{\mathrm{d}x}(x_0, y_0) = \frac{\mathrm{d}v}{\mathrm{d}y}(x_0, y_0) \quad \frac{\mathrm{d}u}{\mathrm{d}y}(x_0, y_0) = -\frac{\mathrm{d}v}{\mathrm{d}x}(x_0, y_0) \tag{A.1}$$

be valid. A proof of this follows. First, we define the differential operators

$$\frac{\partial}{\partial z} = \frac{\partial}{\partial x} + \frac{1}{i}\frac{\partial}{\partial y}$$
$$\frac{\partial}{\partial z^*} = \frac{\partial}{\partial x} - \frac{1}{i}\frac{\partial}{\partial y}.$$

Then,

$$\frac{\partial}{\partial z} z^m z^{*n} = m z^{m-1} z^{*n}$$
$$\frac{\partial}{\partial z^*} z^m z^{*n} = n z^m z^{*n-1}.$$

Let $w = u + iv$ be a function $f(z)$ that satisfies $\frac{\partial w}{\partial z^*} = 0$. That is,

$$\frac{\partial}{\partial z^*}(u + iv) = \frac{\partial (u + iv)}{\partial x} - \frac{1}{i}\frac{\partial (u + iv)}{\partial y}$$
$$= 0.$$

$f(z)$ is a holomorphic function of z in this region and the Cauchy–Riemann equations follow.

If $f(z)$ is holomorphic in the neighbourhood of z_0 but fails to be holomorphic at z_0, then z_0 is called a *singularity* of $f(z)$. For example, consider the function $f(z) = \frac{1}{z}$. Clearly, $z = 0$ represents a singularity of the function, and $f(z)$ is holomorphic at every other point.

A.3 Cauchy–Goursat Theorem

If the function f is holomorphic at all points within and on a simple closed and positively oriented[1] contour C, then the Cauchy–Goursat theorem states that the integral of the function over the contour is

$$\oint_C \mathrm{d}z f(z) = 0. \tag{A.2}$$

An application of the Cauchy–Goursat theorem is demonstrated here by solving the integral $\int_{-\infty+iy_0}^{\infty+iy_0} \mathrm{d}z\, \mathrm{e}^{-z^2}$ (which arises in the derivation of the description of stochastic energy loss with a Gaussian probability density function in Chap. 18). The integral is solved by writing it in the form $\int_{-K+iy_0}^{K+iy_0} \mathrm{d}z\, \mathrm{e}^{-z^2}$ where K is a real constant to be later allowed to go to ∞ as a limit. An appropriate rectangular contour in the complex plane to use is shown in Fig. A.1 and is made up of four individual contours

$$C = C_1 \cup C_2 \cup C_3 \cup C_4. \tag{A.3}$$

---

[1] A positively oriented, or counterclockwise, curve is one in which when one travels on it, the interior is to one's left.

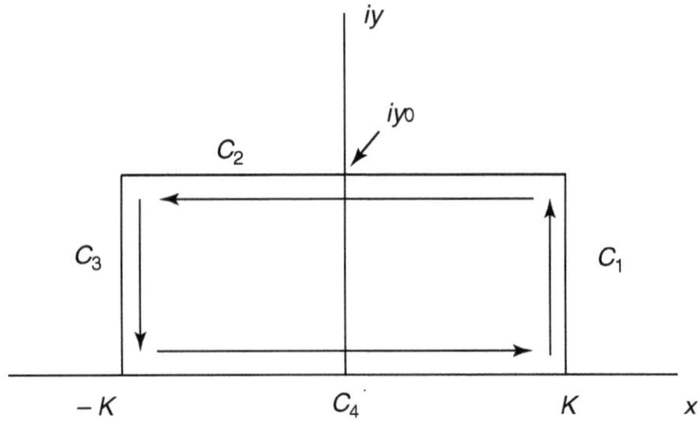

Fig. A.1 Contour in the complex plane used to solve the integral $\int_{-\infty+iy_0}^{\infty+iy_0} \mathrm{d}z\, \mathrm{e}^{-z^2}$.

where the individual contours are

$$\begin{aligned} C_1 &: (K \rightarrow K + iy_0) \\ C_2 &: (K + iy_0 \rightarrow -K + iy_0) \\ C_3 &: (-K + iy_0 \rightarrow -K) \\ C_4 &: (-K \rightarrow +K). \end{aligned} \tag{A.4}$$

Hence, it is the negative of the contour integral over C_2 that we wish to solve for. As e^{-z^2} is holomorphic on and within this contour, the Cauchy–Goursat theorem of (A.2) holds, and we can write the contour integral as the sum of the integrals over the individual components of the total contour which must equal to zero:

$$\begin{aligned} \oint_C \mathrm{d}z\, \mathrm{e}^{-z^2} &= \int_{C_1} \mathrm{d}z\, \mathrm{e}^{-z^2} + \int_{C_2} \mathrm{d}z\, \mathrm{e}^{-z^2} + \int_{C_3} \mathrm{d}z\, \mathrm{e}^{-z^2} + \int_{C_4} \mathrm{d}z\, \mathrm{e}^{-z^2} \\ &= 0. \end{aligned} \tag{A.5}$$

Isolating the integral over C_2 gives

$$\int_{C_2} \mathrm{d}z\, \mathrm{e}^{-z^2} = -\left(\int_{C_1} \mathrm{d}z\, \mathrm{e}^{-z^2} + \int_{C_3} \mathrm{d}z\, \mathrm{e}^{-z^2} + \int_{C_4} \mathrm{d}z\, \mathrm{e}^{-z^2} \right). \tag{A.6}$$

Hence, we must solve the integrals over C_1, C_3 and C_4 for K eventually going to infinity. The integral over C_1 is

$$\begin{aligned} \int_{C_1} \mathrm{d}z\, \mathrm{e}^{-z^2} &= \int_{K}^{K+iy_0} \mathrm{d}z\, \mathrm{e}^{-z^2} \\ &= i \int_0^{y_0} \mathrm{d}y\, \mathrm{e}^{-(K+iy)^2} \\ &= 0 \qquad \text{as } K \rightarrow \infty \end{aligned} \tag{A.7}$$

a result which is due to the e^{-K^2} factor. This will also be the same result for the integration along the contour C_3. The remaining integral is that over the contour C_4, for which the integration variable is real,

$$\begin{aligned}\int_{C_4} dz\, e^{-z^2} &= \int_{-K}^{+K} dx\, e^{-x^2} \\ &= 2\int_0^{\infty} dx\, e^{-x^2} \quad \text{for } K \to \infty. \end{aligned} \tag{A.8}$$

This integral of (A.8) is known as Laplace's integral and is solved by first squaring it:

$$\begin{aligned}\left(\int_0^{\infty} dx\, e^{-x^2}\right)^2 &= \int_0^{\infty} da\, e^{-a^2} \int_0^{\infty} db\, e^{-b^2} \\ &= \int_0^{\infty} da \int_0^{\infty} db\, e^{-\left(a^2+b^2\right)} \end{aligned} \tag{A.9}$$

and converting the resulting double integral into polar co-ordinates and using the fact that $d(r^2) = 2r\, dr$. Following the substitution of variable $y = r^2$, we obtain

$$\begin{aligned}\int_0^{\infty} da \int_0^{\infty} db\, e^{-\left(a^2+b^2\right)} &= \frac{\pi}{2}\int_0^{\infty} dr\, r\, e^{-r^2} \\ &= \pi \end{aligned} \tag{A.10}$$

to arrive at

$$\int_0^{\infty} dx\, e^{-x^2} = \sqrt{\pi}. \tag{A.11}$$

Substituting (A.11) into (A.8) to give

$$\begin{aligned}\int_{C_4} dz\, e^{-z^2} &= 2\sqrt{\pi} \\ &= -\int_{C_2} dz\, e^{-z^2}. \end{aligned} \tag{A.12}$$

Hence,

$$\int_{-\infty+iy_0}^{\infty+iy_0} dz\, e^{-z^2} = -\int_{C_2} dz\, e^{-z^2}; \quad K \to \infty$$
$$= 2\sqrt{\pi}. \tag{A.13}$$

A.4 *Residue Theorem*

However, it is possible for the complex function to fail to be holomorphic at one or more points within a given contour, and the Cauchy–Goursat theorem will no longer hold. Each of these singularities will contribute a number to the net value of the resulting contour integral. As demonstrated below, each of these numbers is referred to as *residues* of the function. From a prior example, the function $\frac{1}{z}$ is analytic everywhere except at $z = 0$. As this is the only singularity (or *pole*) to the function, it is said to be an isolated singularity. In another example, the function $\frac{1}{(1+z^2)}$ has two isolated singularities at $z = \pm i$. On the other hand, the function csc $\left(\frac{\pi}{z}\right)$ has singularities at $z = \frac{1}{n}$ where $n = 0, \pm 1, \pm 2\ldots$. As these singularities lie along the real axis between $z = \pm 1$, then every singularity is isolated except for that of $z = 0$. In this case, every neighbourhood around $z = 0$ contains other singular points of $\csc\left(\frac{\pi}{z}\right)$.

Consider the function $f(z)$ which has an isolated singularity at z_0. There will be a positive number r which defines the region within which $f(z)$ is holomorphic at each point z within the circular domain defined by $0 < |z - z_0| < r$. Within this domain, the function can be expanded in a Laurent series

$$f(z) = \sum_{n=-\infty}^{\infty} a_n (z - z_0)^n. \tag{A.14}$$

The series' coefficients are given by the contour integrals

$$a_n = \frac{1}{2\pi i} \oint_C dz \frac{f(z)}{(z - z_0)^{n+1}} \quad n = 0, \pm 1, \pm 2, \ldots, \tag{A.15}$$

where C is a closed path about the singularity at z_0. In particular, the coefficient for the index of $n = -1$

$$a_{-1} = \frac{1}{2\pi i} \oint_C dz f(z)$$
$$= \mathrm{Res}(f(z)) \tag{A.16}$$

is termed the *residue* of $f(z)$ at z_0, or

$$\mathrm{Res}(f(z)) \equiv \lim_{z \to z_0} (z - z_0) f(z). \quad (A.17)$$

The power of this result becomes apparent when (A.15) is rearranged to give

$$\oint_C \mathrm{d}z f(z) = 2\pi i\, \mathrm{Res}(f(z)). \quad (A.18)$$

This result can be extended to the case of multiple M singularities of the function $f(z)$ within the contour C:

$$\oint_C \mathrm{d}z f(z) = 2\pi i \sum_{m=1}^{M} \mathrm{Res}_m(f(z)). \quad (A.19)$$

An example of the application of this *residue theorem* arises in the derivation of the energy transfer in the Bohr theory of soft collisions in which the integral of the type $\int_{-\infty}^{\infty} \frac{\mathrm{d}x}{1+nx^2}$, where n is an integer, is to be solved. The integral can be rewritten as

$$\frac{1}{n} \int_{-\infty}^{\infty} \frac{\mathrm{d}z}{\left(\frac{1}{n}\right) + z^2} \quad (A.20)$$

for $z = x + iy$. The integrand has multiple singularities at $z = \pm \frac{i}{\sqrt{n}}$. For the example demonstrating the residue theorem considered here, we will consider $n = 1$ for which the singularities are $z = \pm i$. The inequality of $\left|\frac{1}{1+z^2}\right| \leq 1$ suggests that we consider a semi-circular contour of radius R in the upper half plane of the complex plane as shown in Fig. A.2. The contour runs from –R to +R along the real axis

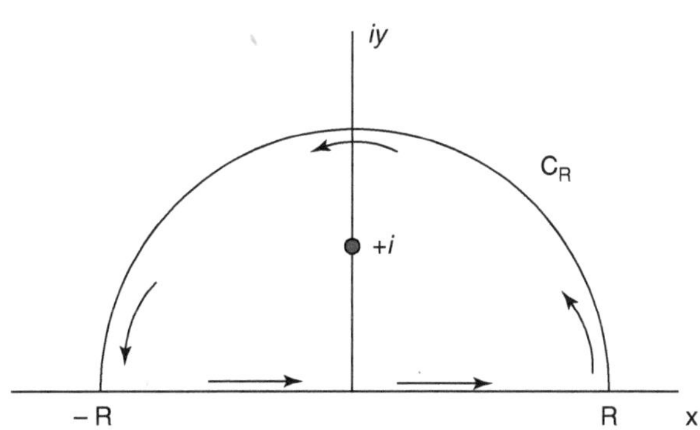

Fig. A.2 Contour in the complex plane used to solve the integral $\int_{-\infty}^{\infty} \frac{\mathrm{d}z}{1+z^2}$.

and then along the semicircle $z = R\,e^{i\theta}$ for θ going from 0 to π (i.e. the contour C_{R}). We have, initially, $1 < \infty$ in order to include the singularity within the contour. Eventually, we will let $R \to \infty$. Integrating in the positive sense over this contour,

$$\int_{-R}^{R} \frac{\mathrm{d}x}{1+x^2} + \int_{C_R} \frac{\mathrm{d}z}{1+z^2} = 2\pi i \,\mathrm{Res}\left(\frac{1}{1+z^2}\right) \tag{A.21}$$

where the residue is

$$\begin{aligned} \mathrm{Res}\left(\frac{1}{1+z^2}\right) &= \lim_{z \to i} (z-i)\frac{1}{1+z^2} \\ &= \lim_{z \to i} (z-i)\frac{1}{(z+i)\,(z-i)} \\ &= \lim_{z \to i} \frac{1}{(z+i)} \\ &= -\frac{i}{2} \end{aligned} \tag{A.22}$$

This gives

$$\int_{-R}^{R} \frac{\mathrm{d}x}{1+x^2} + \int_{C_{\mathrm{R}}} \frac{\mathrm{d}z}{1+z^2} = \pi$$

or,

$$\int_{-R}^{R} \frac{\mathrm{d}x}{1+x^2} = \pi - \int_{C_{\mathrm{R}}} \frac{\mathrm{d}z}{1+z^2} \tag{A.23}$$

That the integral on the right-hand side will reduce to zero as $R \to \infty$ can be demonstrated by considering the inequality

$$\int_{C_{\mathrm{R}}} \frac{\mathrm{d}z}{1+z^2} \leq \frac{\pi R}{1+R^2} \tag{A.24}$$

As the right-hand side is proportional to $\frac{1}{R}$, it thus goes to zero as $R \to \infty$. Hence, the solution for this integral is

$$\int_{-\infty}^{\infty} \frac{\mathrm{d}x}{1+x^2} = \pi. \tag{A.25}$$

A.5 Cauchy Principal Value

The integrals investigated above were examples of the *improper integral* which is defined as an integral in which one or more of the limits is infinite or where the integrand has one or more infinities over the range of integration. The improper integral $\int_{-\infty}^{\infty} \mathrm{d}xf(x)$ is said to be *convergent* if $f(x)$ is continuous for all x and has the value

$$\int_{-\infty}^{\infty} \mathrm{d}xf(x) = \lim_{R\to\infty} \int_{-R}^{\infty} \mathrm{d}xf(x) + \lim_{R\to\infty} \int_{\infty}^{R} \mathrm{d}xf(x), \tag{A.26}$$

if the individual limits exist. The *Cauchy principal value* of $\int_{-\infty}^{\infty} \mathrm{d}xf(x)$ is

$$\mathrm{PV}\int_{-\infty}^{\infty} \mathrm{d}xf(x) = \lim_{R\to\infty} \int_{-R}^{R} \mathrm{d}xf(x) \tag{A.27}$$

provided that the limit on the right-hand side of the equation exists. Thus, if $\int_{-\infty}^{\infty} \mathrm{d}xf(x)$ converges, the value obtained is also the Cauchy principal value. A useful point to note, especially in the earlier integrations performed, is that if $f(x)$ can be written as the ratio of two real polynomials with no factors in common, i.e. $f(x) = \frac{g(x)}{h(x)}$, and $h(x)$ has no real zeros, then the integral $\int_{-\infty}^{\infty} \mathrm{d}xf(x)$ will converge if the degree of the polynomial $h(x)$ is at least two greater than that of $g(x)$. The integral is, as shown earlier, readily solved by determining its Cauchy principal value through the use of residue theory.

The Sokhotski–Plemelj theorem, which uses the Cauchy principal value, is of use in our derivation of the longitudinal dielectric constant in our calculation of the Bethe soft-collision stopping power. We do not provide a proof of it, but the reader is referred to Pipkin (1991) for such. We state simply that

$$\lim_{\varepsilon\to 0} \frac{1}{x \pm i\varepsilon} = \mathrm{PV}\frac{1}{x} \mp i\pi\delta(x). \tag{A.28}$$

A.6 Darboux Inequality

To extend the use of inequalities in contour integration as shown in (A.22) and (A.23), let us consider the general contour integral

$$I = \oint_C \mathrm{d}z f(z) \tag{A.29}$$

The contour C is piecewise regular and $f(z)$ is assumed bound on C. This contour can be considered made up of discrete sides, the number of which goes to infinity or

$$\oint_C \mathrm{d}z f(z) = \lim_{n \to \infty} \sum_{k=1}^{n} (z_k - z_{k-1}) f\left(\frac{z_k + z_{k-1}}{2}\right) \tag{A.30}$$

Let $|f_{\max}|$ be the maximum modulus of $f(z)$ on C. Then,

$$\oint_C \mathrm{d}z f(z) \leq |f_{\max}| \lim_{n \to \infty} \sum_{k=1}^{n} (z_k - z_{k-1}) \tag{A.31}$$

The summation in (A.31) is the perimeter of the polygon inscribed within C and is therefore smaller than the arc length L of the contour. Hence, the inequality of (A.31) can be written as

$$\left| \oint_C \mathrm{d}z f(z) \right| \leq |f_{\max}| L \tag{A.32}$$

This result is known as Darboux inequality.

A.7 Branch Cuts and Branch Points

We will be required to evaluate integrals using, for example, the residue theorem in the region of entities known as branch cuts and branch points which will require special consideration. We first define these entities.

Consider the complex number and its Euler representation $z = x + iy = re^{i\theta}$ where the modulus and argument are

$$\begin{aligned} |z| &= r \\ \arg z &= \tan^{-1}\frac{x}{y} = \theta. \end{aligned} \tag{A.33}$$

arg z is a multivalued function with an infinite number of values differing by integral multiples of 2π. By convention, arg z (i.e. capitalised) is the principal value of arg z for $-\pi \leq \arg z \leq \pi$. We can also write Θ as the principal value of θ over the same range, in which case,

$$\theta = \Theta + 2n\pi; \quad n = 0, \pm 1, \pm 2, \ldots \tag{A.34}$$

Thus, if $\ln r$ is the logarithm of a positive real number, then

$$\begin{aligned} \ln z &= \ln r + i\theta \\ &= \ln r + i(\Theta + 2n\pi); \quad n = 0, \pm 1, \pm 2, \ldots . \end{aligned} \tag{A.35}$$

We can subsequently define the principal value of the logarithm of z for $n = 0$,

$$\ln z = \ln r + i\Theta; \quad (r > 0; -\pi \leq \Theta \leq \pi). \tag{A.36}$$

This can be demonstrated by rewriting this in terms of components

$$\begin{aligned} u(r, \Theta) &= \ln r; \\ v(r, \Theta) &= \Theta \end{aligned}$$

we see that these have continuous first partial derivatives:

$$\begin{aligned} \frac{\partial u(r, \Theta)}{\partial r} &= \frac{1}{r} \\ \frac{\partial v(r, \Theta)}{\partial \Theta} &= 1. \end{aligned}$$

Hence, we can conclude that $\ln z$ is holomorphic within the domain of $r > 0$ and $-\pi \leq \Theta \leq \pi$ and is, hence, the principal value of $\ln z$.

A branch of a multivalued function $f(z)$ is a single value of the function $F(z)$ which is analytic within a domain at the points z for which $F(z)$ is one of the values of $f(z)$. For example, $\ln z$ in the domain of $r > 0$ and $-\pi \leq \Theta \leq \pi$ is a branch of $\ln z$ given by (A.31). This is referred to as the principal branch. The function

$$\ln z = \ln r + i\theta \quad (r > 0; \quad \alpha < \theta < \alpha + 2\pi), \tag{A.37}$$

where θ is arbitrary, is single valued and continuous throughout the specified domain and is a branch of $\ln z$.

Recall the definition of a singularity as being the point at which a function fails to be holomorphic. Every point along the negative real axis (i.e. $\Theta = \pi$), including the origin, is a singularity of the principal branch $\ln z$. The line along $\Theta = \pi$ is referred to as the branch cut for the principal branch: the branch cut is a line or curve of singularities used to define the branch of a multivalued function. Similarly,

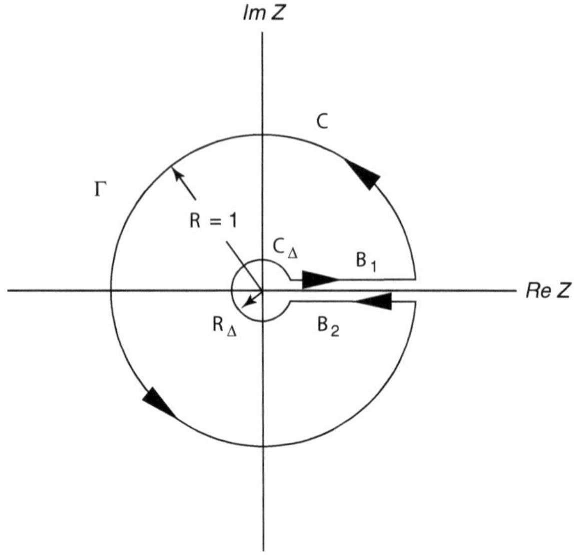

Fig. A.3 Contour for $\oint_C dz\,\sqrt{z}$ where the contour is the unit circle

the line $\theta = \alpha$ will represent a branch cut for (A.37). For both of these functions, the singularity at the origin (i.e. $r = 0$) is common to all branch cuts for $\ln z$ and is referred to as a branch point.

We now consider two applications of the above discussion in performing contour integration. The first example is that of $\oint_C dz\,\sqrt{z}$ where the contour is the unit circle (i.e. $r = 0$; $0 \le \theta \le 2\pi$). As the derivative of $\sqrt{z}$ is not holomorphic at the origin, the Cauchy–Goursat theorem is not applicable. The branch point is the origin, and the branch cut is taken along the real axis. The contour to be integrated over, Γ, is shown in Fig. A.3.

The contour

$$\Gamma = C \cup B_1 \cup C_\Delta \cup B_2 \tag{A.38}$$

is configured so as to exclude the origin in order to allow the Cauchy–Goursat theorem to be applicable:

$$\oint_\Gamma dz\sqrt{z} = 0. \tag{A.39}$$

From (A.38), the contour integral of (A.39) can be decomposed into

$$\oint_\Gamma dz\,\sqrt{z} = \oint_C dz\,\sqrt{z} + \int_{B_1} dz\,\sqrt{z} + \oint_{C_\Delta} dz\,\sqrt{z} + \int_{B_2} dz\,\sqrt{z} \tag{A.40}$$

where it is understood that the radius of the contour C_Δ, r_Δ, will be eventually set equal to zero. Using the Darboux inequality, we have

$$\oint_{C_\Delta} \mathrm{d}z\,\sqrt{z} \leq 2\pi r_\Delta \sqrt{r_\Delta}. \tag{A.41}$$

Clearly, $\oint_{C_\Delta} \mathrm{d}z\,\sqrt{z} \to 0$ as $r_\Delta \to 0$. This leaves us, from (A.39) and (A.40),

$$\begin{aligned}\oint_C \mathrm{d}z\,\sqrt{z} &= -\left(\int_{B_1} \mathrm{d}z\,\sqrt{z} + \int_{B_2} \mathrm{d}z\,\sqrt{z}\right)\\ &= -\left(\int_1^0 \mathrm{d}x\,\sqrt{x}\,\mathrm{e}^{i\pi} + \int_0^1 \mathrm{d}x\,\sqrt{x}\right)\\ &= -2\int_0^1 \mathrm{d}x\sqrt{x}\\ &= -\frac{4}{3}.\end{aligned} \tag{A.42}$$

B Mathematical Transforms

In medical physics, the applications of the Fourier, Laplace and Hankel transforms are well known in signal theory, image analysis, etc. They are also useful tools for solving differential equations and are employed frequently in this book for that purpose.

B.1 Fourier Transform

Consider the periodic function, $f(t)$, with period τ, i.e. $f(t) = f(t + \tau)$. This function can be represented by a complex exponential (Fourier) series

$$\tilde{f}(t) = \sum_{n=-\infty}^{\infty} f_n \mathrm{e}^{in\omega_0 t} \tag{B.1}$$

where $\omega_0 = \frac{2\pi}{\tau}$ and where the Fourier coefficients are

$$f_n = \frac{1}{\tau} \int_{-\frac{\tau}{2}}^{\frac{\tau}{2}} \mathrm{d}t f(t)\, \mathrm{e}^{-in\omega_0 t}. \tag{B.2}$$

Equation (B.2) is obtained by minimising the squared error between the function and its Fourier series representation:

$$\varepsilon = \int_{-\frac{\tau}{2}}^{\frac{\tau}{2}} \mathrm{d}t \left|f(t) - \tilde{f}(t)\right|^2. \tag{B.3}$$

In order to generalise the Fourier series representation so as to enable its applicability to an aperiodic $f(t)$, we substitute (B.2) into (B.1):

$$\tilde{f}(t) = \sum_{n=-\infty}^{\infty} \left(\frac{1}{\tau} \int_{-\frac{\tau}{2}}^{\frac{\tau}{2}} \mathrm{d}t' f(t')\, \mathrm{e}^{-in\omega_0 t'} \right) \mathrm{e}^{in\omega_0 t}. \tag{B.4}$$

In order to represent $f(t)$ over all time, we let $\tau \to \infty$ so that $n\omega_0$ becomes the continuous variable ω, $\frac{1}{\tau}$ becomes the differential $\frac{\mathrm{d}\omega}{2\pi}$, and the summation of (B.4) thus becomes an integral. We then replace $\tilde{f}(t)$ with $f(t)$ and arrive at a representation of the original function:

$$\begin{aligned} f(t) &= \frac{1}{2\pi} \int_{-\infty}^{\infty} \mathrm{d}\omega \left(\int_{-\infty}^{\infty} \mathrm{d}t' f(t')\, \mathrm{e}^{-i\omega t'} \right) \mathrm{e}^{i\omega t} \\ &= \frac{1}{\sqrt{2\pi}} \int_{-\infty}^{\infty} \mathrm{d}\omega \left(\frac{1}{\sqrt{2\pi}} \int_{-\infty}^{\infty} \mathrm{d}t' f(t')\, \mathrm{e}^{-i\omega t'} \right) \mathrm{e}^{i\omega t} \end{aligned} \tag{B.5}$$

The last action of writing $2\pi = \sqrt{2\pi}\sqrt{2\pi}$ is done so as to yield symmetry between the Fourier and inverse Fourier transforms. This normalisation is commonly used in physics, but is not universal. The quantity within the parentheses is defined as the Fourier transform of $f(t)$

$$F(\omega) = \frac{1}{\sqrt{2\pi}} \int_{-\infty}^{\infty} \mathrm{d}t' f(t')\, \mathrm{e}^{-i\omega t'}, \tag{B.6}$$

thus allowing Eq. (B.5) to be written in the form

$$f(t) = \frac{1}{\sqrt{2\pi}} \int_{-\infty}^{\infty} \mathrm{d}\omega\, F(\omega)\, \mathrm{e}^{i\omega t}. \tag{B.7}$$

The $F(\omega)$ of (B.6) is known as the Fourier transform of $f(t)$, and the conversion from the transformed domain back to the original domain is the inverse Fourier transform of (B.7). For convenience, we will frequently write these transforms in the form of operators

$$F(\omega) = F[f(t)] \tag{B.8}$$

and

$$f(t) = F^{-1}[F(\omega)]. \tag{B.9}$$

The useful tool represented by the Fourier transform is that it can change a differential equation to an algebraic equation which is subsequently solved, and following an inverse Fourier transform, the solution to the differential equation is obtained. First, we determine the Fourier transform of the derivative of $f(t)$.

$$F\left[\frac{\mathrm{d}f(t)}{\mathrm{d}t}\right] = \frac{1}{\sqrt{2\pi}} \int_{-\infty}^{\infty} \mathrm{d}t' \frac{\mathrm{d}f(t')}{\mathrm{d}t'}\, \mathrm{e}^{-i\omega t'}. \tag{B.10}$$

The integral is solved by parts

$$\int \mathrm{d}sr = [rs] - \int \mathrm{d}rs$$

where

$$\mathrm{d}s = \mathrm{d}t' \frac{\mathrm{d}f(t')}{\mathrm{d}t'} \qquad r = \frac{\mathrm{e}^{-i\omega t'}}{\sqrt{2\pi}}$$

$$s = f(t') \qquad \mathrm{d}r = -i\omega \frac{\mathrm{e}^{-i\omega t'}}{\sqrt{2\pi}} \mathrm{d}t'.$$

Hence,

$$F\left[\frac{\mathrm{d}f(t)}{\mathrm{d}t}\right] = \left[f(t') \frac{\mathrm{e}^{-i\omega t'}}{\sqrt{2\pi}}\right]_{t'=-\infty}^{t'=\infty} + \frac{i\omega}{\sqrt{2\pi}} \int_{-\infty}^{\infty} \mathrm{d}t' f(t')\, \mathrm{e}^{-i\omega t'}. \tag{B.11}$$

The quantity in the brackets is oscillatory, but if we demand that $f(t) \to 0$ as $t \to \pm\infty$ (which will be the physical case), then that quantity goes to zero, and we are left with the result that

$$F\left[\frac{\mathrm{d}f(t)}{\mathrm{d}t}\right] = i\omega F[f(t)] \tag{B.12}$$

and, by repeated application for higher-order derivatives,

$$F\left[\frac{\mathrm{d}^n f(t)}{\mathrm{d}t^n}\right] = (i\omega)^n F[f(t)]. \tag{B.13}$$

An important series of Fourier transforms of derivatives will be used throughout this book where function is the phasor:

$$f(\mathbf{r}, t) = f_0\, \mathrm{e}^{i(\mathbf{k}\bullet\mathbf{r} - \omega t)}. \tag{B.14}$$

These are

$$\begin{aligned}
F[\nabla f(\mathbf{r}, t)] &= i\mathbf{k} f(\mathbf{k}, \omega) \\
F\left[\nabla^2 f(\mathbf{r}, t)\right] &= -k^2 f(\mathbf{k}, \omega) \\
F\left[\frac{\partial f(\mathbf{r}, t)}{\partial t}\right] &= -i\omega f(\mathbf{k}, \omega) \\
F\left[\frac{\partial^2 f(\mathbf{r}, t)}{\partial t^2}\right] &= -\omega^2 f(\mathbf{k}, \omega) \\
F\left(\nabla \frac{\partial f(\mathbf{r}, t)}{\partial t}\right) &= \omega \mathbf{k} f(\mathbf{k}, \omega)
\end{aligned} \tag{B.15}$$

where

$$f(\mathbf{k}, \omega) = F[f(\mathbf{r}, t)]. \tag{B.16}$$

B.2 Laplace Transform

The Laplace and Fourier transforms are related, and both can be used for solving differential equations. The difference between the two transform types is that the Fourier transform of a function describes it in terms of frequency (or, in general, the reciprocal of the function's argument). On the other hand, the Laplace transform resolves the function in terms of its moments.

Let us reconsider the Fourier transform of (B.6) for which we are interested in the transform of functions which exists only for positive t. We can write the Laplace transform as

$$F(s) = \int_0^\infty dt' f(t') e^{st'} \tag{B.17}$$

where s is a complex parameter

$$s = \sigma + i\omega. \tag{B.18}$$

The normalisation factor of $\sqrt{2\pi}$ is no longer required by this definition. As with the Fourier transform, the Laplace transform of a derivative is multiplicative with the Laplace transform of the original function.

$$L\left[\frac{df(t)}{dt}\right] = sL[f(t)] - f(0) \tag{B.19}$$

where $f(t)$ is assumed to be differentiable and $f(0)$ is set by the boundary condition for $t = 0$. Extending this result,

$$L\left[\frac{d^n f(t)}{dt^n}\right] = s^n F(s) - s^{(n-1)} f(0) - s^{(n-2)} \frac{df(0)}{dt} - \cdots \frac{d^{(n-1)} f(0)}{dt^{(n-1)}} \tag{B.20}$$

These results can be demonstrated using the integration of parts approach used previously for the Fourier transform of a derivative.

The inverse Laplace transform is

$$L^{-1}[F(s)] = \frac{1}{2\pi i} \int_{K-i\infty}^{K+i\infty} ds\, e^{st} f(s). \tag{B.21}$$

The use of the Laplace method to solving a differential equation is, in practice, very similar to that of the Fourier method. It is only followed in one case in this book, that of Vavilov's derivation of the probability distribution function of stochastic energy loss.

B.3 Fourier–Bessel Transform

The Fourier–Bessel transform, also known as the zeroth-order Hankel transform, is used to solve the transport equation in the derivation of Molière's theory of multiple scattering. The νth-order Hankel transform pair is

$$F_\nu(s) \equiv \mathrm{Ha}_\nu[f(r)] = \int_0^\infty dr\, r J_\nu(sr) f(r) \tag{B.22}$$

$$f(r) \equiv \mathrm{Ha}_\nu^{-1}[F_\nu(s)] = \int_0^\infty ds\, s J_\nu(sr) F_\nu(s). \tag{B.23}$$

$J_\nu(x)$ is the Bessel function of the first kind and order ν. It is derived in Appendix C. The transform for $\nu = 0$ is equivalent to the two-dimensional Fourier transform using a radially symmetric integral kernel and is referred to as the Fourier–Bessel transform. For example, consider the two-dimensional Fourier transform

$$\begin{aligned} g(u,w) &= F[f(r)] \\ &= \int_{-\infty}^{\infty}\int_{-\infty}^{\infty} \mathrm{d}x\,\mathrm{d}y f(r)\mathrm{e}^{-i2\pi(ux+wy)}. \end{aligned} \tag{B.24}$$

Now, change to polar co-ordinates

$$x + iy = r\,\mathrm{e}^{i\theta} \tag{B.25}$$

$$u + iw = q\,\mathrm{e}^{i\phi}. \tag{B.26}$$

The transform can then be written as

$$\begin{aligned} g(q) &= \int_0^\infty \mathrm{d}r\, rf(r) \int_0^{2\pi} \mathrm{d}\theta\, \mathrm{e}^{-i2\pi qr(\cos\theta\cos\phi+\sin\theta\sin\phi)} \\ &= \int_0^\infty \mathrm{d}r\, rf(r) \int_0^{2\pi} \mathrm{d}\theta f(r)\, \mathrm{e}^{-i2\pi qr\cos(\theta-\phi)}. \end{aligned} \tag{B.27}$$

Applying the change of variable, $\varphi = \theta - \phi$,

$$\begin{aligned} g(q) &= \int_0^\infty \mathrm{d}r\, rf(r) \int_{-\phi}^{2\pi-\phi} \mathrm{d}\varphi\, \mathrm{e}^{-i2\pi qr\cos\phi} \\ &= \int_0^\infty \mathrm{d}r\, rf(r) \left(\int_0^{2\pi} \mathrm{d}\theta\, \mathrm{e}^{-i2\pi qr\cos\theta} \right) \\ &= 2\pi \int_0^\infty \mathrm{d}r\, rf(r) J_0(2\pi qr). \end{aligned} \tag{B.28}$$

B.4 Convolution Theorem

The convolution theorem is valid for both Fourier and Laplace transforms. As we will employ it in the development of the Vavilov theory of an asymmetric probability distribution function for energy loss, we will present and prove the theorem here in the context of the Laplace transform. Consider two functions, $f(t)$ and $g(t)$, and their corresponding Laplace transforms, $F(s)$ and $G(s)$. The convolution of $f(t)$ and $h(t)$ is

$$
\begin{aligned}
h(t) &= f(t) * g(t) \\
&= \int_0^t \mathrm{d}\tau f(\tau)\, g(t-\tau) \quad \text{or} \\
&= \int_0^t \mathrm{d}\tau g(\tau) f(t-\tau).
\end{aligned}
\tag{B.29}
$$

The product of the Laplace transforms are

$$
\begin{aligned}
F(s)\, G(s) &= \int_0^\infty \mathrm{d}t\, \mathrm{e}^{-st} f(t) \int_0^\infty dt' \mathrm{e}^{-st'} g(t') \\
&= \int_0^\infty \mathrm{d}t' \int_0^\infty \mathrm{d}t \mathrm{e}^{-s(t+t')} f(t) g(t') \\
&= \int_0^\infty \mathrm{d}t' \int_{t'}^\infty \mathrm{d}\xi \mathrm{e}^{-s\xi} f(\xi - t') g(t')
\end{aligned}
\tag{B.30}
$$

where $\xi = t + t'$. Reversing the order of integration,

$$
\begin{aligned}
F(s)G(s) &= \int_0^\infty \mathrm{d}\xi\, \mathrm{e}^{-s\xi} \left(\int_0^\infty \mathrm{d}t' f(\xi - t') g(t') \right) \\
&= L[f(t) * g(t)].
\end{aligned}
\tag{B.31}
$$

C Relevant Functions and Polynomials

Throughout this book, a number of functions and polynomial expressions are either unique or are used frequently. To maintain clarity within the main text of the book, these are derived and summarised in this Appendix.

C.1 Bessel Functions

The Bessel functions and equation are named after Friedrich Bessel (1784–1846), the astronomer and mathematician recognised for, among other accomplishments, having been the first to determine a stellar parallax. However, the eponymous functions and equation described in this Appendix are attributable to Daniel Bernoulli (1700–1782). They are derived from the solution of Laplace's equation

$$\nabla^2\Phi = 0 \tag{C.1}$$

in cylindrical co-ordinates.

The Bessel equation is obtained by first writing the Laplace equation in cylindrical co-ordinates

$$\frac{\partial^2\Phi}{\partial r^2} + \frac{1}{r}\frac{\partial\Phi}{\partial r} + \frac{1}{r^2}\frac{\partial^2\Phi}{\partial\phi^2} + \frac{\partial^2\Phi}{\partial z^2} = 0 \tag{C.2}$$

and solving through the method of the separation of variables

$$\Phi(r.\phi, z) = R(r)\,\Xi(\phi)\,Z(z). \tag{C.3}$$

With this, Laplace's equation can now be written as

$$\Xi(\phi)Z(z)\frac{\mathrm{d}^2R(r)}{\mathrm{d}r^2} + \frac{\Xi(\phi)\,Z(z)}{r}\frac{\mathrm{d}R(r)}{\mathrm{d}r} + \frac{R(r)\,Z(z)}{r^2}\frac{\mathrm{d}^2\Xi(\phi)}{\mathrm{d}\phi^2} + R(r)\Xi(\phi)\frac{\mathrm{d}^2Z(z)}{\mathrm{d}z^2} = 0. \tag{C.4}$$

Dividing through by $R(r)\Xi(\phi)\,Z(z)$ and rearranging,

$$\frac{1}{R(r)}\frac{\mathrm{d}^2R(r)}{\mathrm{d}r^2} + \frac{1}{R(r)r}\frac{\mathrm{d}R(r)}{\mathrm{d}r} + \frac{1}{r^2}\frac{1}{\Xi(\phi)}\frac{\mathrm{d}^2\Xi(\phi)}{\mathrm{d}\phi^2} = -\frac{1}{Z(z)}\frac{\mathrm{d}^2Z(z)}{\mathrm{d}z^2}. \tag{C.5}$$

As the left-hand side of (C.5) is independent of z, the right-hand side must thus equal a constant. For convenience, we write this constant as k^2:

$$\frac{1}{R(r)}\frac{\mathrm{d}^2R(r)}{\mathrm{d}r^2} + \frac{1}{R(r)r}\frac{\mathrm{d}R(r)}{\mathrm{d}r} + \frac{1}{r^2}\frac{1}{\Xi(\phi)}\frac{\mathrm{d}^2\Xi(\phi)}{\mathrm{d}\phi^2} = -k^2. \tag{C.6}$$

Multiplying through by r^2 and, again, rearranging to isolate the angular variable,

$$\frac{r^2}{R(r)}\frac{\mathrm{d}^2R(r)}{\mathrm{d}r^2} + \frac{r}{R(r)}\frac{\mathrm{d}R(r)}{\mathrm{d}r} + r^2k^2 = -\frac{1}{\Xi(\phi)}\frac{\mathrm{d}^2\Xi(\phi)}{\mathrm{d}\phi^2}. \tag{C.7}$$

As the left-hand side of (C.7) is independent of ϕ, then the right-hand side is also constant. For convenience, this result is written as $-\nu^2$.

$$\frac{r^2}{R(r)}\frac{\mathrm{d}^2R(r)}{\mathrm{d}r^2} + \frac{r}{R(r)}\frac{\mathrm{d}R(r)}{\mathrm{d}r} + r^2k^2 = \nu^2. \tag{C.8}$$

Rearranging the result gives the radial equation

$$\frac{\mathrm{d}^2R}{\mathrm{d}r^2}+\frac{1}{r}\frac{\mathrm{d}R}{\mathrm{d}r}+\left(k^2-\frac{\nu^2}{r^2}\right)R=0. \tag{C.9}$$

The parameter ν is initially taken to be an integer. As k is arbitrary, we will consider it to be a positive and real quantity. By using the change of variable $x = kr$ and multiplying through by x^2, (C.9) is written as

$$x^2\frac{\mathrm{d}^2R}{\mathrm{d}x^2}+x\frac{\mathrm{d}R}{\mathrm{d}x}+\left(x^2-\nu^2\right)R=0 \tag{C.10}$$

which is Bessel's equation.

Solutions to Bessel's equation are determined using the Frobenius method, beginning by assuming a power series form

$$R(x)=x^{\alpha}\sum_{j=0}^{\infty}a_jx^j. \tag{C.11}$$

Substituting this, and its first and second derivatives, into (C.10) gives

$$\sum_{j=0}^{\infty}(\alpha+j)(\alpha+j-1)a_jx^{\alpha+j}+\sum_{j=0}^{\infty}(\alpha+j)a_jx^{\alpha+j}+\sum_{j=2}^{\infty}a_{j-2}x^{\alpha+j}-\nu^2\sum_{j=0}^{\infty}a_jx^{\alpha+j}=0. \tag{C.12}$$

For $j = 0$, an indicial equation is obtained:

$$a_0\left(\alpha^2-\nu^2\right)=0. \tag{C.13}$$

For the non-zero coefficient a_0, this result leads to $\alpha = \pm\nu$. Let us first consider the conditions of $\alpha = +\nu$ and $j = 1$. (C.12) gives

$$a_1(2\nu+1)=0 \tag{C.14}$$

which requires that $a_1 = 0$. For values of j of 2 and greater, continuing this procedure results in

$$\left(a_j j(2\nu+j)+a_{j-2}\right)x^{\nu+j}=0 \tag{C.15}$$

which gives the iterative solution for the coefficients

$$a_j=-\frac{a_{j-2}}{j(2\nu+j)}\quad j\geq 2. \tag{C.16}$$

By further iteration, the coefficients with even indices are $a_2 = -\frac{a_0}{4(\nu+1)}$, $a_4 = \frac{a_0}{32(\nu+1)(\nu+2)}$, etc. Thus, in general,

$$a_{2j} = \frac{(-1)^j \Gamma(1+\nu)}{2^{2j} j!\, \Gamma(1+\nu+j)} a_0 \tag{C.17}$$

where the gamma function, defined by $\Gamma(x) = \int_0^\infty dy\, y^{x-1} e^{-y}$, is used to replace the factorials in anticipation of a non-integer ν. The coefficient a_0 is conventionally set to $a_0 = \frac{1}{2^\nu \Gamma(1+\nu)}$. As $\alpha = \pm\nu$, there will be two solutions for $R(x)$, which can now be written as the series

$$J_{\pm\nu}(x) = \left(\frac{x}{2}\right)^{\pm\nu} \sum_{j=0}^{\infty} \frac{(-1)^j}{j!\, \Gamma(j \pm \nu + 1)} \left(\frac{x}{2}\right)^{2j} \tag{C.18}$$

which are referred to as the Bessel functions of the first kind of integer order ν. For $\nu = 0$,

$$J_0(x) = \sum_{j=0}^{\infty} \frac{(-1)^j}{j!\, j!} \left(\frac{x}{2}\right)^{2j} = 1 - \frac{x^2}{2^2 (1!)^2} + \frac{x^4}{2^4 (2!)^2} - \frac{x^6}{2^6 (3!)^2} + \ldots, \tag{C.19}$$

and $\nu = +1$,

$$\begin{aligned} J_1(x) &= \sum_{j=0}^{\infty} \frac{(-1)^j}{j!\, \Gamma(j+2)} \left(\frac{x}{2}\right)^{2j+1} = \sum_{j=0}^{\infty} \frac{(-1)^j}{j!\, (j+1)!} \left(\frac{x}{2}\right)^{2j+1} \\ &= \frac{x}{2} - \frac{x^3}{2^3 1!\, 2!} + \frac{x^5}{2^5 2!\, 3!} - \cdots \end{aligned} \tag{C.20}$$

and $\nu = -1$,

$$\begin{aligned} J_{-1}(x) &= \left(\frac{2}{x}\right) \sum_{j=0}^{\infty} \frac{(-1)^j}{j!\, \Gamma(j)} \left(\frac{x}{2}\right)^{2j} \\ &= \sum_{j=0}^{\infty} \frac{(-1)^j}{j!\, (j-1)!} \left(\frac{x}{2}\right)^{2j-1} \\ &= -\frac{x}{2} + \frac{x^3}{2^3 1!\, 2!} - \frac{x^5}{2^5 2! 3!} + \cdots \end{aligned} \tag{C.21}$$

The derivatives of the Bessel functions will come into use through our derivations. A useful integral form of the zeroth-Bessel function which we will use is

$$J_0(x) = \frac{1}{\pi} \int_0^{\pi} d\theta e^{ix \cos\theta}. \tag{C.22}$$

It further follows that

$$J_{-\nu} = (-1)^{\nu} J_{\nu}(x) \tag{C.23}$$

which demonstrates that the two solutions are not independent for ν being an integer. This is resolved by defining the two solutions to the Bessel equation as $J_\nu(x)$ and the Bessel function of the second kind (also known as the Neumann function):

$$N_{\nu}(x) = \frac{J_{\nu}(x)\left(\cos \nu\pi - (-1)^{\nu}\right)}{\sin \nu\pi}. \tag{C.24}$$

The limiting form of these Bessel functions for small x is

$$J_{\nu}(x) \approx \frac{(x/2)^{\nu}}{\Gamma(1+\nu)} \quad \text{as } x \to 0 \tag{C.25}$$

and

$$N_0(x) \approx \frac{2}{\pi}\left(\ln\left(\frac{x}{2}\right) + \gamma_{\mathrm{EM}} \cdots\right) \quad \text{as } x \to 0 \tag{C.26}$$

$$N_{\nu}(x) \approx -\frac{\Gamma(\nu)}{\pi}\left(\frac{2}{x}\right)^{\nu} k \quad \nu \neq 0 \text{ and as } x \to 0 \tag{C.27}$$

where γ_{EM} is the Euler–Mascheroni constant

$$\gamma_{\mathrm{EM}} = \lim_{n \to \infty}\left(\sum_{k=1}^{n} \frac{1}{k} - \ln n\right) \approx 0.5772\ldots.$$

For large x and a real and positive ν,

$$J_{\nu}(x) \approx \sqrt{\frac{2}{\pi x}} \cos\left(x - \frac{\nu\pi}{2} - \frac{\pi}{4}\right) \quad \text{as } x \to \infty \tag{C.28}$$

and

$$N_{\nu}(x) \approx \sqrt{\frac{2}{\pi x}} \sin\left(x - \frac{\nu\pi}{2} - \frac{\pi}{4}\right). \tag{C.29}$$

The transition between the approximations for small and large x occurs for the argument x of the order of ν.

An additional independent pair of solutions to Bessel's equation is that of the Hankel functions which are also known as the Bessel functions of the third kind and are linear combinations of the Bessel functions of the first and second kind:

$$H_\nu^{\binom{1}{2}}(x) = J_\nu(x) \pm iN_\nu(x). \tag{C.30}$$

The functions $J_\nu(x)$, $N_\nu(x)$ and the Hankel functions all form fundamental sets of solutions to the Bessel equation.

If the separation constant $+k^2$ had been used instead of $-k^2$ in (C.5) and (C.6), then Bessel's equation would be of the form

$$\frac{\mathrm{d}^2R}{\mathrm{d}x^2} + \frac{1}{x}\frac{\mathrm{d}R}{\mathrm{d}x} - \left(1 + \frac{\nu^2}{x^2}\right)R = 0. \tag{C.31}$$

The solutions to this modified Bessel equation will be the Bessel functions with pure imaginary arguments or the modified Bessel functions

$$I_\nu(x) = \frac{1}{i^\nu}J_\nu(ix) \tag{C.32}$$

and

$$K_\nu(x) = \frac{\pi}{2}i^{\nu+1}H_\nu^{(1)}(ix). \tag{C.33}$$

We will often use the integral expressions for the second modified Bessel function in our derivations. The root expression is, for a complex argument z,

$$\begin{aligned} K_\nu(z) &= \frac{\sqrt{\pi}(z/2)^\nu}{\Gamma\left(\nu + \frac{1}{2}\right)} \int_0^\infty \mathrm{d}x\, \mathrm{e}^{-z\cosh x} \sinh^{2\nu} x \\ &= \frac{\sqrt{\pi}(z/2)^\nu}{\Gamma(\nu + (1/2))} \int_1^\infty \mathrm{d}x\, \mathrm{e}^{-zx} \left(x^2 - 1\right)^{\nu - \frac{1}{2}} \end{aligned} \tag{C.34}$$

where $\mathrm{Re}\,\nu > -\frac{1}{2}$ and $|\arg z| < \frac{\pi}{2}$. Another integral expression is

$$K_\nu(yz) = \frac{\Gamma(\nu + (1/2))(2z)^\nu}{\sqrt{\pi}y^\nu} \int_0^\infty \mathrm{d}x \frac{\cos yx}{\left(x^2 + z^2\right)^{\nu + \frac{1}{2}}} \tag{C.35}$$

where $\mathrm{Re}\,\nu > -\frac{1}{2}$, $y > 0$ and $|\arg z| < \frac{\pi}{2}$.

The first derivatives of these functions are

$$\frac{\mathrm{d}K_0(y)}{\mathrm{d}y} = -K_1(y). \tag{C.36}$$

and

$$\frac{\mathrm{d}K_1(y)}{\mathrm{d}y} = -\left(K_0(y) + \frac{K_1(y)}{y}\right). \tag{C.37}$$

The limits of these modified Bessel functions for small and large x will be useful to some of the derivations of this book. These are, for fixed ν zero or positive,

$$I_\nu(x) \approx \frac{(x/2)^\nu}{\Gamma(1+\nu)} \quad x \to 0 \tag{C.38}$$

$$K_0(x) \approx -\left(\ln\left(\frac{x}{2}\right) + \gamma_{\mathrm{EM}}\right) I_0(x) + \frac{(x/2)^2}{(1!)^2} + \left(1+\frac{1}{2}\right)\frac{(x^2/4)^2}{(2!)^2}$$
$$+ \left(1+\frac{1}{2}+\frac{1}{3}\right)\frac{(x^2/4)^3}{(3!)^2} \quad x \to 0 \tag{C.39}$$

$$K_\nu(x) \approx \frac{\Gamma(x)}{2}\left(\frac{2}{x}\right)^\nu \quad x \to 0;\ \nu \neq 0 \tag{C.40}$$

$$I_\nu(x) \approx \frac{\mathrm{e}^x}{\sqrt{2\pi x}}\left(1 - \frac{4\nu^2-1}{8x} + \frac{(4\nu^2-1)(4\nu^2-9)}{2!\,(8x)^2} - \cdots\right)$$

Fixed ν; large x (C.41)

$$K_\nu(x) \approx \sqrt{\frac{\pi}{2x}}\left(1 + \frac{4\nu^2-1}{8x} + \frac{(4\nu^2-1)\,(4\nu^2-9)}{2\,!\,(8x)^2} + \cdots\right)$$

Fixed ν; large x (C.42)

Spherical Bessel and Hankel functions are those solutions to Bessel's equation for non-integer order (i.e. ν is not an integer);

$$j_l(x) = \sqrt{\frac{\pi}{2x}} J_{l+\frac{1}{2}}(x) \tag{C.43}$$

$$n_l(x) = \sqrt{\frac{\pi}{2x}} N_{l+\frac{1}{2}}(x) \tag{C.44}$$

$$h_l^{\binom{1}{2}}(x) = \sqrt{\frac{\pi}{2x}}\left(J_{l+\frac{1}{2}}(x) \pm iN_{l+\frac{1}{2}}(x)\right). \tag{C.45}$$

C.2 Legendre Polynomials

Adrien-Marie Legendre (1752–1833) was the French mathematician who introduced these eponymous polynomials in 1784. Again, they are obtained from the solution to Laplace's equation but now in spherical co-ordinates:

$$\frac{1}{r}\frac{\partial^2(r\Phi)}{\partial r^2}+\frac{1}{r^2\sin\theta}\frac{\partial}{\partial\theta}\sin\theta\left(\frac{\partial\Phi}{\partial\theta}\right)+\frac{1}{r^2\sin^2\theta}\frac{\partial^2\Phi}{\partial\phi^2}=0. \tag{C.46}$$

The equation is solved by again using the separation of variables. Here, the form

$$\Phi(r\ .\ \theta\phi)=\frac{R(r)}{r}P(\theta)\Xi(\phi) \tag{C.47}$$

is used which, following substitution and manipulation, yields

$$r^2\sin^2\theta\frac{1}{R(r)}\frac{\mathrm{d}^2R(r)}{\mathrm{d}r^2}+\frac{1}{P(\theta)r^2\sin\theta}\frac{\mathrm{d}}{\mathrm{d}\theta}\left(\sin\theta\frac{\mathrm{d}P(\theta)}{\mathrm{d}\theta}\right)+\frac{1}{\Xi(\phi)}\frac{\mathrm{d}^2\Xi(\phi)}{\mathrm{d}\phi^2}$$
$$=0. \tag{C.48}$$

The ϕ-dependent term must equal a constant which, conventionally, is set equal to $-m^2$

$$\frac{1}{\Xi(\phi)}\frac{\mathrm{d}^2\Xi(\phi)}{\mathrm{d}\phi^2}=-m^2, \tag{C.49}$$

and which has the solution

$$\Xi(\phi)=\mathrm{e}^{\pm im\phi}, \tag{C.50}$$

where m is an integer in order for $\Xi(\phi)$ to be single valued. We will return to this result in the future derivation of the spherical harmonics. The r- and θ-dependent equations are similarly isolated:

$$\frac{\mathrm{d}^2R(r)}{\mathrm{d}r^2}-\frac{k}{r^2}R(r)=0 \tag{C.51}$$

$$\frac{1}{\sin\theta}\frac{\mathrm{d}}{\mathrm{d}\theta}\left(\sin\theta\frac{\mathrm{d}P(\theta)}{\mathrm{d}\theta}\right)+\left(k-\left(\frac{m}{\sin\theta}\right)^2\right)P(\theta)=0. \tag{C.52}$$

The separation constant k is real and is taken to be equal to $l(l+1)$. (C.52) can be further modified by using the substitution of variable, $x=\cos\theta$,

$$\frac{\mathrm{d}}{\mathrm{d}x}\left(\left(1-x^2\right)\frac{\mathrm{d}P(x)}{\mathrm{d}x}\right)+\left(l(l+1)-\frac{m^2}{1-x^2}\right)P(x)=0. \tag{C.53}$$

This is known as the generalised Legendre equation. For $m \neq 0$, the solution to Equation is the associated Legendre polynomial $P_l^m(x)$ to be discussed later. However, we first consider the case of (C.53) for m set equal to zero:

$$\frac{\mathrm{d}}{\mathrm{d}x}\left(\left(1-x^2\right)\frac{\mathrm{d}P(x)}{\mathrm{d}x}\right)+l(l+1)P(x)=0. \tag{C.54}$$

As $x = \cos\theta$, then $x \in [-1,1]$. Let us write the solution to (C.54) in the form of the power series

$$P(x)=\sum_{j=0}^{\infty}a_jx^j. \tag{C.55}$$

Expanding (C.55),

$$\left(1-x^2\right)\frac{\mathrm{d}^2P(x)}{\mathrm{d}x^2}-2x\frac{\mathrm{d}P(x)}{\mathrm{d}x}+l(l+1)P(x)=0 \tag{C.56}$$

where

$$\frac{\mathrm{d}P(x)}{\mathrm{d}x}=\sum_{j=0}^{\infty}a_jjx^{j-1} \tag{C.57}$$

and

$$\frac{\mathrm{d}^2P(x)}{\mathrm{d}x^2}=\sum_{j=0}^{\infty}a_jj(j-1)x^{j-2}. \tag{C.58}$$

Substituting these two results into (C.56) results in the series

$$\sum_{j=0}^{\infty}\left(a_jj(j-1)x^{j-2}-a_j(j(j+1)-l(l+1))x^j\right)=0. \tag{C.59}$$

As this series must be equal to zero for all values of x, each coefficient of a power of x must also equal to zero. This yields the recursive relationship

$$a_{j+2}=a_j\left(\frac{j(j+1)-l(l+1)}{(j+1)(j+2)}\right), \tag{C.60}$$

i.e. if a_0 is known, then the $a_2, a_4,\ldots$ can be determined, and if a_1 is known, then the $a_3, a_5, a_7\ldots$ can be calculated. Note from (C.60) that as $j \to \infty$, then $a_{j+2} \to a_j$. As the argument $|x| \leq 1$, we will require that $P(x)$ be finite and, hence, that the power series terminate. As j is either zero or a positive integer, (C.60) can only terminate if l is also zero or a positive integer. The resulting solutions are the Legendre polynomials of order l and, by convention, are normalised to equal unity for $x = \pm 1$. The first four Legendre polynomials are

$$\begin{aligned} P_0(x) &= 1 \\ P_1(x) &= x \\ P_2(x) &= \frac{3x^2 - 1}{2} \\ P_3(x) &= \frac{5x^3 - 3x}{2} \\ P_4(x) &= \frac{35x^4 - 30x^2 + 3}{8}. \end{aligned} \tag{C.61}$$

The Legendre polynomials have a variety of useful and interesting properties. They can also be represented by the form

$$P_l(x) = \frac{1}{2^l l!} \frac{\mathrm{d}}{\mathrm{d}x^l} \left(x^2 - 1\right)^l \tag{C.62}$$

which is known as *Rodrigues' formula*.

There is a link between the zeroth-order Bessel function of the first kind (e.g. (C.19)) and the Legendre polynomial. Let us consider (C.54) and use the small angle approximation. In which case, we have, if we replace $l\,(l+1)$ with l^2,

$$\frac{1}{\theta} \frac{\mathrm{d}}{\mathrm{d}\theta} \left(\theta \frac{\mathrm{d}P_l(\theta)}{\mathrm{d}\theta} \right) + l^2 P_l(\theta) = 0. \tag{C.63}$$

That is, for $l \gg 1$ and $\theta \ll 1$, (C.59) which represents the Bessel equation for zeroth-order, we may approximate

$$P_l(\cos\theta) \approx J_0(l\theta). \tag{C.64}$$

Of most practical importance is the fact that they are orthogonal functions and, hence, can be used as a set of basis functions to represent a function over the interval $x \in [-1,1]$. By using Rodrigues' formula, the orthogonality can be demonstrated:

$$\int\limits_{-1}^{1} \mathrm{d}x\, P_l(x) P_n(x) = \frac{2}{2l+1} \delta_{ln} \tag{C.65}$$

where δ_{ln} is the Kronecker delta function

$$\begin{aligned}\delta_{\mathrm{ln}} &= 1 \quad \mathrm{l} = n \\ &= 0 \quad l \neq n.\end{aligned} \tag{C.66}$$

Now, let us return to the original Legendre's equation where $m \neq 0$. That is, both l and m in the equation are arbitrary. By again using a power series solution, it can be shown that l must be a positive integer or zero and that m is also an integer but which can take on only a finite $(2l + 1)$ set of values given by $-l$, $-(l - 1)$, ...,0,...$(l - 1)$,l. The resulting solution is the associated Legendre polynomial defined by

$$P_l^m(x) = (-1)^m \left(1 - x^2\right)^{\frac{m}{2}} \frac{\mathrm{d}^m P_l(x)}{\mathrm{d}x^m}. \tag{C.67}$$

An orthogonal basis set in l is formed by the associated Legendre polynomial function for fixed m,

$$\int_{-1}^{1} \mathrm{d}x \, P_l^m(x) \, P_n^m(x) = \left(\frac{2}{2l + 1}\right) \frac{(l + m)\,!}{(l - m)\,!} \delta_{ln}. \tag{C.68}$$

C.3 Spherical Harmonics

Recall that the Legendre/associated Legendre polynomials were the solution to the component θ of the solution to the Laplace equation in spherical co-ordinates. Both angular products (functions of θ and ϕ) can be combined, and the complete angular part of the solution to the Laplace equation can be

$$Y_{lm}(\theta, \phi) = \sqrt{\frac{2l + 1}{4\pi} \frac{(l - |m|)!}{(l + |m|)!}} P_l^m(x) \, \mathrm{e}^{im\phi} \tag{C.69}$$

where the quantity within the square root is required for normalisation. $Y_{lm}(\theta, \phi)$ is known as the spherical harmonic. It follows from Eq. (C.64) that

$$Y_{l,-m}(\theta, \phi) = (-1)^m Y_{lm}^*(\theta, \phi) \tag{C.70}$$

where the * denotes the complex conjugate. The orthonormalisation property of the spherical harmonic is

$$\int_0^{2\pi} \mathrm{d}\phi \int_{-1}^{1} \mathrm{d}(\cos\theta) \, Y_{l'm'}^*(\theta, \phi) Y_{lm}(\theta, \phi) = \delta_{l'l} \delta_{m'm}. \tag{C.71}$$

The spherical functions allow a general function in two angular variables to be written as the series

$$f(\theta,\phi) = \sum_{l=0}^{\infty} \sum_{m=-l}^{m=l} a_{lm} Y_{lm}(\theta,\phi) \tag{C.72}$$

where the series coefficients are given by the integral

$$a_{lm} = \int_0^{2\pi} \mathrm{d}\phi \int_{-1}^{1} \mathrm{d}(\cos\theta) Y_{lm}^*(\theta,\phi) f(\theta,\phi). \tag{C.73}$$

The first few spherical harmonics are

$$\begin{aligned}
Y_{00} &= \frac{1}{\sqrt{4\pi}} \\
Y_{10} &= \sqrt{\frac{3}{4\pi}} \cos\theta \\
Y_{11} &= -\sqrt{\frac{3}{8\pi}} \sin\theta\, \mathrm{e}^{i\phi} \\
Y_{20} &= \sqrt{\frac{5}{16\pi}} \left(3\cos^2\theta - 1\right) \\
Y_{21} &= \sqrt{\frac{15}{8\pi}} \sin\theta\cos\theta \mathrm{e}^{i\phi} \\
Y_{22} &= \sqrt{\frac{15}{32\pi}} \sin^2\theta \mathrm{e}^{i2\phi}.
\end{aligned} \tag{C.74}$$

The corresponding spherical harmonics with $-m$ can be found using Eq. (C.70).

C.4 Airy Function

The Airy function is named after Sir George Biddell Airy (1801–1892) who held the Lucasian chair of mathematics at Cambridge and was the 7th Astronomer Royal. In this book, the Airy function arises in the derivation of the Gaussian limit to the Vavilov probability distribution function in stochastic collisional energy loss. The Airy function is a solution to Airy's equation

$$\frac{\mathrm{d}^2 y}{\mathrm{d}x^2} - xy = 0. \tag{C.75}$$

This type of equation arises in a variety of areas in physics: a pertinent example being that of Schrödinger's equation describing a charged particle moving in a linearly directed electric field.

An integral solution to Airy's equation can be obtained by using the Laplace method in which the solution is given by the complex integral

$$y = \int_C \mathrm{d}z\, \mathrm{e}^{xz} f(z) \tag{C.76}$$

where z is a complex variable. Equation (C.75) can thus be written in the form

$$\int_C \mathrm{d}z\, (z^2 - x)\mathrm{e}^{xz} f(z) = \int_C \mathrm{d}z\, z^2 \mathrm{e}^{xz} f(z) - \int_C \mathrm{d}z\, x\mathrm{e}^{xz} f(z) = 0. \tag{C.77}$$

The second integral of (C.77) is solved through the integration by parts

$$\begin{aligned} \int_C \mathrm{d}z\, x\mathrm{e}^{xz} f(z) &= \int_C \mathrm{d}z\, \frac{\partial \mathrm{e}^{xz}}{\partial z} f(z) \\ &= \int_C \mathrm{d}z \left(\frac{\partial}{\partial z} (f(z)\mathrm{e}^{xz}) - \frac{\mathrm{d}f(z)}{\mathrm{d}z} \mathrm{e}^{xz} \right) \\ &= f(z)\mathrm{e}^{xz}|_C - \int_C \mathrm{d}z \frac{\mathrm{d}f(z)}{\mathrm{d}z} \mathrm{e}^{xz}. \end{aligned} \tag{C.78}$$

The first term is evaluated at the endpoints of the contour, C. Thus,

$$f(z)\mathrm{e}^{xz}|_C + \int_C \mathrm{d}z\, \mathrm{e}^{xz} \left(z^2 f(z) + \frac{\mathrm{d}f(z)}{\mathrm{d}z} \right) = 0. \tag{C.79}$$

By requiring that $z^2 f(z) + \frac{\mathrm{d}f(z)}{\mathrm{d}z} = 0$, one obtains the solution that

$$f(z) = \mathrm{e}^{-\frac{z^3}{3}}, \tag{C.80}$$

which leads to

$$f(z)\mathrm{e}^{xz}|_C = \mathrm{e}^{xz - \frac{z^3}{3}}\Big|_C = 0. \tag{C.81}$$

The result of (C.81) can be achieved if C is a closed contour. However, as $f(z)\,\mathrm{e}^{xz}$ appears in the integrand of (C.76), this selection of the contour C would give the undesirable solution of the Airy equation of $y = 0$. To circumvent this difficulty,

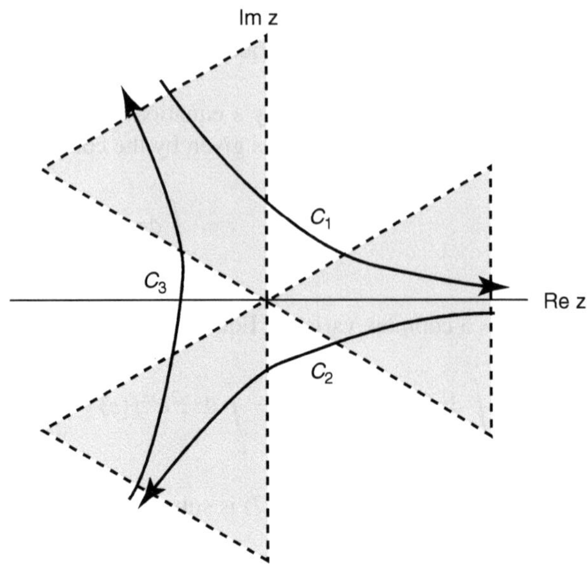

Fig. C.1 Complex plane with sectors for $\cos(3\theta) < 1$ and integration contours

we note that $f(z)\,e^{xz}$ is non-zero for any value of z and consider a contour that starts and terminates in sectors of where $z \to \infty$ for which $f(z)e^{xz} \to 0$ at the boundaries. In other words, the real part of z^3 must be positive in order for $f(z)e^{xz}$ to converge to zero as $z \to \infty$. In addition, we note that $e^{xz-\frac{z^3}{3}} \approx e^{-\frac{z^3}{3}}$ for large z. From Euler's formula, $z = |z|e^{i\theta}$, then

$$\frac{z^3}{3} = \frac{|z|^3}{3}(\cos(3\theta) + i\sin(3\theta)). \tag{C.82}$$

And, thus, the contours must begin and end within the sectors within the complex plane defined by $\cos(3\theta) < 1$, as shown Fig. C.1.

The contours C_1 and C_2 can be neglected for the time being as they will result in solutions that tend to infinity as $x \to \infty$, and only contour C_3 need to be considered. By deforming C_3 so as to fit the imaginary axis (which will still retain the start and end points within the required sectors), one can write the solution to the Airy equation as

$$\mathrm{Ai}(x) = \frac{1}{2\pi i}\int_{-i\infty}^{+i\infty} dz\, e^{xz-\frac{z^3}{3}}. \tag{C.83}$$

From this integral representation, another way of describing the Airy function is given by

$$\mathrm{Ai}(x) = \frac{1}{\pi}\int_0^{\infty} dz\, \cos\left(\frac{z^3}{3} + xz\right). \tag{C.84}$$

Equation (C.84) appears explicitly in the expression of the Gaussian limit to the Vavilov probability distribution function. In that derivation, we are interested in the asymptotic behaviour of the Airy function for large x. This is obtained from the original definition of the Airy function of (C.83)

$$\mathrm{Ai}(x) = \frac{1}{2\pi i} \int\limits_{C_3} \mathrm{d}v\, \mathrm{e}^{xv - \frac{v^3}{3}} \tag{C.85}$$

where C_3 is the contour shown in Fig. C.1 and where $v \equiv \sqrt{r} + is$ for which $s \in (-\infty, +\infty)$. By defining the variable $\lambda = \frac{2x^{\frac{3}{2}}}{3}$, this result can be written as

$$\begin{aligned} \pi \mathrm{e}^{\lambda} \mathrm{Ai}(x) &= \int\limits_{0}^{\infty} \mathrm{d}r \cos\left(\frac{r^3}{3}\right) \mathrm{e}^{-s^2\sqrt{r}} \\ &= \frac{1}{2r^{\frac{1}{4}}} \int\limits_{-\infty}^{+\infty} \mathrm{d}t \cos\left(\frac{t^3}{3r^{\frac{3}{4}}}\right) \mathrm{e}^{-t^2}. \end{aligned} \tag{C.86}$$

By expanding the cosine function into a series and integrating each term sequentially,

$$\pi \mathrm{e}^{\lambda} \mathrm{Ai}(x) \approx \frac{\sqrt{\pi}}{2r^{\frac{1}{4}}} \left(1 - \frac{3 \bullet 5}{1!\, 12^2 r^{\frac{3}{2}}} + \frac{5 \bullet 7 \bullet 9 \bullet 11}{2!\, 12^4 r^3} - \cdots \right) \tag{C.87}$$

which leads to the asymptotic series of the Airy function:

$$\mathrm{Ai}(x) = \frac{1}{2\sqrt{\pi} x^{\frac{1}{4}}} \mathrm{e}^{-\lambda} \left(1 + \frac{3 \bullet 5}{1!\, 216} \frac{1}{\lambda} + \frac{5 \bullet 7 \bullet 9 \bullet 11}{2!\, 216^2} \frac{1}{\lambda^2} + \cdots \right). \tag{C.88}$$

Thus, for large $\lambda = \frac{2x^{\frac{3}{2}}}{3}$, the asymptotic form of the Airy function is

$$\mathrm{Ai}(x) \approx \frac{\mathrm{e}^{-\frac{2x^{\frac{3}{2}}}{3}}}{2\sqrt{\pi} x^{\frac{1}{4}}} \quad \text{as } x \to \infty. \tag{C.89}$$

C.5 Hermite Polynomials and Edgeworth Series

A probability distribution function (*pdf*) can be described by an Edgeworth series. This is an expansion of weighted Hermite, or modified Hermite, polynomials, where the weights are related to the moments of the *pdf*. This type of expansion can be useful in repeated computation of the *pdf*.

C.5.1 Hermite Polynomials

The Hermite polynomial of order n is defined by the contour integral

$$H_n(z) = \frac{n!}{2\pi i} \oint_C \mathrm{d}t \frac{\mathrm{e}^{-t^2+2tz}}{t^{(1+n)}} \tag{C.90}$$

where the contour encloses the origin and the integral is evaluated in the positive sense. It also given by a Rodrigues' formula

$$H_n(x) = (-1)^n \mathrm{e}^{x^2} \frac{\mathrm{d}^n}{\mathrm{d}x^n} \mathrm{e}^{-x^2}. \tag{C.91}$$

The Hermite polynomials are orthogonal with the weighting factor e^{-x^2}, or

$$\int_{-\infty}^{\infty} \mathrm{d}x H_n(x) H_m(x)\, \mathrm{e}^{-x^2} = 2^n n! \sqrt{\pi} \delta_{mn}, \tag{C.92}$$

where δ_{mn} is the Kronecker δ-function. The first nine Hermite polynomials are

$$\begin{aligned}
H_0(x) &= 1 \\
H_1(x) &= 2x \\
H_2(x) &= -2 + 4x^2 \\
H_3(x) &= -12x + 8x^3 \\
H_4(x) &= 12 - 48x^2 + 16x^4 \\
H_5(x) &= 120x - 160x^3 + 32x^5 \\
H_6(x) &= -120 + 720x^2 - 480x^4 + 64x^6 \\
H_7(x) &= -1,680x + 3,360x^3 - 1,344x^5 + 128x^7 \\
H_8(x) &= 1,680 - 13,440x^2 + 13,440x^4 - 3,584x^6 + 256x^8 \\
H_9(x) &= 30,240x - 80,640x^3 + 48,384x^5 - 9,216x^7 + 512x^9.
\end{aligned} \tag{C.93}$$

C.5.2 Use of the Hermite Polynomials in Evaluating Matrix Elements

In Chap. 11, we will find the Hermite polynomial useful in deriving the generalised oscillator strength for a harmonic oscillator. A linear harmonic oscillator has the eigenfunctions

$$|n\rangle = N_n H_n(\zeta) \mathrm{e}^{-\frac{\zeta^2}{2}} \tag{C.94}$$

where $\zeta = \xi\, x$, $\xi = \sqrt{\frac{m\omega_0}{\hbar c^2}}$ and N_n is the normalisation constant for the nth-state. Clearly, ζ is a dimensionless variable. We now present the generating function for $H_n(\zeta)$:

$$S(\zeta, s) = \sum_{n=0}^{\infty} \frac{s^n}{n!} H_n(\zeta) \\ = \mathrm{e}^{-s^2+2s\zeta} \tag{C.95}$$

where s is also a dimensionless variable. It follows from (C.94) that

$$\int_{-\infty}^{\infty} \mathrm{d}x\, \mathrm{e}^{-(\frac{\zeta^2}{2}-i\frac{qx}{\hbar c})} S(\zeta, s) S(\zeta, s') = \sum_{n=0}^{\infty} \sum_{m=0}^{\infty} \frac{s^n s'^m}{n!\, m!} \int_{-\infty}^{\infty} \mathrm{d}x\, H_n(\zeta) H_m(\zeta)\, \mathrm{e}^{-\left(\frac{\zeta^2}{2}-i\frac{qx}{\hbar c}\right)} \\ = \frac{\sqrt{\pi}}{\xi} e^{2ss'+i\frac{q(s+s')}{\xi}-\frac{q^2}{4\xi^2}}. \tag{C.96}$$

We now consider two special cases.
First, consider the case of $q = 0$. Then,

$$\int_{-\infty}^{\infty} \mathrm{d}x\, \mathrm{e}^{-\frac{\zeta^2}{2}} S(\zeta, s)\, S(\zeta, s') = \sum_{n=0}^{\infty} \sum_{m=0}^{\infty} \frac{s^n s'^m}{n! m!} \int_{-\infty}^{\infty} \mathrm{d}x H_n(\zeta) H_m(\zeta) \mathrm{e}^{-\frac{\zeta^2}{2}} \\ = \frac{\sqrt{\pi}}{\xi} \mathrm{e}^{2ss'} \tag{C.97}$$

We next expand the exponential into a series:

$$\frac{\sqrt{\pi}}{\xi} \mathrm{e}^{2ss'} = \frac{\sqrt{\pi}}{\xi} \sum_{n=0}^{\infty} \frac{(2ss')^n}{n!}. \tag{C.98}$$

By then comparing the equal powers of s and s', we find

$$\int_{-\infty}^{\infty} \mathrm{d}x H_n(\zeta) H_y(\zeta) \mathrm{e}^{-\zeta^2} = \frac{\sqrt{\pi}}{\xi} 2^n n!\, \delta_{nm}. \tag{C.99}$$

From (C.94) and (C.99), we now obtain an expression for the normalisation constants:

$$N_n = \sqrt{\frac{\xi}{\sqrt{\pi} 2^n n!}}. \tag{C.100}$$

The second case is for $s' = 0$. Then again expanding the exponential

$$\frac{\sqrt{\pi}}{\xi}\mathrm{e}^{+i\frac{(q/\hbar c)s}{\xi}-\frac{(q/\hbar c)^2}{4\xi^2}} = \frac{\sqrt{\pi}}{\xi}\mathrm{e}^{-\left(\frac{(q/\hbar c)}{2\xi}\right)^2}\sum_{n=0}^{\infty}\frac{s^n}{n!}\left(\frac{i(q/\hbar c)}{\xi}\right)^n \tag{C.101}$$

and comparing powers of s, we obtain

$$\int_{-\infty}^{\infty} \mathrm{d}x H_n(\zeta)H_0(\zeta)\mathrm{e}^{-\left(\frac{\zeta^2}{2}-i\left(\frac{qx}{\hbar c}\right)\right)} = \frac{\sqrt{\pi}}{\xi}\mathrm{e}^{-\left(\frac{(q/\hbar c)}{2\xi}\right)^2}\left(\frac{i(q/\hbar c)}{\xi}\right)^n \tag{C.102}$$

Multiplying through with the product of the normalisation constants N_nN_0, we obtain the matrix element:

$$\langle n|\mathrm{e}^{iqx}|0\rangle = \frac{1}{\sqrt{2^n n!}}\left(\frac{i\frac{q}{\hbar c}}{\xi}\right)^n \mathrm{e}^{-\left(\frac{q/\hbar c}{2\xi}\right)^2}. \tag{C.103}$$

This result will prove useful in our derivation of the generalised oscillator strength in Chap. 11.

C.5.3 Chebyshev–Hermite Polynomials

A modified version of the Hermite polynomial, which is also known as the Chebyshev–Hermite polynomial, is

$$He_n(x) \equiv 2^{-\frac{n}{2}}H_n\left(\frac{x}{\sqrt{2}}\right) \tag{C.104}$$

and where the first few terms are

$$\begin{aligned} He_1(x) &= x \\ He_2(x) &= -1 + x^2 \\ He_3(x) &= -3x + x^3 \\ He_4(x) &= 3 - 6x^2 + x^4. \end{aligned} \tag{C.105}$$

C.5.4 Edgeworth Series

The Edgeworth series provides an approximation to a *pdf* $f(x)$ in terms of the cumulants of that *pdf*. These quantities are defined from the characteristic function of the *pdf* which is the Fourier transform of the *pdf* $f(x)$ which, in the case of a continuous random variable x, is (neglecting the $\frac{1}{\sqrt{2\pi}}$ normalisation factor)

$$\Phi(t) = \int_{-\infty}^{\infty} \mathrm{d}x f(x)\, \mathrm{e}^{itx}. \tag{C.106}$$

The cumulants κ_n are defined through the expansion

$$\ln \Phi(t) = \sum_{n=1}^{\infty} \kappa_n \frac{(it)^n}{n!}. \tag{C.107}$$

Consider a random variable x with the expectation value $E\{x\} = 0$ and variance σ^2. The normalised random variable $\frac{x}{\sigma}$ has the corresponding characteristic function $\phi(t) = \Phi\left(\frac{t}{\sigma}\right)$. From (C.105), it follows that

$$\begin{aligned} \ln \phi(t) &= \ln \Phi\left(\frac{t}{\sigma}\right) \\ &= \sum_{n=2}^{\infty} \frac{\kappa_n}{\sigma^n} \frac{(it)^n}{n!} \end{aligned} \tag{C.108}$$

where the summation begins at $n = 2$ due to $E\{X\} = 0$. As $\kappa_2 = \sigma^2$,

$$\begin{aligned} \ln \phi(t) &= -\frac{t^2}{2} \sum_{n=3}^{\infty} \frac{\kappa_n}{\sigma^n} \frac{(it)^n}{n!} \\ &= -\frac{t^2}{2} \sum_{n=3}^{\infty} \left(\frac{\kappa_n}{\sigma^{2n-2}}\right) \sigma^{n-2} \frac{(it)^n}{n!} \\ &\equiv -\frac{t^2}{2} \sum_{n=3}^{\infty} S_n \sigma^{n-2} \frac{(it)^n}{n!} \end{aligned} \tag{C.109}$$

where the quantity S_n has been defined. We next exponentiate (C.109)

$$\phi(t) = \mathrm{e}^{-\frac{t^2}{2}}\, \mathrm{e}^{\sum_{n=3}^{\infty} S_n \sigma^{n-2} \frac{(it)^n}{n!}}$$

and expand the second exponential in a series in powers of σ

$$\begin{aligned} \mathrm{e}^{\sum_{n=3}^{\infty} S_n \sigma^{n-2} \frac{(it)^n}{n!}} &= \mathrm{e}^{\sum_{m=1}^{\infty} S_{m+2} \sigma^m \frac{(it)^{m+2}}{(m+2)!}} \\ &\cong 1 + \sum_{s=1}^{\infty} P_s(it)\, \sigma^s \end{aligned} \tag{C.110}$$

The coefficient $P_s(it)$ is a polynomial given by

$$P_s(it) = \sum_{\{k_r\}} \prod_{r=1}^{s} \frac{1}{k_r!} \left(\frac{S_{r+2}(it)^{r+2}}{(r+2)!} \right)^{k_r} \tag{C.111}$$

where $\{k_r\}$ are all positive integers satisfying a Diophantine equation

$$\sum_{j=1}^{n} jk^j = n. \tag{C.112}$$

We now derive the Edgeworth series. Assume that a *pdf* $f(x)$ exists for the random variable X which leads to the *pdf* for $\frac{X}{\sigma}$ being

$$g(x) = \sigma f(\sigma x). \tag{C.113}$$

This will also equal the inverse Fourier transform of the corresponding characteristic function φ,

$$g(x) = \frac{1}{\sqrt{2\pi}} \int_{-\infty}^{\infty} \mathrm{d}t\, \mathrm{e}^{-itx} \phi(t). \tag{C.114}$$

Recall from our discussion in Appendix B.1 of the differentiation of Fourier transforms that the Fourier transform of the *nth* derivative of $f(x)$ is $(-it)^n \Phi(t)$ where $\Phi(t)$ is the Fourier transform of $f(x)$. Hence,

$$\frac{\mathrm{d}^n}{\mathrm{d}x^n} f(x) = \frac{1}{\sqrt{2\pi}} \int_{-\infty}^{\infty} \mathrm{d}t (-it)^n \Phi(t). \tag{C.115}$$

Let us now assume that the *pdf* is Gaussian (normal),

$$f(x) = \frac{\mathrm{e}^{-\frac{x^2}{2}}}{\sqrt{2\pi}}. \tag{C.116}$$

The Fourier transform of a Gaussian,

$$\begin{aligned}
\Phi(t) &= \frac{1}{2\pi} \int_{-\infty}^{\infty} \mathrm{d}x\, \mathrm{e}^{-\frac{x^2}{2}} \mathrm{e}^{-ixt} \\
&= \frac{1}{2\pi} \left(\int_{-\infty}^{\infty} \mathrm{d}x\, \mathrm{e}^{-\frac{x^2}{2}} \cos xt - \int_{-\infty}^{\infty} \mathrm{d}x\, \mathrm{e}^{-\frac{x^2}{2}} \sin xt \right) \\
&= \frac{1}{2\pi} \int_{-\infty}^{\infty} \mathrm{d}x\, \mathrm{e}^{-\frac{x^2}{2}} \cos xt \\
&= \frac{1}{\sqrt{2\pi}} \mathrm{e}^{-\frac{t^2}{2}}
\end{aligned} \tag{C.117}$$

where the second integral is equal to zero due to sin xt being odd over the symmetric range of integration. That is, the Fourier transform of a Gaussian function is also Gaussian. Following (C.114) and (C.116), consider the expansion of (C.109) and (C.115). The product of $e^{-\frac{t^2}{2}}$ and every $(it)^n$ creates the nth derivative of the Gaussian function

$$(-1)^n \frac{d^n}{dx^n} f(x) = \int_{-\infty}^{\infty} dt (it)^n e^{-itx} e^{-\frac{t^2}{2}} \tag{C.118}$$

in the expansion of $g(x)$

$$g(x) = f(x) + \sum_{s=1}^{\infty} \sigma^s \left(\sum_{\{k_r\}} \prod_{r=1}^{s} \frac{1}{k_r!} \left(\frac{S_{r+2}(-1)^{r+2}}{(r+2)!} \frac{d^{r+2}}{dx^{r+2}} \right)^{k_r} f(x) \right). \tag{C.119}$$

From the definition of the Hermite and Chebyshev–Hermite polynomials, for a Gaussian *pdf* $f(x)$,

$$\frac{d^n}{dx^n} f(x) = (-1)^n He_n(x) f(x). \tag{C.120}$$

and $b = \sum_{j=1}^{s} k_j$ (C.119) can be written in terms of the Chebyshev–Hermite polynomials

$$g(x) = f(x) \left(1 + \sum_{s=1}^{\infty} \sigma^s \sum_{\{k_r\}} He_{s+2b}(x) \prod_{r=1}^{s} \frac{1}{k_r!} \left(\frac{S_{r+2}}{(r+2)!} \right)^{k_r} \right). \tag{C.121}$$

For example, the Edgeworth series to 4th order for a Gaussian with zero mean is

$$g(x) = \frac{e^{-\frac{x^2}{2\sigma^2}}}{\sqrt{2\pi}\,\sigma} \left(1 + \frac{1}{3!} \left(\frac{\mu_3}{\sigma^3} \right) H_3 \left(\frac{x}{\sigma} \right) + \frac{1}{4!} \left(\frac{\mu_4}{\sigma^4} - 3 \right) H_4 \left(\frac{x}{\sigma} \right) \right). \tag{C.122}$$

where the μ_i are the moments of the *pdf*.

C.6 Confluent Hypergeometric Functions

The *confluent hypergeometric functions* are the result of the exact solution of the Schrödinger equation for the Coulomb potential. These are also solutions to the second-order ordinary differential equation (also known as *Kummer's equation*):

$$\frac{d^2 x}{dy^2} + \left(\frac{b}{x} - 1 \right) \frac{dy}{dx} - \left(\frac{a}{x} \right) y = 0 \tag{C.123}$$

which has a regular singularity at $x = 0$ and an irregular singularity at ∞. There are two independent solutions to this equation. The first is the confluent hypergeometric function of the first kind:

$$M(a,b,x) = \sum_{n=0}^{\infty} \frac{(a)_n}{(b)_n} \frac{x^n}{n!}, \tag{C.124}$$

where $(a)_n$ and $(b)_n$ are Pochhammer symbols defined by

$$(w)_n = \frac{\Gamma(w+n)}{\Gamma(w)}. \tag{C.125}$$

The second solution is the confluent hypergeometric function of the second kind:

$$U(a,b,x) = \frac{\pi}{\sin b\pi} \left(\frac{M(a,b,x)}{\Gamma(1+a-b)\,\Gamma(b)} - x^{1-b} \frac{M(1+a-b,2-b,x)}{\Gamma(a)\,\Gamma(2-b)} \right). \tag{C.126}$$

$M(a,b,x)$ is regular for all finite x and has the asymptotic form for large x:

$$M(a,b,x) \approx \frac{e^{i\pi a}}{x^a \Gamma(1-a)} \left(1 - \left(\frac{a}{x}\right)^2 \right) + \frac{e^x x^{a-1}}{\Gamma(a)} \tag{C.127}$$

for terms up to x^{-1}.

C.7 Exponential and Trigonometric Integrals

The discussion of stochastic energy loss involved the introduction of exponential and trigonometric integrals in the derivation of the Vavilov distribution. These integrals and their relationships are briefly summarised here. Unless otherwise stated, all arguments are complex (i.e. $z = x + iy$). The $E_n(x)$ function is defined as

$$E_n(x) = \int_x^{\infty} \mathrm{d}t \, \frac{e^{-xt}}{t^n}. \tag{C.128}$$

Our interest will be for the case of (C.128) for which $n = 1$,

$$E_1(x) = \int_x^{\infty} \mathrm{d}t \, \frac{e^{-xt}}{t} \tag{C.129}$$

which will be recognised as the incomplete gamma function, $\Gamma(0, 1)$. We now allow a complex argument for

$$\begin{aligned} E_1(z) &= \int_z^\infty \mathrm{d}t\,\frac{\mathrm{e}^{-t}}{t} \quad |\arg z| < \pi \\ &= \mathrm{Re}E_1(x) + i\mathrm{Im}E_1(y). \end{aligned} \tag{C.130}$$

To make $E_1(z)$ a single-valued function, a branch cut is made just below the negative real axis (including the origin) so that $z = x + iy = re^{i\theta}$ where $-\pi < \theta \leq \pi$. The exponential integral is defined as, using a notation by convention,

$$E_1(x) = -\mathrm{Ei}(-x) \tag{C.131}$$

or, in other forms,

$$\begin{aligned} \mathrm{Ei}(x) &= \mathrm{PV}\int_{-\infty}^{x} \mathrm{d}t\,\frac{\mathrm{e}^t}{t} \\ &= -\mathrm{PV}\int_{-x}^{\infty} \mathrm{d}t\,\frac{\mathrm{e}^{-t}}{t} \end{aligned} \tag{C.132}$$

and

$$-\mathrm{Ei}(-x) = \int_x^\infty \mathrm{d}t\,\frac{\mathrm{e}^{-t}}{t} \quad x > 0. \tag{C.133}$$

which is equal to $E_1(x)$ from (C.129). $\mathrm{Ei}(x)$ can be calculated ab initio from its Taylor expansion:

$$\mathrm{Ei}(x) = \gamma_{\mathrm{EM}} + \ln|x| + \sum_{n=1}^{\infty} \frac{x^n}{n\,n!} \tag{C.134}$$

where γ_{EM} is the Euler–Mascheroni constant. This result is convenient for $|x| < 4$, whereas the asymptotic form

$$\mathrm{Ei}(x) \to \mathrm{e}^x \sum_{n=0}^{\infty} \frac{n!}{x^{n+1}} \quad \text{as } x \to \infty \tag{C.135}$$

is accurate only for $50 < |x|$. Recursive relationships are used for the intermediate range of arguments, $4 < |x| < |50|$ (Harris 1957). In complex form,

$$E_1(z) = -\gamma_E - \ln z - \sum_{n=1}^{\infty} \frac{(-1)^n z^n}{n\, n!} \quad \text{for } |\arg z| < \pi. \tag{C.136}$$

Tables of the exponential integral can be found in Abramowitz and Stegun (1972). For example, use will be made of

$$\int_0^z \mathrm{d}t \, \frac{1 - \mathrm{e}^{-t}}{t} = -\mathrm{Ei}(-t) + \ln t + \gamma_{\mathrm{EM}}. \tag{C.137}$$

In order to examine the relationships between the exponential and trigonometric integrals, set $t = iu$. Then (C.130) is

$$\begin{aligned} E_1(x+iy) &= \int_{y-ix}^{\infty} \mathrm{d}u \, \frac{\mathrm{e}^{-iu}}{u} \\ &= \int_{y-ix}^{\infty} \mathrm{d}u \, \frac{\cos u}{u} - i \int_{y-ix}^{\infty} \mathrm{d}u \, \frac{\sin u}{u}. \end{aligned} \tag{C.138}$$

The cosine integral has a variety of forms, but the most commonly used is

$$\mathrm{Ci}\,(z) = -\int_z^{\infty} \mathrm{d}t \, \frac{\cos t}{t}. \tag{C.139}$$

Another expression used commonly for the cosine function is

$$\mathrm{Ci}\,(z) = \gamma_{\mathrm{EM}} + \ln z + \int_0^z \mathrm{d}t \, \frac{\cos t - 1}{t} \quad (|\arg z| < \pi). \tag{C.140}$$

The sine integral is

$$\mathrm{Si}\,(z) = \int_0^z \mathrm{d}t \, \frac{\sin t}{t}. \tag{C.141}$$

Note that the integrand is also the spherical Bessel function of zeroth order. We can move from this definition of the sine integral to a form suitable for the imaginary component of (C.138):

$$\begin{aligned}\mathrm{Si}\,(z) &= \int_0^\infty \mathrm{d}t\,\frac{\sin t}{t} - \int_z^\infty \mathrm{d}t\,\frac{\sin t}{t}\\ &= \frac{\pi}{2} - \int_z^\infty \mathrm{d}t\,\frac{\sin t}{t}.\end{aligned} \tag{C.142}$$

The solution of the first integral follows from our previous investigations of contour integration: we integrate the function $\frac{e^{iz}}{z}$ over a semi-circular contour of radius R with its base on the real axis, except for a small semi-circular contour of radius ρ centred on the origin so as to avoid the pole at the origin. One then lets $R \to \infty$ and $\rho \to 0$. Inserting (C.139) and (C.142) into (C.138) gives

$$\begin{aligned}E_1(x+iy) &= \int_{y-ix}^\infty \mathrm{d}u\,\frac{\cos u}{u} - i\int_{y-ix}^\infty \mathrm{d}u\,\frac{\sin u}{u}\\ &= -\mathrm{Ci}(y-ix) - i\left(\frac{\pi}{2} - \mathrm{Si}(y-ix)\right).\end{aligned} \tag{C.143}$$

Let $y = 0$. Then,

$$E_1(\pm x) = -\mathrm{Ci}(\mp ix) - i\left(\frac{\pi}{2} - \mathrm{Si}(\mp ix)\right). \tag{C.144}$$

Now, from symmetries

$$\begin{aligned}\mathrm{Si}(-ix) &= -\mathrm{Si}(ix)\\ \mathrm{Ci}(-ix) &= \mathrm{Ci}(ix) - i\pi\end{aligned} \tag{C.145}$$

we have

$$\begin{aligned}E_1(+x) &= -\mathrm{Ci}(-ix) - i\left(\frac{\pi}{2} - \mathrm{Si}(-ix)\right)\\ &= \mathrm{Ci}(ix) + i\left(\frac{\pi}{2} - \mathrm{Si}(ix)\right)\end{aligned} \tag{C.146}$$

and

$$E_1(-x) = -\mathrm{Ci}(ix) - i\left(\frac{\pi}{2} - \mathrm{Si}(ix)\right). \tag{C.147}$$

This gives us

$$\mathrm{Ci}(ix) = -\left(\frac{E_1(x) + E_1(-x)}{2}\right)$$
$$\mathrm{Si}(ix) = \frac{\pi}{2} - \left(\frac{E_1(x) - E_1(-x)}{2i}\right). \tag{C.148}$$

C.8 Operators

C.8.1 Introduction

Throughout quantum mechanics, the operator formalism is widely used. That is, if we are able to bring a one-to-one correspondence between a function in one function space Ψ and that in another function space Ψ', then it is stated that Ψ' is obtained by the result of the action of an operator upon Ψ, e.g.

$$\Psi' = \Lambda\Psi \tag{C.149}$$

where Λ is the operator.

In this sub-Appendix, properties of operators which are referred to in the book are summarised.

C.8.2 Linearity and Commutation

A *linear operator* is one in which its action upon the weighted sum of functions is the same as the weighted sum of the modified functions if the weights are constant. In other words, Λ is a linear operator if

$$\Lambda(\alpha_1\Psi_1 + \alpha_2\Psi_2) = \alpha_1(\Lambda\Psi_1) + \alpha_2(\Lambda\Psi_2), \tag{C.150}$$

where the coefficients α_1 and α_2 are constants, holds. From this property, additional linear operators can be defined from algebraic principles. These include the multiplication by a constant α,

$$\alpha(\Lambda\Psi) \equiv (\alpha\Lambda)\Psi, \tag{C.151}$$

the summation of operators,

$$(\Lambda_1 + \Lambda_2)\Psi \equiv \Lambda_1\Psi + \Lambda_2\Psi, \tag{C.152}$$

and the product of two operators,

$$(\Lambda_1\Lambda_2)\Psi \equiv \Lambda_1(\Lambda_2\Psi). \tag{C.153}$$

An important point to note from (C.153) is that operators may not necessarily commute, i.e. $(\Lambda_1\Lambda_2)$ may not necessarily be equal to $(\Lambda_2\Lambda_1)$, a significant difference between ordinary and operator algebras. This is represented by the *commutator* of two operators:

$$[\Lambda_1\Lambda_2] \equiv \Lambda_1\Lambda_2 - \Lambda_2\Lambda_1. \tag{C.154}$$

If

$$\begin{aligned}[\Lambda_1\Lambda_2]\Psi &= \Lambda_1\Lambda_2\Psi - \Lambda_2\Lambda\Psi \\ &= 0,\end{aligned} \tag{C.155}$$

then the two operators Λ_1 and Λ_2 are said to commute with the eigenfunction Ψ with eigenvalue zero. This leads to the important theorem if we were to replace the operators in the commutation process with observables (e.g. spatial position, momentum).

C.8.3 Division

Especially in our discussion of quantum scattering theory, we will have recourse to divide through by an operator. That is, the operator equation $AB = C$ yields $A = \frac{C}{B}$ only if the following conditions are met:

$$AB' = 0 \tag{C.156}$$

which implies that either $A = 0$ or $B' = 0$, and if

$$A \neq 0. \tag{C.157}$$

C.9 Heaviside Step Function

The Heaviside step function will arise in a number of our derivations throughout this book. It is also known as the unit step function. It is defined by

$$\begin{aligned}H(t) &= 0; \quad t < 0 \\ &= \frac{1}{2}; \quad t = 0 \\ &= 1; \quad t = 1.\end{aligned} \tag{C.158}$$

C.10 Dirac δ-Function

The Dirac δ-function is defined through the following two equations:

$$\begin{aligned} \delta(t) &= \infty; \quad t = 0 \\ &= 0; \quad t \neq 0 \end{aligned} \tag{C.159}$$

and

$$\int_{-\infty}^{\infty} \mathrm{d}t\, \delta(t) = 1. \tag{C.160}$$

The Dirac δ-function may also be defined as the derivative of the Heaviside step-function:

$$\delta(t) = \frac{\mathrm{d}H(t)}{\mathrm{d}t} \tag{C.161}$$

where $H(t)$ is given by (C.158). One of the most useful properties of the Dirac δ-function that we will use is the sifting property:

$$\int_{-\infty}^{\infty} \mathrm{d}t f(t)\delta(t-t') = f(t'). \tag{C.162}$$

The Fourier transform of the Dirac δ-function is

$$\begin{aligned} F(\omega) &= \frac{1}{\sqrt{2\pi}} \int_{-\infty}^{\infty} \mathrm{d}t'\, \delta(t')\, \mathrm{e}^{-i\omega t'} \\ &= \frac{1}{\sqrt{2\pi}}. \end{aligned} \tag{C.163}$$

It should be noted that some authors give this result as being equal to $\frac{1}{2\pi}$, reflecting the different normalisations used in the Fourier transform.

From (C.163), it is evident that

$$\begin{aligned} \delta(t) &= F^{-1}[F(\omega)] \\ &= \frac{1}{2\pi} \int_{-\infty}^{\infty} \mathrm{d}t'\, \mathrm{e}^{i\omega t'} \end{aligned} \tag{C.164}$$

which is a result that will be made frequent use of in this book.

C.11 Green's Function

The Green's function will prove a useful tool in some of our derivations and solutions. Let us define a linear differential or integral operator, $\boldsymbol{O}_t$, such that

$$\boldsymbol{O}_t f(t) = g(t) \tag{C.165}$$

where $g(t)$ is a known function and $f(t)$ is a function to be solved for. The Green's function is defined through the equation

$$\boldsymbol{O}_t\, G(t,t') = \delta(t-t') \tag{C.166}$$

where $\delta(t-t')$ is the Dirac δ-function. In other contexts, the Green's function is referred to as the impulse response of a system.

From (C.165) and (C.166) we see that we can write

$$\begin{aligned}\boldsymbol{O}_t f(t) &= \int \mathrm{d}t'\, \delta(t-t')\, g(t') \\ &= \int \mathrm{d}t' \boldsymbol{O}_t\, G(t,t') g(t').\end{aligned} \tag{C.167}$$

The result of (C.165) is possible as $\boldsymbol{O}_t$ is a linear function and acts upon the t variable only. We can thus write

$$f(t) = \int \mathrm{d}t' G(t,t') g(t') + f_0(t) \tag{C.168}$$

where the homogenous solution, $f_0(t)$, is the solution of

$$\boldsymbol{O}_t f_0(t) = 0. \tag{C.169}$$

C.12 Kronecker δ-Function

The Kronecker δ-function is a discrete version of the Dirac δ-function:

$$\begin{aligned}\delta_{jk} &= 1 \quad \text{for} \quad j = k \\ &= 0 \quad \text{for} \quad j \neq k\end{aligned} \tag{C.170}$$

where j and k are integers. It can also be represented as the complex contour integral:

$$\delta_{jk} = \oint_C dz\, z^{j-k-1}. \tag{C.171}$$

where C is the contour over the unit circle.

C.13 Digamma Function

The digamma function is the logarithmic derivative of the gamma function and will be used in the derivation of the Bloch correction factor for the collision stopping power.

It is defined as

$$\psi(z) = \frac{\mathrm{d}}{\mathrm{d}z} \ln \Gamma(z). \tag{C.172}$$

A useful series expansion of this function is

$$\mathrm{Re}(\psi(1 + iy)) = -\gamma_{\mathrm{EM}} + y^2 \sum_{n=1}^{\infty} \frac{1}{n(n^2 + y^2)}. \tag{C.173}$$

For large arguments, it is possible to write

$$\mathrm{Re}(\psi(1 + iy)) = \ln\, y + \frac{1}{12y^2} + \frac{1}{120y^4} + \frac{1}{252y^6} + \cdots. \tag{C.174}$$

D Binary Elastic Collision Kinematics

D.1 Introduction

When we consider energy transfer from moving charged particles to a medium exposed to it, it may seem counterintuitive to consider the fundamental interactions to be binary. But the probability of the transfer of energy from an incident particle to the medium (e.g. through interactions between electrons or positrons with atomic electrons) is relatively small as such collisions require the close spatial interaction between projectile and target. Hence, it is extremely unlikely that those interactions that result in energy transfer to a medium involve the interactions of more than two particles, thus the focus on such binary collisions.

By our definition, an elastic collision is that in which the sum of the pre- and post-collision kinetic energies of the colliding bodies are the same. That is, no energy is channelled into internal degrees of freedom, such as atomic and nuclear excitations or the angular momenta of projectile and target.

We first consider the basic relativistic elastic interaction between two particles in the laboratory frame of reference in which the target particle is at rest. The nature of the interaction potential is of no consequence. Subsequently, we then examine the mapping between the laboratory and centre-of-mass frames of reference.

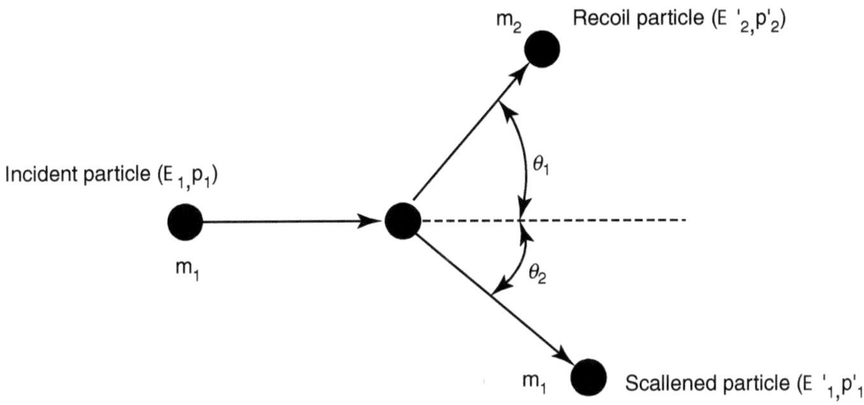

Fig. D.1 The kinematics of two-body elastic scatter

D.2 Relativistic Collision Between Two Particles

Consider the binary case in Fig. D.1 of a projectile with rest mass m_{Proj} and total energy E_{Proj} (i.e. sum of the rest mass and the kinetic energies) colliding with a target with rest mass m_{Tgt} and which is at rest in the laboratory frame of reference.[2] The four-vector momenta of the particles will be used to evaluate the kinematics of the scatter. These are

$$\boldsymbol{p}_{\mathrm{Proj}} = \left(E_{\mathrm{Proj}}, \boldsymbol{p}_{\mathrm{Proj}}\right) \tag{D.1}$$

$$\boldsymbol{p}_{\mathrm{Tgt}} = \left(m_{\mathrm{Tgt}}, 0\right) \tag{D.2}$$

$$\boldsymbol{p}'_{\mathrm{Proj}} = \left(E'_{\mathrm{Proj}}, \mathbf{p}'_{\mathrm{Proj}}\right) \tag{D.3}$$

$$\boldsymbol{p}'_{\mathrm{Proj}} = \left(E'_{\mathrm{Proj}}, \mathbf{p}'_{\mathrm{Proj}}\right) \tag{D.4}$$

where the $\boldsymbol{p}$ are the three-vector momenta. The primed quantities are those post-collision. The kinetic energy of the recoil particle is found by recalling that the pre- and post-collision four-vector momenta will be conserved:

$$\boldsymbol{p}_{\mathrm{Tgt}} + \boldsymbol{p}_{\mathrm{Proj}} = \boldsymbol{p}'_{\mathrm{Tgt}} + \boldsymbol{p}'_{\mathrm{Proj}}. \tag{D.5}$$

[2] In practice, an atomic electron can be considered to be effectively at rest if the projectile's speed greatly exceeds that of the electron.

Recalling the invariance of the four-vector momentum, i.e.

$$\begin{aligned} \boldsymbol{p}_{\text{Proj}}^2 &= E_{\text{Proj}}^2 - p_{\text{Proj}}^2 \\ &= m_{\text{Proj}}^2, \end{aligned} \tag{D.6}$$

and squaring both sides of (D.5) lead to

$$m_{\text{Proj}}^2 = m_{\text{Proj}}^2 + 2m_{\text{Tgt}}^2 + 2\left(\boldsymbol{p}_{\text{Proj}} \bullet \boldsymbol{p}_{\text{Tgt}} - \boldsymbol{p}_{\text{Proj}} \bullet \boldsymbol{p}'_{\text{Tgt}} - \boldsymbol{p}_{\text{Tgt}} \bullet \boldsymbol{p}'_{\text{Tgt}}\right). \tag{D.7}$$

As the dot product of two four-vector momenta is

$$\begin{aligned} \boldsymbol{p}_i \boldsymbol{p}_j &= (E_i, \mathbf{p}_i) \bullet (E_j, \mathbf{p}_j) \\ &= E_i E_j - \mathbf{p}_i \bullet \mathbf{p}_j, \end{aligned} \tag{D.8}$$

then $\boldsymbol{p}_{\text{Proj}} \bullet \boldsymbol{p}'_{\text{Tgt}} = E_{\text{Proj}} E'_{\text{Tgt}} - p_{\text{Proj}} p_{\text{Tgt}} \cos\theta_{\text{Tgt}}$, where θ_{Tgt} is the angle of the recoiling particle's trajectory with respect to the projectile's trajectory, and $\boldsymbol{p}_{\text{Tgt}} \bullet \boldsymbol{p}'_{\text{Tgt}} = m_{\text{Tgt}} E'_{\text{Tgt}}$. Substituting these results into (D.7) yields

$$p_{\text{Proj}} p'_{\text{Tgt}} \cos\theta_{\text{Tgt}} = \left(E'_{\text{Tgt}} - m_{\text{Tgt}}\right)\left(E_{\text{Proj}} + m_{\text{Proj}}\right). \tag{D.9}$$

The kinetic energy of the recoiling target particle, $T'_{\text{Tgt}} = E'_{\text{Tgt}} - m_{\text{Tgt}}$, is solved from (D.9):

$$T'_{\text{Tgt}} = 2m_{\text{Tgt}} \frac{p_{\text{Proj}}^2 \cos^2\theta_{\text{Proj}}}{\left(E_{\text{Tgt}} + m_{\text{Tgt}}\right)^2 - p_{\text{Proj}}^2 \cos^2\theta_{\text{Proj}}}. \tag{D.10}$$

The maximum kinetic energy of the recoiling particle occurs for $\theta_{\text{Tgt}} = 0$ (i.e. the recoil particle is ejected forwards):

$$\begin{aligned} T'_{\text{Tgt,max}} &= 2m_{\text{Tgt}} \frac{p_{\text{Proj}}^2}{\left(E_{\text{Proj}} + m_{\text{Proj}}\right)^2 - p_{\text{Proj}}^2} \\ &= 2m_{\text{Tgt}} \frac{p_{\text{Proj}}^2}{m_{\text{Proj}}^2 + m_{\text{Tgt}}^2 + 2m_{\text{Tgt}}\sqrt{p_{\text{Proj}}^2 + m_{\text{Proj}}^2}}. \end{aligned} \tag{D.11}$$

Let us now consider (D.11) in terms of relative projectile and target masses.

First, consider the case of the projectile and target having equal rest masses, $m_{\text{Proj}} = m_{\text{Tgt}} \equiv m$. This would be the case of, e.g. an electron or positron projectile incident to a stationary electron (or an atomic electron for which case its orbital speed is negligible relative to the incident electron) or, approximately, a neutron incident to a stationary proton. Then, the maximum recoil kinetic energy is

$$T'_{\mathrm{Tgt,max}} = 2m \frac{p^2_{\mathrm{Proj}}}{2m^2 + 2m\sqrt{p^2_{\mathrm{Proj}} + m^2}} \tag{D.12}$$

a result which can be rewritten as

$$\begin{aligned} T'_{\mathrm{Tgt,max}} &= \frac{p^2_{\mathrm{Proj}}}{m + \sqrt{p^2_{\mathrm{Proj}} + m^2}} \\ &= \frac{p^2_{\mathrm{Proj}}}{m + E_{\mathrm{Proj}}} \\ &= \frac{T_{\mathrm{Proj}}\left(T_{\mathrm{Proj}} + 2m\right)}{T_{\mathrm{Proj}} + 2m}. \end{aligned} \tag{D.13}$$

Now, if the projectile's kinetic energy is much higher than the rest masses of the projectile and target, then (D.13) reduces to the extraordinary result of

$$T'_{\mathrm{Tgt,max}} \approx T_{\mathrm{Proj}}, \tag{D.14}$$

i.e. all of the incident kinetic energy can be transferred to the recoil particle. A subtle facet to this collision exists if the identical particles are both the same type such as in electron–electron or proton–proton collisions. In such cases, the projectile and target are indistinguishable, and the inability to differentiate between projectile and target leads to the question of how an observer can tell the difference between when the projectile transfers none or all of its energy to the target. In both cases, one observes a final state with one particle at rest and another with kinetic energy equal to that incident. The fact that the two identical particles are indistinguishable following the collision is addressed by the convention that the maximum kinetic energy of the recoil particle in the relativistic region is set to $\frac{T_{\mathrm{Proj}}}{2}$ and the assumption that the particle with the highest kinetic energy in the final state is the projectile.

Returning to the general case of $m_{\mathrm{Proj}} \neq m_{\mathrm{Tgt}}$, if the projectile has a low incident three-vector momentum, i.e. $p_{\mathrm{Proj}} \ll \left(\frac{m^2_{\mathrm{Proj}}}{m_{\mathrm{Tgt}}}\right)$, then (D.12) can be approximated by

$$T'_{\mathrm{Tgt,max}} \approx 2m_{\mathrm{Tgt}} \left(\frac{p_{\mathrm{Proj}}}{m_{\mathrm{Proj}}}\right)^2 \tag{D.15}$$

which, if the projectile is much more massive than the target (e.g. a proton incident to an atomic electron), becomes

$$T'_{\mathrm{Proj,max}} = 2m_{\mathrm{Tgt}}\gamma^2_{\mathrm{Proj}}\beta^2_{\mathrm{Proj}} \quad \text{for } m_{\mathrm{Proj}} \gg m_{\mathrm{Tgt}}, \quad p_{\mathrm{Proj}} \ll \left(\frac{m^2_{\mathrm{Tgt}}}{m_{\mathrm{Proj}}}\right) \tag{D.16}$$

where β_{Proj} is the projectile speed normalised to the speed of light and $\gamma_{\mathrm{Proj}} = \frac{1}{\sqrt{1-\beta^2_{\mathrm{Proj}}}}$.

Let us now evaluate the kinetic energy of the scattered projectile. From the invariance of the product of the four-vector momenta,

$$\begin{aligned} \boldsymbol{p}_{\mathrm{Tgt}}\boldsymbol{p}_{\mathrm{Proj}} &= \boldsymbol{p}'_{\mathrm{Tgt}}\boldsymbol{p}'_{\mathrm{Proj}} \\ &= \boldsymbol{p}'_{\mathrm{Proj}}\left(\boldsymbol{p}_{\mathrm{Proj}} + \boldsymbol{p}_{\mathrm{Tgt}} - \boldsymbol{p}'_{\mathrm{Proj}}\right) \\ &= \boldsymbol{p}'_{\mathrm{Proj}}\boldsymbol{p}_{\mathrm{Proj}} + \boldsymbol{p}'_{\mathrm{Proj}}\boldsymbol{p}_{\mathrm{Tgt}} - m^2_{\mathrm{Proj}}. \end{aligned} \tag{D.17}$$

Using (D.8), this result is

$$E_{\mathrm{Proj}}m_{\mathrm{Tgt}} = E'_{\mathrm{Proj}}E_{\mathrm{Proj}} - p'_{\mathrm{Proj}}p_{\mathrm{Proj}}\cos\theta_{\mathrm{Porj}} + E'_{\mathrm{Proj}}m_{\mathrm{Tgt}} - m^2_{\mathrm{Proj}}. \tag{D.18}$$

Now, let us assume the example of an ultrarelativistic projectile for which (D.18) is

$$E_{\mathrm{Proj}}m_{\mathrm{Tgt}} \approx E'_{\mathrm{Proj}}E_{\mathrm{Proj}} + E'_{\mathrm{Proj}}m_{\mathrm{Tgt}} - E'_{\mathrm{Proj}}E_{\mathrm{Proj}}\cos\theta_{\mathrm{Proj}}. \tag{D.19}$$

Rearranging this result yields the scattered projectile's total energy:

$$E'_{\mathrm{Proj}}(\theta) = \frac{E_{\mathrm{Proj}}}{1 + \left(\frac{E_{\mathrm{Proj}}}{m_{\mathrm{Tgt}}}\right)\left(1 - \cos\theta_{\mathrm{Tgt}}\right)} \qquad E_{\mathrm{Proj}} \gg m_{\mathrm{Proj}}. \tag{D.20}$$

D.3 Conversion Between Reference Frames: Nonrelativistic Case

While the discussion of Appendix D.2 is of general interest, it is frequently a matter of convenience to consider the interactions between two particles in the centre-of-mass reference frame. In this sub-Appendix, we consider the conversion between the laboratory and centre-of-mass reference frames for nonrelativistic kinematics.

As in Appendix D.2, let us consider a projectile of mass m_{Proj} and velocity v_{Proj} incident to a target of mass m_{Tgt} with a collinear velocity v_{Proj} (we will consider the special case of the projectile at rest later). An elastic collision shall be assumed. We now consider this interaction within the centre-of-mass reference frame which moves with a velocity:

$$\mathbf{V}_{\mathrm{CM}} = \left(\frac{m_{\mathrm{Proj}}}{m_{\mathrm{Proj}} + m_{\mathrm{Tgt}}}\right)\mathbf{v}_{\mathrm{Proj}} + \left(\frac{m_{\mathrm{Tgt}}}{m_{\mathrm{Proj}} + m_{\mathrm{Tgt}}}\right)\mathbf{v}_{\mathrm{Tgt}}. \tag{D.21}$$

Within the centre-of-mass reference frame, the two particles move with velocities described by

$$\upsilon_{\mathrm{Proj}} = \mathbf{v}_{\mathrm{Proj}} - \mathbf{V}_{\mathrm{CM}} \tag{D.22}$$

and

$$\upsilon_{\mathrm{Tgt}} = \mathbf{v}_{\mathrm{Tgt}} - \mathbf{V}_{\mathrm{CM}}. \tag{D.23}$$

By defining the relative velocity $v_{\mathrm{Ref}} = v_{\mathrm{Proj}} - v_{\mathrm{Tgt}}$, then the centre-of-mass velocities are

$$\upsilon_{\mathrm{Proj}} = \left(\frac{m_{\mathrm{Tgt}}}{m_{\mathrm{Proj}} + m_{\mathrm{Tgt}}}\right)\mathbf{v}_{\mathrm{Ref}} \tag{D.24}$$

and

$$\upsilon_{\mathrm{Tgt}} = -\left(\frac{m_{\mathrm{Proj}}}{m_{\mathrm{Proj}} + m_{\mathrm{Tgt}}}\right)\mathbf{v}_{\mathrm{Tgt}}. \tag{D.25}$$

By defining the reduced mass as

$$m_{\mathrm{Red}} = \frac{m_{\mathrm{Proj}} m_{\mathrm{Tgt}}}{m_{\mathrm{Proj}} + m_{\mathrm{Red}}}, \tag{D.26}$$

then the momenta of the two particles in the centre-of-mass reference frame are $\pm\, m_{\mathrm{Red}}\mathbf{v}_{\mathrm{Ref}}$; i.e. they are equal in magnitude but opposite in sign.

Returning to the original examination of the elastic scatter from a target at rest, then clearly $\mathbf{v}_{\mathrm{Tgt}} = \mathbf{0}$. The scattering angle in the centre-of-mass frame is Θ_{Proj}, the angle between the pre- and post-scatter centre-of-mass velocities υ_{Proj} and $\upsilon'_{\mathrm{Proj}}$ which have the respective velocities, in the laboratory reference frame, of $\mathbf{v}_{\mathrm{Proj}}$ and $\mathbf{v}'_{\mathrm{Proj}}$ through the relationships of $\mathbf{v}_{\mathrm{Proj}} = \mathbf{V}_{\mathrm{CM}} + \upsilon_{\mathrm{Proj}}$ and $\mathbf{v}'_{\mathrm{Proj}} = \mathbf{V}_{\mathrm{CM}} + \upsilon'_{\mathrm{Proj}}$. The laboratory scatter angle, θ_{Proj}, is

$$\begin{aligned}\theta_{\text{Proj}} &= \tan^{-1}\left(\frac{v_{\text{Proj}}\sin\Theta_{\text{Proj}}}{V_{\text{CM}} + v_{\text{Proj}}\cos\Theta_{\text{Proj}}}\right)\\ &= \tan^{-1}\left(\frac{\sin\Theta_{\text{Proj}}}{\frac{V_{CM}}{v_{\text{Proj}}} + \cos\Theta_{\text{Proj}}}\right)\\ &= \tan^{-1}\left(\frac{\sin\Theta_{\text{Proj}}}{\frac{m_{\text{Proj}}}{m_{\text{Tgt}}} + \cos\Theta_{\text{Proj}}}\right).\end{aligned} \tag{D.27}$$

It can be seen from (D.27) that for a light projectile incident to a heavy target, $\frac{m_{\text{Proj}}}{m_{\text{Tgt}}} \to 0$ and

$$\theta_{\text{Proj}} \to \Theta_{\text{Proj}}. \tag{D.28}$$

That is, the scattering angles in the laboratory and centre-of-mass reference frames approach each other. It is obvious from (D.27) that the laboratory scattering angle is maximised for

$$\Theta_{\text{Proj,Max}} = -\cos^{-1}\left(\frac{m_{\text{Tgt}}}{m_{\text{Proj}}}\right) \tag{D.29}$$

or

$$\theta_{\text{Proj,Max}} = \sin^{-1}\left(\frac{m_{\text{Proj}}}{m_{\text{Tgt}}}\right). \tag{D.30}$$

It is clear that for $\frac{m_{\text{Proj}}}{m_{\text{Tgt}}} < 1$ the laboratory scattering angle θ_{Proj} will have two values of the centre-of-mass scattering angle Θ_{Proj}.

We next consider the angle of the recoiling target particle following the scatter, θ_{Tgt}. As the target particle was initially at rest in the laboratory frame of reference, then $v_{\text{Proj}} = -\boldsymbol{V}_{\text{CM}}$; the two triangles shown in the Figure are both equilateral leading to $\theta_{\text{Tgt}} = \frac{(\pi - \Theta_{\text{Proj}})}{2}$. From the bisection shown, the recoil particle speed in the laboratory reference frame is

$$v'_{\text{Rec}} = 2V_{\text{CM}}\sin\left(\frac{\Theta}{2}\right), \tag{D.31}$$

leading to the expression for the recoil particle's kinetic energy:

$$T'_{\text{Tgt}} = 2m_{\text{Tgt}}V_{\text{CM}}^2\sin^2\left(\frac{\Theta}{2}\right). \tag{D.32}$$

As, from (D.27), for a stationary target,

$$V_{\text{CM}} = \left(\frac{m_{\text{Proj}}}{m_{\text{Proj}} + m_{\text{Tgt}}} \right) v_{\text{Proj}},$$

then

$$T'_{\text{Tgt}} = 4T_{\text{Proj}} \frac{m_{\text{Proj}} m_{\text{Tgt}}}{\left(m_{\text{Proj}} + m_{\text{Tgt}}\right)^2} \sin^2\left(\frac{\Theta}{2}\right). \tag{D.33}$$

D.4 Conversion Between Reference Frames: Relativistic Case

In the relativistic regime, the total energies and the three-vector momenta of the projectile and target are described in the usual way by

$$E_{\text{Proj}} = m_{\text{Proj}} \gamma_{\text{Proj}} \tag{D.34}$$

$$E_{\text{Tgt}} = m_{\text{Tgt}} \gamma_{\text{Tgt}} \tag{D.35}$$

$$\mathbf{p}_{\text{Proj}} = m_{\text{Proj}} \gamma_{\text{Proj}} \mathbf{v}_{\text{Proj}} \tag{D.36}$$

$$\mathbf{p}_{\text{Tgt}} = m_{\text{Tgt}} \gamma_{\text{Tgt}} \mathbf{v}_{\text{Tgt}}, \tag{D.37}$$

where the relativistic contraction factor is

$$\gamma = \frac{1}{\sqrt{1 - \beta^2}} \tag{D.38}$$

and where β is the particle's speed normalised to the speed of light.

E Dirac *Bra* and *Ket* Notation

In our discussions using quantum mechanics, we will make frequent use of Dirac's *bra* and *ket* notation. This is summarised here.

The notation associates a given quantum state with a *ket vector* or, more commonly, *ket* which is denoted by the symbol $|\rangle$. These *kets* form a linear vector space; i.e. of the two *kets* $|\alpha\rangle$ and $|\beta\rangle$ and the two complex numbers k_1 and kz$_2$, the linear combination

$$|w\rangle = k_1|\alpha\rangle + k_2|\beta\rangle \tag{E.1}$$

is a vector of the *ket* space. Similarly, one can define a *bra vector* or *bra* which is denoted by the symbol $\langle|$. It is taken that these vectors have a one-to-one correspondence which is antilinear. That is, the *bra*

$$\langle w| = k_1^*\langle\alpha| + k_2^*\langle\beta| \tag{E.2}$$

is the complex conjugate of the *ket* of (E.1), and the scalar product of the *bra* $\langle u|$ and *ket* $|w\rangle$ vectors is such that

$$\langle u \mid w\rangle = \langle w \mid u\rangle^*. \tag{E.3}$$

F Dirac's Equation

Dirac's equation and its results will be made frequent use of in our discussions of electron and positron Coulomb scattering in which the intrinsic spin is accounted for. This Appendix provides a very brief review of Dirac's equation and the associated notations that are used in the book. A more comprehensive, indeed definitive, exposition of Dirac's equation is provided by Bjorken and Drell (1964) to which the interested reader is referred.

We begin by considering Schrödinger's equation which is nonrelativistic and is derived from the quantum-mechanical interpretation of energy and momentum as operators, $E \Leftrightarrow i\hbar\frac{\partial}{\partial t}$ and $\mathbf{p} \Leftrightarrow -i\hbar c\nabla$ (note the use of $\hbar c$to ensure that 3-vector momentum has units of energy to be consistent with the units employed by this book). By substituting these operator expressions into the nonrelativistic relationship between energy and momentum, $E = \frac{|\mathbf{p}|^2}{2m}$, we obtain the Schrödinger equation

$$i\hbar\frac{\partial\psi(\mathbf{x},t)}{\partial t} = -\frac{(\hbar c)^2}{2m}\nabla^2\psi(\mathbf{x},t). \tag{F.1}$$

In the 1920s, attempts were made to derive a relativistically invariant analogue to this Schrödinger equation using the relativistic relationship between energy and momentum:

$$E^2 = p^2 + m^2. \tag{F.2}$$

If one were to simply substitute the operator forms of energy and momentum into this relationship, then the Klein–Gordon equation is obtained:

$$\left(\Delta^2 + m^2\right)\psi = 0 \tag{F.3}$$

where we have the operator $\Delta^2 \equiv \hbar^2 \frac{\partial^2}{\partial t^2} - (\hbar c)^2\nabla^2$ has been defined. This result presented immediate problems.

The first was that the energy/momentum relationship of (F.2) allows negative-energy solutions, $E = -\sqrt{p^2 + m^2}$. Although this was a difficulty later confronted and surmounted by Dirac in his interpretation of the negative-energy solution representing the positron, this conundrum was at first assumed to be fatal to this attempt.

The second was that the Klein–Gordon equation of (F.3) is second order in that the time derivative leads to a probability density for the wavefunction that was not positive definite. As a result, the solution represented by the Klein–Gordon equation was rejected, although it was 'rehabilitated' in later times as a descriptor of spinless particles (which are described by scalar or pseudoscalar wavefunctions) such as the pion.

Dirac recognised the need for a first-order time derivative and so wrote a linear *ansatz* of the form

$$E\psi = (\alpha\mathbf{p} + \beta m)\psi \tag{F.4}$$

or, in another form,

$$i\hbar\frac{\partial\psi}{\partial t} = \left(-i\hbar c\sum_{i=1}^{3}\alpha_i\frac{\partial\psi}{\partial x^i}\right) + \beta m\psi \equiv H\psi \tag{F.5}$$

where contravariant notation for the spatial co-ordinates has been used. If this result is to be valid, it must satisfy three conditions:

- Yield the relativistic energy/momentum relationship of (F.2).
- Be Lorentz covariant.
- Produce a continuity equation.

If we consider the first point, the operators α and β are required in order to be consistent with the relativistic relationship between energy and momentum given by (F.2) as can be shown by substitution:

$$E^2 = (\alpha\mathbf{p})^2 + (\beta m)^2 + (\beta m\alpha\mathbf{p} + \alpha\mathbf{p}\beta m). \tag{F.6}$$

Dirac proposed that this be considered as a matrix equation and by equating the terms of (F.3) and (F.6), we find the requirements that the four matrices α_i and β obey the algebra:

$$\begin{aligned}
&\alpha_i^2 = \beta^2 = \mathbf{1}\\
&\alpha_i\alpha_j + \alpha_j\alpha_i = 2\delta_{ij}\mathbf{1}\\
&\alpha_i\beta + \beta\alpha_i = \mathbf{0}.
\end{aligned} \tag{F.7}$$

In order to construct these four matrices, three conditions must be met:

- The matrices must be Hermitian in order for the Hamiltonian to be so (refer to Appendix C.7).
- The eigenvalues of the α and β matrices must be ± 1 as a result of $\alpha_i^2 = \beta^2 = \mathbf{1}$.
- From the anticommutation property that the traces of the matrices must be zero.

These last two conditions allow the dimensions of these matrices to be specified. As the trace of a matrix is also the sum of its eigenvalues, the number of positive and negative eigenvalues must be equal, requiring that the dimensions of the α and β matrices be even. A dimension of 2 is insufficient, and the smallest even-numbered dimension for which these matrices can be realised is 4. Hence, we have

$$\begin{aligned} \alpha_i &= \begin{pmatrix} 0 & \sigma_i \\ \sigma_i & 0 \end{pmatrix} \qquad i = 1, 2, 3 \\ \beta &= \begin{pmatrix} \mathbf{1} & 0 \\ 0 & -\mathbf{1} \end{pmatrix} \\ \beta &= \begin{pmatrix} \mathbf{1} & 0 \\ 0 & -\mathbf{1} \end{pmatrix} \end{aligned} \tag{F.8}$$

where σ_i are the 2×2 Pauli matrices

$$\sigma_1 = \begin{pmatrix} 0 & 1 \\ 1 & 0 \end{pmatrix} \quad \sigma_2 = \begin{pmatrix} 0 & -i \\ i & 0 \end{pmatrix} \quad \sigma_3 = \begin{pmatrix} 1 & 0 \\ 0 & -1 \end{pmatrix} \tag{F.9}$$

and $\mathbf{1}$ is the 2×2 identity matrix.

In order to write the Dirac result in a Lorentz covariant form, we introduce the notation of

$$\begin{aligned} \gamma^0 &\equiv \beta \\ \gamma^i &\equiv \beta\alpha_i = \begin{pmatrix} 0 & \sigma_i \\ -\sigma_i & 0 \end{pmatrix} \qquad i = 1, 2, 3 \\ \gamma_5 &\equiv i\gamma^0\gamma^1\gamma^2\gamma^3 = \sigma_1. \end{aligned} \tag{F.10}$$

We can then write the Dirac equation as

$$\left(\gamma^0 E - \gamma \bullet \mathbf{p} - m\mathbf{1}\right) \psi = 0 \tag{F.11}$$

where $\boldsymbol{p}^\mu = (E, \mathbf{p})$ and the covariant is

$$\begin{aligned} \boldsymbol{p}_\mu &\equiv g_{\mu\nu}\boldsymbol{p}^\nu \\ &= (E, -\mathbf{p}) \end{aligned} \tag{F.12}$$

Equation (F.11) can be rewritten in the form

$$(\gamma^{\mu}\boldsymbol{p}_{\mu} - m\mathbf{1})\psi = 0. \tag{F.13}$$

We now have the $\psi(\boldsymbol{x})$ as a wavefunction with space-time and spinor degrees of freedom:

$$\psi(\boldsymbol{x}) = \begin{pmatrix} \psi_1(\boldsymbol{x}) \\ \psi_2(\boldsymbol{x}) \\ \psi_3(\boldsymbol{x}) \\ \psi_4(\boldsymbol{x}) \end{pmatrix}. \tag{F.14}$$

There are two solutions to Dirac's equation:

$$\begin{aligned} \psi_{+}(\mathbf{x}, t) &= u e^{-ip_{\mu}x^{\mu}} \\ &= u e^{-i\left(\frac{Et}{\hbar} - \frac{\mathbf{p}\bullet\mathbf{x}}{\hbar c}\right)} \end{aligned} \tag{F.15}$$

where u is defined as a Dirac spinor. This solution corresponds to a positive energy. The negative-energy solution is

$$\begin{aligned} \psi_{-}(\mathbf{x}, t) &= v e^{ip_{\mu}x^{\mu}} \\ &= v e^{+i\left(\frac{Et}{\hbar} - \frac{\mathbf{p}\bullet\mathbf{x}}{\hbar c}\right)} \end{aligned} \tag{F.16}$$

$\psi_{-}(\mathbf{x}, t)$ is the wavefunction of a positron or, in the Feynman picture, an electron moving backward in time with momentum $-\mathbf{p}$. The Dirac spinors are

$$u(\boldsymbol{p}, s) = \sqrt{E+m}\begin{pmatrix} \chi_s \\ \frac{\sigma\bullet\mathbf{p}}{\sqrt{E+m}}\chi_s \end{pmatrix} \tag{F.17}$$

$$v(\boldsymbol{p}, s) = \sqrt{E+m}\begin{pmatrix} \frac{\sigma\bullet\mathbf{p}}{\sqrt{E+m}}\hat{\varepsilon}\chi_s^{*} \\ \hat{\varepsilon}\chi_s \end{pmatrix} \tag{F.18}$$

where

$$\hat{\varepsilon} = \begin{pmatrix} 0 & -1 \\ 1 & 0 \end{pmatrix} \tag{F.19}$$

and where χ_s is the (nonrelativistic) Pauli two-component spinor for a spin-½ particle. Note that another variable, s, has been introduced and which describes the spin orientation of the particle. Equations (F.17) and (F.18) describe the particle (electron) and the antiparticle (positron) states, respectively.

G The Harmonic Oscillator: A Review

G.1 The Simple Harmonic Oscillator

The one-dimensional simple harmonic oscillator plays prominent parts in a variety of discussions throughout this book, and a brief review is provided in this Appendix.

Consider a particle originally at equilibrium at the origin $(r = 0)$ that displaced a small amount from the origin. A restorative force, $F(r)$, which is a function of position r will tend to return it to equilibrium. Let us expand the restorative force in a Taylor series

$$F(r) = \sum_{j=0}^{\infty} \frac{1}{j!} \frac{\mathrm{d}^j F(r=0)}{\mathrm{d}r^j} r^j, \tag{G.1}$$

and assume that the displacement is sufficiently small that, for practical purposes, we can truncate the expansion at second order (i.e. $j = 2$)

$$\begin{aligned} F(r) &= F(r=0) + \frac{\mathrm{d}F(r=0)}{\mathrm{d}r} r + \frac{1}{2} \frac{\mathrm{d}^2 F(r=0)}{\mathrm{d}r^2} r^2 \\ &= \frac{\mathrm{d}F(r=0)}{\mathrm{d}r} r + \frac{1}{2} \frac{\mathrm{d}^2 F(r=0)}{\mathrm{d}r^2} r^2 \end{aligned} \tag{G.2}$$

where the last step arises from $r = 0$ being defined as the point of equilibrium. If we truncate (G.2) to the first term, we have

$$\begin{aligned} F(r) &= \frac{\mathrm{d}F(r=0)}{\mathrm{d}r} r \\ &= -kr \end{aligned} \tag{G.3}$$

which, of course, is Hooke's law. Recalling Newton's third law, $F = m\frac{\mathrm{d}^2 r}{\mathrm{d}t^2}$, where m is the mass of the particle, (G.3) is written as

$$\frac{\mathrm{d}^2 r(t)}{\mathrm{d}t^2} + \omega_0^2 r(t) = 0 \tag{G.4}$$

where the natural frequency of the oscillator is

$$\omega_0 = \sqrt{\frac{k}{m}}. \tag{G.5}$$

In physical systems, dissipative forces exist and are usually proportional to the velocity, hence,

$$F_{\mathrm{dis}} = -b\frac{\mathrm{d}r(t)}{\mathrm{d}t} \tag{G.6}$$

where b is a constant. If we include this into the above equation of motion, we have

$$m\frac{\mathrm{d}^2r(t)}{\mathrm{d}t^2} + b\frac{\mathrm{d}r(t)}{\mathrm{d}t} + kr(t) = 0. \tag{G.7}$$

Dividing through by the particle mass and rewriting the result in the nomenclature used in this book, we have

$$\frac{\mathrm{d}^2r(t)}{\mathrm{d}t^2} + \Gamma\frac{\mathrm{d}r(t)}{\mathrm{d}t} + \omega_0^2 r(t) = 0 \tag{G.8}$$

where

$$\Gamma \equiv \frac{b}{m} \tag{G.9}$$

is defined as the damping coefficient and has units of inverse time.

The above result is determined for the case of no external driving forces (i.e. the right-hand side of (G.8) is equal to zero). The general form in this case is

$$\frac{\mathrm{d}^2r(t)}{\mathrm{d}t^2} + \Gamma\frac{\mathrm{d}r(t)}{\mathrm{d}t} + \omega_0^2 r(t) = \frac{F(t)}{m} \tag{G.10}$$

where $F(t)$ is the driving force.

From our knowledge of ordinary differential equations, we know that the solution to (G.10) will be the sum of the complementary solution, $r_{\mathrm{C}}(t)$, which is the solution to the homogeneous equation of (G.8), and the particular solution, $r_{\mathrm{P}}(t)$, which is the solution accounting for the effect of the external driving force:

$$r(t) = r_{\mathrm{C}}(t) + r_{\mathrm{P}}(t). \tag{G.11}$$

As the particular solution is dependent upon the nature of $F(t)$, we defer discussion of it and the means of solving (G.10), using the Green's function. Here, we derive the complementary solution. While there are a variety of means of solving (G.8), we use the time-honoured means of the *ansatz*

$$r(t) = A\mathrm{e}^{Bt} + C\mathrm{e}^{Dt} \tag{G.12}$$

and allowing for the initial conditions of $r(t=0) = 0$ and $\frac{\mathrm{d}r(t=0)}{\mathrm{d}t} = 0$ and solving the auxiliary quadratic equation

$$\xi^2 + \Gamma\xi + \omega_0^2 = 0 \tag{G.13}$$

which gives two roots:

$$\xi_{\frac{1}{2}} = -\frac{\Gamma}{2} \pm \sqrt{\left(\frac{\Gamma}{2}\right)^2 - \omega_0^2}. \tag{G.14}$$

The solution is, thus, of the form

$$r(t) = \mathrm{e}^{-\left(\frac{\Gamma t}{2}\right)} \left(A\mathrm{e}^{\sqrt{\left(\frac{\Gamma}{2}\right)^2 - \omega_0^2}\, t} + C\mathrm{e}^{-\sqrt{\left(\frac{\Gamma}{2}\right)^2 - \omega_0^2}\, t} \right). \tag{G.15}$$

There are three classifications of damped motion that depend upon the relative magnitudes of $\frac{\Gamma}{2}$ and ω_0^2. The classification of use to us in the analysis of soft collisions in Chap. 8 is that of under-damped motion which is defined for $\omega_0^2 > \left(\frac{\Gamma}{2}\right)^2$. In this case, the exponents of (G.15) become imaginary and

$$\begin{aligned} \sqrt{\left(\frac{\Gamma}{2}\right)^2 - \omega_0^2} &= i\sqrt{\omega_0^2 - \left(\frac{\Gamma}{2}\right)^2} \\ &= i\omega_{\mathrm{D}} \end{aligned} \tag{G.16}$$

where the damped frequency, ω_{D}, has been defined. Note that the inclusion of damping has resulted in the inequality, $\omega_{\mathrm{D}} < \omega_0$. Thus, (G.15) is

$$r(t) = \mathrm{e}^{-\left(\frac{\Gamma t}{2}\right)} \left(A\mathrm{e}^{i\omega_{\mathrm{D}} t} + C\mathrm{e}^{-i\omega_{\mathrm{D}} t} \right). \tag{G.17}$$

As the displacement is a real quantity, then we can define $C = A^*$ and write (G.17) in the form

$$\begin{aligned} r(t) &= \frac{A}{2}\mathrm{e}^{-\left(\frac{\Gamma t}{2}\right)} \left(\mathrm{e}^{i(\omega_{\mathrm{D}} t - \delta)} + \mathrm{e}^{-i(\omega_{\mathrm{D}} t - \delta)} \right) \\ &= A_\delta \mathrm{e}^{-\left(\frac{\Gamma t}{2}\right)} \cos(\omega_{\mathrm{D}} t - \delta). \end{aligned} \tag{G.18}$$

G.2 Oscillator Strength

G.2.1 Introductory Comments

In our discussion in Chap. 8 of the soft-collision stopping powers derived from the Bohr and Bethe theories, we will have recourse to the use of the oscillator strength, especially in the discussion of the use of harmonic oscillators as surrogates for bound atomic electrons. In this section, we derive an expression for the oscillator strength that will be of use.

The context will be in the absorption of energy by an oscillator as would be expected here, but the derivations are also applicable to the emission of energy.

G.2.2 Derivation

There are a variety of developments of the derivation of the oscillator strength (Sigmund (2006) provides a very detailed discussion), but Smith (1965) provided a most transparent result in which the oscillator strength is proportional to the product of the energy absorbed and the probability of the absorption for a dipole (E1) transition. As the oscillator strength is to be dimensionless, a succinct representation is the ratio of the squared amplitude matrix element to the square of the reduced de Broglie wavelength:

$$f_{f0} = \frac{|\langle f|x|0\rangle|^2}{\lambda\!\!\!^{-}{}_{f0}^2} \tag{G.19}$$

where

$$\lambda\!\!\!^{-}{}_{f0} = \frac{\hbar c}{\sqrt{2m_\mathrm{e}E_{f0}}}. \tag{G.20}$$

Or, in other words,

$$f_{f0} = 2\frac{m_\mathrm{e}}{(\hbar c)^2}E_{f0}|\langle f|x|0\rangle|^2 \tag{G.21}$$

where E_{f0} is the energy differential between the two states of the transition. Note that x refers to a single dimension; extending this to 3-dimension changes (G.21) to

$$f_{f0} = \frac{2}{3}\frac{m_\mathrm{e}}{(\hbar c)^2}E_{f0}|\langle f|\mathbf{r}|0\rangle|^2. \tag{G.22}$$

G.2.3 Thomas–Reiche–Kuhn Sum Rule

A characteristic of the above results is the expectation that the contributions of the electrons to the overall oscillator strength must be normalised. As noted above, the oscillator strength is a measure of the individual contributions of the Z-electrons within the atom (to either absorption or to emission). Let us consider the hydrogen atom.

We begin with the commutator relationships of

$$[\hat{p}, \mathbf{r}] = -i3\hbar c \tag{G.23}$$

where we have retained our use of the units of energy for the momentum, and

$$\begin{aligned}[\hat{H}, \mathbf{r}] &= \frac{1}{2m_e}[\hat{p}^2, \mathbf{r}] \\ &= -i\frac{\hbar c}{m_e}\hat{p}\end{aligned} \tag{G.24}$$

The $\wedge$ denotes a quantum-mechanical operator. We can write (G.23), through the insertion of a complete set of states, as

$$\begin{aligned}-i3\hbar c &= \langle 0|\hat{p}\mathbf{r} - \mathbf{r}\hat{p}|0\rangle \\ &= \sum_f \left(\langle 0|\hat{p}|f\rangle\langle f|\mathbf{r}|0\rangle - \langle 0|\mathbf{r}|f\rangle\langle f|\hat{p}|0\rangle\right)\end{aligned} \tag{G.25}$$

We can then use (G.24) to give

$$\begin{aligned}\langle f|\hat{p}|0\rangle &= i\frac{m_e}{\hbar c}\langle f|\left[\hat{H}, \mathbf{r}\right]|0\rangle \\ &= i\frac{m_e\left(E_f - E_0\right)}{\hbar c}\langle f|\mathbf{r}|0\rangle \\ &= i\frac{m_e E_{f0}}{\hbar c}\langle f|\mathbf{r}|0\rangle\end{aligned} \tag{G.26}$$

and as

$$\langle 0|\hat{p}|f\rangle = i\frac{m_e\left(E_0 - E_f\right)}{\hbar c}\langle 0|\mathbf{r}|f\rangle. \tag{G.27}$$

Placing (G.26) and (G.27) into (G.25) leads to

$$\sum_f E_{f0}|\langle f|\mathbf{r}|0\rangle|^2 = \frac{3}{2}\frac{(\hbar c)^2}{m_e} \tag{G.28}$$

or, through rearrangement,

$$\frac{2}{3}\frac{m_e}{(\hbar c)^2}\sum_f E_{f0}|\langle f|\mathbf{r}|0\rangle|^2 = 1. \tag{G.29}$$

From the definition of the oscillator strength, we thus have the Thomas–Reiche–Kuhn sum rule:

$$\sum_f f_{f0} = 1. \tag{G.30}$$

This can be generalised to the general case of an atom with Z electrons:

$$\sum_f f_{f0} = Z. \tag{G.31}$$

However, by renormalising the definition of the oscillator strength to the number of electrons from (G.22)

$$f_{f0} = \frac{2}{3} \frac{m_e}{(\hbar c)^2} \frac{E_{f0}}{Z} |\langle f|\mathbf{r}|0\rangle|^2, \tag{G.32}$$

we see that (G.32) holds for all Z.

H Plasma Frequency

A *plasma* is an assembly of charged particles in which the time-averaged charge density is equal to zero. If the plasma is disturbed, powerful restoring forces arise, leading to the charged particles going into oscillations about their equilibrium positions. Consider an infinite plasma slab as shown in Fig. H.1.

In this one-dimensional model, positive and negative charges are displaced from each other over a distance x. The density of charges is considered to be sufficiently small so that there are no collisions. We can use electrostatics to evaluate the time-varying electric field $\mathbf{E}$ if the slab thickness d is sufficiently small. The surface charge densities are

$$\rho_S = \pm \eta e x \quad \mathrm{C/cm^2} \tag{H.1}$$

where the sign reflects the charge sign and η is, again, the charge number density. The straightforward application of Gauss' law gives the electric field within the plasma slab:

$$|\mathbf{E}| = \frac{\eta e x}{\varepsilon_0}. \tag{H.2}$$

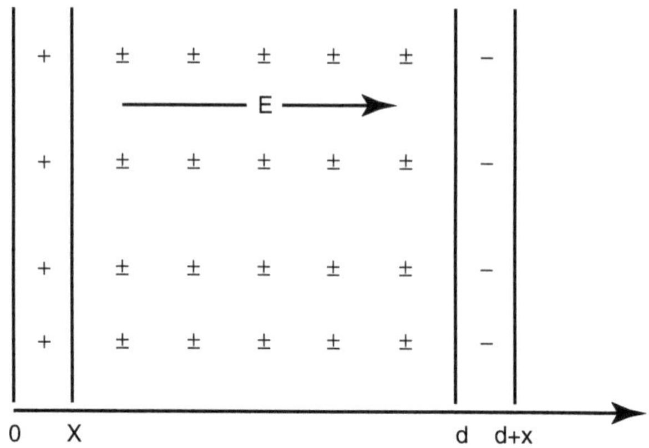

Fig. H.1 Displacement of positive and negative particles in a plasma slab

As the restorative force is linearly proportional to the displacement (Hooke's Law), simple harmonic motion occurs. Let us consider the plasma to be made up of positively charged ions and electrons. Due to the mass differences, we can approximate the ions as being stationary. The equation of motion is

$$m_{\mathrm{e}}\frac{\mathrm{d}^2x}{\mathrm{d}t^2}+\frac{\eta e^2}{\varepsilon_0}x=0. \tag{H.3}$$

This is an undamped system. Recalling the definition of the fine-structure constant, we can rewrite (H.3) as

$$\frac{\mathrm{d}^2x}{\mathrm{d}t^2}+c^2\left(\frac{4\pi\eta\alpha\hbar c}{m_{\mathrm{e}}}\right)x=0 \tag{H.4}$$

where the c^2 has been included to ensure that the coefficient of x has units of inverse time squared when we use our convention of writing the electron rest mass in units of energy. Having solved an equation of this type in Chap. 8, we see that the motion is oscillatory with the plasma frequency:

$$\omega_{\mathrm{P}}=c\sqrt{\frac{4\pi\eta\alpha\hbar c}{m_{\mathrm{e}}}}. \tag{H.5}$$

By inserting all of the physical constants, we can rewrite this as

$$\omega_{\mathrm{P}}=\left(1.59\times10^4\right)\sqrt{\eta}\,\mathrm{s}^{-1} \tag{H.6}$$

where η is the number of electric charges per cm^3.

Bibliography

Belkic D. Principles of quantum scattering theory. Bristol: Institute of Physics Publishing; 2004.
Belloni M, Robinett RW. Quantum mechanical sum rules for two model systems. Am J Physics. 2008;76:798–806.
Carslaw HS. An introduction to the theory of Fourier's series and integrals. New York: Dover Publications; 1950.
Churchill RV, Brown JW, Verhey RF. Complex variables and applications. 3rd ed. New York: McGraw-Hill Inc.; 1974.
Corrington MS. Applications of complex exponential integrals. Math Comput. 1961;15:1–6.
Dyson F. Advanced quantum mechanics. Singapore: World Scientific Publishing; 2007.
Goldberger ML, Watson KM. Collision theory. Mineola: Dover Publications; 2004.
Jackson JD. Classical electrodynamics. 3rd ed. Hoboken: Wiley; 1999.
James F. Statistical methods in experimental physics. 2nd ed. Singapore: World Scientific Publishing Co.; 2006.
McParland BJ. Nuclear medicine dosimetry: Advanced theoretical principles. London: Springer; 2010.
Messiah A. Quantum mechanics. New York: Wiley; 1966.
Vallée O, Soares M. Airy functions and applications to physics. London: Imperial College Press; 2004.

References

Abramowitz M, Stegun IA, editors. Handbook of mathematical functions. New York: Dover Publications; 1972.
Bjorken JD, Drell SD. Relativistic quantum mechanics. New York: McGraw-Hill; 1964.
Harris FE. Tables of the exponential integral Ei(x). Math Tables Aids Comput. 1957;11:9–16.
Pipkin AC. A course on integral equations. Harrisonburg: Springer; 1991.
Sigmund P. Particle penetration and radiation effects. Berlin: Springer; 2006.
Smith CMH. A textbook of nuclear physics. Oxford: Pergamon Press; 1965.

Index

B.J. McParland, *Medical Radiation Dosimetry*,
DOI 10.1007/978-1-4471-5403-7,